Discrete Symmetries and *CP* Violation

Discrete Symmetries and *CP* Violation

From Experiment to Theory

M. S. Sozzi
University of Pisa

OXFORD

UNIVERSITY PRESS

OXFORD
UNIVERSITY PRESS

Great Clarendon Street, Oxford OX2 6DP

Oxford University Press is a department of the University of Oxford.
It furthers the University's objective of excellence in research, scholarship,
and education by publishing worldwide in

Oxford New York

Auckland Cape Town Dar es Salaam Hong Kong Karachi
Kuala Lumpur Madrid Melbourne Mexico City Nairobi
New Delhi Shanghai Taipei Toronto

With offices in

Argentina Austria Brazil Chile Czech Republic France Greece
Guatemala Hungary Italy Japan Poland Portugal Singapore
South Korea Switzerland Thailand Turkey Ukraine Vietnam

Oxford is a registered trade mark of Oxford University Press
in the UK and in certain other countries

Published in the United States
by Oxford University Press Inc., New York

British Library Cataloguing in Publication Data

Data available

Library of Congress Cataloging in Publication Data

Sozzi, M. (Marco)
Discrete symmetries and CP violation: from experiment to theory / M.S. Sozzi.
p. cm.
ISBN 978–0–19–929666–8
1. CP violation (Nuclear physics) 2. Symmetry (Physics) 3. Mirror symmetry.
I. Title.
QC793.3.V5S59 2007
539.7'25—dc22   2007034660

Typeset by Newgen Imaging Systems (P) Ltd., Chennai, India
Printed in Great Britain
on acid free paper by
Biddles Ltd., King's Lynn, Norfolk

ISBN 978–0–19–929666–8

A Edoardo: luce.

PREFACE

As far as I see,
all a priori statements in physics
have their origin in symmetry.
H. Weyl

This work originates from the constructive interference of several considerations.

As usual, lecture notes for courses given over several years to undergraduate and graduate students were the starting point. In teaching the basics of *CP* violation I tried to put some emphasis on 'how' our current knowledge was obtained, and 'why' things were done as they were; in the process I found that none of the several very good books on the subject provided to the student all the aspects I felt to be important.

First, as the subject is not an easy one, sometimes students struggling to find their way among the myriad of phases and amplitudes might be left with the feeling that eventually everything is so nicely 'explained' within the Standard Model of Particle Physics (SM in the following) that no other possibility could actually be conceived: the 'poor researchers' of times past just didn't have a chance to understand what is now clear and pristine. While this is clearly a naive and non-historical view, my objection is mostly to the fact that it tends to dim the critical attitude of the student – a danger to which all researchers are exposed from the great successes of the Standard Model – and which must be resisted for progress to occur. I feel that while learning a subject it is important to keep a broader view initially, to pick connections and analogies with known phenomena; I have often encountered young graduate students, very deeply into the details of their Ph.D. work on some *CP*-violating effect, lacking a solid grasp of the basic connection with their neighbour's work. Older textbooks often presented (quite understandably) a broader treatment of the basics; such books are however long out of print and of course hopelessly outdated in most of their contents.

Second, many textbooks lack any emphasis on the experiments from which our understanding emerged (which can be partially excused when trying to keep the size of a comprehensive book below the 1,000- page mark). Still, to understand how 'the numbers' are obtained, why some of them are easy to get and some are not, why things are measured in one way rather than another, or what distinguishes a successful experiment from another one, is a fundamental part of the formation of any physicist, and several of my experimental colleagues would agree that too often this is the most difficult to convey. Again, young workers in the field (not only those with high theoretical inclinations) sometimes get the feeling that in order to make progress what is needed is just a brilliant theory, and then somebody has to go out and measure all its parameters to check it. I think a thorough appreciation of the experimental principles is really crucial for any new generation of physicists, even more so at the present time in which experiments (at least in what we now call high-energy physics) are evolving towards life-long enterprises: it is not uncommon nowadays to have a young experimental physicist starting work on a specific issue (be it an instrumentation, a simulation or a physics analysis topic) and be stuck with that for many years, without ever getting a chance to learn the principles for designing a new experiment, which might be asked from her or him sometime in the future.

Also, the appreciation of how scientific progress and technological advances are deeply intertwined *in both directions* is something whose importance transcends the formation of a physicist. As V. L. Fitch, one of the protagonists of the discoveries discussed in this work, puts it: 'What always paces the rate of discovery is the development of new instruments – new devices for probing phenomena, either with a new level of precision or by extending the range of observation into realms never before explored'. And already two centuries ago Sir H. Davy stated: 'Nothing tends so much to the advancement of knowledge as the application of a new instrument. The native intellectual powers of men in different times are not so much the causes of the different success of their labours, as the peculiar nature of the means and artificial resources in their possession' (Davy, 1839).

All the above would not have been a sufficient reason for me to write a new book; what really got this started was the feeling that too many students do not get exposed to the *challenging* aspects of experimental physics: if a smart student chooses to pursue a career in theoretical physics this must not be because she/he was not given a chance to appreciate how deep, beautiful and highly stimulating the experimental endeavour can be. The above point was best expressed by V. Telegdi in a nice short essay (Telegdi, 1990), in which he remarked how there is often as much intellectual content in an experiment as in its theoretical interpretation; he concluded that 'we must teach courses in which brilliant experiments of great significance are analyzed in some detail'.

Now, the experimental investigation of tiny symmetry violations in physics is really a field which is full of challenging and careful experiments, often with a long history of improvements and dedication; it encompasses many different branches of physics and even more vastly different approaches; finally, it is a very active field, with several new fundamental results being obtained in the past few years. It appears to the author as an ideal subject to convey the depth and excitement of experimental physics.

The patient reader can now appreciate the ideas behind the conception of this book: while it would be highly pretentious of the author to think of addressing all the above issues with his work, he hopes that this modest attempt can be seen as a step in the directions discussed above, and that the relevance of the important goals will throw some light on the inevitable shortcomings of the actual result.

The book deals mostly with the C, P, T discrete symmetries: other discrete symmetries such as translational symmetry in crystals are outside its scope. While trying to be self-contained, it attempts to take a somewhat novel approach to the subject: more emphasis than usual is put on experimental aspects, trying to provide a somewhat wider picture at the expense of some depth, and to convey the intellectual challenge of the connection between phenomenology and experiment. Excellent research-level books exist for the interested reader to dig more details on specific topics, and the rather extensive bibliography provided might be helpful for this purpose.

In most chapters some relevant experiments are discussed in some detail, trying to provide the average reader more insight than is obtained just by reading the original paper, often written for experts well versed in the approaches and research issues of the time. The focus is not so much on the detection techniques (which grow old) as on the experimental approaches (which do not). The choice of experiments is to some extent arbitrary: it didn't always fall on the first experiment to observe an effect, nor on the last or the most precise, but often on one which was felt to well exemplify some experimental issue. Many more beautiful and successful experiments would have deserved to be discussed in detail, but this would have required several volumes (and several authors too). Also, while the most recent values for experimental measurements are listed, many of them will grow old quickly, and the most up-to-date figures should clearly be looked for in the appropriate place (Yao *et al.*, 2006).

As an experimentalist, I cannot do full justice to the beauty of theory and, needless to say, the balance of topics is due to a personal judgement of their relative importance and affected by my own research activity in high-energy physics: readers who are experts in some field might want to skip the basic explanation of some techniques which are well known to them. Also, the balance is somewhat in favour of more 'established' physics issues, with more recent studies still under active investigation (for which the overall picture is still in the making, and the appraisal of the relevance of individual measurements still lacks the benefit of time) being discussed at a more shallow level.

The level of the book was originally intended to be as accessible as possible, and if this goal was not fully achieved the author is in excellent company, as Okun's masterful (but pretty hard for a novice) work (Okun, 1982) was started with the purpose of being 'readable even in a town bus'. The target audience are undergraduate and graduate students in physics; young researchers working in particle physics and interested in getting a somewhat wider picture on topics related to their own might also find this work of some interest. A generic undergraduate background in particle physics at the level of Perkins (2000) is required to appreciate several topics; also, a general understanding of particle detectors (at the level of standard laboratory courses) is assumed. A basic knowledge of quantum field theory is useful but not mandatory: an appendix provides the minimal knowledge required; some familiarity with the Standard Model of particle physics is required for a full understanding of Chapter 9. Some details lying somewhat outside the main content of the book or requiring some more background for a full understanding are written in smaller print and may be skipped on a first reading. The book might be used for a full graduate course on the subject, but it is probably most useful also as a source book for courses in atomic, nuclear or particle physics, in which only a few chapters might be used.

The plan of the work is as follows: Chapter 1 sets the general discussion on discrete symmetries, and the next three chapters deal with parity, charge conjugation and time reversal in some detail. Chapter 5 discusses the more fundamental *CPT* symmetry and the connection between spin and statistics. *CP* violation is introduced in general terms in Chapter 6, and details of the system in which it was detected, namely neutral flavoured mesons, are presented in Chapter 7. Chapter 8 discusses the kaon system in some detail from a phenomenological point of view, also as a template for heavier flavoured mesons; before dealing with those in Chapter 10, and given their relevance in determining the flavour structure of the SM, the latter is briefly introduced in Chapter 9. Finally, Chapter 11 provides a somewhat wider (and more speculative) view of some related topics which help setting the entire topic in perspective. Sections describing some experiment or set of experiments in detail are interspersed with the main text and clearly identified by shaded boxes; a separate index for these sections is provided.

Some appendices recall some issues formalism and theoretical tools on quantum field theory and scattering, the three-body Dalitz plot and the formal treatment of spin; another one covers some topics concerning the treatment of systematic errors, so central to the whole content of this work.

ACKNOWLEDGEMENTS

It is truly said that all buildings
are composed of countless stones,
and all great rivers are swelled and increased
by many different springs and sources;
in the same way all sciences are extracted
and compiled by different scholars,
one of whom may perhaps be ignorant
of what another has written.
But by the famous writing of ancient authors
all things are known in one place or another.
J. Froissart

First of all I want to thank many colleagues and friends who have unknowingly contributed to some parts of this book with their advice, teaching, and discussions over the years. First among them is Italo Mannelli, who teaches every day by example. I am also indebted to Rosanna Cester, who ignited a spark many years ago which led to a romance with the fascinating subject of *CP* violation.

My colleagues Giovanni Batignani, Luigi Di Lella, Gianluca Lamanna, and Italo Mannelli read and criticized parts of the draft, contributing to making it a better work. It goes without saying that mine is the blame for any remaining errors, hoping for an even number of sign errors in formulæ. The 'naive' questions by countless students, which so often stimulate deeper thoughts, are also gratefully acknowledged.

My editors at OUP, Sonke Adlung and Lynsey Livingston, provided valuable advice while showing considerable patience, and the technical help of Julie Harris and the care of Natasha Forrest were precious. Thanks are due to Dr. Brian H. May (Ph.D) for this kindness and his production, limited in physics but vast in other fields.

Last, but definitely not least, I am indebted to my parents for obvious reasons, and mostly to my family, Giusy and Edoardo, who bore with me during long nights in the attic, when I definitely experienced time dilation: they never left me short of cheers and tea.

Living in the web renaissance I will maintain an updated list of comments, corrections, and updates to this work available for interested readers, hopefully including their welcome suggestions for improvements: these will be hosted at `http://cern.ch/sozzi/CPbook`.

CONTENTS

1

SYMMETRIES AND INVARIANCES

C'est la dissymétrie, qui crée le phénomène.
P. Curie

Impartial, adj. *Unable to perceive*
any promise of personal advantage
from espousing either side of a controversy
or adopting either of two conflicting opinions.
A. Bierce

This flame that burns inside of me
I'm hearin' secret harmonies.
R. Taylor

In this chapter we briefly recall the general theory of transformations in quantum theory. This topic is usually covered in quantum mechanics (QM) and quantum field theory (QFT) textbooks; to be self-contained and to lay down the basic notions for the rest of the book we give a short treatment of the issues that will be relevant for later chapters, at the same time setting our notation.

1.1 Symmetry and invariance

The word $\sigma\upsilon\mu\mu\epsilon\tau\rho\iota\alpha$ in ancient Greek had the meaning of 'well proportioned', 'harmonious', and as such was related to beauty. Ancient science, or 'natural philosophy', was greatly shaped by such concept of symmetry, in a way which we now find somewhat questionable: as one among countless examples just think of the way Kepler obtained his laws, by relating the harmony of musical intervals to the dimension of planetary orbits, while at the same time renouncing with difficulty the 'perfection' of their circular shape.

It is therefore interesting to note how modern science, particularly in the last century, has come back to reserve a very central rôle to the concept of symmetry, although this has evolved to acquire a different and more precise meaning. Much has been written on the fundamental importance of symmetry in modern science: the reason why this concept is so central in physics lies in its generality, and the insensitivity to the details of the specific natural phenomenon considered. When the detailed laws governing a new phenomenon are unknown or only partially understood, symmetry arguments can be fruitful guiding principles to learn more about those, and to restrict the range of viable theories.

Symmetries are associated with transformations of a system: whenever a transformation exists such that the transformed system is indistinguishable from the original one, we say that the system has a symmetry. A simple example is that of rotational symmetry: if we consider a perfect sphere, one which has no marks on its surface nor any inhomogeneity, we can rotate it by an arbitrary angle without observing any change; we say that the sphere is symmetric under rotations. As other solids do not possess the same symmetry, rotational symmetry becomes a useful concept to classify them: indeed, if a given symmetry were be absolutely valid in Nature, there would be no way of discovering this fact, and it would not be discussed at all. As far as we know, few symmetries are really exact in Nature, but even approximate symmetries can be very useful for the advancement of understanding, and actually the mechanism in which such approximate symmetries are broken can give much information on the physics.[1]

Since some natural phenomena exhibit common symmetries, we are led to the idea of physical laws being symmetric: in mathematical terms such laws are expressed in terms of equations, and the symmetry of the laws is reflected in the invariance of the equations under a given mathematical transformation. Symmetry and invariance are thus tightly linked in physics, and represent some of the most fruitful concepts for its development: [The theory] 'possesses certain features, mainly based on invariance properties [...]. The importance of these features can hardly be overestimated, since they offer the most reliable guidance that we have in classifying and interpreting the rapidly growing and already very complex experimental picture' (Wick *et al.*, 1952).

1.2 Transformations and symmetries in quantum theory

We start by discussing transformations in the quantum formalism which is suited to the description of microscopic (and elementary) phenomena.

In quantum theory a transformation maps a state in Hilbert space to another one through a unitary operator U:

$$|\psi\rangle \rightarrow |\psi'\rangle = U|\psi\rangle \tag{1.1}$$

---

[1] Indeed perfect symmetries are not of this world: it is well known that Muslim carpet weavers deliberately introduce small errors when weaving, since Muslim theology states that only God is perfect and human artefacts should not try to share such perfection; for similar reasons elaborate decorations in traditional Japanese gates occasionally have a small detail carved upside down, breaking an otherwise perfect symmetry.

The unitarity condition $U^\dagger U = UU^\dagger = 1$ is required in order to maintain the state normalized (conservation of probability).

A complication arises from the fact that in quantum theory what actually corresponds in a unique way to a physical state is not a state vector (a *ket* in Dirac's terms) $|\psi\rangle$ but rather a *ray* $\hat{\psi} = e^{i\xi}|\psi\rangle$ with arbitrary real ξ; the reason is that all the physical information which can be obtained from a measurement is contained in the probability $|\langle\phi|\psi\rangle|^2$. To each group of transformations acting on a physical system, there corresponds an isomorphic representation which transforms rays accordingly. The transformations for the state vectors with which quantum theory actually deals are only constrained by the corresponding ray transformations, and therefore maintain a large arbitrariness for what concerns the phase factors.

According to Wigner's theorem (Wigner, 1959), for each transformation (1.1) such that

$$|\langle\psi'|\phi'\rangle|^2 = |\langle\psi|\phi\rangle|^2 \tag{1.2}$$

the operators U which give the mapping of the state vectors can always be chosen to be either unitary or *anti-unitary*, a (somewhat poor) name which stands for unitary and anti-linear, i.e. transforming numbers into their complex conjugates; see e.g. Gottfried (1966) on Weinberg (1995) for a proof of Wigner's theorem.

While the linearity of an operator O_L is defined by the property

$$O_L(\alpha|\psi_1\rangle + \beta|\psi_2\rangle) = \alpha\,O_L|\psi_1\rangle + \beta\,O_L|\psi_2\rangle \tag{1.3}$$

(where α, β are numbers), for an anti-linear operator O_A one has instead

$$O_A(\alpha|\psi_1\rangle + \beta|\psi_2\rangle) = \alpha^*O_A|\psi_1\rangle + \beta^*O_A|\psi_2\rangle \tag{1.4}$$

The algebraic manipulation of anti-linear operators requires some care, and different prescriptions for dealing with them can be found in the literature. A simple one consists in remembering that for an anti-linear operator the action on a *ket* $|\psi\rangle$ or on its dual *bra* $\langle\psi|$ has to be explicitly specified, and switching from one to the other involves complex conjugation: writing in general

$$|O\psi\rangle = O|\psi\rangle \quad |O^\dagger\psi\rangle = O^\dagger|\psi\rangle \quad \langle O\psi| = \langle\psi|O^\dagger \quad \langle O^\dagger\psi| = \langle\psi|O$$

for a linear operator O_L one has

$$\langle\psi_1|O_L\psi_2\rangle = \langle\psi_1|(O_L|\psi_2\rangle) = ((\langle\psi_1|O_L)|\psi_2\rangle) = \langle O_L^\dagger\psi_1|\psi_2\rangle = \langle\psi_2|O_L^\dagger\psi_1\rangle^* \tag{1.5}$$

while for an anti-linear operator O_A one has instead

$$\langle\psi_1|O_A\psi_2\rangle = \langle\psi_1|(O_A|\psi_2\rangle) = ((\langle\psi_1|O_A)|\psi_2\rangle)^* = \langle O_A^\dagger\psi_1|\psi_2\rangle^* = \langle\psi_2|O_A^\dagger\psi_1\rangle \tag{1.6}$$

so that when dealing with anti-linear operators parentheses are explicitly used when operators act on their left on a *bra*. Adopting the convention that in absence of parentheses an operator acts on the right, for a linear operator O_L the Hermitian conjugate is defined by

$$\langle\psi_1|O_L^\dagger|\psi_2\rangle = \langle O_L\psi_1|\psi_2\rangle = \langle\psi_2|O_L\psi_1\rangle^* \tag{1.7}$$

while for an anti-linear one O_A one has

$$\langle \psi_1 | O_A^\dagger | \psi_2 \rangle = \langle O_A \psi_1 | \psi_2 \rangle^* = \langle \psi_2 | O_A | \psi_1 \rangle \tag{1.8}$$

Indeed, since only squared modulus quantities have a direct (measurable, operational) meaning, it is easy to see that conservation of probability is achieved not only when

$$\langle \psi_1 | U^\dagger U | \psi_2 \rangle = \langle \psi_1 | \psi_2 \rangle$$

as happens for unitary (linear) operators, but also when

$$\langle \psi_1 | U^\dagger U | \psi_2 \rangle = \langle \psi_1 | \psi_2 \rangle^*$$

as obtained for anti-unitary operators.

The unitary or anti-unitary nature of a transformation operator is not a conventional choice and does not depend on the choice of arbitrary phases for state vectors, but is rather a specific property of the transformation itself: different unitary (or anti-unitary) operators corresponding to the same ray transformation differ from each other at most by phase factors. Continuity implies that any transformation which can be made trivial (i.e. the identity) by varying a continuous parameter must be described by a unitary operator; anti-unitary ones only appear in the case of discrete transformations, such as time reversal, to be discussed in detail in Chapter 4.

Two approaches are usually adopted to deal with transformations (Fonda and Ghirardi, 1970): in the *active* point of view one considers the original physical system and a transformed one, as described by the same observer, while in the *passive* point of view one considers the same (unchanged) physical system as described by one observer and a second one in a transformed reference system. In the active point of view the symmetry can be checked by comparing the behaviour of the transformed system with respect to the original one; in the passive point of view the system is not changed, but the presence of a symmetry implies that both observers describe the system using the same equations. The active point of view is somewhat more natural, since it corresponds more often to transformations which can *actually* be performed on a system (think, for example, of the preparation of a mirror-reflected physical system, as compared to the mirror-reflection of the observer).

In any case, since in quantum physics only the relation of states (state vectors) and observables (operators) has a physical meaning, a transformation can be described formally in different ways, by considering a change of the state vectors or of the operators corresponding to the measured observables, without any physical consequence.

In one approach, the one implicitly chosen in eqn (1.1), a transformation is described in the formalism by the action of an operator U on the state vectors, while all Hermitian operators O corresponding to observables are left unchanged; this is sometimes called the 'Schrödinger' approach (by extension of the 'Schrödinger' picture' for time evolution in quantum mechanics):

$$|\psi\rangle \to |\psi'\rangle = U|\psi\rangle \qquad O \to O \tag{1.9}$$

In the active point of view the system described by $|\psi\rangle$ is transformed into *another* system described by $U|\psi\rangle$, while in the passive point of view the two observers describe the *same* system with two different state vectors $|\psi\rangle$, $U|\psi\rangle$.

Another equivalent approach is possible, in which one defines the effect of a transformation as leaving the state vectors unchanged and modifying instead the operators in the opposite way (sometimes called the 'Heisenberg' approach):

$$|\psi\rangle \to |\psi\rangle \qquad O \to O' = U^{-1}OU = U^\dagger OU \tag{1.10}$$

In both approaches the expression for the transformed matrix element is the same by construction:

$$\langle\psi_1|O|\psi_2\rangle \to \langle\psi_1'|O|\psi_2'\rangle_{(S)} = \langle\psi_1|O'|\psi_2\rangle_{(H)} = \langle\psi_1|U^\dagger OU|\psi_2\rangle$$

Whether the new matrix element is equal or not to the original one depends on whether the transformation is actually a symmetry of the system.

It should be noted that the operator transformation corresponding to the state vector transformation $|\psi\rangle \to U|\psi\rangle$ might actually be considered to be $O \to UOU^\dagger$, in the sense that if *both* the above transformations are applied at the same time, the matrix elements are unchanged. Being a formal identity, such statement has nothing to do with any symmetry property of the system: it just represents the fact that, e.g. in the passive point of view, both observers agree on what each of them measures in his own reference frame (i.e. knowing the transformation laws, observers A and B agree on what A is measuring, and the same for B); the above identity reflects the fact that only the relative relation of state vectors and operators is relevant.

The same change in the measured quantities is obtained by transforming the physical system in one way or the reference system of the observer and the measurement apparatus oppositely. It is clear then that the expression for a transformed matrix element (possibly a physical measured quantity) is the same when only the state vectors are transformed in one way (Schrödinger approach) and when only the operators are transformed in the *opposite* way (Heisenberg approach). As an example, one could define the action of a space rotation as affecting all states, rotated around a given axis by an angle θ (Schrödinger approach) or as rotating all operators by an angle $-\theta$ (Heisenberg approach). The (Heisenberg) operator transformation resulting in a given effect on the (measurable) matrix elements is the one *opposite* to the corresponding (Schrödinger) state transformation. This point is however not very relevant for discrete transformations connecting only two states, such as the inversions considered in this book. In the two approaches either the state vectors or the operators are affected by the transformation, but never both; the two choices are equivalent, and as long as they are not mixed at the same time no trouble arises.

The formalism is closer to that of the classical theory when the Heisenberg approach is used: a rather direct correspondence is obtained when one substitutes for the Poisson brackets between two canonical variables in classical mechanics the commutator between the corresponding quantum-mechanical operators multiplied by $(-i/\hbar)$. The explicit form of the U operator which relates the operators (or state vectors) between the two situations linked by the transformation is obtained in the passive point of view by comparing the measurements of the *same* (subjective) quantity on the same (objective) system, and using the classic correspondence principle (or analogies, for quantities without a classical counterpart) on the quantum expectation values. If one of a set of classical observables $O^{(k)}_{(cl)}$, when measured by the second observer changes as

$$O^{(k)}_{(cl)} \to O^{(k)'}_{(cl)} = \mathcal{F}_k\left[O^{(1)}_{(cl)}, O^{(2)}_{(cl)}, \ldots\right]$$

where $\mathcal{F}_k$ indicates a linear functional dependence on other observables with real coefficients, the requirement for the corresponding quantum operator $O^{(k)}$ is

$$\langle U\psi|O^{(k)}|U\psi\rangle = \langle \psi|U^\dagger O^{(k)} U|\psi\rangle$$

$$= \mathcal{F}_k\left[\langle\psi|O^{(1)}|\psi\rangle, \langle\psi|O^{(2)}|\psi\rangle, \ldots\right] = \langle\psi|\mathcal{F}_k[O^{(1)}, O^{(2)}, \ldots]|\psi\rangle$$

where the first equality is seen to be valid both for linear and anti-linear U using eqn (1.7) or (1.8) and the Hermiticity of $O^{(k)}$. Since the above relation must be valid for any state vector $|\psi\rangle$, the U operators are determined by

$$U^\dagger O^{(k)} U = \mathcal{F}_k\left[O^{(1)}, O^{(2)}, \ldots\right]$$

1.3 Symmetries and conservation laws

The symmetry properties of a physical system are determined in the formalism by the change (or actually the invariance) of the measured quantities when performing a transformation. If a Hermitian operator O commutes with the operator U describing a transformation, $[O, U] = 0$, the measured quantities expressed as expectation values $\langle O\rangle$ are the same after the transformation:

$$\langle O\rangle = \langle\psi|O|\psi\rangle \to \langle\psi|U^\dagger OU|\psi\rangle = \langle\psi|U^\dagger UO|\psi\rangle = \langle O\rangle$$

We also recall that, since the Hamiltonian H drives the time evolution of a system, if O commutes with H, and therefore with the time evolution operator e^{-iHt}, the measured quantities $\langle O\rangle$ are constant in time (for a time-independent operator O):

$$i\hbar\frac{d}{dt}\langle O\rangle = \langle[O, H]\rangle \tag{1.11}$$

The commutativity of O and H allows both operators to be reduced to diagonal form simultaneously, that is to have a complete set of stationary states (energy eigenstates) which are also eigenstates of O. Of course not all observables commuting with H commute among themselves, and in this case those which (in a given representation) are not diagonal do not have constant values (the chosen eigenstates of the Hamiltonian are not eigenstates of the corresponding operators), but still the probability distributions for their expectation values do not vary in time. By considering the time evolution of a system and that of the transformed one, it is easy to see that – among all observables – the Hamiltonian must be invariant for each transformation which is actually a symmetry of the physical system, otherwise the original state and the transformed one could be distinguished by observing their time evolution: a symmetry requires the invariance of the Hamiltonian under the corresponding transformation, i.e. $[U, H] = 0$.

From the above considerations it is seen that, if the unitary operator U corresponding to a transformation commutes with the Hamiltonian, then the energy and time evolution of the system are not affected by such transformation, and on the other hand the matrix elements of U are constant in time. If U is also Hermitian, or if there is a Hermitian operator which only depends on U, then one has an observable which is constant in time, i.e. a conserved quantity. This is the general connection between the symmetries of a system under linear transformations and the existence of conservation laws.

From the passive point of view, we say that a symmetry under a given transformation exists when the same physical laws describing the behaviour of a system for the first observer are also valid for the second one, which uses transformed state vectors (or operators): this means that the same Hamiltonian is used by both. Similarly, from the active point of view a symmetry exists when the time evolution of the transformed system coincides with the transformed of the original system evolved in time.

In collision processes, for a transformation described by a unitary operator U, symmetry implies that the $\mathcal{S}$ matrix (Appendix C), relating free states in the remote past to free states in the remote future, is unchanged, so that $[U, \mathcal{S}] = 0$: this immediately gives selection rules when eigenstates of U are considered: if $U|i, f\rangle = u_{i,f}|i, f\rangle$ then

$$0 = \langle f|[U, \mathcal{S}]|i\rangle = (u_f - u_i)\langle f|\mathcal{S}|i\rangle \tag{1.12}$$

and $i \to f$ transitions are forbidden if $u_f \neq u_i$; moreover

$$\langle Uf|\mathcal{S}|Ui\rangle = \langle f|\mathcal{S}|i\rangle \tag{1.13}$$

Each symmetry is related to the impossibility of measuring a given quantity, that is to a class of 'impossible experiments': translation symmetry is related to the impossibility of determining the absolute position in space for an isolated system (homogeneity of space) and leads to the conservation of momentum; rotational symmetry is related to the impossibility of determining an absolute direction in space (isotropy of space) and leads to the conservation of angular momentum; time displacement symmetry is related to the impossibility of determining an absolute instant of time (homogeneity of time) and leads to the conservation of energy; gauge symmetry is related to the impossibility of determining the relative phase of charged and neutral particle states, and leads to the conservation of electric charge, and so on.

The celebrated Noether's theorem(s) (Noether, 1918) states that for every continuous symmetry (connected to the identity via the change of a continuous parameter) a corresponding conservation law exists, i.e. there is a conserved observable, which is a constant of the motion. The generators of the symmetry group represent the conserved observables, and their commutation relations are unequivocally determined by the group structure.

For continuous transformations there exists a unitary (as opposed to anti-unitary) representation, and in the neighbourhood of the identity, the representation of the group elements can be written as

$$U(\alpha_1, \ldots, \alpha_n) = \exp\left\{\sum_{j=1}^{n} i\alpha_j G_j\right\} \tag{1.14}$$

where α_j are the n real parameters of the group and G_j are operators, the generators of the group; having explicitly factored out a factor i, the unitarity of U implies the Hermiticity of the G_j. Considering for simplicity a one-parameter group (the extension to the case with several parameters is straightforward), if the transformations represented by U are symmetry transformations of the system then the Hamiltonian is invariant under those, which means

$$H \to H' = U^\dagger(\alpha)HU(\alpha) = H$$

or $[U, H] = 0$. For infinitesimal $\alpha \to 0$ ($U \to 1$) this implies

$$(1 - i\alpha G)H(1 + i\alpha G) \simeq H - i\alpha[G, H] = H$$

or $[G, H] = 0$. G commutes with the Hamiltonian and, being Hermitian, represents a constant of the motion.[2]

Symmetry transformations involving space-time (such as Lorentz transformations) require further considerations, as besides the variation *in form* of the operators that induced by the change of their arguments is also involved, see e.g. Leite Lopes (1969).

Note that Noether's theorem says nothing concerning *discrete* symmetries, which therefore do not necessarily lead to conserved quantities.

A special place among symmetry principles is held by the permutation of identical (indistinguishable) particles, according to which only two kinds of particles appear to exist in Nature: bosons, always arranging themselves in systems which are symmetric under exchange of identical particles, and fermions, for which only

---

[2] There are some small caveats to this statement, related to super-selection rules (Chaichian and Hagedorn, 1998), see Section 5.3.

anti-symmetric systems appear instead, leading to the Pauli principle: a system with two fermions in identical states does not exist. Particles with half-integer spin behave as fermions, while those with integer spin behave as bosons: this is the so-called spin-statistics connection. Consistently with observations, we will assume the validity of these principles throughout most of the book, but in Chapter 5 we will consider how their validity is tested experimentally.

1.4 Discrete transformations and inversions

For discrete transformations (not continuously connected to the identity) Noether's theorem does not hold. Among discrete transformations inversions have a particular relevance: these are transformations which give back the same physical system when applied twice (such as mirror-reflection, charge conjugation, etc.).

In general, if for a given quantum field ϕ_i (where i stands for spinor or tensor indexes) the effect of an inversion is described by an unitary operator U_I such that

$$U_I^\dagger \phi_i(x) U_I = \eta_I S_{ij} \phi_j(x') \tag{1.15}$$

(x' being the inverted coordinate), or $\phi \to \eta_I S \phi$ in compact form, where S is a matrix acting on the field components ($S = 1$ for a scalar field) and η_I is a complex number, one has[3]

$$U_I^\dagger \phi_i^\dagger(x) U_I = \eta_I^* S_{ij}^* \phi_j^\dagger(x') \tag{1.16}$$

or[4] $\phi^\dagger \to \eta_I^* \phi^\dagger S^\dagger$.

Because of the unitarity of U_I, and the fact that the fields in the Lagrangian must always appear in terms with the same number of ϕ and $\phi^\dagger$ (ultimately due to charge conservation), the invariance of the free field Lagrangian under the transformation is seen to require $|\eta_I|^2 = 1$, i.e. the $\eta_I = e^{i\xi_I}$ must be phase factors (ξ_I real). Moreover, for Hermitian fields, which might be observables and represent particles identical to their antiparticles, the phase factors must be real, $\eta_I = \pm 1$.

Apart from these general conditions, any further restriction on the values of the phase factors η_I appearing in inversions either involves further physical assumptions or is just a matter

---

[3] Note that in the literature the role of U_I and $U_I^\dagger$ is often interchanged, which only leads to the exchange $\eta_I \leftrightarrow \eta_I^*$ in the expression for the transformation of either the states or the field operators, without any physical consequences.

[4] Note that the meaning of the dagger operation is different for the matrix S, acting in the space of field components, and for the field ϕ which, besides being arranged as a column vector in component space, is an operator in Hilbert space.

of convention (Feinberg and Weinberg, 1959). Considering for simplicity a (non-Hermitian) scalar field, transformations of the type

$$\phi(x) \rightarrow U_M^\dagger \phi(x) U_M = \eta_M \phi(x) \quad \phi^\dagger(x) \rightarrow U_M^\dagger \phi^\dagger(x) U_M = \eta_M^* \phi^\dagger(x) \tag{1.17}$$

where $|\eta_M|^2 = 1$, such as those related to electric charge, baryon number, etc., do not change interaction probabilities (squared matrix elements), so that no measurable difference can occur when using $H' = U_M^\dagger H U_M$ as the Hamiltonian instead of H. If a given inversion symmetry holds, $[U_I, H] = 0$, it will continue to hold for H' when using the corresponding transformed inversion operator U_I'. For inversions U_I which do not commute with U_M but are instead such that

$$U_I U_M = U_M^\dagger U_I \tag{1.18}$$

the transformed operators are

$$U_I' = U_M^\dagger U_I U_M = U_I U_M^2 \tag{1.19}$$

and an appropriate U_M can be chosen in such a way as to cancel the phase factors η_I, leading (for a generic tensor field) to

$$\phi_i(x) \rightarrow U_I^\dagger \phi_i(x) U_I = S_{ij} \phi_j(x') \tag{1.20}$$

instead of eqn (1.15). The conclusion is that, for inversions satisfying eqn (1.18), when the corresponding symmetry holds, the phase factors η_I for non-Hermitian fields are completely unobservable (both absolutely or relatively to each other), as they can be redefined at will without affecting the result of any measurement. The measurable phase factors are therefore only those related to Hermitian fields or to inversions that do not satisfy eqn (1.18): examples are the parity P (Chapter 2) and CPT (Chapter 5) transformations; instead for charge conjugation C (Chapter 3) and time reversal T (Chapter 4) eqn (1.18) is satisfied (in the first case due to the fact that C transforms ϕ into $\phi^\dagger$, in the second one due to the anti-linearity of T), so that the phase factors are arbitrary.

State vectors describing a system which is invariant for the transformation represented by an operator U are usually chosen as eigenstates of this operator, whose eigenvalue is unity within a phase factor. If the effect of the inversion is $|\psi\rangle \leftrightarrow |\psi'\rangle$ then $|\pm\rangle = |\psi\rangle \pm |\psi'\rangle$ are such eigenstates. In general, eigenstates of a unitary operator can have complex eigenvalues, while in the case of anti-unitary operators one can always reduce to the case in which the eigenvalue is real: if $O_A|\psi\rangle = |a|e^{i\xi}|\psi\rangle$ (ξ real), by redefining the phase of the state vector describing the system, i.e. using $|\psi'\rangle = e^{i\xi/2}|\psi\rangle$ instead one gets $O_A|\psi'\rangle = |a||\psi'\rangle$.

A theory is said to be symmetric under a given inversion whenever the phase factors η_I for the fields appearing in it can be chosen in such a way that the inversion operator U_I thus defined commutes with the Hamiltonian; if more than one choice of phase factors leads to operators commuting with the Hamiltonian, any of those can be chosen to represent the inversion. Only if, for *any* possible choice of the phase factors for the different fields, U_I does not commute with H, is the system not symmetric under the corresponding transformation.

Because of the nature of inversions, the unitary operator U_I describing such a transformation must satisfy $U_I^2 = e^{2i\xi_I}$, (ξ_I real), i.e. U_I^2 is the identity to within a phase. Since states themselves are not observable quantities, one cannot directly argue that a double inversion operator must coincide with the identity; the phase factor $e^{2i\xi_I}$ cannot however depend on the state, otherwise when U_I^2 is applied to a linear superposition of states it would not give back an equivalent state. Also in quantum field theory, in which the basic quantities which get transformed are the field operators, when the latter are not observable no constraints on the above phase factor arise; measurable quantities involve field products such as $\phi^\dagger\phi$, and their invariance under double inversions does not constrain U_I^2 at all.

In some cases the phase factor induced by the square of an inversion operator U_I^2 can be shown to be 1, or it can be chosen to be 1 for convention without affecting the physics (see Section 5.3). In this case, the unitarity of U_I implies its Hermiticity, so that U_I itself is a conserved observable if the transformation is actually a symmetry of the physical system; one still has a conserved quantity, despite the non-applicability of Noether's theorem. An important difference with the case of continuous transformations is that in this case the resulting conservation law is *multiplicative* rather than *additive*: the eigenvalue for a system of several (non-interacting) parts which are themselves eigenstates of the transformation is the product (as opposed to the sum) of the individual eigenvalues. The reason is readily seen: in the case of continuous symmetries the conserved observables (Hermitian) are the generators G of the transformation, appearing in $U = \exp[i\alpha G]$, eqn (1.14); considering two eigenstates of G:

$$G|\psi_1\rangle = g_1|\psi_1\rangle \qquad G|\psi_2\rangle = g_2|\psi_2\rangle \qquad (1.21)$$

one has for the combined system state $|\psi_1\psi_2\rangle$

$$U|\psi_1\psi_2\rangle = e^{i\alpha G}|\psi_1\psi_2\rangle = [U|\psi_1\rangle][U|\psi_2\rangle] = \left[e^{i\alpha G}|\psi_1\rangle\right]\left[e^{i\alpha G}|\psi_2\rangle\right]$$

$$= e^{i\alpha g_1}|\psi_1\rangle\, e^{i\alpha g_2}|\psi_2\rangle = e^{i\alpha(g_1+g_2)}|\psi_1\psi_2\rangle \qquad (1.22)$$

which implies

$$G|\psi_1\psi_2\rangle = (g_1 + g_2)|\psi_1\psi_2\rangle \qquad (1.23)$$

On the other hand, in the case of inversions with $U_I^2 = 1$, the conserved observables are the U_I themselves:

$$U_I|\psi_1\rangle = u_1|\psi_1\rangle \qquad U_I|\psi_2\rangle = u_2|\psi_2\rangle \qquad (1.24)$$

$$U_I|\psi_1\psi_2\rangle = [U_I|\psi_1\rangle][U_I|\psi_2\rangle] = u_1|\psi_1\rangle u_2|\psi_2\rangle = u_1 u_2|\psi_1\psi_2\rangle \qquad (1.25)$$

It should be noted that such multiplicative quantum numbers do not have an analogue in classical physics. The impossibility of treating discrete transformations analogously to the continuous ones is also closely related to the absence of any classical analogue for the corresponding conserved quantities.

1.5 Symmetry violations

The existence of a symmetry is always a tentative statement which can only be verified experimentally. The most straightforward way of testing the validity of a discrete symmetry would be to compare the results of two experiments, one of them built in such a way as to be inverted according to the corresponding transformation. This is unfortunately not possible in general, as it is impossible in practice to prepare an entire experimental apparatus inverted as required. One is therefore forced to study microscopic systems, which can be appropriately inverted: while the overall setup of the experiment is still asymmetric (not properly inverted), the disproportion in the time, distance and energy scales involved in the microscopic process and its interaction with the macroscopic apparatus allow one to conclude that any asymmetry observed at the microscopic level actually belongs to the system under study.

Another approach exploits the Hermitian operator associated to the transformation under scrutiny: if the symmetry holds one has a conserved quantum number and a selection rule for transitions among different states; while the values of the conserved quantum numbers are often arbitrary, their *relative* values are usually not, and this allows the experimental test.

It should be clear that one can speak of a symmetry violation when a given symmetry is rather well established for at least a class of physical processes, so that the associated physical quantity is conserved and quantum numbers for particles are well defined, at least in those processes.[5]

One method of testing the validity of a symmetry is based on the search for the *existence* of states with given properties. Whenever a physical state (an eigenstate of the Hamiltonian) is not an eigenstate of a given operator (or a member of an energy degenerate set of such eigenstates) this indicates that there is no valid symmetry associated with this operator.

The existence of two or more states, with different values of a given quantum number, which transform to the same final state, or conversely a physical state (not member of a degenerate pair) which can transform into either of two states with different eigenvalues of such quantum number, proves that the symmetry corresponding to the above quantum number is violated i.e. it is not a *good* quantum

---

[5] Strictly speaking, if a symmetry is violated the symmetry operators themselves cannot be exactly defined (Lee and Wick, 1966).

number (see eqn (1.13)). In this case however one must be quite sure, experimentally, that the the initial and final states under consideration are *really* the same. An example which will be discussed at length in Chapter 6 is the decay of both (distinguishable) physical states of the neutral kaon system into the same two-pion state, which indicates violation of *CP* symmetry.

The symmetries of the theory can be expressed as symmetries of the transition matrix $\mathcal{T}$ (see Appendix C) under the effect of the corresponding unitary operator U, such that the transition between transformed states $|i'\rangle = U|i\rangle$, $|f'\rangle = U|f\rangle$ has the same properties as the one between the original ones $|i\rangle$, $|f\rangle$:

$$\left|\langle f'|\mathcal{T}|i'\rangle\right| = \left|\langle f|U^\dagger \mathcal{T} U|i\rangle\right| = |\langle f|\mathcal{T}|i\rangle| \tag{1.26}$$

As in the general case of eqn (1.13), if the discrete symmetry corresponding to U holds, the conservation of the corresponding quantity means that no transitions between eigenstates of U with different eigenvalues are allowed. If $U|\pm\rangle = \pm|\pm\rangle$ one has

$$\langle -|\mathcal{T}|+\rangle = \langle -|U^\dagger \mathcal{T} U|+\rangle = -\langle -|\mathcal{T}|+\rangle = 0 \tag{1.27}$$

While spontaneous decays of isolated particles are the simplest physical transitions to be studied, more complex processes are also investigated to search for symmetry violations in this way.

Another approach for symmetry tests is the search for asymmetries in a physical process: for example symmetries often impose some constraints on the possible distributions of spin and momenta for the outgoing particles in a reaction among two interacting particles. The parity-violating nature of particle polarization in the production plane, in a reaction with unpolarized initial particles (to be discussed in Chapter 2), illustrates this.

An evidence for the violation of a symmetry is also the observation of a non-zero value for the expectation value of an observable that is *odd* (i.e. changes sign) under the corresponding transformation, although some care might be required in such a case, as will be seen for the case of time reversal in Chapter 4.

Further reading

The literature on symmetries in physics is vast, and even a partial listing is not possible here; two excellent non-technical books are Weyl (1952) and Feynman (1965).

Among older books dealing extensively with discrete symmetries, Sakurai (1964), Low (1967) and Gibson and Pollard (1976) should be noted; others will be referenced in following chapters. Cahn and Goldhaber (1988) is a nice commented collection of reprints of original articles, a few of them dealing with symmetry violations. The history of particle physics is pretty much the story of symmetries, and Pais (1986) is a fascinating account of it.

Quantum transformation theory is discussed, at different levels of detail, in all books on quantum mechanics and quantum field theory, among which Bjorken and Drell (1965), Itzykson and Zuber (1980), Weinberg (1995); an excellent and detailed discussion can be found in Fonda and Ghirardi (1970). Other books such as Scadron (1991) and the excellent books by J. J. Sakurai (1985, 1967) discuss at length these issues, also addressed by older review papers such as Kemmer *et al.* (1959).

An advanced theoretical 'top-down' approach to symmetries (somewhat complementary to the one adopted in this book) can be found in Froggatt and Nielsen (1991).

Problems

1. Show explicitly that space translation symmetry implies conservation of momentum.

2. Rotations in three-dimensional space around different axes do not commute, and the corresponding generators (the components of angular momentum) do not either. Space translations and time translations do commute instead, but still the corresponding generators, the three-momentum p and the Hamiltonian H, do not commute in general, except for systems which are invariant under space translations. Why is this so?

3. Explain why the ray representation of quantum theory does not constrain the existence of eigenvalues for linear transformation operators.

2

PARITY

In this chapter the parity or space inversion transformation and the corresponding symmetry is discussed, and its formal implementation in quantum theory is reviewed in some detail, also as a template for the discussion of other discrete symmetries in following chapters. The intrinsic parity of elementary particles is introduced, and some examples of how it can be determined are discussed. The experimental consequences of parity symmetry for elementary particle processes are then considered, several experimental tests are illustrated and explained. Parity symmetry was the first, among those discussed in this work, which was found not to be universally valid in Nature; because of this and due to its many consequences which can be tested experimentally, the range of experiments in which it has been studied is vast and varied.

2.1 Parity transformation

The effect of a parity transformation is defined as the inversion of the spatial coordinates with respect to the origin

$$x \to x_P = -x \tag{2.1}$$

and in its passive interpretation it corresponds to the reversal of the three spatial axes, under which a right-handed reference system becomes a left-handed one (and vice versa), so that it is also called *space inversion*.

PARITY

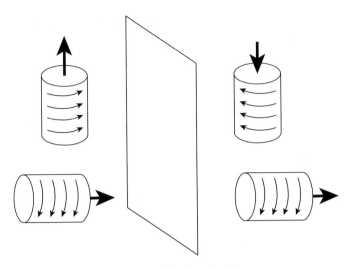

Fig. 2.1. Mirror reflection of axial vectors.

This transformation is equivalent, in the case of our three spatial dimensions, to a *mirror reflection* with respect to an arbitrary plane, followed by a π rotation with respect to an axis orthogonal to this plane. Since rotational symmetry, linked to the isotropy of space, is always assumed to hold, symmetry under mirror reflections in a plane actually corresponds to parity symmetry.[6]

Vectors are defined by their transformation properties with respect to the rotation group, and classified according to their transformation properties with respect to parity: *polar vectors* are reversed by a parity transformation, while *axial vectors* (or *pseudo-vectors*) are not. Correspondingly, under reflection in a plane polar vectors have their component orthogonal to the plane reversed and the components parallel to the plane unaffected, while for axial vectors the opposite occurs (see Fig. 2.1); a π rotation around the axis orthogonal to the reflection plane then turns the reflected image of an axial vector into itself (unaffected by the parity transformation), while a polar vector turns into a reversed copy of itself (inverted by the parity transformation). The above can also be understood by considering that axial vectors can be expressed as vector products of two polar vectors.[7]

Chirality (a term coined by Lord Kelvin, from the Greek $\chi\epsilon\iota\rho$ for hand) is the property of an object not superposable to its mirror image, an example being

[6] The reader who has never pondered about why mirrors do exchange 'right' and 'left' but not 'up' and 'down' is invited to do so.

[7] Axial vectors are actually a grouping of the three independent components of an anti-symmetric tensor A_{ij} of rank 2, and as such a peculiarity of our three-dimensional space: in three dimensions (only) the number of such components is equal to the dimensionality of the space and the axial vector representation is possible: in four dimensions an anti-symmetric tensor $A_{\mu\nu}$ has six independent components, which cannot be represented by a four-component vector.

our hands: 'I call any geometrical figure, or group of points, chiral, and say it has chirality, if its image in a plane mirror, ideally realized, cannot be brought to coincide with itself.' (Thomson, 1904)

While the parity transformation commutes with rotations, it is readily seen to be a qualitatively different transformation from those: it cannot be obtained by any combination of rotations, due to the number of spatial dimensions being odd. More formally, the determinant of the 3×3 transformation matrix M in (vector) space (which is invariant under unitary transformations) is -1 for a parity transformation (and for a mirror reflection), while it is $+1$ for rotations. In general, when space coordinates x_i are transformed by

$$x'_i = M_{ij} x_j \tag{2.2}$$

the components P_i of a pseudo-vector transform according to

$$P'_i = \det(M) M_{ij} P_j \tag{2.3}$$

Examples of the two kinds of vectors are respectively the momentum (polar) vector $\boldsymbol{p}$ and the angular momentum (axial) vector $\boldsymbol{L} = \boldsymbol{x} \times \boldsymbol{p}$: under parity

$$\boldsymbol{p} \to \boldsymbol{p}_P = -\boldsymbol{p} \qquad \boldsymbol{L} \to \boldsymbol{L}_P = +\boldsymbol{L} \tag{2.4}$$

Similarly, while the electric field is a polar vector, the magnetic field is an axial vector (see discussion in Section 2.1.3); this is easily understood by considering a uniform magnetic field parallel to a mirror, as generated by a solenoid (magnetic monopoles seem not to exist): by mirror reflection the electric charges travel along the helix of the solenoid in the opposite screw direction, and according to the right-hand rule the direction of the magnetic field is indeed reversed.[8]

In classical mechanics symmetry under the parity transformation does not result in any constant of the motion, but constrains instead the form of the solutions of the equation of motion: if $\boldsymbol{x}(t)$ is a possible particle trajectory, then the parity inverted function $\boldsymbol{x}_P(t) = -\boldsymbol{x}(t)$ is also possible, if the system has parity symmetry. The fact that parity symmetry is not particularly relevant for phenomena of classical physics is mainly due to the fact that usually arrangements without well-defined parity are considered, and that no constant of motion can appear, since the space-inverted motion cannot evolve in a continuous way from the original one. In quantum physics, when parity symmetry is valid and a system has some point which can be identified as a 'centre of inversion', then states do have a well-defined parity.

---

[8] Never mind that the right hand becomes a left hand under mirror reflection.

2.1.1   Quantum mechanics

The concept of parity is introduced in quantum theory through the correspondence principle: a unitary operator P is introduced[9] to define the parity transformation of state vectors (in the Schrödinger approach)

$$|\psi\rangle \rightarrow |\psi_P\rangle \equiv P|\psi\rangle \tag{2.5}$$

or of operators (in the Heisenberg approach)

$$O \rightarrow O_P \equiv P^\dagger O P \tag{2.6}$$

in such a way that the measured quantities (matrix elements of operators) transform according to the classical interpretation of the transformation, namely

$$\langle \psi_1 | \boldsymbol{x} | \psi_2 \rangle \rightarrow -\langle \psi_1 | \boldsymbol{x} | \psi_2 \rangle$$

for any two state vectors $|\psi_1\rangle, |\psi_2\rangle$; this allows one to define P to within a phase as the operator implementing (2.1).

The effect of parity on the spin operator S (Heisenberg approach) is defined to be

$$S \rightarrow S_P \equiv P^\dagger S P = +S \tag{2.7}$$

in analogy with the behaviour of orbital angular momentum L (for which it follows from its explicit expression in terms of $\boldsymbol{x}$ and $\boldsymbol{p}$). More formally, since rotations and space inversion are commuting operations, P and the generator of an infinitesimal rotation (the total angular momentum J) must commute too, and since $J = L + S$ the above transformation rule is obtained.

Eigenstates of parity are such that

$$P|\psi\rangle = \eta_P|\psi\rangle \tag{2.8}$$

where the (multiplicative) eigenvalue $\eta_P = e^{i\xi_P}$ is a phase factor: $|\eta_P|^2 = 1$ (i.e. real ξ_P) is required, since the probability density $\langle \psi | \psi \rangle$ should be a scalar quantity, unchanged by a parity transformation. The η_P phase factors do not create any difficulty in quantum mechanics, since in absence of particle creation or destruction, all observables contain an η_P^* factor for each η_P. It can be easily seen that a necessary (not sufficient) condition for a vector (pseudo-vector) operator to have non-zero matrix elements between two given states is that these have opposite (same) parity. It should be clear that only systems with zero momentum can possibly be eigenstates of the parity operator, i.e. systems observed in their centre of mass.

---

[9] To streamline notation, we will use P to denote both the parity transformation and the unitary operator (properly U_P) which implements it.

Applying the parity transformation twice the original state must be obtained, so that P^2 must be a phase factor. If (by necessity or by choice) $P^2 = \eta_P^2 = 1$ (i.e. $\xi_P = 0, \pm\pi, \pm 2\pi, \ldots$) one is left with only a sign arbitrariness for P. In this case P is an (Hermitian) observable, with eigenvalues ± 1, usually called the parity of the system; in the following this will usually be taken to be the case (see discussion in Section 5.3).

The existence of a conservation law for parity (without an analogue in classical physics) can be related to the validity of the quantum superposition principle (which also does not have a classical analogue). Its relevance arises from the fact that physical processes are often defined in a convenient way in terms of states of definite parity (eigenstates of P), and ultimately to the experimental fact that the most relevant interactions in Nature (meaning the ones which affect most of the physical processes we are concerned with, in the quantitatively strongest way), indeed appear to be parity symmetric.

If $\varphi(t, x)$ is the wave function for a possible state of a system, then the parity-transformed one $\varphi_P(t, x)$ is also allowed if the Hamiltonian H is parity-invariant, that is if

$$H_P \equiv P^\dagger H P = H \quad \text{or} \quad [H, P] = 0$$

This is the condition for parity symmetry to hold, which is valid if and only if the potential term $V(x)$ in the Hamiltonian is an even function of the coordinates, i.e. if $V(x) = V(-x)$.

In quantum mechanics, the wave function $\varphi_P(t, \boldsymbol{x})$ of a parity-inverted state $|\phi_P\rangle$ is related to the original one $\varphi(t, \boldsymbol{x})$ by

$$\varphi_P(t, \boldsymbol{x}) \equiv \langle \boldsymbol{x} | \varphi_P(t) \rangle = \langle \boldsymbol{x} | P | \varphi(t) \rangle = \eta_P \varphi(t, -x) \tag{2.9}$$

as can be obtained in either the Schrödinger or the Heisenberg approach: the state vector $|\varphi(t)\rangle$ either transforms or remains unchanged as defined above for the two approaches, while $|x\rangle$, being the eigenstate of the position operator x, follows the transformation of the latter, i.e. it is unchanged in the Schrödinger approach and transformed as $|x\rangle \rightarrow P^\dagger |x\rangle$ in the Heisenberg one.

Note also that parity transforms a representation of the Lorentz group, as described by the eigenvalues $(j^{(+)}, j^{(-)})$ of the two three-dimensional operators behaving as angular momenta (see Appendix B), into the representation $(j^{(-)}, j^{(+)})$, and its matrix representation, corresponding to the coordinate transformation described by Λ, namely $\mathcal{D}^{(J)}(\Lambda)$ into $\overline{\mathcal{D}}^{(J)}(\Lambda) = \mathcal{D}^{(J)}(\Lambda^{-1})$ (see e.g. Sakurai (1985) for explicit form of the $\mathcal{D}$ rotation matrices).

If the system is in a parity eigenstate, then

$$\varphi_P(t, x) = \pm \varphi(t, x) \tag{2.10}$$

which means that the wave function is an even or odd function of the coordinates

$$\varphi(t, -\boldsymbol{x}) = \pm \eta_P^* \, \varphi(t, \boldsymbol{x}) \tag{2.11}$$

and one speaks of a state of defined parity (positive or negative). When considering the usual expansion of a generic wave function in terms of spherical harmonics $Y_l^m(\boldsymbol{x})$, one should note that for those

$$Y_l^m(-\boldsymbol{x}) = (-1)^l \, Y_l^m(\boldsymbol{x}) \tag{2.12}$$

i.e. under parity transformation the wave function of a state with angular momentum quantum number l acquires a sign $(-1)^l$ with respect to the $l = 0$ state.

When parity is a valid symmetry the parity eigenvalue is a good quantum number (conserved) and the stationary states can be chosen as parity eigenstates, although this is by no means mandatory (any function can always be expressed in terms of even and odd functions of the coordinates). In this sense 'parity symmetry' and 'parity conservation' are often used interchangeably.

2.1.2   Quantum fields

In quantum field theory the parity transformation is represented by a unitary operator P in Hilbert space, whose effect on a state of definite momentum $\boldsymbol{p}$, spin projection s, and charge q is expected to be

$$P|\boldsymbol{p}, s, q\rangle = \eta_P |-\boldsymbol{p}, s, q\rangle \tag{2.13}$$

where η_P is a phase factor; considering single particle states at rest ($\boldsymbol{p} = 0$) η_P for a given particle can only possibly depend on s; that it actually does not is seen by observing that P commutes with angular momentum, $[P, \boldsymbol{J}] = 0$, so that acting with it on the equation for the raising and lowering angular momentum operators for spin j

$$J_\pm |\boldsymbol{p}, s, q\rangle \equiv (J_1 \pm iJ_2)|\boldsymbol{p}, s, q\rangle = \sqrt{(j \mp s)(j \pm s + 1)}|\boldsymbol{p}, s \pm 1, q\rangle \tag{2.14}$$

gives

$$J_\pm \eta_P(s)|-\boldsymbol{p}, s, q\rangle = \sqrt{(j \mp s)(j \pm s + 1)}\eta_P(s \pm 1)|-\boldsymbol{p}, s \pm 1, q\rangle \tag{2.15}$$

and thus $\eta_P(s) = \eta_P(s \pm 1)$, so that η_P can only depend on the *kind* of particle but not on its state, consistently with the discussion in Chapter 1.

As is evident from eqn (2.13), single-particle states are in general not eigenstates of parity: a state with momentum $\boldsymbol{p}$ (plane wave) and spin j can be labelled by its *helicity* λ, the eigenvalue of the scalar product $\boldsymbol{S} \cdot \boldsymbol{p}$ of spin and momentum, and one has

$$P|\boldsymbol{p}, \lambda\rangle = \eta_P |-\boldsymbol{p}, -\lambda\rangle \tag{2.16}$$

so that parity reverses helicity. A state with a single particle *at rest* is an eigenstate of P with eigenvalue η_P.

It is postulated that the vacuum state is invariant under parity, i.e. that it is an eigenstate of the parity operator with eigenvalue $+1$:

$$P|0\rangle = |0\rangle \tag{2.17}$$

If the P operator acts on the Lagrangian density as $\mathcal{L}(t,x) \rightarrow P^\dagger \mathcal{L}(t,x)P = \mathcal{L}(t,-x)$, leaving unchanged the Lagrangian $L = \int dx \mathcal{L}(t,x)$, the equations of motion and the commutation relations among field operators, it represents a symmetry of the system.

For a non-Hermitian (charged) scalar field the effect of a parity transformation is

$$P^\dagger \phi(t,x)P = \eta_P\, \phi(t,-x) \qquad P^\dagger \phi^\dagger(t,x)P = \eta_P^*\, \phi^\dagger(t,-x) \tag{2.18}$$

In terms of creation and annihilation operators this means

$$P^\dagger a(p)P = \eta_P\, a(-p) \qquad P^\dagger a^\dagger(p)P = \eta_P^*\, a^\dagger(-p) \tag{2.19}$$

$$P^\dagger b(p)P = \eta_P^*\, b(-p) \qquad P^\dagger b^\dagger(p)P = \eta_P\, b^\dagger(-p) \tag{2.20}$$

The fact that P does not interchange creation and annihilation operators is what allows eqn (2.17) to be written.

For a Hermitian field ($\phi^\dagger = \phi$) $a = b$, and η_P must be real and equal to ± 1.

The quantity η_P is the *intrinsic parity* of the particle described by the field ϕ, and is indeed the same factor introduced in the wave function transformation: considering a physical state $|\psi\rangle$ built out of momenta described by the momentum-space wave function $\varphi(p)$ for the Hermitian field ϕ, a parity transformation gives

$$|\psi\rangle \propto \int d\boldsymbol{p}\, \varphi(\boldsymbol{p})a^\dagger(\boldsymbol{p})|0\rangle \rightarrow P|\psi\rangle \propto \int d\boldsymbol{p}\, \varphi(\boldsymbol{p})Pa^\dagger(\boldsymbol{p})P^\dagger|0\rangle$$

$$= \eta_P \int d\boldsymbol{p}\, \varphi(\boldsymbol{p})a^\dagger(-\boldsymbol{p})|0\rangle = \eta_P \int d\boldsymbol{p}\, \varphi(-\boldsymbol{p})a^\dagger(\boldsymbol{p})|0\rangle$$

as the sign of the integration variable can be changed for an integral extending over all space; the wave function in coordinate space, being the Fourier transform of the one in momentum space, transforms in the same way[10]

$$\varphi(t,x) \rightarrow \varphi_P(t,x) = \eta_P \varphi(t,-x) \tag{2.21}$$

so that the intrinsic parity indeed corresponds to the phase factor (or sign) which the wave function at the origin of coordinates acquires under a parity transformation: $\varphi_P(t,\mathbf{0}) = \eta_P \varphi(t,\mathbf{0})$.

---

[10] Note that the similarity of eqn (2.18) and eqn (2.9) is misleading, since the former involves a field operator, and the latter a wave function, being a representation of a state vector.

The concept of intrinsic parity only plays a role in field theory, where particles can be created and destroyed: as already mentioned, in single-particle quantum mechanics the intrinsic parities are identical in the initial and final states for any physical process and therefore need not be taken into account. In quantum field theory, however, the η_P factor in the field transformation can have observable consequences, as it defines the transformation law for an operator which may appear in interaction terms of the Lagrangian together with other *different* fields.

Note that while the intrinsic parity must clearly be the same for all possible states of a given particle in order to have parity symmetry of the free Lagrangian, its definition has an intrinsic sign ambiguity, since P and $-P$ are defined by the same equations and therefore any of them could be chosen as the parity operator. The factor η_P is however not just related to a sign ambiguity in the definition of the P operator, as seen from the fact that under a parity transformation single-particle states get multiplied by η_P, two-particle states by η_P^2, etc.

While in quantum mechanics P^2 can only introduce a phase, in quantum field theory this operator is not even a multiple of the identity, since it can change the relative phases of states with different particle content; this happens because using the correspondence principle one cannot obtain any constraint on the transformation properties of matrix elements between states with different particle numbers, except for the case of Hermitian fields which are measurable and can therefore have a classical analogue (a super-selection rule is in effect, see Section 5.3).

The *orbital parity* for a single particle in an eigenstate of orbital angular momentum $|l\rangle$ (an infinite superposition of plane waves), for which the momentum-space wave function is proportional to a sum of spherical harmonics $Y_l^m(\mathbf{p})$ with the same l, is expressed by the behaviour of the corresponding wave function φ_l:

$$\varphi_l(t,\mathbf{x}) \equiv \langle 0|\phi(t,\mathbf{x})|l\rangle = (-1)^l \varphi_l(t,-\mathbf{x}) \equiv (-1)^l \langle 0|\phi(t,-\mathbf{x})|l\rangle \qquad (2.22)$$

that is, by the wave function being even or odd in the spatial coordinates.

The *total parity* of the single-particle state $|l\rangle$ with defined orbital angular momentum is thus obtained by considering the wave function of the parity-inverted state

$$\varphi_{lP}(t,\mathbf{x}) \equiv \langle 0|\phi(t,\mathbf{x})P|l\rangle = \langle 0|P^\dagger \phi(t,\mathbf{x})P|l\rangle = \eta_P \langle 0|\phi(t,-\mathbf{x})|l\rangle$$

$$= \eta_P \varphi_l(t,-\mathbf{x}) = \eta_P(-1)^l \varphi_l(t,\mathbf{x}) \qquad (2.23)$$

and is seen to be just the product of the intrinsic and orbital parities.

2.1.3 Vector and tensor fields

The electromagnetic (EM) field $A^\mu(x)$ is Hermitian and has a classical counterpart, so that its intrinsic parity is physical; this cannot be determined a priori, but only by considering its interactions with charged particles. Since experimentally electrodynamics appears to respect parity symmetry, one wants to write a parity-invariant Lagrangian to describe it.

If the charge density ρ is defined to be a scalar quantity, $\rho_P(t, x) = +\rho(t, -x)$, then the invariance of the continuity equation under parity requires the current density $j(t, x)$ to be a polar vector: $j_P(t, x) = -j(t, -x)$, or

$$j_P^\mu(t, x) = j_\mu(t, -x) \qquad (2.24)$$

To have invariance of Maxwell's equations the electric field E must then be a polar vector and the magnetic field B must be an axial vector; that is, under parity[11]

$$E(t, x) \rightarrow E_P(t, x) = -E(t, -x) \qquad B(t, x) \rightarrow B_P(t, -x) = +B(t, -x)$$

In this way the Lagrangian for the free electromagnetic field $(1/2) \int dx [E^2(x) + B^2(x)]$ is left unchanged by the parity transformation. Considering the interaction term in the Lagrangian $-j_\mu(x)A^\mu(x)$, by requiring that electromagnetic interactions are parity symmetric (as experimental evidence implies) the transformation of the electromagnetic field operator has to be

$$P^\dagger A^\mu(t, x) P = +A_\mu(t, -x) \qquad (2.25)$$

This can also be seen classically: in order for Lorentz's force equation $m\, dp/dt = q(E + v \times B)$ (m being the mass and q the electric charge of a particle with velocity v and momentum p) to have the same form after a parity transformation, E must be a polar vector and B an axial one, from which the properties for the classical scalar and vector potentials are deduced: $A^0(t, x) \rightarrow A^0(t, -x)$ and $A(t, x) \rightarrow -A(t, -x)$. Since the potentials are the quantities that are carried over to quantum field theory, one can somehow justify the transformation law for field operators.

The photon polarization four-vectors $\epsilon_\mu^{(s)}(k)$ ($s = 0, 1, 2, 3$, with 3 denoting the one whose space part is along the momentum k) can be chosen in such a way that the set of three-vectors $\epsilon^{(1)}(k), \epsilon^{(2)}(k), k/|k|$ form an orthonormal right-handed system, so that the relation

$$\epsilon^{(3)}(k) = k/|k| = \epsilon^{(1)}(k) \times \epsilon^{(2)}(k) \qquad (2.26)$$

holds. This implies

$$-k/|k| = \epsilon^{(1)}(-k) \times \epsilon^{(2)}(-k) = -\epsilon^{(1)}(k) \times \epsilon^{(2)}(k)$$

---

[11] It should be remarked that this is to some extent an arbitrary choice, in that Maxwell's equations would be invariant under a parity transformation (and thus in accord with the experimental validity of parity symmetry in electromagnetic phenomena) also with the opposite transformations for the fields, if electric charge were considered to be a pseudo-scalar quantity.

which can be obtained by choosing[12] e.g.

$$\epsilon^{(1)}(-\mathbf{k}) = -\epsilon^{(1)}(\mathbf{k}) \qquad \epsilon^{(2)}(-\mathbf{k}) = +\epsilon^{(2)}(\mathbf{k}) \qquad (2.27)$$

It should be remembered that the free electromagnetic field is transverse, and only the $1, 2$ polarization directions are physical. Given eqns (2.27), in order to have $A(t, \mathbf{x}) \rightarrow P^{\dagger} A(t, \mathbf{x}) P = -A(t, -\mathbf{x})$ under parity, one writes for the annihilation operators

$$P^{\dagger} a^{(1)}(\mathbf{k}) P = +a^{(1)}(-\mathbf{k}) \qquad P^{\dagger} a^{(2)}(\mathbf{k}) P = -a^{(2)}(-\mathbf{k}) \qquad (2.28)$$

Defining the circularly polarized states (eigenstates of the spin component S_3 along the direction of motion $\mathbf{k}/|\mathbf{k}|$, helicity) as those corresponding to polarization vectors

$$\epsilon^{(R)}(\mathbf{k}) = -\frac{1}{\sqrt{2}}\left[\epsilon^{(1)}(\mathbf{k}) + i\epsilon^{(2)}(\mathbf{k})\right] \qquad \epsilon^{(L)}(\mathbf{k}) = +\frac{1}{\sqrt{2}}\left[\epsilon^{(1)}(\mathbf{k}) - i\epsilon^{(2)}(\mathbf{k})\right]$$

$$(2.29)$$

describing states characterized by their helicity, with the spin aligned along the direction of motion (right-handed) or opposite to it (left-handed) respectively, one has

$$\epsilon^{(R)}(-\mathbf{k}) = \epsilon^{(L)}(\mathbf{k}) \qquad (2.30)$$

In words: a given sense of rotation gives a clockwise screw motion when coupled to a translation in one direction and a counter-clockwise screw motion when coupled to the opposite one. The corresponding annihilation operators are

$$a^{(R)}(\mathbf{k}) = -\frac{1}{\sqrt{2}}\left[a^{(1)}(\mathbf{k}) - i a^{(2)}(\mathbf{k})\right] \qquad a^{(L)}(\mathbf{k}) = +\frac{1}{\sqrt{2}}\left[a^{(1)}(\mathbf{k}) + i a^{(2)}(\mathbf{k})\right]$$

$$(2.31)$$

and one must therefore have

$$P^{\dagger} a^{(R)}(\mathbf{k}) P = -a^{(L)}(-\mathbf{k}) \qquad P^{\dagger} a^{(L)}(\mathbf{k}) P = -a^{(R)}(-\mathbf{k}) \qquad (2.32)$$

From eqns (2.28) and (2.31) it can be verified that the linearly polarized photon states (those annihilated by $a^{(1, 2)}(\mathbf{k})$, which are not eigenstates of S_3), have their spin orientation unchanged by a parity transformation:

$$P|\mathbf{k}, s = 1\rangle = +|-\mathbf{k}, s = 1\rangle \qquad P|\mathbf{k}, s = 2\rangle = -|-\mathbf{k}, s = 2\rangle \qquad (2.33)$$

---

[12] The alternative choice in which the minus sign is swapped among the two definitions (2.27) is equally valid and leads to some sign changes in the definitions for circular polarization vectors and operators, but of course to no observable difference.

while circularly polarized photons (annihilated by $a^{(R,L)}(k)$, which are eigenstates of S_3), transform into each other under parity:

$$P|k,R\rangle = -|-k,L\rangle \qquad P|k,L\rangle = -|-k,R\rangle \qquad (2.34)$$

which of course mean that the spin orientation is unchanged, but the spin projection on the (reversed) direction of motion (helicity) changes sign under parity. The parity transformation changes a right-handed photon propagating with momentum k (free transverse electromagnetic plane wave in an eigenstate of $S_3 = +1$) into a left-handed photon propagating with momentum $-k$ and vice versa, to within a sign in the state vector. Equation (2.25) corresponds to *odd* intrinsic parity for the photon.[13]

For a generic tensor (and vector) field $T^{\mu\nu\rho\cdots}$ the parity transformation gives

$$P^\dagger T^{\mu\nu\rho\cdots}(t,x)P = \pm T_{\mu\nu\rho\ldots}(t,-x) \qquad (2.35)$$

2.1.4 Electromagnetic multipoles

While negative intrinsic parity was assigned to the photon, the analysis of the parity balance in an electromagnetic transition, involving the total parity of a photon, is somewhat complicated by the fact that a massless particle is involved. For photons it is not possible to treat the intrinsic (spin) and orbital angular momenta independently in a consistent way, since the spatial dependence of the wave function is constrained by additional conditions such as $\nabla \cdot A(x) = 0$ (or $k \cdot A(k) = 0$ in the momentum representation), so that not all the components of the vector field (which represent the spin components) are independent. Only the *total* angular momentum is a well-defined concept for photons.

The electromagnetic radiation with total angular momentum $J = 1$ is called *dipole* radiation, the one with $J = 2$ *quadrupole* radiation, and in general one speaks of 2^J-pole radiation. The total angular momentum J can still be considered formally as the composition[14] of the photon spin $S = 1$ and an orbital angular momentum L, which determines the rotation properties of the wave function and can have eigenvalues J or $J \pm 1$; in the first case the radiation is called *magnetic*, and its (total) parity is $\eta_P(MJ) = (-1)(-1)^L = (-1)^{J+1}$ (magnetic dipole M1, quadrupole M2, etc., with $J^P = 1^+, 2^-, \ldots$), while in the second case it is called *electric* and its parity is $\eta_P(EJ) = (-1)(-1)^L = (-1)^J$ (electric dipole E1, quadrupole E2, etc., with $J^P = 1^-, 2^+, \ldots$). The naming originates from the fact that in the long-wavelength limit electric multipole transitions are largely determined by the electric multipoles

---

[13] In an analogous way one could assign negative intrinsic parity to other intermediate gauge bosons with vector interactions such as the gluon, although in this case its non-observability as a free particle makes such assignment of limited use. By no means could one assign a definite intrinsic parity to the weak gauge bosons $W^\pm$ and Z, since weak interactions do not respect parity symmetry.

[14] Just as a convenient approach to classify electromagnetic radiation properties.

TABLE 2.1 Lowest-order EM multipoles

J		L		P
0	Forbidden			
1	Dipole	E1	0, 2	−1
1	Dipole	M1	1	+1
2	Quadrupole	E2	1, 3	+1
2	Quadrupole	M2	2	−1

of the sources, and conversely for the magnetic ones. The parity of a multipole photon is a convenient way of keeping track of the parity changes in EM transitions.

Table 2.1 summarizes the classification of some electromagnetic multipoles in terms of their angular momentum and parity. Note that for $J = 0$ the only possibility appears to be $L = J + 1 = 1$ (electric), with parity $\eta_P = +1$.

The above enumeration of states does not take into account the transversity condition of A: one state, corresponding to the longitudinal polarization case $A(k) = A_L(k) \propto k$, is not allowed and must be discarded. Following Berestetskiǐ *et al.* (1978) we note that such a state, with momentum-space wave function of the generic form $A_L(k) = kf(k)$ (with arbitrary f) behaves as a scalar under rotations,[15] and therefore corresponds to 'spin 0' and a total angular momentum J determined by the spherical harmonic components of f. The parity of this state is also linked to the behaviour of f since from eqn (2.25) $P^\dagger A_L(k)P = - A_L(-k) = - (-k)f(-k) = (-1)^J kf(k)$: it is therefore one of the states with $\eta_P = (-1)^J$ which must be discarded, and this leaves just two states with opposite parities for each $J \neq 0$, and none for $J = 0$; there is no real photon with zero total angular momentum. Also, in terms of angular momentum composition, by choosing as the quantization axis the direction of propagation of the photon, along which the orbital angular momentum is zero, the fact that the helicity of a photon is always ±1 (and never 0) implies that its total angular momentum can never vanish.

It should be noted that the space inversion properties of the electromagnetic field A match those of the electric field E and are opposite to those of the magnetic field B: i.e. according to eqn (2.25) A is a polar vector as E. Since all these three vectors satisfy a wave equation as $[\nabla^2 - (\omega/c)^2]A = 0$, and the wave operator (being of second order) is invariant under parity, if $A(x)$ is a solution then also $A(-x)$ is, and so are the functions of definite parity $A(x) \pm A(-x)$. For an electromagnetic field which is an even or odd function of the space coordinates, under parity

$$A(x) \rightarrow -A(-x) = \mp A(x)$$

$$E(x) \rightarrow -E(-x) = \mp E(x) \qquad B(x) \rightarrow +B(-x) = \pm B(x)$$

[15] For a rotation R, if $k \rightarrow Rk$ then $A_L(k) \rightarrow R\, A_L\, (R^{-1}k) = R\, (R^{-1}k)\, f(R^{-1}k) = k\, f(R^{-1}k)$, and only the argument is changed, and not also the components in an explicit way.

The interaction Hamiltonian of a current with the electromagnetic field is essentially $j \cdot A$, where the electromagnetic current j is a polar vector. Imposing the parity symmetry of electromagnetic interactions, it can be seen immediately that if A changes sign under parity (considering also the change in its argument) as j does, then the above expression is a scalar, so that it has non-zero matrix elements only between states with the same parity; the electromagnetic transition therefore does not change the parity, and when interpreted as the absorption or emission of a photon it means that the photon has parity $+1$. On the other hand, if A does not change sign under parity, the transition induced by the pseudo-scalar term enforces a change of parity, in other words, the emitted (or absorbed) photon has parity -1. The total photon parity is thus seen to be *opposite* to the parity of the electromagnetic field A, or the same as the parity of B. As a consequence, for an electric 2^J-pole transition, it is the *magnetic* field that transforms under parity as the photon:

$$A(x) \to (-1)^{J+1} A(x) \quad E(x) \to (-1)^{J+1} E(x) \quad B(x) \to (-1)^J B(x) \tag{2.36}$$

$$\text{(electric } 2^L\text{-pole transition,} \quad L = J \pm 1) \tag{2.37}$$

and the opposite happens for a magnetic 2^L-pole transition:

$$A(x) \to (-1)^J A(x) \quad E(x) \to (-1)^J E(x) \quad B(x) \to (-1)^{J+1} B(x) \tag{2.38}$$

$$\text{(magnetic } 2^L\text{-pole transition,} \quad L = J) \tag{2.39}$$

The actual functional forms of the electromagnetic fields for the different multipole transitions are expressed in terms of the so-called vector spherical harmonics, eigenfunctions of J^2, L^2 and J_z (see, e.g. Messiah (1961)).

For small energies, that is wavelengths λ of the emitted radiation larger than the size R of the emitting system,[16] only the lowest multipole emission (usually $L = 0$, electric dipole) is important, implying that transitions only occur between states of opposite parity. This is the empirical *Laporte's rule* of atomic physics (Laporte, 1924), known before the advent of quantum mechanics and later explained by Wigner[17] (1927) as a consequence of parity symmetry; its experimental verification indeed confirms the validity of parity symmetry in atomic transitions at the level $(R/\lambda)^2 \sim 10^{-8}$.

2.1.5 Spinor fields

Since the Dirac equation is of first order in the space coordinates, the parity-transformed spinor is not expected simply to acquire a possible sign (or phase) change; even if the potential part of the Hamiltonian is parity symmetric, the kinetic part is not. Indeed, in order to keep the Dirac equation invariant under

---

[16] For a nuclear decay $R \sim 10^{-15}$ m and even for a 10 MeV photon $\lambda \sim 10^{-14}$ m.
[17] The father of discrete symmetries wrote in 1927: 'Only rarely will one be able to use [parity] since it has only two eigenvalues (± 1) and has therefore too little predictive power'.

parity a transformation of the Dirac matrices $\gamma^0 \to \gamma^0$, $\gamma^k \to -\gamma^k$ is necessary, and this requires the four spinor components to transform among themselves. This is expressed by writing

$$P^\dagger \psi(t,\boldsymbol{x})P = \eta_P S_P\, \psi(t,-\boldsymbol{x}) \qquad P^\dagger \overline{\psi}(t,\boldsymbol{x})P = \eta_P^* \,\overline{\psi}(t,-\boldsymbol{x})S_P^\dagger \qquad (2.40)$$

where the 4×4 matrix S_P satisfies the condition

$$S_P^{-1}\gamma^\mu S_P = \gamma_\mu \qquad (2.41)$$

The above transformation on the γ matrices leaves their defining anti-commutation relations (A.2) unchanged, and Pauli's fundamental theorem (Appendix A) therefore ensures that the matrix implementing it exists and is unique up to a multiplicative factor. Such matrix is indeed

$$S_P = \gamma^0 \qquad (2.42)$$

in any representation of the Dirac matrices. In the usual Dirac representation the effect of S_P is that of changing sign to the two lower components of the spinor, which in the non-relativistic limit are the 'large' components for negative-energy states.

The occurrence of a matrix transforming the field components among themselves also means that the required invariance under a double reflection does not necessarily imply $\eta_P^2 = 1$, so that η_P could in general be a complex phase. This is possible since a spinor field is not a measurable quantity, and its sign has no physical meaning.

If for spinor fields one only imposes the weaker condition $P^2 = \pm 1$ (in analogy with the case of rotations, for which a 2π rotation changes the sign of the field and only a 4π rotation restores it), then the phase factors are restricted to $\eta_P = \pm 1, \pm i$ (Yang and Tiomno, 1950). The choice of real or imaginary phase factors would correspond to two different choices for the transformation operators of the orthochronous Lorentz group (see Appendix B); for each such choice there exist two different operators corresponding to the parity transformation, differing by a sign. However, since P^2 is assumed to commute with all observables, it gives origin to a super-selection rule (Section 5.3), forbidding transitions between states with an even and an odd number of fermions with 'imaginary intrinsic parity'.

Anyway, while the above condition is also rather arbitrary, it can be shown (Feinberg and Weinberg, 1959) that under rather general assumptions it is always possible to adopt real η_P factors (that is ± 1) for both bosons and fermions alike (see Section 5.3 for further discussion).

From the action of γ_0 on the free spinors

$$\gamma^0 u(\boldsymbol{p},s) = u(-\boldsymbol{p},s) \qquad \gamma^0 v(\boldsymbol{p},s) = -v(-\boldsymbol{p},s) \qquad (2.43)$$

the corresponding transformation for the operators is obtained

$$P^\dagger a(\boldsymbol{p}, s)P = \eta_P\, a(-\boldsymbol{p}, s) \qquad P^\dagger a^\dagger(\boldsymbol{p}, s)P = \eta_P^*\, a^\dagger(-\boldsymbol{p}, s) \qquad (2.44)$$

$$P^\dagger b(\boldsymbol{p}, s)P = -\eta_P^*\, b(-\boldsymbol{p}, s) \qquad P^\dagger b^\dagger(\boldsymbol{p}, s)P = -\eta_P\, b^\dagger(-\boldsymbol{p}, s) \qquad (2.45)$$

It is important to note that the relative (intrinsic) parity of a spin 1/2 particle f and its antiparticle $\bar{f}$ is always -1, independently of the value of η_P (with $|\eta_P|^2 = 1$): suppressing spin and momentum labels

$$P|f\bar{f}\rangle = Pa^\dagger b^\dagger|0\rangle = -\eta_P^*\eta_P\, a^\dagger b^\dagger|0\rangle = -|f\bar{f}\rangle \qquad (2.46)$$

Since the net fermion number is conserved and fermions cannot be created or annihilated singly, the knowledge of such a relative phase is sufficient to describe any physical process, while the parity of an unpaired fermion is a quantity which never appears in physical quantities.

Another way of seeing the above fact is by observing that the (time-independent) wave function for an electron-positron state (with momenta and spin projections $\boldsymbol{p}, s$ and $\overline{\boldsymbol{p}}, \bar{s}$ respectively) is

$$\varphi(e^-, \boldsymbol{p}, s, \boldsymbol{x}; e^+, \overline{\boldsymbol{p}}, \bar{s}, -\boldsymbol{x}) = \langle 0|\psi(x)\overline{\psi}(-x)|e^-(\boldsymbol{p}, s), e^+(\overline{\boldsymbol{p}}, \bar{s})\rangle$$

$$= u(\boldsymbol{p}, s)\,\overline{v}(\overline{\boldsymbol{p}}, \bar{s})\, e^{i(\boldsymbol{p} - \overline{\boldsymbol{p}})\cdot\boldsymbol{x}}$$

and when the two particles are in a $L = 0$ angular momentum state (with even orbital parity, rather than the above plane waves which are not parity eigenstates), still the wave function contains the product of 'large' and 'small' spinor components, which changes sign for $\boldsymbol{p} \to -\boldsymbol{p}$.

In the single-particle Dirac theory one can consider the relative parity between the vacuum state (filled Dirac sea) and the same state with a single negative energy electron removed (which is interpreted as a positron state); in such theory therefore (with the parity of the vacuum defined to be positive) the intrinsic parity of the positron is equal to that of a negative-energy electron, since parity is multiplicative.

For massless particles with half-integer spin the free field Lagrangian is also invariant if different phase factors are adopted for the states with positive and negative helicities, since in this case such two sets of states can be considered as independent (uncoupled) particles.

Considering the effect of S_P on the complete set $\{\Gamma\}$ of 4×4 matrices obtained from gamma matrices:

$$S_P \{1, \gamma^\mu, \sigma^{\mu\nu}, \gamma^\mu\gamma_5, i\gamma_5\} S_P^{-1} = \{1, -\gamma_\mu, -\sigma_{\mu\nu}, -\gamma_\mu\gamma_5, -i\gamma_5\} \qquad (2.47)$$

one obtains the parity transformation properties of the bilinear covariants of eqns (B.21):

$$S(t, \boldsymbol{x}) \to +S(t, -\boldsymbol{x}) \tag{2.48}$$

$$V^\mu(t, \boldsymbol{x}) \to +V_\mu(t, -\boldsymbol{x}) \tag{2.49}$$

$$T^{\mu\nu}(t, \boldsymbol{x}) \to +T_{\mu\nu}(t, -\boldsymbol{x}) \tag{2.50}$$

$$A^\mu(t, \boldsymbol{x}) \to -A_\mu(t, -\boldsymbol{x}) \tag{2.51}$$

$$P(t, \boldsymbol{x}) \to -P(t, -\boldsymbol{x}) \tag{2.52}$$

Note that for bilinears built from two different fields $\overline{\psi}_2 \Gamma \psi_1$ a factor $\eta_P^*(2)\eta_P(1)$ also appears in the transformed expression so that, if the two fermions had opposite intrinsic parities, it can happen that e.g. $S = \overline{\psi}_2 \psi_1$ actually behaves as a *pseudo-scalar*, etc.

2.2 Parity symmetry

A system is said to be parity symmetric if there exists at least a set of phase factors η_P for all fields such that the Hamiltonian is invariant under parity transformations. If, for each such possible choice, the relative parity of any two particles is the same, then the concept of intrinsic parity has a physical meaning.

An interaction Hamiltonian of the form

$$H_{\text{int}} = \phi_1^\dagger(t, \boldsymbol{x})\overline{\psi}_2(t, \boldsymbol{x})(c + c'\gamma_5)\psi_3(t, \boldsymbol{x}) + \text{H.c.}$$

(where ϕ denotes a scalar or pseudo-scalar field, ψ a spinor field and c, c' are numbers), suited to describe e.g. the decay $\Lambda \to p\pi^-$, transforms under parity into

$$H_{\text{int}} \to \eta_1^* \eta_2^* \eta_3 \, \phi_1^\dagger(t, -\boldsymbol{x})\overline{\psi}_2(t, -\boldsymbol{x})(c - c'\gamma_5)\psi_3(t, -\boldsymbol{x}) + \text{H.c.}$$

If $c' = 0$ then H_{int} is parity invariant with $\eta_1^* \eta_2^* \eta_3 = +1$, while if $c = 0$ invariance is obtained if $\eta_1^* \eta_2^* \eta_3 = -1$. If both $c \neq 0$ and $c' \neq 0$ then it is impossible to choose intrinsic parities in such a way as to leave H_{int} invariant under parity, and in this case one says that the interaction violates parity symmetry.

Similarly, for the generic fermionic interaction Lagrangian (B.29) the interaction is parity symmetric if all the $c_i' = 0$.

The parity symmetry of a theory is therefore expressed by the fact that the Hamiltonian is a scalar quantity; the expectation value of the Hamiltonian must be a scalar quantity and, since the parity of the states cancels in it, pseudo-scalar terms such as $\boldsymbol{S} \cdot \boldsymbol{p}$, $\boldsymbol{S} \times \boldsymbol{p}$ or $\boldsymbol{p}_1 \cdot (\boldsymbol{p}_2 \times \boldsymbol{p}_3)$ (where $\boldsymbol{p}$ is a momentum vector and $\boldsymbol{S}$ a spin vector) are not allowed and cannot appear in cross sections or transition rates. This can be easily understood since any relation involving a product of a polar and an axial vector, such as $\boldsymbol{S} \cdot \boldsymbol{p} = a$ (a being a number) cannot be preserved under parity, since the two vectors transform differently (just as a relation between the components of a vector and a set of numbers is not invariant under generic rotations).

For example, when considering the Hamiltonian describing the scattering of a particle with spin on a spin zero (or unpolarized) target, irrespective of final state polarization, the available vectors are p_i, p_f (initial and final momenta in the centre of mass) and S_i (initial spin), and the only scalar quantity linear in all of them is $p_i \times p_f \cdot S_i$. This is different from zero only for polarization orthogonal to the scattering plane: the cross section therefore cannot depend on the presence of initial state polarization in the scattering plane.

2.2.1   Intrinsic parities of elementary particles

The concept of intrinsic parity is only relevant when considering transitions involving different particles, and it can be defined in a non-ambiguous way only if all their interactions respect parity symmetry.

The assignment of intrinsic parities to particles is a way of keeping track of how orbital parities change when transitions take place among different particles, so it is an experimental issue. This assignment is attempted in such a way as to have conservation of parity in each process, and this turns out to be actually possible for transitions induced by the strong or EM interactions, but not for those induced by weak interactions.

Clearly, if parity were not a symmetry of the most relevant interactions (e.g. electromagnetism) it would hardly have been introduced as a concept at all; the fact that it has been possible to attribute definite parities to states and particles indeed indicates that parity is a fairly good symmetry in Nature.

If parity symmetry was an exact symmetry of Nature, all particles should be eigenstates of parity, or members of degenerate doublets of states transforming into each other by P; indeed a grouping of strongly interacting particles (hadrons) into multiplets with definite spin and intrinsic parity appeared to be possible, since their properties are largely defined by interactions which seem to respect parity symmetry. Such a grouping remains approximately valid even if the weak decays of such particles do not respect such symmetry.

When the intrinsic parity of some particles A, B is known (or chosen by convention), parity-conserving reactions in which such particles are involved can be used to determine the intrinsic parities of other particles (relative to the parity of vacuum), provided they can be created *singly*, e.g. $AB \rightarrow ABC$; this excludes the possibility of determining the intrinsic parity for charged particles (charge conservation) and for fermions (conservation of angular momentum). In general for *truly neutral* particles (particles without any non-zero internal additive quantum numbers, such as π^0, γ), described by Hermitian fields, the intrinsic parity has a physical meaning, as such particles can be created and destroyed singly.

The *relative* parity of two states can instead be determined by considering the processes in which they are involved: for example $AB \rightarrow AC$ allows one to determine the relative intrinsic parity of A and C, if the process is known be parity-conserving. Note that the relative parity of two states can be measured only if both have the same values for all conserved charges, i.e. if they can directly interact or if they can couple to the same field through a parity-conserving interaction (see Section 5.3).

Particles for which the intrinsic parity η_P and the spin J are related by $\eta_P = (-1)^J$ are often said to have *natural* intrinsic parity; these quantum numbers are usually written as J^P, where the superscript actually indicates the value of η_P), so that for these particles $J^P = 0^+, 1^-, 2^+$, etc.

For a system of two particles with intrinsic parities $\eta_P(1), \eta_P(2)$ in a state with angular momentum L the parity is

$$\eta_P = \eta_P(1)\eta_P(2)(-1)^L \tag{2.53}$$

and for a system of three particles with intrinsic parities $\eta_P(1), \eta_P(2), \eta_P(3)$, in a state in which L is the orbital angular momentum of two of the particles in their centre of mass, and l that of the third particle with respect to the centre of mass of the first two, the parity is

$$\eta_P = \eta_P(1)\eta_P(2)\eta_P(3)(-1)^L(-1)^l \tag{2.54}$$

The eigenvalues of total angular momentum and parity for systems of several particles are limited to some combinations:

- Two pions: $J^P = 0^+, 1^-, 2^+, 3^-, \cdots$, because $\eta_P = (-1)^L$.

- Three pions: any combination except $J^P = 0^+$; for $J = 0$ the two angular momenta involved must be equal, $L = l$ and therefore $\eta_P = -1$.

- Two photons: any combination except $J^P = 3^-, 5^-, 7^-, \ldots$ (odd angular momentum and negative parity) because of the Landau-Yang theorem (Section 2.2.2).

- Fermion-antifermion: $\eta_P = (-1)^{L+1}$.

- Boson-antiboson: $\eta_P = (-1)^L$.

Note that for integer spin fields the relative intrinsic parity of particle and antiparticle states is even (same intrinsic parity), while for half-integer spin fields it is odd (opposite intrinsic parity), as will be shown in Chapter 3. In particular this explains why mesons and anti-mesons can be grouped in a single spin-parity multiplet, while baryons and anti-baryons cannot.

Positronium is the bound state of the e^+e^- system, which can decay through the annihilation of e^+ and e^- into photons; this occurs more easily when the two particles are close to each other, therefore favouring an S-wave ($L = 0$). The two spins can be either aligned ($S = 1$, triplet state $^3S_1$, called *ortho-positronium*) or anti-aligned ($S = 0$, singlet state $^1S_0$, called *para-positronium*). Positronium decays into either two or three photons, the latter case having a much lower probability due to an extra factor α in the squared amplitude and to the reduced phase space: $\tau(2\gamma) \simeq 0.12$ ns, $\tau(3\gamma) \simeq 140$ ns.

The state obtained in the two-photon annihilation must be expressed in terms of the centre of mass momentum k of one photon and the two polarization vectors ϵ_1 and ϵ_2. The Hamiltonian describing the annihilation must be linear in the two polarization vectors (as the photon annihilation operators are), and since $k \cdot \epsilon_{1,2} = 0$ the following forms are possible:

$$\epsilon_1 \cdot \epsilon_2 \quad \text{(scalar)}$$

$$\epsilon_1 \times \epsilon_2 \cdot k \quad \text{(pseudo-scalar)}$$

$$(\epsilon_1 \cdot \epsilon_2) \, k \quad \text{(vector)}$$

$$\epsilon_1 \times \epsilon_2, (\epsilon_1 \times \epsilon_2 \cdot k) \, k \quad \text{(pseudo-vector)}$$

Each of the terms above can appear multiplied by some function of $|k|$ in the decay amplitude. Bose–Einstein (B–E) symmetry requires the state to be symmetric under the exchange of the two photons ($\epsilon_1 \leftrightarrow \epsilon_2, k \to -k$), thus allowing only the first term above, and therefore excluding the 2γ decay of the triplet state.

The singlet state must be described by one of the first two expressions depending on its parity (assumed to be conserved in this electromagnetic decay) and thus on the relative intrinsic parity of e^+ and e^-; the first one results in parallel polarization of the two photons and the second one in orthogonal polarization (see discussion in Section 2.2.2 for π^0 decay). Experimental proof of negative relative intrinsic parity of e^+ and e^- was obtained (Wu and Shaknov, 1950) by verifying that the polarization planes of the photons in $e^+e^- \to 2\gamma$ decay are indeed orthogonal, exploiting the strong dependence of the angular distribution for Compton scattering on the photon polarization direction.

Positronium is not an attractive system for parity violation searches, as possible parity-violating effects are strongly hindered by the constraints induced by CP symmetry (see Chapter 6); on the other hand, this system is interesting for testing charge conjugation symmetry, as will be discussed in Chapter 3.

2.2.2 The neutral pion

The first step in the assignment of intrinsic parities to bosons consists in choosing the intrinsic parity of the photon to be negative, as suggested by the polar vector nature of the electric field ($\eta_P = +1$ in eqn (2.25)). Assuming (consistently with observations) that electromagnetic processes conserve parity, the parity of states containing only photons is thus defined.

The decay $\pi^0 \to \gamma\gamma$ of the spinless neutral pion into two spin-1 photons occurs in a time $\sim 10^{-16}$ s, indicating that it is driven by electromagnetic interactions, which are experimentally known to be symmetric under parity with high accuracy. The interaction Hamiltonian describing the decay must therefore be invariant under parity, and it must contain (linearly) the electromagnetic fields of the two photons A_1, A_2 and the scalar (or pseudo-scalar) pion field ϕ; given the polar vector nature of A, and the fact that the only other vector which can be defined (in the π^0 rest frame) is the momentum k of one of the photons, the only possibilities are

$$A_1 \cdot A_2 \, \phi \quad \text{and} \quad A_1 \times A_2 \cdot k \, \phi$$

The first term is a scalar quantity if the pion field ϕ is scalar, and is non-zero for parallel polarization of the two photons; the second requires instead a pseudo-scalar pion field, and corresponds to perpendicular photon polarizations. A joint photon polarization measurement would therefore allow the determination of the parity of the π^0 (Yang, 1950).

The above result can be obtained by considering two-photon decay in full generality. For two photons with momenta k and $-k$, one can define the states (recall eqn (2.31))

$$|RR\rangle \equiv a^{(R)\dagger}(k)a^{(R)\dagger}(-k)|0\rangle \qquad |LL\rangle \equiv a^{(L)\dagger}(k)a^{(L)\dagger}(-k)|0\rangle \qquad (2.55)$$

$$|RL\rangle \equiv a^{(R)\dagger}(k)a^{(L)\dagger}(-k)|0\rangle \qquad |LR\rangle \equiv a^{(L)\dagger}(k)a^{(R)\dagger}(-k)|0\rangle \qquad (2.56)$$

The total angular momentum for a system of two photons cannot be $J = 1$ because of angular momentum conservation and Bose–Einstein symmetry; this is the so-called *Landau–Yang theorem* (Landau, 1948; Yang, 1950): since the helicity of photons can only be ± 1 and orbital angular momentum has zero projection along the direction of motion, the component of total angular momentum along the k direction $J_z = J \cdot k/|k|$ for a state of two photons with $J < 2$ should be 0, i.e. the two photons must have the same helicity, and by rotating the system by an angle π around an axis orthogonal to k they are interchanged, thus leaving the wave function unchanged (B–E symmetry). A state with $(J, J_z) = (1, 0)$ would however change sign under such a rotation and is therefore not allowed.

Another way of understanding this result is to consider that the wave function for the final state (in the centre of mass system) must be constructed out of the two photon polarization vectors $\epsilon_{1,2}$ (or $A_{1,2}$) and the momentum k. Of the three independent vector (spin 1) combinations of these, $\epsilon_1 \times \epsilon_2$ and $(\epsilon_1 \cdot \epsilon_2) k$ are anti-symmetric under the exchange of the two photons ($\epsilon_1 \leftrightarrow \epsilon_2$ and $k \rightarrow -k$) and thus not allowed by B–E symmetry, while $k \times (\epsilon_1 \times \epsilon_2) = \epsilon_1(k \cdot \epsilon_2) - \epsilon_2(k \cdot \epsilon_1) = 0$ because of the transverse nature of the photon.

Given the above, the $|RL\rangle$ and $|LR\rangle$ states must have total angular momentum $J \geq 2$, while $|RR\rangle$ and $|LL\rangle$ can also have $J = 0$. The states $|RL\rangle$ and $|LR\rangle$ are parity eigenstates with eigenvalue $+1$, as can be verified explicitly by using eqn (2.34), while the states $|RR\rangle$ and $|LL\rangle$ are not parity eigenstates, but their linear combinations

$$(|RR\rangle + |LL\rangle)/\sqrt{2} \propto \left[a^{(1)\dagger}(k)a^{(1)\dagger}(-k) - a^{(2)\dagger}(k)a^{(2)\dagger}(-k)\right]|0\rangle$$

$$(|RR\rangle - |LL\rangle)/\sqrt{2} \propto \left[a^{(1)\dagger}(k)a^{(2)\dagger}(-k) + a^{(2)\dagger}(k)a^{(1)\dagger}(-k)\right]|0\rangle$$

do, with eigenvalues $+1$ and -1 respectively.

From the above, one can infer the relative orientation of the photon polarization planes (formed by the direction of polarization and the flight direction) for the two states with zero spin component along the direction of flight ($J_z = 0$):

$$(|RR\rangle + |LL\rangle)/\sqrt{2} \propto (A_1 \cdot A_2)|0\rangle \qquad \text{parallel} \quad (\|) \qquad (2.57)$$

$$(|RR\rangle - |LL\rangle)/\sqrt{2} \propto [(A_1 \times A_2) \cdot k]|0\rangle \qquad \text{orthogonal} \quad (\perp) \qquad (2.58)$$

The parity η_P and helicity J_z eigenvalues, the status of being or not being an eigenstate of a π rotation $R(\pi)$ around an axis orthogonal to the direction of motion (say x), and the relative orientation (parallel or orthogonal) of the photon polarization planes for the different states are

State	η_P	J_z	$R(\pi)$ eigenst.	Polarization planes
$\|RL\rangle$	$+1$	$+2$	no	$\|$ or $\perp$
$\|LR\rangle$	$+1$	-2	no	$\|$ or $\perp$
$(\|RR\rangle + \|LL\rangle)/\sqrt{2}$	$+1$	0	yes	$\|$
$(\|RR\rangle - \|LL\rangle)/\sqrt{2}$	-1	0	yes	$\perp$

Note that a particle with spin J and third component J_z behaves under rotations as the spherical harmonic $Y_J^{J_z}$, and if $J_z = 0$ it is an eigenstate of a π rotation around the x axis WITH eigenvalue $(-1)^J$.

Summarizing, for a two-photon decay of a particle with spin-parity J^P the parity symmetry of electromagnetic interactions enforces the following selection rules:

J^P	Decay	Polarization planes
0^+	allowed	$\|$
0^-	allowed	$\perp$
$1^\pm$	not allowed	
$2^+, 4^+, \ldots$	allowed	$\geq 50\%\ \|,\ \leq 50\%\ \perp$
$2^-, 4^-, \ldots$	allowed	$\perp$
$3^+, 5^+, \ldots$	allowed	$50\%\ \|,\ 50\%\ \perp$
$3^-, 5^-, \ldots$	not allowed	

The cases in which both polarization plane orientations are allowed can be readily understood by considering that e.g. a $J^P = 3^+$ state can only decay to $|RL\rangle$ or $|LR\rangle$ (the other two being prohibited by either parity symmetry or angular momentum conservation and Bose–Einstein symmetry), while a $J^P = 2^+$ state can also decay to the symmetric combination of $|RR\rangle$ and $|LL\rangle$ in addition.

The spinless neutral pion would thus decay into one or the other of the two photon states (2.57) (2.58) depending on its parity. By measuring the circular polarization of one of the two photons, a right or left polarization would be obtained with equal probability, independently of the pion parity. The correlated measurement of the circular polarization state of both photons at the same time would yield no more information on the pion parity, since the probability of finding both photons with right-handed polarization is 1/2, and the same is true for left-handed polarization, while they can never have opposite circular polarizations, but all of this is common to both parity eigenstates. One concludes that information on the neutral pion parity can be obtained from the polarization of the decay photons only by measuring the state of *linear* polarization of both photons.

The determination of the neutral pion parity

The direct measurement of the relative polarization of low energy photons (67 MeV for a π^0 decaying at rest) is experimentally challenging; however, the photon polarization plane is highly correlated to the plane of an e^+e^- pair which it can produce, e.g. when traversing a dense material; unfortunately the angle between the two charged particles is so small (of the order of $m_e/E_\gamma \lesssim 0.01$ rad) that their scattering in the material destroys the correlation. In the rare ($BR \simeq 3{\cdot}10^{-5} \sim O(\alpha^2)$) so-called 'double Dalitz' decay $\pi^0 \to e^+e^-e^+e^-$, which can occur in vacuum, this angle can be larger (~ 0.1 rad) and the correlation, which is expected to be maintained in the decay, could actually be measured.

In an experiment in the early 1960s (Samios *et al.*, 1962), a beam of π^- produced as secondaries from the 390 MeV Cyclotron at Columbia University was slowed down by energy losses (atomic ionization and excitation) in a polyethylene absorber, and then stopped in a hydrogen bubble chamber within a magnetic field. Neutral pions were produced by the rather frequent $\pi^-p \to n\pi^0$ 'charge-exchange' reaction, occurring for $\sim 60\%$ of the negative pions stopping in hydrogen. The produced π^0 are mono-energetic (137.8 MeV) and this constraint helps in the transformation to the centre of mass system. Scanning about 800,000 pictures containing 8 million π^0 decays, 206 double Dalitz decays were detected, out of which 146 had well-measured track momenta (occurring far enough from the edges of the chamber that all tracks had a length within the chamber sufficient for a good measurement of their curvature) and opening angles (large enough to properly define the decay plane), and were used for the measurement.

First, the validity of the theoretical model (based on QED) describing the decay was verified, by comparing the experimental distributions for the invariant mass of the e^+e^- pairs and the energy asymmetry of the two particles with predictions, in order to gain confidence in the assumed correlation of the decay planes with the photon polarizations.

Since the bubble chamber is a uniform detector with 4π acceptance for a decay occurring inside it, all decay configurations are measured with the same efficiency and no correction is required to obtain the true distribution of relative decay plane orientations. Such distribution of events as a function of the angle θ between the two decay planes was fitted to the predicted shape $dN/d\theta = 1 + \alpha\cos(2\theta)$ giving $\alpha = -0.12 \pm 0.15$ in agreement with the theoretically predicted slope $\alpha = -0.18$ for a pseudo-scalar particle and two standard deviations away from the opposite value predicted for a scalar particle. It should be noted that the slope parameter α actually depends on the measured event parameters (track direction and momenta), so that the measured (and predicted) values are actually average values over the different event configurations. Such functional dependence is known from theory and therefore a value of α corresponding to the decay parameters of each event can be computed from the measured track parameters, thus allowing another approach which uses more information: for the purpose of discriminating among two opposite signs of the slope parameter, the event configurations corresponding to larger $|\alpha|$ have more statistical power, so that by properly weighting each event a more powerful discriminating quantity can be evaluated, which is actually the (weighted) average of α itself computed for each event. The result of such analysis favoured the pseudo-scalar assignment by 3.6 standard deviations.

2.2.3   The charged pion

The next step consists in assuming that the charged pions, belonging to the same isospin triplet[18] as the neutral pion, have the same intrinsic parity as the latter: this is consistent with the experimental fact that strong interactions (for which isospin is a good symmetry, so that the three pions behave in the same way) respect parity symmetry. Strictly speaking, however, this is a new arbitrary assumption, as the intrinsic phases (and intrinsic parities) of state vectors for particles with different charges cannot be directly compared due to the presence of a super-selection rule (Wick *et al.*, 1952) (Section 5.3).

   An independent experimental argument for the pseudo-scalar nature of charged pions is the existence of the process $\pi^- d \rightarrow nn$ for stopped pions, (where d is the deuteron), as we now discuss (Ferretti, 1946). The ground state of the deuteron is[19] $^3S_1$ ($L = 0, S = 1, J = 1$), therefore its parity is $\eta_P(p)\eta_P(n)(-1)^L = +1$ if one assumes that the intrinsic parities of the proton and neutron are the same

---

[18]   Isospin or isobaric spin is the internal quantum number, formally analogous to angular momentum, which was introduced to consider the proton and the neutron as two 'states' of a single kind of particle (the nucleon). Its usefulness arises from the fact that strong interactions appear to be insensitive to it, thus resulting in an isospin symmetry (see e.g. Gibson and Pollard (1976)). Such (approximate) symmetry is now understood as being due to the smallness of the masses of the u and d quarks (appearing in ordinary matter) with respect to those of other quarks.

[19]   We recall the spectroscopic notation $^{2S+1}L_J$, where S, L, J are respectively the spin, orbital and total angular momentum and the letters $S, P, D, F, \ldots$ stand for $L = 0, 1, 2, 3, \ldots$ From a relativistic point of view the spin and orbital angular momentum degrees of freedom are coupled, and only their sum J is actually conserved; however such coupling can be small enough for L and S to be considered as good quantum numbers to an excellent approximation.

(with an argument similar to the one above, based on the parity symmetry of strong interactions).[20] A negative pion at rest in matter forms rapidly (in less than a ns) a mesic atom in the ground state, in which the pion orbits around the positive nucleus; the same does not happen for positive pions, which are repelled by the nuclei. If the π^- is captured in an S-wave state ($L = 0$) the total angular momentum is just $J = 1$ from the deuteron spin, since the pion is spinless. The Pauli principle requires the state of two indistinguishable neutrons to be antisymmetric under their interchange, which is equivalent to the exchange of the positions in their centre of mass (the parity transformation, which introduces a factor $\eta_P(n)^2 (-1)^L = (-1)^L$) and of the spins (left unchanged by the parity transformation, as they are axial vectors), which in this case introduces a factor $(-1)^{S+1}$ (as can be checked explicitly from the Clebsch–Gordan coefficients). The above implies that $L + S$ must be even for the two neutrons, and since $L = S = 0$ and $L > 2$ are ruled out by the requirement that $J = 1$, they must be in the $^3P_1$ state ($L = 1, S = 1, J = 1$). The parity of the final state is therefore $(-1)^L = -1$ ($L = 1$), while that of the initial state is $\eta_P(\pi^-)(-1)^L = \eta_P(\pi^-)$ ($L = 0$), thus showing that the charged pion must be a pseudo-scalar particle (the process being driven by parity-conserving strong interactions).

The first direct experimental evidence for the $\pi^- d \rightarrow nn$ reaction was obtained in 1954 using a set-up exploiting the coincidence of two neutron detectors (Chinowsky and Steinberger, 1954).

The above argument relies on the pion capture occurring from an S-wave, and several arguments supported this hypothesis. It is known a priori that only the $L = 0$ wave function has a non-zero probability density at the origin: the range of strong interactions being of the order of the dimensions of the deuteron (a few fm), and the average distance of the electromagnetically bound pion of order $a(m_e/m_\pi) \sim$ 180 fm (where $a \simeq 0.5 \cdot 10^{-10}$ m is the Bohr radius), only S-wave pions have a chance to interact strongly with the nuclei. Moreover, the rate of nn production by pions captured from $L > 0$ states was computed to be much lower than the rate of radiative decay to the $L = 0$ state, and largely insufficient to explain the observed ratio $\sigma(\pi^- d \rightarrow nn)/\sigma(\pi^- d \rightarrow nn\gamma) \simeq 2.5$. Pionic X-rays emitted in the radiative transitions to the $L = 0$ state were also observed. On the other hand, the ratio $\sigma(\pi^- d \rightarrow nn\pi^0)/\sigma(\pi^- d \rightarrow nn\gamma) \simeq 6 \cdot 10^{-4}$ was also measured: the very small phase space available in the reaction with π^0 emission (1 MeV Q-value) favours angular momentum $L = 0$ for the neutrons (for which therefore $S = 0$) and for the neutral pion, therefore giving $J = 0$, while for S-wave capture the total angular momentum of the initial state is equal to the deuteron spin, so that $J = 1$ and the reaction cannot proceed, consistently with the smallness of the observed ratio. The high relative probability for the process in which

---

[20] The deuteron ground state is actually a mixture containing a $^3D_1$ component, which however has the same parity as $^3S_1$, therefore not invalidating the following arguments.

only two neutrons are produced indicates instead that it is an allowed transition. Further support to the S-wave capture hypothesis follows from theoretical computations matching experimental observations, such as that of the short cascade time $O(10^{-12}\,\mathrm{s})$.

Measurements of the $pp \to \pi^+ d$ reaction with polarized beam and target can also determine directly the parity of the π^+ (Wilkin, 1980). As an example of a general remark made earlier, one can see that the intrinsic parity of the charged pion cannot be determined independently from that of the nucleon.

2.2.4 Other states

Among other bosons interacting with pions, kaons were the first to be observed, and played a very important role, actually leading to the discovery of parity violation as will be discussed in detail later. The negative intrinsic parity of the charged K meson (with respect to that of the Λ hyperon, conventionally chosen as $+1$) was determined similarly to that of the π^-, by the existence of the S-wave capture process $K^- {}^4\mathrm{He} \to \mathrm{He}^4_\Lambda \pi^-$, where ${}^4\mathrm{He}_\Lambda$ indicates a *hyper-nucleus* in which one neutron is replaced by a Λ hyperon; since, as will be seen, the weak decays of K or Λ violate parity symmetry, such processes cannot be used to define the intrinsic parities of those particles.

The experimental determination of the intrinsic parities for other particles is discussed in Tripp (1965).

2.2.5 Transition amplitudes

If interactions are parity symmetric, that is if $[P, H] = 0$, the expression of the S matrix in terms of the Hamiltonian H (Appendix C) implies that S also commutes with the parity operator

$$[P, S] = 0 \tag{2.59}$$

and therefore it only connects parity eigenstates with the same parity: parity is conserved. The relation imposed by parity symmetry on the elements of the scattering matrix is

$$S_{fi} = \langle \boldsymbol{p}_f, S_f | S | \boldsymbol{p}_i, S_i \rangle = \langle \boldsymbol{p}_f, S_f | P^\dagger S P | \boldsymbol{p}_i, S_i \rangle$$
$$= \eta_P(f)^* \eta_P(i) \langle -\boldsymbol{p}_f, S_f | S | -\boldsymbol{p}_i, S_i \rangle = \eta_P^*(f) \eta_P(i) S_{f_P i_P} \tag{2.60}$$

where $\boldsymbol{p}_i, \boldsymbol{p}_f$ stand for the whole set of initial and final state momenta and S_i, S_f for the corresponding spins; the subscript P indicates the parity-inverted states (opposite momenta).

The transition matrix (or the Hamiltonian) can always be decomposed as the sum of a parity-even and a parity-odd part

$$\mathcal{T} = \mathcal{T}^{(P+)} + \mathcal{T}^{(P-)} \tag{2.61}$$

$$P^{\dagger} \mathcal{T}^{(P\pm)} P = \pm \mathcal{T}^{(P\pm)} \tag{2.62}$$

$\mathcal{T}^{(P-)}$ can induce transitions among states of opposite parity; note however that, the test of parity symmetry by searching for such 'forbidden' transitions requires initial and final states to be prepared (and detected) in known eigenstates of parity, and the presence of any parity impurities in those must be excluded with sufficient accuracy, which is often difficult and however relies on assumptions on the parity-conserving nature of the processes involved in their preparation. The search for asymmetries in distributions is often an easier approach.

Since parity reverses momenta

$$\mathcal{T}^{(P\pm)}(-\boldsymbol{p}) = \pm \mathcal{T}^{(P\pm)}(\boldsymbol{p}) \tag{2.63}$$

the momentum dependence of the transition amplitudes will reflect the behaviour of the transition matrix. Given a generic transition from an initial state $|i\rangle$ to a final one $|f\rangle$, described in terms of momenta $\boldsymbol{p}$ and spin vectors $\boldsymbol{S}$ by a distribution function $W(\boldsymbol{p}, \boldsymbol{S})$ (differential decay rates or differential cross sections), if both the initial and final states are parity-inverted the same decay distribution is obtained in case the transition matrix is symmetric under parity ($\mathcal{T}^{(P-)} = 0$), so that

$$\langle f_P | \mathcal{T} | i_P \rangle = \langle f | P^{\dagger} \mathcal{T} P | i \rangle = \langle f | \mathcal{T} | i \rangle \tag{2.64}$$

In the case of parity symmetry, W must thus satisfy

$$W(\boldsymbol{p}_1, \boldsymbol{p}_2, \ldots; \boldsymbol{S}_1, \boldsymbol{S}_2, \ldots) = W(-\boldsymbol{p}_1, -\boldsymbol{p}_2, \ldots; \boldsymbol{S}_1, \boldsymbol{S}_2, \ldots) \tag{2.65}$$

It follows that terms which are odd under parity, such as $\boldsymbol{p}_1 \cdot (\boldsymbol{p}_2 \times \boldsymbol{p}_3)$ or $\boldsymbol{S}_1 \cdot \boldsymbol{p}_1$ must have zero expectation values, since for each configuration with a given value for such quantity the probability of producing the configuration with the opposite value is the same. Conversely, the measurement of non-zero expectation values for such P-odd quantities is an indication of parity violation: if parity symmetry is valid, no physical process can depend on a quantity which changes sign under parity, such as $\boldsymbol{S} \cdot \boldsymbol{p}$ (a dependence on $|\boldsymbol{S} \cdot \boldsymbol{p}|$ could be allowed).

It should be noted that a non-zero $\mathcal{T}^{(P-)}$ is not sufficient to produce an asymmetry $\boldsymbol{p} \to -\boldsymbol{p}$, since the observable distributions are squared moduli of amplitudes, and $|\mathcal{T}^{(P-)}(\boldsymbol{p})|^2$ is even in $\boldsymbol{p}$: the presence of both a parity-conserving and a parity-violating amplitude which interfere is required to generate an observable asymmetry:

$$|\langle f | \mathcal{T} | i \rangle|^2 = |\langle f | \mathcal{T}^{(P+)} | i \rangle|^2 + |\langle f | \mathcal{T}^{(P-)} | i \rangle|^2 + 2\mathrm{Re}\left[\langle f | \mathcal{T}^{(P+)} | i \rangle^* \langle f | \mathcal{T}^{(P-)} | i \rangle\right] \tag{2.66}$$

and only the interference term is odd under parity.

Even in presence of both $\mathcal{T}^{(P\pm)}$, it is clear that no parity-violating effect can be observed in the total transition rate (a scalar quantity), since the asymmetry is canceled in the spatial integration of momenta: the measurement of a pseudo-scalar quantity is required to detect parity violation via an asymmetry.

Parity symmetry also reduces the number of independent terms which can be present in the expression of scattering or decay amplitudes. As an example, in the scattering of a spin 1/2 and a spin 0 particle (e.g. πp scattering) the conservation of P and of total angular momentum J imply the conservation of *orbital* angular momentum L (since only two values of L differing by one unit can contribute for a given J, and the corresponding states have opposite parities), which in turn implies that no odd power of $\cos\theta$ (θ being the scattering angle) can appear in the angular distribution (Sakurai, 1964).

2.2.6   Selection rules

Absolute selection rules due to parity conservation do not arise for electromagnetic transitions, even considering a single multipolarity for the emitted radiation, because the photon has spin 1 and can either change the parity of a state or leave it unchanged, depending on the electric or magnetic nature of the transition. When considering the emission or absorption of particles of zero spin, absolute selection rules due to parity symmetry are instead possible.

For a given elementary process, parity symmetry implies that the process and its parity transformed version should occur with equal probabilities, and this imposes some constraints. Considering, for example, the scattering of a polarized particle on an unpolarized target, one can see that the process with the incoming particle having a longitudinal polarization (i.e. along its line of flight) transforms under parity into an identical one (after an uninfluential π rotation around an axis orthogonal to the scattering plane, defined by the incoming and scattered particle momenta) with the longitudinal polarization reversed; if parity symmetry holds, such processes must have the same cross section, leading to the conclusion that the differential cross section for scattering at any given angle is independent of any longitudinal polarization (actually, of any polarization in the scattering plane) of the incoming particle. If transverse polarization (orthogonal to the scattering plane) is considered instead, a parity transformation (followed by a π rotation) turns the original process into itself, and nothing can be said about the properties of the process with opposite transverse polarization.

Note that if the initial state of a reaction has no definite parity (e.g. a plane wave), it can give rise to final states of different parity without violating parity symmetry, and such states can interfere. Only in cases where the reaction is known (by other means) to proceed through an intermediate state of definite parity, are all final states constrained by parity symmetry to have this same parity eigenvalue, so that only even or odd l spherical harmonics are present in the final state amplitude and the differential cross section is symmetric around $\pi/2$ in the centre of mass.

Lack of such a angular symmetry implies that at least two states of opposite parity do contribute to the process.

A related constraint arising from parity symmetry is that in a reaction among unpolarized particles with two particles in the final state, those cannot acquire any net polarization in the production plane, defined by the two outgoing momentum vectors. A reaction induced by a parity-symmetric force can only produce polarization orthogonal to the production plane.

Considering a two-body process $i_1 i_2 \rightarrow f_1 f_2$ with initial and final momenta $p_{i,f}$ (in the centre of mass frame), parity symmetry restricts the polarization to be along the $n \equiv p_i \times p_f$ direction. Reflection with respect to the production plane is equivalent to the parity transformation followed by a π rotation around n, and therefore it is a symmetry operation if parity symmetry (and rotational symmetry) holds; the states of definite momentum (not angular momentum) are the eigenstates of such a transformation. The effect of a parity transformation on the initial state is to multiply it by the product of the intrinsic parities of the two particles and to transform $p_{i,f} \rightarrow -p_{i,f}$, while the rotation restores the momenta and multiplies the state by $e^{i\pi J_n} = (-1)^{J_n}$, where J_n is the sum of the spin components along n. The initial and final states are therefore related by parity symmetry as

$$\eta_P(i_1)\eta_P(i_2)(-1)^{J_n(i)} = \eta_P(f_1)\eta_P(f_2)(-1)^{J_n(f)} \qquad (2.67)$$

so that if the product of intrinsic parities changes a corresponding change in the spin states should also occur (Bohr, 1959).

Another constraint imposed by parity symmetry concerns two-body reactions followed by a subsequent reaction (such as double scattering, or production and decay of a Λ particle): the angular distribution of the final particles of the second reaction must be symmetric with respect to the production plane of the first. The momentum p_1 of the primary incoming particle (assuming the target at rest) and p_2 of the secondary particle define a plane, and each of the particles produced in the second reaction, with momentum p, must be emitted with equal probability above or below this plane, since $p_1 \times p_2 \cdot p$ is a pseudo-scalar quantity and must therefore have zero expectation value if parity is conserved.

2.3 Parity violation

While there is no evidence for parity violation in electromagnetic or strong interactions, weak interactions were found to violate parity symmetry in a *maximal* way, meaning that the amplitudes linking states of opposite parity are as large as those linking states of same parity.

2.3.1 The $\tau - \theta$ puzzle

Around 1955, a few different types of decays of relatively long-lived (lifetime $\sim 10^{-8}$ s) charged strange particles into two and three pions were known, which were labelled as follows:

$$\tau^\pm \to \pi^\pm \pi^+ \pi^- \qquad \tau'^\pm \to \pi^\pm \pi^0 \pi^0 \qquad \theta^\pm \to \pi^\pm \pi^0$$

All three decays appeared to originate from particles (of unknown spin-parity) with the same mass (within $\sim 1\%$ at the time) and lifetime (within $\sim 15\%$), apparently having also the same abundance ratio (independently on their production origin) and the same interactions with heavy nuclei (as measured from the branching ratios of particles which had undergone nuclear scattering); these features were checked using first cosmic rays and then the newly available accelerator beams. But while the first two could be easily described as two different decay modes of the same particle, the third one apparently could not, due to parity symmetry as we will now discuss.

For $\pi^\pm \pi^0$ decay, the orbital angular momentum of the two spinless pions is equal to the spin J of the parent $\theta^\pm$ particle, whose parity eigenvalue is therefore $\eta_P(\theta^\pm) = (-1)^2(-1)^J = (-1)^J$. A neutral 'counterpart' of the $\theta^\pm$ was known, decaying to $\pi^0 \pi^0$, which is only allowed for even spin and positive parity, since identical bosons must be symmetric under exchange, which for spinless particles is equivalent to the parity transformation, so that $\eta_P(\theta^0) = (-1)^J = +1$.

For the three pion decays, denoting with L the relative orbital angular momentum of the two identical ('even') pions, and with l that of the third ('odd') one with respect to the centre of mass of the first two, L must be even due to Bose–Einstein symmetry, and the parity of the system is therefore $\eta_P(\tau, \tau') = (-1)^3(-1)^{L+l} = (-1)^{l+1}$. Since pions are spinless, the total angular momentum J (equal to the spin of the parent particle) is given by the vector sum of L and l Table 2.2 shows the lowest order terms in L, l for several spin-parity assignments of the three-pion state. Generally speaking, high values for L, l are disfavoured, since the angular momentum barrier[21] in those cases would reduce significantly the decay width, increasing the lifetime above the measured value which is consistent with what is expected for a weak decay; this fact would already suggest to consider only $0^-, 1^+$ (and maybe 2^-).

For the three-body decay, it was observed that if the spin and parity of the decaying particle are linked by $\eta_P = (-1)^J$ the following decay configurations are forbidden: (a) the odd pion having zero kinetic energy (then $l = 0$ and $J = L$, even); (b) the odd pion having maximum energy (the two even pions would be relatively at rest, so that $L = 0$ and $J = l$); (c) the three pions being collinear (the rotation properties of the state must then be given by a spherical harmonic of order J, which

---

[21] In non-relativistic quantum mechanics this is the effective potential term $(\hbar^2/2m)[l(l+1)]/r^2$ which appears in the Schrödinger equation for the radial part of the wave function.

TABLE 2.2 Lowest order terms in the spherical wave expansion of the three-pion state with spin-parity J^P up to spin 4; L is the angular momentum of the system of two 'even' pions in their centre of mass frame and l that of the 'odd' pion with respect to such frame. The last row indicates whether a 2π system can be in such a spin-parity state or not.

J^P	0^-	1^+	1^-	2^+	2^-	2^-	3^+	3^+	3^-	4^+	4^+	4^-	4^-	4^-
L	0	0	2	2	2	0	0	2	2	4	2	4	2	0
l	0	1	2	1	0	2	3	1	2	1	3	0	2	4
2π	no	no	yes	yes	no	no	no	no	yes	yes	yes	no	no	no

gives $(-1)^l$ orbital parity). Indeed, in all the above cases the parity of the final state would have to be $\eta_P = (-1)(-1)^l$, and parity symmetry would not allow such a decay. Remembering that the three-body phase space distribution is uniform in the Dalitz plot describing the particle decay configurations (see Appendix D), any non-uniformity directly gives information on the intrinsic properties of the decay, i.e. on the structure of the matrix element.

The analysis of the experimental Dalitz plot distribution of just 71 τ decays from three groups (Orear *et al.*, 1956) did not show any depletion close to the phase space point where the odd pion is at rest and in general no depletion of event distributions was found in the regions of very low or very high kinetic energies for the odd pion, nor in the region corresponding to collinear pions. A more detailed comparison of the Dalitz plot distribution of the events with those expected for various spin-parity assignments, and the check for decay anisotropies in the decay planes distributions which could be present for non-zero spin of the decaying particle showed that any spin-parity assignment compatible with 2π decay was very unlikely, and that $J^P = 0^-$ was instead by far the most favoured hypothesis.[22]

Experiment thus indicated that the spin-parity assignment for the two-pion final state, namely $\eta_P = (-1)^J$, was not the same as for the three-body one, so that – on the grounds that parity must be conserved in the decay – the two modes could not originate from the same initial state: the τ and θ particles could not be the same particle, and it seemed odd that two particles with such similar masses had so different decay modes (since these would correspond to different virtual interactions with other particles). This is the so-called '$\tau - \theta$ puzzle' which dominated the attention of the physicists at the time.

This puzzle led to several proposals for its explanation, such as the possibility that there were two different particles with close mass, with the heavier one decaying

---

[22] Note also that effects due to electromagnetic or strong interaction among the pions in the final state should appear as distortions on the Dalitz plot (e.g. clustering of events in some regions), which were not observed.

into the other with emission of photons (Lee and Orear, 1955), or the one that strange particles always occur in parity doublets (Lee and Yang, 1956*b*).

Perhaps stimulated by M. Block's straight experimenter's question at a conference (Gardner, 1990), T. D. Lee and C. N. Yang critically re-examined the experimental status of parity symmetry (Lee and Yang, 1956*c*). They found that while there was evidence in favour of parity conservation in strong and electromagnetic interactions, such evidence was lacking in the case of weak interactions.[23] Writing a generic decay amplitude as $A = A_{PC} + A_{PV}$, where A_{PC} ($\mathcal{T}^{(P+)}$) is the parity-conserving part and A_{PV} ($\mathcal{T}^{(P-)}$) the parity-violating) part, all of the observables which had been studied so far (decay rates and spectra) are proportional to $|A_{PC}|^2 + |A_{PV}|^2$, and the two amplitudes cannot be distinguished by such measurements; moreover, the parity-conserving or -violating nature of a transition depends on the assignment of the intrinsic parities, which is to some extent arbitrary. For example in beta decay information on parity symmetry cannot be extracted from experiments measuring the electron spectra, due to the fact that the neutrino does not have a measurable mass (Yang and Tiomno, 1950).

Evidence for a violation of parity symmetry could come from the decay of the *same* particle into two states of opposite parity (as in the $\tau - \theta$ puzzle), but one had to be certain that the *same* particle was involved in the two decays. Alternatively, parity violation could be detected from a signal of the presence of amplitudes with different parity properties (A_{PC} and A_{PV}) in a decay, which requires the observation of an interference term among them: $A_{PC}^* A_{PV}$ would give rise to a pseudo-scalar term in the squared modulus of the amplitude (such as $\boldsymbol{p} \cdot \boldsymbol{S}$), which cannot appear as such in a scalar quantity such as a total decay rate, in which integration over $\boldsymbol{p}$ or $\boldsymbol{S}$ would average it to zero. Note that such an interference term would be linear in the (small) parity-violating amplitude, rather than quadratic as a 'forbidden' transition rate. The measurement of the dependence of a physical process on a pseudo-scalar quantity, or of a non-zero average value for such a quantity, is a signal of parity violation, and in their seminal paper Lee and Yang proposed several such possible tests, enlarging the scope from the $\tau - \theta$ puzzle to all of weak-interaction physics.

2.3.2   Beta decay

The most studied process driven by weak interactions at the time was nuclear beta decay $N \rightarrow N' e^- \bar{\nu}$, where N, N' are two nucleons differing by the transformation of a neutron into a proton. Although the triple product of the final state momenta $\boldsymbol{p}_e \times \boldsymbol{p}_N \cdot \boldsymbol{p}_\nu$ in such decay is a pseudo-scalar quantity, it vanishes identically because the three vectors are coplanar, and a non-zero pseudo-scalar quantity can only be

---

[23] Many years later, C. N. Yang would recall that 'neither of us thought this was likely to be the real explanation'.

formed by using a spin vector, thus requiring a polarized state or a polarization measurement, more difficult from an experimental point of view.

The most general expression for the beta decay matrix element is obtained from the generic weak Lagrangian (B.29) in which $S, V^\mu, \dots$ are the bilinear covariants built from the proton and neutron fields $\overline{\psi}_p$ and ψ_n. To lowest order in the weak coupling the transition matrix element is given by

$$\langle N' e \overline{\nu} | H | N \rangle = \sum_i \int d\boldsymbol{x} \, \langle N' | (\overline{\psi}_p \Gamma^i \psi_n) | N \rangle \langle e \overline{\nu} | \left[\overline{\psi}_e (c_i + c_i' \gamma_5) \Gamma_i \psi_\nu \right] | 0 \rangle \qquad (2.68)$$

The leptonic part reduces to ($m_\nu \simeq 0$)

$$\langle e | \overline{\psi}_e | 0 \rangle (c_i + c_i' \gamma_5) \Gamma_i \, v(\boldsymbol{p}_\nu) \, e^{-i \boldsymbol{p}_\nu \cdot \boldsymbol{x}}$$

where the first factor is essentially the electron wave function which, neglecting the distortions due to Coulomb effects, is just given by the free expression $\overline{u}(p_e) e^{-i p_e \cdot x}$, leaving a sum of terms containing

$$\int d\boldsymbol{x} \, e^{-i q \cdot x} \langle p | \overline{\psi}_p \Gamma^i \psi_n | n \rangle$$

where $\boldsymbol{q} = \boldsymbol{p}_e + \boldsymbol{p}_\nu$ is the momentum transfer in the decay; since this is generally small (a few MeV/c) with respect to the inverse mean nuclear radius ($\sim (1 \text{ fm})^{-1} \simeq 200$ MeV/c) over which the integration extends, the first (unity) term in the expansion of the exponential dominates.

In the non-relativistic limit (velocity $\lesssim 0.1c$ for the bound nucleons) only four terms survive (see Appendix B):

$$\langle N' e \overline{\nu} | H | N \rangle \simeq \langle \rho \rangle \left[\overline{u}(p_e)(c_S + c' S \gamma_5) v(p_\nu) + \overline{u}(p_e) \gamma_0 (c_V + c' V \gamma_5) v(p_\nu) \right]$$

$$+ \langle \sigma_k \rangle \left[\overline{u}(p_e)(\sigma_k/2)(c_T + c' T \gamma_5) v(p_\nu) + \overline{u}(p_e) \gamma_k \gamma_5 (c_A + c' A \gamma_5) v(p_\nu) \right]$$

where everything is expressed in terms of only two nuclear matrix elements

$$\langle \rho \rangle \equiv \int d\boldsymbol{x} \, \langle N' | \overline{\psi}_p \psi_n | N \rangle \qquad \langle \sigma_k \rangle \equiv \int d\boldsymbol{x} \, \langle N' | \overline{\psi}_p \sigma_k \psi_n | N \rangle \qquad (2.69)$$

(σ_k being the Pauli matrices), the first one inducing the so-called Fermi transitions, with no spin change between neutron and proton, driven by S and V couplings, and the second one inducing the so-called Gamow–Teller transitions, with possible spin change (as can be seen by the presence of σ_k), driven by A and T couplings; the latter cannot induce spin 0 to spin 0 transitions, though, since $\langle J = 0 | \sigma_k | J = 0 \rangle = 0$ by the Wigner–Eckart theorem, thus such kind of transitions are actually well suited to the study of the magnitude of S and V terms. It should also be remarked that, since the Compton wavelength $\hbar/p$ of the emitted leptons is much larger than the mean nuclear radius, no orbital angular momentum is carried away ($L = 0$); this implies that the electron-neutrino system has total angular momentum equal to the change of nuclear spin.

The outgoing momentum spectra induced by S and V terms only differ in the electron-neutrino correlations: for the scalar interaction the electron and (anti-) neutrino tend to be emitted in opposite directions, while the vector interaction favours parallel momenta. The same difference in correlation (although somewhat reduced) occurs between spectra induced by T and A. Correlation experiments in the 1950s showed that V and A couplings were the dominant ones. Interference terms between S and V, and between A and T (so-called 'Fierz interference terms') give rise to strong energy dependences, not observed in the experimental spectra.

Only some combinations of the c_i and c_i' coefficients appear in the decay probability and angular distribution function for unpolarized particles, namely $|c_i|^2 + |c_i'|^2$ and $c_i^* c_j + c_i'^* c_j'$, so that measurements using only unpolarized particles cannot detect the presence of both c_i and c_i' terms and therefore give information on parity violation.

Eventually, after the fall of parity, the $V - A$ picture of weak interactions emerged.

In the rush of experiments aiming to check the issue raised by Lee and Yang, the first[24] experimental indication of parity violation (Wu *et al.*, 1957) was obtained in the same year, 1956, by measuring the intensities of beta decay electrons from polarized nuclei, emitted in directions parallel and anti-parallel with respect to the nuclear polarization vector $\langle S \rangle$; a difference in the two intensities indicated that the beta decay amplitude depends on the pseudo-scalar quantity $\langle S \rangle \cdot p$ (p being the momentum vector of the emitted electron).

Beta decay asymmetries from polarized nuclei

The first measurement of the beta decay asymmetry from polarized nuclei (Wu *et al.*, 1957), performed in 1956, required that several technical challenges were overcome. The main issue was that of obtaining a polarized nuclear sample, which required low temperature expertise and equipment, at that time only available in few places, such as the laboratory at the National Bureau of Standards in Washington, where the experiment was performed.

The ground state of ^{60}Co has spin-parity $J^P = 5^+$, and it beta decays to an excited state of ^{60}Ni with $J^P = 4^+$ by emitting an electron with maximum energy 316 keV (and an antineutrino), with a half-life of 5.3 years; the excited Ni state in turn gamma decays to its ground state in the sequence of states $4^+ \to 2^+ \to 0^+$ by emitting two photons of 1.2 and 1.3 MeV energy. This beta transition, involving a change of nuclear spin (Gamow–Teller transition), can only be driven by tensor or axial-vector couplings in the generic lepton Lagrangian, while a transition with no change of nuclear spin could also be driven by scalar or vector couplings (Fermi transition); since the latter corresponds to zero total angular momentum being carried away by the electron-neutrino system, it would always lead to

---

[24] Two double scattering experiments with electrons, in 1928 (Cox *et al.*, 1928) and 1930 (Chase, 1930) resulted in asymmetries which show – in retrospect – evidence for parity non-conservation; despite the fact that the authors suggested that 'the source of the asymmetry must be looked for [. . .] in some asymmetry of the electron itself', and that other tentative explanations of the observed effect 'would offer greater difficulties [. . .] than would the acceptance of the hypothesis that we have here a true polarization due to the double scattering of asymmetrical electrons', the significance of such results was not recognized at that time, and no connection to parity was suggested. Times were not yet mature.

isotropic distributions by itself, and any anisotropy could only originate from both the Gamow–Teller part and its interference with the Fermi part, possibly resulting in smaller asymmetry effects than for a pure Gamow–Teller transition alone (as for free neutron decay). If the nucleus is fully polarized, i.e. $|J_z| = +J$, then a Gamow–Teller transition requires the electron and neutrino spins to be aligned with the nuclear polarization (remember that the decay products carry no *orbital* angular momentum).

If parity symmetry is not imposed, the electron angular decay distribution in beta decay has the form

$$\frac{d\Gamma}{d\Omega} \propto 1 + a\mathcal{P}\frac{v}{c}\cos\theta \qquad (2.70)$$

where v is the electron velocity ($\sim 0.6\ c$, on average, for this decay), $\mathcal{P}$ the magnitude of the nuclear polarization vector (see Appendix E) and θ the angle between the nuclear polarization vector and the direction of electron emission; such a distribution results in an asymmetry between θ and $\pi - \theta$ if $a \neq 0$. For a pure Gamow–Teller transition the asymmetry coefficient is

$$a \propto -\frac{2\mathrm{Re}(c_T^* c_T' - c_A^* c_A') + (2Zm_e\alpha/p_e)\mathrm{Im}(c_A^* c_T' + c_A'^* c_T)}{|c_T|^2 + |c_A|^2 + |c_T'|^2 + |c_A'|^2} \qquad (2.71)$$

(where m_e, p_e are the electron mass and momentum, and α the fine structure constant), resulting as it should from the interference of couplings with opposite parities c and c'. The term in eqn (2.71) proportional to the atomic number Z (odd under time reversal) is due to Coulomb effects (Jackson *et al.*, 1957*a*), and an energy-dependent interference term $\propto \mathrm{Re}(c_A'^* c_T' - c_A^* c_T)$, found to be of negligible magnitude, was omitted.

In order to achieve a significant degree of polarization, the nuclei must be cooled, otherwise their thermal motion would prevent any significant alignment of the spins by an external magnetic field: the energies associated with the nuclear spin orientation in a magnetic field **B** are $E \sim \mu_N|\mathbf{B}|$, where the nuclear magneton $\mu_N = e\hbar/(2Mc) \simeq 3.15 \cdot 10^{-8}$ eV/T (M being the proton mass). Even in an intense 20 T magnetic field, such energy corresponds to a temperature of only $\sim 10^{-3}$ K; very low temperatures and/or very strong magnetic fields are required.

Cooling the material in a magnetic field down to the required temperature is technically challenging; adiabatic demagnetization was used in this case. The large atomic magnetic moments of a paramagnetic salt containing ions of the iron or rare earth groups of elements, due to the electrons in the unfilled shells, are easily polarized in a magnetic field while in thermal contact with a liquid helium bath: while at high temperature all the magnetic energy levels (with very small separation) are almost equally occupied according to Boltzmann statistics, the external field splits the levels so that the lower energy ones (corresponding to magnetic moments aligned to the field) are preferentially occupied. The spin system absorbs energy from the magnetic field, losing it as heat delivered to the thermal bath (latent heat of magnetization) and getting more magnetically ordered (lowering the entropy) while maintaining the same temperature. After insulating the material from the bath (by pumping away the helium gas) the magnetic field is then removed adiabatically, that is, in a time which is short with respect to the spin-lattice relaxation time (the characteristic time required for the spin system and the lattice system to reach the same temperature, through phonon exchanges, which can be of the order of 30 min); the level occupation (and the entropy) remains the same and the material therefore cools down. This is analogous to the adiabatic expansion of a gas, and indeed for a paramagnetic system the magnetic field intensity and the magnetization are thermodynamical variables analogous to the pressure and volume of an ideal gas. In this case the entropy S is a function of temperature and magnetic field intensity; if the dipole-dipole interaction among magnetic atoms is negligible (as is indeed the case as long as the temperature is not too low) then S can only depend on the combination $\mu B/kT$, so that the isoentropic transition leads to a final temperature which scales linearly with the ratio of final and initial magnetic fields: $T_f = (B_f/B_i)T_i$ (demagnetization usually does not reach exactly zero field, where the energy levels would become degenerate again, so that the final temperature is slightly higher but the heat capacity of the material, i.e. the amount of heat that can be absorbed while warming up, is increased). The linear relation breaks down when the energy related to the internal magnetic fields among atoms cannot be neglected compared to kT; these internal fields limit the final temperature which can be achieved, so that materials in which magnetic interactions are small are favoured, such as those in which paramagnetic ions are separated by many other atoms.

With this method, using a 2.3 T magnet the temperature was lowered to 0.003 K before starting each measurement. An inductance coil measured the crystal temperature in terms of its magnetic susceptibility; the temperature gradually rose during the measurement, due to the energy exchange between the nuclear system and the electron system, which occurs rather slowly due to the smallness of the hyperfine interaction.

After cooling, a polarizing solenoid was turned on, aligning the nuclear spins along the direction towards the detector, either parallel or anti-parallel to it depending on the sign of the current. The crystal, being strongly anisotropic, has directions in which it does not get strongly magnetized when a field is applied, therefore providing only a small heat release which helps in keeping the Co polarization longer. Indeed the crystals were aligned with their axis along the polarizing field direction (vertical), while their strongest polarizability is in the orthogonal direction along which the demagnetizing field was directed (horizontal); the effective gyromagnetic number for magnetic field parallel or orthogonal to the crystal axis are $g_\parallel = 0.025$ and $g_\perp = 1.84$ respectively.

In paramagnetic salts there are large magnetic fields (O(10–100) T) at the positions of the nuclei due to unpaired electrons and, at temperatures of the order of 0.01 K and below, the nuclear magnetic moments become oriented with respect to such electronic magnetic fields through the same procedure described above, where the paramagnetic salt now plays the role of the thermal bath. Since the large electron magnetic moments are easily aligned, relatively low magnetic fields (less than 0.1 T in this experiment) are sufficient to reach a high nuclear polarization ($\sim$ 65%) with this technique. A suitable salt is cerium magnesium nitrate, $2\,Ce(NO_3)_3 \cdot 3Mg(NO_3)_2 \cdot 24\,H_2O$, in which the water molecules surrounding the Ce^{+++} ion produce an anisotropic electric field in the region of motion of the single electron responsible for the magnetic effects, therefore breaking the degeneracy of the sub-states with different J_z eigenvalues and allowing a significant polarization.

Another problem was that of having a beta source in a very thin surface layer on the cooling crystal, since the low energy electrons emitted cannot penetrate large amounts of matter; for this reason a 50 μm crystalline layer of Co was grown on the surface of a cerium magnesium nitrate crystal and got cooled with it; this method also guaranteed an excellent thermal contact of the radioactive substance with the cooling agent. The sample prepared in this way was actually contained in an enclosure made of the same salt, which has an appreciable specific heat, so that it was possible to maintain the low temperature for a period of time long enough to perform measurements: the time of reheating was about 6 min.

To keep the nuclear polarization long enough, the sample had to be in an evacuated flask immersed in liquid helium (4.2 K down to 1 K at reduced vapour pressure), inside a cryostat containing liquid nitrogen (77 K); the elongated shape of a dewar vessel minimized the heat influx due to conduction along its walls, which was also absorbed by the up-flowing gas (Fig. 2.2). A technical problem was that of also having the electron detector inside the low-temperature enclosure, since beta electrons cannot traverse the cryostat walls. To

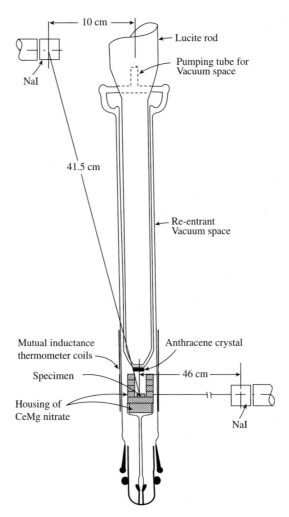

FIG. 2.2. Schematics of the experimental apparatus used for the first measurement of parity violation in beta decay (Wu *et al.*, 1957).

avoid efficiency issues, the same electron detector was used for all measurements; it was an anthracene scintillation counter mounted inside the flask along the polar direction with respect to the magnetic field, 2 cm above the sample. Since a normal photomultiplier does not work well at low temperatures (the photocathode becomes resistive and localized charges appear), nor in high magnetic fields (which perturb the electron trajectories inside the tube), it was placed far from the detector, at room temperature outside the cryostat, and the light flashes were transmitted through a glass window and along a lucite rod. Since lucite is a poor thermal conductor and is opaque to infrared radiation, a rather large diameter pipe could be used, with good light transmission efficiency,[25] without compromising the thermal insulation. The effect of the polarizing magnetic field on the beta electrons was not a problem, since their low energy implies that their trajectories are tight spirals around the magnetic field lines.

As pointed out by Lee and Yang, the study of the angular distribution of the gamma radiation from the excited nucleus produced in the beta decay cannot be used as a test of parity symmetry in weak interactions,[26] since nuclear levels have definite parities (to a very high accuracy dictated by the parity symmetry of strong interactions) and the parity symmetry of electromagnetic interactions implies that emitted photons carry out definite parities, so that the observed distributions must be even functions of p_γ and no pseudo-scalar quantity such as $S \cdot p_\gamma$ can appear.

On the other hand, the anisotropy of the gamma radiation emitted was used to measure the degree of nuclear polarization: the polarization of the Co nucleus is transmitted to the Ni nucleus, and for aligned nuclei the intensity of photon emission depends on the polar angle (on $|\langle S \rangle \cdot p_\gamma|$). Two external NaI scintillation counters were used to detect the photons, one on the equatorial plane and the other close to the polar direction, and their relative rates were seen to change in the few minutes required for warming up, reaching a value which was checked to be independent of the field direction: while the sample warms the nuclear polarization disappears, and such time variation was used to detect the polarization-dependent effects.

The measurements were performed with the polarizing magnetic field directed along opposite directions. Depending on the sign of such field, the electron counts were seen to either increase or decrease in time towards the average value while the sample depolarized, indicating that electrons were being preferably emitted along the direction opposite to the nuclear spin, with a dependence of the rate on the sign of $\langle S \rangle \cdot p_e$, violating parity symmetry.

Possible effects due to residual magnetic fields in the sample were excluded by the absence of any effect when inverting the sign of the demagnetization field, and when using nuclei which were separated from the cooling salt, so that they were not polarized and any residual effect could be due only to a possible residual magnetization of the paramagnetic crystal (which could indeed occur due to its anisotropic polarizability if the magnetic field was not exactly aligned to the crystal axis), or by removing part of the cooling crystal and therefore modifying any residual internal magnetic effect which could affect electrons as they

---

[25] This light guide was machined to have a longitudinal section with the shape of a logarithmic spiral, which has the property that lines from its origin cross the curve at a constant angle, therefore maximizing the collection of light from a point source at its centre, transmitted via total internal reflection.

[26] Indeed, the asymmetry of such distributions was used to test parity symmetry in *electromagnetic* or *strong* interactions.

emerged from the crystal. The back scattering of beta electrons from the source substrate results in a correction which diluted the measured asymmetry.

The large observed asymmetry coefficient $a \, \mathcal{P}(v/c) \simeq -0.4$ implies a large $a \simeq -1$: this magnitude indicates that final state interaction effects, inducing the term proportional to $Z\alpha$ in eqn (2.71), cannot be the cause of the observed effect. As will be discussed in Chapter 4, if time reversal symmetry is valid, the coupling constants c_i, c_i' are real and the choices $(c_A = c_A' = 0, c_T = c_T')$ or $(c_A = -c_A', c_T = c_T' = 0)$ for the couplings could naturally explain both the large magnitude of a and the absence of the interference term in eqn (2.71).

The species of nuclei which could be efficiently polarized were rather few, but the parity non-conservation effect was later confirmed (Postma et al., 1957) using ^{58}Co: this nucleus has a $J^P = 2^+ \rightarrow 2^+ \; \beta^+$ decay (positron emission) with subsequent gamma emission, and a parity-violating asymmetry of opposite sign with respect to the ^{60}Co experiment was measured, as predicted, smaller in magnitude due to the presence of both Fermi and Gamow–Teller terms, which also make the theoretical computation more uncertain.

A second spin vector available in beta decay is the polarization of the emitted electrons: if a longitudinal polarization (along the direction of motion) of the electrons is measured, one has direct access to a pseudo-scalar quantity, $\langle S_e \cdot p_e \rangle$ without the need to use polarized nuclei.

Electron polarization can be measured through its Coulomb scattering on nuclei (Mott scattering) or on free electrons (Møller scattering). In the first case the spin-orbit interaction between the magnetic moment of the incoming electron and the magnetic field it experiences when moving in the electric field of the (unpolarized) nucleus leads to a left-right asymmetry in the scattering cross section, depending on the electron polarization in the direction orthogonal to the scattering plane (transverse polarization). Such asymmetry grows with Z^2, is generally bigger for large-angle (or backwards) scattering, and vanishes in the extreme relativistic limit. An experimental difficulty lies in making sure that only single scattering occurs, since multiple scattering leads to very small polarization, as no effect is present in small angle scattering. This requires using very thin foils, $O(10^{-5}$ cm), as targets. For positrons the spin-dependent asymmetry is much smaller than for electrons, as could be guessed by considering the fact that the region most effective for producing polarization is the neighbourhood of the nucleus, where the inhomogeneity of the electric field over microscopic distances is larger, and that positrons penetrate less in this region due to Coulomb repulsion.

In the second case (scattering on electrons) (Frauenfelder et al., 1957a) a polarized target consisting of a magnetized ferromagnetic foil is used: the scattering cross section at any angle (except zero) is reduced (increased) with respect to the unpolarized case for parallel (anti-parallel) spins along the line of collision. Since extremely high fields are required to magnetize thin foils in the direction normal to its surface, they were magnetized parallel to their surface and tilted with respect to the incident electron direction, so that the polarization had a component along the beam direction, therefore resulting in a dependence on the longitudinal electron polarization without the need to rotate it. The sensitivity is limited by the number of polarizable electrons: in iron, out of 26 electrons only the 2 in the external orbital are aligned at saturation, therefore the analysing power is at most 8%. The spin dependence is maximal in the relativistic limit (a factor 8 in cross section), and is largest for collisions at 90° in the centre of mass, when the transverse component of the polarization does not affect the scattering. On the other hand, the scattering cross section on the orbital electrons is significantly smaller than the Mott one (on nuclei) by a factor $\sim Z$, so that there is a large background from the latter process: the two can be distinguished because in Møller scattering the energy transfer is much larger and both incident and scattered electron can emerge from a sufficiently thin foil and be detected in coincidence. This technique is also applicable to positrons, since Bhabha (e^+e^-) scattering has the same spin dependence as Møller scattering in the extreme relativistic limit, although it vanishes in the non-relativistic limit.

Longitudinal polarization of beta decay electrons

The longitudinal polarization of the emitted beta electrons from an unpolarized ^{60}Co source was first observed in 1957 (Frauenfelder *et al.*, 1957*b*).

By using an electrostatic deflector, i.e. an electric field between two parallel plates shaped as a quarter-circle section of a cylinder, which guides the electron deflecting it by $\pi/2$, the longitudinal polarization was transformed into a transverse one. In the non-relativistic approximation, for an electron gyromagnetic number $g = 2$ the momentum is rotated while the spin direction is left unchanged (see Appendix E), and this can be arranged to occur exactly for a single value of the electron energy (a magnetic deflection would also affect the spin).

The transverse polarization of electrons was measured by scattering them on a high-Z material (a thin gold foil) and measuring the left-right asymmetry for angles between 95° and 140° using Geiger counters. Defining left as $\boldsymbol{p} \times \boldsymbol{p}' \cdot \boldsymbol{p}'' > 0$, where $\boldsymbol{p}$ is the initial electron momentum, $\boldsymbol{p}'$ its momentum after the deflector and before scattering and $\boldsymbol{p}''$ the one after scattering, the scattering ratio for angle θ is

$$A(\theta) = \frac{R_{\text{left}}}{R_{\text{right}}} = \frac{1 + \mathcal{P}_L\, a(\theta)}{1 - \mathcal{P}_L\, a(\theta)}$$

where $\mathcal{P}_L$ is the longitudinal electron polarization, defined as the asymmetry between right-handed (helicity $+1$) and left-handed (helicity -1) electrons

$$\mathcal{P}_L = \frac{I_+ - I_-}{I_+ + I_-}$$

and $a(\theta)$ is the polarization asymmetry factor for scattering angle θ. The left-right asymmetry of the detector arrangement was corrected for by measuring the almost isotropic scattering from aluminium foils. Runs with different target thicknesses and different electron energies gave results indicating non-zero asymmetries at the 3 to 7 standard deviation level. The

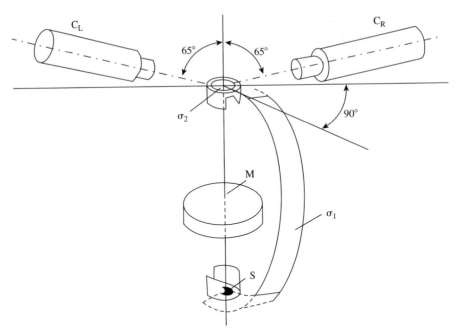

FIG. 2.3. Drawing of the arrangement of scattering foils for the double-scattering experiment (de Shalit *et al.*, 1957). Electrons emitted at different angles from the source S reach the second gold scattering foil σ_2 after a first $\pi/2$ scattering on the aluminium foil σ_1, which is curved along a semi-circumference with diameter equal to the distance between S and σ_2. The lead absorber M prevents electrons from reaching directly σ_2, and the two identical counters C_L, C_R, symmetric with respect to the plane containing S and σ_1, detect the left-right asymmetry in the second scattering.

main difficulty with the above polarization measurement was due to the errors induced by the scattering in the analyser foil, difficult to evaluate accurately.

In another experiment (de Shalit *et al.*, 1957) the longitudinal polarization of electrons was measured through their double scattering. In the non-relativistic regime, the scattering does not affect the electron spin direction, so that a longitudinally polarized electron will become transversely polarized after being scattered by $\pi/2$, and then its polarization can be measured through the left-right asymmetry in a second scattering as described above. A clever geometrical arrangement (see Fig. 2.3) ensured that the electrons reaching the second scattering foil were scattered by $\pi/2$, maintaining the advantage of a large solid angle acceptance.

A 13 standard-deviation left-right asymmetry was measured, dropping to zero when aluminium (lower Z) was used as the second scattering foil, thus showing that instrumental asymmetries are small.

Since for positrons the Mott scattering asymmetry is much smaller, the spin dependence of the e^+e^- (Bhabha) scattering cross section was exploited instead (Frauenfelder *et al.*, 1957*a*) to measure the longitudinal polarization of positrons emitted in beta decay, using as a target a magnetized foil set at an angle with respect to the incident positrons. By detecting both scattered leptons at a given angle and requiring that their energies be about equal (the condition for which the polarization dependence of the cross section is stronger), events strongly affected by multiple Coulomb scattering in the foil were rejected: those are indeed the events for which the kinematical momentum-angle relation between the two leptons (the target electrons can be considered free, as long as the incident energy is much greater than their binding energy) would be distorted and the scattering angle mis-measured. Thin target foils were used and placed in a helium atmosphere in order to further reduce scattering effects. A single electron emitting an energetic *bremsstrahlung* photon could fake a coincidence (the cross section being only one order of magnitude lower than that for Møller scattering) but the missing angular correlation allowed to reject them (thin detectors can be used to reduce the sensitivity to photons).

The asymmetry in the number of events with magnetic field anti-parallel (N_a) or parallel (N_p) to the lepton polarization is

$$A = \frac{N_a - N_p}{N_a + N_p} = \mathcal{P}_f \cos(\alpha) \, \mathcal{P}_L \frac{1 - a(\theta)}{1 + a(\theta)}$$

where $\mathcal{P}_f$ is the fraction of polarized electrons in the foil, inclined at an angle α with respect to the positron direction, $\mathcal{P}_L$ is the positron longitudinal polarization and $a(\theta) = \sigma_p(\theta)/\sigma_a(\theta)$ the ratio of cross sections at angle θ. The measurements on positrons from several β^+ emitting nuclei gave non-zero results. Null checks were done by measuring the counting asymmetry for unpolarized electrons when reversing the magnetic field. The technique was subsequently refined by adding momentum analysis to select monochromatic electrons or positrons and to eliminate background from photons.

The above techniques were used with β^+ nuclear decays (e.g. from ^{22}Na, with a half-life of 2.6 years and emitting a 1.3 MeV positron), and showed that the longitudinal polarization of positrons emitted in nuclear decays was compatible with $+v/c$ (v being the positron velocity), corresponding to positive helicity (right-handed).

In another approach (Goldhaber *et al.*, 1957) the degree and sense of circular polarization of *bremsstrahlung* photons produced by the beta decay electrons was measured. Using an unpolarized ^{90}Y source, which emits high-energy electrons (up to 2.2 MeV) and no photons, the photons originating in the metallic enclosure of the source were detected in a scintillation counter placed after an iron-core electromagnet producing a field aligned along the photon direction.

Part of the beta electron polarization is transferred to the *bremsstrahlung* photon; for a fully polarized electron the circular polarization of the emitted photon varies from zero for very soft radiation to almost 100% at the high-energy end of the spectrum. Actually, as a general rule, for high-energy electrons impinging on matter, all the common processes which can occur tend to preserve the original helicity in the higher-energy components; moreover, fast electrons conveniently emit *bremsstrahlung* photons mostly along their flight direction. Since the Compton cross section depends on the relative spin orientation of incident photon and target electron, by comparing the rates with the magnetic field parallel or anti-parallel to the photon momentum, the asymmetry due to the photon polarization could be measured. The spin-dependent part of the cross section can exceed 1/2 of the spin-independent part at high energies for scattering angles around 40–50°, and can even become comparable to the latter for back scattering (where the cross section is however much lower); for transmission (zero scattering angle) the probability is larger when the photon spin is parallel to the electron spin (variation up to 10%).

The result was that for photons at the high end of the spectrum (energies close to those of the parent electrons) the measured asymmetry indicated a photon polarization close to 100%.

Analogously, polarized positrons annihilating in flight transfer part of their polarization to the resulting photons, most efficiently at high energies. While e^+e^- annihilation into a single photon can occur if a third body is present to satisfy energy–momentum conservation, the dominant annihilation mode is into two photons, and the higher energy one, emitted in the direction of the positron momentum, carries the larger polarization. By directing the positrons on a converter and selecting high energy photons emitted in a narrow forward cone, their Compton scattering allowed to obtain information on the positron polarization (Boehm *et al.*, 1957).

The longitudinal polarization of positrons can also be measured by exploiting their annihilation with polarized electrons. Positrons were directed onto a magnetized iron target thick enough to stop them and the annihilation photons were detected for two opposite orientations of the magnetic field; photons of energy higher than 0.511 MeV originate from annihilation in flight, which has a strong polarization dependence (Page, 1957).

Still another approach, suitable for the measurement of the polarization of slow positrons, is based on the fact that in the zero energy limit only the singlet state with anti-parallel spins can annihilate in two photons. For the two D electrons which are polarized in magnetized iron, the two-photon yield therefore depends on the magnetic field orientation; since D electrons are faster than S ones, annihilations with the former contribute to the wings of the angular correlation distribution of the two photons, and the magnetization dependence of the intensity in these regions is correlated to the positron polarization.[27]

The experiments measuring the helicities of the particles emitted in beta decay, including that of the neutrino (!) in the celebrated experiment by Goldhaber *et al.* (1958*b*), showed that the asymmetry coefficients were of order $\pm v/c$ (v being the velocity of the particle), and eventually led to the $V - A$ structure of the weak interactions.

In order to perform measurements of the kind described, rather than using a sample of oriented nuclei obtained by a strong magnetic field or by any means which selectively populates some J_z sub-states more than others, it is easier to select (*tag*) from an unpolarized sample those nuclei for which the spin is oriented in a particular direction. Due to the asymmetry in beta decay an initially unpolarized nucleus is left partly polarized with respect to the direction in which the decay electron is emitted. If the nucleus does not reach its ground state in the decay but is left instead in an excited state, and a gamma decay follows immediately

---

[27] It is perhaps interesting to note that puzzling asymmetries were observed in 1955, before the discovery of parity violation, in the investigation of ortho-positronium quenching in a magnetic field, and dismissed as instrumental effects, while a posteriori they can be interpreted as the effect of parity-violating longitudinal polarization of positrons.

afterwards, parity violation could also be detected by the existence of a mixture of two gamma transitions with opposite parity and same multipolarity (such as M1 and E1), resulting in a circular polarization of the photon proportional to the cosine of the angle between the photon and the electron directions (another pseudo-scalar quantity $S_\gamma \cdot p_e$).

With this approach the parity-conserving gamma decay acts as an analyser for the parity-violating beta decay, and no polarization of the initial state is necessary, thus allowing more nuclei to be studied and experiments to be performed at room temperature. This method was first used in 1957 on ^{60}Co nuclei (Schopper, 1957).

Experiments with free neutrons were also performed, thus avoiding any complication in theoretical interpretations due to nuclear structure effects: spatial asymmetries in the form of correlations among the neutron spin direction and either the electron or the antineutrino momentum were measured.

Beta decay asymmetries of free polarized neutrons

In one experiment (Burgy et al., 1960), neutrons from the Argonne reactor were collimated by a vertical slit and polarized by grazing reflection on a magnetized Co-Fe mirror. The interaction of slow neutrons with matter is conveniently described by expressing the interference of the incident and scattered waves in terms of an 'index of refraction' (Fermi and Marshall, 1947): $n = p'/p = \sqrt{1 - 2m_n U/p^2}$, p, p' being the neutron momenta outside and inside the material, m_n the neutron mass and U the effective potential for the neutron within the material, which can be of order 100 neV (see e.g. Khriplovich and Lamoreaux (1997)). Neutrons can be totally reflected by the surface of a material; in classical terms this happens when their kinetic energy is smaller than the potential they experience inside the material. The neutron scattering amplitude has a part which depends on the relative spin orientation of the neutron and the atomic electrons, so that n can be different for the two possible relative spin orientations. If a neutron has an angle of incidence which is above the critical angle for one spin state and below it for the opposite state, total reflection is only effective for one of the two, therefore resulting in a polarized reflected beam. While the critical angles depend on the neutron wavelength, and therefore their velocity, it happens that for some materials (such as cobalt) the spin-dependent interaction is strong enough to dominate over other terms, so that the index of refraction for one spin state is always greater than 1 while that for the opposite spin orientation is always less than 1, resulting in a polarizing reflection effect for any wavelength; since the neutrons which are not totally reflected are reflected negligibly (and more so when a thin reflecting layer is stacked on a material with high neutron absorption cross section), very high polarizations (above 95%) can be obtained with this technique. Cobalt is difficult to magnetize, but a Co-Fe alloy with 7% iron is much more easily magnetized.

The above arguments can be applied to neutrons which are slow enough to have a de Broglie wavelength $\lambda = h/p$ of at least a few Å, corresponding to the interatomic spacing in solids; this corresponds to energies below 0.03 eV. The neutrons produced from reactors have kinetic energies of a few MeV, so that they were first slowed down (moderated) in deuterium (hydrogen has a large neutron capture cross section and is therefore not suitable), eventually acquiring a Maxwellian energy distribution at the temperature of the moderator ('thermal' neutrons, with kinetic energy $E_K \simeq 25$ meV and $\lambda \simeq 1.8$ Å, at room temperature, or 'cold' neutrons, with $E_K \simeq 2.5$ meV and $\lambda \sim 6$ Å, at ~ 25 K).

The neutron beam polarization was measured by the reflection probability on a second magnetized mirror. By inserting an unmagnetized steel plate on the beam, which is expected to completely depolarize the beam due to spin precession within its randomly ordered magnetic domains, the reflected intensity on the second magnetized mirror is reduced. Since for complete (100%) initial polarization the intensity should be reduced by 50%, the measured reduction of intensity gives a measurement of the actual polarization, which was $(87 \pm 7)\%$.

In order to maintain the polarization of neutrons along their path a guiding magnetic field was applied; the neutron spin is an adiabatic invariant, so that if the field changes its direction slowly (that is, with an angular velocity which is small compared to the Larmor precession frequency $\omega_B = 2\mu_n B/\hbar$, where μ_n is the neutron magnetic moment and B the magnitude of the guide field) the neutron spin rotates with it.

A proton and an electron detector were mounted orthogonally to the beam on opposite sides of the vacuum chamber, in the horizontal plane. The proton detector consisted of a metallic cathode preceded by an electrostatic field accelerating protons towards it, on which secondary electrons were produced and multiplied by a dynode system; electrons were detected in a crystal scintillator (see Fig. 2.4).

With vertical neutron polarization, the correlation of antineutrino and proton momenta was measured: the detected electron was emitted horizontally towards the detector, so that the vertical (along the polarization direction) components of the antineutrino and proton momenta were opposite. Using a set of angled slits in front of the proton detector only

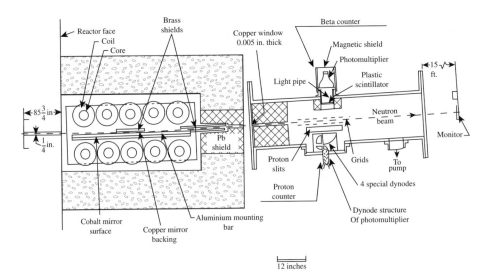

FIG. 2.4. Horizontal section of the experimental apparatus for the measurement of parity non-conservation in the beta decay of free neutrons (Burgy *et al.*, 1960).

protons with downward momenta in a given range were detected; the measured rates were compared with those obtained with a depolarized beam, since reversing the polarization direction by inverting the direction of the magnetic fields, (which would double the measured effect) could also affect detection efficiencies and thus bias the comparison.

The geometrical acceptance of the detector arrangement actually depended on the neutron transverse position in the beam, the electron energy and its direction: a correction factor was computed numerically, taking into account the shape of the beam, the electron energy spectrum and the geometry and energy resolution of the detectors. A significant positive correlation between the neutron spin and the antineutrino momentum was measured.

By rotating the neutron polarization with the guide field, so that it was horizontal in the decay region, the correlation of the electron and proton momenta detected was measured and compared with the one for opposite neutron polarization. For this measurement the angled slits were removed. Since the antineutrino momentum is preferably directed along the neutron spin direction (as measured with the set-up described above) the proton momentum is preferably directed along the opposite direction, i.e. either parallel or anti-parallel to the accelerating field towards the proton detector, depending on the direction of the neutron spin: this required a small acceptance correction to be applied. A very significant negative correlation between the neutron spin and the electron momentum was measured.

The same experimental set-up was also used for a test of symmetry under time reversal (Burgy et al., 1958), discussed in Chapter 4. Briefly, by comparing the proton-electron coincidence rates (with angled slits in front of the proton detector) for neutrons longitudinally polarized with spin parallel or anti-parallel to their momentum, one is comparing a process with its time-reversed counterpart, having all spins and momenta reversed (actually the second process is the time-reversed of the first one after a π rotation around the beam axis, which is assumed to leave physics unchanged).

2.3.3 Decay asymmetries of elementary particles

As also suggested by Lee and Yang (1956c), the study of a double weak decay process allows one to detect a parity-violating polarization term without using any external field: for example, in the weak decay $\pi \rightarrow \mu \nu$ parity violation could manifest itself as a net polarization of the muon along its flight direction, $\langle \mathbf{S}_\mu \cdot \mathbf{p}_\mu \rangle \neq 0$, and in turn such polarization could be detected in the subsequent $\mu \rightarrow e\nu\bar{\nu}$ decay as a parity-violating asymmetry of the electron direction, in the muon centre of mass system, with respect to the muon flight direction (that is, the direction of its possible longitudinal polarization in its rest system).

Muon decay asymmetries

The occurrence of parity violation in muon decay was determined experimentally using stopped muons (Garwin et al., 1957). The use of decay muons from pions stopped in emulsions was problematic, both due to the sub-millimetre range of the muons and because any small asymmetry could be washed out by depolarization effects in the medium, therefore requiring a large number of events to measure any asymmetry (with the consequent danger of a bias in the visual scanning of the emulsions).

Instead of tracking each muon from a parent pion, muons of known direction, originating in vacuum from the decay in flight of 85 MeV pions produced in the cyclotron at Columbia University were used. The key consideration is that the available muons, about 10% of the total beam flux, were naturally polarized since the acceptance channel favoured muons which decayed in the forward direction to be transmitted: for these pion energies, the 30 MeV/c muon momentum in the pion centre of mass results in a large laboratory momentum difference for forward and backward decays in the centre of mass (the recoil velocity in the centre of mass being proportional to the square root of the small recoil kinetic energy), and the momentum spectrum of decaying pions was such that there were many more low energy ones contributing forward-decay muons than there were high energy ones contributing backward-decay muons to the momentum-selection channel used. It should be recalled that if muons have a gyromagnetic number of 2 their polarization is maintained throughout their trajectory, across fringing, deflecting and focusing magnetic fields (see Appendix E).

The beam pions, with a range of 13 cm, were stopped in a 20 cm thick graphite absorber; the range of muons from forward decays was about 21 cm and these crossed a plastic scintillator behind the absorber.

Muons were stopped in a target, which was screened from the residual magnetic field of the accelerator by a metallic shield with high permeability, and their decays were detected by measuring the positrons in a scintillator telescope with a delayed time coincidence of a few μs with respect to the signal indicating that a particle was entering in the target. The target was also made of graphite which, due to its relatively low atomic number, is not expected to depolarize the muons strongly; anyway it should be noted that any possible depolarization would only dilute an existing asymmetry but it could not create a fake one.

The telescope consisted in two counters separated by an absorber, which guaranteed that the detected positrons had at least 25 MeV energy, to reduce the environmental background (see Fig. 2.5). The experiment was performed with a positive beam since μ^- can undergo nuclear capture before decaying.

Since the muons decay at rest, any anisotropy of its decay products' directions can only be ascribed to a net muon polarization, and the experiment aimed at measuring such distribution, which for a spin 1/2 can be of the form $1 + a \cos \theta$, θ being the angle between the positron direction and the original muon flight direction.

To sample the positron distribution at different angles with respect to the muon flight direction the counters would have to be moved, introducing effects due to scattering, energy loss and muon spin precession in the residual magnetic field,[28] non-symmetrical stopping distribution of muons, the variation of photomultiplier gain with the fringing field of the cyclotron, etc., which could generate false forward-backward asymmetries.

The solution was to keep the positron counter fixed in the horizontal plane, and to sample the entire angular distribution by varying the magnitude of a vertical magnetic field $\boldsymbol{B}$ in the target region, thus causing the precession of the muon spin before its decay, with an angular frequency $\omega = ge|\boldsymbol{B}|/(2m_\mu c)$, where g is the muon gyromagnetic number; since $\omega = 14$ kHz

---

[28] Indeed, non-exponential decay distributions for stopped muons, which had sometimes been observed in an irreproducible way (Garwin, 1974), were due to the precession of polarized muons in the fringe fields!

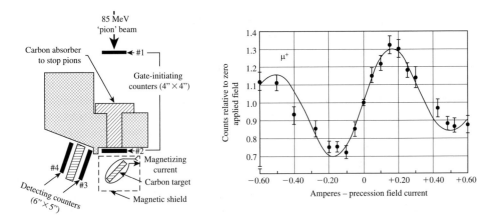

FIG. 2.5. The experiment of Garwin *et al.* (1957) on parity violation in muon decay. Left: sketch of the experimental apparatus; a coil wound around the carbon target provided the uniform vertical (orthogonal to the page) magnetic field. Right: variation of the coincidence rate of counters 3 and 4 with magnetizing current, showing the effect of spin precession; the curve is the one expected for an angular distribution $dN/d\cos\theta = 1 - \cos\theta/3$.

in a field of 1 G, with a relatively weak magnetic field of $\sim$35 G a half-turn rotation was obtained in $\sim$1 μs. The direct measurement of the time dependence of the rate due to the precession was not possible, and the total rate in a fixed time interval of 1.25 μs duration, starting 0.75 μs after the muon entered the target, was measured instead.

A large sinusoidal variation of the positron counting rate with the magnitude of the field was observed, corresponding to $a = -0.30 \pm 0.03$, indicating a large asymmetry of the decay rate with respect to the initial muon direction, and therefore a non-zero $\langle \boldsymbol{S}_\mu \cdot \boldsymbol{p}_e \rangle$. The effect was also confirmed by actually moving the detector at a different angle. By reducing the amount of absorber, thus also allowing pions to reach the target, a higher positron rate was detected but no asymmetry was measured, due to the decaying muons being emitted isotropically from stopping pions.

This measurement provided evidence for parity violation in processes involving elementary particles, which confirmed the result obtained in nuclear beta decays, with quite different systematics.

As a by-product this experiment gave the first measurement of the muon gyromagnetic number $g = 2$, to about 5% accuracy, by the measurement of the precession rate and the fact that the muon polarization when it stopped was consistent with a longitudinal polarization at production which was rotated by the same amount as the momentum, along its path in the magnetic fields. This supported the idea that the muon was indeed a Dirac particle having only electromagnetic and weak interactions.

The same measurement was also performed using $K^+ \to \mu^+\nu$ decays (Coombes *et al.*, 1957).

Another experiment was performed at about the same time in Chicago (Friedman and Telegdi, 1957a), by analysing μ^+ decaying at rest produced from the decay of 40 MeV

π^+ stopped in a stack of unsupported nuclear emulsions.[29] In this case the decay time is not measured, so it is not possible to exploit the muon spin precession as in the previous experiment. It should be noted that in the case of muons produced by moving pions, the longitudinal polarization is reduced by the Lorentz transformation to the laboratory system, with respect to the value it has in the pion rest system; for this reason the largest asymmetry is expected a priori by using pions decaying at rest as in this kind of experiments; on the other hand, the dense material can be expected to induce larger depolarization effects. For each event the angle between the positron emission direction and the initial muon direction, i.e. the direction before any multiple scattering (which could change the muon momentum but not its spin orientation) was measured with an accuracy of about 2°, and a forward-backward asymmetry was formed.

Among the possible sources of muon depolarization[30] which would reduce the measured effect, an important one is the formation of $\mu^+ e^-$ atoms ('muonium'[31]). Since the electrons in the emulsions are not polarized, for a fully polarized muon (say $J_z = +1/2$) half of the muonium would be formed with anti-parallel spins,[32] corresponding to a superposition of the $^1S_0$ ground state and the $^3S_1$ excited state, which are separated by the hyperfine splitting (corresponding to $\omega \sim 4.5$ GHz); the fast spin oscillations of the muon in such state completely depolarize it before decay, thus reducing the measured asymmetry coefficient. Moreover, the other half of muonium, produced in a pure $^3S_1$ state with $J_z = +1$ (parallel spins) would precess at a much higher frequency than a free muon (since the magnetic moment is of order $\mu_B = e\hbar/(2m_e c)$ rather than $e\hbar/(2m_\mu c)$), namely 1.4 MHz in a 1 G magnetic field (the order of magnitude of the intensity of the Earth's magnetic field), therefore losing its polarization much more easily.

Another concern is that any small residual magnetic field (not aligned with the muon direction) can erase an effect by making the polarized muons precess, therefore effectively depolarizing them in the variable time before their decay; for this reason the emulsions were contained inside three coaxial cylindrical magnetic shields to reduce the magnetic field below 4 mG. A disadvantage of the method lies in its inefficiency, since to avoid excessive crowding of the emulsions only a limited exposure time to the pion beam was allowed ($\sim 10^4$ pions/cm^2), resulting in a rather limited number of events.

A forward-backward asymmetry $A = (-0.091 \pm 0.022)$ was measured with 2,000 events (Friedman and Telegdi, 1957b), thus giving further evidence for longitudinal muon polarization and parity violation.

A null check was performed by measuring the asymmetry of muon emission direction with respect to the pion incidence direction, which indeed was found to be compatible with zero as expected for the decay of a spin zero particle at rest.

---

[29] These are photographic emulsions with a particularly high concentration of the heavy elements (silver and bromine) with which the charged particles interact, which have been stripped off the glass on which were prepared, so that the detector is almost fully active.

[30] The slowing down of muons in matter does not affect their polarization in a significant way, since mainly the Coulomb interaction is involved, and magnetic forces are small.

[31] According to the rules this should actually be called 'muium', as 'muonium' is a $\mu^+\mu^-$ atom, but the name stuck.

[32] The capture process is also due to Coulomb forces and does not affect spins.

Another experiment at Brookhaven (Abashian *et al.*, 1957) measured the same asymmetry for 100 MeV π^+ produced in the Cosmotron[33] and stopped in a liquid hydrogen bubble chamber situated after an appropriate amount of absorber so that pions stopped in the chamber. Hydrogen had a small depolarizing effect on muons, and of course a magnetic shield ($|\boldsymbol{B}| < 0.25$ G) surrounded the chamber. An asymmetry coefficient $a = 0.250 \pm 0.045$ was measured, and again the measurement of the isotropic distribution of decay muon directions provided a consistency check.

As in the case of β decay another pseudo-scalar quantity which can be measured experimentally is the longitudinal polarization of the electron or positron in the decay of unpolarized muons. The longitudinal polarization of the $e^\pm$ from unpolarized muon decays was also measured and provided a further indication of parity violation in a purely leptonic process (Culligan *et al.*, 1957).

The experiments described so far only proved that parity symmetry was violated in weak interaction processes involving leptons, and it was not inconceivable at that time that such phenomena could be due to some peculiar property of the neutrino, while the $\tau - \theta$ puzzle (involving hadrons) might still stand: no pseudo-scalar quantity could be measured in the latter, since the two- and three-pion states have different numbers of particles and cannot interfere directly.

The possibility of using the decays of hyperons (another important weak interaction process known at the time) to test parity symmetry was already pointed out in the original paper by Lee and Yang (1956c). Indeed, the combination of strong production and weak decay of hyperons can be exploited to provide a measurement of the expectation value of a pseudo-scalar quantity.

Associated production of Λ by strong interactions occurs e.g. when a π^- beam hits a target via the process $\pi^- p \to \Lambda K^0$. If the Λ is not produced in the forward (pion) direction, a reaction plane is defined by the incident π^- beam momentum $\boldsymbol{p}_b$ and the outgoing Λ momentum $\boldsymbol{p}_\Lambda$, as the plane orthogonal to $\boldsymbol{p}_b \times \boldsymbol{p}_\Lambda$. In the subsequent decay of the Λ, e.g. to $p\pi^-$, parity symmetry implies that there can be no preference for the outgoing proton momentum $\boldsymbol{p}_p$ to be above or below this plane, i.e. the pseudo-scalar quantity $\boldsymbol{p}_b \times \boldsymbol{p}_\Lambda \cdot \boldsymbol{p}_p$ should have zero average value; if this were not the case, the two processes related by a mirror reflection with respect to the production plane would have different probabilities, indicating a parity asymmetry. If parity is conserved in the (strong) production process, any polarization $\mathcal{P}_\Lambda$ of the hyperon must be orthogonal to the production plane ($\mathcal{P}_\Lambda \cdot \boldsymbol{p}_\Lambda$ is a pseudo-scalar quantity, while $\mathcal{P}_\Lambda \cdot \boldsymbol{p}_b \times \boldsymbol{p}_\Lambda$ is scalar), and the above condition for a parity symmetric decay is then equivalent to $\langle \mathcal{P}_\Lambda \cdot \boldsymbol{p}_p \rangle = 0$. The polarization of the

---

[33] This was the name given to Brookhaven's proton synchrotron, the first to ever reach the GeV energy range, recreating in a controlled setting the physics of the cosmic rays, and to provide a proton beam for experiments outside the accelerator. It was built in 1952 and dismantled in 1969.

produced Λ is not generally known, although close to threshold (for sufficiently low beam momenta) it can be assumed that only the states with lowest orbital angular momentum are involved in the production process, and therefore the polarization can be extracted from measurements of the angular production cross section at different energies (Lee *et al.*, 1957*a*).

For the decay of the spin 1/2 hyperon Λ into a spin 1/2 nucleon N and a spinless pion the two final states allowed by angular momentum conservation are $^2S_{1/2}$ ($L = 0$) and $^2P_{1/2}$ ($L = 1$); in case of parity symmetry only the first one is allowed if Λ and N have opposite intrinsic parities, while only the second one is allowed if the two particles have the same intrinsic parities. If parity is not conserved, both states can be present at the same time; in general three real quantities are sufficient to describe the decay, namely the moduli of the decay amplitudes A_S, A_P corresponding to the two final states and their relative phase. These three constants can be expressed (Lee and Yang, 1957) in terms of the decay rate $\Gamma = |A_S|^2 + |A_P|^2$ and of two angular correlation parameters; one of these can be chosen as the asymmetry parameter α describing the anisotropy of the decay angular distribution (when averaging over the nucleon polarization):

$$\frac{d\Gamma}{d\Omega} = \frac{\Gamma}{4\pi}(1 + \alpha\,\mathcal{P}_\Lambda \cdot \hat{\boldsymbol{p}}_N) \tag{2.72}$$

where $\hat{\boldsymbol{p}}_N = \boldsymbol{p}_N/|\boldsymbol{p}_N|$ is the outgoing nucleon momentum versor, in the Λ rest frame. In terms of the two decay amplitudes

$$\alpha = \frac{2\,\mathrm{Re}(A_S^* A_P)}{|A_S|^2 + |A_P|^2} \tag{2.73}$$

and $\alpha \neq 0$ therefore indicates parity violation, i.e. the presence of both amplitudes (α could be accidentally zero in case the two amplitudes are relatively imaginary, though); if $\alpha \neq 0$ the angular distribution of the decay involves an odd power of $\cos\theta = \mathcal{P}_\Lambda \cdot \hat{\boldsymbol{p}}_N)/|\mathcal{P}_\Lambda|$.

It should be noted that if the hyperon is not polarized in the production reaction, no asymmetry can be present even if parity symmetry is violated: indeed in this case no preferred direction is defined in the hyperon centre of mass system. Experiments measuring the decay asymmetry of the emitted nucleon are only sensitive to the product $\alpha|\mathcal{P}_\Lambda|$, and the statistics required for a conclusive experiment depends on the (unknown) hyperon polarization at production.

The relatively short lifetime of the Λ hyperon ($\simeq 2.6 \cdot 10^{-10}$ s) allows experiments to be performed with decay in flight by measuring the decay asymmetry of the final proton

$$\frac{N_+ - N_-}{N_+ + N_-} = 2\alpha|\mathcal{P}_\Lambda| \tag{2.74}$$

where $N_\pm$ are the number of events with $\xi \equiv \boldsymbol{p}_b \times \boldsymbol{p}_\Lambda \cdot \hat{\boldsymbol{p}}_N$ positive or negative. Note that ξ, being the component of $\boldsymbol{p}_N$ orthogonal to the production plane, is not affected by the boost to the laboratory system in which the Λ is moving, and therefore the number of decays 'above' or 'below' the production plane is the same in both systems.

Soon after Lee and Yang's proposal, a large parity-violating asymmetry in the $\Lambda \to p\pi^-$ decay was indeed measured (Crawford et al., 1957; Eisler et al., 1957).

If no average is performed over the polarization states of the emitted nucleon, the angular distribution of the decay (in the hyperon rest frame) is (see e.g. Gasiorowicz (1966))

$$\frac{d\Gamma}{d\Omega} = \frac{\Gamma}{4\pi} \left\{ 1 + \alpha \, \mathcal{P}_\Lambda \cdot \hat{\boldsymbol{p}}_N + \mathcal{P}_N \cdot \left[\alpha + (1 - \gamma)\mathcal{P}_\Lambda \cdot \hat{\boldsymbol{p}}_N \right] \hat{\boldsymbol{p}}_N \right.$$
$$\left. + \beta \, \hat{\boldsymbol{p}}_N \times \mathcal{P}_\Lambda + \gamma \mathcal{P}_\Lambda \right\} \tag{2.75}$$

where

$$\beta = \frac{2 \operatorname{Im}(A_S^* A_P)}{|A_S|^2 + |A_P|^2} \qquad \gamma = \frac{|A_S|^2 - |A_P|^2}{|A_S|^2 + |A_P|^2} \tag{2.76}$$

and the relation $\alpha^2 + \beta^2 + \gamma^2 = 1$ holds.

The polarization of the final nucleon (in its rest system) is

$$\mathcal{P}_N = \frac{(\alpha + \mathcal{P}_\Lambda \cdot \hat{\boldsymbol{p}}_N)\hat{\boldsymbol{p}}_N + \beta \mathcal{P}_\Lambda \times \hat{\boldsymbol{p}}_N + \gamma \hat{\boldsymbol{p}}_N \times (\mathcal{P}_\Lambda \times \hat{\boldsymbol{p}}_N)}{1 + \alpha \mathcal{P}_\Lambda \cdot \hat{\boldsymbol{p}}_N} \tag{2.77}$$

It can be noted from eqn (2.77) that the longitudinal polarization of the emitted nucleon in the decay of an unpolarized hyperon at rest is just given by α: the measurement of such polarization therefore also allows a direct determination of the asymmetry parameter. Since there is no preferred orientation for the hyperon production plane, and therefore for the direction along which it can possibly have a net polarization, by integrating over the orientation of such production planes an unpolarized sample is obtained, irrespective of the polarization in the production process. Measuring the helicity of the decay nucleon in such a configuration thus allows one to extract α without any knowledge of the hyperon polarization (and also in case no net polarization is not present at all).

In the first experiments (Boldt et al., 1958; Birge et al., 1960) to detect a parity-violating proton longitudinal polarization in $\Lambda \to p\pi^-$ decay, the left-right asymmetry in the subsequent elastic scattering of the decay protons was used: this asymmetry is actually sensitive only to *transverse* proton polarization, but for hyperons decaying in flight, the polarization in the laboratory system is rotated with respect to that in the centre of mass, in such a way that a longitudinal polarization is partially turned into transverse (see Appendix E); the same measurement cannot be performed with hyperons decaying at rest.

The magnitude of α can be rather large: $\simeq 0.64$ for $\Lambda \rightarrow p\pi^-$ and $\Lambda \rightarrow n\pi^0$, $\simeq -0.98$ for $\Sigma^+ \rightarrow p\pi^0$, $\simeq -0.41$ for $\Xi^0 \rightarrow \Lambda\pi^0$ and $\simeq -0.46$ for $\Xi^- \rightarrow \Lambda\pi^-$, indicating that A_S and A_P are of comparable magnitude for these decays (however $\alpha \simeq +0.07$ for $\Sigma^+ \rightarrow n\pi^+$ and $\simeq -0.07$ for $\Sigma^- \rightarrow n\pi^-$).

By reversing the previous arguments, if the asymmetry parameter α is known for a weak decay the measurement of the decay angular asymmetry gives information on the parent hyperon polarization: the decay is said to be *self-analysing*. This is particularly interesting when a hyperon decays into another hyperon and a pion, e.g. $\Xi^- \rightarrow \Lambda\pi^-$, since in this case not only the decay asymmetry but also the polarization of the daughter particle (the second hyperon) can be measured from angular distributions, giving in principle (i.e. if the polarization of the parent hyperon is known) experimental access to both the α and β parameters.

The above experiments (and others) eventually led to the resolution of the $\tau - \theta$ puzzle: the two and three pion decays originate from a single particle (now called $K^\pm$) which can actually decay to states of different parity because the weak interactions inducing such decays do not respect parity symmetry.

The measurements of pseudo-scalar quantities did not leave any room for doubt, and the fall of parity symmetry was indeed rapidly accepted as a fact, bringing the Nobel prize to Tsung Dao Lee and Chen Ning Yang in 1957. Nevertheless, this discovery was a major upset for the scientific community, since the parity conservation law was widely expected to be universally valid, somehow akin to the conservation of energy, as due to some property of the 'environment' on which physics is played, rather than something linked to a specific interaction.

While it is now clear that there is no a priori reason for the validity of space (and time) reflection symmetry, the fall of parity had a very strong impact on the physics community, as can be read in the words of eminent physicists: 'Now after the first shock is over, I begin to collect myself. Yes, it was very dramatic.', 'It is good that I did not make a bet . . . I did make a fool of myself.' (W. Pauli, 1964); 'A rather complete theoretical structure has been shattered at the base and we are not sure how the pieces will be put together.' (I. I. Rabi, 1957). In the opinion of the author the best expression of this state of affairs actually predates the experimental discovery by twenty years (Magritte, 1937).

Much has been written on this subject, and the words of physicists deeply involved in the issue are illuminating: 'These people assumed that parity conservation/violation must represent a property of space-time. If parity were violated in τ decay, it should therefore fail for all interactions, whereas the evidence was strong that parity was conserved in nuclear and electromagnetic interactions.' (R. H. Dalitz, 1989); 'Inversion enters into considerations of physics because of another common perception – that right- and left-handedness are matters of convention rather than of substance. [. . .] *Either implicitly or explicitly the dynamic laws were assumed to be invariant under inversion*' (R. G. S. Sachs, 1987). Indeed, since Wigner's explanation of Laporte's rule (Wigner, 1927), the intuitive idea of left-right symmetry was rooted into the quantum description of Nature: 'Thus Lee and Yang's suggestions have led to a great liberation in our thinking on the very structure of physical theory. Once again principle turned out to be prejudice.' (A. Pais, 1988).

However, it should also be noted that other physicists had different preconceptions: when informed of the fall of parity symmetry, P. A. M. Dirac replied: 'If you look carefully, you will see that the concept [of parity] is not once used in my book' (quoted by Telegdi, 1973), and indeed he had stated earlier: 'I do not believe there is any need for physical laws to be invariant under reflections in space and time although the exact laws of nature so far known do have this invariance' (P. A. M. Dirac, 1949). Also: 'The argument against electric dipoles [...] raises directly the question of parity. [...] But there is no compelling reason for excluding this possibility [of parity non-conservation]. It would not be the only asymmetry of particles of ordinary experience, which already exhibit conspicuous asymmetry in respect to electric charge.' (E. M. Purcell, N. F. Ramsey, 1950); 'Even instinctive knowledge of so great a logical force as the principle of symmetry [...] may lead us astray [...]. The instinctive is just as fallible as the distinctly conscious. Its only value is in provinces with which we are very familiar' (E. Mach, 1893).

It should also be mentioned that while the violation of parity symmetry by weak interaction is an unquestionable fact, a strong desire to maintain the validity of such symmetry led to several hypothetical scenarios in which the symmetry is restored at larger energy or distance scales, such as left-right symmetric models (Section 9.5) and the idea of 'mirror matter' (Section 11.4); none of these possibilities has received any significant experimental support, so far.

2.3.4 Atomic systems

Parity tests are also performed with atomic systems: in absence of atomic processes in which the weak interaction manifests uniquely, small effects due to the neutral weak current are studied, see Fortson and Lewis (1984) for a review. The weak interaction is expected to contribute a modification to the potential between an electron and a nucleus at very short distances: the electronic current density of an atom indeed has a helical structure, thus exhibiting a defined chirality. Coulomb repulsion implies that such effects are important in the interaction between the electrons themselves, but they lead to parity mixing among the S and P states, being those in which electrons spend more time close to the nucleus.

Parity violation manifests itself in the presence of a parity-odd amplitude A_{PV}, usually (in processes not driven uniquely by weak interactions) significantly smaller than the parity-even one A_{PC}, so that for two mirror image experiments the transition probability is different:

$$|\langle f_R|T|i_R\rangle|^2 = |A_{PC} + A_{PV}|^2 \qquad |\langle f_L|T|i_L\rangle|^2 = |A_{PC} - A_{PV}|^2 \qquad (2.78)$$

Due to the opposite parity transformation of A_{PC} and A_{PV} the transition probability contains a pseudo-scalar interference term $2\mathrm{Re}(A_{PV}A_{PC}^*)$ whose sign depends on the 'handedness' of the system, resulting in a left-right asymmetry:

$$A = \frac{|\langle f_{(R)}|\mathcal{T}|i_{(R)}\rangle|^2 - |\langle f_{(L)}|\mathcal{T}|i_{(L)}\rangle|^2}{|\langle f_{(R)}|\mathcal{T}|i_{(R)}\rangle|^2 + |\langle f_{(L)}|\mathcal{T}|i_{(L)}\rangle|^2} = 2\frac{\mathrm{Re}(A_{PV}A_{PC}^*)}{|A_{PC}|^2 + |A_{PV}|^2} \simeq 2\mathrm{Re}\left(\frac{A_{PV}}{A_{PC}}\right)$$

$$(2.78)$$

For atomic systems A_{PC} is the electromagnetic amplitude A_{EM} (assumed to be parity-conserving) and A_{PV} is the (parity-violating part of the) weak amplitude A_W; while the order of magnitude of the former is $A_{EM} \sim 1/q^2$ (q being the four-momentum transfer), that for the latter is $A_W \sim 1/(q^2 + M_Z^2)$, with the mass of the neutral Z boson $M_Z \sim 100\,\mathrm{GeV}/c^2$ determining the disparity in the interaction strengths despite the similar magnitude of coupling constants dictated by electroweak unification. The asymmetry is thus of order $A \sim 2q^2/M_Z^2$, and for a simple atomic system $q^2 \sim -\mathbf{q}^2 \sim (\hbar/a_0)^2$ (where a_0 is the Bohr radius) and one expects effects of order $\mathcal{A} \sim 10^{-14}$.

As first pointed out in Bouchiat and Bouchiat (1974), enhancements in heavy atoms can make the effects observable. The size of the expected parity mixing effects is proportional to the matrix element of the parity-violating potential U_{PV} divided by the energy level spacing ΔE. The parity-violating weak potential has very short range, $U_{PV}(\mathbf{r}) \propto Z(v/c)\delta(\mathbf{r})$ (where a factor Z appears in the expression of the effective weak charge of a nucleus, since the contributions of all nucleons add up, and a linear dependence on the electron velocity v is present), so that its matrix element is $\langle \psi|U_{PV}|\psi\rangle \propto |\psi(0)|^2 \sim 1/V$ where V is the volume of the system; this favours nuclei over atoms by a large factor $O(10^{12})$, which reduces to $O(10^6)$ because energy level spacings are $\Delta E \sim \mathrm{eV}$ in atoms and $\Delta E \sim \mathrm{MeV}$ in nuclei. However heavy atoms, for which the region close to the nucleus (where the short-range weak interaction is effective) resembles that of an hydrogenic ion with charge Ze, have a Bohr radius $a_0/Z^{1/3}$. Finally the electron velocity is also proportional to Z, leading to the increase of parity-violating effects in heavy atoms roughly as Z^3 (actually even faster due to relativistic effects).

Atomic species for which theoretical computations are reliable due to the particularly simple structure are used: alkali metals, with a single valence electron, are the favoured choice.

Atomic parity violation experiments look for electric dipole (E1) moments induced by parity-violating forces: permanent electric dipoles also violate time reversal symmetry (see Chapter 4), so that if such symmetry is valid (as assumed here), only *transition* dipole moments can be present, that is dipole moments which are present when the system is in a non-stationary state, such as during absorption or emission processes; these moments are observable through their interference with other atomic moments in radiative transitions.

The transition amplitude interfering with the tiny E1 one induced by parity violation must be a small one, in order for the interference effect to be observable. One possibility is that of a relatively weak M1 atomic transition induced by an oscillating magnetic dipole moment $\tilde{\boldsymbol{d}}_M$; any parity-violating induced electric dipole moment $\tilde{\boldsymbol{d}}_E$ must have a phase difference of $\pi/2$ with respect to $\tilde{\boldsymbol{d}}_M$, so that the quantity $\tilde{\boldsymbol{d}}_M \cdot \tilde{\boldsymbol{d}}_E$, which is odd under time reversal, averages to zero. The phase difference between E1 and M1 radiation causes circular polarization or optical rotation as observable parity-violating helicity effects.

Another possibility is to have an E1 transition induced by a weak static external electric field interfering with the parity-violating E1 one, so that the transition rate depends on the relative orientation of the external field and the polarization of the radiation. With this 'Stark interference' technique the effect can be controlled in magnitude and sign by the external field, which helps in identifying a small signal.

The first class of atomic parity non-conservation experiments looks for *optical rotation* of the polarization plane of light when passing through heavy metal vapours (a high Z medium transparent to light); this is a macroscopic effect, although its expected size is tiny, of the order of 10^{-7} rad per metre of traversed material.

Linearly polarized light can be decomposed as the superposition of oppositely circularly polarized waves: for light of wave vector $\boldsymbol{k} = k\hat{\boldsymbol{z}}$ polarized along $\hat{\boldsymbol{x}}$ at the entrance of a medium ($z = 0$)

$$A(z = 0) \propto \boldsymbol{\epsilon}^{(1)} = \frac{1}{\sqrt{2}} \left[\boldsymbol{\epsilon}^{(L)} - \boldsymbol{\epsilon}^{(R)} \right]$$

If the index of refraction within the medium is different for the two types of circular polarization ($n_L \neq n_R$), then at the exit from the medium ($z = L$):

$$A(z = L) \propto \frac{1}{\sqrt{2}} \left[\boldsymbol{\epsilon}^{(L)} e^{i n_L kL} - \boldsymbol{\epsilon}^{(R)} e^{i n_R kL} \right]$$

$$= e^{i(n_R + n_L)kL/2} \left\{ \boldsymbol{\epsilon}^{(1)} \cos[(n_L - n_R)kL/2] + \boldsymbol{\epsilon}^{(2)} \sin[(n_L - n_R)kL/2] \right\} \quad (2.79)$$

corresponding to a rotation of the polarization plane[34] by an angle $\Delta\phi = \mathrm{Re}(n_L - n_R)kL/2$, greatest close to the absorption lines (where $n_{L,R}$ differ most from 1). Such rotation angle cannot be increased at will by increasing L (the amount of material traversed), since light is also absorbed in the medium.

The optical rotation angle is indeed related to the degree of circular polarization in the transition, or the asymmetry in the microscopic interaction of left and right circularly polarized photons (interaction which determines the difference of the

---

[34] Since the absorption coefficients (the imaginary parts of $n_{R,L}$) also differ for the two circular polarizations, the polarization actually changes from linear to elliptical.

refractive index from 1), that is

$$P = \frac{\text{Re}(n_L - 1) - \text{Re}(n_R - 1)}{\text{Re}(n_L - 1) + \text{Re}(n_R - 1)} \simeq \frac{\text{Re}(n_L - 1) - \text{Re}(n_R - 1)}{2\text{Re}[(n_L + n_R)/2 - 1]}$$

$$= \frac{\Delta\phi}{kL} \frac{1}{\text{Re}[(n_L + n_R)/2 - 1]} \simeq \frac{2\Delta\phi}{\lambda L} \frac{\text{Im}[(n_L + n_R)/2]}{\text{Re}[(n_L + n_R)/2 - 1]}$$

$$(2.80)$$

where $\lambda \simeq 2k\text{Im}[(n_L + n_R)/2]$ is the linear absorption coefficient. The error in the determination of P depends on that of the optical rotation angle $\Delta\phi$, which decreases as $\sim 1/\sqrt{N}$ with the number N of detected photons, and N in turn decreases exponentially with the thickness L as $N = N_0 e^{-L/\lambda}$, so that the statistical error on P is minimum for $L = 2\lambda$, i.e. a thickness of two absorption lengths (which can be of order ~ 1 m in gases).

Optical rotation of light

The first optical rotation experiments in the 1970s exploited the $^4S_{3/2} \to ^2 D_{3/2}$ 876 nm M1 transition in bismuth ($Z = 83$), accessible to tunable lasers available at that time.

A general problem is that the long accumulation times required to measure the tiny parity-violating effect make the experiments sensitive to any time drifts of the detectors and the electronics.

The rotation angle is measured by placing the vapour between two crossed polarizers, so that a slight rotation in the polarization plane can cause a large fractional change in the transmitted light intensity. Denoting with ϕ the (small) angle between the light polarization plane and the plane of minimum transmission, the transmitted light intensity varies as

$$I = I_0 \sin(\phi_0^2 + \phi^2) \simeq I_0(\phi_0^2 + \phi^2)$$

where I_0 is the incident light intensity and ϕ_0^2 is a constant called the *extinction angle*, depending on the quality of the polarizers and of the light beam, which results in non-zero transmission even for $\phi = 0$. Instead of performing the measurement close to $\phi = 0$, where the first derivative of I with respect to ϕ vanishes, the angle ϕ is artificially offset from zero as $\phi = \phi_M + \delta\phi$, where ϕ_M can be chosen at will and $\delta\phi$ is the light rotation angle due to parity violation. The largest fractional intensity change is obtained for $\phi_M = \phi_0$, and for a typical value $\phi_0^2 \sim 10^{-7}$ a rotation of only $\delta\phi \simeq 3 \cdot 10^{-7}$ rad results in a 10^{-3} fractional intensity change (as compared to $\sim 10^{-6}$ if $\phi_M = 0$).

In one experiment in Seattle (Hollister *et al.*, 1981), a water-filled glass cell was placed after the first polarizer and in front of the bismuth vapour. Instead of trying to measure a tiny constant offset $\delta\phi$, the experiment exploited the Faraday effect in the water cell – the rotation of plane polarized light occurring in a medium when a magnetic field is present along the direction of light propagation – to modulate the offset rotation angle ϕ_M with a 10^{-3} rad amplitude at 1 kHz using a coil around the cell. The signal after the second polarizer, being proportional to $(\phi_M + \delta\phi)^2$, exhibits an oscillation at twice this frequency and a 1 kHz term with amplitude $2\phi_M\delta\phi$, so that the detection of this fundamental frequency is the indicator of optical rotation. A second winding on the glass cell was used for the slow correction of the average offset angle to keep the system close to extinction, and to compensate for intensity variations (also due to the bismuth absorption pattern as a function of wavelength).

The bismuth vapour was inside a 1 m long (up to 10 absorption lengths) magnetically shielded tube (magnetic fields had to be kept below 10^{-8} T to avoid spurious effects due to Faraday rotation in bismuth) placed in an oven at ~ 1300 K; a helium 'buffer' gas was used to confine the vapour into the heated part of the tube, to prevent bismuth from condensing on the cool windows.

A solenoid around the bismuth cell could produce a known magnetic field for calibration through the Faraday effect; note that while the parity-violating optical rotation has the same sign for any direction of propagation – therefore being also independent on the relative orientations of the atoms in the sample – the sign of the Faraday rotation reverses for the opposite direction of propagation; typical Faraday rotation effects are O(0.4) rad T^{-1} cm^{-1}. The same coil was also used with small currents to compensate for residual magnetic fields beyond the shielding during the measurement. A schematic of the experimental apparatus is shown in Fig. 2.6.

The light source was a laser diode, emitting most of the light at a single wavelength, which required current regulation to 1 μA and temperature regulation to 10^{-4} K to maintain the desired wavelength stability. The laser current, and therefore its wavelength, was modulated with a triangular shape to sweep over the transition line of interest (of 0.05 nm total width, with

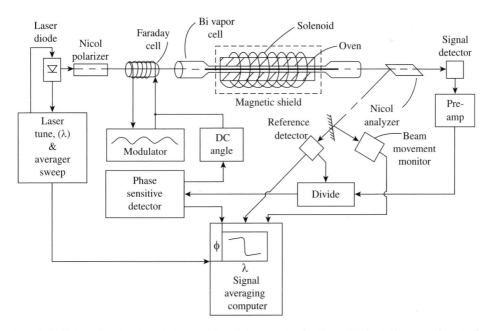

FIG. 2.6. Schematic drawing of the experimental apparatus for the optical rotation experiment of (Hollister *et al.*, 1981).

9 main hyperfine M1 lines) in about 0.1 s. During half of the sweep cycle the coil around the bismuth cell was adjusted to give zero residual Faraday effect to measure the parity-violating rotation (the oven heaters were off during this time to avoid any interference from AC power). During the other half of the cycle a large current in the coil produced the Faraday effect for calibration; this current alternated between two slightly different values, which resulted in rotation angles differing by less than what was expected for the signal, providing in such a way an additional calibration.

The intensity of the light reflected from the face of the second polarizer was continuously monitored and used as a reference for normalizing the signal, to be independent from light intensity variations, including those due to the bismuth absorption. The variation in the effect while crossing each of the hyperfine components of the line helps to distinguish the signal from the background. The curves describing light transmission, Faraday rotation and parity-violating optical rotation as a function of wavelength were compared to the theoretical predictions, resulting in excellent agreement and allowing the extraction of the parity non-conserving rotation effect. The rotation angle is proportional to the path length of light in the vapour, which was measured by the absorption and Faraday rotation curves; the rotation angle per absorption length was measured to be $(-10.4 \pm 1.7) \cdot 10^{-8}$ rad.

Other optical rotation experiments with different atoms also observed the effect, with rotation angles per absorption length in the range 1 to $20 \cdot 10^{-8}$ rad.

In a second class of atomic parity violation experiments, the interference between an E1 transition induced by an external electric field and the one induced by the parity-violating forces is searched for through the measurement of *circular dichroism*, a difference in the light absorption for oppositely circularly polarized light.

An electronic polarization is induced in the heavy atom vapours by the absorption of circularly polarized laser light in the presence of a static electric field: a laser excites a first-order forbidden M1 transition, and the electronic polarization of the excited state in the direction orthogonal to the electric field E and to the photon direction k is measured by the circular polarization of the de-excitation light. The parity-odd quantity is $\pm \langle J \cdot k \times E \rangle \neq 0$ (where the sign is opposite for the two photon circular polarization states); the parity-violating effect is detected by the change in circular polarization when the laser polarization is reversed (see e.g. Bouchiat *et al.* (1984)); moreover, the parity-violating term is also distinguished by reversing E. Again, a time-varying polarization is used for the measurements, together with multiple traversal of the laser light through the vapour by reflection, which adds up the effect being searched for while canceling several systematics. Experiments on caesium and thallium made with this approach measured parity-violating effects consistent with the magnitude expected from weak interactions.

Parity symmetry also implies that elementary particles cannot have permanent electric dipole moments $d_E = \int dx \, x \rho(x)$ (where ρ is the electric charge density), and searches for such quantities are also tests of such symmetry, which will be discussed in more detail in Chapter 4 when dealing with time reversal symmetry.

If parity symmetry is valid the eigenstates of the (free) Hamiltonian are parity eigenstates (in absence of degeneracies); for such states

$$\int d\boldsymbol{x} \, \langle \psi | \boldsymbol{x} | \psi \rangle = - \int d\boldsymbol{x} \, \langle \psi_P | \boldsymbol{x} | \psi_P \rangle = - \int d\boldsymbol{x} \, \langle \psi | \boldsymbol{x} | \psi \rangle = 0$$

where $|\psi_P\rangle = P|\psi\rangle$ is the parity-transformed state.

2.3.5 Electromagnetic interactions

No violation of parity symmetry was found in electromagnetic interactions, and experiments reached the level of sensitivity at which weak interaction effects can be detected for systems which interact electromagnetically.

The analysis of photon polarization in $\pi^0 \to \gamma\gamma$ decay (apparently) allows one to test parity conservation in this electromagnetic process: since the pion is spinless, the two photons must have the same helicity due to angular momentum conservation, without precluding the possibility of a different probability of both helicities being $+1$ with respect to both being -1, so that the photons would have a net polarization. This possibility would, however, violate parity symmetry, as the two situations are related by P. However, the two possible final states (two right-handed photons and two left-handed photons) are also related by CPT (Chapter 5), so that the more fundamental CPT symmetry of EM interactions already constrains the two transition probabilities to be equal (at least to order α^2), spoiling the test of P.

One should note that, in principle, allowing for a possible violation of parity symmetry, the two photons need not be in a parity eigenstate, and could actually be in a mixed state of parallel and perpendicular planes of polarization; in this case (Bernstein and Michel, 1960) the absence of circular polarization of the photons only implies that the two photons are in an eigenstate of TP (where T is time reversal) and the study of parity symmetry would require experiments measuring also the azimuthal (around the relative photon momentum) correlations. The fact that the two photons from π^0 decay have perpendicular polarization planes was actually verified experimentally (Samios et al., 1962).

Electromagnetic transitions among positronium states induced by external electric ($\boldsymbol{E}$) or magnetic ($\boldsymbol{B}$) fields could present parity-violating rate asymmetries corresponding to the reversal of one of the controllable vectors in a pseudo-scalar combination (Conti et al., 1986): examples are asymmetries in $1^3S_1\gamma \to 2^3S_1$ transitions proportional to $\boldsymbol{S}_\gamma \cdot \boldsymbol{E}$ or $\boldsymbol{k} \cdot \boldsymbol{B}$, where $\boldsymbol{k}$ is the external photon momentum and $\boldsymbol{S}_\gamma$ its polarization, or asymmetries in $2^3S_1\gamma \to 2^1S_0$ transitions proportional to $\boldsymbol{E} \cdot \boldsymbol{B}$ or $\boldsymbol{S}_\gamma \cdot \boldsymbol{k}$ The search for forbidden resonant one-photon transitions in positronium is not useful for the purpose since the parity change induced by the photon can be either ± 1 depending on the multipolarity of the radiation.

Parity symmetry violation was detected via asymmetries in the inelastic scattering of longitudinally polarized electrons on unpolarized targets: this process is mainly due to electromagnetic interactions, but weak interactions mediated by Z^0 exchange can also contribute, and the interference between the two amplitudes gives rise to parity-violating effects. Since the asymmetries are proportional to q^2 (Section 2.3.4) high-energy electron scattering is a good place to look for parity violation effects (Zel'dovich, 1959). Experimental results are indeed consistent with no parity violation in EM interactions, and with an effect entirely due to the contribution from weak interactions.

Inelastic electron scattering

The first evidence of parity violation in electron scattering was obtained by studying the deeply inelastic reaction $e^- d \rightarrow e^- X$ (X standing for any other particles) with longitudinally polarized electrons at Stanford (Prescott *et al.*, 1978). The key development was the intense source of polarized electrons, a gallium arsenide crystal cathode: 710 nm circularly polarized laser light incident on the crystal pumped electrons from the valence into the conduction band, and the photon polarization was partially transferred to the emitted electrons (electron polarization $\sim$40%).

The laser light was first polarized linearly by a calcite prism and then transmitted through a Pockels cell, a crystal with no inversion symmetry whose birefringence is proportional to the applied electric field, which produced the circular polarization acting as a voltage controlled quarter-wave plate. This approach allowed a fast inversion of the photon (and thus the electron) polarization by inverting the sign of the electric pulse applied to the cell: this sign was chosen randomly (and recorded) at each accelerator pulse, in order to eliminate spurious asymmetries due to detector variations in time with no relation to changes in beam or detector parameters.

The longitudinally polarized electrons were not depolarized during the acceleration to energies in the 16–22 GeV range, and their polarization was periodically measured by the asymmetry in elastic $e^- e^-$ scattering on a magnetized iron foil. The beam energy and position were stabilized by feedback loops; the intensity was of order 10^{11} electrons per 1.5 μs pulse (at 120 Hz repetition rate).

The electron beam hit a 30 cm thick liquid deuterium target, and a single-arm spectrometer at a fixed angle of 4° with respect to the beam direction accepted scattered particles in a broad momentum range, roughly $\sim$20% lower than the beam momentum; the background from π, μ and K was a few per cent, as that from electron elastic scattering. Electrons were detected by an atmospheric pressure nitrogen Čerenkov counter and a nine radiation lengths thick lead-glass calorimeter, both working in integrating mode due to the high rates (10^3 electrons per pulse), and thus used individually rather than in coincidence. For each accelerator pulse the yield was measured as the ratio of the integrated currents in the detectors over the integrated beam charge given by a monitoring system.

The asymmetry was defined as the difference of the yields for the two polarization directions defined by the sign of the Pockels cell voltage, divided by their sum. A shower counter placed after 15 cm of lead following the calorimeter detected penetrating hadrons and muons, and was used to check that the asymmetry due to the background was indeed consistent with zero. The measured asymmetry was $A = (-1.52 \pm 0.26) \cdot 10^{-4}$ at $q^2 = 1.6 \; (\text{GeV}/c)^2$.

Ancillary measurements allowed one to check for fake asymmetries induced by instrumental effects. By rotating the linear-polarizing prism, the effect of the Pockels cell could be inverted or inhibited: the corresponding asymmetry must vary with the rotation angle in a known way, and any spurious effect linked to the reversal of the cell voltage can thus be separated; such effects were indeed measured to be negligible.

Due to the non-zero anomalous magnetic moment, the electron spin precesses in the deflecting magnets of the beam line by an angle $\gamma\theta(g-2)/2$ (γ being the relativistic factor and $\theta = 24.5°$ the bending angle), proportional to the beam energy. As a result of this spin precession, longitudinal polarization was maintained only for energies which are integer multiples of 3.237 GeV, thus providing a source-independent way of reversing or cancelling the beam polarization. The expected variation of the asymmetry with beam energy was actually observed, thus confirming that the measured asymmetry was indeed due to the electron helicity.

A null measurement with an unpolarized beam was also performed as a cross-check. Fake asymmetries could be generated by any difference in the beam parameters (position, angle, current, energy) if they were correlated with the reversal of the beam helicity: limits on such effects were set by the pulse-by-pulse monitoring of the beam characteristics.

Corrections of a few per cent due to π^- background and *bremsstrahlung* radiation were applied; the main systematic uncertainties were due to the imperfect knowledge of the magnitude of the electron polarization which links the physical asymmetry to the measured one, and to the limits on the correlated beam changes.

The measurement uncertainty of this technique approached the 10^{-8} level in more recent experiments on electron-nucleus scattering.

Just as Mott scattering contains a parity-violating part due to weak interactions, the same is expected to occur for $e^- e^-$ (Møller) scattering, in a much lower range of q^2. The expected effect can be computed with significantly smaller theoretical uncertainties in such a purely leptonic interaction, therefore allowing precision tests of the electro-weak theory.

Møller scattering

Experiment E158 at SLAC was the first to measure a parity-violating effect in the elastic scattering of polarized electrons off unpolarized electrons (Anthony *et al.*, 2004), namely the asymmetry in scattering cross section for incident electrons of opposite longitudinal polarization. This asymmetry is expected to be maximal for a scattering angle of $\pi/2$ in the centre of mass (symmetric decay with both scattered electrons' energies equal to half the incident beam energy).

An intense 50 GeV longitudinally polarized electron beam ($\sim 5 \cdot 10^{11}$ electrons in a ~ 270 ns pulse, with a 120 Hz repetition frequency) was used; the beam was obtained from a gallium arsenide photocathode illuminated by circularly polarized laser light, the laser polarization determining the electron helicity. The laser was operated at a wavelength of 805 nm, for which a photon circular polarization of 99.8% was achieved. The quantum efficiency of the photocathode depends on the orientation of the linear polarization of the incident light at the level of 5–15%, and this was one of the important sources of systematics. The electron beam polarization was $\mathcal{P} \sim 89\%$, periodically measured using Møller scattering off a magnetized foil replacing the target.

The polarized electron beam crossed a 1.5 m long liquid hydrogen target (an electron target with minimum radiation length per electron), in which the hydrogen at a temperature of 20 K circulated at ~ 5 m/s and aluminium meshes enhanced turbulence and mixing to minimize density fluctuations. The entrance and exit windows of the target cell had the same radius (and sign) of curvature so that the effective target thickness was independent of the transverse position of the incident beam. The effective $e^- e^-$ cross section for the apparatus is relatively high (14 μb) at this energy, but Mott scattering is anyway larger; note however that the energy-angle dependence of the cross sections for the two processes is different: σ(Møller) $\propto E^{-1}\theta^{-4}$ and σ(Mott) $\propto E^{-2}\theta^{-4}$, and this is used to separate them (and the *bremsstrahlung*-degraded tail of the primary beam).

An achromatic (no net deflection) set of dipole magnets and collimators after the target was used to put it out of the line-of-sight of the detector. and to guarantee that no photon could reach it with less than two scatterings. A set of quadrupoles followed, which focused Møller scattered electrons (scattering angles between 4.4 and 7.5 mrad) on a concentrated region of the detector.

The main detector was a highly radiation resistant (it had to stand a ~ 100 MJ integrated energy deposition) Čerenkov light calorimeter, made of a 15 radiation lengths deep sandwich of copper and fused silica fibres, covering the annular region between 15 and 35 cm radius, 60 m downstream of the target. At the detector position, Møller electrons of 13–24 GeV energy formed a ring well separated from Mott scattered electrons, which dominated at lower and higher radii and were the major source ($\sim 8\%$) of background. The primary beam and the very intense flux of *bremsstrahlung* photons were confined to the detector-free forward region with radius below 6 cm (the beam size was ~ 5 mm RMS in each transverse coordinate).

A second lead-shielded Čerenkov calorimeter was used to measure the asymmetry of the pion background (which was only $\sim 10^{-3}$ since the nominal momentum transfer was just above the pion mass squared). A luminosity monitor, consisting in a set of 8 ionization chambers at ~ 1 mrad angle, measured the asymmetry for Mott scattered electrons at very forward angles, which is expected to be negligible (5% of the Møller asymmetry) and therefore allowed one to control target density fluctuations.

A rapid flipping of the beam polarization between the two possible helicities resulted in a large cancellation of time-dependent instrumental asymmetries: the electron helicities for a pair of beam pulses were chosen randomly by the sign of the voltage pulse on a Pockels cell in the light source; each such pair was followed by another one with opposite helicities.

Data was collected at beam energies of 45.0 and 48.3 GeV, which correspond to opposite orientation of the electron longitudinal polarization, due to the $g - 2$ precession angles in the magnetic bend of the beam line differing by π. Since the statistical power of the asymmetry measurement (and the asymmetry itself) increases linearly with the beam energy, while the available beam current (and therefore the luminosity) decreases with increasing beam energy at constant power, slightly more data was collected at the higher energy to equalize the statistical errors.

The raw asymmetry A_{raw} was obtained from the integrated detector response for each pulse pair ($\sim 2 \cdot 10^7$ electrons each), and corrected for small fluctuations of the beam trajectory. After correcting for backgrounds the true asymmetry was obtained:

$$A = \frac{\sigma_R - \sigma_L}{\sigma_R + \sigma_L} = \frac{1}{\mathcal{P}} \frac{A_{\mathrm{raw}} - \Delta A_{\mathrm{bkg}}}{1 - f_{\mathrm{bkg}}}$$

where ΔA_{bkg} is the background asymmetry and f_{bkg} the corresponding dilution factor (both were actually sums over several kinds of backgrounds). A correction had to be applied to account for a small non-zero transverse beam polarization inducing an azimuthal modulation of the asymmetry.

Given the expected size of the asymmetry (a few units in 10^{-7}) systematic errors had to be kept at the level of some parts per billion. The most important requirement for the experiment was the minimization of any possible correlation of the beam parameters with the electron helicity (and an excellent control of the residual correlations): the intensity and energy asymmetries had to be below $\sim 10^{-7}$ and $\sim 10^{-8}$ respectively, and the differences in beam position and angle for opposite helicities had to be below 10 nm and 0.4 nrad respectively. These beam parameters were in turn largely determined by those of the laser light.

The laser intensity and the polarity-reversing voltage in the Pockels cell were controlled by a feedback system driven from the measured beam parameters for each pulse: three beam position monitors provided redundant information on the position, angle and energy of the electron beam. The beam intensity was stabilized against slow drifts by controlling a Pockels cell between a pair of crossed polarizers, which time-sliced the 15 μs long laser flash and determined the actual duration of the light pulse; together with temperature and humidity control, this allowed one to achieve a 0.5% pulse stability. A second cell controlled by the feedback system balanced the intensity asymmetry between the two helicity states. One more cell compensated for residual linear polarization of the light: since linear and circular polarization components sum quadratically to 1, a small perturbation which barely changes the magnitude of circular polarization could induce a significant linear polarization, which strongly couples to the anisotropic quantum-efficiency of the cathode resulting in an electron intensity asymmetry. Position and angle differences were compensated by a mirror set on a piezoelectric mount which could translate or tilt it. Besides reducing helicity correlations, the active feedback system also minimized the corrections to be applied to each data set caused by the intrinsic fluctuations of the measured beam parameters: each setting of the updating feedback system was determined by the difference between the current set of measured beam parameters and the previous one, and in the running average the statistical fluctuations cancel in pairs, resulting in a faster convergence to zero.

The correlation coefficients of the detector response with the beam parameters were continuously measured during data taking by varying the latter in a controlled way (slowly with respect to the helicity flip frequency) using corrector coils and RF modules, by an amount large compared to the nominal fluctuations, to extract the first-order derivatives.

Another slow (passive) method for reversing the electron polarization was the toggling of a half-wave plate in the laser light line, which reversed the relation between the Pockels cell voltage polarity and the sign of the electron polarization every 48 h. Finally, a variable-optics system allowed one to invert any light position and angle differences arising from imperfections of the light polarization stage.

All the above inversions allowed one to cancel most of the instrumental symmetries at the required level. The measured parity-violating asymmetry was $A_{PV} = (-175 \pm 30 \pm 20) \cdot 10^{-9}$ (the first error being statistical, the second systematic), therefore establishing parity violation in Møller scattering with a significance of 5 standard deviations at the average $q^2 = 0.026 \, (\text{GeV}/c)^2$, consistent with expectations. The first-order beam-related asymmetry correction was limited to $(-10 \pm 1) \cdot 10^{-9}$ by the approaches described above.

2.3.6 Strong interactions

The natural relative size of the parity-violating effects induced by weak interactions with respect to the strong interactions is of order $G_F m_\pi^2 / \alpha_S \approx 10^{-7}$, where G_F is the weak (Fermi) coupling constant and α_S the strong one. Effects larger than this in hadronic systems would indicate parity violation in the strong interactions themselves and have not been detected, while the sensitivity to detect weak interaction effects was reached. Therefore all the positive results described are actually consistent with parity symmetry of the strong interaction and maximal parity violation of the weak one.

Tests of parity symmetry in strong interactions can be performed by searching for forbidden nuclear transitions among states of opposite (nominal) parity. If parity symmetry is violated the physical states $|i^{(\pm)}\rangle$ do not coincide with the parity eigenstates $|\psi_\pm\rangle$, but for a small violation they will have only a small admixture (of amplitude $\mathcal{F}$) of the state with opposite eigenvalue:

$$|i^{(\pm)}\rangle = |\psi_\pm\rangle + \mathcal{F}|\psi_\mp\rangle$$

A parity conserving interaction will induce a 'forbidden' transition into a final parity eigenstate $|f_\pm\rangle$, e.g.

$$\Gamma(i^{(+)} \to f_-) \propto \mathcal{F}^2 \, |\langle f_-|T|\psi_-\rangle|^2$$

which is proportional to the squared amplitude of parity admixture $\mathcal{F}^2$.

In practice these kinds of tests are limited to the case of states with spin J and parity $\eta_P = (-1)^{J+1}$ (*unnatural* states) which cannot decay to a 0^+ state by alpha emission if parity is conserved. A difficulty in performing these tests is that of populating the unnatural parent state with enough selectivity to be able to discriminate the parity violating transition rate against the background. Also, a practical limit to tests of parity symmetry by the search for forbidden transitions is due to the fact that the emission of an additional undetected soft photon can reverse the parity of the state, so that very good energy resolution is crucial.

An early example of this approach is the study (Wilkinson, 1958) of the nuclear reaction $d + \alpha \to {}^6\text{Li} + \gamma$. The second excited state of ${}^6\text{Li}$ (3.56 MeV above the 1^+

ground state) has spin-parity 0^+, so that it cannot decay to $d\,\alpha$ if parity is conserved $(J^P(d) = 1^+, J^P(\alpha) = 0^+$, the orbital angular momentum would have to be $L = 1$ to match the deuteron spin, and therefore parity would be negative); this excited state can decay to the ground state by gamma emission. The lifetime of the excited ^{6}Li nucleus is smaller than the time required to slow it down after production, and to avoid decays in flight (for which the energy is not known a priori) the inverse reaction was therefore studied, searching for the resonance in which the excited state is formed by bombarding α (^{4}He) with deuterons, and then de-excites by gamma emission. No evidence for the above transition was found.

Positive results were instead obtained with other systems (see e.g. (Neubeck et al., 1974)), in which parity-violating α decays were detected on top of large backgrounds, and found to be consistent with the predicted magnitudes due to weak interaction effects.

Another class of tests of parity conservation in strong interactions are searches for asymmetries in the emission of photons from polarized states (Haas et al., 1959), such as those obtained by capture of polarized neutrons: if the coherent emission of radiation of different parity is allowed by a parity impurity in the initial state, its angular distribution will no longer be isotropic with respect to the initial polarization direction.

Electromagnetic transitions can both keep and change the parity of nuclear states: considering a gamma transition from a polarized initial state $|i\rangle$ to a final state $|f\rangle$ with the same non-zero spin and (predominantly) the same parity, a magnetic dipole transition amplitude A_M will dominate, but any parity impurity in the initial (or final) state will also allow a coherent electric dipole amplitude A_E to be present, and the resulting angular distribution will then be

$$\frac{d\Gamma}{d\Omega} \propto 1 + a\,\cos\theta \qquad (2.81)$$

where θ is the angle between the direction of the initial polarization and that of photon emission, and

$$a = 2\frac{\mathrm{Re}(A_E^* A_M)}{|A_E|^2 + |A_M|^2} \qquad (2.82)$$

In terms of the parity eigenstates $|i_\pm\rangle, |f_\pm\rangle$

$$A_M \propto \langle f_\pm|T|i_\pm\rangle \qquad A_E \propto \mathcal{F}\langle f_\pm|T|i_\mp\rangle$$

so that the angular asymmetry is indeed proportional to the amplitude $\mathcal{F}$ of parity admixture, therefore also allowing one to obtain information on the relative sign of the parity-violating amplitude. For the search of parity-violating effects, the kind of transition discussed above is more favourable than that between states of opposite parities, for which the dominant amplitude will be of electric type and the one arising from the parity impurity of magnetic type, since magnetic transitions are normally smaller than electric transitions of the same order, and the measurable effect in eqn (2.82) is enhanced (over $\mathcal{F}$) by the ratio of the electric to magnetic transition matrix elements (the larger the suppression of the 'normal' transition, the bigger the enhancement factor).

The appearance of odd powers of $\cos\theta$ in angular distributions requires both a parity impurity and initial state polarization: this polarization can be obtained by preparing a polarized gamma source (only possible for initial states fed by relatively long lifetime beta decays, due to the long cooling time required) or by exploiting production reactions with polarized projectiles or targets: intense polarized beams can be produced ($\sim 10^{13}$ s^{-1} protons, $\sim 10^{10}$ s^{-1} cold neutrons, with polarizations above 90%) and in some cases the polarization transfer from the projectile to the decaying state can be as high as 70%.

Clearly the initial state polarization can also arise as a consequence of a parity non-conserving force in a process between unpolarized particles: in this case, however, any observable effect, proportional to such an induced polarization, would be of second order in the parity-violating interaction, and therefore very small.

Parity-violating gamma decay asymmetry

The first measurement of parity violation in a nuclear process (Abov *et al.*, 1964) was obtained by a Moscow group by studying the anisotropy of gamma emission from an excited nucleus, formed by bombardment with polarized neutrons.

From first-order perturbation theory the mixing of excited states of different parity is expected to be larger between states which are close in energy, so that a good option is the study of heavy target nuclei close to the neutron binding energy, where discrete levels spaced only a few eV apart are present. Transitions for which the impurity state is favoured must be considered, such as those between levels with spin differing by one unit and the same parity (in absence of impurity), for which the allowed transition is M1 and the one induced by the parity impurity would be E1, which can better compete in magnitude with the former. Transitions with large enough thermal neutron cross sections are required, and small values of spin are preferred since the statistical factors are larger; finally, the transitions should not be experimentally masked by high energy gamma rays due to other decays. Cadmium is a good candidate: the neutron capture cross section of ^{113}Cd is large and can lead to an excited spin one state of ^{114}Cd; both nuclear states have (mostly) positive parity, and the excited one can decay to the spin zero ground state ($J^P = 1^+ \to 0^+$) with emission of a 9 MeV (mostly M1) photon, or to a spin two first-excited state with emission of a 8 MeV photon.

A transversely polarized neutron beam, of intensity $\sim 6 \cdot 10^6$/s and polarization $\mathcal{P} = 0.7 - 0.85$, was obtained by grazing reflection of reactor neutrons on a 1.5 m long magnetized cobalt mirror. The polarization of the neutrons at the target was horizontal (orthogonal to the beam direction) and, in order to cancel false asymmetries due to the asymmetry of the set-up, measurements with opposite polarization were compared. The neutron polarization could be reversed by inverting all magnetic fields in the apparatus, but this could also affect the performance of the detectors (photomultipliers) therefore introducing a spurious asymmetry, so that a different approach (non-adiabatic spin-flipping) was adopted. An aluminium (non-magnetic) foil was put along the neutron trajectory; this had no effect on neutron spin in 'normal' runs, while in 'spin-reversed' runs the initial direction of polarization (far from the target and the detectors) was inverted magnetically, and an intense current (hundreds of A) was switched on into the foil (orthogonal to the neutron path): this current produced a vertical magnetic field, oppositely directed on the two sides of the foil. The abrupt reversal of this field, occurring over a very short distance, could not be followed by the neutron spin, which therefore remained unaltered; in this way, after crossing the foil the neutrons were precessing around a guide field (the same as in the original setting) *opposite* to their polarization direction, and followed it to reach the target with opposite polarization with respect to 'normal' runs. Note that the large magnetic field gradient in the region of the foil cannot affect the neutron trajectory (Stern–Gerlach effect) as it is along its direction of motion.

The beam passed through a rotating (10 Hz) depolarizer composed of a circular disk with two opposite quadrants covered by a 0.3 mm iron foil: only the neutrons hitting the iron lost their polarization due to the interactions with the randomly aligned magnetic domains, so that polarized and unpolarized neutrons alternated each 25 ms.

The neutrons hit a 0.4 mm thick Cd target, and decay photons from the target were detected by two identical magnetically shielded NaI(Tl) crystals, placed orthogonally to the beam in opposite directions (left and right); the counters were also shielded from background photons by lead, and from scattered neutrons by a lithium-rich material (with high neutron absorption cross section). A sketch of the apparatus is shown in Fig. 2.7. Pulses corresponding to photons in the 8.1–9.4 MeV energy range from polarized and depolarized neutrons were counted separately (and redundantly).

The double ratio of the left and right detector counts N_L and N_R for polarized and depolarized neutrons was compared for opposite neutron polarization directions, in order to exclude effects due to the counter efficiencies and the asymmetry of the apparatus:

$$R_\pm = \frac{(N_L/N_R)_{\text{pol}\,\pm}}{(N_L/N_R)_{\text{unpol}}} = C_\pm \, K_\pm(t) \, \frac{1 \pm a\mathcal{P}\Delta\Omega}{1 \mp a\mathcal{P}\Delta\Omega}$$

where the subscript indicates the sign of neutron polarization, $C_\pm$ is a (time-independent) asymmetry factor, which can be ascribed e.g. to differences in the beam geometry when the depolarizing foil is crossed or not, $K_\pm(t)$ is a time-dependent asymmetry factor, which can account for changes in the neutron flux and time instabilities of the detector and electronics, $\Delta\Omega$ is the acceptance solid angle and a the parity-violating asymmetry coefficient in the angular distribution of gamma rays. The latter can be measured as

$$a = \frac{1}{\mathcal{P}\,\Delta\Omega} \frac{\sqrt{R_+/R_-} - 1}{\sqrt{R_+/R_-} + 1}$$

if $C_+ K_+(t) = C_- K_-(t)$.

The time-dependent factors $K_\pm(t)$ were very close to unity because of the fast switching of polarized to depolarized beam. The equality of the asymmetry factors was checked by periodically repeating the experiment with an unpolarized beam (using a fixed metallic foil along the neutron path), and by detecting intense E1 gamma transitions of Cd, after shifting the energy range of detector sensitivity (4–6 MeV photon energy): in this case any parity-violating

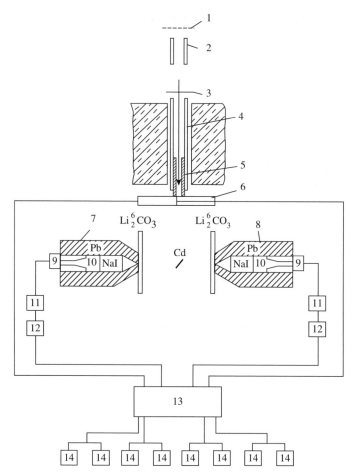

FIG. 2.7. Schematic drawing of the experimental apparatus of Abov *et al.* (1964) for the measurement of the angular asymmetry of nuclear gamma transitions. Neutrons entered from the top of the figure. 1: fixed depolarizing foil; 2: magnet for spin flipping; 3: current-carrying foil for spin flipping; 4: magnetic guide; 5: collimator; 6: rotating depolarizer; 7,8: counters; 9,10: photomultiplier system; 11: amplifier; 12: discriminator; 13: commutator; 14: counter scalers.

contribution would be due to a suppressed M1 transition, therefore reducing any asymmetry. Other nuclei with E1 gamma transitions were also used to check for instrumental asymmetries and possible effects due to circular polarization of the photons. The experimental results were indeed consistent with no asymmetries in all these control experiments.

Other systematic checks were performed by removing the target, therefore measuring any asymmetry due to the $\sim 1.5\%$ background from neutron-induced gamma production in the material around the target, by using graphite as target and measuring the asymmetry due to the $\sim 0.04\%$ neutrons scattered in there, and by using vertical neutron polarization; again, no asymmetries were measured.

The experiment was performed in repeated sets of left-right polarized runs alternated with unpolarized control runs, in 80 min cycles, for a total of 1.5 years. The measured asymmetry was $a = (-3.7 \pm 0.9) \cdot 10^{-4}$, thus giving evidence of a parity violation effect whose magnitude is consistent with being due to weak interaction effects.

The circular polarization of photons emitted by non-polarized nuclei (as any pseudo-scalar quantity) is also given by the interference of parity-conserving and parity-violating transition amplitudes α as in eqn (2.82), and therefore linear in the impurity parameter $\mathcal{F}$. Parity violation effects were also detected with this approach, by exploiting the photon polarization dependence of the Compton cross section in magnetized iron: as for the measurement of electron polarization, one experimental limit of this technique is the limited ($\sim 8\%$ in Fe) maximum analysing power due to the number of polarized electrons in each atom.

Circular polarization of gamma decay radiation

The circular polarization of 482 keV photons emitted in the gamma decay transition of ^{181}Ta to its ground state from an excited one ($J^P = (5/2)^+ \rightarrow (7/2)^+$), obtained by beta decay of ^{181}Hf, was measured (Lobashov et al., 1967).

The interference of a parity-violating E1 transition with the parity conserving M1 transition, strongly suppressed in this particular decay, was the source of the circular photon polarization. Note that for this kind of experiment, as opposed to those measuring the circular polarization of gamma decay photons after a beta decay described earlier, no correlation with the beta decay electron was measured, so that the sample of nuclei had no effective polarization with respect to any measured direction.

The effects expected from weak interactions are of order 10^{-6} to 10^{-7}, so that one would need to accumulate a sample of order 10^{14} events to reach the statistical accuracy required to exclude larger effects due to strong interactions. The available fluxes were not a limitation, but the counting rate of events did, and this suggested to measure the integrated flux of photons instead of individual events: with such an approach no pile-up problems occur and a very intense radioactive source (~ 500 Ci) could be used, which was prepared by neutron irradiation in a reactor.

The photons that underwent Compton scattering in the forward direction in a magnetized iron target were detected by a CsI(Tl) crystal read by a photodiode, which is largely insensitive to stray magnetic fields. Photon energy discrimination was obtained by means of absorbing filters, which removed the low energy ones: a simple gamma spectrum helped in this respect.

The magnet orientation was periodically reversed at 0.5 Hz, and the output photodiode current was resonantly amplified and transformed into a mechanical force which drove a high-Q astronomical pendulum (tuned to 0.5 Hz, period constant to 10^{-7}), acting as a selective filter and allowing the accumulation of the signal over several hours. In this way the periodic current oscillations induced by a dependence of the forward Compton scattering

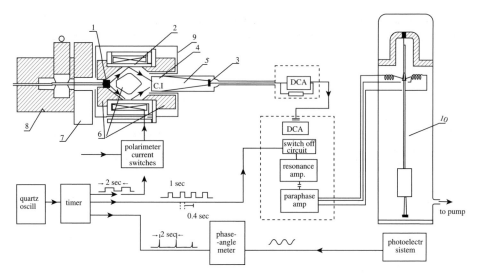

Fig. 2.8. Experimental apparatus for the electron scattering experiment of Lobashov *et al.* (1967). 1: source; 2: polarimeter scattering core; 3: photodiode; 4: CsI(Tl) crystal; 5: light pipe; 6: collimators; 7: shield; 8: container; 9: magnetic shield; 10: pendulum.

cross section on the direction of photon polarization could be distinguished from the random fluctuations due to the Poisson statistics of the number of photons; a schematic of the experiment is shown in Fig. 2.8.

The expected signal is an alternating current of amplitude proportional to the asymmetry

$$A = \frac{I_+ - I_-}{I_+ + I_-}$$

where $I_\pm$ are the current intensities with opposite polarimeter orientations; in turn A is proportional to the photon circular photon polarization.

The amplitude of the oscillations and their phase with respect to the time of current reversal in the polarimeter was measured by a photoelectric system. The phase for a definite polarization was determined experimentally using forward scattered ^{46}Sc gamma photons (practically unpolarized) and *bremsstrahlung* photons produced from beta electrons of ^{198}Au. Null tests were performed using the practically unpolarized unscattered photons from ^{46}Sc.

The effect induced by *bremsstrahlung* photons produced from the beta electrons of Hf or related isotopes had to be subtracted: beta decays with end-point energies above the gamma energy being studied were a source of polarized photons which cannot be simply filtered out by absorbers. This effect was reduced by preparing the sources as tablets diluted in magnesium oxide, therefore decreasing the effective atomic number of the source material. The residual effect, mostly due to internal *bremsstrahlung*, was measured with a very intense (4000 Ci) ^{177}Lu beta source (end point energy 490 keV) followed by filters which removed the soft (200 keV) gamma photons; the measured effect amounted to a correction of $-0.5 \cdot 10^{-5}$ on the measured polarization.

False signals could also be induced by inductive pick-up from the alternating field in the polarimeter, which was reduced to a very low value by screening and then corrected for, or by magnetostrictive changes of the polarimeter dimensions,[35] also found to be negligible after switching off the signal during the 0.4 s required to periodically magnetize the polarimeter.

The asymmetry of the photon intensities for opposite magnetization directions was measured to be $A = (1.4 \pm 0.2) \cdot 10^{-7}$, corresponding to a circular photon polarization $\mathcal{P} = (-6 \pm 1) \cdot 10^{-6}$. The same set-up was used to measure the gamma circular polarization in the $J^P = (9/2)^- \to (7/2)^+$ 396 keV transition to the ground state of ^{175}Lu, with a result $\mathcal{P} = (4 \pm 1) \cdot 10^{-5}$.

Tests of parity symmetry in strong interactions using elementary particles avoid complications related to many-body nuclear physics in their interpretation.

Experiments with polarized pp scattering fall in this category: the signature for parity violation is an asymmetry in the helicity-dependent cross sections for longitudinally polarized protons on an unpolarized target. The asymmetry can be measured either by detecting the scattered protons or by a transmission measurement, the latter approach being possible at high energies where a significant fraction of protons can be scattered by strong interactions in thick targets.

While the incident proton polarization which can be achieved is typically a factor 2 higher than with electrons, in this case a major systematic problem is due to the residual transverse polarization of the beam, which reverses with helicity and couples to the relatively large parity-allowed analysing power causing asymmetries (highly suppressed in the case of electrons). The main effect is due to the variations of the transverse polarization across the beam profile (Adelberger and Haxton, 1985): the coupling of the transverse analysing power of the detector with a 'circulating' transverse polarization on the fringes of the beam, directed clockwise or anti-clockwise, and reversing with the beam helicity, can mimic a parity-violating effect; such a polarization pattern can indeed arise whenever a polarized beam passes through a bending magnet asymmetrically with respect to the bending plane.

Asymmetries of order 10^{-7} were detected in this kind of experiment (Eversheim et al., 1991), in agreement with the predictions of parity-violating effects due to weak interactions, and systematic uncertainties of order $\sim 10^{-8}$ were reached.

With np scattering several parity non-conserving observables have also been studied. After the initial evidence for spin rotation of polarized cold neutrons in the transmission through non-magnetic spinless matter (no strong interaction spin flip being possible, a kind of 'neutron optical activity') (Forte et al., 1980), in recent years, studies have comprised: the measurement of the circular polarization of the

---

[35] Mechanical deformations induced by the magnetic field.

2.2 MeV photon emitted in the capture of unpolarized thermal neutrons by protons, or equivalently the dependence of the deuterium photo-disintegration cross section ($\gamma d \rightarrow np$) on the photon helicity; the dependence of the capture cross section on the neutron helicity, and the measurement of the helicity-correlated transmission probability of polarized neutrons in unpolarized matter, where large ($\sim 10\%$) effects were detected thanks to enhancement factors arising from the existence of closely spaced levels of opposite parity. For a discussion of these beautiful experiments the interested reader is directed to the specialized literature (Gould and Davis, 2002).

Finally we mention that limits for parity symmetry in strong interactions are also obtained by searches for forbidden decays such as $\eta(\eta') \rightarrow \pi^+\pi^-$, $\eta(\eta') \rightarrow \pi^0\pi^0$ (branching ratio limits of a few 10^{-4} for η and 10^{-2} to 10^{-3} for η'), and by limits on electric dipole moments of baryons (p, n, Λ) and leptons (e, μ, τ), to be discussed further in Chapter 4.

Further reading

A nice elementary and wide-ranging illustration of the issues linked to parity violation is Gardner (1990).

The history of the fall of parity is described in several places, and the words of the main protagonists are worth reading (Wu, 1988; Telegdi, 1987; Brown *et al.*, 1989). The reprint book by Cahn and Goldhaber (1988) also dedicates several pages to the early experiments on parity violation.

Readers interested in learning more about atomic parity violation will find a wealth of information in Khriplovich (1991).

Problems

1. Derive the parity transformation of the spherical harmonics, eqn (2.12).

2. How can the optical activity of a material be described by the mirror-reflection invariant Maxwell's equations of electrodynamics?

3. Radiative transitions are favoured between states of *opposite* parity, so that the parity is *changed* by the transition; but such processes are induced by electromagnetic interactions, which *conserve* parity. Explain why this is no paradox.

4. Considering the polarization vectors for linearly and circularly polarized photons and the parity-inverted ones, show that the assignment of negative parity to the photon is consistent.

5. It is well known that many organic molecules important to life are only found in Nature in a single chirality version (e.g. only right-handed helical sugar). Is this evidence of violation of parity symmetry?

6. From the definition of the S matrix (C.4) and the identity $P^\dagger P = 1$ it would appear that relation (2.60) is always true, independently of the validity of parity symmetry. Explain why this is not so.

7. Derive the expression for the time-varying average of the S_z component of the muon spin when a fully polarized muon ($S_z = +1/2$) forms muonium by interacting with unpolarized matter.

8. Writing the most general decay matrix element for $\Lambda \to p\pi^-$ decay (in the centre of mass and in the two-dimensional space of Λ spin projections $S_z(\Lambda) = \pm 1/2$) as $M = A_S + A_P \boldsymbol{\sigma} \cdot \hat{\boldsymbol{p}}_p$, derive the expression (2.72) for the angular distribution of the emitted nucleon.

9. From the composition law for angular momenta it might be expected that if one photon (spin 1) is emitted in a state of orbital angular momentum $L = 1$ a state with total angular momentum $J = 0$ would be possible; explain why single photon emission transitions with zero change of total angular momentum are not observed (or for particle decays e.g. $BR(K^+ \to \pi^+\gamma) < 3.6 \cdot 10^{-7}$).

10. Determine the (dominant) multipolarity of the following radiative decays: η $(J^P = 0^+) \to \pi^+\pi^-\gamma$, ω $(J^P = 1^-) \to \pi^0\gamma$ and $\omega \to \pi^+\pi^-\gamma$.

3

CHARGE CONJUGATION

La sapienza e' figliuola della sperienzia.
L. da Vinci

It does not take an idea so long
to become 'classical' in physics
as it does in the arts.
K. K. Darrow

The black and the white distinctively colouring
Holding the world inside
Now all the world is grey to me.
B. H. May

This chapter deals with charge conjugation symmetry, which emerges naturally out of the mating of quantum theory to special relativity, with the antiparticle concept. Despite not being related to space-time, this symmetry turns out to be closely linked to space-time inversions; its consequences are somewhat less widespread than those of parity symmetry, mainly because only neutral systems can be eigenstates of charge conjugation. Charge conjugation symmetry is maximally violated by weak interactions in a way that is closely linked to parity violation. The large violation of parity and charge conjugation symmetries in weak interactions, the intrinsic difficulties in experimenting with composite antimatter systems and the fact that *CPT* symmetry constrains several properties of matter and antimatter to be equal, all contribute to the fact that investigations shifted towards the study of the combined symmetry *CP* instead.

3.1 Charge conjugation transformation

Charge conjugation is the transformation associated with the exchange of particles and antiparticles, under which all charges (such as electric charge) change sign, while all other quantities (and in particular space-time related ones such as position, momentum and spin) are unaffected (a better name would actually be particle-antiparticle conjugation). Clearly, this concept only acquires meaning when particles are not isolated but are in interaction with an external field.

While classical electrodynamics is intrinsically symmetric with respect to the interchange of positive and negative charges, the differences in the way opposite electric charges manifest themselves in the everyday world (with massive positively charged protons and light negatively charged electrons) are such that charge conjugation as a symmetry only became relevant with the advent of relativistic quantum physics (Kramers, 1937) and the investigations of processes in which particles can be created, namely with the revised interpretation of the Dirac equation which explicitly required the existence of a physical counterpart of ordinary matter, identical by all means to the latter except for the sign of the charge: 'A hole, if there were one, would be a new kind of particle, unknown to experimental physics, having the same mass and opposite charge to an electron. We may call such a particle an antielectron. [...] Presumably the protons will have their own negative-energy states [...] an unoccupied one appearing as an antiproton' (P.A.M. Dirac, 1931).

Indeed, we define *particles* to be the negatively charged electrons and positively charged protons, and call *antiparticles* their oppositely charged counterparts, carrying over this convention to all the other particles for which no significant asymmetry is evident.

In quantum theory the charge conjugation transformation is obtained as usual through a unitary operator C, under which the charge operator changes sign:

$$Q \rightarrow Q_C = C^\dagger Q C = -Q \quad \text{or} \quad \{Q, C\} = 0 \tag{3.1}$$

The effect of such operator (which does not alter the fundamental commutation relations among quantum fields) on a four-vector current is therefore

$$C^\dagger j^\mu(t, \boldsymbol{x}) C = -j^\mu(t, \boldsymbol{x}) \tag{3.2}$$

consistent with the desired transformation for the charge $Q = \int d\boldsymbol{x}\, j^0(t, \boldsymbol{x})$.

It is evident from eqn (3.1) that states with non-zero charge (electric or of any other kind) cannot be eigenstates of the charge conjugation operator C[36].

If charge conjugation is a valid symmetry, physical states are either eigenstates of C or members of degenerate pairs which transform among them under C. The first case only occurs for neutral states; note that in this context 'neutral' does not mean just electrically neutral, but a particle with all internal charges being zero: for example a neutron has no electric charge but carries a non-zero baryon number, therefore it is not neutral in this sense and is not a charge conjugation eigenstate. Since the Hamiltonian describing our (or our detectors') interaction with charged particles is usually *not* charge-symmetric (by and large, we interact with *everything* electromagnetically,[37] we use electric charge to describe and measure

[36] The Hamiltonian commutes with Q for conserved charges (the only relevant ones). Two operators (such as Q and C) which both commute with a third one (H) do not necessarily commute with each other: physical states are eigenstates of the Hamiltonian H and are chosen to be eigenstates of other conserved observables (such as Q rather than C) to remove the remaining degeneration.

[37] Apart from gravity, that is.

particle properties, and therefore we deal with (pairs of) charge eigenstates which are not C eigenstates. Compare this situation with that of parity, with most particles being parity eigenstates and therefore classified to a large extent in terms of their intrinsic parity, while degenerate pairs of 'mirror' particles are not found. For this reason the concept of charge conjugation eigenvalue (or charge conjugation parity, or C-parity, as it is often called) is less useful than that of intrinsic parity for the classification of particles.

Charge conjugation parity is a multiplicative quantity, so that a system composed of several (neutral) particles which are charge conjugation eigenstates, has an eigenvalue given by the product of the single eigenvalues for the individual states.

3.1.1 Scalar and tensor fields

Since in quantum mechanics (with no particle creation and destruction) charge is a parameter and not a dynamical variable, charge conjugation is not associated with a unitary transformation there, and there is no need to actually introduce it.

In quantum field theory instead the concept of charge conjugation arises in a quite natural way with the appearance of antiparticles, see e.g. Weinberg (1995). The concept of antiparticles actually originates in relativistic mechanics, quite independently of quantum theory: the line element of relativity is defined from the Lorentz-invariant $ds^2 \equiv dx_\mu dx^\mu$ through a square root, so that for a particle at rest both signs of $ds = d\tau = \pm dt$ are admissible; the case with a minus sign, corresponding to a particle for which proper time and coordinate time have opposite directions is interpreted as an antiparticle travelling forward in time[38] (Stueckelberg, 1942; Feynman, 1948).

A unitary operator C in Hilbert space is introduced which implements the charge conjugation transformation; its effect on a state with definite momentum p, spin projection s and charge q is expected to be

$$C|p, s, q\rangle = \eta_C |p, s, -q\rangle \tag{3.3}$$

with $\eta_C = e^{i\xi_C}$ a phase factor ($|\eta_C|^2 = 1$, ξ_C real). If C leaves the Lagrangian density $\mathcal{L}(t, x)$ and the commutation relations among field operators unchanged, it is indeed a symmetry of the system.

In QFT charged particles are described by non-Hermitian field operators (required to build a non-vanishing interaction with the electromagnetic field), and the charge conjugation transformation should transform the field into its Hermitian conjugate: the part which creates particles should transform into the part which creates antiparticles and vice versa. For a charged scalar field ϕ, the equation which

---

[38] The concept of backward motion in time appears to be unacceptable mostly because there is no way to act on a system (applying some force on it) causing it to *change* from the forward to the backward time direction (or the other way around).

the charge conjugate field ϕ_C (annihilating antiparticles) must satisfy is consistent with the one for ϕ if

$$\phi_C(t, x) = \eta_C \phi^\dagger(t, x) \tag{3.4}$$

The transformation of the field operators under the effect of C must be such as to give the classical result:

$$\phi(t, x) \rightarrow C^\dagger \phi(t, x) C = \phi_C(t, x) \tag{3.5}$$

so that

$$C^\dagger \phi(t, x) C = \eta_C \phi^\dagger(t, x) \qquad C^\dagger \phi^\dagger(t, x) C = \eta_C^* \phi(t, x) \tag{3.6}$$

which in terms of creation and annihilation operators means

$$C^\dagger a(p) C = \eta_C b(p) \qquad C^\dagger a^\dagger(p) C = \eta_C^* b^\dagger(p) \tag{3.7}$$

$$C^\dagger b(p) C = \eta_C^* a(p) \qquad C^\dagger b^\dagger(p) C = \eta_C a^\dagger(p) \tag{3.8}$$

The charged field ϕ can be written in terms of a pair of Hermitian fields $\phi_{1,2}$ as

$$\phi_1 \equiv (\phi + \phi^\dagger)/\sqrt{2} \qquad \phi_2 \equiv -i(\phi - \phi^\dagger)/\sqrt{2} \tag{3.9}$$

and one has

$$C^\dagger \phi_1(t, x) C = \mathrm{Re}(\eta_C)\phi_1(t, x) + \mathrm{Im}(\eta_C)\phi_2(t, x)$$

$$C^\dagger \phi_2(t, x) C = \mathrm{Im}(\eta_C)\phi_1(t, x) - \mathrm{Re}(\eta_C)\phi_2(t, x)$$

which for $\eta_C = \pm 1$ give

$$C^\dagger a_1(p) C = \pm a_1(p) \qquad C^\dagger a_2(p) C = \mp a_2(p) \tag{3.10}$$

where $a_{1,2}$ are the annihilation operators for the fields $\phi_{1,2}$.

The unitary operator C^2 commutes by construction with all operators; since the phase factors associated with the charge conjugation transformation of a field and its Hermitian conjugate are related by complex conjugation, one has $C^2 = 1$ by construction. Therefore C is Hermitian, and can be an observable, with possible eigenvalues $\eta_C = \pm 1$.

In general, the physical significance of the phase factors η_C is limited by the fact that there can be many equivalent Hamiltonians, describing the same physical system, for which such phase factors are different (Feinberg and Weinberg, 1959). The intrinsic phase associated with the transformation of a non-Hermitian (charged) field under charge conjugation is unphysical (it depends on convention) and cannot be measured, neither in absolute terms nor relative to another field, but the phase factor associated with the transformation of the Hermitian conjugate field (antiparticle) is the complex conjugate of the former.

A neutral particle can be its own antiparticle (π^0 for example[39]) and be described by a Hermitian field such that

$$C^\dagger \phi(t, x) C = \eta_C \, \phi(t, x) \tag{3.11}$$

and in this case the charge conjugate field ϕ_C must also be Hermitian, so that $\eta_C = \pm 1$ has a physical meaning.

The current, when expressed in terms of the field operators, is

$$j^\mu(t, x) \equiv iq : \left[\phi^\dagger(t, x) \partial^\mu \phi(t, x) - \left(\partial^\mu \phi^\dagger(t, x) \right) \phi(t, x) \right] : \tag{3.12}$$

where q is the magnitude of the charge carried by the particle annihilated by the field ϕ (and the colons indicates normal ordering). It can be checked that this current behaves as desired under charge conjugation, and the same is true for the charge

$$Q(t) = iq \int dx : \left[\phi^\dagger(t, x) \frac{\partial \phi(t, x)}{\partial t} - \frac{\partial \phi^\dagger(t, x)}{\partial t} \phi(t, x) \right] : \tag{3.13}$$

It should be noted that the transformation $\phi \leftrightarrow \phi^\dagger$ induced by the charge conjugation operator is not obtained by Hermitian conjugation (that is by $a \leftrightarrow a^\dagger$ and $b \leftrightarrow b^\dagger$, but rather through the transformation $a \leftrightarrow b$. The unitary operator C therefore does not transform *any* operator into its Hermitian conjugate, although its effect on a quantum field is equivalent to exchanging the roles of creation and annihilation operators, inverting the signs of the internal charges without modifying momenta and spins.

For the electromagnetic field, the charge conjugation invariance of the interaction Lagrangian $-j_\mu A^\mu$ (which is imposed in order to have a charge conjugation symmetric description of electromagnetism, as experimentally observed) together with the transformation of the four-vector current (3.2) require

$$C^\dagger A^\mu(t, x) C = -A^\mu(t, x) \tag{3.14}$$

i.e. the (Hermitian) electromagnetic field has negative C-parity. In terms of annihilation and creation operators, eqn (B.12)

$$C^\dagger a^{(s)}(k) C = -a^{(s)}(k) \qquad C^\dagger a^{(s)\dagger}(k) C = -a^{(s)\dagger}(k) \tag{3.15}$$

3.1.2 Spinor fields

Like a non-Hermitian scalar field, a spinor field also contains components describing both particles and antiparticles. When one gets rid of the 'sea' and 'hole'

---

[39] No other distinguishable particle with the same mass and spin is known.

interpretation of Dirac's theory in the second-quantization approach, as done by Majorana (1937), the description of spin 1/2 particles becomes symmetric in the fields ψ and $\overline{\psi}$, which obey the same anti-commutation rules, and can therefore be related by a similarity transformation (see e.g. Sakurai (1967)).

In the Majorana representation of Dirac matrices (in which all the matrices are pure imaginary and the Dirac equation is real) the transformations of a spinor field are similar to those of a scalar field, but in a general representation, in order to obtain for the charge conjugated field a consistent equation, one is forced to introduce a (4×4) matrix S_C which acts in the Dirac space of spinor components, and to define the charge conjugated field as

$$\psi_C(t, \mathbf{x}) = \eta_C S_C \overline{\psi}^T (t, \mathbf{x}) \tag{3.16}$$

where S_C is defined by its property

$$S_C \gamma^\mu S_C^{-1} = -(\gamma^\mu)^T \tag{3.17}$$

Since $-(\gamma^\mu)^T$ satisfy the same anti-commutation relations as the γ^μ, such a matrix S_C is guaranteed to exist by Pauli's fundamental theorem (see Appendix B). C is unitary in Hilbert space so that S_C must be unitary in Dirac space:

$$S_C^\dagger S_C = 1 \tag{3.18}$$

Moreover, one can exchange ψ and ψ_C in the definition, as none of them is more fundamental than the other, and this implies (in any representation in which γ_0 is symmetric, or real):

$$\overline{S}_C^T = \gamma_0 S_C^* \gamma_0 = S_C^{-1} \tag{3.19}$$

(where as usual $\overline{S}_C \equiv \gamma_0 S_C^\dagger \gamma_0$). This property, together with the defining one for S_C (3.17), implies that (with γ_0 is symmetric, or real) the matrix S_C is antisymmetric:

$$S_C^T = -S_C \qquad \overline{S}_C = -S_C^{-1} = -S_C^\dagger \tag{3.20}$$

In the Dirac representation $S_C = i\gamma^2\gamma^0$, so that $\overline{S}_C = S_C$ and therefore

$$S_C^{-1} = S_C^\dagger = S_C^T = -S_C \qquad S_C^2 = -1 \tag{3.21}$$

Restricting in the following to representations of the Dirac matrices in which γ_0 is real, the field transformations are

$$C^\dagger \psi(t, \mathbf{x}) C = \eta_C S_C \overline{\psi}^T \qquad C^\dagger \overline{\psi}(t, \mathbf{x}) C = -\eta_C^* \psi^T S_C^\dagger \tag{3.22}$$

$$C^\dagger a(\mathbf{p}, s) C = \eta_C b(\mathbf{p}, s) \qquad C^\dagger a^\dagger(\mathbf{p}, s) C = \eta_C^* b^\dagger(\mathbf{p}, s) \tag{3.23}$$

$$C^\dagger b(\mathbf{p}, s) C = \eta_C^* a(\mathbf{p}, s) \qquad C^\dagger b^\dagger(\mathbf{p}, s) C = \eta_C a^\dagger(\mathbf{p}, s) \tag{3.24}$$

The positive-energy part of the conjugate field contains $\eta_C\, u_C\, e^{-ipx}$ and must be equal to $\eta_C S_C \bar{v}^T$, resulting in the following transformations for spinors:

$$S_C\, \bar{u}(\boldsymbol{p}, s)^T = v(\boldsymbol{p}, s) \qquad S_C\, \bar{v}(\boldsymbol{p}, s)^T = u(\boldsymbol{p}, s) \qquad (3.25)$$

because $u_C(\boldsymbol{p}, s) = u(\boldsymbol{p}, s)$, since these two spinors satisfy the same Dirac equation $(p^\mu \gamma_\mu - m)u(\boldsymbol{p}, s) = 0$.

The current $j^\mu =\, : \bar{\psi}\gamma^\mu \psi :$ correctly changes sign under charge conjugation

$$: \bar{\psi}\gamma^\mu \psi : \rightarrow\ -: \psi^T(-\gamma^\mu)^T \bar{\psi}^T := -: \bar{\psi}\gamma^\mu \psi : \qquad (3.26)$$

where the last minus sign is due the anti-commutativity of the free fields ψ and $\bar{\psi}$ due to normal ordering (in field theory they are exchanged not only in Dirac space but also in Hilbert space, where they are anti-commuting operators); this important point indicates that charge conjugation can be described in a consistent way only in quantum field theory.

More generally, the bilinear covariants (B.21) built from different spinor fields transform as

$$\bar{\psi}_2 \Gamma \psi_1 \rightarrow \eta_C^*(2)\eta_C(1)\, \bar{\psi}_1 S_C \Gamma^T S_C^\dagger \psi_2 \qquad (3.27)$$

and considering the effect of S_C on the complete set $\{\Gamma\}$ of 4×4 matrices obtained from Dirac matrices

$$S_C\{\mathbf{1}, \gamma_\mu, \sigma_{\mu\nu}, \gamma_\mu \gamma_5, i\gamma_5\}S_C^{-1} = \{\mathbf{1}, -\gamma_\mu, -\sigma_{\mu\nu}, \gamma_\mu \gamma_5, i\gamma_5\}^T \qquad (3.28)$$

one obtains the transformations

$$S_{21}(x) \rightarrow +\eta_C^*(2)\eta_C(1)\, S_{12}(x) \qquad (3.29)$$

$$V_{21}^\mu(x) \rightarrow -\eta_C^*(2)\eta_C(1)\, V_{12}^\mu(x) \qquad (3.30)$$

$$T_{21}^{\mu\nu}(x) \rightarrow -\eta_C^*(2)\eta_C(1)\, T_{12}^{\mu\nu}(x) \qquad (3.31)$$

$$A_{21}^\mu(x) \rightarrow +\eta_C^*(2)\eta_C(1)\, A_{12}^\mu(x) \qquad (3.32)$$

$$P_{21}(x) \rightarrow +\eta_C^*(2)\eta_C(1)\, P_{12}(x) \qquad (3.33)$$

where subscripts have been used to indicate the order in which the fields appear (i.e. $S_{21} =\, : \bar{\psi}_2 \psi_1 :$ etc.); the fact that bilinear covariants are defined as normal-ordered products is crucial.

3.1.3 Majorana spinors

For massless particles with spin there is an intrinsic ambiguity in the definition of particle and antiparticle, since states with different helicities transform independently: the *chirality* of a spinor is defined as the eigenvalue of the γ_5 operator; the chiral components of a spinor are

$$\psi_{R,L} = \frac{1}{2}(1 \pm \gamma_5)\psi \qquad \gamma_5\psi_{R,L} = \pm\psi_{R,L} \tag{3.34}$$

and the Dirac equation in terms of these gives

$$(E - \boldsymbol{\sigma} \cdot \boldsymbol{p})\psi_L = -m\psi_R \qquad (E + \boldsymbol{\sigma} \cdot \boldsymbol{p})\psi_R = -m\psi_L \tag{3.35}$$

which are decoupled if $m = 0$. In this case, remembering that the spin operator is

$$S = \frac{\hbar}{2}\boldsymbol{\Sigma} = \frac{\hbar}{2}(-\gamma^0\gamma_5\boldsymbol{\gamma}) = \frac{\hbar}{2}\begin{pmatrix} 0 & \boldsymbol{\sigma} \\ \boldsymbol{\sigma} & 0 \end{pmatrix} \tag{3.36}$$

the Dirac equation reduces to the Weyl equations:

$$\frac{\boldsymbol{\Sigma} \cdot \boldsymbol{p}}{|\boldsymbol{p}|}\psi_{R,L} = \pm\text{sign}(p^0)\,\psi_{R,L} \tag{3.37}$$

showing that the helicity is positive for right-handed chirality and negative for left-handed chirality, if the energy is positive (and the opposite is true for negative energy states, and therefore for antiparticles). Such relations only make sense for a massless particle, because for a massive particle helicity is not a good quantum number but depends on the reference system; the mass term $m\bar{\psi}\psi$ in the Dirac equation indeed couples ψ_R and ψ_L.

A description of the neutrino using only one of the equations (3.37), which are linear in the momenta and involve only two-component spinors, was proposed by Weyl (1929) but dismissed as not being of physical interest on the grounds that these equations are not invariant under a parity transformation. Indeed $S_P = \gamma_0$ and γ_5 anti-commute, so that under parity $\psi_R \leftrightarrow \psi_L$. Clearly, after the fall of parity this objection was no longer relevant. Indeed, when $m = 0$ not four but only three anti-commuting matrices are necessary to write a *linear* relativistic fermion wave equation, and the 2×2 Pauli matrices are enough, without the need of doubling the spinor components (or coupling particles and antiparticles).

Turning back to charge conjugation, it should be observed that the charge-conjugate of a chiral spinor has the opposite chirality[40]

$$\gamma_5(\psi_R)_C = -(\psi_R)_C \qquad \gamma_5(\psi_L)_C = +(\psi_L)_C \tag{3.38}$$

The important point to note is that, unless some interaction exists which can distinguish a particle from an antiparticle (e.g. electromagnetism if the particle has non-zero electric charge), there is no way to distinguish a (massless, spin 1/2) right-handed chirality particle from the charge-conjugate of a left-handed particle, that

[40] This requires some care to avoid confusion in the notation, since e.g. a (charge-conjugated) spinor with a 'R' subscript is actually *left-handed*.

is ψ_R from $(\psi_L)_C$. Therefore, even if only a single state exists for each helicity, one can either have a description in which the negative-helicity particle is the antiparticle of the positive-helicity one, or a description in which the two different helicity states are different states of the same particle, which is its own antiparticle: the latter is the case of Majorana spinors.

A *Majorana particle* is a self charge-conjugated particle, i.e. one which is equal to its charge-conjugate to within a phase; it is a truly neutral particle, for which all additive charges are zero. For a Majorana spinor therefore

$$\psi^{(M)} = \eta_C S_C \overline{\psi}^{(M)T} \tag{3.39}$$

Note that, if the Majorana representation for gamma matrices is chosen, with an appropriate choice of the phase η_C the Majorana spinor can be made Hermitian; despite this, it remains non-observable, since two fields at different space-time points fail to commute, as can be verified explicitly (Leite Lopes, 1969).

From the transformations (3.29)–(3.33) it is seen that for a Majorana spinor the vector and tensor bilinear covariants are identically zero:

$$: \overline{\psi}^{(M)} \gamma^\mu \psi^{(M)} := 0 \qquad : \overline{\psi}^{(M)} \sigma^{\mu\nu} \psi^{(M)} := 0 \tag{3.40}$$

This means that for a Majorana fermion both charge and magnetic moment are necessarily zero; one can also obtain

$$: \overline{\psi}^{(M)} \sigma^{\mu\nu} \gamma_5 \psi^{(M)} := 0 \tag{3.41}$$

which implies that the electric dipole moment also vanishes (Chapter 4). The above also implies that a Majorana fermion can only interact with scalar, pseudo-scalar or pseudo-vector fields if charge-conjugation symmetry is valid.

A Majorana fermion field transforms under parity transformation as $\psi + \psi_C$, namely

$$P^\dagger \psi^{(M)}(t, \boldsymbol{x}) P = \eta_P S_P \psi(t, -\boldsymbol{x}) - \eta_P^* S_P S_C \overline{\psi}^T(t, -\boldsymbol{x}) \tag{3.42}$$

and to be consistent with the parity transformation for a spinor one needs in this case $\eta_P = -\eta_P^*$, i.e. $\eta_P = \pm i$. Note that from eqns (2.44) and (2.45) it follows that in this case the particle and the antiparticle have *the same* (imaginary) parity, but still the physically significant quantity, which is the parity of a fermion antifermion state (with zero relative orbital angular momentum) is -1 as for Dirac fermions. Indeed the imaginary intrinsic parity for Majorana particles can also be deduced from $\eta_P(\psi)\eta_P(\psi_C) = -1$, as a Majorana spinor coincides with its charge conjugate. This parity assignment implies that e.g. two (neutral) Dirac fermions cannot transform into a (neutral) Dirac fermion and a Majorana fermion if parity is conserved (or if P^2 is a symmetry, see Section 5.3).

So far all known fermions can be described by Dirac spinors; a possible candidate for being a Majorana particle is the neutrino; if this were the case, the neutrino mass term in the Lagrangian proportional to $\overline{\psi}^{(M)}\psi^{(M)}$ would violate the conservation of lepton number (including terms such as $\psi\psi$).

In general, the charge conjugation operation transforms an irreducible representation of Lorentz group (A, B) (see Appendix B) into the representation (B, A). Considering spin j fields ϕ_R, ϕ_L transforming according to the representations $(j, 0)$ and $(0, j)$ respectively, the effect of charge conjugation can be written in general as

$$\phi_R \rightarrow (\phi_C)_R = \eta_C M_C^{(j)} \phi_L^{\dagger T} \qquad \phi_L \rightarrow (\phi_C)_L = \eta_C \overline{M}_C^{(j)} \phi_R^{\dagger T}$$

with $|\eta_C|^2 = 1$ to keep the normalization of $\phi_{R,L}$. An explicit representation for the matrices $M_C^{(j)}, \overline{M}_C^{(j)}$ which shuffle the spin components is given by the matrix elements between eigenstates $|j, s\rangle$ of spin J and the quantized spin component J_z as:

$$\left[M_C^{(j)}\right]_{s's} = \langle js'|M_C^{(j)}|js\rangle = \langle js'|e^{+i\pi J_y/\hbar}|js\rangle = (-1)^{j+s}\delta_{s',-s}$$

from which

$$M_C^{(j)\dagger} M_C^{(j)} = 1 = \overline{M}_C^{(j)\dagger}\overline{M}_C^{(j)} \qquad M_C^{(j)*} M_C^{(j)} = (-1)^{2j} = \overline{M}_C^{(j)*}\overline{M}_C^{(j)}$$

Combining the two chiral components together in a column vector as

$$\Phi \equiv \begin{pmatrix} \phi_R \\ \phi_L \end{pmatrix}$$

the charge conjugation transformation becomes

$$\Phi \rightarrow C^\dagger \Phi C = \Phi_C = \eta_C \mathcal{M}_C^{(j)} \Phi^{\dagger T} \qquad \Phi^\dagger \rightarrow C^\dagger \Phi^\dagger C = \Phi_C^\dagger = \eta_C^* \Phi^T \mathcal{M}_C^{(j)\dagger}$$

where

$$\mathcal{M}_C^{(j)} \equiv \begin{pmatrix} 0 & M_C^{(j)} \\ \overline{M}_C^{(j)} & 0 \end{pmatrix}$$

and from $C^2 = 1$

$$\mathcal{M}_C^T \mathcal{M}_C^\dagger = 1 \quad \text{that is} \quad M_C^{(j)T}\overline{M}_C^{(j)\dagger} = 1$$

which can be satisfied by

$$\overline{M}_C^{(j)} = (-1)^{2j} M_C^{(j)}$$

To make connection with the earlier discussion for spin 1/2, it is sufficient to use the Weyl representation for Dirac matrices, in which

$$M_C^{(1/2)} = i\sigma_2 \sin(\pi/2) = i\sigma_2 \qquad \overline{M}_C^{(1/2)} = -i\sigma_2$$

so that for the four-component Dirac spinor in the Weyl representation

$$M_C^{(1/2)} \equiv \begin{pmatrix} 0 & i\sigma_2 \\ -i\sigma_2 & 0 \end{pmatrix} = i\gamma_2$$

which is consistent with the previous discussion since $S_C = M_C^{(1/2)}\gamma_0^T$ (and in the Weyl representation $S_C = i\gamma_2\gamma_0$ and $\gamma_0^T = \gamma_0$).

By definition the parity transformation exchanges $\phi_R \leftrightarrow \phi_L$, that is

$$P^\dagger \Phi P = \eta_P M_P \Phi \qquad M_P = \begin{pmatrix} 0 & 1 \\ 1 & 0 \end{pmatrix}$$

Considering that the effect of a parity and charge conjugation operation in any order must give the same result, one finds

$$P^\dagger C^\dagger \Phi CP = P^\dagger \Phi_C P = \bar{\eta}_P M_P \Phi_C = \bar{\eta}_P M_P \eta_C M_C \Phi^{\dagger T}$$

$$C^\dagger P^\dagger \Phi PC = C^\dagger \eta_P M_P \Phi C = \eta_C M_C \eta_P^* M_P^{\dagger T} \Phi^{\dagger T}$$

where $\bar{\eta}_P$ denotes the phase factor for the antiparticle; from this, noting that $M_C M_P = (-1)^{2s} M_P M_C$ one obtains the relation between the intrinsic parities for a particle and an antiparticle in the general case:

$$\eta_P \bar{\eta}_P = (-1)^{2s} \tag{3.43}$$

Fermions and antifermions have opposite intrinsic parities, while bosons and antibosons have the same intrinsic parities.

3.2   Charge conjugation symmetry

Charge conjugation symmetry requires that for every non-neutral particle an antiparticle exists which behaves in exactly the same way except for having the sign of all its internal charges reversed. The prediction of the existence of antiparticles just preceded the unexpected discovery of the first element of antimatter, the positron in 1932 (Anderson, 1933); antiprotons were definitely identified only in 1955 (Chamberlain *et al.*, 1955) and antineutrons in 1957 (Cork *et al.*, 1957). Since then for any known particle an antiparticle has also been identified, although for some truly neutral particles, such as the photon, this coincides with the particle itself.

This would seem to suggest a possible way of testing the validity of C symmetry, by verifying the existence of antiparticles and comparing their static properties with those of the corresponding particles. This program cannot work, however, because even if C is violated, another more fundamental symmetry, namely CPT (Chapter 5), guarantees such particle-antiparticle symmetry. Tests of C which are not shadowed by the validity of *CPT* symmetry have to deal with the properties of particle interactions.

Charge conjugation symmetry is expressed as the possibility of choosing the factors η_C for the particles in such a way that the Hamiltonian is invariant under the charge conjugation transformation:

$$H_C \equiv C^\dagger H C = H \quad \text{or} \quad [H, C] = 0 \tag{3.44}$$

As an example, the requirement of charge conjugation invariance of the Lagrangian determines the *C*-parity of neutral (Hermitian) boson fields ϕ coupled to fermion fields; by imposing

$$C^\dagger \left(: \overline{\psi} \Gamma \psi : \phi \right) C = : \overline{\psi} \Gamma \psi : \phi \tag{3.45}$$

(where possible Lorentz indexes of ϕ have been suppressed) from eqn (3.29)–(3.33) one obtains $\eta_C = +1$ for couplings to scalar, pseudo-scalar or pseudo-vector fields, and $\eta_C = -1$ for couplings to vector or tensor fields.

The charge conjugation symmetry of the generic fermionic Lagrangian (B.29) requires (taking into account its Hermiticity) that all the c_i have the same phase, all the c'_i have the same phase, but that the c_i and the c'_i are relatively imaginary (phase difference of $\pm\pi/2$): factorizing an overall phase this means $c_i^* = c_i$ and $c_i'^* = -c_i'$, i.e. all the parity-conserving couplings must be real and all the parity-violating ones must be pure imaginary.

3.2.1 *C*-parity of elementary particles

We recall that the charge conjugation transformation of charged fields contains an arbitrary phase factor η_C; if the system is charge conjugation symmetric, the phase factors η_C for non-Hermitian fields are arbitrary (Feinberg and Weinberg, 1959). Only in the case of neutral particles can the η_C factors be considered as quantum numbers associated to the particle, being the eigenvalues of the C operator.

Choosing by convention the vacuum to have positive *C*-parity,[41] and remembering that *C*-parity is a multiplicative quantum number, from the transformation of the electromagnetic field of eqn (3.14) the *C*-parity of a photon is -1, and that of a state with N photons is $\eta_C = (-1)^N$.

Neutral fields do not interact with photons (although they can decay in states containing photons), but their *C*-parity can be determined by their non-electromagnetic interactions with charged fields (if such interactions respect charge conjugation symmetry). The *C*-parity of the neutral pion is $\eta_C(\pi^0) = +1$, because of its (EM, *C*-conserving) 2γ decay.

---

[41] Despite its possible degeneracies, the vacuum is a C eigenstate (Michel, 1953).

A neutral system of two or more particles can be an eigenstate of charge conjugation, and in this case charge conjugation allows to introduce a generalized version of the spin-statistics theorem. In a system composed of a charged particle and its antiparticle, the two objects are distinct (by their charge) and as such their state is not required to have any special symmetry under exchange. However the two particles transform into each other by charge conjugation, and they can be considered as two identical particles 'labelled' by a charge quantum number, which is reversed by C. In this way one can still require that the two particles are in a symmetric or anti-symmetric state under exchange, provided such exchange is enforced by exchanging the positions (the parity transformation in the centre of mass of the two particles), the spin labels (which introduces a sign depending on the symmetry or anti-symmetry of the total spin state) and also the charge labels (which introduces a sign corresponding to the C-parity of the state). This procedure can be properly justified in field theory by noting that the creation operators for a fermion and an antifermion anti-commute, while those for a boson and antiboson commute. In general, the C-parity of a particle antiparticle state would be $+1$ if such a state were allowed for two identical particles (i.e. symmetric for bosons and anti-symmetric for fermions).

From the above a neutral fermion antifermion system has C-parity eigenvalue

$$\eta_C = (-1)^{L+S} \tag{3.46}$$

(where L, S are the eigenvalues of orbital angular momentum and spin); indeed their exchange must result in a -1 factor in the state due to Fermi–Dirac (F–D) symmetry, and such exchange can be obtained by the swap of the charges (multiplying the state vector by η_C) followed by the swap of the positions (a factor $(-1)^L$), and the swap of the spin states (a factor $(-1)^{S+1}$).

A neutral system of two bosons has C-parity eigenvalue

$$\eta_C = (-1)^{L+S} \tag{3.47}$$

where L is the relative orbital angular momentum and S the total spin. The overall $+1$ factor for exchange is obtained as $\eta_C \times (-1)^L \times (-1)^S$ (spin exchange for bosons).

Note that in the case of two *identical* (neutral) bosons L must be even, so that, for a $\pi^0\pi^0$ pair, the C-parity is always $+1$ (actually, since $\eta_C(\pi^0) = +1$, the C-parity of any system composed only of neutral pions is $+1$, independently of their spatial configuration).

For a neutral system of three bosons (e.g. $\pi^+\pi^-\pi^0$) the C-parity eigenvalue is the same as in (3.47), where now L is the angular momentum of the two particles which are not charge conjugation eigenstates.

For any state of a spinless particle antiparticle pair ($J = L$) the C-parity is $\eta_C = (-1)^J$; for the $J = 0$ state of a particle-antiparticle pair ($L = S$) the C-parity is $\eta_C = +1$.

3.2.2   Transition amplitudes

The symmetry of the theory under charge conjugation can be expressed in terms of the S matrix (see Appendix C) as

$$[C, S] = 0 \qquad (3.48)$$

and denoting with a bar the antiparticle states, when this symmetry is valid

$$S_{fi} \equiv \langle f|S|i \rangle = \langle f|C^\dagger SC|i \rangle = \eta_C(f)^* \eta_C(i) \langle \bar{f}|S|\bar{i} \rangle \equiv \eta_C(f)^* \eta_C(i) S_{\bar{f}\bar{i}} \qquad (3.49)$$

and a similar relation holds for the transition matrix T

$$T_{fi} = \eta_C(f)^* \eta_C(i) T_{\bar{f}\bar{i}} \qquad (3.50)$$

3.2.3   Selection rules

Some consequences of charge conjugation symmetry which can be deduced from what was discussed above are:

- The coupling of a scalar field ϕ with fermions cannot contain both a scalar and a vector term ($S\phi + V^\mu \partial_\mu \phi$, see Appendix B); that of a neutral pseudo-vector field ϕ^μ cannot contain both an axial-vector and a tensor term ($A^\mu \phi_\mu + T^{\mu\nu} \partial_\mu \phi_\nu$) (Pais and Jost, 1952).

- A reaction among neutral bosons is not allowed if the number of vector or tensor couplings to intermediate fermion fields is odd (Leite Lopes, 1969). Among particular cases of this general result:

   (a) The charge conjugation symmetry of QED implies that diagrams with an odd number of external photon lines are identically zero: this is *Furry's theorem* (Furry, 1937).

   (b) Some transitions between states with given spin-parity J^P and states with a given number of photons are not allowed: examples are 1^- (vector) $\rightarrow 2\gamma$ (also forbidden by angular momentum conservation), 1^+ (pseudo-vector) $\rightarrow 3\gamma$, $0^\pm$ (scalar, pseudo-scalar) $\rightarrow 3\gamma$.

   (c) The decay of a C-even neutral boson into any number of neutral π^0 (with $C = +1$) and an odd number of photons is forbidden (Wolfenstein and Ravenhall, 1952).

   (d) Decays such as $\phi \rightarrow \omega\gamma$, $\phi \rightarrow \rho^0\gamma$, $\omega \rightarrow \rho^0\gamma$, $\omega, \rho^0 \rightarrow \eta\pi^0$, $\eta \rightarrow \pi^0 e^+ e^-$ cannot be driven by interactions which are symmetric under charge conjugation.

- Charge-conjugate systems must behave in the same way apart from the charges being opposite: charge-conjugate reactions must have the same differential cross sections, energy spectra, etc. and the same is true for the distribution of charge-conjugate particles in multi-body decays, e.g. for $\eta \rightarrow \pi^+\pi^-\pi^0$ the spectra of π^+ and π^- must be identical if charge conjugation symmetry holds.

• Assuming strong interactions are symmetric under C, P, T separately (and the *CPT* theorem is valid, see Chapter 5), to lowest order in the weak interaction there can be no interference between the parity-conserving and the parity-violating parts of the Hamiltonian (and therefore no parity-violating asymmetry), when the final states are free (non-interacting) particles (Lee Oehme, and Yang 1957*b*) (we will prove this result in Chapter 5). If a parity-violating asymmetry corresponding to a non-zero expectation value for a pseudo-scalar quantity is found, the part of the asymmetry which is independent from the interactions of final state particles can arise only if charge conjugation symmetry is violated (Ioffe *et al.*, 1957); this is indeed how the beta decay experiments which proved parity violation in weak interactions also demonstrated the violation of charge conjugation symmetry, since the interactions among the final particles – despite being present – are too weak to originate asymmetries of the observed magnitude. The case of muon decay asymmetries is even simpler: since in this case the final particles are essentially free (they only interact through weak interactions), parity non-conserving asymmetries cannot arise at all if charge conjugation symmetry is valid (again assuming *CPT* symmetry is valid).

Note that, as mentioned, the equality of lifetimes and other properties of particles and antiparticles does not necessarily require the validity of charge conjugation symmetry, since they follow from the more general CPT symmetry (see Chapter 5); the same is true for total cross sections and total decay widths for charge-conjugate particles.

3.2.4 *G*-parity

The concept of C-parity as charge conjugation eigenvalue is of limited applicability, being defined only for neutral states, but it can be extended to charged particles in case isospin symmetry is also valid.

Considering only processes which respect isospin symmetry, i.e. for which there is independence from the value of the isospin component chosen as quantum number (usually called I_3), a charge-symmetry operator can be defined as

$$R = e^{-i\pi I_2} \tag{3.51}$$

The action of R corresponds to a rotation by an angle π around the 2-axis in isotopic-spin space. For an iso-spinor field (isospin $I = 1/2$, such as nucleons) written as a column vector with components $\psi_{1,2}$ corresponding to $I_3 = \pm 1/2$, one can write in isospin space

$$R^\dagger \begin{pmatrix} \psi_1 \\ \psi_2 \end{pmatrix} R = \mathcal{D}^{(1/2)\dagger}(\theta_2 = \pi) \begin{pmatrix} \psi_1 \\ \psi_2 \end{pmatrix} = \begin{pmatrix} \psi_2 \\ -\psi_1 \end{pmatrix} \tag{3.52}$$

since the rotation matrix (see e.g. Sakurai (1985)) is given by

$$\mathcal{D}^{(1/2)}(\theta_2 = \pi) = e^{-i\pi\sigma_2^{(2)}/2} = -i\sigma_2^{(2)} = \begin{pmatrix} 0 & -1 \\ 1 & 0 \end{pmatrix} \tag{3.53}$$

where $\sigma_2^{(2)}$ is the usual Pauli matrix (the superscript being just a reminder that it is a 2×2 matrix).

For an iso-vector field ($I = 1$, such as pions) with three components $\phi_{1,2,3}$:

$$R^\dagger \begin{pmatrix} \phi_1 \\ \phi_2 \\ \phi_3 \end{pmatrix} R = \mathcal{D}^{(1)\dagger}(\theta_2 = \pi) \begin{pmatrix} \phi_1 \\ \phi_2 \\ \phi_3 \end{pmatrix} = \begin{pmatrix} -\phi_1 \\ \phi_2 \\ -\phi_3 \end{pmatrix} \tag{3.54}$$

since the rotation matrix is given in terms of the three-dimensional Pauli matrix

$$\sigma_2^{(3)} = \begin{pmatrix} 0 & 0 & i \\ 0 & 0 & 0 \\ -i & 0 & 0 \end{pmatrix}$$

as $\mathcal{D}^{(1)\dagger}(\theta_2 = \pi) = e^{-i\pi\sigma_2^{(3)}} = \begin{pmatrix} -1 & 0 & 0 \\ 0 & 1 & 0 \\ 0 & 0 & -1 \end{pmatrix}$

Considering the eigenstates of I_3

$$\phi_\pm = (\phi_1 \pm i\phi_2)/\sqrt{2} \qquad \phi_0 = \phi_3 \tag{3.55}$$

which are also eigenstates of electric charge with eigenvalues ± 1 and 0 respectively (in units of $|e|$), one has

$$R^\dagger \begin{pmatrix} \phi_+ \\ \phi_0 \\ \phi_- \end{pmatrix} R = - \begin{pmatrix} \phi_- \\ \phi_0 \\ \phi_+ \end{pmatrix}$$

Gauge transformations act as $\phi \rightarrow e^{iQ\alpha}\phi$, where Q is the electric charge operator:

$$\begin{pmatrix} \phi_+ \\ \phi_0 \\ \phi_- \end{pmatrix} \rightarrow \begin{pmatrix} e^{+i\alpha}\phi_+ \\ \phi_0 \\ e^{-i\alpha}\phi_- \end{pmatrix} \quad \text{or explicitly} \quad \begin{pmatrix} \phi_1 \\ \phi_2 \\ \phi_3 \end{pmatrix} \rightarrow \begin{pmatrix} \cos\alpha\,\phi_1 - \sin\alpha\,\phi_2 \\ \sin\alpha\,\phi_1 + \cos\alpha\,\phi_2 \\ \phi_3 \end{pmatrix}$$

and are thus related to a rotation around the 3-axis in isospin space. As the charge conjugation operation reverses such rotation, it corresponds to a reflection in a plane containing the 3-axis; in order to have $\phi_+ \leftrightarrow \phi_-$ this is a reflection in the (1,3) plane, that is:

$$C^\dagger \begin{pmatrix} \phi_1 \\ \phi_2 \\ \phi_3 \end{pmatrix} C = \begin{pmatrix} +\phi_1 \\ -\phi_2 \\ +\phi_3 \end{pmatrix} \quad \text{or} \quad C^\dagger \begin{pmatrix} \phi_+ \\ \phi_0 \\ \phi_- \end{pmatrix} C = \begin{pmatrix} \phi_- \\ \phi_0 \\ \phi_+ \end{pmatrix} \tag{3.56}$$

Noting that both the charge symmetry operator R and C invert the charges of a system, the concept of charge conjugation can be generalized to non-neutral states by defining the so-called *G-parity* (for which a much better name would be *isotopic parity*) operator (Michel, 1953; Amati and Vitale, 1955; Lee and Yang, 1956a):

$$G \equiv CR = Ce^{i\pi I_2} \tag{3.57}$$

For an iso-vector system, G-parity corresponds to the inversion of all three Hermitian components of the field (C only inverts the 2 component, transforming $\phi_+ \leftrightarrow \phi_-$, while R only inverts components 1 and 3), which leaves the eigenstates unchanged to within a sign.

$$G^\dagger \begin{pmatrix} \phi_1 \\ \phi_2 \\ \phi_3 \end{pmatrix} G = -\begin{pmatrix} \phi_1 \\ \phi_2 \\ \phi_3 \end{pmatrix} \quad \text{or} \quad G^\dagger \begin{pmatrix} \phi_+ \\ \phi_0 \\ \phi_- \end{pmatrix} G = -\begin{pmatrix} \phi_+ \\ \phi_0 \\ \phi_- \end{pmatrix}$$

amounts to saying that pions are eigenstates of G-parity with eigenvalue $\eta_G(\pi) = -1$. By applying G twice one obtains a state which is equivalent to the original one, so that G^2 can be chosen to be 1 and the eigenvalues of G to be ± 1.

The quantity associated with G is conserved if charge conjugation and isospin symmetries are valid, as appears to be the case for strong interactions; in this case its eigenvalue η_G (called the G-parity of the particle) is a good (conserved) multiplicative quantum number. Electromagnetic interactions violate isospin symmetry and thus also the selection rules based on G-parity.

Any system with integer isospin necessarily includes a state with $I_3 = 0$, which (in absence of other internal charges such as strangeness, etc.) can be an eigenstate of charge conjugation, such that $\eta_G = \eta_C(-1)^I$. For systems that respect charge conjugation symmetry (to a large extent those whose properties are defined by strong or electromagnetic interactions), the eigenvalue of G is obtained by considering the isospin transformation of the I_3 element of the multiplet, and G-parity then results in selection rules (for strong interactions) that are also valid for non-neutral systems; since G (as opposed to C) commutes with all components of isospin all the members of an isospin multiplet share the same G-parity. Since for a neutral ($I_3 = 0$) state $R = (-1)^I$ one has in this case

$$\eta_G = \eta_C(-1)^I \tag{3.58}$$

A $\pi\pi$ state has G-parity $+1$, and in general it can be in a $I = 0, 1, 2$ isospin state with $I + L$ even, that is in an even (odd) angular momentum L state for even (odd) isospin, because of Bose–Einstein symmetry. From eqn (3.58) therefore $\eta_C = (-1)^I$, and this immediately shows that a $\pi^0\pi^0$ state (for which $C = +1$) cannot be in a (isospin anti-symmetric) $I = 1$ state. More generally, a system of an even (odd) number of π^0 can only have even (odd) isospin.

For baryons (e.g. nucleons) and flavoured mesons the particles and antiparticles belong to different isospin multiplets: G (or C) anti-commutes with baryon number and with flavour numbers such as strangeness, etc., so that states containing only particles (and no antiparticles) cannot be G eigenstates.

For a system with n_N nucleons one has $G^2 = (-1)^{n_N}$, and in general when strange particles are also considered, $G^2 = Y(-1)^2$, where the hypercharge $Y = (N + S)/2$ (N being baryon number and S strangeness).

According to (3.52) the nucleon $I = 1/2$ iso-doublet transforms under G-parity as

$$G^\dagger \begin{pmatrix} p \\ n \end{pmatrix} G = \begin{pmatrix} \bar{n} \\ -\bar{p} \end{pmatrix} \tag{3.59}$$

where $C^\dagger(p, n)C = (\bar{p}, \bar{n})$ and a minus sign is introduced[42] in the antinucleon iso-doublet.

---

[42] If the antinucleon iso-doublet were defined to be $(\bar{n}, \bar{p})^T$, always with the $I_3 = +1/2$ component on top but without a minus sign, the transformations under R would be $\bar{p} \to \bar{n}$, $\bar{n} \to -\bar{p}$, opposite to $p \to -n$, $n \to p$ which is valid for nucleons. With the definition $(\bar{n}, -\bar{p})^T$ instead, for which an antiproton state is $(0, -1)^T$, the nucleon and antinucleon iso-doublets transform in the same way.

The implication of the above is that while two-nucleon states of defined isospin $|NN; I, I_3\rangle$ are

$$|NN; 1, +1\rangle = |pp\rangle \qquad |NN; 1, 0\rangle = (|pn\rangle + |np\rangle)/\sqrt{2}$$

$$|NN; 1, -1\rangle = |nn\rangle \qquad |NN; 0, 0\rangle = (|pn\rangle - |np\rangle)/\sqrt{2}$$

(symmetric $I = 1$ and anti-symmetric $I = 0$), the states of a nucleon and an antinucleon (C eigenstates) of defined isospin $|N\overline{N}, ; I, I_3\rangle$ are

$$|N\overline{N}; 1, +1\rangle = +|p\overline{n}\rangle \qquad |N\overline{N}; 1, 0\rangle = (|n\overline{n}\rangle - |p\overline{p}\rangle)/\sqrt{2}$$

$$|N\overline{N}; 1, -1\rangle = -|n\overline{p}\rangle \qquad |N\overline{N}; 0, 0\rangle = -(|n\overline{n}\rangle + |p\overline{p}\rangle)/\sqrt{2}$$

but note that these states have no defined symmetry under particle exchange and can be symmetrized or anti-symmetrized (so that both symmetric and anti-symmetric states exist for both isospin states).

Summarizing:

- A state with N pions has $\eta_G = (-1)^N$.
- A state composed of a K and a $\overline{K}$ (that is $K^0\overline{K}^0$ or K^+K^-) with isospin I ($I = 0, 1$) has $\eta_G = (-1)^{L+I}$ (where L is the orbital angular momentum).
- A nucleon-antinucleon state with isospin I has $\eta_G = (-1)^{L+S+I}$ (where S is the total spin).

As examples of the consequences of G-parity symmetry:

- The same particle cannot decay (by strong interactions) both into states with an even and an odd number of pions.
- The number of pions produced in a nucleon-antinucleon interaction is even (odd) if $L + S + I$ is even (odd). This holds for a isospin eigenstate, which is a mixture of a $p\overline{p}$ state and a $n\overline{n}$ state, so that it actually means that e.g. the annihilation rate of a $p\overline{p}\ {}^1S_0$ state into a given number of pions is the same as the corresponding rate for a $n\overline{n}\ {}^1S_0$ state.

It is worth considering in more detail the case of nucleon-antinucleon annihilation into pions as an example of the application of the symmetries which were introduced so far. First of all, in the initial state only isospin $I = 0, 1$ will be available, with spin $S = 0, 1$ and arbitrary orbital angular momentum L_i.

For the neutral states of two (spinless) pions ($G = +1$), that is $\pi^+\pi^-$ or $\pi^0\pi^0$, with orbital angular momentum L_f, one has $\eta_P = (-1)^{L_f}$; $\eta_C = +1$ for $\pi^0\pi^0$

(only $I = 0$ possible) and $\eta_C = (-1)^I$ for $\pi^+\pi^-$. The possible states with lower L_f and their spin-parities are

2π	J^{PC}	I
S_0	0^{++}	$0,2$
P_1	1^{--}	1 ($\pi^+\pi^-$ only)
D_2	2^{++}	$0,2$
F_3	3^{--}	1 ($\pi^+\pi^-$ only)

Now, the compatibility with a nucleon-antinucleon annihilation (induced by strong interactions) enforces additional constraints (and excludes $I = 2$). Parity conservation requires $(-1)^{L_i+1} = (-1)^{L_f} = (-1)^J$, and therefore when L_i is even (odd) L_f is odd (even) and $I = 1$ ($I = 0$); only even L_f ($I = 0$) is allowed for $\pi^0\pi^0$. This means $J = L_i \pm 1$, or $S = 1$: only the triplet nucleon-antinucleon state contributes. Charge conjugation symmetry requires $(-1)^{L_i+S} = (-1)^{L_f} = (-1)^I$ (or $+1$ for $\pi^0\pi^0$); since $S = 1$ this constraint adds nothing to the previous one.

Considering now the neutral states of 3 pions ($G = -1$), that is $\pi^+\pi^-\pi^0$ or $\pi^0\pi^0\pi^0$, with orbital angular momenta L_f (among the first two particles) and l_f (among the third and the centre of mass of the first two), these have $\eta_P = (-1)^{L_f+l_f+1}$, which for $J = 0$ ($L_f = l_f$) gives $\eta_P = -1$; $\eta_C = +1$ for $\pi^0\pi^0\pi^0$ and $\eta_C = (-1)^{L_f}$ for $\pi^+\pi^-\pi^0$ (note that this can also be expressed in terms of the isospin of the first two pions); from eqn (3.58) $\eta_C = (-1)^{I+1}$, so that $L_f + I$ must be odd. The possible states with lower L_f, l_f and their spin-parities are

3π	J^{PC}	I
$(Ss)_0$	0^{-+}	$1,3$
$(Sp)_1$	1^{++}	$1,3$
$(Ps)_1$	1^{+-}	$0,2$ ($\pi^+\pi^-\pi^0$ only)
$(Pp)_{0,1,2}$	$0,1,2^{--}$	$0,2$ ($\pi^+\pi^-\pi^0$ only)
$(Sd)_2$	2^{-+}	$1,3$
$(Ds)_2$	2^{-+}	$1,3$
$(Pd)_{1,2,3}$	$1,2,3^{+-}$	$0,2$ ($\pi^+\pi^-\pi^0$ only)
$(Dp)_{1,2,3}$	$1,2,3^{++}$	$1,3$
$(Dd)_{0,1,2,3,4}$	$0,1,2,3,4^{-+}$	$1,3$

(in the spectroscopic notation the first (upper case) and second (lower case) letter refer to the eigenvalues of L_f and l_f respectively).

TABLE 3.1 Possible 2π and 3π final states for nucleon-antinucleon annihilation in different states, with their spin-parity J^P, isospin I and G-parity η_G eigenvalues. The entries in the 2π, 3π columns indicate the allowed pion state with lowest angular momentum, if any, or the symmetry which forbids the decay: P, C, G standing for parity, charge conjugation and G-parity symmetry; the last column indicates the minimum number of pions for the annihilation of the $N\overline{N}$ state.

$N\overline{N}$	J^{PC}	I	η_G	$\pi^0\pi^0$	$\pi^+\pi^-$	$\pi^0\pi^0\pi^0$	$\pi^+\pi^-\pi^0$	n_π (min)
$^1S_0$	0^{-+}	0	$+1$	P	P	G	G	4
$^1S_0$	0^{-+}	1	-1	PG	PG	$(Ss)_0$	$(Ss)_0$	3
$^3S_1$	1^{--}	0	-1	CG	G	C	$(Pp)_1$	3
$^3S_1$	1^{--}	1	$+1$	C	P_1	CG	G	2
$^1P_1$	1^{+-}	0	-1	PCG	PG	C	$(Ps)_1$	3
$^1P_1$	1^{+-}	1	$+1$	PC	P	CG	G	4
$^3P_0$	0^{++}	0	$+1$	S_0	S_0	PG	PG	2
$^3P_0$	0^{++}	1	-1	G	G	P	P	5
$^3P_1$	1^{++}	0	$+1$	P	P	G	G	4
$^3P_1$	1^{++}	1	-1	PG	PG	$(Sp)_1$	$(Sp)_1$	3
$^3P_2$	2^{++}	0	$+1$	D_2	D_2	G	G	2
$^3P_2$	2^{++}	1	-1	G	G	$(Dp)_2$	$(Dp)_2$	3

The compatibility with nucleon-antinucleon annihilation excludes $I = 2, 3$; parity symmetry requires $(-1)^{L_i+1} = (-1)^{L_f+l_f+1}$, that is $L_i+L_f+l_f$ even (L_f even for $\pi^0\pi^0\pi^0$, so $L_i + l_f$ even). Charge conjugation symmetry requires $(-1)^{L_i+S} = (-1)^{L_f}$ (L_i+L_f+S even) and therefore (together with the previous condition) l_f+S even (this reduces for $\pi^0\pi^0\pi^0$ to $L_i + l_f$ even, $L_i + S$ even). A full discussion of 3π states and their exchange symmetries, together with the explicit expressions for their wave functions, can be found in Zemach (1964).

Some possible two- and three-pion final states for nucleon-antinucleon annihilation are listed in Table 3.1. Selection rules for other initial and final states can be found, e.g., in Amati and Vitale (1955) or Goebel (1956).

3.3 Charge conjugation violation

As for parity, weak interactions violate charge conjugation symmetry in a maximal way, meaning that the amplitudes linking states of opposite C-parity have the same size as the ones linking states of same C-parity. No violation of this symmetry was found in other interactions.

Except for very simple systems, direct tests of charge conjugation invariance are usually rather difficult to perform in practice, since they require the preparation of antiparticle systems; while such tests are an active area of research, most tests of charge conjugation are done indirectly, by verifying the constraints imposed by such symmetry.

3.3.1   Weak interactions

The first indication that charge conjugation symmetry is violated in Nature came with the discovery of parity violation (Lee, Oehme, and Yang, 1957b); since the two symmetries are tightly linked in weak interactions, most of the early experiments on parity violation also had implications on C violation.

Since the asymmetry parameter of the β angular distribution of polarized ^{60}Co decay (Gamow–Teller transition) is proportional to $\mathrm{Re}(c_T c_T'^* - c_A c_A')$ (see Chapter 2, the small term due to the Coulomb distortion of the wave functions is neglected), the non-zero result implied not only that either c_T' or c_A' (or both) were simultaneously present together with their counterparts c_T and c_A, but also that the primed and unprimed coefficients were not out of phase by $\pi/2$ as would have been required by charge conjugation symmetry. The expression in eqn (2.71) for the beta decay asymmetry shows that if charge conjugation symmetry is valid, the parity asymmetry can indeed arise only as a consequence of a difference in the Coulomb phase shifts for opposite parities (which is odd under time reversal as required by the CPT theorem), but such effects are too small to give effects as large as observed.

Similarly, the measurement of parity violation in hyperon weak decays also allowed to obtain indirect evidence for C violation. Anticipating the discussion of Chapter 4, in absence of final state interactions time reversal symmetry would require the decay amplitudes to be relatively real, so that the decay parameter $\beta = 0$ in eqn (2.76). If such interactions are not neglected the phases corresponding to the (strong) scattering of the final state particles (δ_S, δ_P) will be present; however, since these phases are measured to be relatively small, the above conclusion on the reality of the amplitudes still holds approximately, and this in turn implies that a large value of the α parameter, $|\alpha| > |\sin(\delta_P - \delta_S)|$ (the $\delta_{S,P}$ phase shifts can be measured independently), is evidence for violation of charge conjugation symmetry (which would require the amplitudes to be relatively imaginary).

Opposite polarizations for charge conjugated leptons were observed in the early beta decay experiments on parity violation, and while these were compatible with neutrinos being always left-handed and antineutrinos being always right-handed, no conclusion on charge conjugation could be obtained, since β^- and β^+ decays are not charge conjugated processes (while anti-^{3}He has been produced (Antipov et $al.$, 1971), experiments with antinuclei are very difficult at present).

However, using the polarimetry techniques discussed in Chapter 2, the longitudinal polarizations of positrons and electrons from μ^+ and μ^- decays were also compared, providing information on charge conjugation symmetry.

In one experiment at Berkeley (Macq et $al.$, 1958) decays of unpolarized $\mu^\pm$ at rest were analysed. μ^+ originated from the decay of positive pions at rest: π^+ from a beam of ~ 200 MeV energy were slowed down and stopped in a 2.5 cm thick carbon target; the emitted muons, with zero average polarization also stopped in the target and decayed at rest. μ^- originated instead from the decay in flight of π^- (since π^- interact strongly in matter), and were stopped in carbon where they were almost completely depolarized by the cascade of atomic transitions to the ground state of the μ-mesic atom, and then completely depolarized by the (unshielded) fringing field of the cyclotron.

In both cases the longitudinal polarization of the positrons or electrons emitted in the direction orthogonal to the incoming beam was analysed by measuring the transmission of the (circularly polarized) *bremsstrahlung* photons which they produced, in a magnetized iron cylinder. The magnetization direction was periodically reversed to reduce spurious asymmetries.

In the case of positrons, the spin dependence of the cross section for annihilation in flight $(e^+e^- \to \gamma\gamma)$ in a magnetized Fe plate was also exploited for the determination of their helicity (Buhler *et al.*, 1963). Antiparticles emitted in weak interaction processes were always found to have longitudinal polarization opposite to that of particles.

3.3.2 Electromagnetic interactions

The possibility of charge conjugation symmetry violation in electromagnetic interactions was considered in the 1960s (Bernstein *et al.*, 1965), as a possibility for explaining the observed *CP* violation (Chapter 6) as the combined effect of a *C*- and *P*-violating but *CP*-symmetric weak interaction combined with some other *C*- and *T*-violating but *P*-symmetric interaction. This possibility was ruled out, and no evidence for *C* violation in EM interactions has been found since.

Positronium (Ps), the bound state of an electron and a positron, is the simplest 'atomic' system with charge conjugation eigenstates. The spin-parity and *C*-parity of the different states are

	$^1S_0$	$^3S_1$	$^1P_1$	$^3P_0$	$^3P_1$	$^3P_2$
J	0	1	1	0	1	2
η_P	-1	-1	$+1$	$+1$	$+1$	$+1$
η_C	$+1$	-1	-1	$+1$	$+1$	$+1$

Note that positronium states exist with any given spin-parity, but not always with both *C*-parities. The $^3S_1$ state (ortho-positronium) has negative *C*-parity and therefore if electromagnetic interactions are charge conjugation symmetric it can only decay to an odd number of photons (parity symmetry also implies this), while the $^1S_0$ state (para-positronium) can decay only to an even number of photons (such as 2), and has therefore a shorter lifetime. In general the allowed decays are those for which $(-1)^{L+S} = (-1)^N$, where L, S are the orbital and spin angular momentum of the positronium state, and N the number of photons to which it decays.

Charge conjugation symmetry also implies the absence of a linear Zeeman effect for positronium: for a charge conjugation symmetric Hamiltonian the energy levels are unchanged by the C transformation, which instead reverses the direction of an external magnetic field, so that the energy cannot depend linearly on it.[43] Moreover, the $J_z = \pm 1$ levels of the $^3S_1$ positronium states are degenerate even in the presence of an external magnetic field $\boldsymbol{B}$ (Wolfenstein and Ravenhall, 1952): a π rotation around an axis orthogonal to $\boldsymbol{B}$ interchanges the two levels and reverses the direction of the magnetic field, which can then be restored by the C transformation without affecting the energies (because the Hamiltonian is invariant), from which the result follows. Note in passing that since the magnetic moment of the positron is much larger than that of a proton, the energy differences between the $J = 1$ and $J = 0$ states, which correspond to the hyperfine structure of an hydrogen atom, are relatively large in positronium ($8.5 \cdot 10^{-4}$ eV, or $2 \cdot 10^5$ MHz, to be compared with $5.9 \cdot 10^{-6}$ eV, or 1420 MHz, for hydrogen).

Searches for C-violating (and P-conserving) positronium decays such as $^1S_0 \rightarrow 3\gamma$ and $^3S_1 \rightarrow 4\gamma$ were performed. Such transitions cannot occur as a result of bound state mixing, since there are no other positronium states with opposite C-parity and the same J^P. The contribution to such transitions expected by the C-violating weak interactions are extremely small, many orders of magnitude below the experimental limits.

Charge conjugation tests in positronium decay

Charge conjugation is the only symmetry which forbids the occurrence of the $^1S_0 \rightarrow 3\gamma$ positronium decay (the $^3S_1 \rightarrow 2\gamma$ decay also forbidden by angular momentum conservation, as discussed in Chapter 2). The search for the C-violating decay of the singlet state into three photons is complicated by the presence of allowed $^3S_1 \rightarrow 3\gamma$ decays, which have to be suppressed.

In order to reduce the number of atomic collisions which may decompose positronium it is better to work with gases, thus also reducing the background due to free positron annihilations. The $e^+e^- \rightarrow 3\gamma$ annihilation rate is suppressed with respect to that for $e^+e^- \rightarrow 2\gamma$ by a factor of order α (actually $\simeq \alpha/8$), and the spin-averaged ratio of 3γ (from triplet state collisions) to 2γ (from singlet state collisions) cross sections is three times the ratio of the annihilation rates, i.e. 1/372. When positronium is formed, however, simple counting of the

---

[43] Note that when an external field is present, C may be no longer a good quantum number, since the interactions of the two particles of opposite charge with the external field are in general not the same.

possible relative spin orientations (for unpolarized electrons and positrons) shows that the formation of $^3S_1$ states (mostly decaying to 3γ) is three times more frequent than that of $^1S_0$ states (mostly decaying to 2γ); if the relative abundance of the two positronium states is maintained, the ratio of 3γ to 2γ decays will be essentially 3/4.

In presence of a gas containing molecules with an odd number of electrons (such as NO) the long-lived triplet state can be easily converted to the singlet state by exchanging an electron with the gas and flipping its spin, because the two molecular levels with opposite spin orientations for the unpaired electron have a very small energy difference; since the singlet state decays quickly the opposite reaction is much less important. In this way the amount of ortho-positronium can be drastically reduced (*quenched*). A static magnetic field mixes the triplet and singlet states, and since the latter has a much shorter lifetime this also has the effect of quenching ortho-positronium.

One experiment (Mills and Berko, 1967) exploited the above considerations together with the fact that, independently of the validity of charge conjugation symmetry, Bose–Einstein statistics and rotational symmetry forbid the (*C*-violating) decay $^1S_0 \to 3\gamma$ to occur in the symmetric configuration in which all three photons have the same energy and are emitted in a plane at 120° from each other (in the positronium centre of mass) (Fumi and Wolfenstein, 1953). The experiment detected the 3γ rate from positronium decay, both in this symmetric decay configuration and in others. The ratio of symmetric to non-symmetric decays can depend on the relative amount of triplet to singlet states, which is changed by quenching, only if a *C*-violating para-positronium decay contributes to the non-symmetric configuration.

A gas chamber with thin (0.25 mm) steel walls containing SF_6 gas at a pressure of 12 atm had a positron-emitting ^{64}Cu source (12.8 hours half-life, 1.68 MeV e^+) suspended in its centre and was surrounded by a set of 6 NaI(Tl) scintillator counters enclosed within a lead shielding (Fig. 3.1). The shape of the shielding was symmetric for rotations of multiples of $2\pi/12$, so that each of the detectors (around the source at 13 cm distance) suffered similar background due to scattering in the lead. The total energy spectra were peaked at 1 MeV, corresponding to positron annihilation.

A triple coincidence of counters measured the symmetric decays ($\theta_{12} = 120°$, $\theta_{23} = 120°$, $\theta_{31} = 120°$), while others measured decays corresponding to (60°, 150°, 150°), (90°, 120°, 150°), and (90°, 150°, 120°) configurations, all of them not accessible to 2γ decay. After running with pure SF_6 gas, 1% NO was added in the chamber, reducing the 3γ counts by a factor ~ 30. The number of 3γ counts in a given angular configuration $\Theta \equiv (\theta_{12}, \theta_{23}, \theta_{31})$ is

$$N(\Theta) \propto \left[f_T \Gamma_T(\Theta) + (1 - f_T) b \Gamma_S(\Theta) \right] \varepsilon(\Theta)$$

where f_T and $f_S = 1 - f_T$ are the triplet and singlet annihilation fractions, $\Gamma_{T,S}$ are the corresponding decay rates integrated over the finite counter faces for the given angular configuration, and $b = \Gamma(^1S_0 \to 3\gamma)/\Gamma(^1S_0 \to 2\gamma)$ the *C*-violating ratio of singlet partial decay widths; note that $\Gamma_S \simeq 0$ for the symmetric decay configuration. ε is a configuration-dependent detection efficiency factor; the variation of the differential decay rates over the sizes of the counters (3.8 cm diameter face) was assumed to be small enough for such efficiencies to be the same for triplet and singlet states. For the quenched case the triplet

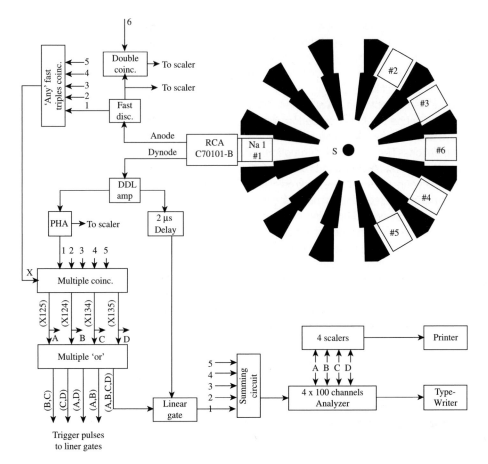

FIG. 3.1. Schematic drawing of the experimental apparatus and electronics for the search of *C*-violating positronium decay of Mills and Berko (1967).

annihilation fraction is reduced to f_T/Q. In the ratio R of counts for the non-symmetric configuration over that for the symmetric one the overall normalization cancels, and in the asymmetry

$$A = \frac{R(\text{quenched}) - R(\text{unquenched})}{R(\text{quenched}) + R(\text{unquenched})}$$

the ratio of detection efficiencies also drops out, so that one can extract

$$b = A \frac{2f_T}{Q-1} \frac{\Gamma_T(\text{non-symmetric})}{\Gamma_S(\text{non-symmetric})}$$

assuming $b \ll 1$; a small correction is required since $\Gamma_S(\text{symmetric})$ is not exactly zero due to the finite angular size of the counters (the 'symmetric' configuration corresponds to a single set of angles, with zero measure).

The quenching factor Q was measured by comparing the 3γ rates in the symmetric configuration with and without the quenching gas, and f_T/Q was measured by comparing the quenched rate in the symmetric configuration with the rate measured when the source is surrounded by metal, in which the triplet fraction is 1/372 as expected for spin-averaged free positron annihilation.

Data was collected in several runs, each consisting of an unquenched and a (ten times longer) quenched data-taking period, leading to the accumulation of $\sim 10^5$ quenched 3γ decays and twice as many unquenched ones, and repeated with two different non-symmetric configurations. The similarity of the photon energy spectra for quenched and unquenched data indicated that backgrounds were small. Corrections were applied to the raw results in order to account for small geometric differences between the angular configurations, detector gain variations with counting rate during the decay of the radioactive source, and accidental coincidences. Besides the normal accidental triple coincidence rate, which can be estimated from the single rates and the coincidence resolving time, accidental triple coincidences could also originate from a single accidental photon together with a true double coincidence, due to one photon from $^1S_0 \to 2\gamma$ decay scattering in the source holder or being preceded by a *bremsstrahlung* photon originating from a β^+ decay positron. To estimate this contribution, the total energy spectra for double coincidences were measured and folded with that for a third counter, obtaining a contribution smaller than 0.5% of the signal. Genuine triple coincidences due to one of the photons from (scattered, non-collinear) $^1S_0 \to 2\gamma$ decay being scattered from one counter to another were estimated using a ^{85}Sr source, emitting single 0.514 MeV photons.

The corrected results were $b = (-11.3 \pm 6.2) \cdot 10^{-6}$ using the (60°, 150°, 150°) non-symmetric configuration and $b = (10.8 \pm 11.9) \cdot 10^{-6}$ for the average of the other two non-symmetric configurations, consistent with charge conjugation symmetry in positronium decay. These led to the limit $b \leq 2.8 \cdot 10^{-6}$ at 68% confidence level.

A more favourable search exploiting the same principle is that for the C-forbidden $^3S_1 \to 4\gamma$ decay with the four photons emitted along the directions of the vertexes of a tetrahedron (four is the minimum even number of photons for a C-violating decay of the triplet state). This configuration is also forbidden by rotational symmetry for the $^1S_0 \to 4\gamma$ decay, which however this time is the allowed mode: instead of searching for a small signal on a large background, this measurement is in principle background free. With sufficiently small source and detectors, the latter placed along the directions indicated above, any plane formed by three of the four counters never contains the source, therefore eliminating the coincidences due to 3γ decays.

Such experiments are based on the separation of the two positronium states by their very different lifetimes: $\tau(^1S_0) \simeq 125$ ps versus $\tau(^3S_1) \simeq 142$ ns; the ortho-positronium lifetime actually measured in experiments is somewhat reduced with respect to the above (vacuum) value and depends on the actual set-up because of the $^3S_1 \to ^1S_0$ transitions induced by spin-flipping atomic collisions.

In one experiment performed at Tokyo University (Yang *et al.*, 1996), the positron source was ^{68}Ge: this isotope decays into ^{68}Ga by orbital electron capture (half-life 271 days), which in turns has a fast (67 min half-life) β^+ decay (1.9 MeV e^+), with small ($\sim 1\%$) background due to gamma emissions. The source was sandwiched between two 0.5 mm thick plastic scintillator sheets used for triggering, and was surrounded by silica aerogel in which positrons stopped. This material is rather efficient for positronium production: about 13% of the stopping positrons (which were 32% of the total in this experiment) resulted in the formation of ortho-positronium (the others annihilating with electrons at rest within ~ 1 ns), with a lifetime $\simeq 121$ ns; N_2 gas flowed in the cases containing the aerogel to reduce ortho-positronium quenching due to collisions with oxygen molecules.

The detector consisted of a set of 32 NaI(Tl) scintillators placed on the surface of an icosidodecahedron[44] surrounding the source in the centre at 26 cm. Each 7.6 cm diameter crystal was shielded with lead to prevent Compton scattered photons from reaching other crystals (suppression better than 10^{-6}). Sixteen pairs of counters are collinear, and fifteen planes of eight counters containing the origin could be formed, thus allowing the measurement of 2γ and 3γ decays. The energy and time resolutions were around 20 keV and 5 ns respectively.

Tight time coincidences among photon detectors in appropriate patterns (no collinear hits) were detected, and backgrounds originating from *bremsstrahlung* photons and 2γ annihilations in which one photon is scattered were rejected by energy cuts. The overall 4γ detection efficiency was estimated to be $\sim 3 \cdot 10^{-5}$ (for singlet state decay). Three 4γ events were observed at times greater than 10 ns when the contamination from the short-lived para-positronium is negligible.

Many possible sources of backgrounds due to the accidental coincidence of two positrons were considered: a first positron crossing the scintillator can trigger the event and its annihilation photons be missed, while the prompt 3γ annihilation of a later positron, together with a *bremsstrahlung* photon or a nuclear gamma decay, can fake a signal; two positrons annihilating in different events with measured photon times in the tails of the distributions can also be mistaken for a signal. Monte Carlo simulation of the above processes resulted in a background estimation of 3.4 ± 0.2 events, leading to the upper limit $\Gamma(^3S_1 \to 4\gamma)/\Gamma$ $(^3S_1 \to 3\gamma) < 2.6 \cdot 10^{-6}$ (at 90% confidence level) for the ratio of C-violating 4γ decay to 3γ decay of ortho-positronium.

A similar test for charge conjugation symmetry is given by the search for the decay of para-positronium into five photons: the upper limit $\Gamma(^1S_0 \to 5\gamma)/\Gamma(^1S_0 \to 2\gamma) < 2.7 \cdot 10^{-7}$ (at 90% confidence level) was obtained at Berkeley (Vetter and Freedman, 2002) by using a high performance detector (Gammasphere) consisting of 110 high-purity germanium crystals, each of them surrounded by a shielding of 6 bismuth germanate (BGO) scintillators, used to detect Compton scattered photons and suppress the corresponding events with a partial shower containment in the crystal, thus improving the energy resolution. The 7 cm diameter crystals are arranged on the surface of a 122-element polyhedron (110 hexagons and 12 pentagons) at a distance of 24.5 cm from the central target, with the BGO shielding detectors around them in a honeycomb pattern. The combination of fine segmentation and good energy resolution (~ 2 keV full-width at half maximum for 30% of 1 MeV photons) allows a strong background rejection of scattered events: for a single candidate 5γ event detected in a ± 20 ns time interval around the trigger, the total estimated contribution was ~ 0.25 events, plus ~ 0.4 events from prompt ortho-positronium decay.

---

[44] This is the 32-face Archimedean polyhedron, that is a solid with faces composed of two or more types of regular polygons meeting in identical vertices.

One-photon positronium transition rates were also considered as tests of C symmetry (Conti *et al.*, 1986): in contrast with the case of parity, the definite C-parity of the photon would allow unambiguous tests using forbidden transitions: examples are $1^3S_1\gamma \to 2^3S_1$ and $n^3S_1 \to 2^1P_1\gamma$.

Besides direct searches for resonant C-violating transitions, asymmetries in the transition rates induced by external electric $\boldsymbol{E}$ or magnetic $\boldsymbol{B}$ fields can also be searched for, corresponding to the reversal of either the field orientation or the photon polarization $\boldsymbol{S}_\gamma$ or direction $\boldsymbol{k}$; these could originate in the interference of a C-violating amplitude with a C-conserving one. Transitions between triplet S states can give C-violating asymmetries proportional to $\boldsymbol{S}_\gamma \cdot \boldsymbol{E}$ or $\boldsymbol{k} \cdot \boldsymbol{B}$, while for those between triplet S and singlet P states the asymmetries can be proportional to $\boldsymbol{S}_\gamma \cdot \boldsymbol{B}$.

Limits on charge conjugation violation in electromagnetic interactions are also obtained from searches for forbidden decays such as $\pi^0 \to 3\gamma$, $\eta \to 3\gamma$, $\eta \to \pi^0 l^+ l^-$, $\omega \to \eta\pi^0$, $\omega \to 3\pi^0$ (and similar decays for η').

Neutral pions are produced either by the decay of stopped charged kaon beams ($K^+ \to \pi^+\pi^0$) or by charge-exchange from a beam of negative pions stopped in a liquid hydrogen target ($\pi^- p \to \pi^0 n$). Kaon decays at rest are well suited for the experiment since the measurement of the direction of the photons and of the π^+ are sufficient to overdetermine the decay kinematics by one constraint ('1C reaction'), useful to reduce the background.

In absence of any selection rule forbidding it, the $\pi^0 \to 3\gamma$ decay would be expected to occur at a rate reduced by $O(\alpha)$ with respect to the 2γ decay. The main difficulties in the search for $\pi^0 \to 3\gamma$ decay are due to backgrounds induced from the abundant $\pi^0 \to 2\gamma$ decays when a third fake photon is detected due to either an accidental particle (intense beams are required) or one of the two photons depositing energy in two separate lumps in the detector, when a secondary photon from the initial electromagnetic cascade interacts far enough from the rest of the shower. The 3γ decay has not been detected, the branching ratio limit at 90% confidence level being $3.1 \cdot 10^{-8}$ (McDonough *et al.*, 1988), consistent with charge conjugation invariance in the electromagnetic decay of the π^0.

Limits on C symmetry are also obtained by null measurements of the charge asymmetries in the decays $\eta \to \pi^+\pi^-\pi^0$ and $\eta \to \pi^+\pi^-\gamma$.

$\eta \to \pi^+\pi^-\pi^0$ asymmetries

In the description of this decay on the Dalitz plane (see Appendix D) the measurement of an asymmetry between the points (u, v) and $(-u, -v)$ implies charge conjugation symmetry violation and requires the presence of final state interactions among the pions (Lee, 1965).

One measurement (Jane *et al.*, 1974) of such asymmetries used a secondary 718 MeV/c π^- beam from the NIMROD proton synchrotron at Rutherford Appleton

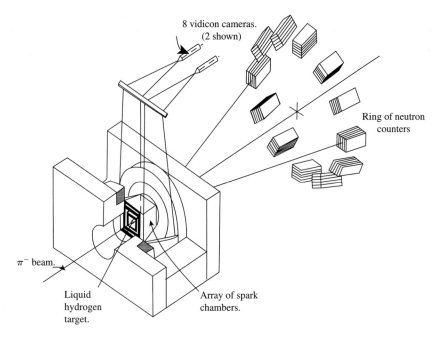

8 vidicon cameras.
(2 shown)

Ring of neutron
counters

π^- beam.

Liquid
hydrogen
target.

Array of spark
chambers.

FIG. 3.2. Experimental arrangement for the measurement of C-violating asymmetries in the $\eta \to \pi^+\pi^-\pi^0$ decay (Jane *et al.*, 1974).

Laboratory, impinging on a liquid hydrogen target, producing η ($J^P = 0^-$, $\eta_C = +1$) through the reaction $\pi^- p \to \eta n$. The beam momentum was measured to 0.5% precision by thin vertical scintillation counters located at a position where it was momentum separated in a magnetic field.

The experimental apparatus (Fig.3.2) had axial symmetry around the beam direction: the target and the pion detectors were within a 0.5 T axial field magnet, which did not deflect the beam on axis. Three concentric sets of four spark chambers, arranged in quadrants around the beam axis and surrounded by scintillator counters, were used to detect charged pions. A set of twelve stacks of scintillator counters arranged in a 2.5 m diameter circle, each preceded by a veto counter, was used to detect neutrons. The magnetic field and the electric field in the chambers were periodically reversed to cancel spurious asymmetries.

The three-pion state of interest was separated from $\pi^+\pi^-$ and $\pi^+\pi^-\gamma$ production by means of missing mass cuts, so that for the $1.7 \cdot 10^5$ $\pi^+\pi^-\pi^0 n$ events the background contamination from the above reactions was below 0.5%. The non-resonant $\pi^+\pi^-\pi^0$ background was estimated by using a different incident beam momentum and by studying the missing mass region aside the η peak, and found to be below $0.5 \cdot 10^{-5}$.

The measured raw asymmetry

$$A = \frac{N_+ - N_-}{N_+ + N_-}$$

where $N_\pm$ are the number of events with positive or negative sign of the energy difference $E(\pi^+) - E(\pi^-)$ in the η centre of mass, was found to depend strongly on the magnetic field orientation (a 5 standard deviation effect for one field orientation). This instrumental asymmetry was studied by measuring it for $\pi^+\pi^- n$ events (for which the isotropy of space forbids any physical asymmetry) and found to be present also in this case, consistent with a geometric distortion of the apparatus due to the magnetic field. The study of the well-measured experimental correlation of the 2π and 3π spurious asymmetries showed that the 2π asymmetry was ten times more sensitive to the distortion than the 3π one; the 3π asymmetry was obtained from the correlation fit as the value corresponding to zero asymmetry for the 2π mode: $A = (2.8 \pm 2.6) \cdot 10^{-3}$, consistent with C symmetry in η decay.

Dalitz plane asymmetries among opposite quadrants and sextants, which are uniquely sensitive to C violation in specific isospin channels, were also measured and found to be compatible with zero.

3.3.3   Strong interactions

In low energy nuclear processes, charge conjugation symmetry does not play any important role, since the creation of particle-antiparticle pairs requires centre of mass energies of at least $2m_\pi$. Tests of C symmetry at higher energies did not show any sign of violation in strong interactions.

Limits on charge conjugation symmetry in strong interactions were obtained by comparing the distributions of kinematical variables for charge conjugated reactions, or in charge conjugated final states produced from self-conjugated initial states such as $\bar{p}p$ (Pais, 1959). For $\bar{p}p$ interactions, the initial state is not in general a C eigenstate but a superposition of $\eta_C = \pm 1$ states of different angular momentum; it is however an eigenstate of CR, where R is a rotation around any axis orthogonal to the p or $\bar{p}$ line of flight in the centre of mass (taken as the z axis), with eigenvalue $(-1)^{J+J_z}$. For unpolarized beam and targets the different angular momentum states do not interfere when summed over spin orientations, and CR symmetry considerations can be applied to the final states.

When two particles a, b are detected in the final state, by comparing abX and $\bar{a}\bar{b}X$ (where X indicates undetected particles) the above symmetry implies the equality of the differential cross sections

$$\frac{d\sigma(\bar{p}p \to abX)}{d\Omega}(E_a, \theta_a; E_b, \theta_b, \phi_b)$$

$$= \frac{d\sigma(\bar{p}p \to \bar{a}\bar{b}X)}{d\Omega}(E_a, \pi - \theta_a; E_b, \pi - \theta_b, \pi - \phi_b) \qquad (3.60)$$

where $E_{a,b}$ are the energies of the measured final state particles, $\theta_{a,b}$ their polar angles (in the centre of mass) with respect to the $\bar{p}$ direction and ϕ_b the azimuthal angle of particle b with respect to the $(\bar{p}a)$ plane ($\phi_a = 0$).

When a single particle is detected in the final state, integrating over some variables, C (or CP) symmetry also implies

$$\frac{d\sigma(\bar{p}p \to aX)}{d\Omega}(E_a, \theta_a) = \frac{d\sigma(\bar{p}p \to \bar{a}X)}{d\Omega}(E_a, \pi - \theta_a)$$

$$\frac{d\sigma(\bar{p}p \to aX)}{d\Omega}(\phi_b) = \frac{d\sigma(\bar{p}p \to \bar{a}X)}{d\Omega}(-\phi_b)$$

$$\frac{d\sigma(\bar{p}p \to aX)}{d\Omega}(E_a) = \frac{d\sigma(\bar{p}p \to \bar{a}X)}{d\Omega}(E_a)$$

The above relations were tested in $\bar{p}p$ reactions by forming momentum and angle asymmetries for several different reactions, and found to be consistent with C symmetry at the level of 1% (e.g. Dobrzynski *et al.* (1966)).

It should be noted that CPT conservation alone already requires the equality of cross sections under the exchange of positive and negative particles in the final states, if final state interactions are absent.

Further reading

The idea of antiparticles was born with Dirac (1931) and Majorana (1937), and the modern concept of charge conjugation symmetry gradually emerged with the advent of quantum field theory; Michel (1987) contains a discussion and many references to the early developments.

Problems

1. Show that under the effect of a charge conjugation transformation the current operator for scalar fields changes sign as required.

2. Evaluate the commutation relation of the parity and charge conjugation operators P, C explicitly in the case of a spinor field, and discuss the conditions under which these operators commute.

3. Show that if $\varphi(p; s)$ is the wave function for an electron with momentum p and spin projection s, the charge-conjugated wave function $S_C\bar{\varphi}(p, s)$ describes a particle with momentum $-p$ and spin projection $-s$. How should this be interpreted, given that charge conjugation is not supposed to affect momentum and spin?

4. Starting from the field transformation properties, derive explicitly the effect of the simultaneous application of the C and P operators on a particle-antiparticle pair (both for spin 0 and spin 1/2 particles).

5. Prove that the charge-conjugate of a positive chirality spinor has negative chirality, then show that Majorana particles carry no charge, i.e. eqn (3.40).

6. From the existence of the indicated decays, determine the C-parity of the following neutral particles: $\eta \rightarrow \gamma\gamma$, ρ^0 (spin 1) $\rightarrow\pi^+\pi^-$, f_0 (spin 0)$\rightarrow$ $\pi^+\pi^-$ and $\rightarrow \pi^0\pi^0$, ω (spin 1)$\rightarrow \pi^+\pi^-\pi^0$.

7. The decay $\eta \rightarrow \pi^0\gamma$ is forbidden by angular momentum conservation, but if the photon is virtual such as in the decay $\eta \rightarrow \pi^0 e^+ e^-$ this is no longer the case; nevertheless such decay does not seem to occur. Why?

8. The ω (spin 1) dominantly decays into 3 pions, indicating it has odd G-parity; nevertheless it also has a $\sim 1.7\%$ branching ratio into 2 pions. Why is this possible? Decays into $3\pi^0$ or $2\pi^0$ are instead not observed, why?

4

TIME REVERSAL

Time is what prevents everything from happening at once.
J. A. Wheeler

Time is an illusion.
Lunchtime doubly so.
D. Adams

Time don't mean a thing
when you're by my side
please stay a while.
J. Deacon

This chapter deals with the time reversal symmetry, which is somehow different
from space inversion, ultimately due to the fact that our description of physics
is in terms of time evolution. The anti-unitarity of the corresponding operator in
the quantum theory leads to some complications, and we devote some pages to
the discussion the formalism to make these clear. The tests of this symmetry are
also somewhat peculiar, and require special care to ensure that other effects faking
asymmetries are properly taken into account.

4.1 Time reversal transformation

The transformation of a system corresponding to the inversion of the time coordinate, the formal substitution $t \rightarrow -t$, is usually called 'time reversal', but a
more appropriate (and less weird) name would actually be *motion reversal* (the
term indeed used by Wigner when he introduced this concept in physics). In our
macroscopic world the very concept of a symmetry with respect to the two directions of time flow seems rather absurd: a very definite direction of time (the 'arrow
of time') is apparent in all real physical processes, which are indeed irreversible.
On the other hand, the laws of classical physics, which determine the properties
of simple systems (e.g. Newton's law), appear to be invariant under the inversion
of the time coordinate. Since the behaviour of macroscopic processes originates

from the sum of very many elementary interactions, this leads to an apparent paradox: macroscopic systems do not share the time reversal symmetry of the microscopic systems of which they are composed.

However it should be noted that the behaviour of a physics process is determined not only by the properties of the basic laws of physics which drive it, but also by the boundary conditions. The macroscopic time irreversibility linked to the increase of entropy of a thermodynamic system indeed originates not in a microscopic time asymmetry of the basic laws which govern it, but from thermodynamical considerations on the number of available microscopic states corresponding to a given macroscopic state of the system. This is readily understood considering one of the simplest irreversible processes, namely the free expansion of a gas: when a gas initially confined in a small volume is allowed to expand into a larger one, the expected final state is to have the gas occupying the entire volume at any later time; conversely, starting from a state with the gas in the larger volume, the state in which all of it is confined in the smaller volume never occurs spontaneously. This is, however, just a probabilistic statement, i.e. the probability of reaching spontaneously the state with the gas in the smaller volume is infinitesimally small, effectively zero for a macroscopic system (Boltzmann, 1895). When the microscopic behaviour is considered, nothing intrinsically irreversible is found in the dynamics of the gas molecules: considering a simpler system with just two particles initially confined in the smaller volume, the probability of finding them both in the same volume in the future is indeed not at all small.

Actually the issue concerns the number of trajectories in phase space leading from a given macroscopic state to another one, either less or more 'probable' than the first; while such trajectories are in one to one correspondence with respect to time inversion (for each trajectory leading to a more probable state there is one, with all momenta reversed, leading to a less probable state), there are trajectories leading to more probable states *in both time directions* (that is, self-conjugate under time reversal), which are by and large the most abundant for a system far from equilibrium (Hurley, 1981). Thus we see that the symmetry of the microscopic laws of physics under time reversal has nothing to do with the macroscopic irreversibility, and it can be effectively studied only with very simple microscopic systems.

In classical physics, time reversal symmetry, or *micro-reversibility*, simply means that for any possible motion of a system, allowed by the dynamical equations, there is another possible one in which the same sequence of states is traversed in the opposite order, i.e. the trajectory in configuration space is reversed. Explicitly, this symmetry implies that a particle, starting at time t_0 from the point (x_0, v_0) in configuration space, moving for a time interval Δt, then reversing its velocity $v(t_0 + \Delta t) \rightarrow -v(t_0 + \Delta t)$, and moving again for a time interval Δt, reaches the point $(x_0, -v_0)$ (see Fig. 4.1).

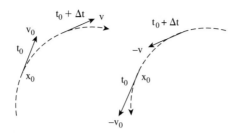

FIG. 4.1. Illustration of time reversal for a point particle in classical mechanics.

Under the time reversal transformation coordinates, momenta and angular momenta transform as

$$x \rightarrow x_T = x \qquad p \rightarrow p_T = -p \qquad L \rightarrow L_T = -L \qquad (4.1)$$

Given a Hamiltonian $H(x, p)$, and a solution $x(t)$ of the corresponding equation of motion, by considering the Hamiltonian with the same functional form as the transformed variables $H_T(x_T, p_T)$ it can be seen that the function $x_T(t) = x(-t)$ is also a solution, since the transformation $t \rightarrow -t$ transforms the equations for the transformed variables into those for the original ones. This is the expression of time reversal symmetry in classical mechanics.

It often happens that the Hamiltonian is quadratic in the momenta and, as such, automatically time reversal symmetric. In electromagnetism this is not the case, as the minimal coupling (Appendix B) introduces terms linear in p in a quadratic Hamiltonian: while the motion of a particle in an electric field is time reversal symmetric, that in a magnetic field is not.[45] The invariance of the Lorentz force equation under time reversal is consistent with the transformation laws for the electric and magnetic fields

$$E(t, x) \rightarrow +E(-t, x) \qquad B(t, x) \rightarrow -B(-t, x) \qquad (4.2)$$

These transformations can be understood from the fact that electric fields can be created by stationary charges, not affected by time reversal, while magnetic fields (in absence of magnetic monopoles) are created by *moving* charges.[46]

Classically the time reversal transformation is not a canonical one, since the Poisson brackets do change sign: it is sometimes called *anti-canonical*. It cannot be stated that a system is not time reversible just because its Hamiltonian function H changes when substituting $x \rightarrow -x$ and $p \rightarrow -p$,

---

[45] It becomes symmetric when the motion reversal of the charged particles which are supposed to generate the magnetic field is also considered, so that the field direction is reversed.

[46] The other possibility, in which only the electric field changes sign under time reversal, cannot be rejected a priori, but it would require the transformation to reverse the sign of stationary electrical charges, which is not a property which we associate with an inversion of time.

since an infinity of Hamiltonians exist which can describe the same system; the invariance of the equations of motion has to be checked explicitly, or the change of the Hamiltonian for the most general anti-canonical transformation leaving x unchanged has to be considered. It can be shown (Morpurgo *et al.*, 1954) that for any conservative system there exists an anti-canonical transformation leaving H unchanged; in order for this transformation to be the one corresponding to time reversal it should also transform x and p appropriately.

As for the other discrete symmetries considered in previous chapters, the concept of time reversal symmetry did not have important consequences in classical physics, despite the fact that all of it is (microscopically) time reversal symmetric; as will be discussed, this symmetry does not lead to conserved quantities. Still, the concept of symmetry under time reversal can be relevant also in classical physics, and led for example to the prediction that crystals with circular symmetry (such as apatite and dolomite) should not exhibit static magnetic moments, in analogy to the static electric fields which are instead present in polar crystals (such as tourmaline) and which give rise to the phenomena of pyroelectricity and piezoelectricity.[47] Static phenomena such as pyromagnetism and piezomagnetism in crystals, not observed for a long time, were thought to be forbidden by time reversal symmetry (Zocher and Török, 1953) according to Neumann's principle, which states that every physical property of a crystal should possess the same symmetry exhibited by the elementary cell of the crystal structure itself (assumed to be T-even): no asymmetry can appear in an effect which does not already exist in the crystal. The above conclusion was later shown to be invalid for magnetic crystals: any microscopic relative motion of the parts of a crystal which does not average to zero destroys the time reversal symmetry of its structure, and this is actually the case for the orderly distribution of spin magnetic moments in a ferromagnetic crystal, which cannot be time symmetric since under time reversal the spontaneous magnetization changes sign. This led to the development of the fruitful concept of magnetic symmetry groups (symmetry groups containing not only translations, rotations, and reflections, but time reversal as well), which increased the number of point groups necessary to faithfully describe crystal properties from the classical 32 to 122 (Birss, 1963).

It should be pointed out that Neumann's principle cannot be applied in space-time (e.g. to transport phenomena, such as current flow in a material) since there is an intrinsically preferred direction in time due to thermodynamic considerations: while the time reversal of a freely moving particle (what would appear when running a movie backwards) does not contradict any physical law, the same is not true for a particle moving steadily in a medium that opposes its motion due to a drag force; after time reversal this would appear as if the medium assisted the motion of the particle, clearly an unphysical situation.

It should also be remarked that the principle does not necessarily work in reverse: the elastic properties of a sugar crystal exhibit parity symmetry independently of its microscopic chiral structure, that is, even if all the molecules have the same handedness; this is because the elasticity properties depend on the strain tensor, which as a second rank symmetric tensor is parity symmetric. Elastic properties are not complicated enough to support parity violation even if it appears microscopically (Froggatt and Nielsen, 1991); this example illustrates a rather general issue: a fundamental asymmetry is not necessarily reflected in all observable quantities.

---

[47] The appearance of electric fields in a crystal when it is heated or subject to mechanical stresses.

4.1.1   Quantum mechanics

Denoting the time reversal operator in Hilbert space by T, so that

$$|\psi\rangle \rightarrow |\psi_T\rangle \equiv T|\psi\rangle \tag{4.3}$$

in analogy to the classical case one considers the time evolution of a state for an infinitesimal time interval δt as described by $|\psi(t + \delta t)\rangle = (1 - i\hbar\delta t)|\psi(t)\rangle$; in the case of time reversal symmetry, the time evolution of the time-reversed state $|\psi_T\rangle$ from time $t + \delta t$ should lead to the time-reversed state at time t, i.e.

$$(1 - iH\delta t)|\psi_T(t + \delta t)\rangle = (1 - iH\delta t)T|\psi(t + \delta t)\rangle$$
$$= (1 - iH\delta t)T(1 - iH\delta t)|\psi(t)\rangle = |\psi_T(t)\rangle$$

which requires $T(-i)H = iHT$. If T were a unitary operator, this means $TH = -HT$, so that given an eigenstate of the Hamiltonian $H|n\rangle = E_n|n\rangle$, the time-reversed state $T|n\rangle$ would have negative energy $-E_n$, leading to an energy spectrum unbounded from below, which cannot be allowed.

Also, it is not possible to hold the correspondence with the classical transformations ($x \rightarrow x$ and $p \rightarrow -p$) and at the same time leave the commutation relations unchanged, if a unitary transformation operator is used for time reversal. In this case, since $[x_i, p_j] = i\hbar\,\delta_{ij}$ is a c-number, one should have $\langle\psi_T|[x_i, p_j]|\phi_T\rangle = \langle\psi|[x_i, p_j]|\phi\rangle$, (where $|\psi_T\rangle$, $|\phi_T\rangle$ are the time-reversed states), but such a relation is inconsistent with the fact that under time reversal $\langle\psi_T|p|\phi_T\rangle = -\langle\psi|p|\phi\rangle$, as dictated by the classical correspondence. Alternatively, in the coordinate representation one has $p_i = -i\hbar\partial/\partial x_i$, but p_i and x_i transform differently under time reversal.

The above difficulties can be overcome if the time reversal operation includes the operation of complex conjugation, so that $i \rightarrow -i$; in this way the canonical commutation relations are left unchanged, while the classical correspondence for the transformation of x and p is maintained. The Hilbert-space transformation operator must therefore be enforced by a *anti-unitary* operator T (Section 1.2), which we recall is also a valid possibility.

Still another way of seeing this is by noting that the time-dependent Schrödinger equation

$$i\hbar\frac{\partial\psi}{\partial t} = H\psi = \left(\frac{-\hbar^2}{2m}\nabla^2 + U(x, t)\right)\psi \tag{4.4}$$

is of first order in the time variable, so that if $\psi(t, x)$ is a solution, then $\psi(-t, x)$ is not.[48] However, if $U(x, t) = U(x) = U^*(x)$ is a real function not depending

---

[48] The choice of the sign $+i\hbar$ in the Schrödinger equation is an arbitrary one, but cannot be changed at will.

explicitly on time, then $\psi^*(-t, x)$ *is* a solution: the complex conjugate Schrödinger equation is

$$-i\hbar\frac{\partial \psi^*(t, x)}{\partial t} = +i\hbar\frac{\partial \psi^*(t, x)}{\partial(-t)} = H^*\psi(t, x) = H\psi(t, x) \qquad (4.5)$$

One can therefore define the wave function of the time-reversed state as

$$\psi_T(t, x) = \psi^*(-t, x) \qquad (4.6)$$

As an example, the wave function for a plane wave travelling in the $+x$ direction $\psi(t, x) = e^{-i(Et - p \cdot x)}$ is transformed into $\psi_T(t, x) = e^{+i(E(-t) - p \cdot x)}$ which correctly describes a plane wave of the same (positive) energy traveling in the $-x$ direction. For spherical harmonics (angular momentum eigenstates) the effect of time reversal amounts to

$$Y_l^m(\theta, \phi) \to \left[Y_l^m(\theta, \phi)\right]^* = (-1)^m Y_l^{-m}(\theta, \phi) \qquad (4.7)$$

as can be checked explicitly from their definition. We anticipate the fact that when spin is introduced the generic Hamiltonian is no longer real (it contains, e.g., σ_2 for spin 1/2) complicating somewhat the picture.

Under the anti-unitary transformation T, the states transform according to eqn (4.3) or (alternatively) one can consider the operators as transforming according to

$$O \to O_T = T^\dagger OT \qquad (4.8)$$

In this approach, adopting the Heisenberg picture for time evolution in which $O = O(t)$ with

$$\frac{dO}{dt} = i\hbar[H, O] \qquad (4.9)$$

the transformation is

$$O(t) \to O_T(t) = T^\dagger O(t)T \qquad (4.10)$$

and if the Hamiltonian H is invariant under time reversal the transformed operator O_T consistently satisfies the equation obtained by applying the time reversal transformation to eqn (4.9), that is

$$\frac{dO_T}{dt} = -\frac{dO_T}{d(-t)} = -i\hbar[H, O_T]$$

which is identical with eqn (4.9) if $O_T(t) = O(-t)$, so that the time evolution of the operator O_T is indeed the same as that of O, just backwards in time.

The two actions of T (the mapping of the states into their motion-reversed counterparts and the complex conjugation of c-numbers) can be split by writing

$$T = U_T K \qquad (4.11)$$

with unitary U_T, where K is the complex conjugation operator: $KzK^{-1} = z^*$ for any complex number z. Since T and K are both anti-unitary, their product $U_T = TK$ is unitary. With this definition one can write:[49]

$$|\psi\rangle \rightarrow |\psi_T\rangle = U_T|\psi\rangle^* \quad \text{or} \quad O \rightarrow O_T = (U_T^\dagger O U_T)^* \tag{4.12}$$

The complex conjugate of a *ket* is defined in a given basis by

$$|\psi\rangle = \sum_n c_n|n\rangle \qquad |\psi\rangle^* = \sum_n c_n^*|n\rangle \tag{4.13}$$

(since the basis vectors $|n\rangle$ are real); the complex conjugate of an operator is defined as that whose matrix elements are the complex conjugates of those of the original operator: $\langle\psi|O^*|\phi\rangle = \langle\psi|O|\phi\rangle^*$ (in a given basis).

It should be remarked that the above definitions depend on the particular choice for the basis vectors: complex conjugation (as opposed to Hermitian conjugation) depends on the representation used. The operator K does not act on the basis kets of the given representation (which are real), and therefore the decomposition (4.11), i.e. the explicit form for U_T, depends on the chosen representation. This is a general property of anti-unitary operators, which are only defined explicitly when a given representation (basis) is specified.

As an example, in the representation in which the position operator x is diagonal (i.e. wave mechanics), the momentum is $p = -i\hbar\nabla$, and its change of sign under time reversal is enforced by K; U_T therefore commutes both with the position operator and the derivatives with respect to position, i.e. it can depend only on spin variables. For a spinless particle one has simply $U_T = 1$, and therefore just complex conjugation of wave functions, eqn (4.6), since indeed the x operator and its eigenstates $|x\rangle$ are not affected by time reversal, so that

$$\psi_T(x) \equiv \langle x|\psi_T\rangle = \langle x|T\psi\rangle = ((\langle x|T)\,\psi)^* = \langle T^\dagger x|\psi\rangle^* = \langle x|\psi\rangle^* \equiv \psi^*(x)$$

On the other hand, in the representation in which the momentum is diagonal the same happens, but the momentum is also inverted:

$$\varphi_T(p) = \varphi^*(-p)$$

Considering instead the spin operator S for a particle with spin 1/2, in order to have $S_T = (U_T^\dagger S U_T)^* = -S$ as required by consistency with eqn (4.1), U_T must commute with σ_2 (the imaginary Pauli matrix) and anti-commute with σ_1, σ_3 (which are real), so that indeed[50] $U_T \propto \sigma_2$. Note also that U_T does not behave as a normal operator under a change of basis (Gottfried, 1966).

---

[49] A second possible approach for describing the time reversal operation (Schwinger, 1951; Kemmer *et al.*, 1959) is that of transforming a *ket* into a *bra* without complex conjugation of numbers, intuitively justified by the fact that initial and final states in a transition are exchanged. In this approach transposition of operators takes the place of complex conjugation, but for observables (represented by Hermitian operators) the two coincide.

[50] The fact that σ_2 appears clearly has nothing to do with the y direction, but rather depends on the standard choice of Pauli matrices (A.3) for the representation of the algebra of the $SU(2)$ group, in which this matrix is imaginary.

A valid choice for U_T must be such that

$$U_T x^* U_T^\dagger = x \qquad U_T p^* U_T^\dagger = -p \qquad (4.14)$$

in such a way as to have

$$x_T = (U_T^\dagger x U_T)^* = x \qquad p_T = (U_T^\dagger p U_T)^* = -p$$

Expanding the states in a given basis (4.13)

$$\langle \psi_T | x | \psi_T \rangle = \sum_{n,m} c_m c_n^* \langle m | U_T^\dagger x U_T | n \rangle = \sum_{n,m} c_m c_n^* \langle m | (x^*) | n \rangle$$

$$= \sum_{n,m} c_m c_n^* \langle m | x | n \rangle^* = \sum_{n,m} c_m c_n^* \langle n | x^\dagger | m \rangle = \sum_{n,m} c_m c_n^* \langle n | x | m \rangle = \langle \psi | x | \psi \rangle$$

as desired. Equation (4.14) defines U_T to within a phase. Also, given a (Hermitian) Hamiltonian H, the Hamiltonian H^* has the same spectrum of eigenvalues, and is therefore linked to the original one by a unitary transformation $H^* = U_T^\dagger H U_T$ (which defines U_T up to a phase). A system is said to be symmetric under time reversal if a (unitary) operator U_T exists, for which the Hamiltonian H satisfies

$$H_T = (U_T^\dagger H U_T)^* = H \qquad (4.15)$$

and which has the required properties (4.14) to describe the time reversal operation.

For a spinless state $U_T = 1$, so that the symmetry condition (4.15) is equivalent to the reality of the Hamiltonian. In this case

$$T | \psi(t) \rangle = \eta_T | \psi(-t) \rangle^* \qquad (4.16)$$

with $|\eta_T|^2 = 1$.

Note that the η_T factor is not a kind of 'temporal parity' quantum number, since $|\psi(t)\rangle$ is not an eigenstate of T. Indeed the association of an eigenvalue of time reversal to a state would be inconsistent: if $T|\psi\rangle = \tau|\psi\rangle$, since $e^{i\alpha}|\psi\rangle$ (real α) describes the same state it should have the same eigenvalue, while $T e^{i\alpha}|\psi\rangle = e^{-i\alpha} \tau|\psi\rangle$. This is the reason why one never refers to T symmetry as 'T conservation'. In any case, for anti-unitary operators a phase factor can always be transformed to 1 with an appropriate choice of phase: if $T|\psi\rangle = e^{i\xi}|\psi\rangle$, for the equivalent state $|\psi'\rangle \equiv e^{-i\xi/2}|\psi\rangle$ one has $T|\psi'\rangle = |\psi'\rangle$. η_T therefore does not represent a physical quantity.

A consequence of time reversal symmetry is that non-degenerate wave functions can always be chosen to be real: this is readily seen, as in this case if $\psi(x)$ is a stationary solution (an eigenfunction of energy), then also $\psi^*(x)$ is one, belonging to the same energy eigenvalue, and in absence of degeneracy it must therefore be equal to $\psi(x)$ to within a phase factor independent from x, so the wavefunctions are actually real (except for such a constant phase). A plane wave $e^{ik\cdot x}$ is complex but degenerate with $e^{-ik\cdot x}$, and the same is true for the wave function of a state of defined angular momentum (a spherical harmonic) which is degenerate with respect to the sign of the m quantum number (except for $L = 0$, and Y_0^0 is actually real).

As a consequence of the anti-unitarity of T the following identity holds for a generic operator O (independently from the validity of time reversal symmetry):

$$\langle \psi | O | \phi \rangle = \langle \psi | T^\dagger T O T^\dagger T | \phi \rangle = \langle \psi_T | T O T^\dagger | \phi_T \rangle^* = \langle \phi_T | T O^\dagger T^\dagger | \psi_T \rangle \quad (4.17)$$

For an operator $O^{(T\pm)}$ such that $O_T^{(T\pm)} = T^\dagger O^{(T\pm)} T = \pm O^{(T\pm)}$ (called 'T-even' or sometimes 'real' operator for the plus sign and 'T-odd' or 'imaginary' for the minus sign), then the above identity implies

$$\langle \psi | O^{(T\pm)} | \phi \rangle = \pm \langle \psi_T | O^{(T\pm)} | \phi_T \rangle^* \quad (4.18)$$

which for $|\psi\rangle = |\phi\rangle$ and a Hermitian operator ($O = O^\dagger$) gives a relation among the expectation values in time-reversed states

$$\langle O^{(T\pm)} \rangle_T = \pm \langle O^{(T\pm)} \rangle \quad (4.19)$$

According to a theorem by Wigner, for a T-odd Hermitian operator $O^{(T-)}$ the sum of expectation values $\langle O^{(T-)} \rangle$ corresponding to all eigenstates of the Hamiltonian belonging to the same eigenvalue E is zero, if the Hamiltonian itself is invariant under time reversal (Hamilton, 1959). Indeed, if $|E_\alpha\rangle$ are the eigenstates of H with eigenvalue E (the index α labelling additional quantum numbers for degenerate states), the states $|E_{\alpha T}\rangle = T|E_\alpha\rangle$ are also eigenstates of H belonging to the same eigenvalue (the Hamiltonian commutes with T); if $|E_{\alpha T}\rangle \neq |E_\alpha\rangle$, for a T-odd Hermitian operator $O^{(T-)}$:

$$\langle E_\alpha | O^{(T-)} | E_\alpha \rangle = -\langle E_\alpha | (T^\dagger O^{(T-)} T) | E_\alpha \rangle$$
$$= -\langle E_{\alpha T} | O^{(T-)} | E_{\alpha T} \rangle^* = -\langle E_{\alpha T} | O^{(T-)} | E_{\alpha T} \rangle$$

and in the sum over all eigenstates of H with eigenvalue E both $|E_\alpha\rangle$ and $|E_{\alpha T}\rangle$ appear, making the sum zero. Because of the above, when time reversal is a valid symmetry, the sum of the expectation values of momenta, angular momentum, spin, etc. over the states of a system with the same energy is zero. The same is true for the sum over all the eigenstates of a generic Hermitian operator commuting with T belonging to the same eigenvalue.

In relativistic quantum mechanics, the spinor wave function corresponding to the time-reversed state is

$$\varphi_T(t, x) = \eta_T S_T \varphi^*(-t, x) \quad (4.20)$$

where η_T is a phase factor and the unitary 4×4 matrix S_T acts in the space of the field components, transforming Dirac matrices as

$$S_T \gamma^\mu S_T^\dagger = \gamma_\mu^* \tag{4.21}$$

and in the Dirac representation (in which $\gamma_\mu^* = \gamma^{\mu T}$) has the form

$$S_T = i\gamma^1 \gamma^3 = -i\gamma_5 S_C \tag{4.22}$$

to within a phase (S_C being the matrix involved in charge conjugation, defined in eqn (3.17)), with

$$S_T^\dagger = S_T^{-1} = -S_T^* = -S_T^T = S_T \tag{4.23}$$

As the set of matrices γ_μ^* satisfy the same commutation relations as γ_μ, by Pauli's fundamental theorem a unitary matrix exists which connects the two, and this is indeed S_T.

4.2 Of phase factors

The odd nature of $\boldsymbol{J}$ under time reversal

$$T^\dagger \boldsymbol{J} T = -\boldsymbol{J} \tag{4.24}$$

implies that for eigenstates $|j, m\rangle$ of $\boldsymbol{J}^2$ and J_z one has $J_z T |j, m\rangle = -mT |j, m\rangle$. When $|j, m\rangle$ is an energy eigenstate (that is when rotational symmetry holds) time reversal symmetry requires that $T |j, m\rangle$ is also one and in absence of further accidental degeneracies this must be precisely $|j, -m\rangle$ to within a phase[51] $\eta_T(m)$. By applying time reversal to the equation for the raising and lowering angular momentum operators, eqn (2.14) with $s = m$ one obtains

$$-\eta_T(m)(J_1 \mp iJ_2)|j, -m\rangle = \eta_T(m \pm 1)\sqrt{(j \mp m)(j \pm m + 1)}|j, -m \mp 1\rangle \tag{4.25}$$

i.e. $T^\dagger J_\pm T = -J_\mp$, so that

$$\eta_T(m) = -\eta_T(m \pm 1) \tag{4.26}$$

and one can choose

$$\eta_T(m) = \bar{\eta}_T (-1)^{j+m} \tag{4.27}$$

---

[51] In presence of degeneracies the subspaces corresponding to different eigenvalues of the additional conserved quantity can be considered separately.

where $\bar{\eta}_T$ is arbitrary and independent from m, and can only depend on the particle type: such a phase was seen to be irrelevant and eliminable for time reversal, so that $\bar{\eta}_T = 1$ will be used in the following:[52]

$$T|j,m\rangle = (-1)^{j+m}|j,-m\rangle \tag{4.28}$$

As an example, for spin 1/2 ($\boldsymbol{J} = \hbar\boldsymbol{\sigma}/2$) the choice

$$U_T = e^{i\pi J_y} = i\sigma_2$$

satisfies the condition to be the unitary operator associated with time reversal: $(U_T^\dagger \boldsymbol{J} U_T)^* = -\boldsymbol{J}$, so that working in the basis of $|j,m\rangle$ eigenstates

$$T|1/2,m\rangle = U_T|1/2,m\rangle^* = U_T|1/2,m\rangle = e^{i\pi J_y}|1/2,m\rangle = (-1)^{1/2+m}|1/2,-m\rangle$$

The above can be generalized to obtain the phase factor for a generic angular momentum eigenstate $|J,M\rangle$ written as the combination of a spinor part (spin j, with m as quantum number for the component along the quantization axis) and a spatial part (spherical harmonics corresponding to an orbital angular momentum l with quantum number m_l for the quantized component) with appropriate Clebsch–Gordan coefficients. The latter (in the usual convention) have the property

$$\langle j_1, j_2; -m_1, -m_2 | J, -M \rangle = (-1)^{j_1+j_2-J} \langle j_1, j_2; m_1, m_2 | J, M \rangle \tag{4.29}$$

and by choosing an appropriate phase for the total wave function by writing

$$|J,M\rangle = i^l \sum_{m_l} \langle l,j; m_l, m | J, M \rangle Y_l^{m_l} | j, m \rangle \qquad (m = M - m_l)$$

one obtains $T|J,M\rangle = (-1)^{M+2j-J}|J,-M\rangle$ and since $J - j$ is always an integer this is

$$\boxed{T|J,M\rangle = (-1)^{J+M}|J,-M\rangle} \tag{4.30}$$

The same choice of phase can be used in combining more angular momenta. In general, if a *non-degenerate* eigenstate of angular momentum is a linear combination of states of given J, M normalized as above, then the coefficients may all be chosen to be real: if $|\psi(J,M)\rangle = \sum_\alpha a_\alpha |J,M,\alpha\rangle$ is such state, then $|\psi(J,-M)\rangle = \sum_\alpha a_\alpha |J,-M,\alpha\rangle$ must coincide with the state obtained by time reversal, namely $(-1)^{J+M} T|\psi(J,M)\rangle = \sum_\alpha a_\alpha^* |J,M,\alpha\rangle$.

The anti-linearity of T implies that it cannot possibly be an observable. Its square T^2 is however linear and unitary: it is an observable and its possible eigenvalues are ± 1. The effect of T^2 must be just the multiplication by a phase factor[53] $\tilde{\eta}$ ($|\tilde{\eta}|^2 = 1$),

---

[52] The choice $\eta_T(m) = (-1)^m = i^{2m}$ would result in imaginary phase factors for half-integer spins; the introduction of j at the exponent gives real phase factors while still satisfying eqn (4.26), since j and m are always both integers or half-integers. This choice is admissible since the relative phase factors for states with different j is not measurable.

[53] See Weinberg (1995) for more 'exotic' – but apparently not physically relevant – possibilities.

and since it commutes with T, then $\tilde{\eta}T - T\tilde{\eta} = (\tilde{\eta} - \tilde{\eta}^*)T = 0$, so that $\tilde{\eta} = \tilde{\eta}^* = \pm 1$. More generally, by applying time reversal twice one must get a phase factor, or a diagonal matrix of phase factors $M(\xi) = \text{diag}(e^{i\xi_i})$ (with real ξ_i; $i = 1, \ldots, n$ where n is the spin dimensionality of the wave function considered). Since $T = U_T K$, one has $T^2 = U_T U_T^* = U_T (U_T^T)^{-1} = M(\xi)$, i.e. $U_T = M(\xi)U_T^T = M(\xi)U_T M(\xi)$, which implies $e^{i(\xi_i + \xi_j)} = 1$ for each i, j, and therefore $e^{i\xi_i} = \pm 1$. Multiplication of U_T by a phase factor does not alter the value of $T^2 = U_T U_T^*$. Moreover, if $T' = U_T' K$ is the time reversal operator in another representation, obtained from the first by a unitary transformation U, then $(T')^2 = U_T' K U_T' K = U_T' U_T'^* = U^\dagger U_T U_T^* U = U^\dagger T^2 U = T^2$. One can see therefore that the eigenvalue of T^2 is an intrinsic property of the system.

Considering the expression for the time reversal operation in the spin-diagonal representation, from eqn (4.30):

$$T^2|J, M\rangle = (-1)^{2J}|J, M\rangle \tag{4.31}$$

so that the eigenvalue of T^2 distinguishes states of integer and semi-integer spin. As seen, this property is independent from phase conventions and the choice of basis. T^2 coincides with the operator for 2π rotation around any axis, and describes therefore a symmetry operation for any system which is symmetric under proper Lorentz transformations.

This is consistent with the previous discussion: for a system of n spin 1/2 particles in a state defined by the eigenvalues of one component of spin, $|m_1, m_2, \ldots, m_n\rangle$, the time reversal operator is the direct product of operators acting on each of the particles according to eqn (4.28), so that

$$T|m_1, m_2, \ldots, m_n\rangle = (-1)^{n/2}(-1)^{\sum_i m_i}| - m_1, -m_2, \ldots, -m_n\rangle \tag{4.32}$$

$$T^2|m_1, m_2, \ldots, m_n\rangle = (-1)^n|m_1, m_2, \ldots, m_n\rangle \tag{4.33}$$

which is $+1$ for an even and -1 for an odd number of fermions.

Considering an energy eigenstate for a system of n particles of half-integer spin $|E, n\rangle$, if the Hamiltonian is symmetric under time reversal the state $T|E, n\rangle$ is also an eigenstate belonging to the same energy eigenvalue. If $T^2 = -1$ (n odd), this state vector cannot represent the same state as $|E, n\rangle$, because this would require $T|E, n\rangle = \eta|E, n\rangle$ (with $|\eta|^2 = 1$), and therefore $T^2|E, n\rangle = |\eta|^2|E, n\rangle = |E, n\rangle$ contradicting the hypothesis. Indeed $|E, n\rangle$ and $|T(E, n)\rangle$ are orthogonal:

$$\langle T(E, n)|E, n\rangle = \langle T(E, n)|T^\dagger T|E, n\rangle = \langle T^2(E, n)|T(E, n)\rangle^*$$

$$= \langle T(E, n)|T^2(E, n)\rangle = -\langle T(E, n)|E, n\rangle = 0$$

The two states are therefore linearly independent, and all energy levels are at least degenerate of order 2. This is the *Kramers' degeneracy* (Kramers, 1930) (and its existence is called Kramers' theorem), which had been known for many years before it was explained by Wigner (1932) as being a consequence of time reversal invariance.[54] A consequence of the above is that for a generic system with an odd number of electrons placed in an external electric field, each energy level is always at least twice degenerate, whatever the arbitrary shape of the field is: for a spin 1/2 fermion a twofold degeneracy between the two angular momentum states is trivial in a rotationally invariant system, but Kramers' degeneracy guarantees that the two states remain degenerate even in a non rotationally symmetric (but T-symmetric) environment, such as an arbitrary electrostatic field. Kramers' degeneracy is lifted in presence of a magnetic field, which is not time reversal invariant: this fact allows the measurement of local magnetic fields in crystals and the study of electronic states in paramagnetic crystals (Sakurai, 1985). A related consequence of time reversal symmetry is that a photon cannot couple states which are conjugated under T in a system with an even number of electrons (Stedman, 1983).

4.2.1 Scalar and vector fields

In quantum field theory an anti-linear implementation of time reversal is also required (Bell, 1955), since the field (operator) transformation law

$$\frac{d\phi}{dx_\mu} = \frac{i}{\hbar}[P^\mu, \phi] \tag{4.34}$$

(P^μ being the four-momentum operator) cannot be left unchanged for $t \to -t$ with a linear transformation, if the Hamiltonian must not change sign (for parity this problem did not arise, since $P \to -P$ is an admissible transformation).

The time reversal transformation is thus represented by an anti-unitary operator T, whose effect on a state of definite momentum p, spin projection s and charge q is expected to be

$$T|p, s, q\rangle = \eta_T| - p, -s, q\rangle \tag{4.35}$$

with $\eta_T = e^{i\xi_T}$ a phase factor ($|\eta_T|^2 = 1$, ξ_T real).

This time, however, one cannot set $T^\dagger \phi(t, x)T - \phi(t, x)$ to be equal to the variation in form of the corresponding classical field: in analogy to the case of quantum mechanics, in which the multi-component spinorial wave function transforms as in eqn (4.20) one could think of defining the time reversal transformation as $\phi(t, x) \to S_T \phi^\dagger(-t, x)$, but this is not an admissible transformation to describe the effect of time reversal: in the second-quantized theory $\phi^\dagger$ is the Hermitian conjugate not only in the space of field components, but also in Hilbert space, and therefore such a transformation would transform, e.g., a particle at rest into an antiparticle.

This situation is different from the case of charge conjugation, where one also has $\phi \leftrightarrow \phi^\dagger$ but obtained by transforming particle and antiparticle creation and annihilation operators among themselves, eqns (3.7), (3.8). Considering the expression

---

[54] The attentive reader might feel a sense of *déjà vu* at this point.

(B.1) for the charged scalar field in terms of creation and annihilation operators, one can obtain from it the expression $\phi^\dagger$ for the charge-conjugated field, containing the terms $a^\dagger(p)e^{ipx}$ and $b(p)e^{-ipx}$ (the sign in the exponent being inverted by complex conjugation), using a unitary operator C in Hilbert space transforming $a \leftrightarrow b$ (which is an equivalence transformation if also external fields are appropriately transformed). For time reversal the sign in the exponent $e^{\pm iEt}$ is further inverted by $t \to -t$ (the sign in the exponent for space variables is not a problem since the infinite integration over dp allows one to redefine $p \to -p$), so that in order to get $\phi^\dagger$ from the original operator by the transformation $\phi \to T^\dagger\phi T$, the Hilbert space operator T should necessarily transform $a \leftrightarrow a^\dagger$ and $b \leftrightarrow b^\dagger$, which is not admissible as an equivalence transformation.

One therefore defines the time-reversed field to be equal to the Hermitian conjugate of the one which the classical correspondence would suggest:

$$\phi(t,x) \to \phi_T(t,x) = T^\dagger \phi(t,x) T = \left[U_T^\dagger \phi(t,x) U_T \right]^* \propto \phi(-t,x)$$

In this way, for a scalar field

$$T^\dagger \phi(t,x)T = \eta_T \phi(-t,x) \qquad T^\dagger \phi^\dagger(t,x)T = \eta_T^* \phi^\dagger(-t,x) \tag{4.36}$$

and no Hermitian conjugation appears.

The transformations of creation and annihilation operators are

$$T^\dagger a(p)T = \eta_T\, a(-p) \qquad T^\dagger b(p)T = \eta_T\, b(-p) \tag{4.37}$$

For a Hermitian field one must have $\eta_T = \pm 1$.

We see that the T operator time-reverses field operators and complex conjugates c-numbers, but does *not* transform field operators into their Hermitian conjugates: the complex conjugation introduced by time reversal in a scalar product is arranged to arise from the transformation of state vectors.

For the electromagnetic field, the requirement that the current $j^\mu(t,x)$ transforms according to the classical correspondence into $j_\mu(-t,x)$ implies that in order to have time reversal symmetry (as indicated by experimental evidence) of the interaction Hamiltonian $H_{int} = j_\mu A^\mu$ one must have

$$T^\dagger A^\mu(t,x)T = A_\mu(-t,x) \tag{4.38}$$

i.e. $\eta_T = +1$ for the electromagnetic field. This is consistent with the transformations of classical electric and magnetic fields, eqn (4.2).

In terms of creation and annihilation operators for linear polarization, remembering eqn (2.27):

$$T^\dagger a^{(1)}(k)T = +a^{(1)}(-k) \qquad T^\dagger a^{(2)}(k)T = -a^{(2)}(-k) \tag{4.39}$$

$$T^\dagger a_R(k)T = +a_R(-k) \qquad T^\dagger a_L(k)T = +a_L(-k) \tag{4.40}$$

so that helicity is unchanged under time reversal (both the momentum and the spin are reversed) as desired.

4.2.2 Spinor fields

For a spinor field

$$T^\dagger \psi(t, x)T = \eta_T S_T \psi(-t, x) \qquad T^\dagger \overline{\psi}(t, x)T = \eta_T^* \overline{\psi}(-t, x)S_T^\dagger \qquad (4.41)$$

($|\eta_T|^2 = 1$), with S_T defined in (4.21). Moreover (now ψ is an operator)

$$(T^2)^\dagger \psi(t, x)T^2 = T^\dagger [\eta_T S_T \psi(-t, x)]T = S_T^* S_T \psi(t, x) = -\psi(t, x) \qquad (4.42)$$

as expected for a spin 1/2 field. Spinors transform according to

$$S_T u(\boldsymbol{p}, s) = [u(-\boldsymbol{p}, -s)]^* e^{i\xi_a(s)} \qquad S_T v(\boldsymbol{p}, s) = [v(-\boldsymbol{p}, -s)]^* e^{i\xi_b(s)} \qquad (4.43)$$

where $\xi_{a,b}(s)$ are spin-dependent phases. Since $S_T^2 = 1$, one has

$$u(\boldsymbol{p}, s) = S_T^2 u(\boldsymbol{p}, s) = S_T [u(-\boldsymbol{p}, -s)]^* e^{i\xi_a(s)}$$

$$= -[S_T u(-\boldsymbol{p}, -s)]^* e^{i\xi_a(s)} = -u(\boldsymbol{p}, s)e^{-i\xi_a(-s)} e^{i\xi_a(s)}$$

(and analogously for v) from which

$$\xi_{a,b}(s) = \pm\pi + \xi_{a,b}(-s) \qquad (4.44)$$

The phases $\xi_{a,b}$ can be chosen to be equal for particles and antiparticles: $e^{i\xi_a(s)} = e^{i\xi_b(s)}$. Writing $T = U_T K$ (with unitary U_T), in order to satisfy eqn (4.41), which explicitly is

$$\int d\boldsymbol{p} \sum_s \left[T^\dagger a(\boldsymbol{p}, s)T \ [u(\boldsymbol{p}, s)]^* e^{iEt - i\boldsymbol{p}\cdot x} + T^\dagger b^\dagger(\boldsymbol{p}, s)T \ [v(\boldsymbol{p}, s)]^* e^{-iEt + i\boldsymbol{p}\cdot x} \right]$$

$$= \eta_T \int d\boldsymbol{p} \sum_s \left[a(\boldsymbol{p}, s) \, S_T \, u(\boldsymbol{p}, s)e^{-iE(-t) + i\boldsymbol{p}\cdot x} \right.$$

$$\left. + b^\dagger(\boldsymbol{p}, s) \, S_T \, v(\boldsymbol{p}, s)e^{iE(-t) - i\boldsymbol{p}\cdot x} \right]$$

one sets

$$T^\dagger a(\boldsymbol{p}, s)T = U_T^\dagger a(\boldsymbol{p}, s) U_T = -\eta_T \, a(-\boldsymbol{p}, -s)e^{i\xi_a(s)} \qquad (4.45)$$

$$T^\dagger b^\dagger(\boldsymbol{p}, s)T = U_T^\dagger b^\dagger(\boldsymbol{p}, s) U_T = -\eta_T \, b^\dagger(-\boldsymbol{p}, -s)e^{i\xi_b(s)} \qquad (4.46)$$

For the wave function φ (which is a c-number function), the usual transformation of QM (4.20) holds. It is worth spelling out in detail the derivation of such a transformation in the case of an anti-unitary operator: for a single particle state with energy E, momentum $\boldsymbol{p}$ and spin projection s:

$$
\begin{aligned}
\varphi(t,\boldsymbol{x};E,\boldsymbol{p},s) &\equiv \langle 0|\psi(t,\boldsymbol{x})|\boldsymbol{p},s\rangle = \langle K0|K\psi(t,\boldsymbol{x})|\boldsymbol{p},s\rangle^* \\
&= \langle 0|U^\dagger UK\psi(t,\boldsymbol{x})K^{-1}U^\dagger|UK(\boldsymbol{p},s)\rangle^* = \langle 0|T\psi(t,\boldsymbol{x})T^\dagger|T(\boldsymbol{p},s)\rangle^* \\
&= \eta_T^* S_T^* \langle 0|\psi(-t,\boldsymbol{x})|(\boldsymbol{p},s)_T\rangle^* = \eta_T^* S_T^* \langle 0|\psi(-t,\boldsymbol{x})Ta^\dagger(\boldsymbol{p},s)T^\dagger|0\rangle^* \\
&= -\eta_T^* S_T^* e^{i\xi_a(s)} \langle 0|\psi(-t,\boldsymbol{x})a^\dagger(-\boldsymbol{p},-s)|0\rangle^* \\
&= -\eta_T^* S_T^* e^{i\xi_a(s)} \langle 0|\psi(-t,\boldsymbol{x})|-\boldsymbol{p},-s\rangle^* \\
&= \eta_T^* S_T e^{i\xi_a(s)} \varphi(-t,\boldsymbol{x};E,-\boldsymbol{p},-s)^*
\end{aligned}
$$

The fact that the quantized field and the wave function transform differently is a consequence of the anti-unitary nature of the time reversal operator.

The transformation under time reversal of bilinear covariants built from spinor fields can be obtained by considering the effect of S_T on the complete set $\{\Gamma\}$ of 4×4 matrices

$$S_T\{\mathbf{1},\gamma^\mu,\sigma^{\mu\nu},\gamma^\mu\gamma_5,i\gamma_5\}S_T^{-1} = \{\mathbf{1},\gamma_\mu^*,-\sigma_{\mu\nu}^*,\gamma_\mu^*\gamma_5^*,-i\gamma_5^*\} \tag{4.47}$$

$$S(t,\boldsymbol{x}) \rightarrow +S(-t,\boldsymbol{x}) \tag{4.48}$$

$$V^\mu(t,\boldsymbol{x}) \rightarrow +V_\mu(-t,\boldsymbol{x}) \tag{4.49}$$

$$T^{\mu\nu}(t,\boldsymbol{x}) \rightarrow -T_{\mu\nu}(-t,\boldsymbol{x}) \tag{4.50}$$

$$A^\mu(t,\boldsymbol{x}) \rightarrow +A_\mu(-t,\boldsymbol{x}) \tag{4.51}$$

$$P(t,\boldsymbol{x}) \rightarrow -P(-t,\boldsymbol{x}) \tag{4.52}$$

For bilinears built from two different fields $\overline{\psi}_2\Gamma\psi_1$ a factor $\eta_T^*(2)\eta_T(1)$ also appears in the transformed expression.

4.3   Time reversal symmetry

While in classical mechanics the time-reversed motion of an elementary system which is symmetric under time reversal is always possible, in quantum mechanics the situation is quite different. Considering the decay of a particle with spin, the time-reversed reaction, obtained from the decay products with inverted momenta and spins, does not result in the original particle with inverted spin, because the asymptotic states (incoherent) of incident particles are not a coherent superposition of ingoing spherical waves, which would be the true time-reversed state of the decaying one. Only for a single particle is the time-reversed state a simple one in quantum mechanics (being essentially obtained by a π rotation around any axis orthogonal to the particle spin). Time-reversed states in quantum mechanics are usually complicated and very 'improbable' even for microscopic systems that are time reversal symmetric (Lee, 1988): the classical difficulty (of statistical nature) in preparing the time-reversed version of a state with a large number of particles often becomes a real impossibility in quantum mechanics. For this reason tests of time reversal symmetry are usually performed in an indirect way, by checking its consequences rather than by studying a time-reversed state in the laboratory.

As mentioned, the anti-unitary nature of T implies that this operator does not have observable eigenvalues which can be used to label states. Therefore time reversal does not enforce selection rules which can be put to a test.

While the experimental measurement of a quantity which is odd under space inversion directly implies violation of parity symmetry, the same is not true for a quantity that is odd under time reversal, both because the exchange of initial and final states is more complicated and also because, in case of higher order contributions in the Hamiltonian, the iterated time evolution factors $\exp(-i \int H dt)$ do not commute with an anti-linear operator such as T, even when the Hamiltonian H itself does, see eqn (C.25).

In contrast with the case of parity symmetry, it can be shown (Arash *et al.*, 1985) that for time reversal no null experiment can be performed; that is, without making assumptions on the dynamics, there exists no single observable which must vanish if time reversal symmetry is valid. The reason is that, even in the simplest case of elastic scattering in which the forward and time-reversed reactions are the same, the constraints imposed by time reversal symmetry relate a physical observable for the forward reaction to a *different* observable for the backward (time-reversed) reaction.

The above implies in turn that measuring a non-zero value for any single observable cannot be taken as a proof of time reversal symmetry violation without introducing further assumptions, and tests of time reversal performed by measurements of expectation values of T-odd operators always require consideration of the fact that non-vanishing contributions can arise from time reversal symmetric processes (this is why such quantities are appropriately called 'T-odd' rather than 'T-violating' quantities), and that the precision attainable in such tests is limited also by the knowledge of such effects.

The only exception to the above is the case of total cross section or forward scattering, i.e. transmission experiments, for which null tests of time reversal symmetry can be found in specific cases (Conzett, 1993).

As an example of the consequence of time reversal symmetry we note that the non-derivative Yukawa coupling of a neutral spinless meson (described by $\phi = \phi^\dagger$)

to a spinor field can be written in general as

$$H_{\text{int}} = c_S \, \phi \overline{\psi} \psi + c_P \, i \phi \overline{\psi} \gamma_5 \psi \tag{4.53}$$

and Hermiticity requires $c_{S,P}$ to be real. The opposite transformation properties of the two terms under time reversal (under which $\phi \to \eta_T \phi$, with real η_T to maintain Hermiticity) imply that time reversal symmetry forbids both of them to appear together (Feinberg, 1957): in this case T symmetry imposes the same constraints as P symmetry would (derivative couplings such as $\partial_\mu \phi \overline{\psi} \gamma^\mu \psi$ and $\partial_\mu \phi \overline{\psi} \gamma^\mu \gamma_5 \psi$ could also appear together with the second term of (4.53), possibly violating P). As will be discussed in Section 4.3.4, similar constraints do apply for couplings to the EM field, where $\overline{\psi} \gamma_5 \sigma_{\mu\nu} \psi F^{\mu\nu}$ cannot appear together with any combination of $\overline{\psi} \gamma^\mu \psi A_\mu$ and $\overline{\psi} \sigma_{\mu\nu} \psi F^{\mu\nu}$, if time reversal symmetry holds.

The time reversal symmetry of the generic fermionic Lagrangian (B.29) requires $c_i^* = c_i$ and $c_i'^* = c_i'$, i.e. real coupling constants.

4.3.1   Transition amplitudes

In case of time reversal symmetry for the interaction Hamiltonian

$$T^\dagger H_{\text{int}}^{(S)} T = H_{\text{int}}^{(S)} \tag{4.54}$$

in the Schrödinger picture, and in the interaction picture

$$T^\dagger H_{\text{int}}^{(I)}(t) T = H_{\text{int}}^{(I)}(-t) \tag{4.55}$$

Introducing the time evolution operator $U^{(I)}$ and applying time reversal to eqn (C.10) one gets

$$-i\hbar \frac{\partial}{\partial t} T^\dagger U^{(I)}(t, t_0) T = H_{\text{int}}^{(I)}(-t) T^\dagger U^{(I)}(t, t_0) T$$

from which it can be concluded that

$$T^\dagger U^{(I)}(t, t_0) T = U^{(I)}(-t, -t_0) \tag{4.56}$$

One has therefore

$$T^\dagger U^{(I)}(0, -\infty) T = U^{(I)}(0, \infty) = U^{(I)\dagger}(\infty, 0) \tag{4.57}$$

For the scattering matrix $\mathcal{S} \equiv U^{(I)}(\infty, -\infty)$ the condition of time reversal symmetry is thus

$$\mathcal{S}_T \equiv T^\dagger \mathcal{S} T = U^{(I)}(-\infty, \infty) = \mathcal{S}^{-1} = \mathcal{S}^\dagger \qquad (4.58)$$

Time reversal symmetry does not imply that the scattering matrix is left unchanged, but rather that it is transformed into its inverse (compare with eqn (2.59) for parity). This is consistent with the fact that no conserved quantum number is associated with T symmetry, since $\mathcal{S}$ does not commute with T.

Analogously, for the transition matrix $\mathcal{T}$ time reversal symmetry implies

$$\mathcal{T}_T \equiv T^\dagger \mathcal{T} T = \mathcal{T}^\dagger \qquad (4.59)$$

The above results can also be understood by recalling that time reversal is equivalent to the inversion of all T-odd quantities such as momenta and spins (sometimes called 'naive' time reversal or 'kinematical' time reversal, the effect of the unitary operator U_T) plus the exchange of initial and final states; considering the definitions of asymptotic 'in' and 'out' states (see Appendix C) the effect of time reversal as in (4.57) is

$$T|\psi_{(\text{in})}\rangle = \eta_T(\psi)|\psi_{T\,(\text{out})}\rangle \qquad (4.60)$$

where the subscript T indicates the effect of 'naive' time reversal on the states, and an 'in' state is transformed into the *corresponding* 'out' state only if time reversal symmetry is valid. The exchange of 'in' and 'out' states, which is of course only relevant when these states contain interacting particles, indeed implies that in case of time reversal symmetry

$$\mathcal{S}_{fi} = \langle f_{(\text{out})}|i_{(\text{in})}\rangle = \langle f_{(\text{out})}|T^\dagger T|i_{(\text{in})}\rangle = \langle Tf_{(\text{out})}|Ti_{(\text{in})}\rangle^*$$
$$= \eta_T(f)\eta_T^*(i)\langle f_{T\,(\text{in})}|i_{T\,(\text{out})}\rangle^* = \eta_T(f)\eta_T^*(i)\langle i_{T\,(\text{out})}|f_{T\,(\text{in})}\rangle = \eta_T(f)\eta_T^*(i)\mathcal{S}_{i_T f_T}$$
$$(4.61)$$

(where $\eta_T(i)$ stands for the product of the η_T factors for each particle in the initial state, etc.) or more simply in terms of the $\mathcal{S}$ operator

$$\langle f|\mathcal{S}|i\rangle = \langle i|\mathcal{S}^\dagger|f\rangle^* = \langle i|\mathcal{S}_T|f\rangle^* = \langle i|T^\dagger \mathcal{S} T|f\rangle^* = \eta_T(f)\eta_T^*(i)\langle i_T|\mathcal{S}|f_T\rangle$$
$$(4.62)$$

An analogous relation holds for the transition matrix elements $A(i \rightarrow f) = T_{fi} = \langle f|T|i \rangle$:

$$A(i \rightarrow f) = \eta_T(f)\eta_T^*(i)A(f_T \rightarrow i_T) \tag{4.63}$$

which is called the *reciprocity relation*. This relation implies the equality of the rates for the $i \rightarrow f$ and $f_T \rightarrow i_T$ reactions (*not* for the reaction with time-reversed states $i_T \rightarrow f_T$, nor for the inverse reaction $f \rightarrow i$ among the same states).

For states which are unchanged under 'naive' time reversal, $|i_T, f_T\rangle = |i, f\rangle$ (e.g. spinless particles or spin-averaged states; the inversion of momenta is irrelevant when dealing with decay amplitudes which are integrated over phase space) eqn (4.63) becomes

$$A(i \rightarrow f) = \eta_T(f)\eta_T^*(i)A(f \rightarrow i) \tag{4.64}$$

and in this case with an appropriate choice of phases time reversal symmetry implies that the transition matrix (and the S matrix) is symmetric (Hamilton, 1959), actually leading to the equality of rates for the $i \rightarrow f$ and $f \rightarrow i$ processes.

For a generic reaction $i \rightarrow f$, time reversal symmetry in eqn (4.63) is explicitly

$$\langle \boldsymbol{p}_f, \boldsymbol{S}_f|T|\boldsymbol{p}_i, \boldsymbol{S}_i\rangle = \langle -\boldsymbol{p}_i, -\boldsymbol{S}_i|T|-\boldsymbol{p}_f, -\boldsymbol{S}_f\rangle \tag{4.65}$$

where $\boldsymbol{p}_{i,f}$ and $\boldsymbol{S}_{i,f}$ stand for all initial and final momenta and spins. In its generality eqn (4.65) states that the probability of an initial state i to be scattered into a final state f is the same as that of an initial state identical to f, but with all momenta and spins reversed, to be scattered into a final state identical to i with reversed momenta and spins, which is what is expected from time reversal symmetry.

The *principle of detailed balance* is the equality for the transition probabilities with initial and final states interchanged:

$$|\langle \boldsymbol{p}_f, \boldsymbol{S}_f|T|\boldsymbol{p}_i, \boldsymbol{S}_i\rangle|^2 = |\langle \boldsymbol{p}_i, \boldsymbol{S}_i|T|\boldsymbol{p}_f, \boldsymbol{S}_f\rangle|^2 \tag{4.66}$$

For a $2 \rightarrow 2$ reaction, if the spins of initial and final particles lie in the reaction plane this result follows from time reversal symmetry (reciprocity) and the symmetry for a rotation by π around an axis orthogonal to such a plane.

It should be noted that the principle of detailed balance is always (approximately) valid for weak interactions: if the first order of the perturbative expansion in H_{int} (Born approximation) is a good approximation, the transition matrix coincides with the interaction Hamiltonian and is therefore Hermitian ($T^\dagger = T$), so that

$$\langle f|T|i \rangle \simeq \langle f|H_{int}|i \rangle = \langle f|H_{int}^\dagger|i \rangle = \langle i|H_{int}|f \rangle^* \simeq \langle i|T|f \rangle^*$$

from which the equality of the transition rates $i \rightarrow f$ and $f \rightarrow i$ follows, *independently* of the validity of time reversal symmetry.

Conversely, in this case, the reciprocity relation (4.63) can relate directly a process to itself (after appropriate spin and momentum inversion) rather than to its time-reversed counterpart, so that time reversal symmetry implies (for a Hermitian transition matrix)

$$\langle f|T|i\rangle = \eta_T(f)\eta_T^*(i)\langle f_T|T^\dagger|i_T\rangle^* = \eta_T(f)\eta_T^*(i)\langle f_T|T|i_T\rangle^*$$

from which the equality for the properties of the processes $i \to f$ and $i_T \to f_T$ follows.

Note that this possibility, of obtaining from the symmetry under an anti-unitary operator a relation between the process $i \to f$ and the corresponding one between suitably modified states (here $i_T \to f_T$) *in the same direction* (as opposed to $f_T \to i_T$), is always present when dealing with weakly interacting initial and final states (see Chapter 5).

The Born approximation is not valid for strong interactions; still for some types of reactions detailed balance follows directly from the unitarity of the S matrix, independently of time reversal symmetry (Henley and Jacobsohn, 1959). When the S matrix splits in disconnected blocks, and in each of those blocks two sets of states $|a_i\rangle, |b_i\rangle$ can be found for which the matrix elements have the form

$$\langle a_j|S|b_k\rangle = A_{jk}M \qquad \langle b_k|S|a_j\rangle = \widehat{A}_{kj}\widehat{M}$$

where the dependence on the individual j, k states is only contained in coefficients which differ just in phase, $|A_{jk}| = |\widehat{A}_{kj}|$, then from the unitarity of S (which works for each disconnected block), by inserting a complete set of states spanning the entire block:

$$1 = \langle a_j|S^\dagger S|a_j\rangle = \sum_{j'}|\langle a_j'|S|a_j\rangle|^2 + \sum_k |\widehat{A}_{kj}|^2|\widehat{M}|^2$$

$$1 = \langle a_j|SS^\dagger|a_j\rangle = \sum_{j'}|\langle a_j|S|a_j'\rangle|^2 + \sum_k |A_{jk}|^2|M|^2$$

Summing over j and subtracting, it is found that $|M|^2 = |\widehat{M}|^2$, so that

$$|\langle a_j|S|b_k\rangle| = |\langle b_k|S|a_j\rangle|$$

Note that the above always applies when the S matrix breaks into 2×2 blocks, since in this case the most general form for each block is

$$S = \begin{pmatrix} \cos\theta\, e^{i\phi} & \sin\theta\, e^{+i\xi} \\ \sin\theta\, e^{-i\xi} & \cos\theta\, e^{-i\phi} \end{pmatrix}$$

for which indeed $|\langle a|S|b\rangle| = |\langle b|S|a\rangle|$, and time reversal symmetry only enforces the equality of the phases of the two matrix elements for the opposite reactions, which are already equal in modulus.

For generic spin orientations, the parity and time reversal symmetries together enforce the *principle of semi-detailed balance*:

$$\sum_s |\langle \boldsymbol{p}_f, \boldsymbol{S}_f | \mathcal{T} | \boldsymbol{p}_i, \boldsymbol{S}_i \rangle|^2 = \sum_s |\langle \boldsymbol{p}_f, -\boldsymbol{S}_f | \mathcal{T} | \boldsymbol{p}_i, -\boldsymbol{S}_i \rangle|^2 = \sum_s |\langle \boldsymbol{p}_i, \boldsymbol{S}_i | \mathcal{T} | \boldsymbol{p}_f, \boldsymbol{S}_f \rangle|^2$$

$$(4.67)$$

where the sums run over all the polarization states of the particles.

From the principle of semi-detailed balance a relation between differential cross sections in the centre of mass system follows. For a reaction $ab \rightarrow cd$:

$$(2s_a + 1)(2s_b + 1)|\boldsymbol{p}_i|^2 \left(\frac{d\sigma_{i \rightarrow f}}{d\Omega}\right)_{CM} = (2s_c + 1)(2s_d + 1)|\boldsymbol{p}_f|^2 \left(\frac{d\sigma_{f \rightarrow i}}{d\Omega}\right)_{CM}$$

$$(4.68)$$

where $s_{a,b,c,d}$ are the particle spins, $|\boldsymbol{p}_i| = |\boldsymbol{p}_a| = |\boldsymbol{p}_b|$, $|\boldsymbol{p}_f| = |\boldsymbol{p}_c| = |\boldsymbol{p}_d|$ their initial and final centre of mass momenta, and the cross sections are averaged over spin polarization states. In the case of indistinguishable particles the number of accessible final states, and thus the cross section, is reduced (a factor 1/2 for binary reactions), but such reduction does not appear in the expression for the differential cross section, since at any angle there is a double contribution due to the interchange of the identical particles.

It should also be noted that for elastic scattering reactions, which are their own reciprocal, time reversal symmetry (and parity symmetry as well) does not give any restriction if spin measurements are not performed, because the scalar S matrix can only depend on p_i^2, p_f^2, and $\boldsymbol{p}_i \cdot \boldsymbol{p}_f$, which are all invariant under T (and P).

In general, the reality conditions imposed by time reversal symmetry also limit the terms which can appear in the general expression for a scattering amplitude, see, e.g., Sakurai (1964).

4.3.2   Phase relations

As discussed above, time reversal symmetry does not enforce the equality of the $i \rightarrow f$ and $i_T \rightarrow f_T$ transition rates (the subscript T indicates 'naive' time reversal, the inversion of spin and momenta), but rather the equality (to within a phase factor) of the $i \rightarrow f$ and $f_T \rightarrow i_T$ transition rates, and only in case of weakly interacting particles (when the first-order Born approximation is valid) the Hermiticity of $\mathcal{T}$ allows to deduce the equality to the $i_T \rightarrow f_T$ transition rate.

When the particles involved in a reaction cannot be considered as free because they interact (strongly) through a part of the Hamiltonian, even when only a weak part of H_{int} is active for a transition one does not have a 'small' transition matrix element, and the first-order approximation is not valid. However, by separating the strong and weak parts of $\mathcal{T}$ (as discussed in Appendix C), for transitions $i \rightarrow f$

which can be driven only by the weak part $\mathcal{T}_W$ ($\langle f|\mathcal{S}_S|i\rangle = 0$), some interesting relations between the transition amplitudes can be obtained.

First recall the unitarity condition (C.36), which written in terms of matrix elements is

$$\langle f|\mathcal{T}_W|i\rangle = \sum_{n,m}\langle f|\mathcal{S}_S|m\rangle\langle m|\mathcal{T}_W^\dagger|n\rangle\langle n|\mathcal{S}_S|i\rangle$$

$$= \sum_{n,m}\langle f|\mathcal{S}_S|m\rangle\langle n|\mathcal{T}_W|m\rangle^*\langle n|\mathcal{S}_S|i\rangle \qquad (4.69)$$

where the sums extend over complete sets of states. The transition rate will be proportional to

$$|\langle f|\mathcal{T}|i\rangle|^2 = |\langle f|\mathcal{T}_W|i\rangle|^2 = \sum_{n,m}\sum_{n',m'}\langle f|\mathcal{S}_S|m\rangle\langle m'|\mathcal{S}_S^\dagger|f\rangle$$

$$\langle n|\mathcal{T}_W|m\rangle^*\langle n'|\mathcal{T}_W|m'\rangle\langle n|\mathcal{S}_S|i\rangle\langle i|\mathcal{S}_S^\dagger|n'\rangle$$

Considering the sum of transition rates for all the initial and final states belonging to sets of states I, F which are complete for $\mathcal{S}_S$ (that is, if $|n\rangle \in I$ and $\langle n|\mathcal{S}_S|m\rangle \neq 0$ then also $|m\rangle \in I$, therefore sets of states which are unchanged by $\mathcal{S}_S$, which does not induce transitions leading out of them), one has

$$\sum_{i\in I}\langle n|\mathcal{S}_S|i\rangle\langle i|\mathcal{S}_S^\dagger|n'\rangle = \delta_{n,n'} \qquad (4.70)$$

if $n, n' \in I$ (in this case the sums over states in I are equivalent to sums over *all* states), and zero otherwise. Therefore

$$\sum_{i\in I}\sum_{f\in F}|\langle f|\mathcal{T}_W|i\rangle|^2 = \sum_{n\in I}\sum_{m\in F}|\langle n|\mathcal{T}_W|m\rangle|^2 \qquad (4.71)$$

and one obtains the equality of the summed transition rates for $I \to F$ and $F \to I$ (semi-detailed balance):

$$\sum_{i\in I}\sum_{f\in F}|A(i \to f)|^2 = \sum_{i\in I}\sum_{f\in F}|A(f \to i)|^2 \qquad (4.72)$$

Note that this is just a consequence of unitarity and of the smallness of $\mathcal{T}_W$, independent from the validity of time reversal symmetry. Examples of sets of states disconnected by the strong interactions are semi-leptonic states with different leptons: $he\nu$ and $h\mu\nu$ (where h indicates generic hadrons), or states with two and three pions, which have different eigenvalues of G-parity (Section 3.2.4), conserved by strong interactions (not by EM ones, though).

If the strong part of the interaction is time reversal symmetric, the completeness of the sets of states I, F for $\mathcal{S}_S$ (as defined above) is also valid for the time-reversed sets of states I_T, F_T: if $|i\rangle \in I$ implies $\mathcal{S}_S|i\rangle \in I$, then using $\mathcal{S}_S T = T\mathcal{S}_S^\dagger$ one can see that $|i_T\rangle \in I_T$ also implies $\mathcal{S}_S|i_T\rangle \in I_T$. Therefore, if in addition $\mathcal{T}_W$ is also T-symmetric (so that time reversal symmetry holds overall) then unitarity implies from eqn (4.69) the equality of the (summed) transition rates for the reactions $I \to F$ and $I_T \to F_T$ (at first order in $\mathcal{T}_W$).

In the simplest case, the complete sets of states I, F just contain a single state each, an eigenstate of $\mathcal{S}_S$ (below any inelastic threshold), which would be stable if it were not for $\mathcal{T}_W$. Recalling that the effect of interactions is just to introduce a phase shift:

$$\mathcal{S}_S|i,f\rangle = e^{2i\delta_{i,f}}|i,f\rangle \qquad \langle n|\mathcal{S}_S|i\rangle = e^{2i\delta_i}\delta_{ni} \qquad \langle m|\mathcal{S}_S|f\rangle = e^{2i\delta_f}\delta_{mf} \quad (4.73)$$

where[55] the phase shifts $\delta_{i,f}$ are real due to the unitarity of $\mathcal{S}_S$. In this case time reversal symmetry (of $\mathcal{S}_S$ and $\mathcal{T}_W$) enforces (at first order in the weak interaction $\mathcal{T}_W$), besides the equality of $i \to f$ and $f_T \to i_T$ transition rates, the following relation among the *amplitudes*, obtained from eqn (4.69):

$$\langle f|\mathcal{T}_W|i\rangle = \eta_T(f)\eta_T^*(i)e^{2i(\delta_i+\delta_f)}\langle f_T|\mathcal{T}_W|i_T\rangle^* \qquad (4.74)$$

which relates the $i \to f$ and $i_T \to f_T$ processes (both realizable in the laboratory). Factorizing the strong phase shifts as

$$\langle f|\mathcal{T}_W|i\rangle = Me^{i(\delta_i+\delta_f)} \qquad (4.75)$$

one obtains

$$\langle f_T|\mathcal{T}_W|i_T\rangle = M^*e^{i(\delta_i+\delta_f)} \qquad (4.76)$$

which shows that the *modulus* of the transition matrix element is equal for the $i \to f$ and $i_T \to f_T$ transitions.

For states which are not affected by 'naive' time reversal eqn (4.74) actually *determines the phases* of the transition matrix elements, which are in this case just given by the scattering phase shifts induced by the $\mathcal{S}_S$ matrix on the particles in the initial and final states:

$$A(i \to f) = \pm e^{i(\delta_i+\delta_f)}|A(i \to f)| \qquad (4.77)$$

---

[55] In eqns (4.73) δ_{ni} and δ_{mf} are Kronecker deltas, not phase shifts!

where the product of the phases $\eta_T(f)\eta_T^*(i)$ gives the arbitrariness in sign.[56]. This result is known as the (Aizu–)Fermi–Watson theorem (Aizu, 1954; Watson, 1954; Fermi, 1955): if time reversal symmetry is valid the phases of weak transition amplitudes are just given by the strong phase shifts due to the scattering of the initial and final particles; such phases can in principle be measured by studying the physical scattering of those particles. Once more, time reversal symmetry is seen to enforce a reality condition: the transition matrix elements are real apart from the above phase shifts.

For particle decays, the initial state $|i\rangle$ is a single-particle state, with no phase shift due to $\mathcal{S}_S$ ($\delta_i = 0$); this is the reason why the above phase shifts are usually called 'final state interaction' (FSI) phases.

The Fermi–Watson theorem is very important at low energies (compared to the scale of hadron masses), but when considering the decays of heavy particles it can rarely be used, since elastic unitarity is usually not valid due to the creation of additional particles in the final state.

Note finally that in general, from the unitarity equation for $\mathcal{T}$ (6.31), if time reversal invariance is valid and $|i_T, f_T\rangle = |i, f\rangle$ then

$$A^*(f \to i) - A(i \to f) = A^*(i_T \to f_T) - A(i \to f)$$

$$= A^*(i \to f) - A(i \to f) = i \sum_n A^*(f \to n)A(i \to n)$$

$$(4.78)$$

so that a transition amplitude can be complex due to the presence of (real) intermediate physical states on the mass shell (conserving energy and momenta) which can be reached (at second order or above in the interaction); this reflects the fact that the absorptive part of the amplitudes is imaginary.

4.3.3 T-odd observables

As a consequence of time reversal symmetry it might be expected that observables which change sign under such transformation should have vanishing expectation values, and that the Hamiltonian should not contain T-odd terms such as $S \cdot (p_1 \times p_2)$, $S_1 \cdot (S_2 \times S_3)$, $p_1 \cdot (p_2 \times p_3)$, $p \cdot (S_1 \times S_2)$ (where p are momenta and S spins). As might be guessed from (4.78), the above is actually only true in the limit of vanishing initial and final state interactions: as an example, time reversal symmetry implies that the muon polarization in the reaction $\nu n \to \mu^- p$ (with unpolarized neutrinos) can only lie in the production plane, since $S_\mu \cdot (p_\mu \times p_p)$ is T-odd, if EM interactions between μ^- and p are neglected.

In presence of interactions which cannot be neglected among the particles participating in a reaction the above statement has to be qualified better, since such

---

[56] With the choice of phases of eqn (4.30) the η_T factors are always real.

interactions are time reversal symmetric but not symmetric under 'naive' time reversal.

The unitarity constraint (C.20) on the transition matrix for a generic transition $i \to f$ gives

$$|\mathcal{T}_{if}|^2 = |\mathcal{T}_{fi}|^2 - 2\mathrm{Im}(\mathcal{I}_{fi}\mathcal{T}_{fi}) - |\mathcal{I}_{fi}|^2 \tag{4.79}$$

where $\mathcal{I}_{fi} = (2\pi) \sum_n \delta(E_n - E_i) T_{nf}^* T_{ni}$ is the absorptive part of the $i \to f$ process (real intermediate states); from this the expression for a T-odd quantity is obtained

$$|\mathcal{T}_{f_T i_T}|^2 - |\mathcal{T}_{fi}|^2 = |\mathcal{T}_{f_T i_T}|^2 - |\mathcal{T}_{if}|^2 - 2\mathrm{Im}(\mathcal{I}_{fi}\mathcal{T}_{fi}) - |\mathcal{I}_{fi}|^2 \tag{4.80}$$

Time reversal symmetry would imply $|\mathcal{T}_{f_T i_T}| = |\mathcal{T}_{if}|$, eqn (4.63), but the last two terms in eqn (4.80) still make the T-odd quantity different from zero in presence of an absorptive part due to final state interactions ($\mathcal{I}_{fi} \neq 0$).

Following Gasiorowicz (1966) we now sketch the general approach to address P-odd or T-odd observables. When a weak decay $i \to f$ is considered, the expectation value of a generic observable $O(p_f, s_i, s_f)$ built out of the momenta and spin of the initial and final particles (in the centre of mass p_i is a constant) is in general

$$\langle O \rangle = N \frac{1}{2s_i + 1} \sum_{s_i} \sum_{p_f, s_f} O(p_f, s_i, s_f) \, |\langle p_f, s_f | \mathcal{T}_W | s_i \rangle|^2$$

where N is a normalization factor and $\mathcal{T}_W$ is the weak part of the transition matrix (see Appendix C). In the presence of final state interactions the matrix element is complex even if T symmetry holds; inserting a complete set of angular momentum eigenstates $|p_f, j, m\rangle$ (eigenstates of the strong S matrix) and factorizing out the corresponding scattering phase shift δ_j for the strongly interacting particles in the final state one writes

$$\langle p_f, s_f | \mathcal{T}_W | s_i \rangle \equiv \sum_j e^{i\delta_j} M(j, p, s) \tag{4.81}$$

(where $\{p, s\}$ is a shorthand notation for $\{p_f, s_f, s_i\}$) so that

$$\langle O \rangle \propto \sum_{p,s} \sum_{j,j'} O(p, s) \, e^{i(\delta_j - \delta_{j'})} M(j, p, s) M^*(j', p, s)$$

Splitting P-even and P-odd parts

$$\mathcal{T}_W = \sum_i \left(c_i O_i^{(P+)} + c_i' O_i^{(P-)} \right) \tag{4.82}$$

$$P^\dagger O_i^{(P\pm)} P = \pm O_i^{(P\pm)} \tag{4.83}$$

parity symmetry would require that either $c_i = 0$ or $c_i' = 0$ for all i. Under time reversal

$$\mathcal{T}_W \to T^\dagger \mathcal{T}_W T = \sum_i \left(c_i^* T^\dagger O_i^{(P+)} T + c_i'^* T^\dagger O_i^{(P-)} T \right)$$

which should be equal to $T_W^\dagger$ in case of T symmetry. Recalling eqn (C.36) and choosing O such that $T^\dagger O_i^{(P\pm)} T = S_S^\dagger O_i^{(P\pm)} S_S^\dagger$, a possible T violation is described by the complexity of the c_i, c_i' coefficients.

The above decomposition corresponds to a similar one for M defined in eqn (4.81), with the properties:

$$M_i^{(P\pm)}(j,p,s) = \pm M_i^{(P\pm)}(j,-p,s) \tag{4.84}$$

$$M_i^{(P\pm)*}(j,p,s) = M_i^{(P\pm)}(j,-p,-s) \tag{4.85}$$

A generic observable $O^{(P\pm,T\pm)}$ can be even or odd under parity and time reversal (as indicated by the superscripts); some examples are

$$O^{(P+,T+)} = p_1 \cdot p_2, \; S_1 \cdot S_2 \qquad\qquad O^{(P-,T+)} = S \cdot p$$

$$O^{(P+,T-)} = S_1 \cdot p_1 \times p_2, \; S_1 \cdot S_2 \times S_3 \qquad O^{(P-,T-)} = p_1 \cdot p_2 \times p_3$$

The expectation values, expressed in terms of the $M_{i,j}^{(\pm)}$, will contain different terms with products of the c_i, c_i'; expressing M^* in terms of M for opposite momenta and spin through eqn (4.85), some terms vanish in the sum (integration) over momentum, because of the symmetry or antisymmetry of O under momentum inversion.

As an example, for a P-even, T-odd observable, using

$$O^{(P+,T-)}(-p,-s) = -O^{(P+,T-)}(p,s)$$

the terms containing $c_i c_j'^*$ vanish, showing that no parity-violating (interference) term can be measured by a P-even observable, and one can write

$$\langle O^{(P+,T-)} \rangle \propto \sum_{j,j'} e^{i(\delta_j - \delta_{j'})} \left\{ \sum_{i,j} c_i c_j^* \sum_{p,s} O^{(P+,T-)}(p,s) \right.$$

$$\frac{1}{2}\left[M_i^{(P+)}(j,p,s)M_j^{(P+)}(j',-p,-s) - M_i^{(P+)}(j,-p,-s)M_j^{(P+)}(j',p,s) \right]$$

$$\left. + \sum_{i,j} c_i' c_j'^* \cdots \right\} = \frac{1}{4} \sum_{j,j'} \sum_{i,j} \left[\cos(\delta_j - \delta_{j'}) \right.$$

$$+ i\sin(\delta_j - \delta_{j'}) \right] \left[(c_i c_j^* + c_i^* c_j) + (c_i c_j^* - c_i^* c_j) \right]$$

$$\sum_{p,s} O^{(P+,T-)}(p,s) \left[M_i^{(P+)}(j,p,s)M_j^{(P+)}(j',-p,-s) \right.$$

$$\left. - M_i^{(P+)}(j,-p,-s)M_j^{(P+)}(j',p,s) \right] + \cdots$$

The last term in square brackets is antisymmetric under the simultaneous interchange $i \leftrightarrow j$ and $j \leftrightarrow j'$, so that the only non-vanishing terms are those containing $\cos(\delta_j - \delta_{j'})(c_i c_j^* - c_i^* c_j)$ and $\sin(\delta_j - \delta_{j'})(c_i c_j^* + c_i^* c_j)$. This means that in absence of strong interactions among final particles

TABLE 4.1 Symmetry violations implied by the measurement of non-zero *P*-odd or *T*-odd observables, in the case of final state interactions (FSI) being present or not present. As an example: the expectation values of a *P*-odd and *T*-even observable contain a term violating both parity and charge conjugation, multiplied by $\cos(\delta_j - \delta_{j'})$ (thus present even if FSI are absent), and a term violating both parity and time reversal, multiplied by $\sin(\delta_j - \delta_{j'})$ (thus vanishing in absence of FSI).

	$\cos(\delta_j - \delta_{j'}) \times$ Present also if no FSI	$\sin(\delta_j - \delta_{j'}) \times$ Only present if FSI
$(P+, T+)$	non-violating	T, C-violating
$(P-, T+)$	P, C-violating	P, T-violating
$(P+, T-)$	T, C-violating	non-violating
$(P-, T-)$	P, T-violating	P, C-violating

$(\delta_j = \delta_{j'} = 0)$ $\langle O^{(P+,T-)} \rangle$ can be non-zero only if time reversal is violated ($c_i \neq c_i^*$), but if such interactions are present a non-zero expectation value can be present even if time reversal symmetry holds.

   The above result is of general validity if the C-violating entries are read as PT-violating (equivalent if CPT symmetry holds).

   The expectation values of the various kinds of observables O contain the terms shown in Table 4.1.

   Non-zero expectation values for P-odd observables $O^{(P-,T\pm)}$ always indicate the presence of a parity-violating term (independently from the presence of FSI), while for T-odd observables the situation is different, and only in absence of FSI (when the terms in the second column vanish) such non-zero expectation values imply a true violation of time reversal symmetry. If independent information on FSI (such as an upper limit on the phase difference) is available the observation of a large non-zero expectation value for a $O^{(P-,T-)}$ observable can be considered as an indication of time reversal violation if the term in the second column above is not sufficient to produce the observed effect. Also, if one could measure accurately the dependence of an expectation value on the value of the phase shift difference (e.g. if their energy dependence were precisely known), the two terms could be disentangled and information on the validity of time reversal symmetry could be extracted from any of the above expectation values; unfortunately this is not usually possible.

The discussion above just reflects the general issue of null tests of time reversal symmetry, linked to the fact that the expectation values of a given T-odd operator are usually different observables for a reaction and its inverse.

An important example is the polarization-asymmetry equality: in the scattering of spin 1/2 particles time reversal symmetry enforces the equality of the polarization $\mathcal{P}$ of the (initially unpolarized) scattered beam from an unpolarized target, and the spatial scattering asymmetry for a fully polarized beam by an unpolarized target (Bell and Mandl, 1958). This can be seen in a rather simple way: in the scattering of unpolarized spin 1/2 particles from an unpolarized target, denoting with $\sigma(\theta;f,i)$ the cross section for scattering at an angle θ with initial spin state i and final spin state f ($i, f = \pm$ stand for spin components $m_{i,f} = \pm 1/2$ along the quantization axis, chosen orthogonal to the scattering plane), the polarization of the scattered beam is

$$P(\theta) = \frac{[\sigma(\theta;++) + \sigma(\theta;+-)] - [\sigma(\theta;-+) + \sigma(\theta;--)]}{[\sigma(\theta;++) + \sigma(\theta;+-)] + [\sigma(\theta;-+) + \sigma(\theta;--)]} \qquad (4.86)$$

while the asymmetry in the scattering of a fully polarized $m_i = +1/2$ beam (polarization not observed) is

$$A(\theta) = \frac{[\sigma(\theta;++) + \sigma(\theta;-+)] - [\sigma(-\theta;++) + \sigma(-\theta;-+)]}{[\sigma(\theta;++) + \sigma(\theta;-+)] + [\sigma(-\theta;++) + \sigma(-\theta;-+)]} \qquad (4.87)$$

and since time reversal symmetry implies $\sigma(\theta;f,i) = \sigma(-\theta;-i,-f)$, after applying a π rotation around the scattered particle direction (assumed to be a valid symmetry) one obtains

$$A(\theta) = P(\theta) \qquad (4.88)$$

The above applies to transverse polarization: if parity symmetry is violated in the scattering, longitudinal polarization can also be produced, which however does not affect the asymmetry. Note also that in the case of a spin 0 target, the polarization-asymmetry equality follows directly also from parity symmetry alone, independently from the validity of time reversal symmetry.

4.3.4 Electric dipole moments

It was recognized long ago that searches for permanent electric dipole moments (EDM) are very interesting as experimental tests of time reversal symmetry (Ramsey, 1958), since EDMs are T-odd (Landau, 1957a).

The Pauli interaction between a spin 1/2 fermion and the electromagnetic field due to the anomalous magnetic moment is described by the interaction term

$$H_{\text{int}}^{(M)} = \frac{d_M}{2} \overline{\psi} \sigma_{\mu\nu} \psi F^{\mu\nu} \qquad (4.89)$$

where d_M is a constant, and $F^{\mu\nu} = \partial^\mu A^\nu - \partial^\nu A^\mu$ is the electromagnetic tensor. Remember that the non-anomalous (i.e. $g = 2$) magnetic moment interaction is included in the usual $j_\mu A^\mu$ interaction term. The term (4.89) is invariant under parity, as from eqn (2.50) $\overline{\psi}\sigma_{\mu\nu}\psi \to \overline{\psi}\sigma^{\mu\nu}\psi$, and $F^{\mu\nu} \to F_{\mu\nu}$; it is also invariant under time reversal, as from eqn (4.50) $\overline{\psi}\sigma_{\mu\nu}\psi \to T^\dagger \overline{\psi}\sigma_{\mu\nu}\psi T = -\overline{\psi}\sigma^{\mu\nu}\psi$ and $T^\dagger F^{\mu\nu}T = -F_{\mu\nu}$.

This term reduces to $-d_M \langle \boldsymbol{\sigma} \rangle \cdot \boldsymbol{B}$ in the non-relativistic limit ($\boldsymbol{\sigma}$ being the vector of Pauli matrices and $\boldsymbol{B}$ the external magnetic field), and therefore

$$d_M = (1+\kappa)\frac{e\hbar}{2m} \tag{4.90}$$

where m is the fermion mass and $\kappa = (g-2)/2$ the dimensionless anomalous magnetic moment ($\kappa \simeq \alpha/2\pi \simeq 1 \cdot 10^{-3}$ for the electron, $\kappa \simeq 1.79$ for the proton and $\kappa \simeq -1.91$ for the neutron); the magnetic moment is

$$\boldsymbol{\mu} \equiv \boldsymbol{d}_M = d_M \frac{\boldsymbol{S}}{|\boldsymbol{S}|} = 2d_M \frac{\boldsymbol{S}}{\hbar} = g\frac{e}{2m}\boldsymbol{S} \tag{4.91}$$

or in general

$$\boldsymbol{\mu} = \frac{1}{2}\int d\boldsymbol{x}\, \boldsymbol{x} \times \boldsymbol{j}(\boldsymbol{x}) \tag{4.92}$$

where $\boldsymbol{j}$ is the space component of the electromagnetic current operator $j^\mu = (\rho, \boldsymbol{j})$; its interaction with a magnetic field is indeed

$$H_{\text{int}}^{(M)} = -\boldsymbol{\mu} \cdot \boldsymbol{B} \tag{4.93}$$

(invariant under parity and time reversal as it should be, as $\boldsymbol{\mu}$ and $\boldsymbol{B}$ transform in the same way under such operations).

An electric dipole moment $\boldsymbol{d}_E$ can originate from an asymmetry of its charge density distribution $\rho(\boldsymbol{x})$:

$$\boldsymbol{d}_E = \int d\boldsymbol{x}\, \boldsymbol{x}\, \rho(\boldsymbol{x}) \tag{4.94}$$

and its interaction with an electric field $\boldsymbol{E}$ is

$$H_{\text{int}}^{(E)} = -\boldsymbol{d}_E \cdot \boldsymbol{E} \tag{4.95}$$

For this interaction to be invariant under parity and time reversal $\boldsymbol{d}_E$ must transform as $\boldsymbol{E}$, namely as a polar vector which does not change sign under time reversal.

For a spin 1/2 particle the electric dipole interaction is obtained from the magnetic one (4.89) by duality, that is by exchanging $\boldsymbol{B} \leftrightarrow \boldsymbol{E}$, which corresponds to $F^{\mu\nu} \to -\tilde{F}^{\mu\nu}$ (where $\tilde{F}^{\mu\nu} = \epsilon^{\mu\nu\rho\sigma}F_{\rho\sigma}/2$):

$$H_{\text{int}}^{(E)} = -\frac{d_E}{2}\overline{\psi}\sigma_{\mu\nu}\psi \tilde{F}^{\mu\nu} = i\frac{d_E}{2}\overline{\psi}\gamma_5\sigma_{\mu\nu}\psi F^{\mu\nu} \tag{4.96}$$

where d_E is a constant. Since $T^\dagger \overline{\psi} \gamma_5 \sigma_{\mu\nu} \psi T = -\gamma_5 \overline{\psi} \sigma^{\mu\nu} \psi$ (see eqn (4.50)) and $T^\dagger \tilde{F}^{\mu\nu} T = +\tilde{F}_{\mu\nu}$, this interaction is not symmetric under time reversal unless d_E is pure imaginary, which is not possible in order to have H_{int} Hermitian; it is also not invariant under parity, since $\overline{\psi} \gamma_5 \sigma_{\mu\nu} \psi \rightarrow -\overline{\psi} \gamma_5 \sigma^{\mu\nu} \psi$ from eqn (2.50) and $\tilde{F}^{\mu\nu} \rightarrow \tilde{F}_{\mu\nu}$.

The non-relativistic limit of the interaction (4.96) is indeed $-d_E \langle \boldsymbol{\sigma} \rangle \cdot \boldsymbol{E}$, and indeed EDMs can be detected by the presence of a term linear in the external electric field in the expression for the energy of a system. From this last expression the non-invariance under time reversal is also evident, as $\boldsymbol{\sigma} \rightarrow -\boldsymbol{\sigma}$ while $\boldsymbol{E} \rightarrow \boldsymbol{E}$ (and under parity as well, with $\boldsymbol{E}$ acquiring a minus sign and $\boldsymbol{\sigma}$ not).

The above followed from the fact that in the above expressions the electric dipole moment (a polar vector) turned out to be proportional to spin (an axial vector); while this was obtained above for spin 1/2 particles, it is a general result, as we now show.

The existence of a static electric dipole moment for a non-degenerate system violates both parity and time reversal symmetries: for a state with spin $J > 0$ and no accidental degeneracies (i.e. no degeneracy besides that due to the $2j + 1$ possible orientations of its angular momentum, an 'elementary' particle), the expectation value of a vector operator such as the electric dipole moment $\boldsymbol{d}_E$ for eigenstates of angular momentum J (eigenstates of the free Hamiltonian) can be expressed through the Wigner–Eckart theorem (see, e.g., Sakurai (1985)) as

$$\langle j, m | \boldsymbol{d}_E | j, m \rangle = \langle j, m | \boldsymbol{J} | j, m \rangle \frac{\langle j | \boldsymbol{d}_E \cdot \boldsymbol{J} | j \rangle}{j(j+1)} \tag{4.97}$$

where the second term on the right-hand side is independent from m. As discussed, time reversal symmetry and absence of degeneracy imply that $T | j, m \rangle$ is equal to $| j, -m \rangle$ to within a phase, eqn (4.30); by using the time reversal transformation properties of a real $\boldsymbol{d}_E$ ($T^\dagger \boldsymbol{d}_E T = \boldsymbol{d}_E$) and $\boldsymbol{J}$ (4.24)

$$\langle j, -m | \boldsymbol{d}_E | j, -m \rangle = -\langle j, -m | \boldsymbol{J} | j, -m \rangle \frac{\langle j | \boldsymbol{d}_E \cdot \boldsymbol{J} | j \rangle}{j(j+1)} \tag{4.98}$$

and therefore, exchanging $m \rightarrow -m$ and comparing with eqn (4.97), one must conclude that $\langle j, m | \boldsymbol{d}_E | j, m \rangle = 0$.

For a particle with no spin one can note instead that the same state (to within a phase) is obtained by rotation of π around any axis, which changes sign to the vector $\boldsymbol{d}_E$ for an axis orthogonal to it, so that rotational symmetry alone implies in this case $\langle \boldsymbol{d}_E \rangle = 0$.

In less rigorous but more intuitive terms, for a non-degenerate system a perma-nent electric dipole moment must be directed along its spin vector $\boldsymbol{S}$, being the only preferred direction in space; if one could specify another (independent) preferred direction, the system would be degenerate because additional quantum numbers would be required to specify its state (if this were the case for a neutron or an

electron, the Pauli principle would allow nuclei and atoms with twice as many constituents as normal). But under the time reversal transformation the spin (pseudo-)
vector S changes sign while the vector d_E does not, in order for the interaction term
$d_E \cdot E$ to respect time reversal symmetry, showing that the EDM must vanish. As
seen in Chapter 2, the same conclusion can be obtained by considering symmetry
under the parity transformation, which also transforms in the opposite way the spin
axial vector and the electric dipole moment vector.

Still another way of seeing the above result is in connection with Kramers'
degeneracy discussed in Section 4.2: for a particle with an EDM the degeneracy among its $2j + 1$ spin states would be entirely lifted by placing it in a
T-symmetric static electric field, thus contradicting a consequence of time reversal
symmetry.

The energy W of an atom in external electric (E) and magnetic (B) fields is
unchanged by a π rotation so that $W(E, B) = W(-E, -B)$, while time reversal
symmetry would require $W(E, B) = W(E, -B)$, so that $W(E, B) = W(-E, B)$;
any term $-d_E \cdot E$ is therefore not admissible in this case.[57]

More generally, it can be shown that both parity and time reversal symmetry separately imply that non-
degenerate systems can only have even electric and odd magnetic multipole moments. The multipole
moments are defined as the expectation values of the corresponding tensor operators $O^{(J,M)}$ in a state
of angular momentum j with $m = j$:

$$d_J = \langle \alpha, j, j | O^{(J,0)} | \alpha, j, j \rangle$$

where α labels other quantum numbers (a 2^J pole can occur only for a system with spin $j \geq J/2$).
Time reversal symmetry and the Hermiticity of $O^{(J,M)}$ give

$$\langle \alpha, j, j | O^{(J,0)} | \alpha, j, j \rangle = \langle \alpha, j, -j | TO^{(J,0)} T^\dagger | \alpha, j, -j \rangle$$

and using the Wigner–Eckart theorem and eqn (4.29)

$$TO^{(J,0)} T^\dagger = (-1)^J O^{(J,0)}$$

For electric multipoles O_E, originating from charge distributions

$$TO_E^{(J,M)} T^\dagger = (-1)^M O_E^{(J,-M)} \tag{4.99}$$

while for magnetic multipoles O_M, originating from currents and magnetizations

$$TO_M^{(J,M)} T^\dagger = (-1)^{M+1} O_M^{(J,-M)} \tag{4.100}$$

from which the stated result follows.

Despite the fact that unstable particles are not transformed into themselves by the time reversal
operation, the conclusion also holds in this case (Bell, 1962).

---

[57] Parity symmetry would also require $W(E, B) = W(-E, B)$, and thus $W(E, B) = W(E, -B)$,
so that a magnetic dipole interaction term $-d_M \cdot B$ is allowed, provided d_M is an axial vector (which
is indeed the case if it is proportional to spin).

TABLE 4.2 Transformation properties of magnetic ($\mu \equiv d_M$) and electric (d_E) dipole moments under the different discrete transformations.

	P	C	CP	T
μ	μ	$-\mu$	$-\mu$	$-\mu^\dagger$
d_E	$-d_E$	$-d_E$	d_E	$d_E^\dagger$

The existence of a permanent EDM for an elementary system would be therefore an indication of the failure of P and T symmetries,[58] (and therefore also CP and CT if the CPT theorem is valid), as summarized in Table 4.2.

Before discussing EDM searches as tests of time reversal symmetry, one point has to be clarified. A composite system can have an electric dipole moment due to its internal structure, without violating P and T symmetries: it is well known that polar molecules (such as water) can have large 'electric dipole moments': these are actually *induced* electric dipole moments which apply to the situation in which the external electric field is strong enough to orient the molecule, and have nothing to do with time reversal violation, being originated by the presence of degeneracies as sketched below.

For a bi-atomic molecule, any elementary EDM should be directed along the molecular axis, while since the orbital angular momentum L is orthogonal to such axis, any state of definite L has zero EDM. Similar considerations can be made for polyatomic molecules (Khriplovich and Lamoreaux, 1997): in general the non-degenerate states are symmetric or antisymmetric superpositions of energy eigenstates with opposite orientations, which have no net EDM.

For a system with a pair of levels with opposite parities $|+\rangle$ and $|-\rangle$ which are degenerate (or almost degenerate, with energies E_+, E_- such that $|E_+ - E_-| < kT$), the application of an external electric field E can induce (parity-conserving) transitions between the two states, mixing them into new energy eigenstates containing parity impurities of order $\Delta \equiv |\langle -|T|+\rangle||E|$ with energy eigenvalues

$$E = \frac{1}{2}(E_+ + E_-) \pm \sqrt{(E_+ - E_-)^2/4 + \Delta^2} \simeq \frac{1}{2}(E_+ + E_-) \pm \Delta \qquad (4.101)$$

for $|E_+ - E_-| \ll |\Delta|$. For exact initial degeneracy the energy difference of the two levels is linear in the electric field magnitude $|E|$: degenerate systems can exhibit electric dipole moments. Note however that degeneracy is never exact in practice, and atomic ground states are expected to be non-degenerate; for measurements

---

[58] It should be noted that this conclusion could be invalidated if magnetic monopoles existed (Ramsey, 1958): the circulation of monopoles could generate an electric dipole moment with the proper polar vector nature, being given by the product of angular momentum and the pseudo-scalar magnetic pole strength.

performed at very low temperatures in very weak external fields the energy levels
are actually

$$E_1 = E_+ + \frac{\Delta^2}{E_+ - E_-} + \cdots \qquad E_2 = E_- - \frac{\Delta^2}{E_+ - E_-} + \cdots$$

and the energy shift is actually proportional to Δ^2, i.e. quadratic in E; such induced
EDMs do not violate P or T symmetries and are actually enhanced by the small
energy difference between states of opposite parity.

The 'true' EDM is the one defined in the weak-field limit, which refers to a pure
state.[59]

For neutrons the absence of further degeneracies is supported by the fact that they
obey the Pauli principle; moreover, Kramers' theorem requires that the eigenvalues
are at least twofold degenerate for a fermion system described by a time reversal
symmetric Hamiltonian, and since a free neutron (as any spin 1/2 particle) has only
two eigenstates, these remain degenerate also in the presence of an electric field.

On dimensional grounds EDMs are given by $|d_E| \sim Q l f$, where Q is electric
charge, l a length scale for the charge displacement and f a factor parametrizing
the violation of T. For an effect linked to weak interactions the natural length scale
is $l \sim G_F m \sim (G_F m^2) \times (1/m)$ with a suitable mass m, and setting for a nucleon
$|Q| = |e|$ and $m = m_N$, so that $l \sim 10^{-5} \times \lambda_N$ with the Compton wavelength
$\lambda_N = 1/m_N \simeq 2 \cdot 10^{-14}$ cm ($\ll r_N \simeq 10^{-13}$ cm). Taking for f the amount of CP
(and T) violation in weak interactions measured in kaon decays ($f \sim 2 \cdot 10^{-3}$) one
obtains as an order of magnitude estimate $|d_E| \sim 5 \cdot 10^{-22} |e|$ cm. Experimental
limits (and theoretical predictions) for the neutron EDM are much lower than this
order of magnitude estimate.

4.4   Time reversal violation

After the fall of parity, physicists started enquiring about the validity of the other
discrete symmetries: while charge conjugation fell together with parity, tests of
time reversal symmetry at the per cent level showed no evidence for violation.

As mentioned, the non-Hermiticity of the transition matrix implies that no *single*
experiment tests time reversal symmetry, and one should rather compare 'reciprocal'
experiments: this generally restricts the experimental tests to reactions with only
two particles in the final state.

It is important to remember that in searches for time reversal violation fake
asymmetries can be induced by final state interaction effects, as discussed. While

---

[59] It should be noted (Sachs, 1987) that, as is usually the case with time reversal, some care is
required in the attribution of T asymmetric effects to microscopic T violation: Ohm's law $j = \sigma E$,
where j is the T-odd current, E the T-even electric field and σ the (T-even) electric conductivity, is
not invariant for time reversal, just because it arises from a dissipative process.

the spurious effects might in principle be computed if the underlying theory of the T-conserving process is well understood, their size is usually much larger than that of the true T-violating effects being searched for, and the uncertainties in such computations set a limit on the sensitivity of this kind of tests. As will be seen in the following, searches for T violation usually involve the measurement of some asymmetry when some quantity (an external electric field, a polarization, etc.) is reversed; as a general result, a measurement is immune from T violation mimicry by final state interactions if (and only if) such reversal is equivalent to a time reversal operation, that is if it results in a process which looks like the original one when time is run backwards (Schäfer and Adelberger, 1991).

Among the measurements which can provide evidence of violations of time reversal symmetry are:

- A non-zero expectation value for a T-odd observable in a stationary state, or in the final state of a decay into weakly interacting particles (that is, when final state interactions can be neglected).

- A difference in the expectation values of two observables related by time reversal, in a reaction and its inverse.

- A violation of the reciprocity relation (4.63).

- A difference in the transition probability among two states which are invariant under 'naive' time reversal ($i_T = i, f_T = f$): $P(i \to f) \neq P(f \to i)$ at a given time.

The first (and so far unique) direct measurement of time reversal violation in weak interactions was obtained from the study of virtual transitions of neutral kaons, and will be discussed in Section 8.4.

4.4.1 Measurements of T-odd correlations

The spin-momentum correlation $\langle S \cdot p \rangle \neq 0$ measured in the beta decay experiments (Wu $et\ al.$, 1957; Garwin $et\ al.$, 1957) is of type $O^{(P-,T+)}$ in the classification of Section 4.3.3. Independently of the explicit form of the interaction Hamiltonian, the measurement of such non-zero correlation showed that the symmetry PT is violated: this is described by an anti-unitary operator, and its validity would require, since in this case final state interactions (Coulomb interaction between the electron and the proton) are relatively small, the equality of rates for the transitions $i \to f$ and $i_{PT} \to f_{PT}$ (from eqn (4.74) in the case $\mathcal{T} \simeq \mathcal{T}_W$); the PT-inverted state would have the opposite value of $\langle S \cdot p \rangle$, since spin is inverted and momentum is not, and electrons could not show an asymmetry in their emission direction parallel or anti-parallel with respect to the nuclear spin. Since the experiments showed that P is violated, the violation of PT (equivalent to violation of C if CPT symmetry is valid) when final state interactions are small does not allow one to conclude anything concerning the validity of T symmetry for weak interactions. In order to perform tests of time reversal symmetry in nuclear beta decay, one has to measure either the nuclear recoil momentum (giving the neutrino momentum) or the electron polarization, as discussed originally in Jackson *et al.* (1957b) (see also the discussion of the free polarized neutron decay experiment in Chapter 2).

We stress once more that the measurements of expectation values of T-odd quantities are good tests of time reversal symmetry only for weakly interacting particles, since FSI can generate non-zero values even if such symmetry is valid.

It should be noted that scalar T-odd correlations built out of momentum and spin vectors always involve at least three vector quantities; since spins are usually more difficult to measure than momenta, the simplest observable is $\mathbf{S} \cdot \mathbf{p}_1 \times \mathbf{p}_2$ (a polarization transverse to the plane containing the two momenta $\mathbf{p}_1, \mathbf{p}_2$). In the simplest case of the decay of a spinless particle into two particles with spin, such as $\pi \to \mu\nu$ decay, the above correlation vanishes identically due to Lorentz and rotational symmetry: the measured quantity must be a Lorentz invariant (a symmetry violation must be the same in any frame), and the expression $\mathbf{p}_\pi \times \mathbf{p}_\mu \cdot \mathbf{S}_\mu = \mathbf{p}_\nu \times \mathbf{p}_\mu \cdot \mathbf{S}_\mu$ (since $\mathbf{p}_\pi = \mathbf{p}_\mu + \mathbf{p}_\nu$), when boosted to the centre of mass frame (with a boost lying in the plane defined by $\mathbf{p}_\pi$ and $\mathbf{p}_\mu$, which does not affect the transverse polarization (see Appendix E) becomes $\mathbf{p}_\nu^{(CM)} \times \mathbf{p}_\mu^{(CM)} \cdot \mathbf{S}_\mu$ which vanishes because $\mathbf{p}_\nu^{(CM)}$ and $\mathbf{p}_\mu^{(CM)}$ are parallel.[60] One needs at least a three-body decay of a spinless particle to have a non-zero T-odd correlation of this kind.

An example is the measurement of the charged lepton polarization transverse to the decay plane in the weak semi-leptonic decay of kaons $K \to \pi \ell \nu$ ($\ell = e, \mu$, often denoted as $K_{\ell 3}$). The charged lepton polarization can be described by three components:

- **Longitudinal**, parallel to the charged lepton momentum:
  $$\mathcal{P}_L = \mathcal{P} \cdot \hat{\mathbf{u}}_L = \mathcal{P} \cdot \mathbf{p}_\ell / |\mathbf{p}_\ell|;$$

- **Transverse**, perpendicular to the decay plane:
  $$\mathcal{P}_T = \mathcal{P} \cdot \hat{\mathbf{u}}_T = \mathcal{P} \cdot (\mathbf{p}_\pi \times \mathbf{p}_\ell) / |\mathbf{p}_\pi \times \mathbf{p}_\ell|;$$

- **Normal** to the charged lepton momentum in the decay plane:
  $$\mathcal{P}_N = \mathcal{P} \cdot \hat{\mathbf{u}}_N = \mathcal{P} \cdot [\mathbf{p}_\ell \times (\mathbf{p}_\pi \times \mathbf{p}_\ell)] / |\mathbf{p}_\ell \times (\mathbf{p}_\pi \times \mathbf{p}_\ell)|.$$

Here $(\hat{\mathbf{u}}_L, \hat{\mathbf{u}}_T, \hat{\mathbf{u}}_N)$ form a right-handed orthonormal set of vectors, and the transverse polarization $\mathcal{P}_T$ is readily seen to be the T-odd quantity.

---

[60] One can also see that the covariant quantity corresponding to the above triple product must be a Lorentz scalar built with the completely antisymmetric tensor $\epsilon_{\mu\nu\delta\sigma}$, but with only two independent momenta and one spin four-vector no such quantity can be formed.

The amplitude for the $K \to \pi \ell \nu$ decay can be written in the SM as a sum of two terms, containing two arbitrary functions (form factors) $f_{+,-}$, one of them proportional to m_ℓ/m_K, and therefore negligible for $\ell = e$; $\mathcal{P}_T$ is proportional to the T-violating phase difference between the two.

Precision measurements of the transverse muon polarization have been performed both for $K_L \to \pi^- \mu^+ \nu$ ($K_{L\mu 3}$) (Morse *et al.*, 1980) and for $K^+ \to \pi^0 \mu^+ \nu$ ($K^+_{\mu 3}$) (Abe, 2006b) decays. The second case is more favourable for the search of small violations of time reversal symmetry, since the interactions among the final state particles are much weaker: their effect is estimated to contribute a transverse polarization below $\sim 10^{-5}$ in the SM, and experiments have not yet reached such level of sensitivity (the true T-violating contribution due to the known CP violation in the SM is at the level of 10^{-7}). On the contrary, the experiments performed on K_L decay already reached a sensitivity of a few 10^{-2} on $\mathrm{Im}(f_-/f_+)$, comparable to the value expected from electromagnetic FSI in the SM (~ 0.008), so that this decay is no longer competitive for the test of T symmetry in weak interactions.

Muon transverse polarization in semi-leptonic K decays

The muon polarization can be conveniently measured by the (parity-violating) asymmetry in the angular distribution of the decay electron: such distribution has the form $dN/d\theta = N_0(1 + \mathcal{A}\mathcal{P}_\mu \cos\theta)$, where $\mathcal{P}_\mu$ is the muon polarization, θ the electron emission angle with respect to the muon polarization direction, and $\mathcal{A}$ the analysing power ($\mathcal{A} = 1/3$ in the ideal case). Positive muons are used uniquely, since the interactions of muons with the material used to slow them leads to much stronger depolarization effects for μ^-.

The most recent experiment is E246 at KEK in Japan (Abe, 2006b), which used the decay of stopped K^+: in this way, at the price of having a smaller flux and reduced acceptance, an important constraint on the decay kinematics is available ($p_K = 0$), which allows a precise measurement of the decay directions and an effective suppression of the background.

The detector, shown schematically in Fig. 4.2, consisted of a segmented crystal electromagnetic calorimeter for measuring the π^0 energy and direction, a tracking spectrometer for measuring the μ^+ momentum and direction, and a muon polarimeter. All detectors had a twelve fold azimuthal symmetry around the direction of the incoming beam, and the averaging over such sectors helped in reducing systematic errors; the use of stopped kaons allows the detector to be off the beam axis, thus reducing the backgrounds.

A 660 MeV/c ($\pm 0.3\%$) positive hadron beam, produced by 12 GeV protons impinging on a platinum target, travelled along the axis of the experiment. Incoming K^+ (about 10^5 per second) were tagged by a differential Čerenkov counter to separate them from pions, still six times more abundant than kaons in the beam after electrostatic separation. They were then slowed in a Al-BeO degrader and stopped in a target; the use of a low Z material resulted in high stopping power with minimal multiple scattering, and reduced the interactions of photons from the target.

The beam momentum was chosen to give the maximum kaon stopping rate in the active target ($\sim 40\%$ efficiency), consisting of a cylindrical array of 256 scintillating fibers parallel to

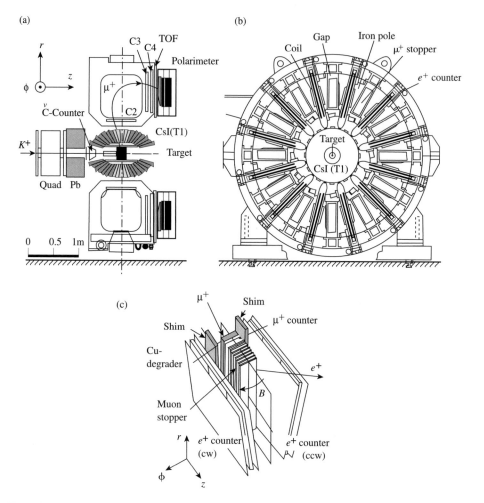

FIG. 4.2. Schematic drawing of the KEK-E246 experimental apparatus (Abe, 2006*b*). (a) side view, (b) end view, (c) one sector of the polarimeter.

the beam direction (9 cm overall diameter, 20 cm fiducial length along beam direction), which measured the transverse coordinates of the stopping point. It was surrounded by 32 plastic scintillator rings coaxial with the beam (6 mm length each) which measured the longitudinal coordinate of the decay muon. Stopped kaons decayed in the target, to $\pi^0\mu^+\nu$ in 3.3% of the cases, and kaon decays in flight were suppressed by requiring a minimum time delay of 2 ns between their arrival in the target and the detection of a decay particle.

The target was surrounded by an electromagnetic calorimeter, consisting of a 1 m diameter barrel of 13.5 radiation length thick CsI(Tl) scintillating crystals with projective geometry, covering $\sim70\%$ of the solid angle and providing a 4.3% photon energy resolution at 100 MeV. Muons emitted around the radial direction $\hat{r}$ (orthogonal to the beam axis) could pass through twelve azimuthal holes in the EM calorimeter, crossing one of twelve lon-gitudinal plastic scintillator counters used for time-of-flight measurement, and were then momentum analysed in a spectrometer, with a superconducting magnet providing a 0.9 T toroidal magnetic field and twelve sets (one per sector) of three multi-wire proportional chambers. The toroidal field did not rotate the azimuthal component of the muon polariza-tion (parallel to the field). This 'longitudinal field' approach gives a higher statistical sensitivity with respect to the alternative one used in other experiments, in which the polarization component of interest is allowed to precess in a 'transverse field' and its phase is measured from the decay distribution. While the method used is more sensitive to instrumental mis-alignments, the comparison of two opposite asymmetries allows a cancellation of induced spurious effects.

Muons were bent by approximately 90° in the forward direction, passed time-of-flight scin-tillator counters and reached a sector of the polarimeter, consisting of a copper degrader to slow the particles and make them stop in a stack of pure aluminium plates (stopping efficiency $\sim 80\%$). The degrader was wedge-shaped to compensate for the momentum dispersion: higher momentum muons hit the degrader at larger radius, where it was thicker, so that particles of different momenta stopped approximately at the same depth in the aluminium.

Relativistic effects in Coulomb scattering with nuclei can depolarize muons: the depo-larizing power per unit length is proportional to the square of the multiple scattering angle $-d\mathcal{P}/dl \propto Z^2/A$, while the energy loss is $-dE/dl \propto Z/A$ so that the ratio of depolarization to energy loss is $\propto Z$ and a light material was used. Muon depolarization can also occur due to local internal magnetic fields, muonium formation and nuclear interactions; a nonmagnetic metal is required (muonium formation is absent in a pure metal), and pure (99.9%) alu-minium is chosen because depolarization effects are known to be small from spin rotation experiments (measurements of the damping rate of the parity-violating forward/backward asymmetry in the decay of stopped muons precessing in a magnetic field).

For events whose decay plane contains the beam axis, a transverse muon polarization $\mathcal{P}_T$ appears as an azimuthal component of the polarization around such axis; any such component originates a clockwise-counterclockwise asymmetry in the distribution of decay positrons:

$$A_T = \frac{N^{(\text{cw})} - N^{(\text{ccw})}}{N^{(\text{cw})} + N^{(\text{ccw})}} = \mathcal{P}_T \, \mathcal{A} \, \langle \cos \theta_T \rangle$$

where θ_T the angle between the normal to the decay plane and the azimuthal direction $\hat{\phi}$, and $\mathcal{A}$ the analysing power.

Plastic scintillator sheets parallel to the beam axis on both sides of the aluminium stack detected the decay positrons and measured their times to allow reducing the accidental background, while several veto counters around the polarimeter helped to identify good stopping muons; the time interval between 20 ns and 6 μs after a muon stop was used. Each positron counter was used for two adjacent sectors of the polarimeter, so that the effect of inefficiencies cancelled on average in the asymmetry.

Iron plates guided the fringe magnetic field of the spectrometer in the polarimeter (130 G average) so that it was properly directed along the azimuthal direction. The guide field was not exactly azimuthal at all positions in the polarimeter, so that locally $\mathcal{P}_N$ and $\mathcal{P}_L$ components were mixed with $\mathcal{P}_T$ during precession: the symmetry of the field around the median plane of the magnetic gap was therefore very important for making any instrumental asymmetry due to such effect vanish on average.

The analysing power $\mathcal{A}$ was affected by the finite size of the scintillator counters, the scattering and absorption in the muon stopper and the spin precession around non-azimuthal magnetic fields. It was measured on the (large) component of the muon polarization in the decay plane $\mathcal{P}_N$: such component (about 0.6 when averaged over the accepted phase space volume) was measured with the same apparatus using the events with a π^0 emitted in a radial direction (so that $\mathcal{P}_N$ is in the azimuthal direction), by forming the asymmetry

$$A_N = \frac{N^{(L)} - N^{(R)}}{N^{(L)} + N^{(R)}} = \mathcal{P}_N \, \mathcal{A}$$

where L, R superscripts label the left or right π^0 direction with respect to the plane containing the radially emitted muon and the detector axis (i.e. the sign of $\boldsymbol{p}_\pi \cdot \boldsymbol{p}_K \times \boldsymbol{p}_\mu$). The average analysing power was $\mathcal{A} \simeq 0.27$, determined from the ratio of asymmetry to polarization from Monte Carlo simulation.

$K^+_{\mu 3}$ decays were selected based on the charged particle momentum, its time of flight between target and polarimeter, and the invariant mass of the two photons in the electromagnetic calorimeter; the residual background due to $\pi^+ \to \mu^+ \nu$ decays was below 10%. Background in the detector and in the polarimeter diluted the $\mathcal{P}_T$ measurement by $\sim 10\%$ but could not create any spurious asymmetry. The overall acceptance (about 8%) was optimized taking into account the sensitivity of $\mathcal{P}_T$ to the T-violating parameter $\mathrm{Im}(f_-/f_+)$, which varies by an order of magnitude over the phase space.

The two classes of events with π^0 emitted parallel or anti-parallel to the beam, give opposite transverse polarization asymmetries (see Fig. 4.3), and by forming the difference of the asymmetries for forward (polar angle $\theta < 70°$) and backward ($\theta > 110°$) π^0 events the effect due to $\mathcal{P}_T$ adds up, resulting in a doubled sensitivity

$$A_T^{\text{(forward)}} - A_T^{\text{(backward)}} \simeq 2\mathcal{P}_T \, \mathcal{A} \, \langle \cos \theta_T \rangle$$

while most of the spurious instrumental asymmetries cancel, such as those due to the imperfect alignment of the magnetic field in the polarimeter with respect to the detector. With this approach there is no need to invert the direction of the magnetic field (which was not possible in this experiment and is anyway difficult to do in a highly symmetrical way) to keep systematics under control.

The sum of the forward and backward asymmetries provides instead a null check, since its difference with respect to zero can only be induced by the asymmetry of the instrumental apparatus. The average over the twelve azimuthal sectors also reduces spurious asymmetries. The intrinsic left-right asymmetry in the polarimeter, due to its finite dimension in the azimuthal direction, was eliminated by averaging the asymmetries computed for events corresponding to the same azimuthal offset in the polarimeter, performing the analysis in bins. The distribution of the decay plane angle was also checked to be symmetric with respect to the polarimeter axis, since any asymmetry would indicate an admixture of $\mathcal{P}_N$ (in-plane) polarization.

The analysis was performed both on fully reconstructed π^0 events (6.3 million) and on events in which only one photon with energy above 70 MeV (sufficient to determine the π^0

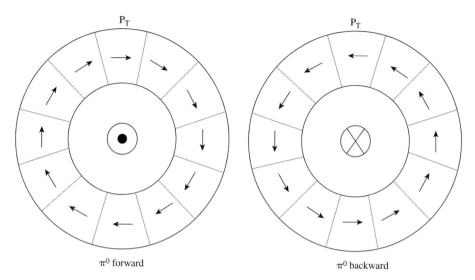

π^0 forward π^0 backward

FIG. 4.3. Comparison of the transverse muon polarization $\mathcal{P}_T$ for $K^+_{\mu3}$ events with forward and backward emitted π^0 in the KEK-E246 experiment (Abe, 2006b) (the muon is emitted in the radial direction).

direction to some accuracy) was detected (5.5 million); two independent analysis on both data sets gave consistent results.

The left-over systematic effects were due to the imperfect cancellation of effects in the forward/backward asymmetry difference: residual misalignments of detector elements, the magnetic field not being perfectly azimuthal, and the coupling of any difference in the azimuthal dependence of stopping muons for forward and backward events (induced, e.g., by muon multiple scattering in the degrader) with asymmetries in the polarimeter geometrical asymmetry, the latter two being the largest single effects.

The final result from 5 years of data taking was $\mathcal{P}_T = (-1.7 \pm 2.3 \pm 1.1) \cdot 10^{-3}$. This result shows no evidence of time reversal violation; the experimental error is still two order of magnitudes larger than the intrinsic limit of this measurement due to final state interaction effects.

With the same set-up a zero result at the level of 10^{-2} was also obtained for the transverse muon polarization in the $K^+ \rightarrow \mu^+\nu\gamma$ decay (Anisimovsky, 2003): $\mathcal{P}_T = (-0.64 \pm 1.85) \cdot 10^{-2}$, for which final state interaction effects are expected to be below 10^{-3} in the SM.

Muon decay $\mu \rightarrow e\nu\bar{\nu}$, a purely leptonic process, has been studied in great detail by many experiments. Since the two neutrinos are not observed, the electron polarization is conveniently analysed with respect to the plane defined by the electron momentum $\boldsymbol{p}_e$ and the muon polarization $\boldsymbol{S}_\mu$. As final state interactions are weak, the measurement of the electron (positron) polarization transverse to such decay plane, corresponding to the T-odd quantity $\langle \boldsymbol{S}_e \cdot \boldsymbol{p}_e \times \boldsymbol{S}_\mu \rangle$, is a good test of time reversal symmetry. Such polarization was measured to be consistent with zero (Burkard et al., 1985) in μ^+ decay, $\mathcal{P}_T = 0.007 \pm 0.023$, by exploiting the spin dependence of the positron annihilation cross section on polarized electrons in a magnetized foil.

Other T-odd triple product correlations were measured in weak decays: in the decay of a spinless particle to two vector mesons $A \to B + C$ the quantity $\xi = \boldsymbol{p}_B \cdot (\boldsymbol{S}_B \times \boldsymbol{S}_C)$ can be formed, and a non-zero average value would show up statistically as an asymmetry in the number of events corresponding to $\xi > 0$ and $\xi < 0$. Again, this quantity can be shown to be different from zero if the amplitudes contributing to the decay (corresponding to different helicity states) are not relatively real and interfere. If the secondary particles are unstable, their polarization is measured by the asymmetry of the angular distribution of their decay products. In general ξ will be different from zero due to the final state interactions of particles B and C; still, if CPT symmetry (Chapter 5) is assumed, the difference of the measured T-odd correlations for the decays of particles and antiparticles could be used to provide a true T-violating asymmetry independent of FSI.

As an example, in the decay $B^0 \to \phi K^{0*}$ with the subsequent decays $\phi \to K^+ K^-$ and $K^{0*} \to K^+ \pi^-$, two T-odd triple products were measured in a B-factory experiment (Chen et $al.$, 2005) to be different from zero at the level of 2.5 standard deviations, but they appear to be equal within errors for the decays of B^0 and $\overline{B}^0$ mesons.

T-odd triple products can also be formed with momentum variables only: $\boldsymbol{p}_1 \cdot (\boldsymbol{p}_2 \times \boldsymbol{p}_3)$. Not involving spin, these correlations are easier to measure, especially in high-energy processes, but due to momentum conservation they vanish identically for a three-body decay; at least four distinguishable particles are needed in the final state of a decay to build a non-zero triple product of momenta; this generally requires studying somewhat rare decay modes. One example (Gervais et $al.$, 1966) is the quantity $\boldsymbol{p}_\pi \cdot (\boldsymbol{p}_\ell \times \boldsymbol{p}_\gamma)$ in the radiative semi-leptonic decay of kaons $K^\pm \to \pi^0 \ell^\pm \nu \gamma$ (branching ratio $\sim 2 \cdot 10^{-4}$ for $\ell = e$, $\lesssim 6 \cdot 10^{-5}$ for $\ell = \mu$). The current experimental value (Bolotov et $al.$, 2005) for the asymmetry in the above T-odd quantity is $A_T = (-1.5 \pm 2.1) \cdot 10^{-2}$ for K^- decays, in the phase space region corresponding to photon energies above 10 MeV and angles (with respect to the electron) between $26°$ and $53°$ in the centre of mass frame (radiative decays always require qualifications such as these, since the cross section diverges for soft and collinear photons, in the region where the radiative decays cannot possibly be distinguished from non-radiative ones).

Time reversal symmetry is also studied with triple correlations involving one or two spin vectors in beta decay: the differential decay rate can be written as (Jackson et $al.$, 1957b)

$$\frac{d\Gamma}{dE_e \, d\Omega_e \, d\Omega_\nu} \propto \left[1 + a \frac{\boldsymbol{p}_e \cdot \boldsymbol{p}_\nu}{E_e E_\nu} + b \frac{m_e}{E_e} + \mathcal{P}_n \cdot \left(A \frac{\boldsymbol{p}_e}{E_e} + B \frac{\boldsymbol{p}_\nu}{E_\nu} + D \frac{\boldsymbol{p}_e \times \boldsymbol{p}_\nu}{E_e E_\nu} \right) \right]$$

$$(4.102)$$

(where p, E are the momenta and energies of the decay products, $\mathcal{P}_n$ is the neutron polarization and a, A, B, D are constants). The terms with coefficients $A \simeq -0.1$ and $B \simeq 1$ are parity-violating, while the last term is T-odd and can be used to test time reversal symmetry, after taking into account the effect of final state interactions; the D coefficient is indeed proportional to the relative phase of the vector and axial vector form factors (constants, in the limit of zero momentum transfer): $D \propto \mathrm{Im}(g_A/g_V)$. Since (EM) FSI contribute additional T-odd terms which are proportional to Zm_e/p_e, where Z is the charge of the residual nucleus, the study of light nuclei with high energy release is preferred. Within the SM, the effect of FSI is estimated to give contributions of order 10^{-5} in D for a free neutron (the true T-violating effect consistent with the known amount of CP violation is of order 10^{-12}), while recent experiments reached a precision of a few 10^{-4} (Yao *et al.*, 2006).

If the electron polarization is measured, other nuclear T-odd correlations can be formed, such as the triple product of nuclear spin, electron spin, and electron momentum; this is a P-odd quantity, and as such it is linked to weak interaction effects, as opposed to the D term in eqn (4.102) which is P-even and can probe effects linked to EM interactions. The expression for the beta decay rate when the electron polarization is observed contains (among others) the terms (Jackson *et al.*, 1957b)

$$\frac{d\Gamma}{dE_e \, d\Omega_e \, d\Omega_\nu} \propto 1 + \mathcal{P}_n \cdot \left(A \frac{p_e}{E_e} + D \frac{p_e \times p_\nu}{E_e E_\nu} + R \frac{p_e \times \mathcal{P}_e}{E_e} \right) \qquad (4.103)$$

with A the parity-violating asymmetry parameter and D, R the T-odd parameters (P-even the first, P-odd the second). P-odd, T-odd correlations such as that related to the R term are generally more difficult to measure experimentally because two polarization measurements are required.

Transverse polarization in beta decay

In one experiment at PSI (Sromicki *et al.*, 1996) the beta decay of ^{8}Li was studied: due to the low charge ($Z = 4$) and relatively high decay energy ($E_{max} = 13.1$ MeV), FSI effects are particularly small for this nucleus (below $1 \cdot 10^{-3}$ on R). The ^{8}Li $\to ^8$Be beta decay ($J^P = 2^+ \to 2^+$, isospin $I = 1 \to 0$) is dominated by Gamow–Teller type of transitions, therefore eliminating some nuclear structure uncertainties in the interpretation of the results.

The nuclei were polarized by bombarding a ^{7}Li target with a beam of polarized 10 MeV deuterons, through the reaction $d\,^7$Li $\to p\,^8$Li: the target (of a few mm size) was irradiated for 0.33 s and then beta decays (with a half-life of 0.84 s) were measured for 1 s. The polarization could be easily reversed (at a 0.15 Hz rate) by inverting the deuteron polarization direction, therefore providing an important cancellation of systematic effects induced by detector drifts. Working at relatively low beam energies limited the amount of background reactions present: ^{6}He and ^{17}F beta decays were present, but they could be easily removed by a detector threshold since the electron end-point energies are lower than for ^{8}Li, leading to a contamination level below 1%; the electron energy spectrum was also measured and found to be in good agreement with the one expected for beta decay.

From a beam polarization of about 50% a vertical polarization around 11% was obtained for the target. Such polarization was maintained by keeping the cooled (77 K) target in a 7 mT magnetic field, and continuously monitored by measuring the up-down asymmetry of beta decay electrons in two semiconductor Si detectors placed at $\approx 45°$, exploiting the $d\,^{12}C \rightarrow p\,^{13}C$ reaction. The relaxation time of the target polarization was measured to be 3.5 s (much longer than the nucleus lifetime), and slow drifts (below 1%/day) due to temperature effects were also detected.

The electron transverse polarization was measured by the up-down asymmetry in Mott scattering on a lead foil. A significant analysing power is obtained for large angle scattering (where cross sections are low), and a polarimeter with approximate axial symmetry around the target polarization axis was developed: this geometry not only had a large acceptance, both for electrons emitted from the source and scattered ones, but also allowed the suppression of some systematic effects. The polarimeter (see Fig. 4.4) comprised two 90° segments downstream of the target and two 60° segments upstream, each with a curved lead scattering foil and a set of counters accepting electrons scattering at angles 124°–161°. Scattering foils of different vertical size (5–8 cm) and thickness (24–30 μm) were used: while the rate increases with foil thickness, depolarization effects (more important for low energy electrons) prevented the use of thick foils. The average analysing power of the polarimeter was $\mathcal{A} = -10\%$ (known to 8% relative precision).

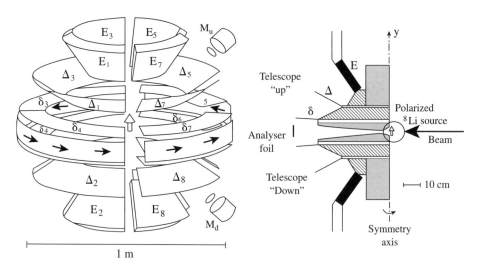

FIG. 4.4. Sketch of the setup used for the measurement of the transverse electron polarization in beta decay (Sromicki *et al.*, 1996). The beam enters horizontally, the target and polarized ^{8}Li source is indicated by the thick vertical arrow (parallel to the polarization). Electrons emitted in the horizontal plane are scattered by the foil and detected in the δ, Δ and E layers of the scintillator hodoscope. Left: the polarimeter; the $M_{u,d}$ counters are used for polarization monitoring. Right: vertical cross section of the apparatus.

The three-layer plastic scintillator hodoscopes were shielded from the target by brass collimators in such a way that a particle emitted from the latter had to undergo at least two scatterings to reach the detector (lead, while more effective in absorbing electrons and *bremsstrahlung* photons, would have induced a higher back-scattering rate); the signal to noise ratio obtained was 14:1 (7% background). The third layer of scintillator detectors was thick (4 cm, compared to 2 mm for the first two layers) and used for energy measurement with a resolution better than 17%.

Since the measured asymmetry is proportional to the value of the R parameter of interest multiplied by the magnitude of the target polarization and the polarimeter analysing power, a 10^{-5} precision on the asymmetry was required to reach a $\sim 10^{-3}$ error on R, and systematic effects had to be limited to the level of a few 10^{-6} in the asymmetry. Any geometrical imperfections or misalignments of the polarimeter detectors leading to spurious up-down asymmetries are independent of the target polarization, so that by averaging the results for the two orientations of nuclear spin they could be cancelled. A polarimeter misalignment can produce transverse polarization components which reverse with the target polarization through a T-even $\mathcal{P}_n \cdot \mathcal{P}_e$ correlation in the decay rate; this correlation is however small, and the asymmetry originating from different regions of the scattering foil are opposite in sign, leading to a negligible effect.

A non-uniform vertical illumination of the scattering foil arises from the parity-violating beta decay asymmetry, leading to different detector acceptances for the upper and lower hodoscopes even in the case of perfect alignment; this effect reverses with the target polarization, therefore simulating a time reversal violation which is not cancelled by polarization reversal. The linear dependence of the effective cross section times acceptance on the polar angle of electron emission was measured with a scattering foil of small vertical size (1 cm) displaced in steps from the median plane; by integrating the acceptance-weighted asymmetric decay distribution over the size of the scattering foil the spurious asymmetry effect was determined with an error determined by the uncertainty of the above linear slope.

By forming the double ratio of the up-down counts for both orientations of the target polarization

$$r = \sqrt{\frac{N_+^{(up)}/N_+^{(dn)}}{N_-^{(up)}/N_-^{(dn)}}} \tag{4.104}$$

(where the subscript $\pm$ indicates target polarization up or down) the asymmetry was extracted

$$A_r = \frac{r-1}{r+1} = -\mathcal{A} \mathcal{P}_n R$$

while the detector efficiencies, solid angles, and normalization cancelled. The target polarization was obtained from the up-down asymmetry in emission direction measured by the monitor counters at polar angle θ

$$A_M = \frac{r_M - 1}{r_M + 1} = a \mathcal{P}_n \cos\theta$$

where a is the beta decay asymmetry parameter.

Periodic measurements performed without the scattering foil gave information on the intensity and asymmetry of background radiation, which were then weighted and subtracted from the data; note that an asymmetry is expected in the background, since its rate depends on the asymmetric electron angular distribution which follows polarization reversal. The data was recorded in 33 ms wide time bins after the activation of the source, and the data in the first bin, affected by higher background and less steady detector conditions, was discarded.

The counts were recorded separately for each of the four sectors of the polarimeter, which experienced different background conditions. For a single target polarization direction, taking into account background counts $B^{(up)}, B^{(dn)}$:

$$R = \frac{1}{\mathcal{P}_n \mathcal{A}} \frac{(N^{(up)} - B^{(up)}) - (N^{(dn)} - B^{(dn)})}{(N^{(up)} - B^{(up)}) + (N^{(dn)} - B^{(dn)})} = \frac{R_{raw} - R_{bkg} \mathcal{P}_{nbkg}/\mathcal{P}_n}{1 - I}$$

where R_{raw}, R_{bkg} are the 'raw' (uncorrected) R value and the R value due to background measured with no scattering foil, with corresponding measured polarization $\mathcal{P}_{nbkg}$, and $I = (B^{(up)} + B^{(dn)})/(N^{(up)} + N^{(dn)})$ the background intensity ratio. The correction applied to the raw R value is

$$\frac{I}{1 - I}\left(R_{raw} - R_{bkg}\mathcal{P}_{nbkg}/\mathcal{P}_n\right)$$

which goes to zero as the background $I \to 0$ and when the raw and background asymmetries are the same. The described procedure was actually applied to the double ratio of event counts (4.104). Care was taken not to introduce correlations through the use of the same background rate measurements for different measurements.

The data was analysed in different ways, and the measurements performed as a function of electron energy, time after activation or time of run gave statistically consistent results (taking in account their correlation) after background subtraction.

Detector gain variations, monitored by a calibration system, were small and corrected for. The accidental coincidence rates in the three-layered detectors, measured by counting with delayed signals, were below 0.1% of the event rates, and the corresponding asymmetries were consistent with the background asymmetry; they were subtracted as for the background. Effects arising from the dead time of the readout electronics were equalized by gating all the ADCs with mixed pulses from all detectors. The result of the three-year experiment was $R = (-0.2 \pm 4.0) \cdot 10^{-3}$, consistent with time reversal symmetry.

In non-leptonic hyperon decays, such as $\Lambda \to p\pi^-$, in absence of final state interactions time reversal symmetry would require the parameter β defined in eqn (2.76) to vanish, which would imply the absence of net polarization for the emitted baryon in the direction orthogonal to the plane formed by the parent hyperon spin and the final baryon momentum, eqn (2.77) (Gatto, 1957; Lee and Yang, 1957). Since FSI cannot usually be neglected in such decays, one writes

$$A_S = a_S\, e^{i\delta_S} \quad A_P = a_P\, e^{i\delta_P}$$

where according to the Fermi–Watson theorem the $\delta_{S,P}$ phase shifts are those corresponding to the scattering of the hadrons in the final state. Time reversal symmetry therefore requires the quantities a_S, a_P to be relatively real, so that

$$\beta/\alpha = -\tan(\delta_S - \delta_P) \tag{4.105}$$

In the case of $\Lambda \to p\pi^-$ decay the left-hand side of eqn (4.105) can be determined by analysing the proton polarization in the decay of polarized Λ, through the left-right asymmetry of the proton scattering with respect to the $\boldsymbol{p}_p \times \hat{\boldsymbol{n}}$ direction ($\hat{\boldsymbol{n}}$ is orthogonal to the Λ production plane). The right-hand side of eqn (4.105) is measured in $\pi^- p$ scattering experiments at a centre of mass energy corresponding to the Λ energy; in general the phase shifts depend on the orbital angular momentum L and isospin I of the state, but $L = 0, I = 0$ dominate in this case. The measurements (Overseth and Roth, 1967) are consistent with time reversal symmetry. One such scattering phase is present for each final state, and while in the case of $\pi^- p \to \Sigma^- K^+$ production followed by $\Sigma^- \to n\pi^-$ the final state is an eigenstate of isospin (a good quantum number for strong interactions), in the case of $\pi^- p \to \Lambda K^0$ production followed by $\Lambda \to p\pi^-$ or $\Lambda \to n\pi^0$, the final states are mixtures of two isospin states, so that two phase shifts appear for each of A_S and A_P.

More information could be obtained from the measurement of the polarization of the emitted nucleon, and an interesting approach (Gatto, 1957), used experimentally, is that of exploiting $K^- p \to \Xi^- K^+$ production and the cascade decay $\Xi^- \to \Lambda \pi^0$, in which the Λ takes the role of the emitted baryon, with the advantage that its subsequent weak decay (e.g. to $p\pi^-$) can be used as the polarization analyser, if the angular asymmetry parameter $\alpha(\Lambda)$ is known.

4.4.2 Electromagnetic interactions

The possibility of time reversal symmetry violation for electromagnetic interactions was first discussed in Bernstein *et al.* (1965), where it was noted that if such interactions are introduced by the so-called 'minimal coupling', that is simply replacing $\partial_\mu \to \partial_\mu - iQA_\mu$ in the free Lagrangian (Q is the electric charge operator), then time reversal symmetry simply follows from that of the free Lagrangian itself, since ∂_μ and $\partial_\mu - iQA_\mu$ transform in the same way under time reversal.

Electromagnetic interactions are described by the coupling of the current to the electromagnetic field: $H_{\text{int}} = j^\mu A_\mu$; for a complex system such as a nucleon (which is not a Dirac particle) the current does not have the simple expression $Q\bar{\psi}\gamma^\mu\psi$, and must be written as the most general expression compatible with Lorentz (and parity, since no parity violation has been detected in EM interactions) symmetry. The matrix element of the current between nucleon states is written in momentum space as

$$\langle N_f | j^\mu | N_i \rangle = Q\,\bar{u}^{(s_f)}(\boldsymbol{p}_f)\Gamma^\mu(p_f, p_i)u^{(s_i)}(\boldsymbol{p}_i)e^{i(p_f - p_i)x} \tag{4.106}$$

where $|N_i\rangle$, $|N_f\rangle$ are the initial and final nucleon states, $u^{(s_i)}(\boldsymbol{p}_i)$, $u^{(s_f)}(\boldsymbol{p}_f)$ the corresponding spinors, p_i, p_f and s_i, s_f the four momenta and spins. $\Gamma^\mu(p_f, p_i)$ is a 4×4 matrix depending on p_i and p_f such that $\overline{\psi}\Gamma^\mu\psi$ transforms as a four-vector; note that the Hermiticity of j^μ imposes the constraint $\Gamma^\mu(p_f, p_i) = \gamma^0\Gamma^{\mu\dagger}(p_i, p_f)\gamma^0$.

Γ^μ must be formed by the available vectors p_i^μ, p_f^μ, and γ^μ; writing $p = p_i + p_f$, $q = p_i - p_f$ the following Hermitian terms are available for $\Gamma^\mu(p_f, p_i)$ if parity symmetry is imposed: p^μ, iq^μ, γ^μ, $\sigma^{\mu\nu}p_\nu$, $i\sigma^{\mu\nu}q_\nu$. The term in p^μ, sandwiched between spinors, can be re-expressed in terms of the γ^μ and $i\sigma^{\mu\nu}q_\nu$ ones (Gordon decomposition), and analogously the term $\sigma^{\mu\nu}p_\nu$ is equivalent to the iq^μ one, so that the most general (parity-conserving) form is

$$\langle N_f|j^\mu|N_i\rangle = Q\,\overline{u}^{(s_f)}(\boldsymbol{p}_f)\left[\gamma^\mu F_1 + i\sigma^{\mu\nu}q_\nu F_2 + iq^\mu F_3\right]u^{(s_i)}(\boldsymbol{p}_i)e^{-iqx} \quad (4.107)$$

The form factors F_i written above are real functions (because of the Hermiticity condition) of the only independent scalar quantity, namely the squared four-momentum transfer $q^2 = (p_i - p_f)^2$ (since $Pq = 0$ and $P^2 = 4m_N^2 - q^2$, where $p_i^2 = p_f^2 = m_N$ is the nucleon mass, when restricting to the case of real nucleons, on the mass shell). Setting $p_i = p_f$ and integrating the time component of eqn (4.107) over all space results in the constraint $F_1(0) = 1$ from the conservation of charge, while $F_2(0) = (g - 2)/4m_N = \kappa/2m_N$ gives the anomalous magnetic moment (so that the magnetic dipole moment is $d_M = Q[F_1(0)/2m_N + F_2(0)]$, see eqn (4.90)).

From the transformation properties of bilinear covariants built from gamma matrices, eqns (4.48)–(4.52), it is readily seen that the third term in the right-hand side of eqn (4.107) results in an interaction with the EM field (see eqn (4.38)) which changes sign under time reversal: $iq^\mu A_\mu \to -iq_\mu A^\mu$, unless F_3 is imaginary (which cannot be for Hermiticity); time reversal symmetry therefore requires this term to be absent: $F_3 = 0$. However this fact cannot be used for a symmetry test because the conservation of the electromagnetic current $\partial_\mu j^\mu = 0$ already implies

$$0 = \langle N_f|\partial_\mu j^\mu|N_i\rangle = i\langle N_f|[P_\mu j^\mu]|N_i\rangle = -iq_\mu\langle N_f|j^\mu|N_i\rangle$$

$$= Q\,q^2 F_3\overline{u}^{(s_f)}(\boldsymbol{p}_f)u^{(s_i)}(\boldsymbol{p}_i)e^{-iqx}$$

since the term containing F_1 is zero because the spinors satisfy the Dirac equation (B.16), and the one containing F_2 vanishes identically due to the antisymmetry of $\sigma^{\mu\nu}$. Current conservation already requires the F_3 term to be absent.[61]

---

[61] This limitation is not present for systems of spin $> 1/2$; see Khriplovich (1991) for the general case.

If parity symmetry is not imposed, three more terms can appear in the expression for the matrix element of the electromagnetic current:

$$\langle N_f | j^\mu | N_i \rangle = Q \bar{u}^{(s_f)}(p_f) \left[\gamma^\mu F_1 + i\sigma^{\mu\nu} q_\nu F_2 + i q^\mu F_3 \right.$$
$$\left. + \gamma^\mu \gamma_5 G_1 + \sigma^{\mu\nu} \gamma_5 q_\nu G_2 + i q^\mu \gamma_5 G_3 \right] u^{(s_i)}(p_i) e^{-iqx} \qquad (4.108)$$

Current conservation requires $F_3 = 0$ and links G_1 to G_3 into $G_A(\gamma^\mu q^2 + 2m_N q^\mu)\gamma_5$, with $G_A(0)$ being called the (parity-violating) anapole moment.[62] This leaves four independent terms.[63] The term containing G_2 is unconstrained: it violates parity and time reversal symmetry, and gives the electric dipole interaction, eqn (4.96), with $G_2(0) = d_E/Q$.

For nucleons which are not on the mass shell ($p_{i,f}^2 \neq m_N^2$, i.e. not free), the matrix element contains in general more form factors, which are functions of the independent scalars p_i^2, p_f^2, and q^2, and the above restrictions on T-odd terms do not apply. This suggests that tests of time reversal in nuclear electromagnetic reactions could be performed in energy regions where contributions due to intermediate resonances are significant.

Other experiments search for T-violating effects induced by non-zero relative phases between two interfering nuclear transition amplitudes of comparable strength, e.g. mixed E2 and M1 gamma decays. Since parity symmetry is known to hold to a very good accuracy, electric 2^J-pole and magnetic 2^{J-1}-pole transitions can add coherently; at first order in the fine structure constant, when the transition matrix is equal to the Hamiltonian (Hermitian), which contains the multipole operators $O^{(J,M)}$ introduced in Section 4.3.4:

$$\langle \alpha', j', m' | O^{(J,M)} | \alpha, j, m \rangle = (-1)^{j+m}(-1)^{j'+m'} \langle \alpha', j', -m' | TO^{(J,M)} T^\dagger | \alpha, j, -m \rangle^* \qquad (4.109)$$

(α, α' stand for other quantum numbers). Using the Wigner–Eckart theorem, eqn (4.30), the reality of the Clebsch–Gordan condition and the properties of electric and magnetic multipoles under a time reversal transformation, eqn (4.99), (4.100), one obtains for the matrix elements of $O^{(J)}$ independent from m, m':

$$\langle \alpha', j' | O_E^{(J,M)} | \alpha, j \rangle = (-1)^J \langle \alpha', j' | O_E^{(J,M)} | \alpha, j \rangle^* \qquad (4.110)$$

$$\langle \alpha', j' | O_M^{(J,M)} | \alpha, j \rangle = (-1)^{J+1} \langle \alpha', j' | O_M^{(J,M)} | \alpha, j \rangle^* \qquad (4.111)$$

Combining this result with the parity selection rule mentioned above, one concludes that the electromagnetic transition matrix elements which may interfere can all be chosen to have the same phase (within a sign) (Lloyd, 1951); any relative phase

---

[62] Note that this is the only possible electromagnetic moment for a Majorana fermion, because of eqn (3.40).

[63] The expression in eqn (4.108) is usually considered for the weak current, which is known to be parity-violating; in this case however only the vector (and not the axial-vector) part of the current is conserved, so that the only constraint is $F_3 = 0$.

different from 0 or π between two amplitudes is evidence for time reversal violation (barring final state interaction effects). The realization of the above constraint
opened the way to practical tests of time reversal symmetry.

The effect of the phase difference was searched for in non-zero T-odd correlations in nuclear gamma decay, which require the polarization measurement of the
emitted photon and/or the initial and final nuclei (Jacobsohn and Henley, 1959).
The initial nucleus can be polarized by orientation in a magnetic field, a previous
beta or gamma decay, the capture of polarized neutrons, etc., while the final nuclear
polarization is usually measured by a subsequent gamma decay occurring in a short
time (with respect to the spin relaxation time of the state) after the transition under
investigation.

In this respect the tests exploiting resonant absorption of nuclear gamma rays are
particularly significant (Henley, 1969): the hyperfine gamma resonant absorption
cross section in an absorber oriented in a magnetic field depends on the linear
polarization of the incident photon, which can thus be measured by comparing the
rates when the direction of the magnetic field is reversed. The photon polarization
is obtained by placing another magnetized absorber between the source and the
target, which selectively absorbs one polarization component. With this method
the T-odd correlation $\langle (\boldsymbol{k} \cdot \boldsymbol{B} \times \boldsymbol{\epsilon})(\boldsymbol{k} \cdot \boldsymbol{B})(\boldsymbol{\epsilon} \cdot \boldsymbol{B}) \rangle$ is measured, where $\boldsymbol{k}, \boldsymbol{\epsilon}$ are the
photon momentum and polarization vectors respectively, and $\boldsymbol{B}$ is the magnetic
field in the absorber.

The study of the transmission probability of photons through magnetized foils
can be performed by using resonant photons exploiting the Mössbauer effect, so
that inelastic processes are excluded and initial and final states are guaranteed to
be identical.

With clever choices of foil configurations, some differences in the transmission
probability for opposite magnetic field orientations can be found which are insensitive to any final state interaction effect and therefore probe directly time reversal
(Schäfer and Adelberger, 1991). The limit on the sensitivity of such experiments
is given by the error in the reversal of the magnetic fields and on the equality of
the foils.

4.4.3   Searches for electric dipole moments

Since parity violation is established, searches for electric dipole moments (EDM) of
elementary systems (which violate both P and T) now represent an important class
of tests of time reversal symmetry; indeed the most stringent limits on time reversal
violation in electromagnetic interactions are given by this kind of measurements.
The interest in EDM searches was revamped with the discovery in 1964 of the
violation of CP symmetry (see Chapter 8) to which time reversal is linked by the
CPT theorem (Chapter 5); actually 'the neutron EDM has ruled out more theories
(put forward to explain [CP violation]) than any experiment in the history of physics'
(Golub and Lamoreaux, 1994).

No static EDM has been detected in any system, so far. An exhaustive and detailed discussion of both theoretical and experimental aspects can be found in Khriplovich and Lamoreaux (1997). For EDM, there are no intrinsic phenomena which can mimic the effect being searched for,[64] and a non-zero measurement would be direct evidence for violation of time reversal symmetry.

Experiments are usually performed on neutral systems, since the intense electric fields required to put in evidence the interaction energy of a tiny electric dipole moment would accelerate charged particles out of the experimental apparatus. Even in matter, where collisions lead to time-averaged uniform velocities for charged particles in an electric field, the particles move in a zero average electric field, since the forces acting on the particles by the medium are themselves of electric nature, and produce an electric field which exactly compensates the external one (Garwin and Lederman, 1959), so that no EDM can be detected for non-accelerated charged particles.

Working with spin 1/2 systems, in which the only possible electromagnetic moments are dipoles, represents a simplification: for such systems, in absence of time reversal violation there can be no energy shift between the two $J_z = \pm 1/2$ states, even in presence of external electric fields, since the ground state must remain twofold degenerate because of Kramers' theorem; for example the neutron precession frequency is not affected by an electric field. In absence of quadrupole moments no interaction with gradients of the applied electric field is present.

The neutron, easily produced, easily polarized and with a long lifetime, is an ideal candidate for an EDM search. The first limits on the electric dipole moment of the neutron were obtained using the resonance technique with separated oscillatory fields (Ramsey, 1949) introduced to measure accurately magnetic dipole moments. This is the technique which, with several improvements, has been used since for neutron EDM measurements, and is briefly reviewed here starting with its application in beam experiments (see Ramsey (1956) for an extensive discussion).

In a beam resonance experiment, a collimated beam of electrically neutral particles (or atoms, molecules) directed towards a detector crosses two regions with equal and oppositely oriented inhomogeneous magnetic fields; such fields deflect the particles in opposite directions by interacting with their magnetic moments, so that they are refocused onto the detector independently from their velocity. If the state of the particles (and the direction of their magnetic moment) changes between the two inhomogeneous field regions, they will fail to be refocused on the detector, resulting in a drop in the transmitted beam intensity. Such a change of state is induced by letting the particles traverse a region with uniform magnetic

---

[64] Actually (Khriplovich and Pospelov, 1990), a macroscopically time-asymmetric phenomenon such as the leakage current due to the presence of an electric field can couple to nuclear spin through a P-odd interaction with its anapole moment (of weak origin) faking an EDM interaction, but this effect is negligible at the level of precision of present experiments.

field B_0 (which splits the energies of the different magnetic levels through the magnetic dipole interaction) in which a second weaker magnetic field B_1, rotating in the plane orthogonal to the first, is also present. In a semi-classical picture, when only the constant field B_0 (chosen to be along the z axis) is present, the magnetic moment $\mu = \gamma_{GM} S$ (where the gyromagnetic ratio $\gamma_{GM} = g(Q/2m)$, Q, m and S being the electric charge, mass and spin of the particle, and $g \simeq 2$ for a Dirac particle) precesses around it at the Larmor frequency $\omega_0 = \gamma_{GM} |B_0|$ keeping the angle ϕ between μ and B_0 constant. When the orthogonal field B_1 (rotating in the x, y plane) is also present, the magnetic moment will also tend to precess around it, therefore changing the value of ϕ, but the net torque will average to zero unless the frequency of the rotating field equals ω_0, when a cumulative effect will result in a net change of ϕ; such a change of alignment with respect to the external field B_0 can lead to a drop in the transmission rate when the frequency ω_1 of the external field B_1 exactly matches ω_0.

This is the traditional magnetic resonance method, which can be formulated from a quantum viewpoint by writing the time-dependent transition probability between two energy eigenstates, induced by an oscillatory perturbation (Ramsey, 1956). In practice an oscillating magnetic field B_1 is used, which can be expressed as the superposition of two counter-rotating ones with the same frequency and half the amplitude:

$$|B_1|\hat{y} \cos(\omega_1 t) = |B_1|[\hat{x} \sin(\omega_1 t) + \hat{y} \cos(\omega_1 t)]/2$$
$$+ |B_1|[-\hat{x} \sin(\omega_1 t) + \hat{y} \cos(\omega_1 t)]/2$$

and the effect of the component rotating oppositely to the precession at ω_0 is negligible close to resonance. Also, the net spin rotation of the particle depends on the time it spends in the region with B_1, and therefore on its velocity: as a result a widening of the resonant shape is obtained when averaging over the velocity distribution of the beam. A better resolution (a sharpening of the resonance) is obtained if the length of the region in which transitions are induced is increased, but a limitation arises from the difficulty in keeping B_0 uniform over such region: changes in the magnitude of the field give varying resonance frequencies leading to a broadening of the integrated resonance pattern.

Ramsey's method (Ramsey, 1949) is an improvement on the above approach, in which the oscillating field B_1 is only present in two short regions at the beginning and at the end of the zone with constant field. Along the lines of the semi-classical discussion above, the magnitude of the rotating (oscillating) field B_1 in the first region is chosen such that it rotates a magnetic moment initially parallel to B_0 ($\phi = 0$) by an angle $\pi/2$, into the plane of B_1 ($\phi = \pi/2$), orthogonal to B_0 and the instantaneous direction of B_1. The spin then precesses around B_0 at the frequency ω_0 as the particle travels along the beam, until it reaches the region with the second oscillating field, which is phase coherent with the first one. Here, if the frequency ω_1 of $\boldsymbol{B}_1$ exactly matches ω_0, the spin of the particle will again be orthogonal to $\boldsymbol{B}_1$, therefore experiencing a second $\pi/2$ rotation, resulting in a net reversal $\Delta\phi = \pi$; note that this occurs independently of the arrival time of the particle in the second region, and therefore of its velocity. If instead $\omega_1 \neq \omega_0$, there will be a phase difference between the spin vector and the field in the second region, so that the rotation will be only partial, or even opposite. Note that if the two frequencies differ by such an amount that the relative phase shift is an integer multiple of 2π there is also complete spin reversal, but this condition fails to be satisfied for a particle of different velocity (a slower particle gets a larger phase shift than a faster one) and therefore the reorientation is once more only partial. Again, the visibility of the resonance structure is reduced by the broadening due to the velocity distribution of the particles, and the spin rotation in the oscillating field regions is also somewhat velocity dependent, due to the amount of time the particle spends in this (relatively small) region. The amplitude of the resonance curve (or how much the transmission gets close to zero) depends on the initial polarization of the particle and its depolarization on inhomogeneities of the magnetic fields, on the accuracy of the spin rotation, on the phase coherence of the two pulses, etc., and all the above factors ultimately contribute to the sensitivity of the method. Note that with this approach any non-uniformity of the field $\boldsymbol{B}_0$ in the precession region simply results in an effective resonance frequency which is given by a weighted average of those corresponding to the different values of the magnetic field in the different points.

A variation of the above method consists in having the two oscillating fields coherent but phase shifted, and comparing the intensity drop when this phase difference is switched between $-\pi/2$ and $+\pi/2$: in this case the change in transition probability as a function of frequency is more pronounced, and it changes sign at the resonance frequency, so that the steepness of the resonance curve at that point gives a linear sensitivity to frequency changes (instead of a quadratic one at a minimum) helping to locate ω_0 more accurately; moreover, a phase reversal introduces less RF pick-up with respect to the case in which the intensity is compared with oscillators on and off.

The above process can be very conveniently described by transforming to a frame of reference rotating with the field $\boldsymbol{B}_1$ (Rabi et al., 1954), as we now sketch. The equation of motion in a stationary reference frame is

$$\frac{d\boldsymbol{S}}{dt} = \gamma_{GM} \, \boldsymbol{S} \times \boldsymbol{B} \tag{4.112}$$

but in a coordinate system rotating with angular velocity $\boldsymbol{\omega}$ one has

$$\left.\frac{d\boldsymbol{S}}{dt}\right|_{\text{rot}} = \frac{d\boldsymbol{S}}{dt} - \boldsymbol{\omega} \times \boldsymbol{S} = \gamma_{GM} \, \boldsymbol{S} \times \boldsymbol{B}_{\text{rot}} \tag{4.113}$$

($\boldsymbol{S}$ is the spin vector as measured by the stationary observer), where $\boldsymbol{B}_{\text{rot}} = \boldsymbol{B} + \boldsymbol{\omega}/\gamma_{GM}$ is the effective field in the rotating system.

In the case at hand, in the frame rotating with angular velocity $-\omega_1$, with the component of the oscillating field $\boldsymbol{B}_1$ with the same sense of rotation as the precession (x axis along such field component), there is a constant field of magnitude $|\boldsymbol{B}_0| - \omega/\gamma_{GM}$ along z, a constant field of magnitude $|\boldsymbol{B}_1|/2$ along x and a field of the same magnitude counter-rotating with frequency $2\omega_1$, whose effects can usually be ignored. In this approximation, all the fields are constant in the rotating frame, and the spin precesses around the effective field direction, making an angle

$$\theta = \arctan\left(\frac{|\boldsymbol{B}_1|/2}{|\boldsymbol{B}_0| - \omega/\gamma_{GM}}\right) = \arctan\left(\frac{\gamma|\boldsymbol{B}_1|/2}{\omega_0 - \omega}\right) \tag{4.114}$$

with the z axis, at a frequency $\gamma_{GM}|\boldsymbol{B}_1|/2$; at the resonance frequency $\theta = \pi/2$ and a magnetic moment initially parallel to $\boldsymbol{B}_0$ can be reversed completely. When drifting in the region between the two oscillatory fields, the magnetic field in the rotating frame is zero at the resonance, so that the spin does not rotate with respect to the (fixed) direction of $\boldsymbol{B}_1$, independently from the particle velocity.

The above method is readily applied to the measurement of an electric dipole moment, by also having an intense electric field $\boldsymbol{E}_0$ in the drift region and observing the change in the resonance frequency associated with a reversal of $\boldsymbol{E}_0$ with respect to $\boldsymbol{B}_0$, which corresponds to an energy shift correlated to the P- and T-odd quantity $\boldsymbol{E}_0 \cdot \boldsymbol{B}_0$

The interaction of a neutron in the magnetic and electric fields is described by the sum of (4.93) and (4.95)

$$H_{\text{int}} = -(d_M \boldsymbol{B}_0 + d_E \boldsymbol{E}_0) \cdot \frac{\boldsymbol{S}}{|\boldsymbol{S}|} \tag{4.115}$$

(with $\boldsymbol{\mu} \equiv \boldsymbol{d}_M$). The neutron precession frequency becomes $\omega_0 \pm 2d_E|\boldsymbol{E}|/\hbar$, depending on whether $\boldsymbol{E}_0$ is parallel or anti-parallel to $\boldsymbol{B}_0$. The possibility of turning the T-violating effect on or off at will by controlling the external electric field $\boldsymbol{E}_0$ represents an important strong point in this kind of experiment.

The sensitivity of an EDM experiment scales linearly with the intensity of the source of particles and with the magnitude of the applied electric field. For $|\boldsymbol{E}_0| \sim 100$ kV/cm (about the maximum electric field which can be applied in vacuum) and $d_E \sim 10^{-26}\,e$ cm, the energy difference induced by the electric dipole interaction is $\Delta E \sim 4 \cdot 10^{-21}$ eV, and the frequency difference $\sim 10^{-6}$ Hz; by comparison, since the magnetic dipole moment of the neutron is $|\boldsymbol{\mu}_n| = 6 \cdot 10^{-12}$ eV/G, such a frequency change corresponds to a variation of $3 \cdot 10^{-10}$ G in the precession magnetic field (to be compared, e.g., with the earth's average magnetic field of ~ 0.5 G). Magnetic field shielding and control are clearly essential features of these experiments: shielding factors in excess of 10^6 can be achieved. In modern experiments the magnetic field in the precession region needs to be stable over the time interval in which the field reversal is made at the level of the nG (10^{-9} of the Earth's field). The amount of time the particles spend in the spin precession region also determines the EDM sensitivity: for a beam experiment such time is of the order of some ms (beam velocities ~ 300 m/s).

One of the most crucial aspects of EDM searches with this approach is to make sure that there are no spurious magnetic fields linked to the system used for reversing the electric field: the coupling of such fields to the magnetic moment would result in a frequency shift indistinguishable from that due to an EDM. One source of this kind of effects are leakage currents flowing through insulators, which generate magnetic fields correlated with the electric field and have to be limited as much as possible.

A more fundamental effect of similar nature is due to the fact that, according to special relativity, in the reference frame of a particle moving with velocity v with respect to an electric field E_0 a motional magnetic field B_m appears too, given by

$$B_m = (v/c) \times E_0 \qquad\qquad (4.116)$$

If the electric field E_0 is almost parallel to B_0 (small angle θ among the two), the effective magnetic field experienced by the particle has magnitude

$$|B_{\text{eff}}| \simeq |B_0| + \theta|B_m| + \frac{1}{2}|B_m|^2/|B_0|$$

(for $|B_m| \ll |B_0|$), resulting in a corresponding shift of the precession frequency correlated with the electric field;[65] even for a perfect geometry ($\theta = 0$), a shift quadratic in $|E_0|$ is present, so that any difference in the magnitude of the electric field before and after reversal results in a fake signal. This effect, being proportional to the velocity of the particles, is the ultimate limiting factor for beam EDM experiments.

Beam searches for neutron EDM

The first beam EDM experiment was performed by Ramsey and collaborators at the Oak Ridge reactor in 1949–1951, but the results were only published in full in 1957 (Smith *et al.*, 1957), after the fall of parity.

A sketch of the experimental set-up is shown in Fig. 4.5. An intense beam of transversely polarized neutrons ($T \sim 500$ K) was obtained by total reflection of the reactor neutrons on a magnetized (2,500 G) iron reflecting sheet (a new technique specially implemented for this experiment, see Chapter 2) and travelled through a region containing the uniform magnetic and electric fields in about 0.5 ms. The 250 G magnetic field (corresponding to a 750 kHz neutron precession frequency) was obtained by magnetizing two 180 cm long steel bars with permanent magnets; a 70 kV/cm electric field was set between two 135 cm long nickel-plated copper plates enclosed in vacuum (while the ferromagnetic nickel disturbed the magnetic field somewhat, it was used to avoid sparks). Before and after the field region were the two

---

[65] Note that the interaction term due to the motional magnetic field, $-\mu \cdot v \times E_0$, is T-even, as it should be.

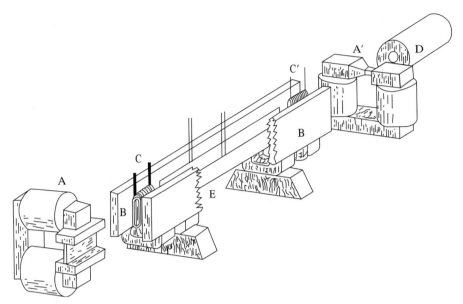

FIG. 4.5. Sketch of the setup used in the first neutron EDM measurement (Smith *et al.*, 1957). A: magnetized iron mirror polarizer; A′: magnetized iron transmission analyzer; B: pole faces for the static magnetic field, with permanent magnets attached on bottom; C, C′: RF spin-rotation coils; D: neutron detector; E: plates for electric field generation.

spin-rotating RF coils, 5 cm long and separated by 159 cm, powered by a stable oscillator. The neutrons were detected by a proportional counter containing BF_3 gas (exploiting the $n \, ^{10}B \rightarrow ^7 Li \alpha$ reaction, with a large cross section, $\sigma \simeq 4$ b, for thermal neutrons) at the end of the beam line. The counter was preceded by a small saturated iron block placed between the poles of an electromagnet: part of the neutron scattering cross section (and therefore the neutron transmission probability), due to the magnetic dipole interaction with the atoms, depends on the relative spin alignment, so that the detector worked as a polarization analyser (a high degree of magnetic saturation is required, randomly oriented magnetic domains cause the neutron spin direction to precess around the local magnetic field direction, therefore depolarizing them). Note that the direction of the magnetic field in the polarizing mirror was orthogonal to that of the static precession field (and of the analysing iron block), and twisted iron strips between them were used to rotate the neutron spin: it can be shown (Ramsey, 1956) that if a magnetic field varies slowly with respect to the precession frequency ($|\dot{B} \times B|/|B|^2 \ll \gamma |B|$), a dipole will slowly turn, following the change of such field ('adiabatic spin rotation').

The frequency dependent counting rate, normalized to that of a monitor counter in the neutron beam, to eliminate the dependence on reactor power fluctuations, was measured for two angle settings of the polarizing mirror with the electric field off, and found to be in excellent agreement with the predictions; then the frequency was set to a point on one of the steep sides of the curve close to the resonance, and rates were measured as the electric field polarity was reversed. With about 4 million total counts, interspersed with calibration measurements in which the slope of the resonance curve at the working point was measured, a result $d_E(n) = (-0.1 \pm 2.4) \cdot 10^{-20}$ e cm was obtained, with a corresponding upper limit $d_E(n) < 5 \cdot 10^{-20}$ e cm.

Other neutron beam EDM searches were performed over the following 25 years, the most precise one at the Institut Laue Langevin (ILL) reactor in Grenoble (Dress *et al.*, 1977).

The neutrons were polarized by reflection on iron mirrors and precessed in a 180 cm long region with a 17 G magnetic field (51 kHz precession frequency) and about 100 kV/cm reversible electric field. The high flux of neutrons ($\sim 5 \cdot 10^6$/s) required fast detectors, ^{6}Li-loaded glass scintillators read by two small (fast) photomultipliers, which had to be far from the magnetic field regions. The neutron transmission polarimetry method does not work for slow neutrons, with de Broglie wavelengths larger than $2d$ (d being the maximum inter-atomic spacing, $\simeq 2$ Å in iron), since in this case there is no orientation of the magnetized micro-crystals for which the Bragg condition $n\lambda = 2d \sin \theta$ is satisfied, and therefore strong reflection does not occur. The polarization was therefore analysed in the same way as it was produced, namely by reflection on magnetized mirrors.

An important feature of this experiment was its placement on a surplus naval-gun mount, allowing the entire set-up (except neutron source and detector) to be rotated by π. In this way the direction of the neutron beam with respect to the apparatus could be reversed, therefore cancelling the most important spurious effect due to the motional magnetic field ($\boldsymbol{B}_m \rightarrow -\boldsymbol{B}_m$ for $v \rightarrow -v$ in eqn (4.116)).

The resonance frequency was measured by comparing the neutron rate changing the phase shift between the two RF coils from $-\pi/2$ to $+\pi/2$ once per second; after 100 such measurements the electric field was reversed (with appropriate dead time introduced to avoid collecting data during periods of unstable conditions) and another set of measurements collected. This approach averaged out high frequency noise drifts and allowed to follow the slow drifts of the magnetic field, temperature, and reactor intensity variations; the electric field magnitude was steadily increased under computer control at every cycle, until a spark was detected, when it was reduced, thus providing an automatic optimization. About every day the set-up was rotated, and null measurements included periodic measurements with zero electric field.

Among the spurious effects correlated with the electric field direction were: leakage currents, which were reduced to the pA level; the permanent local magnetization induced by sparks, for which data surrounding each (recorded) spark were symmetrically rejected; the interference of the high-voltage reversal system with the ambient magnetic field, which required further shielding and periodic reversal of power cables; the electrostatic force between the electric field plates inducing mechanical stresses distorting the magnetic pole faces differently for different magnitude of the electric field.

By combining the data points in different running weighted averages, several time drifts could be eliminated. The final result of many months of data-taking was $d_E(n) = (0.4 \pm 1.5) \cdot 10^{-24}$ e cm, corresponding to a limit $d_E(n) < 3 \cdot 10^{-24}$ e cm (at 90% confidence level).

Another class of EDM searches is based on the same principle outlined above for beam experiments, but uses stored cold particles. In this case the two spin rotations are obtained by switching on a RF field for a time Δt such that $\gamma |\boldsymbol{B}_1| \Delta t/2 = \pi/2$

(a so-called '$\pi/2$ pulse'), and then repeating the process after a precession time ΔT. In this case the two (coherent) spin-rotating pulses are separated in time rather than in space. The particles are prepared in a given spin state and after the above procedure are extracted and spin-analysed to measure their spin-flip probability as a function of the frequency of the field B_1.

EDM experiments with stored particles have some advantages over beam experiments. First, the spin rotation pulses are generated by the same coil, so that both the rotation time and the precession time are the same for all the particles, independently of their velocity. Second, the storage times, which determine the maximum amount of time the particles can precess, can be much larger: for atoms this is determined by the spin relaxation time, the time scale over which the spins lose their orientation due to e.g. collisions with walls or other atoms, up to ~ 100 ms for paramagnetic atoms or 1000 s for diamagnetic atoms (in which the nucleus has a smaller magnetic moment which is shielded by atomic electrons). For stored neutrons the time is limited to ~ 100 s by the interaction with the cell walls and ultimately by the neutron lifetime. Finally, in a storage experiment the average velocity of the particles is zero, so the motional magnetic field effect mentioned earlier is reduced in this case (Khriplovich and Lamoreaux, 1997).

The accurate measurement of the field in the region where the particles precess is performed in modern experiments by using lighter polarized atoms (expected to have much smaller EDMs, if any) placed in the same volume as the sample under study (co-magnetometer), measuring the spin-dependent transmission (or scattering) of circularly polarized light. The reduction of magnetic field gradients in the storage volume is also important in storage experiments, since when coupled to the random movements of the particles in the storage volume they can result in a loss of spin orientation.

Recent experiments exploit ultra-cold neutrons (UCN), that is neutrons with kinetic energies smaller than the effective potential at the boundary with a material, which are reflected for any incidence angle and can therefore be stored in closed vessels (Golub and Lamoreaux, 1994). Due to the small penetration depth during total reflection, $O(10^{-5}$ mm), the interaction time of UCN with the walls of the vessel is small and such interactions do not cause appreciable losses on the scale of the neutron lifetime. UCN have velocities of order 5m/s (wavelength ~ 500 Å) corresponding to effective temperatures of order 2 mK, thus requiring cryogenic moderators (liquid H, D or He); further neutron energy reduction is also obtained by extracting them vertically from the source, since they lose about 110 neV per metre due to the gravitational potential. Note that the small magnitude of the velocity further contributes to making the motional field effect unimportant.

As in beam experiments, the highest possible electric fields and density of particles are required to enhance the effect and reduce the statistical uncertainties. In storage experiments the electric fields are applied over larger regions (the storage cell), and their maximum magnitude before sparking occurs is somewhat reduced with respect to beam experiments (depending very much on the composition and treatment of the cell surfaces). The intensities of UCN sources are generally rather low when compared with higher energy ones, but the possibility of storing the neutrons, to allow measurements with long durations, ultimately gives UCN the advantage. When spurious effects correlated with electric field reversal are kept under control, the experimental limits for this technique ultimately arise from the accuracy to which the magnetic field in the precession region is known.

ILL storage EDM experiment

A storage experiment at the ILL in Grenoble (Baker *et al.*, 2006) used UCN with a range of kinetic energies up to 350 neV (density $\sim$ 50/cm^3), which were polarized by transmission through a thin (1 μm) foil of iron evaporated onto aluminium, magnetized close to saturation (1 kG), which totally reflected one spin component while letting the other pass through. The storage cell of 22 l volume was a hollow quartz cylinder 20 cm high and closed by two 60 cm diameter aluminum electrodes, in which a highly uniform 10 mG magnetic field parallel to the axis was maintained (see Fig. 4.6). A four-layer magnetic shielding suppressed external magnetic fields by a factor $\sim 10^4$.

The cell was filled for 20 s with UCN and then closed; a first 2 s oscillating magnetic field pulse rotated the neutron spins by $\pi/2$, which were then allowed to precess (30 Hz frequency) for about 2 min before the second spin-rotation pulse. The cell was then opened again and the neutrons could exit, those of the appropriate polarization falling through the magnetized foil to be counted in a ^{3}He-doped proportional counter (exploiting the reaction $n\,^3\text{He} \rightarrow\,^3\text{H}\,p$). After some counting time, an RF spin-flip coil above the magnetized foil was switched on, resulting in a resonance at some point along the region in which the magnetic field is varying (between the cell and the foil), which rotated the spin of the remaining neutrons, allowing them to reach the detector and be counted too.

The frequency of the spin-rotation RF pulses was cycled through four points in the steep region around the half-maximum of the central fringe, close to the resonance, providing a measurement of the neutron polarization and the resonance frequency; this approach averaged out neutron flux fluctuations. The measurement on a batch of about 13,000 neutrons took 210 s, and was repeated continuously, each time slightly changing the oscillating field frequency, to scan the resonance pattern. The duration of the precession time is determined by the survival time of the neutrons in the cell, which was about 110 s depending on the presence of hydrogen in the cell surfaces. About once per hour the direction of the 4.5 kV/cm electric field (as large as possible to keep a low sparking rate) was switched between parallel and anti-parallel to the magnetic field to measure the EDM signal.

The major limitation of previous storage EDM experiments were the unavoidable small magnetic field fluctuations (of order a few nG), inducing shifts in the precession frequency which could not be corrected for accurately. To overcome this problem, the experiment implemented for the first time a co-magnetometer, reducing in this way the corresponding systematic error by a factor more than 20 with respect to previous experiments. Polarized

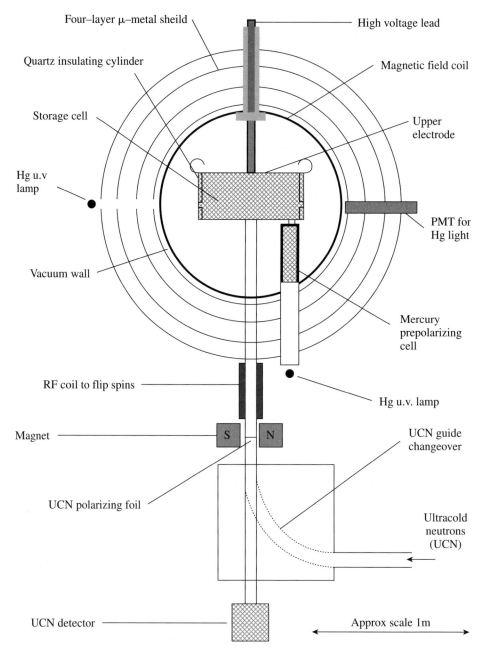

Four–layer μ–metal sheild

Quartz insulating cylinder

Storage cell

Hg u.v lamp

Vacuum wall

RF coil to flip spins

Magnet

UCN polarizing foil

UCN detector

High voltage lead

Magnetic field coil

Upper electrode

PMT for Hg light

Mercury prepolarizing cell

Hg u.v. lamp

UCN guide changeover

Ultracold neutrons (UCN)

S   N

Approx scale 1m

FIG. 4.6. Schematic drawing of the stored UCN EDM measurement apparatus at ILL (Baker *et al.*, 2006).

atoms were stored together with the neutrons in the same volume, allowing an accurate measurement of the magnetic field in a region and a time interval which closely match those in which the neutrons precess. ^{199}Hg atoms were used, which have a $^1S_0$ electronic configuration so that their polarization is determined by the nuclear spin 1/2 due to an odd neutron; Kramers' degeneracy guarantees that the only two existing ground state levels remain degenerate and the Larmor frequency is not affected by the presence of electric fields. This atom is suitable to be used as a magnetometer also because a very low experimental limit on a possible EDM is available for it (smaller than the neutron EDM sensitivity of the experiment), so that their precession rate is sufficiently independent from the electric field. Other atomic species used as magnetometers have a large electric polarizability so that the ground state Zeeman levels split proportionally to $|E_0|^2$, and an exact reversal of the electric field is required to avoid spurious effects.

The mercury atoms (density $\sim 3 \cdot 10^{10}/cm^3$) obtained from heated powder were polarized by optical pumping with circularly polarized laser light and then stored together with the neutrons; this required a careful choice of the material used for the cell walls, to guarantee long spin relaxation times. The atoms were spin-rotated by $\pi/2$ with an oscillating magnetic field pulse to have polarization in the plane orthogonal to the magnetic field; this rotation occurred before that of the neutrons, and did not affect them in a significant way, since the mercury precession frequency of 8 Hz is quite different from the 30 Hz neutron frequency. The mercury precession was continuously monitored by measuring the transmission of circularly polarized light through the cell: the absorption of the horizontal light beam is minimal when the mercury polarization is parallel to that of the photons. The optimal density of mercury atoms to have the largest modulation effect corresponded to a 20% light absorption. From the 100 Hz digitized data for the damped (a few minutes depolarization times) oscillations of the transmitted light oscillations, the average precession frequency was determined with a precision of a few parts in 10^{-8}, corresponding to the determination of the average magnetic field with an error of about 0.5 nG.

A vertical shift between the average position of the (ultra-cold) neutrons and (room temperature) mercury atoms in the storage cell was present due to the gravitational field: while the mercury density was rather uniform in the cell, the neutron density changed by about 10% over the height of the cell, resulting in an imperfect sampling of the magnetic field measurement with respect to the position of the neutrons; indeed, to minimize the effective distance between the centres of gravity of the two distributions (~ 0.3 cm) the cell had the shorter axis vertical.

The efficiency of the polarizer was not 100% due to the magnetic effects of surface oxide layers and roughness, and this reduced somewhat the amplitude of the resonance pattern. Another dilution factor arose from some spin relaxation of the neutrons in the storage cell due to residual magnetic field inhomogeneities. Leakage currents, generating with their circulation around the cell spurious magnetic fields correlated to the direction of the electric field, were of order 1 nA, and the related systematic effect was negligible. Other instrumental effects such as variations in the magnetic field due to changes in the magnetic shield magnetization caused by sparks or stresses induced by the electric field were automatically cancelled by the co-magnetometer normalization. As pointed out, the effect of motional magnetic fields was negligible, since the only coherent velocity of the neutrons could be due to the slight warming of the neutrons over the time of the measurement, leading to an overall upward displacement of their centre of mass, and to any residual net circulation of the neutrons arising from some asymmetry in filling; both effects were very small, since the upward thermal drift velocity was of order 10^{-2} cm/s and the neutron filling occurred from the centre of the cell, and anyway the precession only started after several seconds. Moreover, in order to induce an effect in the measurement, the above motions should couple to a component of the electric field not parallel to the magnetic field, which could only originate from distortions induced by charges accumulating on the insulators and was also very small.

The result from about 500 days worth of data with this technique, combined with earlier data in which the magnetic field was monitored by three external magnetometers (at about 40 cm from the cell), was $d_E(n) = (+0.6 \pm 1.5 \pm 0.8) \cdot 10^{-26}$ e cm, corresponding to $|d_E(n)| < 3.0 \cdot 10^{-26}$ e cm (90% confidence level). The direction of the magnetic field was also periodically reversed, thus reversing any true EDM shift, and the results obtained in the two conditions were compatible.

The largest correction applied to the measurement, $(1.1 \pm 0.5) \cdot 10^{-26}$ e cm, was due to the residual vertical magnetic dipole field in the region of the cell door, which coupled to the average vertical displacement between the neutrons and the Hg atoms; the largest contributions to the systematic error arose from the limits on the knowledge of this and other possible dipole fields in the cavity.

The above result provides the current best limit on the neutron EDM (about six orders of magnitude larger than the value expected in the SM). A feeling for the smallness of this number is obtained considering that such an EDM corresponds to a unit charge being displaced from the centre of a neutron by a distance 10^{-12} of its size (such as less than 10 μm if the neutron were the size of the Earth). Further progress depends on the development of more precise co-magnetometer systems, such as those using super-fluid helium, in which the strong spin dependence of the n^3He cross section is exploited to monitor the field magnitude.

Permanent EDMs for a composite (molecular, atomic, nuclear) non-degenerate system could originate from either the elementary EDMs of their constituents or from T- (and P-) violating interactions among those. As nucleon EDMs can arise from quark EDMs or their interactions, atomic EDMs can be generated by nuclear or electron EDMs, or by the T-odd interaction among them. The contribution from nuclei arises from their non point-like nature, as the interaction of P- and T-odd nuclear multipoles interacting with electrons and leading to the mixing of closely spaced opposite parity states; a nuclear EDM does not contribute directly since the average electric field is zero at its position. The largest effect contributing to atomic EDMs is usually the electron EDM, which can indeed be measured with high precision by exploiting the fact that in some heavy atoms it can induce large effects. Note also that, while for the neutron EDM the connection with the T-violating parameters of an underlying fundamental theory (at the quark level) is difficult because of the trouble dealing with low-energy strong interactions, the

EDM of an electron is easier to compute in terms of the parameters of a given model, although it is usually predicted to be smaller than for the neutron.

A theorem by Schiff (1963) states that a system of point-like non-relativistic particles interacting electrostatically adjusts its charge distribution in an external electric field in such a way that any electric dipole moment of its constituents is completely screened (i.e. the average local static field is zero (Garwin and Lederman, 1959)). Following Khriplovich and Lamoreaux (1997), for a system of particles with charges Q_k and electric dipole moments d_{Ek}, an additional dipole moment can arise because of the mixing of states of opposite parity induced by a P-odd interaction term with the internal electric field $E_{(int)}$ through

$$H_{int} = -\sum_k d_{Ek} \cdot E_{(int)}(r_k) = \sum_k \frac{d_{Ek}}{Q_k} \cdot \nabla_k U(r) = \frac{i}{\hbar} \sum_k \frac{d_{Ek}}{Q_k} \cdot [p_k, H_0]$$

where H_0, U are the unperturbed (no dipole moments) Hamiltonian and potential energy. Such mixing is quantified by the ratio of the matrix element of the interaction and the energy level spacing, giving origin to the perturbed states

$$|\tilde{n}\rangle = |n\rangle + \sum_m \frac{\langle m|H_{int}|n\rangle}{E_n - E_m}|m\rangle = \left(1 + \frac{i}{\hbar} \sum_k \frac{d_{Ek}}{Q_k} \cdot p_k\right)|n\rangle$$

for which the electric dipole moment is

$$\langle \tilde{n}| \sum_k Q_k r_k |\tilde{n}\rangle = -\langle n| \sum_k d_{Ek} |n\rangle$$

which exactly cancels the one due to the individual electric dipoles.

Indeed, the effective electric field experienced by a nucleon in a diamagnetic atom (such as ^{3}He, ^{199}Hg, ^{225}Ra) can be $\sim 10^{-3}$ of the externally applied field (the cancellation is not exact due to hyperfine interactions with the nuclear spin).

Schiff's theorem however breaks down dramatically for heavy atoms, due to the importance of relativistic effects (e.g. magnetic forces) and the finite size of the nucleus: for paramagnetic atoms, with an unpaired electron (such as Tl, Cs) this can actually lead to an *enhancement* of the effect due to an electron EDM (even by 2–3 orders of magnitude), scaling as Z^3. Enhancement factors for polar molecules (such as TlF, PbO) can even approach 10^7.

The size of the enhancement factor in heavy paramagnetic atoms can be roughly understood as follows: the interaction of the electron EDM $d_E(e)$ with the internal electric field $E_{(int)}$ in the atom is a relativistic effect violating Schiff's theorem, and as such proportional to $(v/c)^2$ (v being the electron velocity). This interaction leads to a mixing of energy levels of opposite parity which in turn originates an atomic dipole moment whose order of magnitude is $d_E \sim ea_0\mathcal{F}$ (e being the electron charge and a_0 Bohr's radius), where $\mathcal{F} \sim \langle m|d_E(e) \cdot E_{(int)}|n\rangle/(E_m - E_n)$. The interaction is stronger at short distances ($r \sim a_0/Z$), where the electron feels the unshielded field of the nucleus $|E_{int}| \sim Ze/r^2$. In this region (of volume $r^3 \sim a_0^3/Z^3$) the wave function is such that $|\psi|^2 \sim Z/a_0^3$ (see, e.g., Landau and Lifshitz (1965)) and since the electron velocity is $v/c \sim Z\alpha$ and the level spacing $\Delta E \sim e^2/a_0$, the resulting estimate is

$$d_E = Kd_E(e) \sim ea_0 \frac{Ze}{r^2} Z^2 \alpha^2 \frac{Z}{a_0^3} r^3 \frac{a_0}{e^2} d_E(e) = Z^3 \alpha^2 d_E(e)$$

that is an enhancement $\propto Z^3$ over the magnitude of the electron EDM.

Some EDM searches in heavy atoms look for the changes in the equilibrium atomic polarization when an electric field is applied. Circularly polarized laser light illuminates an ensemble of atoms, pumping them into an excited state with given angular momentum projection, while a magnetic field parallel to the light direction keeps the polarization direction. After a sufficient atomic polarization is obtained, this magnetic field is removed and replaced by suitably oriented electric (E_0) and magnetic (B_0) fields, around which the atomic spins precess; this results in a rotating polarization, which is detected by measuring the scattering or the transmission of circularly polarized light from a second (weaker) laser, whose polarization is modulated so that a synchronous signal can be analysed. The effect of an EDM would be a variation in the precession frequency correlated with the direction of the applied electric field, proportional to $\langle J \cdot (E_0 \times P) \rangle$, where J is the angular momentum of the probe photons and P the atomic polarization vector.

The duration of the measurements is limited by the depolarization rate due to atomic collisions with the cell walls or with the 'buffer gas' used to improve the high voltage characteristics; such rates depend on the atomic species and the experimental arrangement (order $\sim 1/\text{ms}$). Techniques to reduce spurious effects involve the use of multiple cells with oppositely directed electric fields, the elimination of any variable magnetic fields, the reduction of spatial inhomogeneities of the polarization and the equalization of the magnitude of the electric field for the two opposite orientations (Khriplovich and Lamoreaux, 1997).

Beam-type experiments are also performed with atoms, using elegant arrangements such as the use of pairs of beams, each one periodically reversed, crossing opposite electric fields (to cancel effects due to motional magnetic fields) and propagating vertically in order to insure that atoms with different velocities have the same trajectories, and containing a second atomic species used as a co-magnetometer. The most recent Tl experiment at Berkeley (Regan et al., 2002) using such techniques gave the best current limit on the electron EDM: $d_E(e) = (6.9 \pm 7.4) \cdot 10^{-28} \, e \, \text{cm}$, corresponding to $|d_E(e)| < 1.6 \cdot 10^{-27} \, e \, \text{cm}$ (90% confidence limit), some 10 orders of magnitude larger than the order of magnitude estimate in the SM due to the measured CP-violating effects.

While there is no electron EDM enhancement for diamagnetic atoms, these present several advantages: the EDM effect is associated with the nuclear spin direction, and since nuclear magnetic moments are $\sim 10^{-3}$ smaller than electronic ones, these atoms are much less affected by spurious magnetic effects such as those due to motion, leakage currents or imperfections of the setup. Moreover, diamagnetic atoms with nuclear spin $S_N = 1/2$ do not couple to electric field gradients on collisions with cell walls, and can therefore have very long spin relaxation times (several minutes), so that ultimately, despite the large suppression of electron EDM effects, experiments with diamagnetic atoms can be competitive with those using paramagnetic atoms. EDM searches in diamagnetic atoms were performed on the two heavy stable isotopes with $^1S_0$ electronic configuration and $S_N = 1/2$ which are suited for room temperature optical pumping: ^{129}Xe and ^{199}Hg.

Polar molecules such as TlF, in which very closely spaced rotational levels can be mixed by a T-violating process, leading to very large enhancements of the effect of an electron EDM, have also been studied.

Limits on the electric dipole moments of other elementary particles have been obtained with a variety of techniques (Yao *et al.*, 2006). Here we briefly discuss that of the muon, which gives us the opportunity to mention another very elegant experiment.

The muon EDM

The muon EDM was studied as a by-product in the storage ring experiments which measured the muon anomalous magnetic moment (Bailey *et al.*, 1978). The $g - 2$ experiments form a class of fascinating investigations which would themselves deserve a detailed description, which cannot be attempted here.

The CERN storage ring experiment (Bailey *et al.*, 1979) used highly relativistic muons ($\gamma \simeq 29$) in order to achieve a longer observation time of their spin precession ($\tau_{LAB} \simeq 64\,\mu s$). Pions produced from 22 GeV/c proton collisions on a copper target were momentum selected and directed tangentially into the 14 m diameter storage ring; the latter stored muons of a given momentum, which originated from forward decay of pions and had a (parity-violating) longitudinal polarization $\geq 95\%$.

For spin and momentum both orthogonal to a uniform vertical magnetic field $\boldsymbol{B}$, the spin precession frequency $\omega_S = g(e/2m)|\boldsymbol{B}|$ and the orbit frequency $\omega_c = (e/m)|\boldsymbol{B}|/\gamma$ (g is the gyromagnetic number, e, m the muon charge and mass, γ the relativistic factor) are equal if $g = 2$, so that the change of beam polarization $S \cdot p$ with time is a measure of $\kappa = (g-2)/2$. The precession frequency of the spin relative to the momentum, which is orthogonal to the magnetic and electric fields $\boldsymbol{B}, \boldsymbol{E}$, is

$$\boldsymbol{\omega} = -\frac{e}{mc}\left[\kappa\boldsymbol{B} + \left(\frac{1}{\gamma^2 - 1} - \kappa\right)\boldsymbol{v} \times \boldsymbol{E} + 2d_E(\boldsymbol{E} + \boldsymbol{v} \times \boldsymbol{B})\right] \qquad (4.117)$$

where d_E is the electric dipole moment. The storage muon momentum chosen was 3.094 GeV/c, such that the second term inside the square bracket of eqn (4.117) vanishes ('magic' γ). Electric fields being negligible compared with the 1.47 T vertical magnetic field, the precession due to an EDM was given by the motional electric field appearing in the muon rest frame, that is the last term of eqn (4.117), around an axis radial to the orbit, so that

$$\omega = \omega_a + \omega_{EDM} = -\frac{e}{mc}\left(\kappa\mathbf{B} + 2d_E\,\mathbf{v} \times \mathbf{B}\right)$$

The effect of an EDM is to tilt the spin precession direction to an angle $\delta \simeq |\omega_{EDM}|/|\omega_a|$ from the vertical (see Fig. 4.7), so that the muon polarization acquires a vertically oscillating component with the same frequency as the precession of the horizontal component (the $g-2$ precession frequency is also increased to $|\omega_a|\sqrt{1+\delta^2}$).

The muon polarization direction was observed through the anisotropic (because of parity violation) distribution of the decay electrons, by selecting high energy electrons (forward-going in the muon rest frame) with a detector threshold. These electrons were detected by a set of twenty-two plastic scintillator/lead sandwich detectors placed along the inner circumference of the ring (see Fig. 4.7); five of these detectors were preceded by two plastic scintillator sheets, covering the upper and lower part, which were used to identify whether the electron was above or below the median plane of the ring. Taking into account the vertical position of the decaying muons, a hit in the upper or lower detector was highly correlated to upward or downward electron directions, so that the up-down electron asymmetry in the split counters for the chosen energy threshold was $A_e \simeq 0.16\,\delta$.

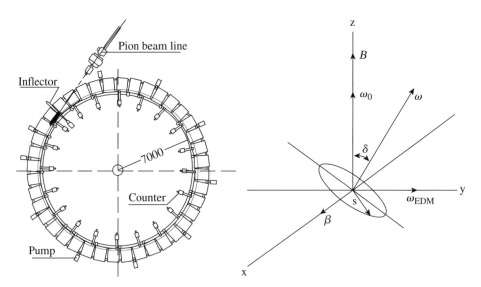

FIG. 4.7. Left: plan view of the CERN muon storage ring with the 22 electron detectors. Right: sketch of the relevant directions for the EDM measurement (Bailey *et al.*, 1978): the precession plane of the spin s relative to the velocity $\boldsymbol{\beta}$ is tilted by an angle δ relative to the horizontal (orthogonal to the magnetic field $\mathbf{B}$, as the $(g-2)$ precession around ω_a combines with the EDM precession around ω_{EDM} into an overall precession around ω.

The time evolution of the upward (downward) going electrons is

$$N = \frac{N_0}{2} e^{-t/\tau} \left[1 - A_\mu \cos(\omega t + \phi) \pm A_e \sin(\omega t + \phi) \right]$$

where A_μ is the asymmetry due to the $g-2$ precession, around 15% for the chosen electron energy threshold. For $A_e \neq 0$ a phase shift occurs between the two oscillating distributions, $\Delta = 2 \arctan(A_e/A_\mu) \simeq 0.32 \, \delta/A_\mu$, so that tilts of the precession plane of a few mrad with respect to the horizontal could be detected as phase shifts of the same magnitude, extracted from a fit with fixed $\omega = \omega_a$, which also allowed for background and other small distortions of the data.

Spurious phase shifts could be introduced by a different energy response of the upper and lower half of the counters, but these were checked by measuring the $(g-2)$ asymmetry separately for the two, and a negligible difference was found. The upper and lower scintillator counters were also swapped to avoid biases due to any efficiency difference. The sensitivity of the measurement to the vertical misalignment of the muons with respect to the median plane of the scintillator pair was checked by artificially displacing the detector: an effect arises from the correlation of the electron arrival time on the detector with its vertical distance from the median plane, since electrons arriving later (outward emitted ones, the deflection being towards the centre of the ring) have a larger vertical spread compared to the ones arriving earlier, and a net phase difference arises if the two detectors are not symmetrical with respect to the median plane (about 8 mrad for each 1 mm displacement). The position of the detector was set by equalizing the rates in the upper and lower counters, and was found to be consistent with the average muon position as computed from the measured value of the radial component of the magnetic field, and also checked by using a stop placed at different heights within the ring.

The result from 11 million $\mu^\pm$ decays detected was $d_E(\mu) = (3.7 \pm 3.4) \cdot 10^{-19} \, e$ cm, corresponding to $|d_E(\mu)| < 1.05 \cdot 10^{-18} \, e$ cm (95% confidence level).

The best limit on the muon EDM is $|d_E(\mu)| < 2.8 \cdot 10^{-19} \, e$ cm (95% confidence level), obtained by the BNL $g-2$ collaboration (McNabb, 2005), essentially with the technique described above.

A new experimental approach was proposed for the measurement of the muon EDM in a storage ring (Farley et al., 2004): by choosing γ such that $\gamma^2 - 1 \ll \kappa$ the second term in the square bracket of eqn (4.117) becomes practically independent of κ, and for a suitable value of an applied radial electric field $|\mathbf{E}| = \kappa |\mathbf{B}| c\beta\gamma^2$ the $g-2$ precession can be completely cancelled. In this way any electric dipole coupling will have its full effect of inducing a rotation around the radial direction: a particle initially polarized longitudinally will develop a spin component orthogonal to the plane of the trajectory, which increases linearly in time and can become rather large. It is possible that with this method significant improvements in the sensitivity can be achieved.

Limits on the EDM of the Λ hyperon were also determined by studying its spin precession. In their production by high energy (hundreds of GeV) proton collisions,

Λ are polarized (transversely with respect to the production plane), and their polarization is known from the measurement of the (parity-violating) decay asymmetry into $p\pi^-$. Starting from Λ with spin orthogonal to their velocity v, a static magnetic field B produced by conventional magnets can give large spin precession angles over the relatively long flight distances (metres) allowed by the long average decay lengths of the high-energy hyperons. The precession would be perturbed by the interaction of an EDM with the electric field $\gamma v \times B$ which appears in the Λ rest frame moving in the magnetic field: this interaction would originate a component of the polarization along the direction of the magnetic field B, otherwise absent. From the analysis of 3 million decays of Λ with momenta between 60 and 250 GeV/c, a Fermilab experiment obtained (Pondrom *et al.*, 1981) $|d_E(\Lambda)| < 1.5 \cdot 10^{-16}\, e$ cm.

4.4.4   Strong interactions

Since time reversal symmetry has been tested to be valid in electromagnetic interactions to a good extent, any significant test of this symmetry in strong interactions requires using either strong-interacting probes (e.g. nucleons) or photons (real or virtual) with wavelengths which are comparable or smaller than the size of the nucleus.

Limits on time reversal violation in strong interactions are usually obtained by checking the validity of the reciprocity relation (Henley, 1969). This requires some care, since in some cases reciprocity can be valid independently of time reversal symmetry. As mentioned, this happens when the first-order (Born) approximation is valid for a process (not the case for strong interactions) but also for reactions in which few states are available (e.g. close to the energy threshold), so that the most useful tests of time reversal symmetry through reciprocity are those in which many competing channels are open on the energy shell. For 'statistical' nuclear reactions, in which the reaction energy widths are much larger than the average energy spacing between compound nuclear levels, an enhancement in the sensitivity to time reversal violating amplitudes can be present (Ericson, 1966). On the other hand it is convenient to choose reactions dominated by a single amplitude, in order to avoid possible accidental cancellations of the T-violating phases; two-body reactions with spinless particles are optimal in this sense.

To avoid the need of absolute cross section measurements (and therefore fluxes, target thicknesses, etc.), the ratios of differential cross sections at two different scattering angles can be compared for the forward and backward reactions, or the cross sections for the two reactions can be normalized at some point. Since any possible time reversal violating amplitude is expected to be small, it is argued that cross section minima, where the relative effect of such amplitude can be more relevant, are the most sensitive regions for the tests.

An example of reciprocity test is the study of the reactions $p\,^{27}\mathrm{Al} \leftrightarrow \alpha\,^{24}\mathrm{Mg}$ (Blanke *et al.*, 1983). By using protons and α particles of MeV energies, for which the reaction proceeds via the formation of several compound nuclear states, the forward and backward reactions are normalized at a cross section maximum and compared at a minimum; regions in which the cross section is rather constant with energy are chosen for the comparison, in order to be insensitive against small energy shifts. By measuring the cross sections at scattering angles close to 180°, one ensures that a single helicity amplitude contributes even in presence of the spin 1/2 proton, since orbital angular momentum does not contribute along the direction of motion (the same would be true at 0°); note that the actual scattering angles in the laboratory must be different for the forward and backward reactions, to have the same scattering angle in the centre of mass.

In these experiments great care must be taken to control the energy calibration, stability, and reproducibility of the results, for example by a careful analysis of the energy structure of the cross section and the scattering angle, the precise knowledge of the beam, target, and detector geometries and by monitoring any left-right angular asymmetries. The possible presence of background (clearly different in the forward and backward reactions) must also be addressed and eliminated as much as possible (or measured and subtracted); in any case the two reciprocal reactions always have ineliminable differences at some level: for example *bremsstrahlung* produced by p and α particles is necessarily different, and this can in principle mimic T violation effects for a finite target thickness.

No violation of the reciprocity condition was found in detailed balance experiments, excluding the presence of time reversal non-invariant amplitudes in the reactions studied at the per mille level.

Tests of time reversal symmetry in elastic scattering reactions involve polarization measurements. The equality of final state polarization for unpolarized beams and asymmetry for polarized beams was seen to follow from time reversal symmetry, and as such it was tested in early experiments.

In spin 1/2–spin 0 scattering the only T-odd term which can be formed is $S\cdot(p_i - p_f)$, where S is the spin vector of the spin 1/2 particle and $p_{i,f}$ the initial and final momenta of one of the particles in the centre of mass. Parity symmetry however already requires $S\cdot p_{i,f}$ to vanish, and writing the scattering matrix in terms of the available vectors S, $p_{i,f}$ this will not contain T-odd terms, therefore leading to the conclusion that – if P symmetry is valid – tests of time reversal are not possible in this case. If parity symmetry is valid (true to a good extent for strong interactions), polarization-asymmetry tests with nucleons require target nuclei of non-zero spin. Conversely, under the same assumption, scattering on spin 0 nuclei (such as carbon) can be used as polarizer and/or analyser with $\mathcal{P} = \mathcal{A}$ without complications.

The simplest interesting case is pp scattering, and the use of nucleons instead of nuclei is an advantage because no nuclear structure complications are involved.

The only T-odd term allowed by parity symmetry (Wolfenstein and Ashkin, 1952) is in this case $S_1 \cdot (p_i - p_f)S_2 \cdot (p_i + p_f) + S_1 \cdot (p_i + p_f)S_2 \cdot (p_i - p_f)$, since the term $(S_1 \times S_2) \cdot (p_i \times p_f)$, being antisymmetric in the exchange of the two protons, is not allowed for identical particles. The measurement of such a term requires a polarized H_2 target and either a polarized beam or the measurement of the polarization of the scattered proton.

A double scattering experiment, in which the first target is carbon (spin 0) and the second one is a nucleus with non-zero spin (e.g. hydrogen) results in an asymmetry $\mathcal{P}(C)\,\mathcal{A}(H_2)$, $\mathcal{P}(C)$ being the beam polarization due to the first scattering on carbon and $\mathcal{A}(H_2)$ the analysing power of hydrogen; by interchanging the two targets the asymmetry becomes $\mathcal{P}(H_2)\,\mathcal{A}(C)$, and since $\mathcal{P}(C) = \mathcal{A}(C)$ the ratio of the two asymmetries allows one to test the equality of $\mathcal{A}(H_2)$ and $\mathcal{P}(H_2)$. Care must be taken to have the same centre of mass energy for the H_2 scattering in the two cases, by using degraders or by slightly changing the scattering angle in the two reactions. Early tests of this kind (e.g. Hillman *et al.* (1958)) verified the polarization-asymmetry equality in strong interactions at the percent level.

A similar test is based on the comparison of the final polarization in pp triple scattering. Two configurations with given initial and final polarizations are compared, which are related by time reversal (and rotation) in the centre of mass (see Fig. 4.8); when boosting to the laboratory frame the incident spin directions are unchanged while the scattering angle is modified ($\theta^* \rightarrow \theta$), as well as the angle between final spin and momenta. Time reversal symmetry requires the final polarization along the direction $\phi_f + \theta$ in the first reaction to be equal to that along the direction $\phi_i - \theta$ in the second one.

In one experiment (Limon *et al.*, 1968), a 430 MeV proton beam scattered on a beryllium target acquiring 54% polarization normal to the scattering plane; a magnetic field along the beam direction rotated such polarization by 90° into the scattering plane, and then the $(g - 2)$ precession in a bending magnet rotated it at an angle $\phi_i = 45°$ with respect to the momentum. The protons were then scattered by $\theta = 30°$ (65° in the centre of mass) on hydrogen, and precession in a second bending magnet rotated the final transverse polarization by $\pi/2 - \phi_i - \theta = 15°$ to make the component at 75° normal and analyse it by the up-down asymmetry in a third scattering on carbon. In a second configuration protons were scattered in the opposite direction $\theta = -30°$ and the second precession field was such that a rotation by $\pi/2 - \phi_i - \theta = 75°$ transformed the component at 15° to normal. The difference of the measured asymmetries in the two configurations is proportional to the difference in the two final state polarizations, and was measured to be consistent with zero at two standard deviations.

The analysis of the propagation of polarized neutrons through polarized media can also give information on the validity of time reversal symmetry (Kabir, 1988*b*). In general the neutron-nucleus forward scattering amplitude can depend on the

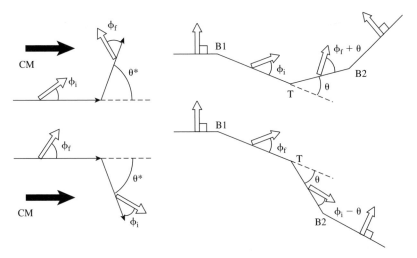

FIG. 4.8. The two *pp* scattering configurations related by time reversal to be compared (Limon *et al.*, 1968); single arrows represent momenta and double arrows polarization. Left: centre of mass (CM) frame (the direction of the CM momentum in the laboratory is indicated by the thick arrow); right: laboratory frame (B1,B2 are bending magnets, T is the target).

neutron spin S_n and the target spin S_T as

$$f(0) = f_0 + f_M \, S_n \cdot S_T + f_P \, S_n \cdot p_n + f_T \, S_n \cdot (p_n \times S_T) \qquad (4.118)$$

where p_n is the neutron momentum and the f_i coefficients are in general polynomials in $S_T \cdot p_n$. f_0 is the spin-independent part of the forward scattering amplitude, and f_M represents the spin-spin interaction. f_P is a *P*-odd and *T*-even term resulting in the different transmission probability for polarized neutrons with spin parallel or anti-parallel to the nuclear polarization, detected experimentally. f_T is a *P*-odd, *T*-odd term leading to triple correlations: its real part would cause the neutron spin to precess around the direction $p_n \times S_T$, while its imaginary part would cause differential absorption of neutrons depending on their polarization along the same direction. A f_T term linear in $S_T \cdot p_n$ would give a fivefold correlation which can be measured with polarized neutrons and a spin-aligned target.

An analogue of the polarization-asymmetry theorem also exists for neutron scattering (Kabir, 1988a): denoting by $\mathcal{P}_j$ the *j* component of the polarization acquired by an unpolarized neutron beam transmitted by a material, and by $\mathcal{A}_j$ the asymmetry in the transmission for a neutron beam fully polarized along the direction *j* (parallel or anti-parallel to it), one has

$$P_j = \frac{\sigma(j,k) + \sigma(j,-k) - [\sigma(-j,k) + \sigma(-j,-k)]}{\sigma(j,k) + \sigma(j,-k) + [\sigma(-j,k) + \sigma(-j,-k)]} \tag{4.119}$$

$$A_j = \frac{\sigma(k,j) + \sigma(-k,j) - [\sigma(k,-j) + \sigma(-k,-j)]}{\sigma(k,j) + \sigma(-k,j) + [\sigma(k,-j) + \sigma(-k,-j)]} \tag{4.120}$$

where $\sigma(\alpha,\beta)$ is the probability of having a neutron transmitted with spin up along α when the incident beam is fully polarized along β. Time reversal symmetry (reciprocity) requires

$$\sigma(\alpha,\beta) = \sigma(\beta_T,\alpha_T) \tag{4.121}$$

where α_T, β_T are the time-reversed directions; under motion reversal all three vectors p_n, S_n, S_T are inverted, and then a further π rotation around $p_n \times S_T$ (assumed to be a valid symmetry) restores the directions of p_n, S_T and the components of S_n lying in the plane of the above two vectors, so that

$$\sigma(j,-j) = \sigma(-j,j) \quad (\hat{j} \perp p_n \times S_T)$$

On the other hand, the component of S_n orthogonal to the (p_n, S_T) plane is unchanged by the rotation, so that

$$\sigma(-j,-j) = \sigma(j,j) \quad (\hat{j} \parallel p_n \times S_T)$$

Using the above equations in eqns (4.119), (4.120) one obtains

$$P_j = \pm A_j \tag{4.122}$$

where the sign is positive if the direction $\hat{j}$ lies in the plane containing the neutron momentum p_n and the target spin S_T, and negative otherwise. This result also allows one to determine whether a non-vanishing transverse polarization P_T (orthogonal to the neutron direction and target spin) generated in an unpolarized beam is due to spurious T-invariant effects: in this case the polarization should be opposite to the corresponding asymmetry, while if it arises uniquely from the selective absorption of neutrons with spin parallel or anti-parallel to the direction $p_n \times S_T$ it should be equal to the corresponding asymmetry.

Experiments are performed which test the symmetry of the elastic scattering S matrix by measuring the relative cross sections for spin-up and spin-down polarized neutrons on a spin-aligned target as a function of the angle between the neutron momentum and the target spin.

In one experiment (Huffman *et al.*, 1997) the forward transmission of a 5.9 MeV neutron beam with 67% polarization traversing a cryogenically cooled (160 mK) and spin-aligned ^{165}Ho target was measured as a function of the neutron spin direction, toggled every 100 ms, and the spin alignment direction of the target, which was periodically rotated. Holmium was chosen because it is mono-isotopic and the large internal hyperfine fields present in single crystals are such that at low temperatures (below 10 K) the nuclei become oriented forming a spiral structure with cylindrical symmetry around a crystal axis. The cross sections were measured to be equal at the 10^{-5} level, consistent with time reversal symmetry.

Further reading

An excellent (if somewhat dated) book on the subtleties of time reversal symmetry is Sachs (1987). The tight relationship of time reversal symmetry with CP symmetry enforced by the CPT theorem (Chapter 5) is such that T is usually discussed in the literature together with CP, so that all books on CP violation are good sources of material on T. A detailed discussion of the theoretical and experimental aspects of EDM searches can be found in Khriplovich and Lamoreaux (1997).

Problems

1. Verify the sign introduced by a time reversal transformation for a generic angular momentum state $|J, M\rangle$ built from a spin and an orbital part, i.e. eqn (4.30).

2. Obtain the reciprocity relation (4.63) from the explicit expression of the S matrix in terms of the interaction Hamiltonian, eq (C.25), for a T-symmetric H_{int}.

3. If $\varphi(x, t)$ is the wave function for a spinless particle corresponding to a plane wave, show explicitly that $\varphi^*(x, -t)$ is the one corresponding to a motion-reversed state. Do the same for the two-component Pauli spinor $\chi(\hat{n})$ corresponding to an eigenstate of the spin component along $\hat{n}$ (that is $S \cdot \hat{n}$) with eigenvalue $+\hbar/2$, showing that $i\sigma_2\chi^*(\hat{n})$ is indeed the time-reversed spinor.

4. For a spinless particle bound to a fixed centre by an arbitrary real potential $V(x)$, show that time reversal symmetry requires $\langle L \rangle = 0$ for orbital angular momentum.

5. Verify that the electric dipole interaction (4.96) can be obtained from the magnetic dipole one using a duality transformation.

6. Evaluate the commutation relation of the parity and time reversal operators P, T explicitly in the case of a spinor field, and determine the choice of the arbitrary phase η_P for which these operators commute.

7. Write an explicit expression in terms of creation and annihilation operators for the T operator enforcing the trasformation (4.39).

8. Consider the (weak) scattering reaction $\nu n \rightarrow \mu^- p$: assuming time reversal symmetry, what can be said about the polarization of the outgoing muon? Does the conclusion depend on a possible neutron polarization?

9. Consider the decay $K^+ \rightarrow \mu^+ \nu \gamma$: what are the constraints induced by time reversal symmetry on the polarization of the decay muon? Are such constraints valid independently of the strength of the interaction which drives the decay?

5

CPT AND SPIN STATISTICS

A mathematician may say anything he pleases,
but a physicist must be at least partially sane.
J. W. Gibbs

Alice laughed. 'There's no use trying' she said: 'One can't believe impossible things'.
'I daresay you haven't had much practice,' said the Queen.
'When I was your age I always did it for half-an-hour a day.
Why, sometimes I've believed as many as six
impossible things before breakfast.'
L. Carroll

There's no escape from my authority, I tell you
I am the one the only one
I am the god of kingdom come.
Brian H. May

This chapter deals with two 'sacred' principles in the current framework of physics, which are really fundamental cornerstones of the present theories: the *CPT* theorem and the spin-statistics connection. The validity of these principles is well verified experimentally and they are usually assumed to be universally valid; for the very same reasons a steady, if minor, experimental activity on their verification continues, since any evidence of their violation would have very profound significance.

5.1 *CPT* symmetry

The product *CPT* of charge conjugation, parity and time reversal transformations is an anti-unitary transformation, just as the factor T which it includes. Its effect on the (scalar, spinor, vector) fields is

$$(CPT)^\dagger \phi(x)(CPT) = \eta_{CPT}\, \phi^\dagger(-x) \tag{5.1}$$

$$(CPT)^\dagger \psi(x)(CPT) = \eta_{CPT}\, \gamma_5\gamma^0 \overline{\psi}^T(-x) \tag{5.2}$$

$$(CPT)^\dagger A^\mu(x)(CPT) = -\eta_{CPT}\, A^\mu(-x) \tag{5.3}$$

where the phase factor η_{CPT} depending on the field (with $|\eta_{CPT}|^2 = 1$) is not completely arbitrary (just as η_P), and is necessarily $+1$ for a scalar field. A Majorana particle (Section 3.1.3) is an eigenstate of *CPT* for which $\eta_{CPT}^2 = (-1)^{2S}$, where S is the particle spin (Carruthers, 1968).

The (normal-ordered) bilinear quantities built from fermion fields transform under *CPT* as

$$S_{21}(x) \rightarrow +\eta_{CPT}^*(2)\eta_{CPT}(1)\, S_{12}(-x) \tag{5.4}$$

$$V_{21}^\mu(x) \rightarrow -\eta_{CPT}^*(2)\eta_{CPT}(1)\, V_{12}^\mu(-x) \tag{5.5}$$

$$T_{21}^{\mu\nu}(x) \rightarrow +\eta_{CPT}^*(2)\eta_{CPT}(1)\, T_{12}^{\mu\nu}(-x) \tag{5.6}$$

$$A_{21}^\mu(x) \rightarrow -\eta_{CPT}^*(2)\eta_{CPT}(1)\, A_{12}^\mu(-x) \tag{5.7}$$

$$P_{21}(x) \rightarrow +\eta_{CPT}^*(2)\eta_{CPT}(1)\, P_{12}(-x) \tag{5.8}$$

so that, e.g., the electromagnetic current transforms as

$$j^\mu(t,\boldsymbol{x}) =: \overline{\psi}(t,\boldsymbol{x})\gamma^\mu\psi(t,\boldsymbol{x}) : \rightarrow - : \overline{\psi}(-t,-\boldsymbol{x})\gamma^\mu\psi(-t,-\boldsymbol{x}) := -j^\mu(-t,-\boldsymbol{x}) \tag{5.9}$$

Intuitively, since space-time has an even number of dimensions, we could naively think that the operation *PT* is equivalent to a (four-dimensional) rotation under which all the components of a four-vector are reversed, and which could be obtained as a Lorentz transformation (corresponding to a valid symmetry). However this does not work, as can be checked, e.g. , for the EM current j^μ: its space component (the current density vector) reverses under both P and T, while the time component (charge density) is unaffected by both, so that under *PT* the four-current transforms into itself $j^\mu \rightarrow j^\mu$. The reason lies in the pseudo-Euclidean nature of space-time, in which time and space coordinates behave differently: while momentum is easily reversed, we know that reversing the sign of the energy is related to antiparticles. C actually reverses the four current, so that it is the *CPT* transformation which has indeed the same effect on j^μ as an inversion of all four space-time axes. From the discussion above, the effect of a four-dimensional inversion of the axis is to change the sign of the electric charge, so that for each solution of the equations of motion a new solution appears, which hints at the deep connection of antiparticles to relativity.

During the consolidation phase of quantum field theory it was recognized that the effect of space-time inversion was deeply linked to the effect of charge conjugation, and this was formalized in the so-called *CPT* theorem (Schwinger, 1951; Bell, 1955; Pauli, 1955; Lüders, 1957). The *CPT* theorem is one of the most far-reaching theorems in physics, is deeply rooted in the theoretical framework used to describe Nature at the microscopic level and provides the justification of the Stueckelberg–Feynman picture of antiparticles as negative-energy particles travelling backwards in space-time.

5.1.1 The *CPT* theorem

The theorem states that the product of charge conjugation, parity, and time reversal transformations is a valid symmetry. More precisely: any quantum field theory based on a Hermitian, local (no action at a distance), normal-ordered Lagrangian which is invariant under Lorentz transformations, and for which the usual field commutation or anti-commutation rules hold, is also invariant under the transformation corresponding to the product of *C*, *P*, and *T*, taken in any order, irrespectively of its symmetry under the three inversions separately. This means that it is always possible to choose the phases which appear in the *C*, *P*, *T* transformation laws for the fields in such a way that the product of those operators is a symmetry of the theory. The *CPT* transformation is seen to possess a higher significance than the three component transformations. Note also that *CPT* is the only combination of *C*, *P*, and *T* which is at this time observed to be an exact symmetry of Nature.

Several proofs of the *CPT* theorem exist, based on slightly different hypotheses, but in any case its validity follows from very general assumptions; it is actually not easy to have consistent theories in which the *CPT* theorem is not automatically valid. We will just sketch a simple proof of the theorem (not the most general one) for Lagrangian quantum field theory.

The effect of the *PT* transformation, also called *strong reflection* (*SR*), is the inversion of all the space-time coordinates: $x \to -x$, and space-time derivatives are also inverted: $\partial_\mu \to -\partial_\mu$. Charge conjugation *C* exchanges the fields with their Hermitian conjugates. From the transformation laws for the fields and bilinear covariants one can see that the successive application of *C*, *P*, and *T* transforms even-rank tensors, including bilinear covariants built from normal-ordered products of spinor fields or terms involving derivatives, into their Hermitian conjugates, provided the normal (anti-)commutation rules hold. Similarly, odd-rank tensors are transformed into the negatives of their Hermitian conjugates; *c*-numbers are transformed into their complex conjugates (because of *T*). Since a generic local Lorentz-invariant Lagrangian density $\mathcal{L}(x)$ is a Lorentz scalar, all tensor indexes are contracted and the number of minus signs is thus even, $\mathcal{L}(x)$ transforms into its Hermitian conjugate, remaining it unchanged if it is Hermitian:

$$(CPT)^\dagger \mathcal{L}(t, \boldsymbol{x})(CPT) = +\mathcal{L}(-t, -\boldsymbol{x}) \tag{5.10}$$

The Hamiltonian density shares the same property, and the Hamiltonian (the integral over space-time of the Hamiltonian density) is unchanged under *CPT*, which is therefore a symmetry of the theory.

The hypothesis that $\mathcal{L}(x)$ is local means that it is built from terms containing the fields evaluated at the same space-time points, or derivatives of *finite* order; non-local theories have Lagrangian densities with terms of the form $\phi(x)\phi(y)$, which indeed contain an infinite number of derivatives, as one can write

$$\phi(x)\phi(y) = \phi(x)e^{(y-x)^\mu \partial_\mu}\phi(x)$$

and in this case it is no longer obvious that the Lagrangian contains a contraction of an even number of tensor indexes, so that the above proof does not hold.

More sophisticated proofs obtained in axiomatic field theory only require relativistic invariance and the local field concept (Streater and Wightman, 1964). The locality assumption can be exchanged for the more general one of micro-causality (Jost, 1957), requiring that the fields commute or anti-commute at space-time separations[66]: in physical terms this means that the field at a given space-time point can have no influence whatsoever on that at another point, separated from the first by a space-like interval.

Axiomatic proofs (Greenberg, 2003), exhibiting all the generality of the *CPT* theorem, start from the hypothesis of symmetry under proper orthochronous Lorentz transformations, and only require weak local commutativity, positiveness of the total energy and existence of a unique vacuum (the spin-statistics connection already follows from these assumptions).

As mentioned above, if space-time were Euclidean, the strong reflection $x \to -x$ (which in four dimensions would have determinant $+1$), could be obtained with continuity from the identity, and therefore symmetry under *SR* would follow directly from Lorentz invariance. In Minkowski space-time, however, strong reflection is not an orthochronous transformation and cannot be obtained in this way. Nevertheless, it is possible to obtain such transformation with *complex* Lorentz transformations (e.g. infinitesimal transformations such as $x_3 \to x_3 + i\epsilon x_0$, $x_0 \to x_0 + i\epsilon x_3$, from which the finite transformations $x_3 \to ix_0 \to -x_3$ and $x_0 \to ix_3 \to -x_0$ are obtained). The vacuum matrix elements of field products, from which all observables are computed, can be analytically continued into complex space, where they are invariant under the complex Lorentz group (and therefore also space-time inversion); by reordering the fields in the products (taking into account their commutativity or anti-commutativity) and going back to the real axis, one can obtain the relation between the original and transformed matrix elements in a way which is exactly the one enforced by *CPT*.

In all proofs of the *CPT* theorem Lorentz symmetry is the basic hypothesis, and indeed a theorem (Greenberg, 2002) states that if *CPT* symmetry is violated then Lorentz symmetry must be violated too; for this reason the tests of *CPT* and Lorentz symmetries are closely linked.

Speculations on the violation of *CPT* symmetry recently arose in the discussion of possible failures of conventional quantum mechanics in quantum gravity, leading to a violation of unitarity, and in the context of string theory, since strings would be extended objects with nonlocal interactions and could possibly evade the *CPT* theorem (see e.g. Mavromatos (2005) for a review).

5.1.2   Transition amplitudes

By applying C, P, and T a single particle state is transformed into an antiparticle state with reversed spin; due to the anti-unitary nature of the *CPT* operator, some care is required in spelling out the consequences of *CPT* symmetry, just as in the

---

[66] Actually, one only needs such property to hold for the *vacuum matrix elements* of field products, which is called *weak local commutativity*.

case of time reversal. If CPT symmetry is valid, the S and T matrices transform under the (anti-linear) CPT transformation as

$$S \to S_{CPT} = (CPT)^\dagger S (CPT) = S^\dagger \tag{5.11}$$

$$T \to T_{CPT} = (CPT)^\dagger T (CPT) = T^\dagger \tag{5.12}$$

Denoting with a bar the C-conjugate states and with subscripts PT the parity- and (naive) time-inverted ones (only spins inverted), this implies for the transition amplitudes $A(i \to f) \equiv T_{fi} = \langle f | T | i \rangle$:

$$A(i \to f) = \eta_{CPT}(f) \eta^*_{CPT}(i) A(\bar{f}_{PT} \to \bar{i}_{PT}) \tag{5.13}$$

Frequently it is possible to deal with states which are unchanged by PT (spinless particles or spin-averaged transition amplitudes)

$$|i_{PT}\rangle = |i\rangle \quad |f_{PT}\rangle = |f\rangle \tag{5.14}$$

so that the subscript can be dropped and eqn (5.13) links the amplitudes between C-conjugate states $i \to f$ and $\bar{f} \to \bar{i}$; in this case, for self-conjugated states CPT enforces the equality of the $i \to f$ and $f \to i$ processes (as time reversal symmetry, under the same conditions, see eqn (4.64)).

As in the case of time reversal (Chapter 4), if the transition matrix can be approximated at first order by the (Hermitian) interaction Hamiltonian, then CPT symmetry (5.12) requires

$$A(i \to f) = \eta_{CPT}(f) \, \eta^*_{CPT}(i) \, A^*(\bar{i}_{PT} \to \bar{f}_{PT}) \tag{5.15}$$

and the transition amplitudes between self-conjugated states respecting (5.14) can be chosen to be real. The above is actually (approximately) valid when the initial and final states are weakly interacting, independently from the Hermiticity of T (Feinberg, 1960): using the unitarity condition (C.20)

$$\langle f | (T^\dagger - T) | i \rangle = i \, (2\pi) \sum_n \delta(E_f - E_n) \langle f | T^\dagger | n \rangle \langle n | T | i \rangle \tag{5.16}$$

and if the states $|i\rangle$ and $|f\rangle$ contain only weakly interacting particles, then the term on the right-hand side of eqn (5.16) is necessarily smaller than those in the left-hand side, as it contains a higher power of a small coupling constant, and can thus be neglected; this gives

$$\langle f | (T^\dagger - T) | i \rangle \simeq 0 \quad \langle f | T^\dagger | i \rangle = \frac{1}{2} \langle f | (T + T^\dagger) | i \rangle = \langle f | T | i \rangle$$

and the transition matrix elements are expressed in terms of the matrix elements of the Hermitian operator $\mathcal{T} + \mathcal{T}^\dagger$, so that finally *CPT* symmetry gives

$$\langle f|\mathcal{T}|i\rangle = \langle f|(CPT)^\dagger \mathcal{T}^\dagger (CPT)|i\rangle = \langle \overline{f}_{PT}|\mathcal{T}^\dagger|\overline{i}_{PT}\rangle^* = \langle \overline{f}_{PT}|\mathcal{T}|\overline{i}_{PT}\rangle^* \quad (5.17)$$

(the same kind of argument applies to time reversal).

In any case a first-order approximation is always legitimate at least for the weak part of the transition matrix $\mathcal{T}_W$ (Appendix C), and the derivation of the Fermi–Watson theorem in Chapter 4 can be repeated for the case of *CPT* symmetry: for states (5.14) which are unchanged by *PT* and which are eigenstates of the strong interaction (eqns 4.73), it is possible to go beyond eqn (5.13) obtaining the relation

$$A(i \to f) = \eta_{CPT}(f)\eta^*_{CPT}(i)\, e^{2i(\delta_i + \delta_f)} A^*(\overline{i} \to \overline{f}) \quad (5.18)$$

where $\delta_{i,f}$ are the strong phase shifts: the *CP*-conjugate decay amplitude is the complex conjugate of the original one, apart from the strong phase shifts.

5.1.3 Consequences of *CPT* symmetry

A first point to be noted is that *CPT* symmetry requires that if one of the C, P, T symmetries is violated then at least another one must be violated: it is in this sense (and only in this sense) that, e.g., *CP* violation is said to be 'equivalent' to T violation; indeed this is why the known violation of *CP* in weak interactions (Chapter 6) strongly hinted at the fact that T might also be violated there, even before this was directly proved to be the case.

Given the great generality of the *CPT* theorem, the consequences of *CPT* symmetry are of very wide applicability; in general, many of the statements which can be made about the relation between particles and antiparticles do not really depend on the exact validity of charge-conjugation symmetry, but rather stem directly from the more fundamental *CPT* symmetry.

Here follow the most important consequences which follow from the requirement of *CPT* symmetry.

■ Antiparticles must exist, even if charge conjugation is not an exact symmetry of Nature.

■ Particles and antiparticles have equal masses (if they are stable or cannot decay into other single-particle states) (Lüders and Zumino, 1957).
Stable particles are eigenstates of the Hamiltonian H, so that considering particle $|a\rangle$ and antiparticle states $|\overline{a}\rangle$ at rest, for which

$$CPT|a\rangle = \eta_{CPT}|\overline{a}\rangle \qquad H|a\rangle = E(a)|a\rangle \qquad H|\overline{a}\rangle = E(\overline{a})|\overline{a}\rangle$$

if H commutes with *CPT* one has $E(a) = E(\overline{a})$.

For unstable particles *CPT* symmetry analogously implies the equality of the expectation value of H for the states $|a\rangle$ and $|\bar{a}\rangle$; if the particles do not couple to other single-particle states the Hamiltonian is diagonal in the subspace of single-particle states, and the diagonal elements $E(a)$ and $E(\bar{a})$ are directly related to the masses.

■ Particles and antiparticles have the same lifetimes (total decay widths) (Lüders and Zumino, 1957).
Consider a Hamiltonian $H_0 + H_{\text{int}}$, and a particle $|a\rangle$ which can decay into states $|f\rangle$ due to the effect of H_{int}. The lifetime τ_a is

$$\tau_a^{-1} \propto \sum_f |\langle f|T|a\rangle|^2$$

where the sum extends to all possible final states. Given the equality of the masses for particles and antiparticles, the phase space factor is the same for them and the only concern is the T matrix element; for this

$$\sum_f |\langle f|T|a\rangle|^2 = \sum_f \langle a|T^\dagger|f\rangle\langle f|T|a\rangle = \langle a|T^\dagger T|a\rangle$$

due to the completeness of the sum over states f. *CPT* symmetry implies that this is equal to

$$\langle a|(CPT)^\dagger T T^\dagger (CPT)|a\rangle = \langle \bar{a}_{PT}|TT^\dagger|\bar{a}_{PT}\rangle = \langle \bar{a}_{PT}|T^\dagger T|\bar{a}_{PT}\rangle$$

since $TT^\dagger = T^\dagger T$ from unitarity, eqn (C.9). Inserting a sum over a complete set of states this becomes

$$\sum_f \langle \bar{a}_{PT}|T^\dagger|f\rangle\langle f|T|\bar{a}_{PT}\rangle = \sum_f \langle f|T|\bar{a}_{PT}\rangle^*\langle f|T|\bar{a}_{PT}\rangle = \sum_f |\langle f|T|\bar{a}_{PT}\rangle|^2$$

since the effect of PT is just to invert the spin orientation of a, on which the total decay rate cannot depend due to rotational invariance, the result follows. The validity of the above proof hinged on the sum over a complete set of final states, therefore *CPT* symmetry does not imply the equality of any *partial* decay width for particles and antiparticles (required instead by the C or CP symmetries).

■ *Partial* decay widths corresponding to sums over all states which are coupled among them by the strong part $\mathcal{S}_S$ of the scattering matrix are equal if *CPT* symmetry is valid, since the sum over states which gives the equality of the decay widths is only required to run over those states which would be mixed together in forming eigenstates of $\mathcal{S}_S$.
The proof follows along the same lines used to derive eqn (4.71) in the case of time reversal invariance: consider a particle $|i\rangle$, which would be stable under

the effect of the strong interactions (an eigenstate of $\mathcal{S}_S$), and which can decay through weak interactions ($\mathcal{T}_W$) into several different final states $|f\rangle$. If the $\mathcal{S}_S$ matrix is block-diagonal, i.e. there are sets F_α of final states which do not couple to others through strong interactions:

$$\langle f|\mathcal{S}_S|f'\rangle = 0 \qquad \text{if} \qquad f \in F_\alpha, f' \in F_\beta \, (\alpha \neq \beta) \qquad (5.19)$$

then the $\mathcal{S}_S$ matrix is unitary over each of those sets, and by summing over all the states in them

$$\sum_{\bar{f} \in \bar{F}_\alpha} |\langle \bar{f}|T|\bar{i}\rangle|^2 = \sum_{\bar{f} \in \bar{F}_\alpha} |\langle \bar{f}|\mathcal{T}_W|\bar{i}\rangle|^2$$

$$= \sum_{\bar{f} \in \bar{F}_\alpha} |\langle \bar{f}|(CPT)^\dagger \mathcal{T}_W^\dagger (CPT)|\bar{i}\rangle|^2 = \sum_{f \in F_\alpha} |\langle f|\mathcal{T}_W^\dagger|i\rangle|^2$$

Using eqn (C.36) and inserting a sum over a complete set of states this becomes

$$\sum_{f \in F_\alpha} |\langle f|\mathcal{S}_S^\dagger \mathcal{T}_W \mathcal{S}_S^\dagger|i\rangle|^2 = \sum_{f \in F_\alpha} \left|\sum_n \langle f|\mathcal{S}_S^\dagger|n\rangle \langle n|\mathcal{T}_W|i\rangle\right|^2$$

$$= \sum_{f \in F_\alpha} \sum_{n,m} \langle f|\mathcal{S}_S^\dagger|n\rangle \langle m|\mathcal{S}_S|f\rangle \langle n|\mathcal{T}_W|i\rangle \langle m|\mathcal{T}_W|i\rangle^*$$

where the sum over states makes the inversion of spins introduced by *PT* irrelevant. The term

$$\sum_{n,m} \sum_{f \in F_\alpha} \langle m|\mathcal{S}_S|f\rangle \langle f|\mathcal{S}_S^\dagger|n\rangle = \begin{cases} \delta_{n,m} & \text{if } n, m \in F_\alpha \\ 0 & \text{otherwise} \end{cases}$$

since in the first case the sum over f is equivalent to a sum over all states, as the contribution from states not belonging to F_α vanishes due to (5.19). Therefore

$$\sum_{\bar{f} \in \bar{F}_\alpha} |\langle \bar{f}|T|\bar{i}\rangle|^2 = \sum_{n \in F_\alpha} |\langle n|\mathcal{T}_W|i\rangle|^2$$

and the result is proved.

In practical cases final state interactions induced by weak interactions are neglected, and the above enforces the equality of the partial rates for semi-inclusive decay modes, summed over all the final states in a set which does not couple to another set through strong or EM interactions. In particular, if states containing photons (induced by electromagnetic interactions) are neglected, this enforces the equality of meson decay widths into leptonic and semi-leptonic modes

$$\Gamma(M^+ \rightarrow \ell^+ \nu_\ell) = \Gamma(M^- \rightarrow \ell^- \overline{\nu}_\ell)$$

$$\Gamma(M^+ \rightarrow h^0 \ell^+ \nu_\ell) = \Gamma(M^- \rightarrow \overline{h}^0 \ell^- \overline{\nu}_\ell)$$

$$\Gamma(M^0 \rightarrow h^- \ell^+ \nu_\ell) = \Gamma(\overline{M}^0 \rightarrow h^+ \ell^- \overline{\nu}_\ell)$$

where $\ell^\pm$ are charged leptons, and h^+, h^0 and $h^-, \overline{h}^0$ are C-conjugated hadrons (a single hadron is necessarily an eigenstate of strong interactions). Other specific selection rules may give rise to sets of disconnected states: as an example, when neglecting electromagnetic final state interactions, the partial widths for semi-leptonic decays of K mesons with any given number of pions in the final state are equal by *CPT*: the conservation of G-parity (Section 3.2.4) in strong interactions decouples the $2\pi \ell \nu$ and $3\pi \ell \nu$ modes (two pions cannot scatter into three by strong interactions); for a single pion the result was stated in the second and third of the above equations, and more than three pions are not allowed by phase space. In general the sum runs over all final states sharing the same quantum numbers for all the observables which are conserved by the strong part of the interaction.

The equalities of partial rates enforced by *CPT* must be taken into account when considering tests of C (or CP) symmetry, which are thus somewhat hindered.

■ In the absence of final state interactions there can be no interference effect between parity-conserving and parity-violating amplitudes for a decay induced by the weak interaction at first order, if charge conjugation symmetry holds (as stated without proof in Chapter 3); this applies to the decay rate and other observables which are even under time reversal (Lee, Oehme, and Yang, 1957b; Coester, 1957).

The transition operator $\mathcal{T}$ can be split in two parts with opposite behaviour under parity, eqn (2.61); charge conjugation (and *CPT*) symmetry links the effect of time reversal to that of parity: *CPT* and C symmetry together (*PT* symmetry) require

$$\langle \beta | \mathcal{T} | \alpha \rangle^* = \langle \alpha | \mathcal{T}^\dagger | \beta \rangle = \langle \alpha | (CPT)^\dagger \mathcal{T}(CPT) | \beta \rangle$$
$$= \langle \alpha_T | (CP)^\dagger \mathcal{T}(CP) | \beta_T \rangle^* = \langle \alpha_T | P^\dagger \mathcal{T} P | \beta_T \rangle^*$$

so that if $\mathcal{T}$ is Hermitian (as it happens at first order in absence of FSI)

$$\langle \beta | \mathcal{T}^{(\pm)} | \alpha \rangle^* = \pm \langle \beta_T | \mathcal{T}^{(\pm)} | \alpha_T \rangle \qquad (5.20)$$

for states unaffected by naive time reversal the transition matrix elements are real (imaginary) for the parity-conserving (violating) part of the interaction, and they cannot interfere in the integrated decay rate.

More generally, the expectation value of an operator O in a state $|i\rangle$ is written as

$$\langle O \rangle = \sum_{f_\alpha, f_\beta} \langle f_\alpha | T | i \rangle \langle f_\beta | T | i \rangle^* \langle f_\beta | O | f_\alpha \rangle \tag{5.21}$$

where the states $|f_\alpha, f_\beta\rangle$ correspond to specific spin and momentum config-urations of the particles in the final state. Inserting eqn (4.82) in the above expression, an interference term appears

$$\sum_{f_\alpha, f_\beta} \left[\langle f_\alpha | T^{(+)} | i \rangle \langle f_\beta | T^{(-)} | i \rangle^* + \langle f_\alpha | T^{(-)} | i \rangle \langle f_\beta | T^{(+)} | i \rangle^* \right] \langle f_\beta | O | f_\alpha \rangle$$

$$\tag{5.22}$$

If $O = O^{(T+)}$ is T-even ($T^\dagger O^{(T+)} T = +O^{(T+)}$) then

$$\langle f_\beta | O^{(T+)} | f_\alpha \rangle = \langle f_{\beta T} | O^{(T+)} | f_{\alpha T} \rangle^* = \langle f_{\alpha T} | O^{(T+)} | f_{\beta T} \rangle$$

(O is Hermitian); this equation and eqn (5.20) allow one to write

$$\sum_{f_\alpha, f_\beta} \langle f_\alpha | T^{(+)} | i \rangle \langle f_\beta | T^{(-)} | i \rangle^* \langle f_\beta | O^{(T+)} | f_\alpha \rangle$$

$$= - \sum_{f_\alpha, f_\beta} \langle f_{\alpha T} | T^{(+)} | i_T \rangle^* \langle f_{\beta T} | T^{(-)} | i_T \rangle \langle f_{\alpha T} | O^{(T+)} | f_{\beta T} \rangle$$

For the decay of a spinless particle $|i_T\rangle = |i\rangle$; for a particle with spin $|i_T\rangle$ differs from $|i\rangle$ by the spin direction, but rotational invariance allows to exchange the two. Finally one can see that the first term in (5.22) cancels the second after a relabelling, and the interference term vanishes.

This result was already included in the general discussion of P and T-odd observables in Chapter 4: indeed Table 4.1 shows that with no FSI a non-zero T-even, P-odd quantity (which must arise from the interference of a parity-conserving and a parity-violating amplitude) violates C; no parity-violating asymmetry can be observed unless either charge conjugation symmetry is also violated or final state interactions are present.

■ The previous results are not limited to particle decays, but can be trivially extended to the equality of total *reaction rates* from an initial set of particles and the corresponding set of antiparticles with spins reversed. For example the *CPT* relation (5.13) requires the equality of the elastic amplitudes, and therefore cross sections:

$$|A(i \to i)|^2 = |A(\bar{i}_{PT} \to \bar{i}_{PT})|^2$$

while using the unitarity relation (C.7)

$$\sum_f |A(i \rightarrow f)|^2 = \sum_f |A(f \rightarrow i)|^2 = \sum_f |A(\bar{i}_{PT} \rightarrow \bar{f}_{PT})|^2 \qquad (5.23)$$

since the sum over f includes both states and anti-states. Total cross sections are equal for particles and antiparticles (clearly *not* for particles and antiparticles on *matter*).

- The electromagnetic properties of particles and antiparticles are equal and opposite, as can be seen by considering the fact that the (Hermitian) charge density ρ transforms under *CPT* as

$$(CPT)^{\dagger} \rho(x)(CPT) = -\rho(-x)$$

It should be noted that several of the consequences of *CPT* symmetry would also follow from *C* or *CP* symmetry, but since these are already known to be violated, the tests of the above equalities are indeed tests of *CPT* (moreover the size of *CPT* violation, if any, must be at least smaller than the measured *CP* violation).

5.2 Experimental tests of *CPT*

As discussed, *CPT* symmetry is a very fundamental property of the way we understand and describe Nature, and most theories do incorporate it right from the start. Nevertheless, its validity must ultimately rest on experiment, so that many tests of *CPT* symmetry have been and are being performed.

Some caution is necessary when comparing different limits on *CPT* violation, because in absence of a specific model it is not possible to rank the experimental tests in terms of sensitivity, as different phenomena might be affected by different types of *CPT* violation.

5.2.1 Particle-antiparticle comparisons

Precise limits on the possible failure of *CPT* are mostly obtained by the accurate comparison of the masses, lifetimes, electric charges and anomalous magnetic moments of particles and antiparticles, and by the detailed analysis of the behaviour of neutral flavoured meson systems. The latter will be discussed in Chapters 8 and 10, while some examples of the former are presented here.

Table 5.1 lists some limits for the equality of static properties of particles and antiparticles (Yao *et al.*, 2006); strong limits are usually obtained for stable particles which allow longer measurements (the lifetime of an unstable state τ sets a natural time limit for the observation, corresponding to an energy resolution $\Delta E \sim h/\tau$).

TABLE 5.1 Experimental limits for some tests of *CPT* symmetry based on the comparison of static properties for particles and antiparticles (Yao *et al.*, 2006). m, Q, g, and τ
indicate mass, charge, spin factor, and lifetime respectively; since no violation of *CPT*
has been detected, the reference quantities in the second column are the averages of the
measured values for particle and antiparticle, unless noted otherwise.

Quantity	Value	Notes
$m(e^+) - m(e^-)$	$< 8 \cdot 10^{-9}\, m_e$	Ps spectroscopy
$Q(e^+) + Q(e^-)$	$< 4 \cdot 10^{-8}\, e$	Ps spectroscopy $+ \omega_c$
$g(e^+) - g(e^-)$	$(-0.5 \pm 2.1) \cdot 10^{-12}\, g_e$	Penning trap
$m(\overline{p}) - m(p)$	$< 1 \cdot 10^{-8}\, m_p$	$\overline{p}\,^4\mathrm{He}^+$ atoms
$|Q(\overline{p})/m(\overline{p})| - Q(p)/m(p)$	$(-9 \pm 9) \cdot 10^{-11}\, |e|/m_p$	Penning trap
$\tau(\mu^+) - \tau(\mu^-)$	$(2 \pm 8) \cdot 10^{-5}\, \tau_\mu$	Pulsed beam and H_2
$\tau(\pi^+) - \tau(\pi^-)$	$(5.5 \pm 7.1) \cdot 10^{-4}\, \tau_\pi$	Beams in vacuum
$\tau(K^+) - \tau(K^-)$	$(0.11 \pm 0.09)\%\, \tau_K$	Beams in vacuum
$|m(K^0) - m(\overline{K}\,^0)|$	$< 10^{-18}\, m_K$	See Chapter 8

Since the different figures cannot be directly compared without a specific model of
CPT violation, tests in a variety of systems are interesting.

A limit on the difference between electron and positron masses is obtained
indirectly by comparing the measurement of the $2^3 S_1 - 1^3 S_1$ energy level difference
in positronium (Fee *et al.*, 1993) with the QED predictions, which assume the
equality of such masses: the Rydberg constant for positronium (Ps)

$$R(\mathrm{Ps}) = \frac{m(e^+)\, m(e^-)}{m(e^+) + m(e^-)} \frac{c^2 \alpha^2}{2hc} \tag{5.24}$$

is half of R_∞, the one for the hydrogen atom in the limit of infinite proton mass

$$R_\infty = m(e^-) \frac{c^2 \alpha^2}{2hc} \tag{5.25}$$

only if $m(e^+) = m(e^-)$. For Ps the ratio of binding energy to rest mass is $\mathrm{O}(10^{-5})$,
requiring relativistic approaches to compute its energy levels; despite this complication, correction terms of order 10^{-8} are evaluated for the above energy
difference.

An atom can be excited by two-photon absorption from an initial state to a
final state with the same parity: the relatively long lifetimes of the first two triplet S
states correspond to a very small natural line-width $\mathrm{O}(10^{-9})$. By using two counter-
propagating laser beams the Doppler shifts seen by a moving atom cancel in the
sum of the frequencies to first order, thus making the absorption line sharper and

allowing more atoms to be excited; the measurement of the transition frequency to a precision of $2.6 \cdot 10^{-9}$ reflects in the error on the electron and positron mass ratio quoted in Table 5.1; this limit actually assumes that the magnitudes of the e^+ and e^- electric charges are the same.

By combining the above measurement with experimental data on quantities with a different functional dependence on charges and masses, a limit on the charge difference of e^+ and e^- can be also obtained (Hughes and Deutch, 1992): the cyclotron frequency of a particle of charge Q and mass m in a magnetic field $\boldsymbol{B}$ is

$$\omega_c = \frac{|Q\boldsymbol{B}|}{m} \tag{5.26}$$

and by comparing the ratio of cyclotron frequencies for e^+ and e^- in the same field

$$\frac{\omega_c(e^+)}{\omega_c(e^-)} = \frac{|Q(e^+)|\, m(e^-)}{|Q(e^-)|\, m(e^+)}$$

with the ratio of the Rydberg constants for positronium and hydrogen

$$\frac{R(\text{Ps})}{R_\infty} = \frac{m(e^+)\, m(e^-)}{m(e^-)[m(e^+) + m(e^-)]} \frac{Q^2(e^+)}{Q^2(e^-)}$$

the mass ratio can be eliminated, obtaining a limit on the electron to positron charge ratio; writing

$$\frac{|Q(e^+)|}{|Q(e^-)|} = 1 + \Delta Q \qquad \frac{\omega_c(e^+)}{\omega_c(e^-)} = 1 + \Delta\omega_c \qquad \frac{R(\text{Ps})}{R_\infty} = 1/2 + \Delta R$$

one obtains

$$\Delta Q = \frac{1}{5}(\Delta\omega_c + 4\Delta R) \tag{5.27}$$

at first order. The $\sim 10^{-7}$ error on the measured ratio of e^+ and e^- cyclotron frequencies in the same trap, and the one from the positronium experiment discussed above, give the *CPT* limit on the electron-positron charge equality in Table 5.1.

Note also that, if electric charge conservation (and the absence of charged particles with very small charge) were assumed, the experimental limit on the neutrality of the photon, $Q(\gamma) < 5 \cdot 10^{-30}$ (obtained from the dispersion of light from pulsars (Raffelt, 1994)), would give a more stringent limit on the equality of particle and antiparticle charges by the observation of their annihilation into photons.

As the only stable anti-hadron, the antiproton is a potentially interesting system for the study of antiparticle properties. Antihydrogen appears to be the best system for sensitive tests of *CPT* symmetry: first of all hydrogen, as the simplest atomic structure, is quite well understood, and its energy levels (which should be equal for antihydrogen, if *CPT* symmetry holds) are predicted with very high accuracy; moreover, very precise spectroscopic techniques can be used for its study, offering in principle an unmatched potential for comparisons.

Other neutral antiparticle systems are experimentally disfavoured for various reasons: while positronium is even simpler than hydrogen, it is a self-charge-conjugate system and has a short lifetime; protonium ($p\bar{p}$) shares the same kind of problems, with an even shorter lifetime. On the other hand muonium (μ^+e^-) is not self-charge-conjugate, and its charge-conjugate cannot be easily prepared, as negative muons in matter undergo nuclear capture.

Cold antihydrogen ($\bar{p}e^+$) samples have been produced recently at CERN in the ATHENA (Amoretti *et al.*, 2002) and ATRAP (Gabrielse *et al.*, 2002) experiments, using nested Penning traps, in which low energy $\bar{p}$ and e^+ separately trapped and cooled to O(10 K) were allowed to mix. When antihydrogen spectroscopy becomes possible, an approach similar to the one described above could be used to obtain a limit on the proton antiproton charge ratio, by using data on the cyclotron frequencies of p and $\bar{p}$, and the measurement of the extremely small line-width of the $1S - 2S$ two-photon transition ($2.5 \cdot 10^{15}$ Hz, measured to $2 \cdot 10^{-14}$ accuracy in hydrogen) in presence of a magnetic field; alternatively, the measurement of the 1.4 GHz microwave transition between the ground state hyperfine levels, which is proportional to the antiproton magnetic moment, or the measurement of the Lamb shift in a relativistic antiproton beam, also appear as possible tests of *CPT* (Eades and Hartmann, 1999).

The present limit on the p-$\bar{p}$ mass difference (Hori *et al.*, 2003) is obtained by the study of energy levels in antiprotonic atoms, in which a $\bar{p}$ substitutes one of the electrons after having knocked it out in a collision. While such kinds of atoms have usually very short lifetimes before the $\bar{p}$ annihilates, in the peculiar case of helium, the atom lifetime was accidentally discovered to be relatively long ($\sim$3 μs) for some range of levels, due to a combination of the shielding effect of the remaining electron and the level structure; this allowed the possibility of laser excitation and spectroscopic measurements on such atoms, effectively opening a new field of research (see Yamazaki *et al.* (2002) for a review).

Penning trap measurements

Precise measurements of cyclotron frequencies are obtained for single particles stored inside Penning traps, the so-called 'geonium atoms' (particle bound to an apparatus on Earth by a trapping field). Measurements on single trapped particles are free from the unavoidable and complicated space charge effects which are present with clouds of charged particles. This field has a long history of precision experiments, with constant improvements obtained by developing sophisticated techniques: a comprehensive discussion is found e.g. in Brown and Gabrielse (1986).

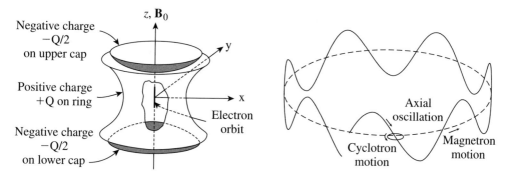

FIG. 5.1. Left: schematic of a Penning trap (Van Dyck *et al.*, 1986). Right:orbit of a charged particle inside the trap, showing the fast cyclotron motion around the guiding centre moving by magnetron motion, and the axial oscillation (Brown and Gabrielse, 1986).

Basically, a Penning trap is an electrode configuration with axial symmetry, consisting of two end caps in the shape of hyperbolas of revolution which produce an electric quadrupole field, and a central ring electrode; a strong uniform magnetic field B_0 is superimposed along the axis of the trap. Typical dimensions are a few cm, and the whole trap is in a cryogenic environment at liquid He temperature of 4.2 K.

The magnetic field bounds radially the motion of a charged particle with mass m and charge Q in the trap, and the electric field bounds it axially: the particle makes a fast cyclotron motion with small radius (at frequency ω_c), around a slowly rotating centre (magnetron motion, at frequency ω_m), with a harmonic oscillation (axial motion, at frequency ω_z) superimposed (Fig. 5.1). The frequencies corresponding to cyclotron, magnetron and axial motion for a trapped proton (electron) can be of order $\omega_c \sim 100$ MHz (100 GHz), $\omega_m \sim 500$ kHz (10 kHz) and $\omega_z \sim 10$ MHz (50 MHz); in general $\omega_c \gg \omega_z \gg \omega_m$. For typical trap parameters (a few T magnetic field, few tens of V over distances of a fraction of cm) the equilibrium orbits have radii of order 10^{-5} mm for the cyclotron motion and 10^{-3} mm for the magnetron motion, and the amplitude of the axial motion is of order $10^{-2} - 10^{-3}$ mm: the total motion thus occupies a very small volume in which the fields are mostly homogeneous, limiting the widening of the line-widths due to non-linear effects.

In the ideal case the three frequencies are properties of the trap, independent of the location of the particle inside it. The use of an intense magnetic field is required to make the system insensitive to other small fluctuating fields present in the laboratory and to separate the cyclotron energy levels such that the thermal blackbody radiation from the electrodes does not excite the particle out of the lowest quantum state.

The frequency ω_z of the axial harmonic motion is determined by the electric trapping potential

$$\omega_z = \sqrt{\frac{QV_0}{md^2}}$$

where V_0 is the potential difference between the end-cap and ring electrodes and d a characteristic linear dimension of the trap. The presence of the trapping electric field, with a non-zero radial component, slightly shifts the cyclotron frequency from its free value

$$\omega_c^{(0)} = \frac{|QB_0|}{m} \rightarrow \omega_c = \omega_c^{(0)} - \frac{\omega_z^2}{2\omega_c}$$

This cyclotron frequency shift is practically insensitive to imperfections of the trap geometry; even for an imperfect trap however the following relation holds

$$\omega_c^2 + \omega_z^2 + \omega_m^2 = \omega_c^{(0)2} \tag{5.28}$$

so that the free cyclotron frequency can always be determined from the measurement of the three frequencies

The particles are initially captured in large orbits, then synchrotron radiation emission rapidly shrinks the cyclotron radius for electrons (while for heavier particles an external damping circuit is used). The radius of the magnetron motion instead *increases* when energy is lost, and thus tends to bring the particle outside the trap: energy has therefore to be supplied in order to effectively shrink it. This is achieved by introducing an inhomogeneous oscillatory electric field which couples the magnetron motion to the axial motion, which is in contact with a thermal bath.

The axial motion is driven by applying an alternating voltage between the ring and one end-cap electrode (to increase its oscillation amplitude), and monitored by the oscillating current due to the image charges induced on the other end-cap. This approach is possible because ω_z is in the radio-frequency range: a phase sensitive current detector measures the above current, which is proportional to the number of trapped particles. By increasing the drive amplitude the particles can be selectively pushed outside the trap, until only a single one is left. If the drive frequency coincided with the one ($\sim\omega_z$) at which the detector is tuned, the noise induced by the former would swamp the small signal due to the axial motion: for this reason the trapping electric potential is weakly modulated at a frequency $\omega_t \ll \omega_z$, so that the axial motion can be driven at the sideband frequencies $\omega_z \pm \omega_t$ instead, eliminating the signal pick-up.

For a trapped proton or ion, the cyclotron oscillation (at RF frequencies) can be detected with a method analogous to the above, by segmenting the ring electrode azimuthally and using one part of it to drive the motion and the other to detect the induced signal; in this case, however, the elimination of the feed-through signal from drive to detector must be reduced with different techniques, as the modulation of the magnetic field is difficult to implement. Note that the energy damping due to the induced signal is also effective in reducing the radius of the cyclotron motion.

For a trapped electron ω_c is much higher (in the high microwave range, and decreasing it by reducing the magnitude of the axial magnetic field is not a good option, as mentioned above), so that a different technique is used to detect it. An additional weak inhomogeneous magnetic field B_b is introduced, which along the axis of symmetry of the trap ($r = \sqrt{x^2 + y^2} = 0$, see Fig. 5.1) is parallel to the main magnetic field B_0 (along the z axis) and increases quadratically with the distance from the median plane $z = 0$:

$$B_b(r, z) = \beta[(z^2 - r^2/2)\hat{B}_0 - (\hat{B}_0 \cdot z)r] \qquad (|B_b| \ll |B_0|) \tag{5.29}$$

The interaction term of this field with a magnetic moment effectively confines it in the axial direction, forming a 'magnetic bottle'. This additional confinement is much weaker than the one provided by the electric field, and is only used for detection purposes: the axial restoring force due to the weak magnetic field makes the axial oscillation frequency slightly dependent on a magnetic moment along the trap axis, such as those are associated to both the cyclotron and the magnetron motions: along the axis

$$\omega_z \rightarrow \omega_z - \frac{\beta \mu_z}{m \omega_z}$$

By observing (small) frequency shifts in the axial frequency, information on the state of such magnetic moments is obtained ('continuous Stern–Gerlach effect'). The resonance is thus measured by exciting the cyclotron motion with microwaves, introduced in the trap for a given time, and then measuring the related magnetic moment (actually the cyclotron orbit radius) after a fixed time interval by the shift it induces on the axial frequency. The magnetron frequency can be measured in the same way, and can provide information on the trap imperfections by comparing it with its theoretical value.

A trapped particle with spin 1/2 also interacts with the magnetic field through its magnetic moment (4.91), and the interaction Hamiltonian (4.93) can be written as

$$H_{int}^{(M)} = -\frac{g}{2} \omega_c^{(0)} |\mathbf{S}|$$

where $\mathbf{S} = \hbar \boldsymbol{\sigma} / 2$ is the spin vector. The two energy levels corresponding to $|\mathbf{S}| = \pm \hbar/2$ thus have an energy difference

$$\hbar \omega_s = \frac{g}{2} \hbar \omega_c^{(0)}$$

and for the electron or the positron $g \simeq 2 + \alpha/\pi$, so that the spin precession frequency ω_s is just slightly higher than the free cyclotron frequency $\omega_c^{(0)}$. The small difference is the anomaly frequency

$$\omega_a^{(0)} = \omega_s - \omega_c^{(0)}$$

corresponding to a transition to the next upper level of cyclotron motion with a simultaneous spin flip from up to down or vice versa. The g factor and $\kappa = (g - 2)/2$ can thus be defined as ratios of two measurable frequencies:

$$g = 2\omega_s/\omega_c^{(0)} \qquad \kappa = \omega_a^{(0)}/\omega_c^{(0)}$$

In terms of the modified frequencies which are measured inside a trap (even with imperfections)

$$\kappa = \frac{\omega_a - \omega_c^{(0)} + \omega_c}{\omega_c^{(0)}}$$

where $\omega_a = \omega_s - \omega_c$, and $\omega_c^{(0)}$ can be obtained from (5.28).

The anomaly frequency can be measured directly, rather than as a difference of two much larger terms, thus obtaining a much smaller error. As the particle moves along a cyclotron orbit centered on the axis, it experiences (as long as it is outside the $z = 0$ plane) a magnetic field rotating at the cyclotron frequency in the (xy) plane, corresponding to the radial component of the bottle field B_b of eqn (5.29). If the particle is forced to oscillate in the axial direction, the amplitude of such field also oscillates at the same frequency and, recalling the discussion on EDM measurements in Section 4.4.3, such an oscillating amplitude can be decomposed in a co-rotating and a counter-rotating component with respect to the cyclotron motion. If the axial oscillation frequency coincides with ω_a, then the former component rotates at $\omega_c + \omega_a = \omega_s$, thus inducing a spin flip with a significant probability.[67] The presence of the trapping electric field, and the consequent movement of the origin of the cyclotron motion, does not alter essentially the above discussion. After driving the particle at a given frequency for some fixed amount of time, the spin state is measured by the shift of the axial frequency in the magnetic bottle (of order a few Hz), and the flip probability is determined with repeated measurements before moving to a new frequency, producing a line-shape from which the resonance frequency ω_a is determined, and finally $\omega_a^{(0)}$ and g are obtained.

Precise measurements of the electron and positron g factors were obtained with these techniques, and their comparison provides a test of *CPT*. Such a comparison (Van Dyck *et al.*, 1987) was performed at Seattle in the same trap, with only the trapping potential reversed; the measured 0.4 ppm difference in cyclotron frequencies limited any possible difference in the magnetic fields. Electrons were obtained from a field emission point inside the trap, while positrons originated from a ^{22}Na β^+ source and were moderated before being inserted in the trap (a single e^+ stored for more than 100 days in the trap was used for several measurements). The radial oscillating magnetic field inducing the transitions was obtained by splitting the upper and lower portions of the trap electrodes, in which two effective counter-circulating current loops at frequency ω_a were generated. The limit is quoted in Table 5.1.[68]

The measurement of cyclotron frequencies in a Penning trap also provided a sensitive *CPT* test, from the comparison of such frequencies for proton and antiproton at CERN (Gabrielse *et al.*, 1999), which is equivalent to a comparison of the charge to mass ratios Q/m. A single H^- ion (*eep*) was used instead of a proton: this was inside the trap simultaneously to a $\bar{p}$ (which has the same sign of Q/m), thus avoiding the need to perform measurements with reversed trapping potentials which – if not exactly opposite in the two cases – would result in the particles being stored at slightly different locations in the trap and not experiencing the same magnetic field. Moreover, the measurements can be alternated much more rapidly in this way, and the residual effects due to small unavoidable magnetic field drifts in time (due to environmental effects or temperature changes) are reduced. Note that the conversion factor between the cyclotron frequencies of H^- and p is known to higher accuracy than required.

The ~89 MHz cyclotron frequencies, differing by ~100 kHz, could be individually excited, so that while one particle was being measured with a small cyclotron orbit the other one was kept in a large orbit ($r > 1.6$ mm), not perturbing the measurement of the former.

---

[67] Contrary to the case of EDM experiments, the co-rotating field cannot produce a finite spin rotation, since the thermal random oscillations of the axial motion coupled with the inhomogeneous magnetic field cause a spread of ω_s, and the spin flip probability saturates at 1/2, if the drive is applied for a long time interval.

[68] If this limit is re-expressed as the ratio of a difference in interaction energies to the rest mass of the electron: $\hbar|\omega_a(e^+) - \omega_a(e^-)|/2m_ec^2$, its numerical value becomes even more impressive (Dehmelt *et al.*, 1999), reminding that direct comparisons of the figures for *CPT* limits are often not too meaningful.

The free cyclotron frequencies $\omega_c^{(0)}$ were determined from the measurement of the frequencies for the three motions in the trap (with widely different accuracy) via eqn (5.28), obtaining the limit in Table 5.1.

Lifetime differences for unstable particle and antiparticles are also used to obtain limits on *CPT* symmetry violation. While the lifetime of a positively charged particle can be measured accurately from the time distribution of its decay products when it is at rest, after being stopped in a material, this approach is usually not possible for a negatively charged particle, which is attracted by and interacts with the positive nuclei in the surrounding matter; the time scale for this process is usually much shorter than the lifetime to be measured, completely masking the latter.

Since to make an accurate comparison of the lifetimes for positive and negative particles these have to be measured in the same conditions, the use of particle beams propagating in vacuum is the most standard approach. Exceptions arise when the formation of metastable atoms in matter is possible, as for negative muons forming muonium; in this case an approach based on time measurements after a pulsed beam is stopped is possible (Bardin *et al.*, 1984).

For in-flight measurements with particle beams of alternating polarities, an accurate knowledge of momentum (or relative momentum differences for the two polarities) is crucial for transforming a decay path into a decay time. Either the number of decay products for a fixed path length or the number of particles surviving after a given distance can be measured: in the former case an accurate equalization of solid angle acceptance and efficiency for the decay particles (usually different for particle and antiparticle) is required. In the second method the path length used for the measurement is determined by the required accuracy, and increases with increasing momentum due to relativistic time dilation: measuring the fraction R of particles surviving after a distance Δl

$$R = \frac{N(\Delta l)}{N(0)} = e^{-m\Delta l/p\tau}$$

where p, m, τ are the momentum, mass, and lifetime of the particle, the relative error on the lifetime measurement is

$$\frac{\sigma_\tau}{\tau} = \frac{1}{\sqrt{N(0)}} \sqrt{\frac{N(0) - N(\Delta l)}{N(\Delta l)}} \frac{p\tau}{m\Delta l} \qquad (5.30)$$

which for a fixed initial flux $N(0)$ is minimum for a path length of $m\Delta l/p\tau = 1.59$ decay lengths, and decreases by increasing flux and path length.

For the measurement of the ratio of lifetimes the errors related to the path length Δl and the particle mass cancel in the ratio (even without assuming *CPT* symmetry, the mass equality is usually experimentally verified at a sufficient accuracy).

The beams' properties have to be carefully controlled and equalized, in terms of composition (other kinds of particles in the beam), momentum spectrum, and angles; magnet polarities are reversed to switch from particles to antiparticles along the same beam line, and such reversals have to be performed at a sufficiently high rate to be insensitive to detector and beam line drifts. As it is important that particles are not lost from the beam (except due to decays) the use of well-collimated beams and the minimization of the amount of material crossed are important issues, as any difference in beams' divergences and interaction lengths could introduce biases. Positive identification of the particles of interest reduces the sensitivity to beam backgrounds, and the discrimination against decay products remaining in the beam (being counted as undecayed particles) could be obtained in the same way.

With the above provisos, the ratio of particle and antiparticle beam intensities as a function of distance gives the information on the lifetime ratio; in this way the lifetimes of $\pi^\pm$ (Ayres *et al.*, 1971) and $K^\pm$ (Lobkowicz *et al.*, 1966) were compared, with the results shown in Table 5.1.

5.2.2   Interaction tests

As for time reversal, tests of *CPT* symmetry involving interactions are difficult since the states which are related by such symmetry are time-reversed versions of each other, and the preparation of the time-reversed state might not be possible. Again, when a decay only occurs due to a weak interaction which can be treated in first order, the principle of detailed balance can be used instead to relate the transition rate for $|i\rangle \to |f\rangle$ to that for $|i_{CPT}\rangle \to |f_{CPT}\rangle$, and for simple cases 'naively' *CPT*-inverted states (charge-conjugated and spin-inverted) might be prepared. The comparison of two-body decays such as $\pi^+ \to \mu^+ \nu$ and $\pi^- \to \mu^- \bar{\nu}$ provides an example (Stapp, 1957): *CPT* symmetry requires the equality of the decay rates and equal but opposite polarizations for μ^+ and μ^- in decays at rest; on the contrary, *C* symmetry would also require the equality of the decay rates with equal polarizations.

Positronium, being an atomic system whose states are nominally *C* and *CP* eigenstates, is a rather unique system for the symmetry tests which it allows, as discussed in Chapter 3. In particular Ps decay correlations allow to test *CPT* symmetry in electromagnetic interactions, analogously to the way *T* symmetry is tested in other systems, without complications due to strong interactions. In the (allowed) 3γ decays of *polarized* $^3S_1$ positronium, the angular correlation

$$O_{CPT} = \mathcal{P} \cdot (k_1 \times k_2) \tag{5.31}$$

can be measured, where $\mathcal{P}$ is the positronium polarization vector and $k_{1,2}$ the momenta of the two most energetic annihilation photons ($|k_1| > |k_2|$). This is a *CPT*-odd quantity, since both triplet positronium and the final state are *C* eigenstates with eigenvalue -1 (*C* is conserved in the decay) and only spins change direction under *PT*; it represents a correlation between spin and decay plane orientation. As in the case of time reversal (Chapter 4), the measurement of a non-zero expectation value for a *CPT*-odd quantity does not imply directly a violation of *CPT*, as final state interaction effects could also generate it. The electromagnetic FSI of the photons (due to the creation of virtual charged particle pairs) are very small ($O(10^{-9})$ in the coefficient of $\langle O_{CPT} \rangle$) (Bernreuther *et al.*, 1988), so that the measurement of a larger effect would indicate the existence of a new kind of interaction (or *CPT* violation).

Positronium decay correlation tests

The first measurement of the angular correlation in eqn (5.31) was performed at Ann Arbor (Arbic *et al.*, 1988).

200 keV positrons from a ^{22}Na source are longitudinally polarized due to parity violation in the β^+ decay, with a helicity $\langle \mathbf{S} \cdot \hat{\mathbf{p}} \rangle = \langle v/c \rangle (\hbar/2)$, where v is the e^+ velocity. The net polarization along a fixed direction is related to the averaged angular distribution of the particles and in particular is reduced in presence of back-scattered particles (with opposite spin direction), so that the source holder was made of a low Z material (beryllium) to limit this effect. Passing through a low Z absorber plane, the lower energy positrons (smaller helicity) were eliminated and those emitted at larger angles were preferentially absorbed, thus increasing the polarization along the direction orthogonal to the absorber itself. The positrons impinged on a tungsten moderator consisting of seven vanes made by 0.025 mm thick W foils, with their surfaces parallel to the beam direction, so that they were hit at glancing angles; positrons lost energy in collisions, thermalized and annihilated, but a small $O(10^{-3})$ fraction of them could reach the surface and be expelled from it with energies $\simeq 2$ eV, determined by the negative work function of the moderator at a rate $\sim 10^5$/s.

The polarization of the positrons exiting the moderator was measured by a different apparatus, exploiting the fact that in a magnetic field (0.65 T) the singlet and triplet states with spin projection quantum number $m = 0$ mix, thus decreasing the lifetime of the $m = 0$ triplet state, which in this case depends on the positron polarization along the magnetic field direction. The ratio of the amount of (perturbed, lifetime ~ 15 ns) $m = 0$ triplet states to (unperturbed, lifetime ~ 140 ns) $m = \pm 1$ triplet states was measured by counting events in two different time windows, and allowed the determination of the positron polarization: $\mathcal{P} \sim 0.4$.

The positrons were then electrostatically collected, focused, and accelerated to 450 eV, and then passed through a Wien filter, whose crossed electric and magnetic fields selected the positron velocity and rotated its spin to be orthogonal to the momentum; by inverting the direction of both fields the final spin direction could also be reversed. An electrostatic mirror then deflected the positrons by 90° (without rotating their spin) to be again longitudinally polarized, and directed them towards a positronium formation cavity (see Fig. 5.2).

Here the positrons hit a channel electron multiplier array (CEMA), on which positronium was formed, and which signalled the e^+ arrival time while also allowing to monitor the spatial distribution of the beam. Polarized positrons and unpolarized electrons form polarized

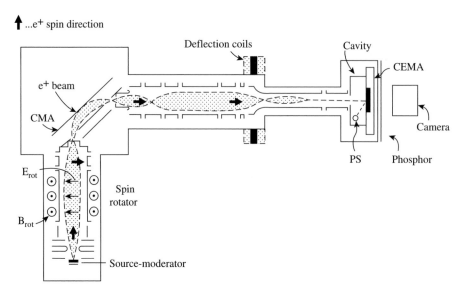

FIG. 5.2. Sketch of the beam set-up for the experiment of (Arbic *et al.*, 1988) to search of a *CPT*-odd correlation in positronium decay. The polarized e^+ pass through the spin rotator and the cylindrical deflector (CMA), then travel towards the Ps formation cavity and detector on the right.

positronium: a spin up ($m = +1/2$) e^+ forms triplet Ps with ($m = 1$) in half of the cases, and an equal mixture of singlet and triplet Ps with ($m = 0$) in the other half, so that triplet Ps forms in 3/4 of the cases, with a polarization which is 2/3 of that of the positrons; this result holds provided there are no large (~ 0.1 T) magnetic fields, which would mix the triplet and singlet ($m = 0$) states, thus depleting ('quenching') up to 40% of the former, due to the much higher decay rate of the latter. The polarization is retained in collisions with the cavity walls during the lifetime of the ortho-positronium ($\simeq 135$ ns).

Three NaI crystal detectors surrounded by lead shielding (Fig. 5.3) were placed around the Ps cavity, in the plane orthogonal to the e^+ beam, and two of them were required to fire in coincidence: one detecting the highest energy photon (400–500 keV range) and one of the two opposite ones detecting the next to highest energy one (300–400 keV). Depending on which of the two symmetric detectors fired, the decay plane normal was either parallel or anti-parallel to the positron beam direction, effectively flipping randomly the sign of the correlation O_{CPT}. Ortho-positronium decays were selected by the delayed coincidence of the photon detectors with respect to the positron arrival signal, thus eliminating the prompt signals due to para-positronium decay and positron annihilation.

In order to average out effects due to detector drifts or instabilities, the polarization of the positron beam was reversed every 4 minutes and the two symmetric photon detectors were interchanged daily, thus also averaging out effects due to their different efficiencies and resolutions.

The asymmetry of events with the normal to the decay plane parallel or anti-parallel to the positron spin was measured and averaged for the two orientations of the beam polarization,

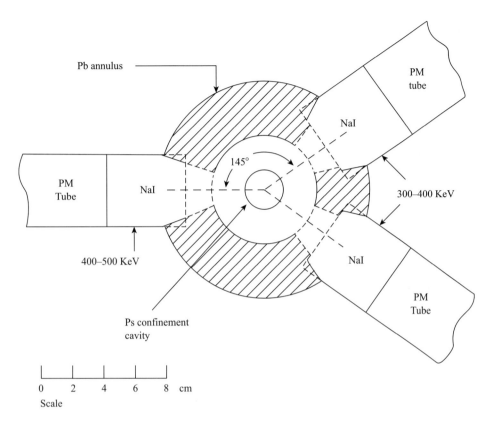

Fig. 5.3. Sketch of the detector set-up in the experiment of Arbic *et al.* (1988) for the search of a *CPT*-odd correlation in positronium decay. The slow e^+ beam enters perpendicularly to the drawing; the left detector detects the highest-energy photon and one of the two on the right the second highest one.

cancelling instrumental effects. The relation of the measured asymmetry to the coefficient of the angular correlation was determined by several factors including: the detector acceptance and geometry, the Ps polarization, the 8% background due to badly measured two-photon annihilation events. From a total of $3.5 \cdot 10^5$ events the null result $C_{CPT} = 0.020 \pm 0.023$ was obtained for the coefficient of the angular correlation O_{CPT}.

The largest systematic effect was due to the correlation of the e^+ beam position with the polarization direction due to the asymmetry of the magnetic fringing fields of the spin rotator, which caused acceptance differences which did not average out. The error was estimated by artificially enhancing the beam position differences (up to 2 mm), measuring the resulting effect and linearly interpolating to the actual ~ 0.1 mm difference in running conditions.

By using a 4π-acceptance, high-resolution detector (Gammasphere, see Chapter 3) detecting all three annihilation photons, a more precise measurement was obtained at Berkeley (Vetter and Freedman, 2003), which also has a reduced sensitivity to detector asymmetries. Two different β^+ positron sources were used, and their different polarization allowed to cross-check the consistency of the results. The positrons emitted in an hemisphere could form positronium in an aerogel target; the source and target assembly was rotated at different angles, to cancel possible detector asymmetries. The measured asymmetries for each orientation of the decay plane were compared to the Monte Carlo simulation containing a *CPT*-odd asymmetry and then averaged; with $2.7 \cdot 10^7$ events the measured coefficient was $C_{CPT} = 0.0071 \pm 0.0062$.

Other angular correlation measurements can be measured which are sensitive to the violation of different discrete symmetries, such as $\mathcal{P} \cdot \mathbf{k}$ (*P*-odd and *CPT*-odd), $(\mathcal{P} \cdot \mathbf{k}_1)(\mathcal{P} \cdot \mathbf{k}_1 \times \mathbf{k}_2)$ (*P*-odd and *T*-odd) or, by measuring the circular polarization $\mathbf{S}$ of one decay photon: $\mathcal{P} \cdot \mathbf{S}_1 \times \mathbf{k}_2$ (*P*-odd and *T*-odd).

5.2.3 Longitudinal polarizations

The measurement of a circular photon polarization in the $\gamma\gamma$ decay of a π^0 (or η) would indicate *CPT* symmetry violation (Okun, 2003), provided the electromagnetic decay is C symmetric, since the product $\mathbf{S} \cdot \mathbf{p}$ is *P*-odd and *T*-even. Such a polarization could arise from the interference of two terms in the Lagrangian having opposite *CPT* character between the pseudo-scalar neutral pion field $\phi^\dagger = \phi$ and the EM field, such as $c_S \, \phi F_{\mu\nu} \tilde{F}^{\mu\nu}$ and $c_P \, \phi F_{\mu\nu} F^{\mu\nu}$ (where $c_{S,P}$ are real because of the Hermiticity of the Lagrangian), but a longitudinal polarization term $\langle \mathbf{S}_\gamma \cdot \mathbf{p}_\gamma \rangle \propto \mathrm{Im}(c_S^* c_P)$ could arise from the difference of the absorptive parts of a two-loop diagram at a higher order. A similar argument applies to the longitudinal polarization of muons in the rare $\eta \to \mu^+ \mu^-$ decay.

5.3 Super-selection rules

In quantum mechanics, it is usually assumed that the correspondence between rays in Hilbert space and physical states is one to one, which means that every ray corresponds to a physically realizable state, and from this the superposition principle follows (an arbitrary linear superposition of state vectors is a possible state vector). This is equivalent to the *irreducibility postulate* (Fonda and Ghirardi, 1970), according to which it is always possible to find, among the family of all the observables of a given system, a set such that there exists no operator (except multiples of the identity) commuting with all its members: such a set completely characterizes the Hilbert space of the system. It is also assumed that *all* Hermitian operators correspond to observable quantities.

   In quantum field theory, dealing with different types of particles which (sometimes) cannot transform into each other, the above is no longer true: there exist rays which do not correspond to physically realizable states. One example is a superposition of two rays corresponding to states of different electric charge: this would be a state for which a measurement of electric charge would not always give the same result, which was never observed. The only physically realizable states are thus limited to those which are simultaneous eigenstates of a given set of observables O_X, which define what are called *super-selection* rules (Wick et al., 1952). Such observables are characterized by the fact that their generators commute with *all* the observables of the system, i.e. the observables O_X can always be measured without disturbing the system. Physical states are always eigenstates of such quantities, which are therefore always good quantum numbers, and belong to each complete set of commuting observables used to define the states. Examples are (as far as we know) electric charge, baryon number, and lepton number; if only strong interactions existed flavour would also be an observable of this kind.

Any observable can be written as

$$O_X = \sum_{x,\alpha} x|x,\alpha\rangle\langle x,\alpha| = \sum_x xP(x)$$

where α stands for the eigenvalues of all the other observables; $P(x)$ is the projector for the subspace corresponding to the eigenvalue x of O_X. Since O_X, and therefore $P(x)$, commutes with all observables, for any observable O

$$\langle x,\alpha|O|x',\alpha\rangle = \langle x,\alpha|P(x)O|x',\alpha\rangle = \langle x,\alpha|OP(x)|x',\alpha\rangle = \delta_{x,x'}\langle x,\alpha|O|x',\alpha\rangle$$

A super-selection rule splits the Hilbert space in disconnected subspaces for different eigenvalues of the corresponding quantity: no observable can have non-zero matrix elements between two state vectors of different subspaces, there can be no transition between state vectors belonging to two different subspaces (i.e. an absolute selection rule exists), and no physically meaningful superposition can be built with such state vectors. Conversely, Hermitian operators which do not commute with all the observables defining the super-selection rules, and which therefore can have non-zero matrix elements between states with to different eigenvalues of O_X, cannot be observables.

Also, the relative phases of state vectors belonging to different subspaces, defined by the eigenvalues of an observables corresponding to a super-selection rule, are unphysical and cannot be measured; this is a generalization of the multiplication of each state by an arbitrary phase factor which is intrinsic to the ray nature; there is actually one such arbitrary phase factor for each disconnected subspace: not only $|\psi\rangle$ and $e^{i\xi}|\psi\rangle$ (the same ray, with arbitrary real ξ) represent the same state, but also $|\psi_1\rangle + |\psi_2\rangle$ and $|\psi_1\rangle + e^{i\xi}|\psi_2\rangle$ (*different* rays) correspond to the same (non-physically realizable) state if $|\psi_1\rangle$ and $|\psi_2\rangle$ belong to different subspaces.

For an inversion described by the operator I (e.g. P, T, C) its square (which is a unitary operator, even when I is anti-unitary) must leave a system unchanged and therefore behaves as in eqn (1.17); if I^2 is a symmetry it commutes with all observables, originating a super-selection rule. For any two states

$$I^2|\psi^{(1)}\rangle = \tilde{\eta}_I(1)|\psi^{(1)}\rangle \quad I^2|\psi^{(2)}\rangle = \tilde{\eta}_I(2)|\psi^{(2)}\rangle$$

where $\tilde{\eta}_I(1, 2)$ are phase factors. The matrix elements of any observable O between $|\psi^{(1)}\rangle$ and $|\psi^{(2)}\rangle$ satisfy

$$\langle\psi^{(2)}|O|\psi^{(1)}\rangle = \tilde{\eta}_I(1)^*\langle\psi^{(2)}|OI^2|\psi^{(1)}\rangle$$
$$= \tilde{\eta}_I(1)^*\langle\psi^{(2)}|I^2 O|\psi^{(1)}\rangle = \tilde{\eta}_I(1)^*\tilde{\eta}_I(2)\langle\psi^{(2)}|O|\psi^{(1)}\rangle$$

and therefore vanish if $\tilde{\eta}_I(1) \neq \tilde{\eta}_I(2)$.

In this case, when considering the superposition state $|\Phi\rangle = e^{i\xi_1}|\psi^{(1)}\rangle + e^{i\xi_2}|\psi^{(2)}\rangle$, no measurement of any observable O in such a state can give information on the relative phase, since

$$\langle\Phi|O|\Phi\rangle = \langle\psi^{(1)}|O|\psi^{(1)}\rangle + \langle\psi^{(2)}|O|\psi^{(2)}\rangle$$
$$+ e^{-i(\xi_1-\xi_2)}\langle\psi^{(1)}|O|\psi^{(2)}\rangle + e^{i(\xi_1-\xi_2)}\langle\psi^{(2)}|O|\psi^{(1)}\rangle$$

but the last two terms, containing the relative phase, are zero as required by the discussion above.

As an example of the application of the above considerations, the super-selection rule given by the square of the parity operator P^2 could be considered (Leite Lopes, 1969): this requires that there can be no transitions between states containing even and odd numbers of fermions with imaginary parity, described by Majorana fields (Chapter 3). Indeed if

$$P^2|\psi\rangle = +|\psi\rangle \qquad P^2|\psi'\rangle = -|\psi'\rangle$$

then

$$\langle\psi + \psi'|T|\psi + \psi'\rangle = \langle\psi + \psi'|U_P^{\dagger 2}TU_P^2|\psi + \psi'\rangle = \langle\psi - \psi'|T|\psi - \psi'\rangle$$

so that $\langle\psi|T|\psi'\rangle = 0$.

An analogous result could be obtained for the square of the time reversal operator T^2, which is unitary and has eigenvalues ± 1. It follows that, if T^2 is a symmetry of the theory, a super-selection rule exists between the subspaces of states with $T^2 = \pm 1$, which are then disconnected. As seen from eqn (4.31), these correspond to states with integer or half-integer total angular momentum respectively, and there can be no physical observables connecting such states: spinor quantities, such as a field ψ, or Hermitian combinations of them such as $\psi + \psi^\dagger$ and $i(\psi - \psi^\dagger)$, which connect such subspaces, are not observables.

T^2 actually coincides with the operator for a rotation by 2π about any axis, so that the above super-selection rule exists independently of the validity of time reversal symmetry, following of invariance under proper, orthochronous Lorentz transformations: this requires all observables to be relativistically covariant, i.e. four-tensors transforming as

$$U^\dagger O_{\alpha,\beta,\dots}U = \Lambda_\alpha^{\alpha'}\Lambda_\beta^{\beta'}\cdots O_{\alpha,\beta,\dots}$$

where U is the unitary operator corresponding to Lorentz transformations, Λ_β^α the corresponding transformation matrix in space-time, and a possible phase factor (restricted to ± 1 for Hermitian O) was fixed to be $+1$, since proper orthochronous Lorentz transformations include the identity. For the Lorentz transformation corresponding to a 2π rotation around the z axis, $\Lambda^{\mu}_{\nu} = 1$ and since the corresponding operator is $U_R = e^{-i2\pi J_z}$

$$\langle \psi^{(1)} | O | \psi^{(2)} \rangle = \left\langle \psi^{(1)} \left| U_R^{\dagger} O U_R \right| \psi^{(2)} \right\rangle = e^{-i2\pi(m_2 - m_1)} \langle \psi^{(1)} | O | \psi^{(2)} \rangle$$

so that $\langle \psi^{(1)} | O | \psi^{(2)} \rangle = 0$ unless $m_1 - m_2$ is an integer, i.e. m_1 and m_2 have both integer or half-integer values.

Consequences of super-selection rules were mentioned in earlier chapters, one being the fact that the intrinsic parities η_P (which can be measurable, see Chapter 1) cannot be directly compared for particles with different electric charge (or other conserved internal quantum numbers); of course, when some of those are chosen by convention, then others might result to be fixed, in order to have an assignment consistent with parity conservation.

In general, every absolutely conserved charge allows one to redefine the phases of the states in each of the subspaces corresponding to its eigenvalues: considering electric charge Q and the baryon and lepton numbers B, L, this means that a phase factor

$$\eta_X = e^{i\epsilon_Q Q} \, e^{i\epsilon_B B} \, e^{i\epsilon_L L} \tag{5.32}$$

(with arbitrary, real $\epsilon_{Q,B,L}$) can be used to redefine, e.g., the parity operator as $P \to \eta_X P$, which is conserved if P is conserved but has different eigenvalues: the choice of appropriate $\epsilon_{Q,B,L}$ allows to redefine the intrinsic parities for particles with different eigenvalues of the conserved operators Q, B, L.

At this point the condition under which the phase factors for an inversion I can be chosen to be real may be stated (Feinberg and Weinberg, 1959): if I^2 belongs to a continuous symmetry group of phase transformations (i.e. its effect on any state is just the multiplication by an appropriate phase factor such as in eqn (5.32)), then it can always be redefined to be just $+1$, so that in this case the eigenvalues of I can always be chosen to be $\eta_I = \pm 1$, also for those fields and inversion operators for which the phase factors are not entirely arbitrary; this appears indeed to be the case for the parity operator, at least in absence of Majorana particles.

5.4   Spin and statistics

The symmetry arising from the interchange of identical particles belonging to the same system is one of the most striking consequences of quantum theory. The essential difference between classical and quantum mechanics in this respect is that the limitation on the simultaneous measurability of the position and the momentum of a particle invalidates the concept of a trajectory: an individual particle cannot be 'followed' in its motion, so that there is no other way to distinguish it from another one except by considering its specific physical attributes (such as charge, mass, spin, etc.). Two particles which are identical with respect to such attributes, i.e. which have all the same quantum numbers, completely lose their individuality in quantum mechanics, and therefore the system in which they are interchanged is indistinguishable from the original one, resulting in a symmetry.

All observables commute with the unitary permutation operator $\mathcal{P}$ exchanging two identical particles; this is true in particular for the Hamiltonian, so that the symmetry character of a state vector under such permutation is preserved in time. The same situation occurs with more than two identical particles, and many permutation operators can be defined, all of them commuting with all the observables.

It can be assumed that $\mathcal{P}$ is a multiple of the identity[69], and since $\mathcal{P}^2 = +1$ is always a possible choice, one has $\mathcal{P} = \pm 1$ (see e.g. (Fonda and Ghirardi, 1970)), thus distinguishing the particles into bosons, obeying Bose–Einstein (B–E) statistics and forming symmetric states of identical particles, and fermions, obeying Fermi–Dirac (F–D) statistics and forming antisymmetric states; this is called the *symmetrization postulate* (Messiah and Greenberg, 1964): no other (mixed) symmetry states of identical particles exist.

The Pauli exclusion principle, stating that in a many-fermion system there cannot be two fermions with all their quantum numbers equal (Pauli, 1925) is well known to give a straightforward explanation of the structure of the periodic table of elements and of the stability of matter. More generally the connection between the statistics obeyed by a particle and its spin is expressed by the *spin-statistics theorem*, according to which particles of integer spin are bosons and those of half-integer spin are fermions. This is a postulate in quantum mechanics, which in quantum field theory follows from general requirements (Fierz, 1939; Pauli, 1940): loosely speaking Bose–Einstein symmetry is obtained when quantizing a field with commutators, and Fermi–Dirac symmetry when anti-commutators are used; the requirement of non-negative energy forbids the quantization of spinor fields with commutators, and that of non-negative probability density forbids the quantization of integer-spin fields with anti-commutators[70].

Several proofs (with different sets of hypotheses and levels of mathematical rigour) exist of the spin-statistics theorem (Streater and Wightman, 1964; Duck and Sudarshan, 1997), the most elementary one (Sudarshan, 1975) reducing it to a consequence of rotational symmetry in three spatial dimensions, with only implicit reliance on relativistic quantum theory.

A nice discussion of the link between C, P, T and the spin-statistics connection was given by Feynman (1987), but while appealing parallels can be made with the

---

[69] So that the irreducibility postulate (Section 5.3) is valid for every subspace defined by super-selection rules.

[70] Note that the EM field is measurable, and as such it is required to commute with itself at space-like separation.

topological properties of rotations[71] and the fact that spinors (in contrast to integer-spin fields) acquire a minus sign under a rotation by 2π (the rotational properties of spinors are not linked to the Fermi–Dirac statistics of the particle which they describe, but rather to the properties of three-dimensional space), it seems that an intuitive, elementary understanding of this theorem still eludes us.

Composite systems behave as fermions if they contain an odd number of fermions, and as bosons otherwise (Ehrenfest and Oppenheimer, 1931); this is sometimes called the *conservation of statistics* theorem, and is valid provided the interaction between composite particles is negligible compared to their internal excitation energy (so that the composite particles can be considered as single particles ignoring their internal structure). This is confirmed by Bose–Einstein condensation of atoms being observed only in case the atomic nucleus has an even number of neutrons (so that the total number of fermions in the atom is even).

There are thus two quite separate issues which can be questioned: the first is the fact that for assemblies of identical particles, among the many possible states distinguished by permutations of particles in pairs, only two appear to be realized in Nature, namely the one which is completely symmetric under any such permutation and the one which is completely antisymmetric (the symmetrization postulate). This is a much stronger condition than what is implied by the indistinguishability of identical particles, and indeed quantum mechanics would allow states with different symmetries from the above two: the question about the physical existence of such states is a legitimate one.

The second independent issue is whether states in which half-integer (integer) spin particles are in an exchange-symmetric (antisymmetric) state exist (the spin-statistics connection).

Starting with Gentile (1940) there have been several theoretical attempts to build consistent theories with symmetries different from those corresponding to bosons and fermions: statistics differing in a discrete way from these (parastatistics) are easily ruled out by experiment, while theories with *small* deviations from the usual statistics encounter fundamental difficulties[72] in the framework of local quantum field theory (Greenberg and Mohapatra, 1989a). While ingenious new schemes are nevertheless being considered, the lack of a consistent model incorporating violations of the above principles should not stop the experimental tests of their validity: 'If something in fundamental physics *can* be tested, then it absolutely *must* be tested' (Okun 1939).

---

[71] It is well known that for an extended body the spatial relations with the environment are not left unchanged by a 2π rotation, and only a 4π rotation brings back the system to its original state; this can be explicitly verified by rotating a solid object attached to a fixed frame with strings: the 2π rotation introduces a twist in the strings which is undone by a further 2π rotation (see e.g. Huang (1992)).

[72] It might be interesting to note that this is true only if space has three or more dimensions.

5.4.1   Consequences of symmetrization and spin-statistics

The most fundamental consequence of the validity of the symmetrization principle is the stability of matter and the existence of the periodic table of elements: indeed Pauli's principle was introduced to justify the fact that in an atom the electrons in higher energy orbits do not fall all down into the fundamental state.

The spin-statistics connection was already used in several places; it leads in general to relations between internal and orbital quantum numbers for pairs of particles:

- For a state of two identical particles X with orbital angular momentum L and total spin S

$$(-1)^L = (-1)^S \tag{5.33}$$

- For a (neutral) $X\overline{X}$ state (where $\overline{X}$ is the antiparticle of X) with orbital angular momentum L and total spin S the charge conjugation parity eigenvalue is

$$\eta_C(X\overline{X}) = (-1)^{L+S} \tag{5.34}$$

- For a pair of identical strongly interacting particles of isospin I_X

$$(-1)^L = (-1)^{S+I+2I_X} \tag{5.35}$$

   I being the total isospin of the pair.

- For a $X\overline{X}$ state of strongly interacting particles (where $\overline{X}$ now stands for the G-parity conjugate of X), the G-parity eigenvalue (Section 3.2.4) is

$$\eta_G(X\overline{X}) = (-1)^{L+S+I} \tag{5.36}$$

While the above relations have been expressed in non-relativistic terms, their validity is more general, and they can be cast in the helicity formalism (see e.g. Gibson and Pollard (1976)).

5.4.2   Experimental tests of symmetrization and spin-statistics

Overwhelming experimental evidence that electron and nucleons are fermions and photons are bosons is provided by the stability of matter and by the study of black-body radiation respectively, and more generally by the high accuracy to which quantum electrodynamics has been verified. Possible tests for other kinds of particles, which are not easily produced in large amounts, are reviewed in Messiah and

Greenberg (1964) and Greenberg and Mohapatra (1989*b*). To date all the experimental results are consistent with an exact validity of the symmetrization postulate and the spin-statistics connection.

Two important points should be noted which have consequences for possible experimental tests of symmetry properties.

The first is that for a system composed of only two identical particles the only possible irreducible representations of permutation symmetry are the symmetric and anti-symmetric states, so that experiments with two particles can say nothing about the validity of the symmetrization postulate: three or more identical particles are required to test it.

The second point was briefly mentioned above: since the Hamiltonian treats symmetrically all identical particles (otherwise they would not be identical), it commutes with any permutation of those and as such it cannot induce transitions from a state with a given permutation symmetry to one with a different one. In other terms within the framework of ordinary quantum mechanics a super-selection rule (Section 5.3) exists between states with different behaviour under exchange symmetry, which therefore cannot mix[73]. This implies, for example, that even if the Pauli principle were violated, searches for spontaneous transitions of an atomic electron into an already occupied state would give negative results (Amado and Primakoff, 1980), as an electron obeying the Pauli principle cannot start violating it.

States with different symmetry properties can only appear together in incoherent mixtures, as superpositions of states with different symmetry are not allowed: the absence of any observable interference term between states which obey the symmetrization postulate and states which do not makes it difficult to obtain significant tests of such postulate. Limits on the violations of the usual statistics are conventionally expressed by stating that two identical particles have a probability $1 - \beta^2/2$ of being found in a normal symmetry state and $\beta^2/2$ of being found in a forbidden state.

Some tests of the validity of the Pauli principle by searching forbidden transitions were performed: the objection described above against such kind of tests does not apply when 'new' particles particles are involved, which could be created in an 'anomalous' state not respecting the Pauli principle. This possibility was tested by an experiment (Ramberg and Snow, 1990) which looked for anomalous X-rays with energies corresponding to the cascade of an electron into the filled 1S shell: external electrons were injected into a copper strip traversed by a 30 A current for about two months. No significant signal was detected above background,

---

[73] We cannot avoid noting that the original proof of this rule (Messiah and Greenberg, 1964) was based on the validity of *CP* symmetry.

leading to a limit $\beta^2/2 \leq 1.7 \cdot 10^{-26}$ at 95% confidence level. Analogous limits for the rate of emission of γ rays arising from a de-excitation of a p-shell nucleon into the s-shell forming a new particle state have been obtained, for example by the large Kamiokande detector (3,000 tons of water) (Suzuki *et al.*, 1993), corresponding to $\beta^2 < 2.3 \cdot 10^{-57}$.

Note that the existence of hypothetical particles which are quite similar but not exactly identical to others (so that symmetrization is not required for them) would pose other problems: if, e.g., 'anomalous' electrons existed (even in a tiny amount) which are somehow slightly different from ordinary ones, they would be pair-produced copiously, thus dramatically altering the known properties of quantum electrodynamics[74].

Another class of tests of the violation of the Pauli principle is based on searches for anomalous atoms or nucleons which are not totally antisymmetric under permutations: if this principle is violated any substance should contain a fraction of order β^2 of anomalous atoms.

The simplest system in which violations of the Pauli principle can be searched for is the helium atom, with a ground state with two electrons in the symmetric triplet state $^3S_1$: since the chemical properties are (largely) determined by the outer valence electrons, such an atom would behave chemically like hydrogen, but it could be distinguished from it by its spin and magnetic properties (e.g. a non-vanishing magnetic moment). Theoretical computations of the energy levels for 'wrong-symmetry' states are considered to be reliable for helium, and laser spectroscopy searches for forbidden transitions corresponding to such states have been carried out (Deilamian *et al.*, 1995), leading to $\beta^2/2 \leq 5 \cdot 10^{-6}$.

Atoms with completely filled shells behave as inert gases: a beryllium atom with all four electrons in the 1S shell would behave as helium, and experimental limits on its presence in the atmosphere and in metallic samples were set by mass spectrometry. In general searches for anomalous atoms of the element Z can be performed in natural samples of the element $Z - 1$, with a high sensitivity in case of pairs for which the abundance of Z is much larger than that of $Z - 1$, which is the case, e.g., for Ne and F.

Violation of the Pauli principle at the quark level would give origin to a 70-plet of baryons with quarks in the S wave (Greenberg and Mohapatra, 1989*b*), including an octet with $J^P = 3/2^+$ and a decuplet with $J^P = 1/2^+$, which are not observed.

While searches for the violation of the Pauli principle are usually null experiments in which a signal is searched with no background, in the case of B–E statistics the signature being looked for is usually a small change in the properties of a

---

[74] Searches for X-ray transitions of the kind described were performed a long time ago when the issue of the identity of β particles and electrons was not completely settled yet (Goldhaber and Scharff-Goldhaber, 1948).

many-particle system. For this reason significant tests on integer-spin particles are more difficult in general.

Photons are the only fundamental bosons which can be readily produced in states with large number density, and indeed the existence of finite energy in a static EM field (photon energy $\to$ 0 and number density $\to \infty$) implies no bounds on the photon occupation number N (easily of order 10^{40} in a laboratory).

In the case of photons, the Landau–Yang theorem (Section 2.2.2) forbidding the two-photon decay of a spin 1 particle offers a possible test of B–E symmetry, and indeed the experimental limit on the non-observed decay $Z \to \gamma\gamma$ of the heaviest known vector boson (with the rationale that violations of B–E symmetry could be expected to appear more prominently at higher energies, if they are linked to unknown high-mass new physics[75]) was used to set a limit on the validity of such symmetry (Ignatiev *et al.*, 1996).

Search for antisymmetric two-photon states

The Landau–Yang theorem was used to set a limit for the two-photon transition between an atomic state with angular momentum $J = 0$ and an excited level with $J = 1$ induced by two photons of the same frequency (DeMille *et al.*, 1999). With two counter-propagating laser beams of the same energy, Bose–Einstein symmetry requires the transition amplitude to be even under exchange of the identical photons, while all the gauge-invariant expressions bilinear in the photon polarization vectors $\epsilon_{1,2}$ are odd under such exchange (see Section 2.2.2). In this case, however, the possibility exists of absorbing two photons from the same beam, and in this case the amplitude $(\epsilon_1 \cdot \epsilon_2)\mathbf{k}$ (where $\mathbf{k} = \mathbf{k}_1 = \mathbf{k}_2$ is the photon momentum) is not odd under exchange; this amplitude is, however, odd under parity, and thus vanishes for transitions between states of the same total parity, since EM interactions are symmetric under parity[76]. Other forms for the amplitude involving higher powers of the photon momentum, such as $[(\epsilon_1 \times \epsilon_2) \cdot \mathbf{k}]\mathbf{k}$ correspond to higher order multipoles (Section 2.1.4) which are highly suppressed for atomic transitions.

A two-photon transition can be considered as two single-photon transitions proceeding through an intermediate virtual state: if the two photons have different energies two different possible amplitudes exist, corresponding to either photon being absorbed first; these two amplitudes must be added (with a plus sign) because photons are bosons. For a $J = 0 \to J = 1$ transition the relevant amplitude is of the form $\epsilon_1 \times \epsilon_2$, which is indeed antisymmetric and requires the two photons to have different polarizations.

The excitation of the transition between the $^1S_0$ state (valence electron configuration $6s^2$) and the $^3S_1$ state ($5d\,6d$) was searched for by using two 549 nm photons: the corresponding forbidden amplitude is expected to be largely enhanced by the presence of the intermediate

---

[75] Note however that the Z, being neutral, does not couple to photons directly but only through (suppressed) loop diagrams or quartic couplings.

[76] The effects induced by parity-violating weak interactions are negligibly small in this context.

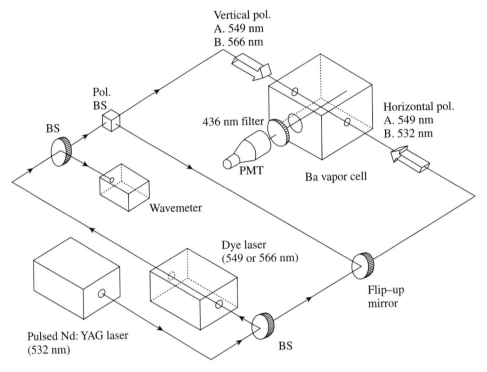

FIG. 5.4. Schematic of the apparatus for the search of exchange-antisymmetric two-photon states (DeMille *et al.*, 1999); BS: beam splitter.

$^1P_1$ state (6s 6p) close to half the energy difference, with large dipole matrix elements with both the initial and final states.

The experimental arrangement is shown in Fig. 5.4: light with wavelength around 549 nm from a pulsed dye laser, pumped by a Nd:YAG laser at 532 nm, was split into two beams with orthogonal linear polarization, which impinged on the cell containing Ba vapour (at $\simeq$900 K temperature) from opposite directions.

The transitions were detected by observing fluorescence due to the decay of the excited $^3S_1$ state in a 25 ns time window after the laser pulses (of $\simeq$7 ns duration), using a photomultiplier with filters: the signature of B–E symmetry violation would appear as an excess signal in a narrow range of laser frequencies.

By removing a mirror the cell was illuminated on one side directly by the Nd:YAG laser, and by tuning the dye laser to 566 nm the same transition could be excited with two non-degenerate photons, providing a calibration. The intensity ratio of forbidden (equal energy photons) to allowed transitions depends on the laser intensities and spectral widths: the intensities were determined by measuring the pulse energy ($\simeq$1.5 mJ for the degenerate transition search and 1 (532 nm) and 100 (566 nm) μJ for calibration to avoid saturation) and beam area ($\simeq$5 mm^2). The laser spectral widths ($\simeq$3 GHz) were measured with a scanning Fabry–Pérot interferometer: using an allowed degenerate two-photon transition the bandwidth of the dye laser was obtained, and then the one of the Nd:YAG laser was determined by deconvolution at the calibration transition. In the limit in which the single state mentioned above dominates the sum over intermediate states for the two-photon transition (which was verified to be a good approximation by measuring the lifetime and branching ratios of all decays from the excited state) the matrix elements cancel in the ratio of forbidden to allowed transition rates, which thus depend only on known atomic and photon energies.

The results of three data-taking runs showed small but statistically significant peaks above background at the predicted position of the degenerate transition: such peaks are consistent with the predicted effect due to the finite bandwidth of the dye laser, when the transition is excited by two photons of different frequencies; ignoring such effects a conservative limit $\beta^2/2 < 1.2 \cdot 10^{-7}$ (at 90% CL) was obtained.

Experiments studying the spectra of molecules containing identical spin-0 nuclei are analogous to those on helium atoms mentioned above, in that they offer a virtually zero background search. The total wave function of the molecule can be factorized in its electronic, vibrational, rotational, and nuclear spin parts: the spin part is clearly symmetric under the exchange of the two spin-0 nuclei, and the same is true for the vibrational part, which only depends on the magnitude of the inter-nuclear distance; the electronic and rotational parts must then be both even (symmetric) or odd (anti-symmetric), in order for the total wave function to be symmetric under particle exchange as required by Bose-Einstein symmetry. The position of absorption lines corresponding to transitions involving forbidden states can often be computed with good accuracy, and by scanning the corresponding frequency region with a tunable laser the light absorption spectrum is measured. To increase the sensitivity and reduce background due to the variation of laser intensity during the scan the laser frequency is modulated and the output signal from the photo-detector is passed through a lock-in amplifier and demodulated.

Sensitive experiments were performed on the vibrational spectrum of the $^{12}C^{16}O_2$ molecule, looking for transitions between states which are anti-symmetric under the exchange of the two identical ^{16}O nuclei (Mazzotti et al., 2001). This molecule has strong roto-vibrational bands in the infrared region, which can be exploited with adequate sources of laser light; its ground state electronic wave function is symmetric under the exchange of nuclei, so that well-isolated forbidden lines corresponding to transitions from states with odd values of the rotational quantum number (antisymmetric) around 4.25 μm were searched for. By comparing to an observed calibration line the limit on the presence of forbidden lines was translated into an upper limit $\beta^2/2 < 1.7 \cdot 10^{-11}$.

Another constraint on the violation of Bose-Einstein statistics for pions based on the measurement of direct CP violation will be discussed in Chapter 8.

Further reading

Up-to-date information on possible violations of *CPT* and Lorentz symmetries in many systems can be found in Kostelecký (2005). A recent detailed review of possible *CPT* tests on antihydrogen is Holzscheiter *et al.* (2004).

A review of tests of symmetry under particle interchange can be found in Hilborn and Tino (2000).

6

CP SYMMETRY

The most exciting phrase to hear in science,
the one that heralds new discoveries,
is not 'Eureka!' but 'That's funny ...'
I. Asimov

Perhaps looking-glass milk isn't good to drink.
L. Carroll

But now you've found another partner
And left me like a broken toy.
B. H. May

CP symmetry is perhaps the discrete symmetry which has drawn more experimental and theoretical attention for several reasons: besides its fundamental significance and its connection to time reversal symmetry through the *CPT* theorem, it is the most striking case of a fundamental symmetry which is violated in Nature by a very small amount, and its detected manifestations have been rather limited, even within the realm of phenomena governed by the weak interactions.

In this chapter *CP* symmetry is addressed in general terms and some of its consequences and experimental tests are considered. The discovery of *CP* violation (*CPV*) in neutral kaon decays is illustrated, briefly anticipating some results from the detailed discussion of this system in following chapters.

6.1 *CP* symmetry: found and lost

When the first experimental indications appeared that parity symmetry was not universally valid in Nature, but was violated in processes governed by weak interactions, nobody really questioned the validity of the results, because the evidence was overwhelming and very direct (things would be quite different when *CP* symmetry fell 7 years later). However, as seen in Chapter 2, this had a rather shocking impact on physics, and accepting that Nature actually distinguishes left from right at a very fundamental level was a difficult step. Such uneasiness was however quickly mitigated by the suggestion that the left-right symmetry might just be implemented in Nature in a slightly more subtle way: L. D. Landau (1957*a*,1957*b*) remarked that to have such a symmetry it might also be necessary to invert the particle charges, so that the mirror image of a particle world would actually be an antiparticle world, in which the same laws of physics would continue to be valid, thus restoring the symmetry: 'At first glance it appears that non-conservation of parity would imply an asymmetry of space with respect to inversion. Such an asymmetry, in view of the complete isotropy of space [...] would be more than strange. In my opinion a simple denial of parity-conservation would place theoretical physics in an unhappy situation'. Landau proposed that the 'combined inversion' operator *CP*, the product of parity and charge conjugation, is actually what implements left-right symmetry in Nature: 'The invariance of the interactions with respect to combined inversion leaves space completely symmetrical, and only the electrical charges will be asymmetrical'. It should be noted that, remarkably, this possibility had been considered even before the validity of parity symmetry was questioned.[77]

The suggestion that the 'proper' way of having a mirror reflection in Nature would also require charge conjugation, so that left-right symmetry would indeed be somehow restored as *CP* symmetry, did not clash with the known description of the parity-conserving electromagnetic phenomena, since those are separately symmetric under charge conjugation (always proportional to the product of two charges, or to $\alpha \propto e^2$), so that the ordinary parity symmetry would still hold for them.

The restoration of left-right symmetry by *CP* can be understood by considering how we could communicate our definition of right and left to an extraterrestrial civilization living on a distant galaxy (Feynman, 1965); supposing we can only exchange information with them and not material particles (the use of circularly polarized radio-waves to transmit messages is not allowed either). Before 1956 this was thought to be impossible, since left and right were assumed to have only a conventional meaning. After the fall of parity, since the same laws of physics which hold on Earth are assumed to hold in the far away galaxy, we could think of telling the extraterrestrials to perform a weak interaction experiment and observe the resulting parity asymmetry, e.g. checking in which direction electrons are preferably emitted from polarized nuclei and thus defining unambiguously which of the two verses of rotation associated with nuclear spin should be called clockwise and which anti-clockwise, which defines right and left. However, if the extraterrestrials were living in an antimatter world, they would perform the experiment with antinuclei and come up with the opposite definition for right and left. With *CP* symmetry, any space asymmetry due to parity violation would just be an expression of a matter-antimatter asymmetry, and the significance of right and left would be just

---

[77] 'The disturbing possibility remains that *C* and [*P*] are both only approximate and *C*[*P*] is the only exact symmetry law. [...] This possibility, however, seems rather remote at the moment' (Wick *et al.*, 1952).

as fundamental as that of positive and negative electric charge. Conversely, the absolute distinction of matter and antimatter can only be obtained if an absolute definition of right and left is possible: C violation allows to state that the two behave differently, but is not sufficient to tell us which is which, in a non-conventional way. In this sense CP is seen to be more fundamental than the individual P and C symmetries.

For some years this seemed to be a quite satisfactory state of affairs, and indeed the V-A theory of weak interactions incorporated maximal P and C violation by construction and was CP symmetric. As an example, a weak decay such as $\pi^+ \to \mu^+ \nu$ gives left-handed (helicity -1) muons, while its parity-reversed process in which right-handed muons (helicity $+1$) are emitted never occurs; similarly, the C-conjugated process $\pi^- \to \mu^- \bar{\nu}$ with left-handed muons is also not observed in nature. Both P and C symmetries are maximally violated by the fact that only one of the two (parity-inverted and charge-conjugated) processes occurs in Nature, while the other one is completely absent. Nevertheless, the CP-conjugated decay $\pi^- \to \mu^- \bar{\nu}$ with right-handed muons does occur, with exactly the same rate and (parity-violating) asymmetry as the original process: the combined symmetry CP is not violated in this decay.

From the above discussion it should be clear that, when in 1964 also the combined CP symmetry was shown to be violated, this represented an even more fundamental blow to the basic principles which were assumed to lie at the heart of Nature's behaviour.

6.1.1 Kaons again

The same kind of particles which led to the discovery of parity violation through the $\theta - \tau$ puzzle (Section 2.3.1) had something more in reserve. Kaons, as the lightest particles containing a quark from the second family, can be considered as the 'minimal flavour laboratory', and indeed showed that the introduction of more flavours brings much more than a dull proliferation of new particles: the physics of flavour is a very rich subject which introduces many qualitatively new phenomena not present with a single quark family.

This time *neutral* kaons were involved: like all neutral mesons with non-zero flavour quantum number, these exhibit a rather subtle phenomenology which we briefly outline here and explore in much more detail in Chapter 8.

Like most 'early' particles, kaons, the first containing a 'new' kind of matter not normally present on Earth, were discovered in cosmic rays. In 1944, the study of photographs from a cloud chamber exposed to cosmic rays on the French alps indicated the existence of a charged particle with mass between that of the electron and the proton (Leprince-Ringuet and L'Héritier, 1944). A single event, interpreted as an elastic scattering on an electron, was attributed to a particle of mass 506 ± 61 MeV/c^2 (the K^+ mass is 494 MeV/c^2).

Three years later, some unstable 'V-particles' (named after the characteristic shape of the events in the photographs) were detected at Manchester[78] (Rochester and Butler, 1947), again in cosmic rays using cloud chambers. In order to study the penetration of particles, cloud chambers were usually filled with stacked iron plates, while in this case a chamber with a single thick plate was used, allowing the experimenters to detect two unusual events: one now interpreted as a $K^0 \to \pi^+\pi^-$ decay (a 'V', on 15 October 1946) and one a $K^+ \to \mu^+\nu$ decay (a 'kink', on 23 May 1947). The masses of the neutral and charged particle were roughly compatible among them, around 500 MeV/c^2.

After many further observations of such new particles, some 'strangeness' in their behaviour was put in evidence: they were copiously produced, with cross sections comparable to those for pion production (e.g. about 1 mb for $\pi^- p \to K^0 \Lambda$, a few per cent of the total $\pi^- p$ interaction cross section) and therefore consistent with a mechanism due to *strong* interactions, but then lived for a disproportionately long time (e.g. 10^{-10} s for the neutral K), much longer than the typical lifetimes of strongly interacting particles (order of magnitude 10^{-23} s), and more akin to particles affected by *weak* interactions.

A solution to this puzzle emerged (Pais, 1952; Gell-Mann, 1953; Nakano and Nishijima, 1953; Sachs, 1955) with the introduction of an additive 'strangeness' quantum number,[79] assumed to be conserved by the strong interactions responsible for their production but *not* by the weak interactions responsible for their decays. In Gell-Mann's scheme the new quantum number was additive, being +1 for the K^+ and K^0 mesons and -1 for the Λ and Σ baryons (and 0 for pions and nucleons); this could explain some results which were not accounted for by the earlier proposal of a multiplicative quantum number. Pair ('associate') production of strange particles of opposite strangeness by strong interactions was thus possible, e.g. $\pi^- p \to K^+ \Sigma^-$ or $K^0 \Lambda$, while single production e.g. $\pi^- p \to K^0 n$ was forbidden. The lightest strange mesons and baryons, K and Λ, could only decay into non-strange particles by violating strangeness conservation, and this had to be a weak process.

With the additive quantum number scheme, antiparticles have strangeness opposite to that of the corresponding particle: this required in particular that *two* neutral K particles had to exist, with strangeness ± 1; at a seminar in Chicago in 1954, when M. Gell-Mann described this scheme, E. Fermi asked how the two neutral particles, which could decay into the very same final states, would still retain any (distinguishable) individuality.

---

[78] This was the same year the charged pion was discovered; the neutral pion would enter the scene only in 1950.

[79] While this name might not be the best example of a carefully chosen one in physics, it should be compared with the 'attribute' proposed in Sachs (1955); things can always be worse.

These considerations led to the very remarkable observation by Gell-Mann and Pais (1955) that *neutral strange mesons* could be particles in a rather weird class of their own. Their argument went as follows.

For laws of Nature which are invariant under *CP*,[80] two classes of neutral particles are usually considered: those which are self-*CP*-conjugate (the photon, the π^0, etc.), and those which have an antiparticle which is somehow distinguished from it by a conserved quantum number (e.g. the neutron, by baryon number). The authors noted that the neutral K meson would belong to the latter class as long as strong interactions are considered, so that a second neutral K antimeson should exist, but that the strangeness quantum number distinguishing K^0 and $\overline{K}^0$ (+1 for K^0 and -1 for $\overline{K}^0$) *is not conserved* by weak interactions (putting them in the first class as far as weak processes are concerned). While e.g. a neutron and an antineutron are distinct at all times, and their respective decay products are different, the $\pi^+\pi^-$ states obtained from K^0 decays cannot be distinguished from those obtained from $\overline{K}^0$; the two mesons would therefore couple through processes such as $K^0 \rightarrow \pi^+\pi^- \rightarrow \overline{K}^0$, and in general $K^0 \leftrightarrow \overline{K}^0$ transitions would be allowed through the common (virtual) decay states. Even if such transitions are a second order effect in weak interactions, the degeneracy of K^0 and $\overline{K}^0$ (enforced by *CPT* symmetry) could make such processes significant.

A non self-conjugated particle is described by a non-Hermitian (complex) field, so that (with an appropriate choice of phases):

$$CP|K^0\rangle = |\overline{K}^0\rangle \qquad CP|\overline{K}^0\rangle = |K^0\rangle \qquad (6.1)$$

and if *CP* is a valid symmetry it can be used to characterize the physical states, which should therefore be

$$|K_1\rangle = \frac{1}{\sqrt{2}}\left(|K^0\rangle + |\overline{K}^0\rangle\right) \qquad |K_2\rangle = \frac{1}{\sqrt{2}}\left(|K^0\rangle - |\overline{K}^0\rangle\right) \qquad (6.2)$$

for which

$$CP|K_1\rangle = +|K_1\rangle \qquad CP|K_2\rangle = -|K_2\rangle \qquad (6.3)$$

Two neutral physical states (6.2), both linear superpositions of those of defined strangeness, were thus predicted to exist. If all interactions respected *CP* symmetry, the K_1 would only decay to states with $CP = +1$, and the K_2 to states with $CP = -1$, so that they could exhibit different lifetimes; since their virtual transitions would also involve different states, their masses could also be different (not being a

---

[80] As this was before the fall of parity, the authors actually considered charge-conjugation (*C*) invariance, but the same argument stands when only the combined *CP* symmetry is valid, as L. D. Landau pointed out.

particle-antiparticle pair, K_1 and K_2 are not constrained by *CPT* symmetry to have equal masses and lifetimes).

The search for the second neutral K meson was started at the newly built Brookhaven Cosmotron. The 3 GeV proton beam was directed onto a copper target, a neutral beam was extracted at an angle of 68° and travelled past a sweeping magnet, to eliminate charged particles, and a lead collimator, to reduce the amount of photons. The largest cloud chamber ever built (90 cm diameter) was placed at 6 m from the target (corresponding to $\sim$100 lifetimes for the Λ and the known neutral K meson); the large size of the chamber increased the chances to have long-lived particles stopping inside it, and indeed 23 'V events' were detected out of 1,200 photographs (Lande *et al.*, 1956). For almost all events the Q values were incompatible with $\pi^+\pi^-$ decay, the plane formed by the two visible decay tracks did not include the direction from the target, and the presence of an undetected neutral particle was necessary to balance energy and momentum, indicating a decay into at least three particles. After excluding alternative explanations, these events were attributed to $\pi e\nu$ and $\pi\mu\nu$ (and occasionally 3π) decays of a neutral particle of mass comparable to that of the known neutral K meson, and lifetime between 10^{-6} and 10^{-9} s.

With further experiments, a confirmation of the Gell-Mann–Pais scheme emerged, with two neutral K mesons, now called short-lived (K_S) and long-lived (K_L), with lifetimes

$$\tau_S \equiv \tau(K_S) = 0.89 \cdot 10^{-10}\,\text{s} \qquad \tau_L \equiv \tau(K_L) = 5.17 \cdot 10^{-8}\,\text{s}$$

The *huge* lifetime difference (a factor $\approx$600) between the two states originates from phase space volume: the only hadrons lighter than the K are pions, and the ratio of the kaon mass to the pion mass is small enough that decays into two or three pions have a relatively large difference in Q-values. This lifetime difference, while accidental in origin, has very important experimental consequences for the richness of the investigations which are possible on the neutral K system, as will be discussed in Chapter 8. Starting from an arbitrary mixture the two physical states can be separated by just waiting long enough: far enough from the production point a beam will contain exclusively the long-lived K_L.

A second prediction of the Gell-Mann–Pais theory was the following: since the strangeness eigenstates K^0 and $\overline{K}^0$ are coupled through weak interactions, their time evolution should not be a simple exponential (they are not physical states with definite lifetime), but should rather exhibit *strangeness oscillations*. The two strangeness eigenstates are not really two distinct 'particles' but rather the 'components' of a two-state system, since they can transform into each other spontaneously. By expressing a state which at time $t = 0$ is a K^0 in terms of the physical states K_S and K_L, which have the usual (e^{-iEt}) time evolution, then projecting back the state at the generic time t on a component of definite strangeness (K^0 or $\overline{K}^0$), the probability of measuring positive or negative strangeness as a function of time is found to be

$$P[K^0(t=0) \to K^0](t) = \frac{1}{4}\left[e^{-\Gamma_S t} + e^{-\Gamma_L t} + 2e^{-(\Gamma_S+\Gamma_L)t/2}\cos(\Delta mt)\right] \quad (6.4)$$

$$P[K^0(t=0) \to \overline{K}^0](t) = \frac{1}{4}\left[e^{-\Gamma_S t} + e^{-\Gamma_L t} - 2e^{-(\Gamma_S+\Gamma_L)t/2}\cos(\Delta mt)\right] \quad (6.5)$$

where $\Gamma_{S,L}$ are the total decay widths of the physical states $K_{S,L}$, and $\Delta m = m(K_L) - m(K_S)$ is their mass difference.

The time evolution of the strangeness eigenstates is not a simple sum of two exponentials but also exhibits an interference oscillatory term, showing that the $K^0, \overline{K}^0$ states are indeed not physical states, and that strangeness is not conserved in time. Note that when ignoring the strangeness content of the final state (i.e. summing over both strangeness states) the oscillatory term vanishes. Indeed the observation of strangeness oscillations allowed a measurement of the mass difference $\Delta m \simeq 3.5 \cdot 10^{-6}$ eV. We note in passing that the *sign* of Δm cannot be extracted from the above $K_L - K_S$ interference term (the cosine is an even function).

The phenomenon of strangeness oscillations was first observed in 1957 (Lande et al., 1957) when, starting from an almost pure K^0 beam (positive strangeness) produced by $pn \to p\Lambda K^0$ just above threshold (in order to conserve strangeness and baryon number the energy threshold for $\overline{K}^0$ production is higher), an event containing $\Sigma^- ppn\pi^+$ (negative strangeness) was found, produced by a $\overline{K}^0 \text{He}$ interaction.

It was later found that weak decays appear to respect the '$\Delta S = \Delta Q$ rule' (Feynman and Gell-Mann, 1958) linking the change in strangeness and in electric charge of the hadrons in the initial and final states[81]: as an example, while the decay $\Sigma^- \to ne^-\overline{\nu}_e$ occurs (branching ratio $\sim 1 \cdot 10^{-3}$), the decay $\Sigma^+ \to ne^+\nu_e$ does not[82] (branching ratio $< 5 \cdot 10^{-6}$). For semi-leptonic decays of neutral kaons this implies that $K^0 \to \pi^- e^+ \nu_e$ and $\overline{K}^0 \to \pi^+ e^- \overline{\nu}_e$ are possible, while the opposite decays (e.g. that in which a K^0 produces a negative electron) do not occur. This fact provides a way to determine the strangeness of a neutral kaon at the time of its decay (if it decays in the above modes, which have large branching ratios), thus allowing an easy experimental detection of strangeness oscillations, as shown in Fig. 6.1.

A very important empirical fact is that for neutral kaons the oscillation frequency Δm is comparable in magnitude to the inverse of the decay lifetime $\Gamma_L + \Gamma_S \simeq \Gamma_S$, so that the oscillations are neither too fast nor too slow to be observed before the

---

[81] This is readily understood today in terms of the valence quark content of the decaying particle: K^0 ($d\overline{s}$) and $\overline{K}^0$ ($\overline{d}s$).

[82] Note that neither the initial nor the final states are the *CP* conjugates of the previous process.

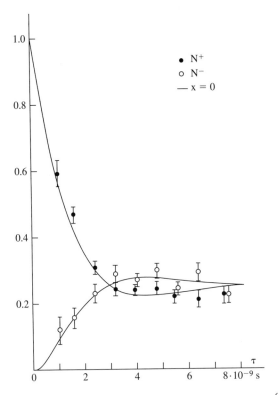

FIG. 6.1. Strangeness oscillations in neutral K decays, starting with a pure K^0 beam ($S = +1$) at production (proper time $\tau = 0$) obtained from $K^+p \rightarrow K^0p\pi^+$. The points show the relative rates of $\pi^-e^+\nu$ decays (N^+, filled circles, $S = +1$) and $\pi^+e^-\bar{\nu}$ decays (N^-, empty circles, $S = -1$) (Niebergall *et al.*, 1974). The lines indicate the expectation if the '$\Delta S = \Delta Q$' rule holds.

kaons decay in the third term of eqns (6.4), (6.5): with the same experimental time resolution which allows to detect the exponential decay of the K_S the strangeness oscillations are visible, as $\Delta m\,t \sim \Delta m/\Gamma_S \sim 1/2 \sim 0.1 \times 2\pi$.

A third consequence of the Gell-Mann–Pais theory, 'an even more bizarre man-ifestation of the mixing of K^0 and $\overline{K}^0$' (J.D. Jackson), was spelled out by Pais and Piccioni (1955). Strong interactions of kaons with matter are not independent of their strangeness (matter is not charge-conjugation symmetric): a $\overline{K}^0$ has more final states available than a K^0 when interacting with a baryon, e.g. besides the charge-exchange process $\overline{K}^0 n \rightarrow K^- p$ (which has the counterpart $K^0 p \rightarrow K^+ n$) also $\overline{K}^0 p \rightarrow \Lambda\pi^+$ is possible, while an analogous reaction does not exist for $\overline{K}^0$, due to the non-existence of a baryon with positive strangeness; the $\overline{K}^0$ interaction cross section is thus higher than that of the K^0. Hadronic interactions are therefore expected to distinguish the two strangeness components of a physical state, so that starting from an equal mixture of flavour eigenstates an asymmetric one will evolve spontaneously, since $\overline{K}^0$ are preferably absorbed. Quite analogously to what happens to circularly polarized light traversing an optically active medium, or to a polarized electron beam in a Stern–Gerlach experiment, when a pure K_L (K_2) beam (50%–50% strangeness mixture) traverses matter the differential interaction probability of K^0 and $\overline{K}^0$ leads to the *regeneration* of a K_S (K_1) component[83]. This phenomenon occurs through scattering, but also simply by transmission of the beam through matter ('coherent regeneration'), because of the difference in the forward-scattering amplitudes.[84]

The occurrence of regeneration was indeed verified experimentally by O. Piccioni and collaborators at Berkeley in 1960 (Muller *et al.*, 1960), who exposed a propane bubble chamber to a beam of neutral kaons, produced by a secondary π^- beam obtained from the 6 GeV Bevatron accelerator. The chamber was at a distance of 7 m from the kaon production target, corresponding to more than 100 K_S lifetimes; the beam therefore contained only K_L for all practical purposes. A thick (15 cm) slab of metal was inserted inside the chamber, and 2-track 'V-events' (the signature of $K_S \rightarrow \pi^+\pi^-$ decays) with the same momentum as that of the incident K_L were searched for, right after the slab (within 2 K_S lifetimes); a significant number of these were indeed detected, with a characteristic peak in the forward direction, by scanning 200,000 pictures.

From the investigations described above the picture emerged of a system with two *different, distinguishable* neutral K particles: this was not at all a straightforward conceptual step (compare e.g. to the π^+, π^-, π^0 system). Also note that, even if the physical states are superpositions of K^0 and $\overline{K}^0$, these two states do not lose their individuality completely: despite having the same charge, mass and lifetime (because of *CPT* symmetry), in the interaction with other hadrons the flavour eigenstates K^0 and $\overline{K}^0$ are clearly distinguished, and the same happens for some of their decay modes: the Hamiltonians describing such interactions and decays distinguish the two flavour eigenstates; it is just the *free* Hamiltonian, describing their propagation in vacuum, which does not.

---

[83] Note that this is quantum mechanics at its best: one is not just dealing with some counter-intuitive reappearance of a (non-classical) property of a particle such as spin orientation, but rather to the *creation* of a particle by quantum-mechanical interference. While the phenomena are completely analogous within the formalism of quantum mechanics, one cannot avoid being somewhat dazzled: 'There is nothing else quite like it in nature' (Feynman, 1970); 'This is one of the greatest achievements of theoretical physics. [...] Especially interesting is the fact that we have taken the principle of superposition to its ultimately logical conclusion' (Feynman, 1961).

[84] Note that this is the only instance in which a forward scattered beam can be distinguished from the original incident beam.

It is worth reading the words in a paper from that time: 'It is by no means certain that, if the complex ensemble of phenomena concerning the neutral K mesons were known without the benefit of the Gell-Mann Pais theory, we could, even today, correctly interpret the behavior of these particles. That their theory, published in 1955, actually preceded most of the experimental evidence known at present, is one of the most astonishing and gratifying successes in the history of the elementary particles' (Good *et al.*, 1961).

6.1.2   *CP violation*

As usual, the CP eigenvalues of the two neutral K were determined by considering the CP-parities of the final states in the decays and trying to find an assignment which satisfies CP conservation. While semi-leptonic states ($\pi \ell \nu$) are clearly not CP eigenstates, the two- and three-pion neutral states can be. Kaons are pseudo-scalar particles ($J^P = 0^-$), like as pions.[85]

Due to angular momentum conservation, the two pions (spin $S = 0$) in a $K \rightarrow 2\pi$ decay must be in a state of orbital angular momentum $L = 0$, so that the parity of the final state is $\eta_P = (-1)^2(-1)^L = +1$ and the C-parity is $\eta_C = (-1)^L = +1$. The CP-parity of the 2π final state is therefore $+1$; recalling from Section 3.2.4 that $\eta_{CP} = (-1)^{L+I}$ for a two-pion state, the isospin of the pair is necessarily even ($I = 0, 2$).

For the $K \rightarrow 3\pi$ decay the orbital angular momentum L of two pions in their centre of mass system must be equal to that (indicated as l) of the third one with respect to said system, in order to give a total angular momentum zero. The parity is thus $\eta_P = (-1)^3(-1)^{L+l} = -1$, and the C-parity is $+1$ for the $3\pi^0$ state and $(-1)^L$ for the $\pi^+\pi^-\pi^0$ state. The CP-parity of the 3π final state is therefore -1 for $3\pi^0$ and $(-1)^{L+1}$ for $\pi^+\pi^-\pi^0$; for the latter state $L = 0$ ($CP = -1$) is expected to be favoured, due to the angular momentum repulsion. Recalling that for a three-pion state $I + L$ must be odd, in this case one has $\eta_{CP} = (-1)^I$.

Since the short-lived neutral kaon was seen to decay into two pions, while the long-lived one apparently only decayed into three-body final states, the K_S was identified with the $CP = +1$ eigenstate (K_1), and the K_L with the $CP = -1$ eigenstate (K_2).

The discovery of *CP* violation

In 1963, a bubble chamber experiment on K_S regeneration at the Brookhaven Cosmotron by R. Adair and collaborators (Leipuner *et al.*, 1963) showed an anomalous excess of 2π events with the K_L beam traversing liquid hydrogen: the effect was statistically very significant,

---

[85] Although, strictly speaking, it should be recalled that a parity assignment is only meaningful when weak interactions are ignored.

and the authors stated that 'The possibility of interpreting the events as two-pion decays of K_2, which would be allowed if CP invariance were violated, is excluded by the result of observation of 411 K_2 decays in cloud chambers [in two previous experiments], none of which were consistent with two-pion decays'; they suggested instead that a new kind of long-range interaction between protons and K might explain the observations, although the article ends stating that: '... it is necessary to emphasize that we cannot, at this time, completely exclude the possibility [...] that the striking character of the data results from a combination of real effects underestimated by us together with strong statistical fluctuations'.[86] The effect in the data was just too large for the correct conclusion to be drawn.

A proposal was submitted at Brookhaven in April 1963 by the Princeton group of J. Cronin, V. Fitch, and R. Turlay to measure this anomaly with much higher statistics at the Alternating Gradient Synchrotron (AGS), using the newly developed spark chambers; such detector was much faster than bubble chambers and resulted in a five times better mass resolution. A secondary goal of the experiment, mentioned in the proposal, was to improve the limit on the CP-violating $K_L \to 2\pi$ decay mode; some (very few) attentive people had actually remarked that existing tests of CP symmetry for weak interactions were limited, as such interactions were tested only at lowest order (except by neutral K decays) so that CP symmetry in one class of processes (such as beta decay) gave no information on the properties of other ones (see e.g. Sachs and Treiman (1962)). In one month the experiment was approved, and the next month the three physicists, joined by J. Christenson, started a 40-day data-taking period.

A K_L beam was obtained from 30 GeV protons impinging on an beryllium target inside the accelerator, and was directed at an angle of 30° towards a crammed and hot experimental area at the centre of the machine dubbed 'inner Mongolia' by the physicists working there. The beam crossed a 3.8 cm lead absorber to reduce the amount of photons and a sweeping magnet placed between two collimators to remove charged particles, before entering a 'helium bag' (poor man's vacuum) to reduce spurious regeneration, after 17 m from the production target; at the average K momentum of ~ 1.1 GeV/c the K_S decay length is $\beta\gamma c\tau_S \simeq$ 6.5 cm, so practically no short-lived kaons were left in the beam. The detector was a two-arm spectrometer, the angle between the two arms being optimized for the detection of $\pi^+\pi^-$ pairs and the measurement of their momenta: each arm had two sets of spark chambers with a 0.45 T m magnetic field integral in between (Fig. 6.2). The chambers were triggered by scintillator counters and water Čerenkov counters used to discriminate against $e^\pm$; the good track resolution and the possibility of triggering the detector were the key advantages over previous experiments.

Events were reconstructed by measuring the invariant mass and the transverse momentum p_T of the two detected charged tracks: track pairs from three-body decays are not expected to give any peak at the K^0 mass, and they also have a non-zero p_T in general when one particle is undetected. The detector was calibrated by observing the $\pi^+\pi^-$ decays of K_S regenerated in a tungsten regenerator followed by an anti-coincidence counter, which was placed at different longitudinal positions in steps of 28 cm to simulate the spatial distribution of K_L decays. As the $K_S \to \pi^+\pi^-$ coherently regenerated have the same momentum as the original K_L, they closely simulate $K_L \to \pi^+\pi^-$ decay; the data showed standard

---

[86] A fair statement which could apply to most experiments.

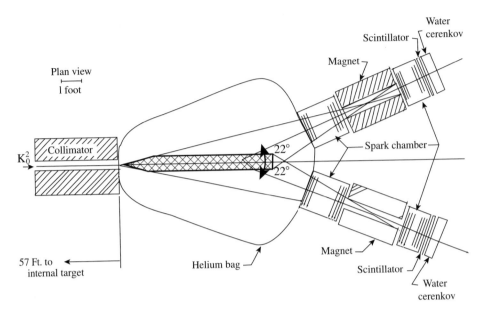

FIG. 6.2. Sketch of the apparatus of Christenson, Cronin, Fitch, and Turlay at Brookhaven (Christenson *et al.*, 1964). The cross-hatched area is the fiducial region for decays.

deviations of 3.6 MeV/c^2 and 3.4 mrad on the reconstructed invariant mass and kaon direction.

After several runs with regenerators, some data was taken without, and still a clear forward peak (with a Gaussian width of 4 mrad) for invariant mass corresponding to $m(K^0)$ was found (Fig. 6.3): this peak only appeared for the interval of invariant masses around the neutral kaon mass, and the events had invariant mass and angle distributions with resolutions in agreement with those for the decays of regenerated K_S, indeed appearing indistinguishable from those. Data taken with a liquid hydrogen target instead of helium gas showed a similar forward peak and, after subtracting the expected background from regeneration, the yield was consistent with the previous case (but much smaller than the 'anomalous regeneration' result would have indicated).

The signal seemed to indicate the occasional decay of K_L into two charged pions, violating *CP* conservation. The puzzling result was extensively checked for six months, looking hard for conventional explanations. K_S regeneration in the helium gas was estimated to be a factor $\sim 10^6$ too small to account for the data, and the spatial distribution of the $\pi\pi$ decays was consistent with that of other K_L decay modes, not concentrated at any particular position; moreover, if regeneration was responsible for the excess events, a much higher rate would have been observed in the liquid hydrogen runs. Three-body decays of K_L with a missing particle could not reproduce the experimental distributions by any reasonable mechanism. Finally, the possibility that $\pi^+\pi^-$ were accompanied by an undetected soft photon, possibly resulting in a *CP* $= -1$ final state, required some weird unknown mechanism forcing the photon energy to be always

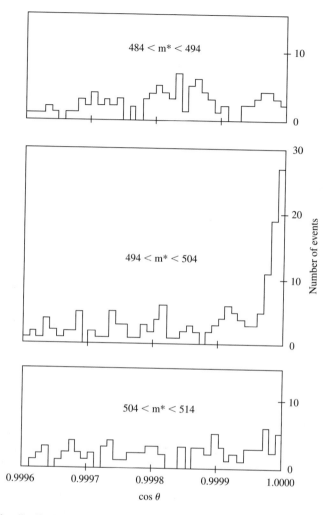

FIG. 6.3. Angular distribution ($\cos\theta = 1$ is the forward direction) of the reconstructed momentum of $\pi^+\pi^-$ events for three ranges of invariant mass m^*, below (top), above (bottom) and around (centre) the neutral kaon mass. A forward peak is evident only for 2-pion invariant masses consistent with $m(K^0)$ (Christenson *et al.*, 1964).

below 1 MeV, in order to fit the data (such decays do indeed occur, although at a much smaller rate).

The paper announcing the fall of *CP* symmetry was eventually published in July 1964 (Christenson *et al.*, 1964), reporting the decay ratio

$$\frac{\Gamma(K_L \to \pi^+\pi^-)}{\Gamma(K_L \to \text{charged particles})} = (2.0 \pm 0.4) \cdot 10^{-3} \qquad (6.6)$$

based on 45 ± 9 events after background subtraction, out of a total acceptance-corrected sample of 22,700 K_L decays.

A second result was reported immediately after, by another Brookhaven experiment (Abashian *et al.*, 1964) designed to study semi-leptonic kaon decays: by using a set of spark chambers in two different configurations to detect K_L decays in vacuum, a forward peak of eleven events was observed for the proper invariant mass range. Although this was not significant enough by itself, it confirmed the previous result, as indeed other experiments did soon after.

Contrary to the case of parity violation, *CP* violation was not immediately accepted by the whole physics community: the evidence of a symmetry violation from the existence of a relatively rare process (the 'Princeton effect') was somewhat less immediate than the observation of an up-down asymmetry in a decay, and the fact that the effect was small left open the possibility that something had been overlooked. But most importantly, this time the left-right symmetry of Nature was receiving a final blow: 'What shook all concerned now was that with *CP* gone there was nothing elegant to replace it with.' (Pais, 1986). Many ingenious possible explanations were put forward to save CP symmetry, from mundane ones such as the fact that the detected final state was not really $\pi^+\pi^-$ (or the initial one not really a K_L), to more exotic ones such as the existence of some asymmetry in the environment (so that the effective vacuum is not a *CP* eigenstate), to which the very sensitive neutral kaon system would react and tentatively proposed as the effect of a new field of 'cosmological' origin (the effect was also tentatively ascribed to the regeneration on a fly trapped inside the helium bag!), up to really dramatic ones such as giving up the validity of the superposition principle of quantum mechanics in attempts to make the existence of $K_L \to \pi\pi$ decay innocuous (see e.g. Kabir (1968) or Zumino (1966) to get a feeling). They were all proved wrong.

A conclusive piece of evidence actually came just one year later, when the interference between the 2π state obtained from K_S decays and that obtained from the newly discovered K_L decays was detected.

In presence of *CP* violation, not only do the physical states K_S and K_L interfere in decays to strangeness eigenstates (leading to strangeness oscillations) but they do so also in decays to *CP* eigenstates: for a state which is produced at $t = 0$ as a $K^0(\overline{K}^0)$ and propagates freely in vacuum, the decay rate to $\pi^+\pi^-$ is

$$\Gamma[K^0(\overline{K}^0)(t = 0) \to \pi\pi](t) \propto e^{-\Gamma_S t} + |\eta_{\pi\pi}|^2 e^{-\Gamma_L t}$$
$$\pm 2|\eta_{\pi\pi}|e^{-(\Gamma_S+\Gamma_L)t/2}\cos(\Delta m t - \phi_{\pi\pi}) \quad (6.7)$$

where

$$\eta_{\pi\pi} = |\eta_{\pi\pi}|e^{i\phi_{\pi\pi}} = \frac{A(K_L \to \pi\pi)}{A(K_S \to \pi\pi)}$$

is the *CP*-violating ratio of $K \rightarrow \pi\pi$ decay amplitudes, and the $+ (-)$ sign applies to an initial K^0 ($\overline{K}^0$) state. The oscillating term in eqn (6.7) describes the interference of K_S and K_L in vacuum, which vanishes with increasing time (distance from the production point) as the short-lived component decays away: when the K_S decay amplitude has been attenuated appropriately relative to the K_L one, that is for $e^{-t/2\tau_s} \sim |\eta_{\pi\pi}|$ or $t \sim 12\tau_S$, the interference effects are large.

The interference term in eqn (6.7), of opposite sign for initial K^0 and $\overline{K}^0$, is an indication of *CP* violation: if *CP* symmetry holds two *CP*-conjugate processes should look the same, and, for decays into a self *CP*-conjugate final state f_{CP}, indeed $K^0 \rightarrow f_{CP}$ and $\overline{K}^0 \rightarrow f_{CP}$, are related in this way. More generally, in absence of *CP* violation there can be no $K_S - K_L$ interference in the decay rate of an arbitrary neutral kaon beam to any set of states which transforms into itself under *CP*, a result analogous to the one discussed in Chapter 5 for *C* symmetry.

The same kind of interference is more easily detected beyond a regenerator, where a significant K_S component can be obtained without the difficulty of performing an experiment too close to the production target, as would be required with the former approach at the kaon energies available in the 1960s. In this case an initially pure K_L beam develops a coherently regenerated K_S component at the exit of a regenerator (defined to be $t = 0$)

$$|\psi(t = 0)\rangle \propto |K_L(t = 0)\rangle + \rho|K_S(t = 0)\rangle \qquad (6.8)$$

where the complex quantity $\rho = |\rho|e^{i\phi_\rho}$ describes the coherent regeneration process (it is defined as the ratio of the K_S to the K_L amplitudes at the exit of the regenerator). As a consequence the $\pi\pi$ yield has three components

$$\Gamma[K^0(\overline{K}^0)(t = 0) \rightarrow \pi\pi](t) \propto |\rho|^2 e^{-\Gamma_S t} + |\eta_{\pi\pi}|^2 e^{-\Gamma_L t}$$
$$\pm 2|\eta_{\pi\pi}||\rho|e^{-(\Gamma_S + \Gamma_L)t/2} \cos[\Delta m t - (\phi_{\pi\pi} - \phi_\rho)] \qquad (6.9)$$

exhibiting a $K_S - K_L$ interference term, which is large when ρ and $\eta_{\pi\pi}$ have comparable magnitudes.

The proof of *CP* violation

An experiment by V. Fitch and collaborators (Fitch *et al.*, 1965) showed evidence for $K_S - K_L$ interference in $\pi^+\pi^-$ decays by comparing the yields with a thick solid regenerator ($|\rho| \gg |\eta_{\pi\pi}|$), in air ($|\rho| \sim 0$) and with a diffuse regenerator (such that $|\rho| \sim |\eta_{\pi\pi}|$) .

The experiment was carried out at the AGS, with a neutral beam extracted at 30° and directed towards a spark chamber spectrometer placed 27 m after the K_L production target. At the average kaon momentum the beginning of the fiducial region was at a distance corresponding to ~ 270 K_S lifetimes from the target.

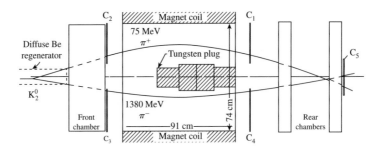

FIG. 6.4. Schematic plan view of the detector for the first $\pi\pi$ interference experiment by Fitch *et al.* (1965). C1–C5 are scintillator counters.

The spectrometer had sets of chambers in front and behind a dipole magnet with a vertical 1.2 T m field integral, inside which a thick tungsten absorber stopped particles travelling along the beam direction (see Fig. 6.4). Charged particles bent by the magnetic field could bypass the absorber and reach the second set of chambers; for a two-body decay the trajectories of the two particles (with equal transverse momentum) crossed beyond the magnet (in the horizontal projection) along the projected kaon trajectory, resulting in a good resolution for the measured kaon direction and an insensitivity of the reconstructed invariant mass to measurement errors in the front chambers. The last chamber plates were preceded by three radiation lengths of lead inducing electromagnetic showering of $e^\pm$ to help identifying $\pi e \nu$ decays. Five plastic scintillator counters were used for triggering, the last one at the centre of the rear spark chamber discriminating against neutron-induced background.

Both the invariant mass and the kaon direction with respect to the beam line were reconstructed in the hypothesis of a $\pi^+\pi^-$ decay; the accepted kaon momentum was 1.55 ± 0.30 GeV/c. As the beam travels inside the diffuse regenerator a K_S component builds up and decays, and at depths much larger than the K_S decay length the equilibrium regeneration amplitude is

$$\rho \simeq \frac{i2\pi N_d}{|p_K|}[f(0) - \bar{f}(0)]\frac{\beta\gamma\,\tau_S}{1 - i\Delta m\tau_S} \qquad (6.10)$$

In this formula N_d is the nuclear density of the regenerator, p_K the kaon momentum, and $f(0), \bar{f}(0)$ are the forward-scattering amplitudes for $K^0, \overline{K}^0$ respectively (β, γ are the relativistic factors). This depends on the value of Δm, as it represents the sum of events regenerated at different depths, which therefore propagate inside the regenerator (with a given effective mass) for different lengths. The K_S flux is therefore constant; since $\Gamma_L t \ll 1$ the K_L flux is also constant, and the $\pi\pi$ decay rate is just

$$\Gamma[K^0(\overline{K}^0)(t = 0) \to \pi\pi](t) \propto |\rho|^2 + |\eta_{\pi\pi}|^2 \pm 2|\eta_{\pi\pi}||\rho|\cos(\phi_{\pi\pi} - \phi_\rho)$$

Since the regeneration amplitude ρ depends linearly on the density, by measuring the $\pi^+\pi^-$ yield in a given region of space for three different densities, the three quantities $|\eta_{\pi\pi}|$, $|\rho|$, and $\cos(\phi_{\pi\pi} - \phi_\rho)$ can be determined.

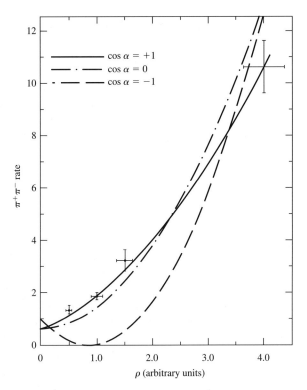

FIG. 6.5. Measured $\pi^+\pi^-$ event rate as a function of regeneration amplitude in Fitch *et al.* (1967). The lines indicate best fits for the cases $\alpha \equiv \phi_{\pi\pi} - \phi_\rho = 0, \pi/2, \pi$.

The regenerator material was beryllium, chosen to minimize multiple scattering of the charged pions, due to its low atomic number. The largest interference effects are obtained for $|\rho| \sim |\eta_{\pi\pi}|$, which for Be corresponds to a density $\sim 0.1\,\mathrm{g/cm^3}$. A low-density diffuse regenerator was obtained by using a stack of 0.6 mm thick plates spaced by 1 cm, in a 1 m long arrangement: this effectively appeared as an homogeneous medium to kaons with 7.5 cm decay length and 90 cm strangeness oscillation length. To retain the applicability of the equilibrium formula (6.10) the yield of decays from the last 45 cm of the diffuse regenerator ($\sim 7 K_S$ lifetimes after its front face) was compared to that obtained for the same length in vacuum and to the one expected from regeneration. The latter was determined by measuring the $\pi^+\pi^-$ yield behind a dense regenerator, an 8 cm thick beryllium slab; in this case the terms containing $\eta_{\pi\pi}$ in (6.9) can be neglected and at the end of the regenerator

$$\Gamma[K^0(\overline{K}^0)(t=0) \to \pi\pi](t) \propto |\rho|^2 \left(\frac{N_D}{N_d}\right)^2 \left[1 + e^{-l/\beta\gamma\tau_S} - 2e^{-l/2\beta\gamma\tau_S}\cos(\Delta m\,l)\right]$$

where N_D is the nuclear density of the dense regenerator and l its length, so that $|\rho|$ can be obtained. The data was corrected for kaon (and pion) attenuation and interference inside the dense regenerator, in the anti-coincidence counter and behind it; this regenerator was moved in 10 cm steps along the beam, in order to average out space-dependent effects within the fiducial region.

The experiment showed that the two terms due to regenerated K_S and *CP*-violating K_L decays did interfere: the yield of $\pi^+\pi^-$ events with the diffuse regenerator was not the sum of those due to K_L decays and regenerated K_S decays (the latter determined by the measurement of the regeneration amplitude with the dense regenerator), but rather up to twice that value, requiring a large constructive interference term between the two amplitudes to fit the data. By using different spacings for the beryllium slabs, data was also collected with different densities for the diffuse regenerator, clearly showing the constructive interference (Fig. 6.5).

This result proved that the $\pi\pi$ final states from K_S and K_L decays were indeed the same, so that the conclusion that *CP* was violated could not be escaped by assuming that some extra particle went undetected or in general that a final state different from $\pi^+\pi^-$ was involved, nor by requiring that the initial state was an incoherent K_S obtained e.g. in the decay of a K_L with some undetected light particle.

CP violation was thus established.

In the following years it was detected also in the more experimentally challenging $K_L \rightarrow \pi^0\pi^0$ decay mode, first at the 28 GeV CERN PS (Gaillard *et al.*, 1967) and at the Princeton–Pennsylvania 4 GeV proton synchrotron (Cronin *et al.*, 1967), in the former case by comparing the yields with and without a regenerator. For this decay the large $K_L \rightarrow 3\pi^0$ background is more difficult to control, and the experiments exploited either the higher photon energy obtained from a two-body decay (at Princeton) or the constraints obtained by measuring the photon directions from their electromagnetic showers in spark chambers (at CERN): the energy of a π^0 can be obtained from the measurement of the opening angle ϕ and ratio of energies E_1/E_2 for the two decay photons, and this energy can be compared with the one that is actually measured from the sum of photon energies, thus providing constraints for each π^0. Together with the measured $\pi^0\pi^0$ invariant mass and the angle of the final state with respect to the beam direction these constraints can be used to suppress the background.[87]

Despite earlier claims to the contrary, the ratio of the *CP*-violating to *CP*-conserving amplitude for $K \rightarrow \pi^0\pi^0$ decays appeared to be consistent with that for the $\pi^+\pi^-$ decay (Bartlett, 1968), as shown by improved experiments which made use of segmented electromagnetic calorimeters, allowing a proper rejection of the contribution due to $K_L \rightarrow 3\pi^0$ background. It was noted that $K_L \rightarrow \pi^+\pi^-$ could be due to a violation of Bose–Einstein statistics for pions, allowing an antisymmetric $I = 1$ state, while fully respecting *CP* symmetry (Bludman, 1964). The same possibility is not present for $\pi^0\pi^0$, as an anti-symmetric state of two identical pions

---

[87] However at that time the quality of the photon detector measurement was not enough to allow exploiting all of the above constraints without introducing significant biases; the history of kaon physics is a history of improvements in calorimetry.

cannot be formed independently of pion statistics; the observation of $K_L \rightarrow \pi^0 \pi^0$ thus proved that *CP* violation exists, removing the above loophole.

Another manifestation of *CP* violation was later detected as a charge asymmetry in semi-leptonic decays $K_L \rightarrow \pi^+ \ell^- \bar{\nu}_\ell$ vs. $K_L \rightarrow \pi^- \ell^+ \nu_\ell$ (where $\ell = e, \mu$), first at Brookhaven (Bennett *et al.*, 1967) and at Stanford (Dorfan *et al.*, 1967); in these cases backgrounds were not a problem, while the key feature was the control of systematic biases at a sufficient level to extract the small (3 per mille) asymmetry.

The interference between K_S and K_L after a regenerator (6.9) was further exploited to measure the mass difference Δm and the difference between the *CP*-violating phase ϕ_{+-} and the regeneration phase at CERN (Alff-Steinberger *et al.*, 1966; Bott-Bodenhausen *et al.*, 1966). The interference was also studied in vacuum (i.e. without a regenerator): in this way for a known Δm the phases could be measured directly because the initial state (a strangeness eigenstate) is known a priori. The need for intense beams implied that the shielding from the intense background radiation was a serious issue, as one has to consider events originating in the region where K_S are still present, close enough to the production target: the interference term is large when the K_S and K_L decay rates into $\pi\pi$ decays are comparable, that is (see eqn (6.7)) at proper time $\sim 2 \log |\eta_{\pi\pi}|/(\Gamma_L - \Gamma_S) \sim 12 \tau_S$ although earlier times also contribute in a significant way to the results because the higher statistics compensates for a smaller interference term. High kaon momenta are chosen to profit from relativistic time dilation and move this region away from the target, and as a consequence one has to deal with an incoherent mixture of K^0 and $\overline{K}^0$ (rather than with an almost pure K^0 beam which is possible at lower energies) and the interference gets somewhat diluted (see Chapter 8); note, however, that the measurement of the zero-crossing point at which $\Delta m\,t - \phi_{+-} = \pi/2$ is independent of the knowledge of the coefficient of the interference term. In the first spark-chamber CERN experiment (Böhm *et al.*, 1969) the decays of ~ 6.5 GeV/*c* kaons were detected at 2.5–6 m (6 to 16 τ_S) distance from the target.

The same experiment was then repeated with a higher precision at CERN (Geweniger *et al.*, 1974) as the first application of the newly invented multi-wire proportional chambers (MWPC), which allowed rates two order of magnitudes higher than spark chambers and could be self-triggered, thus avoiding the need for scintillator plates introducing material before the spectrometer and degrading its resolution. The *CP*-violating phase for the more challenging neutral mode was then also measured, starting with (Chollet *et al.*, 1970) at CERN.

In the words of the protagonists, the observation of *CP* violation was 'a purely experimental discovery, a discovery for which there were no precursive indications, either theoretical or experimental' (Fitch, 1981), and was 'made with a simple apparatus, designed to ask the right question of nature, and received a clear response' (Cronin, 1981). For this discovery Jim Cronin and Val Fitch shared the 1980 Nobel prize.

Moreover, in this respect it may be important to note that, although most discoveries do not occur in a vacuum but are quite influenced by the environment and by the concurring experimental and theoretical investigations, they are still paced by the technical advances: 'If you know anything at all about physics, you will conclude that [the *CP* violation] experiment could not have been done a couple of years earlier at all. Because we did not have the technical tools. Such an experiment is not seasoned by brilliant people at the nuclear laboratories to measure that. It came with a certain season, at a certain time, at a certain place. And even if you have the tools, it does not mean that you will take the decision to use them [There is a] kind of progress in physics, step by step, by the development of tools, and by increasing the accelerator energies.' (Telegdi, 1987).

It is important to note that the sign of the interference term in eqn (6.7) provides an *absolute* way of distinguishing matter and antimatter: independently of any convention, the $K \to \pi\pi$ decay yield in vacuum as a function of proper decay time starting with an initial $\overline{K}^0$ develops to be *smaller* that for an initial K^0, in the proper time region 12–17 τ_S. Since pure K^0 can be obtained from the charge-exchange scattering of K^+ (even on an antimatter target), and conversely $\overline{K}^0$ from K^-, this provides an absolute distinction of positive and negative charges. Analogously, the sign of the oscillating term in the interference behind a regenerator, eqn (6.9) would be opposite for a regenerator made of antimatter, thus providing a way to distinguish it from matter (Sakurai and Wattenberg, 1967): at small times the interference is constructive for matter and destructive for antimatter.

The existence of *CP* symmetry violation truly allows one to define in an absolute way the meaning of right and left; recalling the discussion in Section 6.1, *CP* violation gives a way of communicating in an unambiguous way our definition of right and left to an extraterrestrial civilization: we could then tell the extraterrestrials about which sign of charge we call 'positive' (e.g. the charge of the electron which is most frequently obtained in the semi-leptonic decays of a long-lived neutral kaon), and thus make sure that they agree with what we call 'right'[88]. Also in this sense *CP* violation is more fundamental than *C* (or *P*) violation.

6.2 Consequences of *CP* symmetry

To learn more about *CP* violation it is necessary to study it in many phenomena, so we now consider its signatures by discussing the constraints that *CP* symmetry would enforce on physical processes.

As remarked in the discussion of *C* symmetry, direct tests of *CP* cannot be performed by directly comparing *CP*-conjugate experiments, and one is rather forced to examine microscopic processes in which charge-conjugate particles appear; we

---

[88] In case one is tempted to speculate about extraterrestrials having an opposite definition of 'past' and 'future' (to recover the right-left ambiguity through *CPT*), it should be considered that the absoluteness of these concepts is guaranteed by the existence of a macroscopic arrow of time. If that were also reversed in the two worlds, no communication would be possible between the two, although this possibility represents an interesting subject for science fiction (Piper, 1977).

note in passing that this excludes (apart from what was discussed in Chapter 5) atomic and nuclear processes as probes of *CP* symmetry, since the energies involved in those systems are not sufficient for the production of antiparticles (*CP* violation studies in nuclear physics (Gudkov *et al.*, 1992) are thus really tests of time reversal symmetry, linked to *CP* by the *CPT* theorem).

If *CP* symmetry held exactly in Nature, *neutral* physical states (with well defined masses and lifetimes) should be either *CP* eigenstates or *degenerate* pairs of states linked by the *CP* transformation. As discussed in Chapter 3, the fact that *C* anti-commutes with the electric charge makes *CP*-parity not a very useful quantum number for the classification of particles, except for neutral ones. Again, as in the case of *C*, the fact that *CP* symmetry requires antiparticles to exist and have the same static properties as the corresponding particles (no such property is *P*-odd, so *CP* is equivalent to *C* for this purpose) is irrelevant, since the more fundamental *CPT* symmetry already enforces the same (see Chapter 5). On the other hand the general validity of *CPT* symmetry implies that the constraints enforced by *CP* are usually considered to be the same as those imposed by *T* symmetry, such as in the case of interaction terms which are forbidden to appear in the Lagrangian (Feinberg, 1957).

6.2.1 *CP* and complex phases

In general, *CP* symmetry holds if the Hamiltonian is invariant under the *CP* transformation:

$$H_{CP} \equiv (CP)^{\dagger} H (CP) = H \quad \text{or} \quad [H, CP] = 0 \qquad (6.11)$$

The link of *CP* symmetry to *T* symmetry enforced by the *CPT* theorem (which we assume to be valid in the following) means that *CP* violation is described by complex terms appearing in the Lagrangian density; in general, writing this as a sum of operators O_i with coefficients a_i

$$\mathcal{L} = \sum_i a_i O_i + \text{H.c.} \qquad (6.12)$$

where the operators are such that $(CP)^{\dagger} O_i (CP) = O_i^{\dagger}$, *CP* symmetry holds if the constants a_i are real. As an example, considering the Yukawa couplings in the SM, Hermiticity requires that they appear in pairs as

$$\mathcal{L}^{(Y)} = g \, \overline{\psi}_L \phi \psi_R + g^* \, \overline{\psi}_R \phi^{\dagger} \psi_L$$

(where $\psi_{L,R}$ are the left, right quark fields and ϕ the Higgs scalar field); this Lagrangian density transforms under *CP* into

$$\mathcal{L}^{(Y)} \to \mathcal{L}^{(Y)}_{CP} = g \, \overline{\psi}_R \phi^{\dagger} \psi_L + g^* \, \overline{\psi}_L \phi \psi_R$$

which is identical to the original if $g = g^*$.

It should be noted that not all the possible complex phases have a physical meaning, as some of them could be eliminated by just redefining the unphysical phase of the fields. Indeed, the phase factors η_{CP} associated to the *CP* transformation of a field, e.g. for a scalar field

$$(CP)^\dagger \phi(t, \boldsymbol{x})(CP) = \eta_{CP}\, \phi^\dagger(t, -\boldsymbol{x}) \tag{6.13}$$

are not physical (for non-Hermitian fields) just as for charge conjugation. *CP* violation is only present when there exist no phase convention in which all the complex parameters can be eliminated from the Lagrangian density.

The complex parameters in the Lagrangian density translate into phase factors in transition amplitudes, which have opposite values for two *CP*-conjugated processes; since *CP* violation has been detected only in phenomena driven by weak interactions, such phases are often called 'weak phases'.

The fact that *CP* symmetry is linked to the phases in the transition amplitudes introduces a complication, since all the quantities which can be observed experimentally are expressed as squared matrix elements, and therefore such phases do not always lead to observable effects. In particular it is clear that the absolute phase of a transition amplitude is not an observable quantity (depending on a convention it can be redefined at will, since it disappears when taking the squared modulus), and only phase *differences* between amplitudes can possibly be observed: such phase differences can be independent of the phase convention, as they cannot always be redefined away by changing the phases of the initial or final states. Indeed at least two interfering amplitudes are required to observe a phase-related effect and to study *CP* symmetry in a given process: in this case

$$A(i \to f) = |A_1|e^{i\phi_1} + |A_2|e^{i\phi_2} \tag{6.14}$$

$$|A(i \to f)|^2 = |A_1|^2 + |A_2|^2 + 2|A_1||A_2|\cos(\phi_1 - \phi_2) \tag{6.15}$$

Note the analogy to the situation of parity violation, which could not be detected by just looking at *scalar* quantities such as transition rates, and required pseudo-scalar observable to be put in evidence, obtained by the interference of a parity-conserving and a parity-violating amplitude. Similarly, *CP* (*T*) violation is not detected by considering only single real quantities, and the interference of two amplitudes with a non-zero relative phase is required instead (unfortunately there's no way of directly measuring a complex quantity, since all observable quantities are real).

The difficulties with the elusive nature of phases in quantum theory do not end here, though: in presence of two interfering amplitudes with different phases, the observables of a process (e.g. the magnitude of a cross section, or an angular distribution) can indeed depend on the value of such phase difference as shown by eqn (6.15), but how can one assess that any specific feature of the observed process actually arises from a phase difference? To detect the third term in eqn (6.15) just from the measurement of the transition rate would require knowledge of the first two terms with sufficient precision, which is not possible without further insight. A better way is to compare the process with its *CP*-conjugated one, in which any *CP*-violating phase is opposite in sign; however the squared modulus of the total amplitude is insensitive to the *sign* of the phase difference: a cosine term appears in eqn (6.15) which is not affected by $\phi_i \to -\phi_i$. Still something more is required to translate the *CP*-violating phases into observables.

The expression in eqn (6.15) cannot give a *CP*-odd quantity, that is one which only vanishes when the phase difference goes to zero; in order to obtain such an expression, linear in $\sin(\phi_1 - \phi_2)$ in the squared modulus, an imaginary part in the amplitude is required, which could interfere with $i|A_1||A_2|\sin(\phi_1 - \phi_2)$.

Such additional phases can arise due to final state interactions. Indeed, even in absence of *CP* violation complex phases can appear in transition amplitudes, due to contributions from intermediate states on the mass shell (beyond the Born approximation), i.e. absorptive parts in amplitudes to which coupled channels can contribute. Usually the most important effects of this type are due to the strong interactions among the particles involved in a reaction, and therefore the phases involved are called 'strong phases' or 'scattering phases'. These phases, contrary to the *CP*-violating ones discussed above, have the same sign for the process in which all the particles involved are replaced by their antiparticles, since the interactions producing them are *CP*-conserving (see Chapter 5). The presence of such scattering phases is what makes the observation of the presence of the *CP*-violating phases possible at all, since the transformation of the two kinds of phases when passing to the *CP*-conjugated process is different. This is readily understood by considering the simplest observable difference between two *CP*-conjugated processes, namely their rate difference. Denoting with a bar the *CP*-conjugate state:

$$A(i \to f) = |A_1|e^{i\phi_i}e^{i\delta_1} + |A_2|e^{i\phi_2}e^{i\delta_2} \tag{6.16}$$

$$A(\bar{i} \to \bar{f}) = |A_1|e^{-i\phi_i}e^{i\delta_1} + |A_2|e^{-i\phi_2}e^{i\delta_2} \tag{6.17}$$

where the scattering phases δ_i do not change sign for the *CP*-conjugate process. The difference in transition rates is now proportional to

$$|A(\bar{i} \to \bar{f})|^2 - |A(i \to f)|^2 = 2|A_1||A_2|\sin(\phi_1 - \phi_2)\sin(\delta_1 - \delta_2) \tag{6.18}$$

which indeed vanishes if the two interfering amplitudes have the same phase $\phi_1 = \phi_2$, but also if the two scattering phases are equal (or absent). The interplay of the weak and strong phases is different for the two *CP*-conjugate amplitudes, and final state interactions are indeed seen to be an essential ingredient for the observability of the *CP*-violating phases (see Fig. 6.6). They are however a necessary nuisance, since being induced by strong interaction effects they cannot be reliably computed, so that it is usually difficult to extract information on the weak phases from the measurement of *CP* asymmetries: the size of the *CP*-violating asymmetry necessarily depends on the strong interaction dynamics.

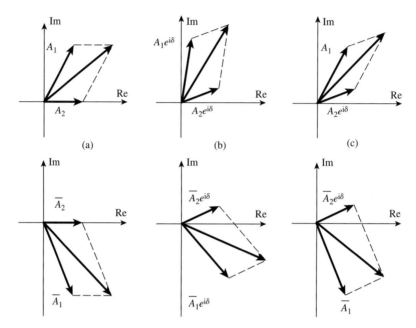

FIG. 6.6. Schematic drawing of the effect of strong and weak phases on decay rate differences: the arrows represent two complex amplitudes A_1, A_2 with different (weak) phases (top), and the corresponding amplitudes $\bar{A}_1, \bar{A}_2$ for the *CP*-conjugate process. (a) The opposite weak phases do not result in an observable rate difference (the length of the arrow representing the total amplitude). (b) If both amplitudes acquire equal strong phases the pictures are just rigidly rotated and still no rate difference arises. (c) If the two amplitudes get different strong phases the two processes have different rates.

Note also the important fact that the asymmetry (6.18), being proportional to the product of two interfering amplitudes, is only sizeable when such amplitudes have comparable magnitude, which might often not be the case: decays in which a tree amplitude is suppressed by some approximate selection rule are the best candidates to exhibit large asymmetries.

One way to observe *CP* violation in decays for which a single elementary amplitude dominates is to exploit *cascade* decays to the same final state (Bander *et al.*, 1979): if $i \to f_1$ and $i \to f_2$ are two possible decay modes of i with comparable amplitudes, and if f_1 and f_2 in turn contain unstable particles and share some common decay mode f, then the decay $i \to f$ can proceed in two ways:

$$A(i \to f) = A(i \to f_1)A(f_1 \to f) + A(i \to f_2)A(f_2 \to f) \qquad (6.19)$$

and again interference is possible, even if each of the individual decays is dominated by a single amplitude. If the scattering phases for the f_1 and f_2 states are different, any *CP*-violating difference in the weak phases of $A(i \rightarrow f_1)$ and $A(i \rightarrow f_2)$ can be detected in the comparison with the *CP*-conjugate decay of $\bar{i}$, for which

$$A(\bar{i} \rightarrow \bar{f}) = e^{2i\delta_1} A^*(i \rightarrow f_1) A(f_1 \rightarrow f) + e^{2i\delta_2} A^*(i \rightarrow f_2) A(f_2 \rightarrow f)$$

(the $f_1, f_2 \rightarrow f$ decays were assumed to be *CP*-conserving); this possibility is relevant for heavy meson decays.

In the case of neutral flavoured mesons (for which all *conserved* charges are zero), a peculiar situation arises: despite $i \neq \bar{i}$, transitions $i \leftrightarrow \bar{i}$ are possible, that is *state mixing* occurs. Indeed in general two *CP*-conjugate states should mix if no absolute conservation law forbids this (as when they can decay into the same final state). The mixed states which diagonalize the Hamiltonian might be *CP* eigenstates or not (if *CP* symmetry is violated).

In this case, even considering a *single* final state f (accessible to both i and $\bar{i}$), there are always two possible decay amplitudes: $A(i \rightarrow f)$ and $A(i \rightarrow \bar{i} \rightarrow f)$ can be different and interfere, possibly resulting in *CP* asymmetries, even when $A(i \rightarrow f)$ and $A(\bar{i} \rightarrow f)$ contain a single elementary amplitude. This possibility is clearly not present for charged particle decays for which i and $\bar{i}$ are always distinct, and will be discussed in more detail in Section 7.2.1.

6.2.2 *CP*-odd quantities

Classifying the *CP*-violating observables as even or odd with respect to naive time reversal, that is the inversion of spin and momenta (Section 4.3.3), it can be seen that the quantities which are even under such operation are proportional to $\sin(\phi_1 - \phi_2) \sin(\delta_1 - \delta_2)$, that is they require the presence of *CP*-conserving scattering phases to be different from zero, and therefore vanish in the Born approximation. The quantities which are instead odd under naive time reversal are proportional to $\sin(\phi_1 - \phi_2) \cos(\delta_1 - \delta_2) + \sin(\delta_1 - \delta_2) \cos(\phi_1 - \phi_2)$, and don't necessarily require the presence of scattering phases (first-term), but on the other hand can also be non-zero without any *CP* violation (second term). This discussion parallels that of *T*-odd quantities in Chapter 4, as *CPT* symmetry is assumed.

When considering transition amplitudes involving fermions, complex phases can appear in quantities such as triple products of momenta and spins: consider the Lorentz-invariant quantity

$$O_T = \epsilon_{\alpha\beta\gamma\delta} \, p^\alpha q^\beta r^\gamma s^\delta \tag{6.20}$$

where p, q, r, s are four-vectors related to the particles in the final state of a decay, such as four-momenta or relativistic spin vectors (see Appendix E), and $\epsilon_{\alpha\beta\gamma\delta}$ is the completely antisymmetric tensor in four dimensions. This can be written in a given frame as

$$O_T = p^0 \boldsymbol{q} \cdot (\boldsymbol{r} \times \boldsymbol{s}) - q^0 \boldsymbol{r} \cdot (\boldsymbol{s} \times \boldsymbol{p}) + r^0 \boldsymbol{s} \cdot (\boldsymbol{p} \times \boldsymbol{q}) - s^0 \boldsymbol{p} \cdot (\boldsymbol{q} \times \boldsymbol{r})$$

Considering now the case in which p, q, r, s are the momenta of the four particles in a four-body decay, and restricting to the centre of mass frame: $\boldsymbol{p} = -\boldsymbol{q} - \boldsymbol{r} - \boldsymbol{s}$, so that the above reduces to

$$O_T = \left[E_p + E_q + E_r + E_s \right] \boldsymbol{q} \cdot (\boldsymbol{r} \times \boldsymbol{s})$$

which is (a constant times) the triple product of three momenta.

In the computation of the matrix element for a decay, this (T-odd) quantity can arise from the trace of four gamma matrices and one γ_5, which contains a factor i as compared to other (real) expressions:

$$\mathrm{Tr}(\gamma^\mu \gamma^\nu \gamma^\rho \gamma^\delta \gamma_5) = 4i\epsilon^{\mu\nu\rho\delta} \qquad (6.21)$$

thus introducing a complex term in the amplitude, which can be used to make a CP-violating phase observable as discussed above. Note that in this case the value of the T-odd and CP-odd observable is opposite for the two CP-conjugate decays.

As discussed in Chapter 4, while in the above example of a T-odd correlation built from momentum vectors a minimum of four particles are required[89] in the final state for this quantity to be non-zero (with three particles, momentum conservation alone implies that one of the momenta can be expressed in terms of the other two and the triple product vanishes), if the final state particles have spin, T-odd quantities involving polarization vectors can be formed also in decays involving fewer particles.

In general, the CP-odd quantities that are even under 'naive' time reversal require the presence of a CP-conserving scattering phase which can arise from the absorptive part of an amplitude or from the presence of resonances in a decay chain, while those that are odd under 'naive' time reversal, obtained from the real part of the amplitudes, can get a CP-conserving phase from the above trace containing γ_5 (all such quantities are odd under true time reversal, consistently with CPT symmetry). Note that, even when final state interactions are present, which can induce non-zero values for triple product correlations with no CP violation, the comparison of the measured value of a CP-odd correlation for a process and its CP-conjugate provides an unambiguous test of CP violation.

6.2.3 Transition amplitudes

The constraint enforced by CP symmetry on $\mathcal{T}$ is

$$\mathcal{T}_{CP} \equiv (CP)^\dagger \mathcal{T} (CP) = \mathcal{T} \qquad (6.22)$$

---

[89] The particles should also be all distinguishable, otherwise the product is necessarily symmetric in the quantities referring to two identical particles and therefore vanishes identically.

which for transition amplitudes $A(i \to f) \equiv \mathcal{T}_{fi} = \langle f|T|i \rangle$ implies

$$A(i \to f) = \eta^*_{CP}(f)\eta_{CP}(i)A(\bar{i} \to \bar{f}) \tag{6.23}$$

where $\eta_{CP}(i,f)$ are the phase factors appearing in the *CP* transformation of the states:

$$CP|i\rangle = \eta_{CP}(i)|\bar{i}\rangle \qquad CP|f\rangle = \eta_{CP}(f)|\bar{f}\rangle \tag{6.24}$$

$$CP|\bar{i}\rangle = \eta^*_{CP}(i)|i\rangle \qquad CP|\bar{f}\rangle = \eta^*_{CP}(f)|f\rangle \tag{6.25}$$

with $|\eta_{CP}(i,f)|^2 = 1$.

Considering for simplicity the weak transition matrix element between two eigenstates of *PT* (e.g. spinless particles or spin-averaged states) and the corresponding charge conjugated states, when they are not connected by the strong interactions, *CPT* symmetry imposes the equality of the total cross sections and of the transition rates summed over subspaces F disconnected from the point of view of the strong interactions (see Chapter 5), for particles and antiparticles.

$$\sum_{f \in F} |A(i \to f)|^2 = \sum_{\bar{f} \in \bar{F}} |A(\bar{i} \to \bar{f})|^2$$

irrespective of the validity of the C and *CP* symmetries and independently of the strength of the interaction.

However only *CP* symmetry (C also would) guarantees the equality of the transition rates for any specific state

$$|A(i \to f)|^2 = |A(\bar{i} \to \bar{f})|^2 \tag{6.26}$$

and obviously the combined effect of the T (4.64) and *CPT* (5.13) symmetries would lead to the same result:

$$|A(i \to f)|^2 = |A(f \to i)|^2 = |A(\bar{i} \to \bar{f})|^2 \tag{6.27}$$

Unitarity requires that summing over *all* states

$$\sum_f |A(i \to f)|^2 = \sum_f |A(f \to i)|^2 = 1 \tag{6.28}$$

so that for a system with only two available states $1, 2$, unitarity alone (C.7) implies

$$|A(1 \to 1)|^2 + |A(1 \to 2)|^2 = |A(1 \to 1)|^2 + |A(2 \to 1)|^2 \tag{6.29}$$

and together with *CPT* symmetry

$$|A(1 \to 2)|^2 = |A(2 \to 1)|^2 = |A(\bar{1} \to \bar{2})|^2 \tag{6.30}$$

which is what *CP* symmetry would require. This means that at least two different final states are necessary in a decay in order to possibly have any rate difference: if a particle has a single possible decay mode, the decay rate for the antiparticle is the same as for the particle (indeed in this case the decay rate coincides with the total decay rate).

CPT symmetry implies that the absorptive parts of amplitudes which result from the scattering of a state into itself cannot contribute to partial decay width differences, so that the class of such differences does not represent an exhaustive determination of *CP* violation, as they could all be zero even if *CP* symmetry is violated.

The unitarity condition (C.20) is

$$A^*(f \to i) - A(i \to f) = i(2\pi) \sum_n \delta(E_n - E_i)A^*(f \to n)A(i \to n) \qquad (6.31)$$

where the sum is over a complete set of states $|n\rangle$. *CP* symmetry would require

$$|A(f \to i)|^2 = |A(\bar{f} \to \bar{i})|^2 \qquad (6.32)$$

which if *CPT* symmetry (5.13) is valid gives

$$|A(f \to i)|^2 = |A(i \to f)|^2 \qquad (6.33)$$

Any violation of the condition (6.33) is constrained by unitarity, since from eqn (6.31)

$$|A(f \to i)|^2 - |A(i \to f)|^2$$
$$= -2\,\mathrm{Im}[(2\pi) \sum_n \delta(E_n - E_i)A^*(f \to n)A(i \to n)A^*(i \to f)]$$
$$+ \left|(2\pi) \sum_n \delta(E_n - E_i)A^*(f \to n)A(i \to n)\right|^2$$

and if the rate of $i \to f$ transitions is determined by a small parameter g (e.g. the coupling constant of weak interactions), so that $|A(i \to f)|^2 = O(g^k)$, the *CP*-violating decay rate differences between particle and antiparticle are at least of order g^{k+1}: such differences are therefore zero in the Born approximation, and only arise in amplitudes corresponding to loop diagrams. Moreover, the above expression also shows explicitly that the intermediate states in such loops must correspond to physical states $|n\rangle$ in order to possibly contribute to a rate difference: even if such intermediate states have *CP*-violating complex couplings, they can only contribute if their masses are small enough to possibly appear on the mass shell in the transition (in this case the amplitudes have absorptive parts, and the same conclusions as those of Section 6.2.1 are reached, where it was shown that strong phase shifts are necessary to get rate differences).[90]

Let us also recall from Chapter 5 that for decay rates summed over all final states corresponding to a given eigenvalue a of a quantity A which is conserved by the part of the interaction which introduces the loop corrections, the particle and antiparticle rates (into states with eigenvalues a and $\bar{a}$) are necessarily equal by *CPT* and unitarity; a non-zero rate difference requires that the interaction responsible for the loop contributions violates the conservation of A.

To fully appreciate the conditions imposed by *CP* symmetry on decays, the presence of the phase factors (6.24) in the *CP* transformation of states should be considered. Using the abbreviated notation

$$A_f = A(i \to f) \qquad A_{\bar{f}} = A(i \to \bar{f}) \qquad \bar{A}_f = A(\bar{i} \to f) \qquad \bar{A}_{\bar{f}} = A(\bar{i} \to \bar{f})$$

the condition of *CP* symmetry for $\mathcal{T}$, eqn (6.22), requires

$$A_f = \eta_{CP}(i)\eta_{CP}^*(f)\bar{A}_{\bar{f}} \qquad A_{\bar{f}} = \eta_{CP}(i)\eta_{CP}(f)\bar{A}_f \qquad (6.34)$$

so that the transition amplitudes are required to be equal in modulus

$$|\bar{A}_{\bar{f}}| = |A_f| \qquad |A_{\bar{f}}| = |\bar{A}_f| \qquad (6.35)$$

There are, however, other constraints imposed by *CP* symmetry when more than a single final state is considered for the same decaying particle: relations similar to those in eqn (6.34) hold for any other pair of *CP*-conjugate states $|g\rangle, |\bar{g}\rangle$ allowed for the decay of $|i\rangle, |\bar{i}\rangle$, in which another arbitrary phase factor $\eta_{CP}(g)$ appears but obviously the *same* phase factor $\eta_{CP}(i)$. Considering these relations together, *CP* symmetry is also seen to impose the relation

$$\frac{\bar{A}_f \bar{A}_{\bar{f}}}{A_f A_{\bar{f}}} = \frac{\bar{A}_g \bar{A}_{\bar{g}}}{A_g A_{\bar{g}}} \qquad (6.36)$$

among complex quantities. For decays into *CP* eigenstates $(\bar{f} = f, \bar{g} = g)$

$$CP|f\rangle = \eta_{CP}(f)|f\rangle \qquad CP|g\rangle = \eta_{CP}(g)|g\rangle$$

where $\eta_{CP}(f, g) = \pm 1$ (without loss of generality) are the *CP*-parities of the final states; *CP* symmetry gives

$$A_f = \eta_{CP}(f)\eta_{CP}(i)\bar{A}_f \qquad (6.37)$$

---

[90] Note, however, that when mixing between particles and antiparticles is possible, *CP* violation can occur also without intermediate physical states.

and a similar relation for A_g, so that the constraint (6.36) becomes

$$\eta_{CP}(f)\frac{\overline{A_f}}{A_f} = \eta_{CP}(g)\frac{\overline{A_g}}{A_g} \tag{6.38}$$

which requires the complex quantity

$$A_f\overline{A_g} - \eta_{CP}(f)\,\eta_{CP}(g)\,A_g\overline{A_f} \tag{6.39}$$

to vanish. This quantity is not observable unless the two final states are linked in such a way that a physical decay exists involving both; an example is given by the decays of neutral kaons into $\pi\pi$ states with isospin $I = 0, 2$, since the detectors select final state particles of definite charge ($\pi^+\pi^-$ and $\pi^0\pi^0$) and thus detect final states which are coherent mixtures of different isospin states.

6.3   *CP* violation signatures

CP violation can manifest itself in several ways, either directly or (if the validity of *CPT* symmetry is assumed) as the 'equivalent' violation of time reversal symmetry. Here is a summary of signatures which are studied to look for *CP* violation; we anticipate that while *CP* symmetry is known to be violated by weak interactions, other interactions are consistent with it.

- **Decay selection rules**: a *CP* eigenstate decaying into a final state with a different *CP* eigenvalue exhibits a clear violation of *CP* symmetry; since the *CP* quantum numbers are usually attributed to a state according to its observed decay modes, in an attempt to get a consistent *CP*-conserving assignment, *CP* violation is detected in this way when a particle is seen to decay in two states with opposite *CP* eigenvalues.[91] This is how *CP* violation was discovered, with the K_L state decaying *both* into 2π and 3π states.

- **Partial decay widths**: *CP* symmetry requires the equality of *any* partial decay width for particles and antiparticles:[92] considering a single decay state or a subset of states which is not independent of others with respect to strong or electromagnetic interactions (e.g. the $\pi^+\pi^-$ state in the decay of a neutral kaon, which is not independent from the $\pi^0\pi^0$ state since strong interactions mix them into two isospin eigenstates), *CP* violation allows the two partial decay rates for particle and antiparticle to be different even if *CPT* symmetry is valid. Since *CPT* symmetry enforces the equality of the lifetimes for

---

[91] Compare with the $\tau - \theta$ puzzle in the case of parity.

[92] Clearly, C symmetry would also require such equality, since space inversion is irrelevant for quantities such as decay widths which are integrated over space variables.

CP-conjugate states, a comparison of the branching ratios is sufficient for the test.

The constraint imposed by *CPT* symmetry, which spoils the significance of this comparison as a test of *CP* symmetry, becomes less relevant for heavier particle decays, for which many more final states are available (and therefore belong to each of the above subsets).

It should be recalled that final state interactions are required for a rate difference to be observable, otherwise the *CP*-conjugate decay amplitudes only differ by a phase which cannot be measured.

- **Decay differences**: differences in the decay energy spectra (for decays into more than two particles) or angular distributions for a particle and its antiparticle (after taking into account the inversion of the momenta due to P) are an indication of *CP* violation: FSI are required, otherwise the squared moduli of decay amplitudes are equal at each point of phase space. With respect to decay rate comparisons this signature has the advantage that it does not require any knowledge of the relative flux normalization.

- **Decay asymmetries**: differences in the energy spectra or angular distributions between charge conjugate particles in the decay of a self-*CP*-conjugate state are an indication of *CP* violation, and can also be present in absence of FSI (an example would be a difference in the π^+ and π^- spectra in the decay $\eta \to \pi^+\pi^-\pi^0$ or $K_L \to \pi^+\pi^-\gamma$).

- **Particle production asymmetries**: statistical asymmetries between particle and antiparticle production rates or distributions in reactions with an initial state which is a *CP* eigenstate (such as e^+e^- or $\bar{p}p$ in a state of definite angular momentum) indicate *CP* violation.

- **Interference effects**: considering the decay of a generic meson state, given by the coherent superposition of physical eigenstates, the amplitude for the decay into a given final state (or set of final states) has an interference term (which leads to non-exponential decay) containing the product of the decay amplitudes of the two physical states into the same final state; for a final *CP* eigenstate such a non-exponential decay is a signal of *CP* violation. This is the case if the set of states under consideration is mapped into itself by *CP* (that is, a state with any number of γ, π^0, $\pi^+\pi^-$ pairs, etc.); if the decay in a partial region of phase space is considered (e.g. because the experimental set-up implies that some kinematic configurations cannot be detected) such region must be symmetric under the exchange of *CP*-conjugate particles (such as $\pi^+ \leftrightarrow \pi^-$).

- **Static properties of elementary systems**: non-zero values of T-odd quantities (such as electric dipole moments) for non-degenerate systems translate into *CP* violation if the *CPT* symmetry is valid.

- **Triple products**: in decays to final states with at least four distinguishable particles, or at least three distinguishable particles when considering polarization variables, T-odd triple products such as $\boldsymbol{p}_a \cdot (\boldsymbol{p}_b \times \boldsymbol{p}_c)$ or $\boldsymbol{S}_a \cdot (\boldsymbol{p}_a \times \boldsymbol{p}_b)$ can be formed; non-zero measured values for such correlations can also be induced by FSI with no T violation (see Chapter 4), but any difference between the magnitude of the triple products for particle and antiparticle decays indicates *CP* violation.

6.3.1   Hyperon decays

Comparisons of the non-leptonic weak decays of hyperons and antihyperons into a baryon and a pion, $Y \to B\pi$ (Y can be Λ, Σ, Ξ), can be used to test *CP* symmetry. This system is useful as an illustration of the above general discussion on the signatures of *CP* violation.

A first *CP*-violating quantity is clearly the rate difference between the two decays:

$$\Delta_{CP} = \frac{\Gamma - \overline{\Gamma}}{\Gamma + \overline{\Gamma}} \tag{6.40}$$

where $\overline{\Gamma}$ stands for the partial decay width for the *CP*-conjugate ($\overline{Y} \to \overline{B}\overline{\pi}$) decay.

Other *CP*-violating quantities can be formed by considering the decay asymmetries. The parametrization of these decays was discussed in Chapter 2: we recall that due to parity violation they can give both S-wave and P-wave final states; the parameter α (2.73) describes the asymmetry in the angular distribution of the emitted baryon with respect to the polarization direction of the parent hyperon, as well as the longitudinal polarization of such a baryon in the decay of an unpolarized hyperon, while β and γ (2.76) describe the transverse polarization of the baryon. A non-zero value for any of the above parameters indicates P violation, while β is also T-odd (and can also be different from zero due to FSI).

CP symmetry would require the decay parameters for antihyperon decays to be equal in magnitude and opposite to those for hyperon decays (Pais, 1959): if *CP* symmetry holds the transition amplitudes are related by eqn (6.23)

$$\langle f|T|i \rangle = \langle f|(CP)^\dagger T(CP)|i \rangle = (-1)^{L+1} \eta_C^*(f) \eta_C(i) \langle \overline{f}|T|\overline{i} \rangle \tag{6.41}$$

where $L = 0, 1$ is the orbital angular momentum in the final state (the pion is pseudo-scalar and the intrinsic parities of the parent hyperon and final baryon are chosen to be equal, so that the S-wave amplitude is parity-changing and the P-wave one is parity-conserving). The relative sign of the two interfering amplitudes is therefore opposite in hyperon and antihyperon decays, and from eqns (2.73), (2.76) the stated result follows.[93] This can also be seen by considering explicitly the effect of a CP transformation on a $Y \rightarrow B\pi$ decay configuration: by inverting the momenta and not the spins, P maps the angle θ between the direction of the emitted baryon and the hyperon polarization into $\pi - \theta$, and after transforming to the antihyperon decay (by C) a configuration is obtained which must have the same rate as the original one if CP holds.

Two more CP asymmetries can thus be written as

$$A_{CP} = \frac{\alpha + \overline{\alpha}}{\alpha - \overline{\alpha}} \qquad B_{CP} = \frac{\beta + \overline{\beta}}{\beta - \overline{\beta}} \tag{6.42}$$

where bars denote antihyperon parameters.

According to the general discussion above, for the CP asymmetries to be different from zero at least two interfering amplitudes are required, with different weak phases. Due to their different symmetry properties, amplitudes corresponding to opposite parities cannot interfere in the total decay rate, which is integrated over the whole phase space: the rate asymmetry Δ_{CP} cannot be generated by the interference of the S-wave and P-wave amplitudes. In several cases, however, more amplitudes are present which can interfere, since the final states can be in different isospin eigenstates: each isospin state can have a different strong phase for both the S and the P wave, and the amplitudes for different isospin states can also interfere in the total decay rate. Exceptions are the decays $\Sigma^- \rightarrow n\pi^-$ and $\Xi \rightarrow \Lambda\pi$, for which a single isospin final state is allowed: for these decays Δ_{CP} vanishes, independently of the validity of CP symmetry (the same is not true for $\Sigma^+ \rightarrow n\pi^+$ (Okubo, 1958), which is not the CP-conjugate decay of the one above). The A_{CP}, B_{CP} asymmetries can also be different from zero for decays with a single allowed isospin state, since the two angular momentum amplitudes for a given decay angle can interfere.

Since Δ_{CP} and A_{CP} are CP-odd but (naive) T-even, they require the presence of final state interactions (different for the two interfering amplitudes) in order not to vanish. B_{CP} is instead a CP-odd and T-odd quantity, and as such it can be different from zero independently of the presence of FSI; this is relevant, since the strong phases, which according to the Fermi–Watson theorem (Chapter 4) are given by the scattering phase shifts for the particles in the final state, are known to be rather small ($\lesssim 10°$) in this case, thus suppressing the magnitude of any CP-violating effect. B_{CP} actually increases when FSI are small, since in the limit in which they are absent $\beta = +\overline{\beta}$. The hierarchy in the magnitudes of the asymmetries is generally $|B_{CP}| \gg |A_{CP}| \gg |\Delta_{CP}|$. General expressions for the asymmetries in presence of several amplitudes can be found in (Donoghue et al., 1986b).

---

[93] Note that the arbitrary phase factor $\eta_C^*(f)\eta_C(i)$ appearing in eqn (6.41) is irrelevant, being the same for all decay amplitudes, and can be chosen to be $+1$.

Hyperon decay asymmetry measurements

Quite generally, the CP-violating asymmetry B_{CP} is expected to be the largest one, due to its dependence on the (small) strong phase differences, as discussed above. The measurement of β, however, requires that of the polarization of the emitted baryon from a polarized hyperon decay, which is experimentally challenging. A_{CP} is much more easily measured, since it is related to the angular distribution of the emitted baryon with respect to the hyperon polarization direction as in eqn (2.72), not requiring a polarization measurement. Δ_{CP}, besides being generally smaller in magnitude, would require for its measurement a precise knowledge of the relative flux of hyperon and antihyperon samples, which is experimentally difficult to achieve without introducing significant experimental biases.

The hyperon polarization direction is fixed by the parity-conserving nature of its (strong) production process, which requires it to be orthogonal to the production plane: the measurement of the hyperon direction (together with the beam direction) is sufficient to determine this plane, even in inclusive production where the accompanying particles are undetected. The measurement of the angular distribution of the emitted baryon only determines the product $\alpha \mathcal{P}_Y$ of the asymmetry parameter and the parent hyperon polarization, so that a comparison of the α and $\bar{\alpha}$ parameters requires the knowledge of such polarization for the two cases.

An early experiment at the CERN ISR (Chauvat et al., 1985) studied inclusive Λ and $\overline{\Lambda}$ production in pp and $\bar{p}p$ collisions assuming equal polarization in the two cases, but such assumption is not expected to hold at high accuracy, not being based on any conservation law. Exploiting the exclusive reaction $\bar{p}p \rightarrow Y\overline{Y}$, the C symmetry of strong interactions guarantees that hyperons and antihyperons are produced with the same polarization, so that A_{CP} can be directly obtained from the comparison of the angular decay distributions, without requiring the knowledge of the polarization, which in general depends from the interaction energy and production angle and cannot be reliably computed (as it arises from strong interactions).

Indeed comparisons of particle and antiparticle properties are better performed starting from self-conjugate initial states, such as $\bar{p}p$ (or e^+e^-, but in this case production cross sections are much lower), since the CP symmetry of strong (or EM) interactions guarantees the final state to have the same CP properties as the initial one, thus removing possible spurious asymmetries (Donoghue et al., 1986a). As an example, while T-odd (and CP-odd if CPT is assumed) triple products suffer from the possible contamination from T-conserving FSI (see Chapter 4), which are difficult to compute, for a self-conjugate system some asymmetries can be formed which cannot be faked by FSI effects (sums of CP-conjugated T-odd correlations, as discussed in Chapter 4). Consider the self-conjugated reaction

$$p\bar{p} \rightarrow Y\overline{Y} \rightarrow B\pi \overline{B}\overline{\pi}$$

and the T-odd triple product (all momenta in the centre of mass frame)

$$O = \boldsymbol{p}_p \cdot (\boldsymbol{p}_B \times \boldsymbol{p}_\pi)$$

The corresponding *CP*-conjugate quantity is

$$\overline{O} = \boldsymbol{p}_{\overline{p}} \cdot (\boldsymbol{p}_{\overline{B}} \times \boldsymbol{p}_{\overline{\pi}}) = -\boldsymbol{p}_p \cdot (\boldsymbol{p}_{\overline{B}} \times \boldsymbol{p}_{\overline{\pi}})$$

so that a *CP*-violating quantity can be written as

$$O + \overline{O} = \boldsymbol{p}_p \cdot (\boldsymbol{p}_B \times \boldsymbol{p}_\pi - \boldsymbol{p}_{\overline{B}} \times \boldsymbol{p}_{\overline{\pi}}) = -\boldsymbol{p}_p \times \boldsymbol{p}_Y \cdot (\boldsymbol{p}_B + \boldsymbol{p}_{\overline{B}})$$

which measures the asymmetry in the distribution of the final state baryons (irrespective of their charge) above or below the production plane, defined by the normal direction $\boldsymbol{p}_p \times \boldsymbol{p}_Y$.

This approach was used to put limits on *CP* violation at the 2% level in the reaction $\overline{p}p \to \Lambda\overline{\Lambda} \to p\pi^-\overline{p}\pi^+$, with 1.5–2 GeV/*c* antiprotons incident on a CH$_2$ target at the CERN LEAR (Barnes *et al.*, 1996).

The HyperCP experiment at Fermilab (Holmstrom *et al.*, 2004) adopted a similar approach to analyse the decays $\Xi^- \to \Lambda\pi^- \to p\pi^-\pi^-$ and $\overline{\Xi}^+ \to \overline{\Lambda}\pi^+ \to \overline{p}\pi^+\pi^+$ for hyperons inclusively produced at an average angle of 0°: the parity symmetry of strong interactions constrains the hyperon polarization to be zero in this case (unpolarized beam and target, undefined production plane). The angular distribution of the Λ is isotropic in the Ξ^- rest frame, since the parent hyperon is unpolarized; the *p* angular distribution in the Λ rest frame is described by eqn (2.72) with an asymmetry parameter $\alpha(\Lambda)$, and the magnitude of the Λ polarization in this case is just given by $\alpha(\Xi)$, so that

$$\frac{d\Gamma}{d\theta} = \frac{\Gamma}{2}[1 + \alpha(\Xi)\alpha(\Lambda)\cos\theta] \tag{6.43}$$

where θ is the polar angle corresponding to the direction of the emitted proton, measured in the Λ rest frame with respect to the direction of the Λ in the Ξ rest frame (so-called helicity frame). *CP* symmetry would require the angular distribution for $\overline{\Xi}^+$ decays to be the same as in eqn (6.43), since both $\alpha(\Lambda)$ and $\alpha(\Xi)$ would be opposite in that case; any difference between the two angular distributions is evidence of *CP* violation expressed by

$$A_{CP}(\Xi\Lambda) = \frac{\alpha(\Lambda)\alpha(\Xi) - \overline{\alpha}(\Lambda)\overline{\alpha}(\Xi)}{\alpha(\Lambda)\alpha(\Xi) + \overline{\alpha}(\Lambda)\overline{\alpha}(\Xi)} \simeq A_{CP}(\Lambda) + A_{CP}(\Xi)$$

Hyperons were produced by 800 GeV/*c* protons on a copper target, followed by a 6 m long curved magnetized ($\sim$ 1.7 T) collimator selecting particles of mean momentum 160 GeV/*c*: this relatively low value was a compromise between the requirements of a reasonably large acceptance for the decay and a relatively small difference between Ξ^- and $\overline{\Xi}^+$ yields. The collimator also worked as a beam dump for the non-interacting protons.

A magnetic spectrometer with nine multi-wire chambers and two analysing dipole magnets ($\sim$ 4.7 T m field integral) measured the momenta of charged particles produced in a 13 m long evacuated decay region. Two scintillator hodoscopes triggered on events compatible with $\Lambda \to p\pi^-$ (or the *CP*-conjugate) kinematics, and an iron-scintillator hadronic calorimeter measured the *p* (or $\overline{p}$) energy, which was required to be above a threshold to reduce the high rate of events due to beam interactions in the spectrometer (see Fig. 6.7).

Data-taking periods for Ξ^- and $\overline{\Xi}^+$, with opposite magnet polarities and different target lengths (three times longer for Ξ^- production, to equalize the event rates within $\sim$ 5%) alternated four times per day, to minimize effects due to slow drifts in detector performances.

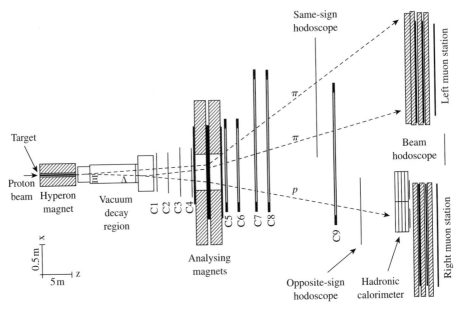

F<small>IG.</small> 6.7. Plan view of the HyperCP experimental setup for the measurement of *CP* asymmetries in hyperon decays (Holmstrom *et al.*, 2004). The Ξ hyperons (of both signs) are directed upward (positive y) at an angle of 19.5 mrad and would bend to the left in the analyzing magnets. C1-9 are multi-wire chambers.

The asymmetry was extracted from the ratio of p and $\bar{p}$ angular distributions, thus eliminating any sensitivity to common biases. The acceptance of the detector does not introduce any bias if it is constant in time, which requires in particular the magnitude of the magnetic fields to be accurately equalized in both running modes: this was continuously monitored by Hall probes and measured to be the same at the $\sim 3 \cdot 10^{-4}$ level; the uncertainty in the knowledge of such magnetic fields eventually induced the major systematic error in the result. Differences in the momentum and position-angle distributions for Ξ^- and $\overline{\Xi}^+$ can also induce acceptance differences, so that the events were weighted at the analysis stage to equalize such distributions. The result of the above procedures was that no acceptance or efficiency corrections based on simulation had to be applied.

It should also be noted that the orientation of the helicity frame in which the asymmetry is computed varies from one event to the other, so that any effect of localized detector efficiency variations is diluted in the measurement, being only weakly correlated to any particular region in the decay distribution. The power of this approach was demonstrated by the good agreement of the helicity frame distributions for two additional data sets with Ξ^- polarized in opposite directions (obtained with proton targeting angles of ± 2.5 mrad in the horizontal plane), and therefore illuminating differently the detector.

The raw result was corrected for the measured asymmetry of the $\sim 0.4\%$ background, while effects due to different interactions of the *CP*-conjugated particles in the spectrometer material, minimized by the presence of helium bags (2.3% interaction lengths total), or any small residual Ξ polarization due to (different) imperfect targeting angles were checked to be negligible.

From the analysis of 120 million Ξ^- and 40 million $\overline{\Xi}^+$ decays an asymmetry $A_{CP}(\Xi\Lambda) = (0.0 \pm 5.1 \pm 4.4) \cdot 10^{-4}$ was measured, consistent with no *CP* violation in hyperon decays (Holmstrom *et al.*, 2004).

It should be remarked that by studying the decays of *polarized* Ξ^- $\overline{\Xi}^+$ pairs, the *CP*-violating quantity B_{CP} in the $\Xi \to \Lambda\pi$ decay would be in principle more easily accessible, due to the self-analysing nature of the subsequent Λ decay, which allows a measurement of β from just the momenta of the final state particles (Donoghue *et al.*, 1986a). Considering the polarization of the emitted baryon (Λ in this case) as given by eqn (2.77), a *CP*-odd asymmetry of the T-odd correlation $\hat{n} \cdot \mathcal{P}_\Lambda \times \hat{p}_\Lambda$ can be formed (where $\mathcal{P}_\Lambda$ is the Λ polarization and $\hat{n}$ is the normal to the Ξ production plane along which the Ξ polarization lies) since $\mathcal{P}_\Lambda$ is correlated to the final proton direction:

$$\hat{n} \cdot (\hat{p}_p \times \hat{p}_\Lambda - \hat{p}_{\overline{p}} \times \hat{p}_{\overline{\Lambda}}) \propto |\mathcal{P}_\Xi|\alpha_\Lambda(\beta + \overline{\beta})_\Xi$$

where all momenta are evaluated in the Ξ rest frame(s). This kind of measurement would however, require Ξ^- and $\overline{\Xi}^+$ with identical (or precisely known) polarization: while such particles are polarized when inclusively produced in proton collisions, the magnitude of their polarization in realistic conditions is relatively small ($\sim 10\%$), thus diluting the sensitivity and requiring prohibitively large samples; such polarization is not expected to be the same for both *CP*-conjugate particles (except when produced in exclusive pairs by $\overline{p}p$). Note also that in such a measurement the asymmetry is more directly correlated to positions in the detector, if the production plane has a preferred orientation for the particles accepted by the experimental setup, leading to stringent requirements on detector symmetry to avoid large biases.

6.3.2   Other particle decays

The angular asymmetry in $\pi \to \mu \to e$ decays due to parity violation in weak interactions must be the same for π^+ and π^- if *CP* symmetry is valid. Because of the different interaction properties of pions of opposite charge with matter, the comparison of the two decays must be performed in vacuum. Any effect is expected to be very suppressed by the smallness of the (required) final state interactions, which can only be due to weak interactions (10^{-7} asymmetry expected in the SM). An experimental limit was obtained from the comparison of the measured amplitude of spin precession due to the anomalous magnetic moment for μ^+ and μ^- in a magnetic field (Kaplan, 1988), which are consistent at the 1–2% level.

As will be discussed in Chapter 9, the mechanism with which *CP* violation is accommodated in the SM predicts small effects for decays within the same quark family; limits on *CP* violation for such processes can be set by studying the decays of the η meson, which is a *CP* eigenstate and easily produced, while being massive enough ($m_\eta \simeq 548$ MeV/c^2) to have hadronic as well as EM decays.

The η has a relatively small decay width ($\Gamma \simeq 1.2$ keV) because its major decay mode into 3π is forbidden by *G*-parity conservation (Section 3.2.4), and only occurs because isospin symmetry is broken; decays into 2π and 4π violate parity and *CP* symmetries.

The CMD-2 experiment at the Novosibirsk VEPP-2M collider studied the decays of η produced in the reaction $e^+e^- \to \phi \to \eta\gamma$, and placed limits on the occurrence of the $\eta \to 2\pi$ decays from a sample of about 20 million ϕ decays (Akhmetshin *et al.*, 1999a, 1999b) $BR(\eta \to \pi^+\pi^-) < 3.3 \cdot 10^{-4}$ and $BR(\eta \to \pi^0\pi^0) < 3.3 \cdot 10^{-4}$ at 90% confidence level; the limits of the above measurements are due to the background from $\pi\pi\gamma$ production and 2π decays of other states.

The Crystal Ball collaboration at the Brookhaven AGS used an electrostatically separated π^- beam of ~ 720 MeV/c to produce η close to threshold by the process $\pi^-p \to \eta n$, and searched for the $\eta \to 4\pi^0$ decay. Despite the small phase space available, this decay had the advantage of being essentially free of background in the above experimental conditions: the two-body production constrains the η kinematics and close to threshold the cross section is large (~ 2 mb), thus allowing a sensitive search with low beam intensities, reducing accidental effects. With no events found, a 90% CL limit was set (Prakhov *et al.*, 2000): $BR(\eta \to 4\pi^0) \leq 6.9 \cdot 10^{-7}$.

CP-violating decays of the η' meson ($m'_\eta \simeq 958$ MeV/c^2) were also investigated, but in this case the sensitivity is intrinsically smaller because of the larger (~ 160 times) decay width.

A non-zero longitudinal polarization in the decay of a spin zero particle into a fermion-antifermion pair is a signature of *CP* violation, as can be seen by the fact that *CP* transforms the final state into itself with opposite helicities, so that if *CP* symmetry holds these two configurations must have the same rate and no net longitudinal polarization can occur.

For the $\eta \to \mu^+\mu^-$ decay (BR $\simeq 6 \cdot 10^{-6}$), such a longitudinal polarization would also violate *CPT*, since the η has *C*-parity $+1$ (because of its $\gamma\gamma$ decay) and $\eta_C(\mu^+\mu^-) = (-1)^{L+S} = +1$ (because the η is spinless, independently from the parity $\eta_P(\mu^+\mu^-) = (-1)^{L+1}$).

6.3.3 Positronium decay correlations

Since positronium states are *CP* eigenstates ($CP = -1$ $(+1)$ for singlet (triplet) states), if *CP* is a valid symmetry no parity-violating effects due to the mixing of the singlet and triplet states can occur; in this case singlet states (for which the parity is fixed by the total angular momentum) cannot exhibit any parity mixing at all. Neglecting radiative corrections, only triplet states with total angular momentum 1 (that is $^3S_1$ and $^3P_1$) can be parity mixed (Bernreuther and Nachtmann, 1981), although the effects induced by the known weak interactions are extremely small (mixing amplitudes $O(10^{-14})$). Moreover, if CP symmetry holds no parity-violating interference effects can occur for decays into states which are eigenstates of C (as for Ps annihilation into any number of photons): for a P-odd observable O ($P^\dagger OP = -P$) one has

$$\langle \beta_P | O | \alpha_P \rangle = -\langle \beta | O | \alpha \rangle$$

where the subscript P indicates the inversion of momenta. On the other hand for C eigenstates CP symmetry implies

$$\langle \beta | \mathcal{T} | \alpha \rangle = \langle \beta | (CP)^\dagger \mathcal{T} (CP) | \alpha \rangle = \eta_C^*(\beta) \eta_C(\alpha) \langle \beta_P | \mathcal{T} | \alpha_P \rangle$$

so that the expectation value of O, as in (5.21), vanishes identically ($|i_P\rangle = |i\rangle$ for a particle at rest, and the η_C phase factors cancel). Thus CP symmetry forbids optical activity and similar effects in positronium, although this is no longer true in presence of an external magnetic field (which breaks C symmetry), and in such case parity violation tests would be possible, in principle.

From the above discussion, the observation of parity-violating interference terms in Ps decays into photons would constitute evidence for CP violation: the measurement of P-odd correlations in the 3γ decays of polarized $^3S_1$ positronium, such as $\mathcal{P} \cdot \boldsymbol{p}$ (where $\mathcal{P}$ is the Ps polarization vector and $\boldsymbol{p}$ a photon momentum vector) is therefore a test of CP. Note that, as usual, while the above quantity is also CPT-odd, its non-zero expectation value would not automatically imply a violation of CPT symmetry. Note also that the parity-violating neutral current interaction of weak interactions (Z exchange), being CP-conserving, cannot generate a non-zero expectation value for such quantity, which is therefore a 'clean' observable, highly sensitive to other possible sources of CP violation.[94]

6.3.4 Production asymmetries

Starting from an initial state $|i\rangle$ which is a CP eigenstate, the comparison of CP-conjugated final states obtained from the reactions $i \to f$ and $i \to \bar{f}$ tests the CP symmetry of the reaction: any interference between two amplitudes with opposite CP behaviour can originate a difference between their parameters. This is similar to the case of C symmetry tests (Chapter 3), but in this case the reactions with reversed momenta are compared.

Such asymmetries can be fruitfully studied at e^+e^- colliders, both for real particles in the final state, as for lepton pair production, and for virtual particles, as

---

[94] Even if weak charged-current CP violation existed in the lepton sector, as allowed by the non-zero neutrino masses, its effect would be highly suppressed (Bernreuther et al., 1988).

for $e^+e^- \rightarrow b\bar{b}g$ (b being a bottom quark and g a gluon); in the latter case the information on the elementary process involving quarks and gluons are extracted from the properties of the particles in which they hadronize, using statistical approaches.

The approach to the extraction of a small parameter by measuring the change it induces in a differential cross section is based on the definition of the optimal observable as described in (Atwood and Soni, 1992), which is of general applicability. The differential cross section for a process is written as

$$\frac{d\sigma}{d\phi} = \left(\frac{d\sigma}{d\phi}\right)_0 + \lambda \left(\frac{d\sigma}{d\phi}\right)_1 \tag{6.44}$$

where ϕ represents the relevant phase-space variables under consideration, and λ is the small parameter to be determined. Experimentally an observable O is measured as the average value of some function $f_O(\phi)$

$$\langle O \rangle = \int f_O(\phi) \frac{d\sigma}{d\phi} d\phi \tag{6.45}$$

and the comparison of the measured value with the one expected for $\lambda = 0$, namely

$$\langle O \rangle_0 = \int f_O(\phi) \left(\frac{d\sigma}{d\phi}\right)_0 d\phi \tag{6.46}$$

is used to extract information on λ:

$$\Delta O = \langle O \rangle - \langle O \rangle_0 = \lambda \int f_O(\phi) \left(\frac{d\sigma}{d\phi}\right)_1 d\phi \tag{6.47}$$

Since $\langle O \rangle_0$ is assumed to be exactly known, the error on ΔO is just given by that on the quantity $\langle O \rangle$: for a measurement with N events this is

$$\delta_O = \frac{1}{\sqrt{N}} \sqrt{\sigma \int f_O^2(\phi) \frac{d\sigma}{d\phi} d\phi} \quad \text{where} \quad \sigma = \int \frac{d\sigma}{d\phi} d\phi$$

σ is the measured total cross section (note that also this quantity is subject to statistical fluctuations). The statistical significance with which the presence of a non-zero λ parameter can be determined is therefore

$$\frac{\Delta O}{\delta_O} = \lambda \sqrt{N} \sqrt{R} \quad \text{with} \quad R = \frac{\left[\int f_O(\phi) (d\sigma/d\phi)_1 d\phi\right]^2}{\sigma \int f_O^2(\phi) (d\sigma/d\phi)_0 d\phi}$$

($\lambda \simeq 0$ used in the denominator) and the optimal observable is the one for which R is maximum. The function f_O is conveniently split as

$$f_O(\phi) = A \left[\frac{(d\sigma/d\phi)_1}{(d\sigma/d\phi)_0} + \widehat{f}(\phi) \right] \tag{6.48}$$

$$A = \frac{\int f_O \, (d\sigma/d\phi)_1 \, d\phi}{\int \left[(d\sigma/d\phi)_1^2 \, / \, (d\sigma/d\phi)_0 \right] d\phi} \tag{6.49}$$

where by construction

$$\int \widehat{f}(\phi) \left(\frac{d\sigma}{d\phi} \right)_1 d\phi = 0 \tag{6.50}$$

and since the statistical significance is independent of A, in the following $A = 1$ is set. Using (6.48) and expanding to first order in λ

$$R = \frac{\frac{1}{\sigma} \left[\int \frac{(d\sigma/d\phi)_1^2}{(d\sigma/d\phi)_0} d\phi \right]^2}{\int \frac{(d\sigma/d\phi)_1^2}{(d\sigma/d\phi)_0} d\phi + \int \widehat{f} \, (d\sigma/d\phi)_0 \, d\phi}$$

and since both terms in the denominator are semi-positive defined, R is maximal for $\widehat{f} = 0$. The optimal observable for the determination of λ is therefore

$$f_O^{\text{(opt)}} = \frac{(d\sigma/d\phi)_1}{(d\sigma/d\phi)_0} \tag{6.51}$$

An interaction term of the form (4.96) can be considered as an addition to the basic lepton-photon vertex $\gamma \ell^+ \ell^-$ ($\ell = \mu, \tau$), with $d_E(q^2)$ being a (complex) electric dipole form factor, depending on the q^2 of the photon (the real part of $d_E(0)$ is the electric dipole moment). The cross section for the reaction $e^+ e^- \to \ell^+ \ell^-$ for unpolarized $e^\pm$ contains (Bernreuther *et al.*, 1993)

$$\sigma \propto |M_0|^2 + \text{Re}(d_E)|M_R|^2 + \text{Im}(d_E)|M_I|^2 \tag{6.52}$$

where $|M_0|^2$ corresponds to the standard QED contribution and $|M_{R,I}|^2$ are functions of spin and momenta. Here the form factor d_E is actually a function of the centre of mass energy s (fixed for a $e^+ e^-$ collider). The term proportional to $\text{Re}(d_E)$ is *T*-odd; the form factor can also have an imaginary part arising from absorptive parts in the process which produces this effective coupling.

In the case of the process $e^+ e^- \to \tau^+ \tau^- \to f^+ f^-$, in which the charged particles in the final state are observed, the matrix elements for their production and subsequent decay factorize due to the long lifetime of the τ, and since decays are independent of the electric dipole form factor, the dependence on d_E is retained in the cross section to the observed final states. Information on the τ polarization can be obtained from the analysis of their decays, even when only some of the final state particles are observed.

The production of $\tau^+\tau^-$ pairs was studied at e^+e^- colliders, recently with high statistics in *B*-factory experiments. The Belle experiment (Inami *et al.*, 2003) analysed $\tau^+\tau^-$ pairs and extracted a value of d_E for τ using the optimal observable method (Atwood and Soni, 1992): for each detected event the value of the terms appearing in the squared matrix element (6.52) is computed by averaging over all the decay configurations compatible with the observed quantities (due to the undetected neutrinos) to evaluate the quantity $|M_R|^2/|M_I|^2$. From this quantity the value of $\mathrm{Re}(d_E)$ is obtained, using the correlation among the two quantities extracted from a Monte Carlo simulation which includes detector acceptance effects. The result from the analysis of 27 million lepton pairs was $\mathrm{Re}(d_E^{(\tau)}) \in (-2.2, 4.5) \cdot 10^{-17}\, e\, \mathrm{cm}$ at 95% confidence level.

Analogously, by studying the above reaction at a centre of mass energy corresponding to the Z resonance, the LEP experiments obtained limits on the weak dipole form factor describing the $Z\ell^+\ell^-$ vertex (Escribano and Massó, 1997).

The study of top quark production is interesting for *CP* symmetry tests: the top quark is so massive that it decays before hadronizing, so that its production and decay processes are not masked by non-perturbative physics. For the same reason, the spin of the top quark is an important observable (just as for leptons) which can be analysed in its decay by studying the charged lepton produced in the semi-leptonic decay of the W ($t \to b\ell^+\nu$). A large sample of top quark pairs could indeed be obtained at future $pp, e^+e^-, \mu^+\mu^-, \gamma\gamma$ colliders, and SM *CP*-violating effects are expected to be small (Atwood *et al.*, 2001a).

If the $t \to bW^+$ decay dominates top quark decays, partial rate asymmetries are expected to be small, as they would vanish due to *CPT* symmetry in the limit of a single decay mode; other decay modes would have larger asymmetries, but their rates are expected to be so as small as to make measurements difficult. Decay asymmetries which are not suppressed by the above argument are those obtained by integrating only over a part of the phase space; one example is the lepton average energy asymmetry in $t \to b\ell\nu$ decays

$$A_{CP}^{(E)} = \frac{\langle E(\ell^+) \rangle - \langle E(\ell^-) \rangle}{\langle E(\ell^+) \rangle + \langle E(\ell^-) \rangle}$$

Such partially integrated rate asymmetries in semi-leptonic decays turn out to be proportional to the lepton mass, so that the $\ell = \tau$ channel is usually considered; asymmetries involving the lepton polarization in semi-leptonic decays are expected to be larger (in a model-independent way); cross section asymmetries involving top quark polarization have also been considered, such as

$$A_S = \frac{\sigma_{LL} - \sigma_{RR}}{\sigma}$$

where the subscripts refer to the t and $\bar{t}$ helicities, which could be measured from the angular distributions of the decay products: in semi-leptonic decays the correlation of the top-quark polarization and the lepton direction is maximal (Atwood *et al.*, 2001*a*).

6.3.5 Neutrinos

The properties of neutrinos, which are only affected by the weak interaction, are experimentally difficult to determine. On the other hand, this very same fact makes them a very interesting system to extract information on the fundamental parameters of such interaction, without undue complications from strong or EM effects; moreover, the relationship with the theory is much more direct, since neutrinos are closer to being measurable free states than quarks, which are permanently confined within hadrons in a complicated and poorly known way.

Following the discovery of neutrino flavour oscillations (and therefore, indirectly, of non-degenerate neutrino masses) the issue of *CP* violation in the lepton sector is being actively considered While searches for *CP* violation in the neutrino sector are clearly something for the future, it is worth mentioning some of the approaches in which they could be performed, referring the interested reader to the growing literature (e.g. Lipari (2001)) for more insight.

CP violation effects can be present both in the time evolution of neutrinos in flavour space (oscillations) and in their interactions with matter (detection); the latter is clearly problematic to study, due to the macroscopic *CP* asymmetry of the detection system. *CP* violation in neutrino oscillations can be experimentally measured by comparing the probability of a neutrino, prepared in a given flavour state, to be detected after propagation in vacuum as a given different flavour, with the same probability for an antineutrino:

$$P(\bar{\nu}_\alpha \to \bar{\nu}_\beta) \neq P(\nu_\alpha \to \nu_\beta)$$

indicates *CP* violation, if *CPT* symmetry is valid.

As usual, *CPT* symmetry imposes some constraints on the relations among the above probabilities: among these is the fact that *CP* violation effects in neutrino oscillations, implying differences in transition probabilities between different flavour states, can be studied only in appearance experiments, since the probability of a neutrino to remain in the same flavour eigenstate, measured in disappearance experiments, is equal to that for an antineutrino, if *CPT* symmetry is valid. Neutrino mixing implies that lepton flavour conservation is not exact; nevertheless, neutrino-antineutrino mixing is not expected to play any role: either neutrinos are their own antiparticle (Majorana neutrinos) or, if they are not (Dirac neutrinos), then total lepton number conservation (regardless of flavour) is valid, therefore preventing any $\nu - \bar{\nu}$ mixing.

The measurable CP violation effects in neutrino oscillations arise from the interference of two neutrino mixing amplitudes with different weak phases and different oscillation phases (due to the mass differences). Unfortunately, while FSI are not an issue, there is nevertheless a catch in this case: the CP-asymmetric matter effects in long-baseline experiments add more complications, so that the simple CP-odd asymmetries between oscillation probabilities are no longer a sufficient signal for CP violation.

The sensitivity to CP-violating asymmetries using intense conventional ν_μ neutrino beams would be limited by the intrinsic contamination of $\bar{\nu}_\mu$ ($\sim 1\%$) and ν_e ($\sim 0.5\%$), although this can be accurately predicted for a primary proton energy below the kaon production threshold.

The best and least model-dependent approach to the search for CP violation in neutrino oscillations appears to be the direct search for a CP asymmetry between $\nu_e \rightarrow \nu_\mu$ transitions and their CP-conjugates. This could be performed using future intense muon storage facilities ('neutrino factories'), which would provide very large fluxes of neutrinos with only two different flavours and opposite helicities ($\nu_\mu, \bar{\nu}_e$ or $\bar{\nu}_\mu, \nu_e$), by searching for the appearance of wrong sign muons and comparing the results for runs with opposite muon charges. Although precise estimates of the size of expected effects are not possible at the moment, the search of CP violation in neutrinos is clearly an interesting field of research, as its detection would establish that such phenomenon is not restricted to the confined quarks.

Further reading

Many excellent books and review articles exist on the issue of CP violation, and more will undoubtedly appear in the future, as the investigation of this phenomenon has been central in elementary particle physics for several decades, with a very intense research activity, both theoretical and experimental, and remains high on the current agenda of fundamental science. For the same reason any discussion of the subject (including the present one) grows old rather quickly as new information is gathered. It is not possible to list all the books on the subject, so only a few recent ones will be mentioned such as Branco, Lavoura, and Silva (1999), Bigi and Sanda (2000) (both leaning on the theoretical side), and Kleinknecht (2003) (with some discussion of experimental issues).

An excellent account of the early period in the study of CP violation in neutral kaons is given in Kabir (1968): despite being pretty much outdated (and having been written just too early, before all the dust had settled and some misleading results could be corrected), it still makes very interesting reading. The recollections of the main protagonists, from their Nobel prize acceptance speeches (Fitch, 1981; Cronin, 1981), give a nice insight on the whole issue.

A book focusing exclusively on the phenomenology of neutral kaons, which also deals with early experiments is Belušević (1999); several original papers concerned with the discovery of their peculiar behaviour are collected and briefly discussed in Goldhaber *et al.* (1957).

Detailed and rather complete discussions on *CP* violation searches can be found in the proceedings of recent schools devoted to the subject, such as Beyer (2002) and particularly Giorgi *et al.* (2006). Valuable information can be obtained from the proceedings of the topical conferences, such as Costantini *et al.* (2002); Perret (2003), and following editions.

Problem

1. The phenomenon of regeneration is related to a difference in the behaviour between particle and antiparticle (K^0 and $\overline{K}^0$): does this imply a violation of *CPT* symmetry?

2. If the pion had been much lighter or 20% heavier than it is, what would have changed in the history of *P* and *CP* symmetry violation?

3. What would the π^+ angular distribution $d\Gamma/d\Omega$ be for the decay $K^+ \to \pi^+\pi^0$ if the K^+ had spin 1? Compare it to the actual case of spin 0.

4. From a purely statistical point of view, which would be the optimal arrangement of neutral kaon beams for a comparison of 2π decays of K_S and K_L?

5. Suppose that in the decay $K^+ \to \pi^0\mu^+\nu$ it was found that $\langle S_\mu \cdot (p_\mu \times p_\pi) \rangle > 0$; explain how you would tell an extraterrestrial how to define unambiguously right and left.

6. After the discovery of $K_L \to 2\pi$ decay, many possible models were devised in which the violation of *CP* symmetry was restricted to a specific class of interactions, such as the parity-conserving ones, or those inducing isospin variations $\Delta I > 1/2$, or the electromagnetic interactions of hadrons. Which experimental results can be used to dismiss such possibilities?

7

FLAVOURED NEUTRAL MESONS

Not only is the universe stranger than we imagine,
it is stranger than we can imagine.
A. Eddington

For every complex question there is a simple answer –
and it's wrong.
H. L. Mencken

Where are the answers that we're all searching for?
There's nothing in this world to be sure of anymore.
F. Mercury

In this chapter we discuss the general phenomenology of flavoured neutral mesons, in the decays of which *CP* symmetry violation was observed. We collect some formulæ which will be useful in the following chapters. While the formalism applies equally well to all species of flavoured neutral mesons, the relevant approximations are often different in each case, making a fully generic treatment not always useful; for this reason some specific formulæ will only be introduced in Chapters 8 and 10.

7.1   Flavoured neutral meson mixing

Flavoured neutral mesons, those with a net non-zero strangeness, charm or beauty content, are a peculiar and interesting system in that particle and antiparticle are distinguished only by the flavour quantum number, which however is *not conserved* by the weak interactions responsible for their decay. Four families of mesons of this kind are known to exist: neutral kaons (Chapter 8), neutral D mesons and neutral B_d and B_s mesons (Chapter 10). While flavoured mesons with different spin-parity assignments are known, we will focus on those of lowest mass in each family, being the pseudo-scalar mesons; all higher mass states rapidly decay into these via strong interactions.

Mesons, being bosons, can in principle be created and destroyed singly,[95] and therefore the two *CP*-conjugate neutral states are subject to *mixing* through weak

---

95 Although not by the flavour-conserving strong interactions.

interaction effects, as they share common decay modes and therefore exhibit virtual transitions among them; note that this cannot happen for neutral fermions, since a number conservation law (most likely) holds.[96]

7.1.1 Flavour eigenstates

A generic flavoured neutral meson M^0 (K^0, D^0, B^0, where the latter stands for both B_d^0 and B_s^0) with non-zero eigenvalue of flavour F (strangeness, charm, or beauty) and its antiparticle $\overline{M}^0$ are defined by

$$F|M^0\rangle = +|M^0\rangle \qquad F|\overline{M}^0\rangle = -|\overline{M}^0\rangle \tag{7.1}$$

Being pseudo-scalar particles:

$$P|M^0\rangle = -|M^0\rangle \qquad P|\overline{M}^0\rangle = -|\overline{M}^0\rangle \tag{7.2}$$

Charge conjugation transforms M^0 into $\overline{M}^0$ and vice versa, introducing a phase factor η_C:

$$C|M^0\rangle = \eta_C|\overline{M}^0\rangle \qquad C|\overline{M}^0\rangle = \eta_C^*|M^0\rangle \tag{7.3}$$

with $|\eta_C|^2 = 1$. Defining

$$CP|M^0\rangle = \eta_{CP}|\overline{M}^0\rangle \qquad CP|\overline{M}^0\rangle = \eta_{CP}^*|M^0\rangle \tag{7.4}$$

the CP-parity phases $\eta_{CP} = e^{i\xi_{CP}}$ ($|\eta_{CP}|^2 = 1, \xi_{CP}$ real) are therefore[97] $\eta_{CP} = -\eta_C$. Since the mesons we are considering are spinless

$$T|M^0\rangle = \eta_T|M^0\rangle \qquad (\text{with } |\eta_T|^2 = 1) \quad \text{and}$$

$$T|\overline{M}^0\rangle = T\eta_C^*C|M^0\rangle = \eta_C TC|M^0\rangle = \eta_C CT|M^0\rangle = \eta_C C\eta_T|M^0\rangle = \eta_C^2\eta_T|\overline{M}^0\rangle$$

More symmetric expressions are obtained by defining $\eta_T = \tilde{\eta}_T\eta_C^*$:

$$T|M^0\rangle = \tilde{\eta}_T\eta_C^*|M^0\rangle \qquad T|\overline{M}^0\rangle = \tilde{\eta}_T\eta_C|\overline{M}^0\rangle \tag{7.5}$$

For what concerns CPT:

$$CPT|M^0\rangle = \eta_{CPT}|\overline{M}^0\rangle \qquad CPT|\overline{M}^0\rangle = \eta_{CPT}|M^0\rangle \tag{7.6}$$

---

[96] In the case of Majorana fermions there's no distinction at all between particle and antiparticle.
[97] We note that attempts to define the phase factors in terms of more 'fundamental' quantities, such as the phase factors for the quark and antiquark which compose the meson, are not fruitful: in general a composite system has an arbitrary phase due to the wave function, on top of the phases due to its constituents.

with $\eta_{CPT} = e^{i\xi_{CPT}}$ ($|\eta_{CPT}|^2 = 1$); with the definitions (7.5) and $\eta_{CPT} = -\tilde{\eta}_T$ is obtained. Note also that with the above choice

$$T|M^0\rangle = e^{i(\xi_{CPT}-\xi_{CP})}|M^0\rangle \qquad T|\overline{M}^0\rangle = e^{i(\xi_{CPT}+\xi_{CP})}|\overline{M}^0\rangle \qquad (7.7)$$

where both ξ_{CP} and ξ_{CPT} are arbitrary real numbers.

Let us recall that while strong and EM interactions are invariant under the CP and CPT transformations for *any* choice of the above phases, it is sufficient to find a specific set of phases such that the constraints imposed by CP or CPT symmetry are satisfied for such symmetries to be valid overall: the freedom of phase choice could be given up for a good cause, but such a sacrifice turns out not to be sufficient to save CP symmetry. Still, this freedom must not be forgotten, as will be discussed in Section 7.1.5.

7.1.2   *CP* eigenstates

The full Hamiltonian does not commute with flavour (since weak interactions change it), so that the flavour eigenstates $|M^0\rangle$ and $|\overline{M}^0\rangle$ are not expected to be the physical states (eigenstates of the full Hamiltonian) with defined life-times. Another conserved quantum number must be found to characterize the physical states; while charge conjugation C is not conserved as well by weak interactions, the weak Hamiltonian might be expected to commute with CP, and CP eigenstates are therefore good candidates as physical states. If CP symmetry holds, the physical states could indeed be chosen as the eigenstates of CP, namely

$$|M_\pm\rangle = \frac{1}{\sqrt{2}}\left(e^{-i\xi_{CP}/2}|M^0\rangle \pm e^{i\xi_{CP}/2}|\overline{M}^0\rangle\right) \qquad (7.8)$$

for which

$$CP|M_\pm\rangle = \pm|M_\pm\rangle \qquad CPT|M_\pm\rangle = \pm\eta_{CPT}|M_\pm\rangle \qquad (7.9)$$

The flavour eigenstates are written in terms of the CP eigenstates as

$$|M^0\rangle = \frac{e^{i\xi_{CP}/2}}{\sqrt{2}}(|M_+\rangle + |M_-\rangle) \qquad |\overline{M}^0\rangle = \frac{e^{-i\xi_{CP}/2}}{\sqrt{2}}(|M_+\rangle - |M_-\rangle) \qquad (7.10)$$

Note that $|M_+\rangle$ and $|M_-\rangle$ are not CP-conjugate states as $|M^0\rangle$ and $|\overline{M}^0\rangle$, but rather two self-conjugated states, which therefore can have different masses (and lifetimes) without violating CPT symmetry.

A convenient choice of the arbitrary relative phase between $|M^0\rangle$ and $|\overline{M}^0\rangle$ is the one which makes the phase factor in (7.4) $\eta_{CP} = +1$ ($\xi_{CP} = 0$), so that

$$CP|M^0\rangle = +|\overline{M}^0\rangle \qquad CP|\overline{M}^0\rangle = +|M^0\rangle \qquad (7.11)$$

and the CP eigenstates are then

$$|M_+\rangle = \frac{1}{\sqrt{2}}\left[|M^0\rangle + |\overline{M}^0\rangle\right] \qquad |M_-\rangle = \frac{1}{\sqrt{2}}\left[|M^0\rangle - |\overline{M}^0\rangle\right] \qquad (7.12)$$

This choice will often be made in the following, but it is important to remember that an arbitrariness exists for the above phase.

Note that such phase arbitrariness allows one to assign the CP eigenvalues as desired: if CP symmetry held, so that the physical states coincide with $|M_\pm\rangle$, it would be possible (albeit quite useless) to assign negative CP-parity to the state decaying into a CP-even final state, instead of the more natural opposite choice (this would amount to saying that the weak interaction Hamiltonian is CP-odd and *always* connects CP eigenstates of opposite CP-parity): the relative CP-parity of states with different flavour content is arbitrary. Of course, once a choice is made for such relative CP-parity, the fact that it can be maintained consistently for all known processes is what defines CP invariance.

7.1.3 Effective Hamiltonian

Since mixing couples M^0 and $\overline{M}^0$ together, and these can decay into other states, the exact solution of the time evolution in the general case is prohibitively complicated, as it deals with states which are linear combinations of both mesons and all other states which can couple to them. The problem is greatly simplified in the Weisskopf–Wigner approach (Weisskopf and Wigner, 1930a, 1930b), in which initial states that are linear combinations of M^0 and $\overline{M}^0$ only are considered, and the time evolution of the coefficients describing these two components are studied for times which are much larger than the typical strong-interaction scale. In this approach (Kabir, 1968; Nachtmann, 1989) the weak interactions are considered as a perturbation to the strong ones: the time evolution of the eigenstates of the latter is evaluated up to second order in weak interactions and then an effective Hamiltonian is considered which would give the same evolution in leading order of perturbation theory. The time evolution of a generic state is described[98] in the $M^0 - \overline{M}^0$ subspace by the effective non-Hermitian Hamiltonian $\mathcal{H}$:

$$i\hbar\frac{d}{dt}|\psi\rangle = \mathcal{H}|\psi\rangle \qquad (7.13)$$

---

[98] Admittedly exotic alternative possibilities are also considered in the literature, such as those in which conventional quantum theory fails and pure states can evolve into statistical mixtures (Ellis *et al.*, 1996).

$\mathcal{H}$ can be split into a Hermitian and an anti-Hermitian part

$$\mathcal{H} = M - \frac{i}{2}\Gamma \tag{7.14}$$

where both M (*mass matrix*) and Γ (*decay matrix*) are Hermitian

$$M = \frac{1}{2}\left(\mathcal{H} + \mathcal{H}^\dagger\right) = M^\dagger \qquad \Gamma = i\left(\mathcal{H} - \mathcal{H}^\dagger\right) = \Gamma^\dagger \tag{7.15}$$

Writing a generic state $|\psi\rangle = a(t)|M^0\rangle + b(t)|\overline{M}^0\rangle$ as a column vector

$$|\psi\rangle = \begin{pmatrix} a(t) \\ b(t) \end{pmatrix}$$

the effective Hamiltonian is ($1 = M^0, 2 = \overline{M}^0$)

$$\mathcal{H} = \begin{pmatrix} M_{11} - i\Gamma_{11}/2 & M_{12} - i\Gamma_{12}/2 \\ M_{21} - i\Gamma_{21}/2 & M_{22} - i\Gamma_{22}/2 \end{pmatrix}$$

The non-Hermiticity of $\mathcal{H}$ reflects the non-conservation of the norm in the $M^0 - \overline{M}^0$ subspace (mesons decay into other states, outside this subspace); from (7.13)

$$\frac{d}{dt}\langle\psi|\psi\rangle = \frac{i}{\hbar}\langle\psi|(\mathcal{H}^\dagger - \mathcal{H})|\psi\rangle = -\frac{1}{\hbar}\langle\psi|\Gamma|\psi\rangle$$

which must be negative (the mesons decay), so that Γ is positive definite: Γ_{11}, Γ_{22} and $\det(\Gamma)$ are positive.

The effective Hamiltonian $\mathcal{H}$ describes (at first order) the effects of the true Hamiltonian at second order in perturbation theory; the latter can be split into an unperturbed part H_0 and a perturbation H_{int}:

$$\mathcal{H} = H_0 + H_{\text{int}} + \sum_n \frac{H_{\text{int}}|n\rangle\langle n|H_{\text{int}}}{m_0 - E_n} \tag{7.16}$$

where m_0 is the unperturbed mass of $M^0, \overline{M}^0$ and E_n the energy of the intermediate state n (in the rest system of the meson).

The states $|M^0\rangle, |\overline{M}^0\rangle$ are eigenstates of H_0, which does not induce decay nor mixing and which therefore only contributes to the diagonal elements of M. The explicit expressions are

$$M_{11} = M_{11}^* = m_0 + \langle M^0 | H_{int} | M^0 \rangle + \sum_n \mathcal{P} \frac{|\langle n | H_{int} | M^0 \rangle|^2}{m_0 - E_n} \tag{7.17}$$

$$M_{22} = M_{22}^* = m_0 + \langle \overline{M}^0 | H_{int} | \overline{M}^0 \rangle + \sum_n \mathcal{P} \frac{|\langle n | H_{int} | \overline{M}^0 \rangle|^2}{m_0 - E_n} \tag{7.18}$$

$$M_{12} = M_{21}^* = \langle M^0 | H_{int} | \overline{M}^0 \rangle + \sum_n \mathcal{P} \frac{\langle M^0 | H_{int} | n \rangle \langle n | H_{int} | \overline{M}^0 \rangle}{m_0 - E_n} \tag{7.19}$$

$$\Gamma_{11} = \Gamma_{11}^* = 2\pi \sum_n \delta(m_0 - E_n) |\langle n | H_{int} | M^0 \rangle|^2 \tag{7.20}$$

$$\Gamma_{22} = \Gamma_{22}^* = 2\pi \sum_n \delta(m_0 - E_n) |\langle n | H_{int} | \overline{M}^0 \rangle|^2 \tag{7.21}$$

$$\Gamma_{12} = \Gamma_{21}^* = 2\pi \sum_n \delta(m_0 - E_n) \langle M^0 | H_{int} | n \rangle \langle n | H_{int} | \overline{M}^0 \rangle \tag{7.22}$$

where $\mathcal{P}$ indicates the principal part of what is really an integral over final states. M_{12} is the dispersive part of the transition amplitude between M^0 and $\overline{M}^0$ (virtual intermediate states) while Γ_{12} is the absorptive part (real intermediate states); both can be complex. Note that $\mathbf{\Gamma}$ is entirely determined by matrix elements of H_{int} for energy-conserving transitions (physical decay amplitudes) so that the contributions due to the different decay channels n can in principle be determined experimentally.

H_0 can be identified with the Hamiltonian for strong and EM interactions; if H_{int} is identified with the weak interaction Hamiltonian H_W then some of the terms in the general expressions above are absent, as for example $\langle M^0 | H_W | \overline{M}^0 \rangle = 0$ since weak interactions (in first order) only change flavour by one unit; in absence of other new interactions, the mixing of neutral flavoured meson states is the only measurable effect of *second order* weak interactions.

Note that despite the Hermiticity of $\mathbf{M}$ and $\mathbf{\Gamma}$ implying $\mathcal{H}_{21} = M_{12}^* - i\Gamma_{12}^*/2$, $\mathcal{H}_{12}$ is independent of $\mathcal{H}_{21}$ unless further constraints are imposed.

The physical states $|M_{a,b}\rangle$ are the eigenstates of $\mathcal{H}$, with complex eigenvalues:

$$\mathcal{H} |M_{a,b}\rangle = \lambda_{a,b} |M_{a,b}\rangle \qquad \lambda_{a,b} \equiv m_{a,b} - i\Gamma_{a,b}/2 \tag{7.23}$$

with masses $m_{a,b}$ and total decay widths $\Gamma_{a,b}$, so that

$$|M_{a,b}(t)\rangle = e^{-im_{a,b}t} e^{-\Gamma_{a,b}t/2} |M_{a,b}(0)\rangle \tag{7.24}$$

The physical states can be labelled by the property which better distinguishes them: the lifetime for neutral kaons (short-lived K_S and long-lived K_L), or the mass for neutral B mesons (heavy B_H and light B_L).

Note that since $\mathcal{H}$ is not a normal matrix ($[\mathcal{H}, \mathcal{H}^\dagger] \neq 0$), the matrix D which diagonalizes it is not unitary, and the eigenstates $|M_{a,b}\rangle$ of $\mathcal{H}$ are not orthogonal. For this reason $\langle M_a|\psi\rangle$ cannot be interpreted as the probability of finding a state $|M_a\rangle$ inside $|\psi\rangle$ (it is non-zero also for $|\psi\rangle = |M_b\rangle$). Note also that D does not diagonalize M nor Γ (which do not commute), and the real and imaginary parts of the eigenvalues $\lambda_{a,b}$ are not the eigenvalues of M and Γ. For the above reason two different sets of eigenstates of $\mathcal{H}$ have to be introduced, with the consequence that decay rates of $M_{a,b}$ are properly defined only if the $M^0, \overline{M}^0$ from which they originated are specified (Enz and Lewis, 1965). This turns out to be irrelevant in practice because CP violation is a small effect.

The physical states are written in full generality as

$$|M_a\rangle = p_a|M^0\rangle + q_a|\overline{M}^0\rangle \qquad |M_b\rangle = p_b|M^0\rangle - q_b|\overline{M}^0\rangle \qquad (7.25)$$

normalized by $|p_a|^2 + |q_a|^2 = 1 = |p_b|^2 + |q_b|^2$ from which

$$|M^0\rangle = \frac{1}{q_a p_b + q_b p_a} [q_b|M_a\rangle + q_a|M_b\rangle] \qquad (7.26)$$

$$|\overline{M}^0\rangle = \frac{1}{q_a p_b + q_b p_a} [p_b|M_a\rangle - p_a|M_b\rangle] \qquad (7.27)$$

Out of the four complex parameters $p_{a,b}$, $q_{a,b}$ two real parameters are eliminated by the above normalizations and three more are unphysical as they correspond to the arbitrary phases of $|M_a\rangle$, $|M_b\rangle$ and to the relative phase between $|M^0\rangle$ and $|\overline{M}^0\rangle$, leaving three independent real parameters.

Note that, while both the flavour and CP eigenstates are orthogonal: $\langle M^0|\overline{M}^0\rangle = 0$, $\langle M_+|M_-\rangle = 0$, the same is not true in general for the physical states:

$$\zeta \equiv \langle M_a|M_b\rangle = p_a^* p_b - q_a^* q_b \qquad (7.28)$$

Diagonalizing the matrix $\mathcal{H}$:

$$\lambda_a = \mathcal{H}_{11} + (q_a/p_a)\mathcal{H}_{12} \qquad \lambda_b = \mathcal{H}_{11} - (q_b/p_b)\mathcal{H}_{12} \qquad (7.29)$$

$$\lambda_a + \lambda_b = \mathcal{H}_{11} + \mathcal{H}_{22} \qquad (7.30)$$

$$\frac{q_a}{p_a}\frac{q_b}{p_b} = \frac{\mathcal{H}_{21}}{\mathcal{H}_{12}} \qquad (7.31)$$

and we *define* $\Delta\lambda \equiv \lambda_b - \lambda_a$ so that

$$\Delta m \equiv m_b - m_a \qquad \Delta\Gamma \equiv \Gamma_b - \Gamma_a \qquad (7.32)$$

We also define

$$\bar{\Gamma} = \frac{1}{2}(\Gamma_a + \Gamma_b) \tag{7.33}$$

$$\frac{q}{p} \equiv \sqrt{\frac{q_a \, q_b}{p_a \, p_b}} = \sqrt{\frac{\mathcal{H}_{21}}{\mathcal{H}_{12}}} \tag{7.34}$$

It should be noted that the sign of Δm and $\Delta \Gamma$ is arbitrary until it is linked to some other property of the physical states, and the sign of q/p acquires a meaning only in relation to the signs of Δm and $\Delta \Gamma$. Different conventions are found in the literature for these differences and for the signs in (7.25), therefore some care is required when comparing formulæ by different authors. In some cases for example Δm is *defined* to be always positive for any system, or the convention for Δm and $\Delta \Gamma$ is opposite. Here we define the mass and the width differences consistently and independently of the actual values which these quantities have in Nature for a specific system: they are the differences between the physical state which, for $q \simeq p$ and with the phase convention (7.11), is almost a CP-odd state ($|M_b\rangle \simeq |M_-\rangle$), and the one which in the same limit is almost CP-even ($|M_a\rangle \simeq |M_+\rangle$). For the K system $\Delta m_K \equiv m_L - m_S$ agrees with this convention, while $\Delta \Gamma_K \equiv \Gamma_S - \Gamma_L$ is usually defined in the opposite way, to make it a positive quantity.

The link between the eigenvalues and the elements of the $\mathcal{H}$ matrix is given by

$$\mathcal{H}_{11} = \frac{\lambda_a + \lambda_b}{2} + \frac{\Delta\lambda}{2} \frac{q_a/p_a - q_b/p_b}{q_a/p_a + q_b/p_b} \tag{7.35}$$

$$\mathcal{H}_{12} = -\Delta\lambda \frac{1}{q_a/p_a + q_b/p_b} \tag{7.36}$$

$$\mathcal{H}_{21} = -\Delta\lambda \frac{(q_a/p_a)(q_b/p_b)}{q_a/p_a + q_b/p_b} \tag{7.37}$$

$$\mathcal{H}_{22} = \frac{\lambda_a + \lambda_b}{2} - \frac{\Delta\lambda}{2} \frac{q_a/p_a - q_b/p_b}{q_a/p_a + q_b/p_b} \tag{7.38}$$

$$\lambda_{a,b} = \frac{\mathcal{H}_{11} + \mathcal{H}_{22}}{2} \pm \frac{1}{2}\sqrt{(\mathcal{H}_{11} - \mathcal{H}_{22})^2 + 4\mathcal{H}_{12}\mathcal{H}_{21}} \tag{7.39}$$

The choice of which of the two eigenvalues $\lambda_{a,b}$ corresponds to the plus sign in eqn (7.39) is arbitrary and corresponds to the choice of sign in

$$\Delta\lambda = \mp\sqrt{(\mathcal{H}_{11} - \mathcal{H}_{22})^2 + 4\mathcal{H}_{12}\mathcal{H}_{21}} \tag{7.40}$$

$$\frac{q}{p} = \pm\sqrt{\frac{\mathcal{H}_{21}}{\mathcal{H}_{12}}} = \pm\sqrt{\frac{M_{12}^* - i\Gamma_{12}^*/2}{M_{12} - i\Gamma_{12}/2}} \tag{7.41}$$

For example, with a plus sign in (7.41) (corresponding to a minus sign in (7.40) and a plus sign in (7.39) for λ_a), if the mass difference Δm as defined above turns out to be positive it means that the real part of the square root in (7.40) is a negative quantity. This is the choice which we will make in the following, but it should be remembered that a different one would result in opposite signs appearing in some formulæ, while the physical results (in terms of measurable quantities) are of course the same in any convention.

Interesting analogies exist between this two-state system and the behaviour of mechanical and electrical classical systems described by two normal modes (Rosner and Slezak, 2001).

7.1.4 Symmetries

The constraints enforced by the CPT, T and CP symmetries on the effective Hamiltonian can be obtained by using the explicit expressions (7.17)–(7.22) in terms of (Hermitian) Hamiltonians H_0, H_{int} and the results of the previous chapters; for example CP symmetry, eqn (6.11) requires

$$\Gamma_{21} = 2\pi \sum_n \delta(m_0 - E_n) \langle \overline{M}^0|(CP)^\dagger H_{int}(CP)|n\rangle \langle n|(CP)^\dagger H_{int}(CP)|M^0\rangle$$

$$= 2\pi \sum_n \delta(m_0 - E_n) \langle M^0|\eta_{CP} H_{int}\eta_{CP}(n)|\overline{n}\rangle \langle \overline{n}|\eta_{CP}(n)^* H_{int}\eta_{CP}|\overline{M}^0\rangle$$

$$= e^{2i\xi_{CP}}\Gamma_{12}$$

as the sum over intermediate states includes both $|n\rangle$ and $|\overline{n}\rangle$; if the phase factors are ignored ($\xi_{CP} = 0$) this requires $\Gamma_{21} = \Gamma_{12} = \Gamma_{21}^*$ to be real.

As a second example, setting again for simplicity all phase factors to 1, T symmetry, eqn (4.63) requires

$$\langle \overline{M}^0|\mathcal{H}|M^0\rangle = \langle M^0|\mathcal{H}|\overline{M}^0\rangle = \langle \overline{M}^0|\mathcal{H}^\dagger|M^0\rangle^*$$

as the spinless states $|M^0\rangle$, $|\overline{M}^0\rangle$ are invariant under naive time reversal and $\mathcal{H}$ is not Hermitian; this means $M_{12} = M_{21}$, $\Gamma_{12} = \Gamma_{21}$, and all the matrix elements are real (the diagonal elements are always real due to the Hermiticity of M and Γ).

In general the constraints imposed by the above discrete symmetries are:

■ **CPT symmetry**

$$M_{11} = M_{22} \qquad \Gamma_{11} = \Gamma_{22} \qquad (7.42)$$

which imply $\mathcal{H}_{11} = \mathcal{H}_{22}$, with no constraint on the off-diagonal elements of $\mathcal{H}$. The equality of the real and imaginary parts of these matrix elements corresponds to the equality of the masses and total decay widths (lifetimes) of M^0 and $\overline{M}^0$, as required by CPT symmetry (Chapter 5). The above suggests the complex dimensionless parameter

$$\delta_{CPT} \equiv \frac{\mathcal{H}_{22} - \mathcal{H}_{11}}{\Delta\lambda} = \frac{q_b/p_b - q_a/p_a}{q_b/p_b + q_a/p_a} \qquad (7.43)$$

as characterizing CPT symmetry violation. CPT symmetry thus requires the ratio of the mixing parameters q, p to be the same for the two physical states:

$$\frac{q_a}{p_a} = \frac{q_b}{p_b} = \frac{q}{p} \qquad (7.44)$$

In this case it is convenient to adopt a phase convention in which $p_a = p_b = p$, $q_a = q_b = q$:

$$|M_a\rangle = p|M^0\rangle + q|\overline{M}^0\rangle \qquad |M_b\rangle = p|M^0\rangle - q|\overline{M}^0\rangle \qquad (7.45)$$

Recalling eqn (7.26), CPT symmetry requires the composition of each flavour eigenstate to be symmetric in terms of the two physical states; conversely, the magnitude of any flavour asymmetry in the composition of the physical states must be the same for both physical states.

The general expressions of the previous section for the matrix elements of $\mathcal{H}$ simplify in this case to

$$\mathcal{H}_{11} = \mathcal{H}_{22} = \frac{\lambda_a + \lambda_b}{2} \qquad \mathcal{H}_{12} = -\frac{\Delta\lambda}{2}\frac{p}{q} \qquad \mathcal{H}_{21} = -\frac{\Delta\lambda}{2}\frac{q}{p} \qquad (7.46)$$

and (recall the choice of plus sign in (7.41))

$$\lambda_{a,b} = \mathcal{H}_{11} \pm \frac{q}{p}\mathcal{H}_{12} = \mathcal{H}_{11} \pm \sqrt{\mathcal{H}_{12}\mathcal{H}_{21}} \qquad (7.47)$$

$$\Delta\lambda = -2\frac{q}{p}\mathcal{H}_{12} = -2\sqrt{\mathcal{H}_{12}\mathcal{H}_{21}} \qquad (7.48)$$

or in terms of real observables

$$m_{a,b} = M_{11} \pm \mathrm{Re}\sqrt{M_{12}M_{21}} \qquad \Gamma_{a,b} = \Gamma_{11} \mp 2\,\mathrm{Im}\sqrt{M_{12}M_{21}} \qquad (7.49)$$

$$(\Delta m)^2 + (\Delta\Gamma)^2/4 = 4|\mathcal{H}_{12}||\mathcal{H}_{21}| \qquad (7.50)$$

$$(\Delta m)^2 - (\Delta\Gamma)^2/4 = 4|M_{12}|^2 - |\Gamma_{12}|^2 \qquad (7.51)$$

$$\Delta m\,\Delta\Gamma = 4\,\mathrm{Re}(M_{12}\Gamma_{12}^*) \qquad (7.52)$$

The following relation also holds in this case:

$$\zeta \equiv \langle M_a|M_b\rangle = |p|^2 - |q|^2 = \frac{4\,\mathrm{Im}(M_{12}^*\Gamma_{12})}{4|M_{12}|^2 + |\Gamma_{12}|^2 + (\Delta m)^2 + (\Delta\Gamma)^2/4} \qquad (7.53)$$

The above simplified expressions have been written down explicitly since they are those which will be mostly used, as *CPT* symmetry will be often be assumed to hold.

■ **T symmetry**

$$M_{21} = e^{2i\xi_{CP}} M_{12} \qquad \Gamma_{21} = e^{2i\xi_{CP}} \Gamma_{12} \tag{7.54}$$

which can be written as $\mathcal{H}_{21} = e^{2i\xi_{CP}} \mathcal{H}_{12}$.

Note that only the phase ξ_{CP} appears above, as ξ_{CPT} of eqn (7.7) cancels. Since ξ_{CP} is not a physical quantity by itself, the actual meaning of this constraint is better expressed by saying that the phases appearing in the matrix elements of M and Γ are the same:

$$\text{Im}\left(M_{12}^* \Gamma_{12}\right) = 0 \quad \text{or} \quad |\mathcal{H}_{12}| = |\mathcal{H}_{21}| \tag{7.55}$$

This is what is expected from the anti-linearity of T giving $\langle f|\mathcal{M}|i\rangle = \langle f|\mathcal{M}^\dagger|i\rangle^*$; apart from a global phase all the matrix elements of Hermitian matrices should be real if T symmetry holds. A real dimensionless T-violating parameter can be defined as

$$\delta_T \equiv \frac{|\mathcal{H}_{12}| - |\mathcal{H}_{21}|}{|\mathcal{H}_{12}| + |\mathcal{H}_{21}|} = \frac{|p_b/q_b| - |q_a/p_a|}{|p_b/q_b| + |q_a/p_a|} \tag{7.56}$$

If *CPT* symmetry holds in addition, this parameter becomes

$$\delta_T = \frac{1 - |q/p|^2}{1 + |q/p|^2} = |p|^2 - |q|^2 = \zeta \tag{7.57}$$

$$\text{and} \quad |p|^2 = \frac{1}{2} + \frac{\delta_T}{2} \quad |q|^2 = \frac{1}{2} - \frac{\delta_T}{2} \tag{7.58}$$

■ **CP symmetry**

$$M_{11} = M_{22} \qquad \Gamma_{11} = \Gamma_{22} \tag{7.59}$$

$$M_{21} = e^{2i\xi_{CP}} M_{12} \qquad \Gamma_{21} = e^{2i\xi_{CP}} \Gamma_{12} \tag{7.60}$$

which are the same constraints imposed by the *CPT* and *T* symmetries together. Both parameters δ_{CPT} and δ_T, introduced in (7.43) and (7.56), are therefore also *CP*-violating quantities; if *CP* symmetry is valid both of them are zero and

$$\frac{q_b}{p_b} = \frac{q_a}{p_a} = \frac{q}{p} \quad \text{with} \quad \left|\frac{q}{p}\right| = 1 \tag{7.61}$$

The following relations hold in case of *CP* symmetry:

$$\lambda_{a,b} = \mathcal{H}_{11} \pm \eta_{CP}\mathcal{H}_{12} \tag{7.62}$$

$$\Delta\lambda = -2\eta_{CP}\mathcal{H}_{12} = -2\eta_{CP}^*\mathcal{H}_{21} \tag{7.63}$$

$$\zeta \equiv \langle M_a|M_b\rangle = 0 \tag{7.64}$$

or in terms of real observable quantities

$$m_{a,b} = M_{11} \pm \mathrm{Re}(\eta_{CP}\mathcal{H}_{12}) \qquad \Gamma_{a,b} = \Gamma_{11} \mp 2\,\mathrm{Im}(\eta_{CP}\mathcal{H}_{12}) \tag{7.65}$$

$$\Delta m = -2\,\mathrm{Re}(\eta_{CP}\mathcal{H}_{12}) \qquad \Delta\Gamma = 4\,\mathrm{Im}(\eta_{CP}\mathcal{H}_{12}) \tag{7.66}$$

The eight real parameters which determine the matrix $\mathcal{H}$ can be expressed in terms of the two complex eigenvalues $\lambda_{a,b}$ (four real parameters), δ_{CPT} and δ_T (three real parameters); the remaining parameter represents the unobservable global phase of $\mathcal{H}$. If *CPT* symmetry holds the real physical parameters reduce to five, and if *CP* symmetry holds to four.

More insight into the meaning of the different elements of the effective Hamiltonian might be obtained by writing it as

$$\mathcal{H} = \begin{pmatrix} m_0 + h_{CPT} & \delta m + h_{CP} \\ \delta m - h_{CP} & m_0 - h_{CPT} \end{pmatrix} \tag{7.67}$$

In absence of flavour-changing interactions the matrix is diagonal with $\mathcal{H}_{11} = m(M^0) = m(\overline{M}^0) = \mathcal{H}_{22}$, and the two equal-mass flavour eigenstates are decoupled. If the *CP* and *CPT* symmetries hold ($h_{CP} = 0 = h_{CPT}$) the eigenvalues of $\mathcal{H}$ are $m_0 \pm \delta m$, so that the coupling of the flavour eigenstates just introduces a mass shift $m_0 - m(M^0)$ and a mass splitting $2\delta m$; δm is complex in presence of decays (real intermediate states). *CP* violation is introduced by $h_{CP} \neq 0$ (for $\Gamma = 0$ the matrix $\mathcal{H}$ would be Hermitian, with orthogonal eigenstates and no *CPV*); as the mass degeneracy between the two eigenstates of $\mathcal{H}$ (the physical states) is already removed by the *CP*-conserving term δm, *CPV* does not change the eigenvalues at first order.

To measure any *CPT*-violating difference $2h_{CPT}$ in the diagonal terms another kind of interaction which is different for M^0 and $\overline{M}^0$ can be used, e.g. strong interactions with matter (see the discussion on regeneration in Chapter 6). Due to the flavour-conserving nature of strong interactions, which does not induce $M^0 - \overline{M}^0$ transitions, the measurement of the *CP*-violating difference $2h_{CP}$ between the off-diagonal terms cannot be made in this way, and only their moduli can be measured using strong interactions. Weak interactions could be exploited instead, but still the phase of an amplitude inducing $M^0 - \overline{M}^0$ transitions is known only if time reversal symmetry holds: if a decay mode exists for which this symmetry is known to be valid (by other means), it can be used to measure the phase difference of the off-diagonal terms of $\mathcal{H}$.

Note that, while when CP symmetry holds the physical states can be chosen to be CP eigenstates, the same is not true for T or CPT symmetry: not being stable states, the physical states $|M_{a,b}\rangle$ *are not* eigenstates of T or CPT, even if the corresponding symmetry holds.

Also note that the physical states are orthogonal ($\zeta = 0$) if and only if both $\delta_T = 0$ and $\mathrm{Im}(\delta_{CPT}) = 0$, which is certainly the case if CP symmetry holds, but can also happen in presence of CP and CPT violation with $\mathrm{Re}(\delta_{CPT}) \neq 0$ (Branco, Lavoura, and Silva, 1999).

7.1.5   Rephasing invariance

A complication in the study of CP violation arises due to its connection, through the CPT theorem, to T violation and the presence of complex phases; phases are however rather elusive quantities in quantum physics, as they are arbitrary and irrelevant for any single state, and pinpointing which are physically relevant might not be immediate: 'Anyone who has played with these invariances knows that it is an orgy of relative phases' (Pais, 1986).

The relative phase of $|M^0\rangle$ and $|\overline{M}^0\rangle$ is arbitrary, since such states are properly defined when only strong (and EM) interactions are present, but in that case flavour conservation holds exactly, originating a super-selection rule (see Section 5.3): no strong (or EM) process can induce $M^0 \leftrightarrow \overline{M}^0$ transitions and thus allow a determination of the relative phase. On the contrary, the flavour-violating weak transitions do link M^0 to $\overline{M}^0$, but they also violate C symmetry, and cannot be used to get information on such phase either. It can be concluded that no physical quantity can depend on this relative phase, and invariance under rephasing

$$|M^0\rangle \to |M^{0(\chi)}\rangle = e^{i\chi}|M^0\rangle \qquad |\overline{M}^0\rangle \to |\overline{M}^{0(\chi)}\rangle = e^{-i\chi}|\overline{M}^0\rangle \qquad (7.68)$$

can be used as a check on each equation among physical quantities. The above relative phase is changed by a transformation induced by the flavour operator F, conserved by strong and EM interactions:

$$|\psi\rangle \to |\psi^{(\chi)}\rangle = e^{i\chi F}|\psi\rangle \qquad (7.69)$$

Under rephasing (7.68):

$$\mathcal{H}_{ii}^{(\chi)} = \mathcal{H}_{ii} \qquad \mathcal{H}_{12}^{(\chi)} = e^{-2i\chi}\mathcal{H}_{12} \qquad (7.70)$$

$$(q/p)^{(\chi)} = e^{2i\chi}(q/p) \qquad (7.71)$$

$$A_f^{(\chi)} \equiv A(M^{0(\chi)} \to f) = e^{i\chi} A(M^0 \to f) \equiv e^{i\chi} A_f \tag{7.72}$$

$$\overline{A}_f^{(\chi)} \equiv A(\overline{M}^{0(\chi)} \to f) = e^{-i\chi} A(\overline{M}^0 \to f) \equiv e^{-i\chi} \overline{A}_f \tag{7.73}$$

$$\left(\overline{A}_f / A_f\right)^{(\chi)} = e^{-2i\chi} \left(\overline{A}_f / A_f\right) \tag{7.74}$$

For some specific change of phases only some quantities might be affected: for $\chi = \pi$ the flavour eigenstates acquire a minus sign, but q/p and the matrix elements of $\mathcal{H}$ do not. Note in particular that while e.g. the quantity $\sqrt{\mathcal{H}_{12}\mathcal{H}_{21}}$ appearing in (7.40) is rephasing invariant, the phase of the quantity q/p is not, and is thus devoid of physical significance. This justifies the fact that the CP symmetry constraint (7.61) only involves the moduli of the mixing coefficients p, q, and not their phases: indeed if $|q| = |p|$ then q/p is a pure phase, which can always be made $+1$ by an appropriate choice of the relative phase of $|M^0\rangle$ and $|\overline{M}^0\rangle$ in (7.45), so that $|M_{a,b}\rangle = |M_\pm\rangle$ and the physical states are CP eigenstates. Conversely, if $|q| \neq |p|$ CP symmetry is violated, as the condition for $|M_{a,b}\rangle$ to be CP eigenstates is $q/p = \pm\eta_{CP}$, a pure phase (see (7.34) and (7.59)).

Analogously, the T (and CP) constraint (7.55) on the elements of $\mathcal{H}$ only requires M_{12} and Γ_{12} to be *relatively* real, as their absolute phase is not physical (if their relative phase vanishes a rephasing can make them both real): while T violation is linked to the appearance of any term not proportional to σ_1 (or 1) in (7.67), it is not expressed by $\mathcal{H}_{12} - \mathcal{H}_{21} \neq 0$ but rather by the rephasing invariant condition

$$|\mathcal{H}_{12}|^2 - |\mathcal{H}_{21}|^2 = 4 \,\mathrm{Im}\left(M_{12}^* \Gamma_{12}\right) \neq 0 \tag{7.75}$$

describing a phase mismatch between the off-diagonal elements of M and Γ resulting in a difference *in modulus* between the transitions $M^0 \to \overline{M}^0$ and $\overline{M}^0 \to M^0$. In other words, it makes no sense to discuss CP violation entirely within a closed $M^0 - \overline{M}^0$ system with no decays: only the phase difference of the mass and decay matrix elements has physical meaning.

Note also that if $\eta_{CP} \neq 1$ ($\xi_{CP} \neq 0$) in (7.4), by rephasing with $\chi = -\xi_{CP}/2$ the case $\eta_{CP} = 1$ can be recovered, so that the choice of (7.11) can always be made.

The CP operator is also defined to within a phase factor, and by using

$$CP^{(\phi)} = e^{i\phi F} (CP) e^{-i\phi F}$$

the relation between the flavour eigenstates M^0 and $\overline{M}^0$ can be left unchanged, despite the flavour-induced transformation (7.68), by choosing $\phi = -\chi$:

$$CP^{(\phi=\chi)}|M^{0(\chi)}\rangle = \eta_{CP}|\overline{M}^{0(\chi)}\rangle$$

Both the P and CPT operators are unaffected by the above phase transformation.

Note that also the physical states $|M_{a,b}\rangle$ have an arbitrary relative phase, but a change of this phase does not affect the ratios q_a/p_a and q_b/p_b, which on the other hand change in the same way under the rephasing of $|M^0\rangle, |\overline{M}^0\rangle$; for this reason the phase of $\gamma = (q_a p_b/p_a q_b)$ is significant, and indeed the measurable CPT-violating complex parameter introduced in (7.43) is

$$\delta_{CPT} = \frac{1 - \gamma}{1 + \gamma} \tag{7.76}$$

The scalar product of physical states ζ (7.28) is

$$\zeta = \frac{2\mathrm{Im}(M_{12}^*\Gamma_{12})}{\Delta m^2 + |\Gamma_{12}|^2} = \frac{\mathrm{Im}(M_{12}^*\Gamma_{12})}{|M_{12}|^2 + |\Gamma_{12}|^2/4 + (\Delta m^2 + \Delta\Gamma^2/4)/4}$$

and since $|M_{12}|^2 \geq \Delta m^2/4$, $|\Gamma_{12}|^2 \geq \Delta\Gamma^2/4$, a small non-orthogonality of the physical states (which does not imply necessarily small CPV, while the converse is true) means that $\mathrm{Im}(M_{12}^*\Gamma_{12}) \ll |M_{12}|^2 + |\Gamma_{12}|^2/4$. Note that ζ is invariant under rephasing of $|M^0\rangle, |\overline{M}^0\rangle$, but not of $|M_a\rangle, |M_b\rangle$, while its modulus is rephasing invariant and is indeed small, $|\zeta|^2 \ll 1$, as all CP and CPT violation effects are small). Choosing ζ to be real, in the case of CPT symmetry

$$\zeta \simeq \frac{\mathrm{Im}(M_{12}^*\Gamma_{12})/2}{|M_{12}|^2 + |\Gamma_{12}|^2/4} \tag{7.77}$$

at first order in ζ, and the following approximate relations hold for the mass and width differences in terms of this small quantity

$$(\Delta m)^2 \simeq 4|M_{12}|^2 \left[1 - \zeta^2 \left(1 + \left| \frac{\Gamma_{12}}{2M_{12}} \right|^2 \right) \right] \tag{7.78}$$

$$(\Delta\Gamma)^2 \simeq 4|\Gamma_{12}|^2 \left[1 - \zeta^2 \left(1 + \left| \frac{2M_{12}}{\Gamma_{12}} \right|^2 \right) \right] \tag{7.79}$$

Note that the expressions (7.47) and (7.48) for the eigenvalues of $\mathcal{H}$ and their differences are explicitly invariant under rephasing, just as $(q/p)\mathcal{H}_{12} = \pm\sqrt{\mathcal{H}_{12}\mathcal{H}_{21}}$ is. Recalling eqn (7.41) CP symmetry requires M_{12} and Γ_{12} to share the same phase (modulo π), so that q/p is a pure unimodular phase factor in this case, and $|q/p| = 1$ is the rephasing-invariant condition for CP symmetry.

Since the (small) breaking of CP symmetry depends on the relative phase of M_{12} and Γ_{12}, it is appropriate to write

$$M_{12} = \overline{M}_{12}\, e^{i\xi_M} \qquad \Gamma_{12} = \overline{\Gamma}_{12}\, e^{i\xi_M} e^{i\Delta\xi} \tag{7.80}$$

where $\overline{M}_{12}$ and $\overline{\Gamma}_{12}$ are signed real quantities ($\overline{M}_{12} = \pm|M_{12}|$, $\overline{\Gamma}_{12} = \pm|\Gamma_{12}|$) chosen to have the phases $\xi_M, \xi_M + \Delta\xi \in [-\pi/2, \pi/2]$ and $\Delta\xi \simeq 0$ (rather than

$\Delta\xi \simeq \pi$). In terms of these phase angles

$$\left|\frac{q}{p}\right|^2 = \sqrt{\frac{|\mathcal{M}_{21}|^2}{|\mathcal{M}_{12}|^2}} = \sqrt{\frac{|M_{12}|^2 + |\Gamma_{12}|^2/4 - \overline{M}_{12}\overline{\Gamma}_{12}\sin\Delta\xi}{|M_{12}|^2 + |\Gamma_{12}|^2/4 + \overline{M}_{12}\overline{\Gamma}_{12}\sin\Delta\xi}} \tag{7.81}$$

$$\frac{q}{p} = \pm e^{-i\xi_M}\sqrt{\frac{|M_{12}|^2 + e^{-2i\Delta\xi}|\Gamma_{12}|^2/4}{|M_{12}|^2 + |\Gamma_{12}|^2/4 + \overline{M}_{12}\overline{\Gamma}_{12}\sin\Delta\xi}} \tag{7.82}$$

and at first order in the small phase difference $\Delta\xi$

$$\frac{q}{p} \simeq \pm e^{-i\xi_M}\sqrt{1 - \Delta\xi\frac{\overline{M}_{12}\overline{\Gamma}_{12} + 2i|\Gamma_{12}|^2/4}{|M_{12}|^2 + |\Gamma_{12}|^2/4}} + O(\Delta\xi^2) \tag{7.83}$$

Moreover

$$\sqrt{\mathcal{M}_{12}\mathcal{M}_{21}} = \left(\overline{M}_{12} - i\frac{\overline{\Gamma}_{12}}{2}\right) + O(\Delta\xi^2) \tag{7.84}$$

These expressions will be used for the K and B systems in Chapters 8 and 10.

7.1.6 Unitarity

Since the physical states (7.25) are not orthogonal, from eqn (7.23) and the Hermiticity of M and Γ

$$\langle M_\alpha|(M - i\Gamma/2)|M_\beta\rangle = \lambda_\beta\langle M_\alpha|M_\beta\rangle$$

$$\langle M_\alpha|(M + i\Gamma/2)|M_\beta\rangle = \lambda_\alpha^*\langle M_\alpha|M_\beta\rangle$$

where $\alpha, \beta = a, b$. Subtracting the above relations and recalling (7.23)

$$\langle M_{a,b}|\Gamma|M_{a,b}\rangle = \Gamma_{a,b}$$

$$i\langle M_a|\Gamma|M_b\rangle = [-\Delta m + i\overline{\Gamma}]\langle M_a|M_b\rangle$$

and using eqns (7.20)–(7.22) and writing $A_f^{(\alpha)} \equiv A(M^{(\alpha)} \to f)$

$$A_f^{(\alpha)*}A_f^{(\beta)} = 2\pi\delta(m_0 - E_f)\langle M_\alpha|H_{\text{int}}|f\rangle\langle f|H_{\text{int}}|M_\beta\rangle$$

(apart from factors of 2π, phase space and energy-conservation delta functions)

$$\Gamma_{a,b} = \sum_f \left|A_f^{(a,b)}\right|^2 \tag{7.85}$$

$$[i\Delta m + \overline{\Gamma}]\langle M_a|M_b\rangle = \sum_f A_f^{(a)*}A_f^{(b)} \tag{7.86}$$

Equation (7.86) is called the Bell–Steinberger relation (Bell and Steinberger, 1966), and equates two terms which are *CP*-violating: the left-hand one because of $\zeta \neq 0$, and the right-hand one because it depends on the existence of common decay modes for M_a and M_b (flavour-specific decay modes cancel pairwise in the sum if *CP* symmetry holds). The Bell–Steinberger relation is a consequence of unitarity (implicit in the Hermiticity of M and Γ), as can be seen from an alternative derivation: the time evolution of a generic state $|\psi(t=0)\rangle = c_a|M_a\rangle + c_b|M_b\rangle$ gives

$$\langle\psi|\psi\rangle = |c_a|^2 e^{-\Gamma_a t} + |c_b|^2 e^{-\Gamma_b t} + 2\mathrm{Re}[c_a^* c_b e^{+i(\lambda_a^* - \lambda_b)t}\langle M_a|M_b\rangle]$$

$$\frac{d}{dt}\langle\psi|\psi\rangle\bigg|_{t=0} = -\Gamma_a|c_a|^2 - \Gamma_b|c_b|^2 + 2\mathrm{Re}[ic_a^* c_b(\lambda_a^* - \lambda_b)\langle M_a|M_b\rangle]$$

The latter expression must correspond to the decay rate into all final states

$$\frac{d}{dt}\langle\psi|\psi\rangle\bigg|_{t=0} = -\sum_f \left|c_a A_f^{(a)} + c_b A_f^{(b)}\right|^2$$

and by equalizing the imaginary parts the Bell–Steinberger equation is retrieved.
  Exploiting the Schwarz inequality on (7.86)

$$\left|\Gamma - i\Delta m\right||\zeta| \leq \sum_f \left|A_f^{(a)}\right|\left|A_f^{(b)}\right| \leq \left(\sum_f |A_f^{(a)}|\right)\left(\sum_f |A_f^{(b)}|\right) = \sqrt{\Gamma_a \Gamma_b}$$

from which

$$|\zeta|^2 \leq \frac{\Gamma_a \Gamma_b}{(\Delta m)^2 + \overline{\Gamma}^2} \tag{7.87}$$

which is called the Lee–Wolfenstein inequality (Lee and Wolfenstein, 1965) (first written in (Lee, Oehme, and Yang, 1957b)): remarkably, a bound can be placed on the *CP*-violating quantity ζ just from the knowledge of the *CP*-conserving quantities on the right-hand side of eqn (7.87).

7.2 Flavoured neutral meson decays

M^0 and $\overline{M}^0$ cannot be observed directly, but only through their decays into a variety of final states: these allow one to study *CP* violation, also in cases where this phenomenon does not actually occur in the decay process itself, in which case it is just a probe of the composition of the decaying meson.

7.2.1 Types of *CP* violation

Flavoured neutral mesons can exhibit a rich phenomenology of *CP* violation, and indeed are the only system in which this phenomenon has been observed so far. *CP* violation in these systems can be classified in terms of where the complex phases appear, keeping in mind the fact that the arbitrariness in the phases of the states allows some freedom in shifting them from one place to another:

- **Indirect *CP* violation**: in which all the uneliminable phases can be defined to appear in the amplitudes for $M^0 - \overline{M}^0$ mixing, describing a process with flavour change $\Delta F = 2$;

- **Direct *CP* violation**: in which some of the uneliminable phases necessarily appear in the physical decay amplitudes $M^0, \overline{M}^0 \to f$, usually[99] describing a process with $\Delta F = 1$.

  Of course both types of *CPV* can also be present at the same time.

  Another way of classifying *CP* violation in neutral flavoured meson decays is based on the way in which its effects appear:

- ***CP* violation in the mixing**: this occurs when physical states do not coincide with *CP* eigenstates, that is when in eqn (7.45)

$$|q| \neq |p| \tag{7.88}$$

so that the composition of the physical states is not flavour symmetric. This is due to the virtual transition $M^0 \to \overline{M}^0$, induced by the flavour-changing part of the Hamiltonian, having a different probability with respect to the *CP*-conjugate transition $\overline{M}^0 \to M^0$, because of the presence of two amplitudes with different phases in such process. Clearly, this type of *CPV* is of the indirect type.

- ***CP* violation in the decays**: this occurs when the physical decay amplitudes for *CP* conjugate processes into final states f and $\bar{f}$ are different in modulus, that is when eqns (6.35) are not satisfied

$$|\overline{A}_{\bar{f}}| \neq |A_f| \tag{7.89}$$

Again, this requires the presence of at least two interfering decay amplitudes, with different weak and strong phases. Clearly, this type of *CPV* is of the direct type, and is the only one which is also possible for charged particles, which are forbidden to mix by charge conservation.

---

[99] In a decay such as $K^{*0} \to K^0 \pi^0$ there is no change of strangeness.

■ *CP* **violation in the interference of mixing and decay**, also called 'mixing-induced' *CP* violation: for neutral flavoured mesons, a third possibility exists (Carter and Sanda, 1980) to observe *CPV* with final states f which can be reached by both flavour eigenstates (these include all *CP* eigenstates, but not only those), even if the amplitudes for both flavour oscillations and physical decays have the same magnitude for *CP*-conjugate states, that is if $|p| = |q|$ and $|\overline{A}_f| = |A_f|$.

Considering the decay of a flavour eigenstate M^0 into a final state f (accessible to both M^0 and $\overline{M}^0$), the latter can be reached not only via the decay amplitude $A(M^0 \to f)$, but also through the process in which M^0 first oscillates to $\overline{M}^0$ and then the latter decays to f via $A(\overline{M}^0 \to f)$. In the second case the two processes (flavour mixing oscillation and decay) act together, so that any relative phase between the corresponding amplitudes is relevant. The processes in which the meson does or does not oscillate before decaying cannot be distinguished, and can therefore interfere in the overall amplitude: in this case, even if $A(M^0 \to f)$ does not contain two interfering elementary amplitudes, so that it equals *in modulus* the *CP*-conjugate $A(\overline{M}^0 \to f)$, its phase is no longer irrelevant, as it interferes with the flavour oscillation amplitude. Two interfering amplitudes are therefore always present, which can be identified as the meson and antimeson decay amplitudes for the evolving coherent mixture of M^0 and $\overline{M}^0$. The overall decay amplitude is

$$A(M^0 \to f) + A(M^0 \to \overline{M}^0)A(\overline{M}^0 \to f) \qquad (7.90)$$

and denoting as θ_M the phase of the mixing amplitude and as $\theta_f, \overline{\theta}_f$ those of the $M^0, \overline{M}^0$ decay amplitudes to f, the two terms in (7.90) have a relative phase $\theta_f - \overline{\theta}_f - \theta_M$.

For the *CP*-conjugate process, in which an initial $\overline{M}^0$ is considered, the moduli of the amplitudes in the expression corresponding to (7.90) are constrained to be the same by the assumption of *CP* symmetry in both mixing and decay, and the phase difference is $\overline{\theta}_f - \theta_f - \overline{\theta}_M$, where $\overline{\theta}_M$ is the phase of the $\overline{M}^0 \to M^0$ amplitude. As will be seen, it is always possible to equalize *either* the mixing phases ($\theta_M = \overline{\theta}_M$) or the decay phases ($\theta_f = \overline{\theta}_f$) for the *CP* conjugate processes, but not both at the same time unless *CP* symmetry holds; this means that the decays of an initial M^0 and an initial $\overline{M}^0$ into the final state f have overall phases which can differ not only by a sign, and could therefore result in a difference of decay probabilities. Final state interactions are not required for a phase difference to arise in this kind of process, while flavour mixing (even *CP*-conserving) is clearly a necessary ingredient.

This type of *CPV*, arising from the phase mismatch between the mixing and decay amplitudes, cannot be unambiguously classified as direct or indirect, since absolute phases have no meaning and a complex phase can be attributed to either process at will. Note however that any *difference* in *CP* violation between *different* final states would be an unambiguous evidence of direct *CPV*, since the single phase in the mixing amplitude is the same for all decay final states, and cannot induce any such difference by itself.

When dealing with decays into a state f which can be reached by both flavour eigenstates, it is convenient to introduce the complex quantity[100]

$$\lambda_f = \frac{q}{p}\frac{\overline{A}_f}{A_f} \tag{7.91}$$

Contrary to its factors, λ_f is invariant with respect to the choice of the arbitrary relative phase between $|M^0\rangle$ and $|\overline{M}^0\rangle$, and its phase is thus physical; this is easily seen by writing explicitly

$$\lambda_f = \frac{\langle \overline{M}^0|M_a\rangle\langle f|\mathcal{T}|\overline{M}^0\rangle}{\langle M^0|M_a\rangle\langle f|\mathcal{T}|M^0\rangle} \tag{7.92}$$

or from eqns (7.70). For a final state which is a *CP* eigenstate ($CP|f\rangle = |\bar{f}\rangle = \eta_{CP}(f)|f\rangle$, $\eta_{CP}(f) = \pm 1$), λ_f can be written as

$$\lambda_f = \eta_{CP}(f)\frac{q}{p}\frac{\overline{A}_{\bar{f}}}{A_f} \tag{7.93}$$

which is more transparent as A_f and $\overline{A}_{\bar{f}}$ are the amplitudes connected by a *CP* transformation.

Writing the phase involved in the mixing process as

$$\frac{q}{p} = \left|\frac{q}{p}\right| e^{-2i\phi_M} \tag{7.94}$$

($\phi_M = (\overline{\theta}_M - \theta_M)/4$), an explicit expression is

$$\text{Im}(\lambda_f) = |\lambda_f|\sin(\overline{\theta}_f - \theta_f - 2\phi_M) \tag{7.95}$$

If the decay $M^0 \to f$ is driven by a single elementary amplitude (no *CPV* in the decay possible, $|A_f| = |\overline{A}_f|$), then if the conditions for the validity of the

---

[100] This should not be confused with the eigenvalue of the $\mathcal{H}$ matrix of eqn (7.23).

CPT version of the Fermi–Watson theorem (5.18) are satisfied the phases of the $M^0, \overline{M}^0 \to f$ decay amplitudes can be written as

$$\theta_f = \delta_f + \phi_f \qquad \overline{\theta}_f = \delta_f - \phi_f \qquad (7.96)$$

in terms of the strong (*CP*-conserving) and weak (*CP*-violating) phases δ_f, ϕ_f; both the moduli of the amplitude and the strong phases cancel in λ_f, leaving

$$\lambda_f = \frac{q}{p} e^{-2i\phi_f} = \left| \frac{q}{p} \right| = e^{-2i(\phi_M + \phi_f)} \qquad (7.97)$$

From this expression it is once more evident that if for a different final state g the corresponding parameter λ_g is different from λ_f (apart from a sign), then $\phi_f \neq \phi_g$ and direct *CPV* must be present.

All three kinds of *CP* violation discussed above can be conveniently expressed in terms of λ_f: the condition imposed by *CP* symmetry is

$$\lambda_f = \frac{1}{\lambda_{\overline{f}}} \qquad (7.98)$$

Indeed *CP* symmetry relates the decay amplitudes according to eqn (6.23) and allows $|M_a\rangle$ to be a *CP* eigenstate (eigenvalue ± 1), so that

$$q = \langle \overline{M}^0 | M_a \rangle = \langle \overline{M}^0 | (CP)^\dagger (CP) | M_a \rangle = \pm \eta_{CP} \langle M^0 | M_a \rangle = \pm \eta_{CP} \, p$$

and the result follows. For final states which are *CP* eigenstates ($\overline{f} = f$) the *CP* symmetry condition (7.98) reduces to $\lambda_f = \pm 1$.

Note that if both (7.88) and (7.89) hold, then (7.98) is automatically satisfied in modulus: $|\lambda_f| = 1/|\lambda_{\overline{f}}|$, but not necessarily in phase, as it is still possible that $\lambda_f \neq \lambda_{\overline{f}}^*$, while *CP* symmetry would require these parameters to have opposite or zero phases (necessarily zero for final *CP*-eigenstates); the signature of *CP* violation in the interference of mixing and decay is thus

$$\arg(\lambda_f) + \arg(\lambda_{\overline{f}}) \neq 0 \qquad (7.99)$$

For a *CP* eigenstate *CPV* in either mixing or decay is indicated by

$$|\lambda_f| \neq 1 \qquad (7.100)$$

while *CPV* in the interference of mixing and decay corresponds to

$$\text{Im}(\lambda_f) \neq 0 \qquad (7.101)$$

For decays into CP eigenstates with eigenvalue $\eta_{CP}(f) = \pm 1$ (accessible to both M^0 and $\overline{M}^0$), if CP symmetry holds

$$\lambda_f = \eta_{CP}(f) \left(\frac{q}{p} \eta_{CP}^* \right) = \pm \eta_{CP}(f) \tag{7.102}$$

Finally, considering two final states f, g, CP symmetry requires

$$\lambda_f \lambda_{\bar{f}} = \lambda_g \lambda_{\bar{g}} \tag{7.103}$$

which for final CP eigenstates reduces to

$$\eta_{CP}(f)\lambda_f = \eta_{CP}(g)\lambda_g \tag{7.104}$$

and we get back condition (6.39). For any two states the λ parameters must be either equal or opposite, depending on whether the CP-parities of the two states are equal or opposite; a difference between the $\eta_{CP}(f)\lambda_f$ values for any two final states is a signature of CP violation in the decay process (direct CPV) which does not require the presence of strong phases, since

$$\eta_{CP}(f)\lambda_f - \eta_{CP}(g)\lambda_g \propto \frac{\overline{A}_{\bar{f}}}{A_f} - \frac{\overline{A}_{\bar{g}}}{A_g} \tag{7.105}$$

Note that the *production* process of the $M^0, \overline{M}^0$ mesons from an initial state i can also interfere with the mixing to give CP violation, described by the (phase-convention independent) parameters

$$\zeta_i = \frac{A(i \to \overline{M}^0)}{A(i \to M^0)} \frac{p}{q} \qquad \zeta_{\bar{i}} = \frac{A(\bar{i} \to \overline{M}^0)}{A(\bar{i} \to M^0)} \frac{p}{q} \tag{7.106}$$

with CPV indicated by

$$\arg(\zeta_i) + \arg(\zeta_{\bar{i}}) \neq 0 \tag{7.107}$$

7.2.2 Flavour-specific decays

Not all final states allow the observation of all three kinds of CP violation. We now discuss CPV effects displayed in the decay rate asymmetry for mesons which at some initial time are flavour eigenstates: the knowledge of the initial flavour eigenvalue can be obtained exploiting associate production, by identifying (some of) the particles produced together with the meson or, for exclusive coherent $M^0\overline{M}^0$ pair production in a known state, by detecting the decay of the companion meson, as in this case the time evolution of the pair is correlated at all times (Section 7.5).

   Consider now decays into *flavour-specific* final states, that is states which, due to some selection rule, can be reached directly only by M^0 and not by $\overline{M}^0$ or vice versa; such states are clearly not eigenstates of CP, one example being semi-leptonic final states such as $\pi^{\pm}\ell^{\mp}\nu$ (where ℓ is a charged lepton). If $A(M^0 \to f) \neq 0$ and $A(\overline{M}^0 \to \bar{f}) \neq 0$ but

$$A(\overline{M}^0 \to f) = 0 = A(M^0 \to \bar{f})$$

then the fact that the same p, q coefficients appear in the expressions (7.45) for the physical states (a consequence of CPT symmetry) implies

$$A(M_a \to f) = A(M_b \to f) \qquad A(M_a \to \bar{f}) = -A(M_b \to \bar{f}) \qquad (7.108)$$

since the flavour eigenstates $M^0, \overline{M}^0$ contained in the physical states $M_{a,b}$ cannot interfere, as they decay to different final states. Still, if CP is violated the decay rates of physical states into f and $\bar{f}$ can be different:

$$\Gamma(M_{a,b} \to \bar{f}) \neq \Gamma(M_{a,b} \to f) \qquad (7.109)$$

In this case CPV in the interference of mixing and decay clearly cannot occur, as only one of the two flavour eigenstates can feed the final state. CP violation in the mixing and in the decay are both possible, but in some important cases the decay is dominated by a single amplitude, and/or there are no different strong scattering phases as required to observe CPV in the decay; this is the case for the semi-leptonic decays mentioned above, for which CPT symmetry for weakly interacting states gives, through eq (5.17):[101]

$$A(M^0 \to f) = \eta^*_{CPT}\,\eta_{CPT}(f)\,A^*(\overline{M}^0 \to \bar{f}) \qquad (7.110)$$

and thus no CPV in the decay, by eqn (7.89). Only CPV in the mixing can then occur, which still allows (7.109), because of the different amount of each flavour eigenstate in the composition of the physical states ($|p| \neq |q|$). However in this case

$$\frac{A(M_{a,b} \to \bar{f})}{A(M_{a,b} \to f)} = \pm\eta_{CPT}(f)\frac{q}{p}\eta^*_{CPT}\frac{A^*_f}{A_f} \qquad (7.111)$$

and the phase η_{CPT} can be chosen such that the last factor is unity (changing q/p appropriately at the same time) and only the mixing parameters q, p appear in the amplitude ratio; this implies that the rate asymmetry

$$\Delta_f^{(a,b)} = \frac{\Gamma(M_{a,b} \to \bar{f}) - \Gamma(M_{a,b} \to f)}{\Gamma(M_{a,b} \to \bar{f}) + \Gamma(M_{a,b} \to f)} \qquad (7.112)$$

must be the same (to within a sign) for *any* (flavour-specific) final state f, as it can always be ascribed to a property of the decaying states (independent from f).

---

[101] The PT subscript, corresponding to the reversal of spin, has been omitted as we deal with spinless mesons, and is anyway irrelevant because of rotational symmetry.

7.2.3 Non flavour-specific decays

For final states f which are accessible to both M^0 and $\overline{M}^0$, all three types of CP violation can in principle occur. Such final states need not be CP eigenstates: for example doubly Cabibbo suppressed decays $D^0 \to K^+\pi^-$ are allowed, besides the favoured $\overline{D}^0 \to K^+\pi^-$; still the simplest case is that of CP eigenstates.

In this case note that if *both* physical states can reach the same CP eigenstate then CP is necessarily violated (if it were conserved the two physical states would be CP eigenstates with opposite eigenvalues, and one of them would be forbidden to decay into f), so that a non-zero ratio of decay amplitudes

$$\frac{A(M_b \to f)}{A(M_a \to f)} = \frac{1 - \lambda_f}{1 + \lambda_f} \tag{7.113}$$

(or its reciprocal) is a signature of CPV.

Under the same conditions leading to eqn (7.110) one would get a relation between decay amplitudes into the same state

$$A(M^0 \to f) = \eta^*_{CPT}\, \eta_{CPT}(f)\, A^*(\overline{M}^0 \to f) \tag{7.114}$$

so that

$$\frac{A(M_b \to f)}{A(M_a \to f)} = \frac{1 + \eta_{CPT}(f)(q/p)(\eta^*_{CPT}A^*_f/A_f)}{1 - \eta_{CPT}(f)(q/p)(\eta^*_{CPT}A^*_f/A_f)} \tag{7.115}$$

could be written, with an appropriate choice of η_{CPT}, exclusively in terms of the mixing parameters, and would be independent of the specific final state f (except for $\eta_{CPT}(f)$). This is not the general case, however, and the amplitude ratio (7.113) is usually also sensitive to CPV in the decay.

In any case, while the observation of decays into a *single* final state can prove the existence of CP violation, it is usually not enough to identify whether it belongs to the mixing or the decay process (or both); in other words, it is always possible to redefine the relative phase of the flavour eigenstates to make any *one* decay amplitude real; however, if a process gets contributions from more amplitudes, which have a non-zero CP-violating relative phase, the rephasing freedom cannot be used to make *both* of them real, and by observing decays into another final state, in which such amplitudes appear in a different combination, CPV in the decay can be put into evidence, independently from the phase arbitrariness.

7.2.4 Decays of physical states

When the physical particles $M_{a,b}$ can be distinguished in a practical way, it is possible to perform experiments on them separately, as in the case of K mesons (Chapter 8).

The decays of one of them into *CP* eigenstates offer a direct way to learn about *CP* violation, which is actually how this phenomenon was discovered (Chapter 6): if decay into a given *CP* eigenstate f is observed for one state (say M_a), the observation of the decay of the other one (M_b) into f indicates *CPV*, parametrized by a non-zero ratio of decay amplitudes

$$\eta_f = \frac{A(M_b \rightarrow f)}{A(M_a \rightarrow f)} \qquad (7.116)$$

which can originate from any kind of *CP* violation, in the asymmetric mixing process which makes the physical states impure mixtures of *CP* eigenstates, in the decay process into f being different between M^0 and $\overline{M}^0$, or in the interference between the two.

7.3 Time evolution of flavour eigenstates

Since the flavour quantum number F is not conserved by the (weak part of the) Hamiltonian, the physical states are not flavour eigenstates and any state of definite flavour will evolve into a coherent superposition of $|M^0\rangle$ and $|\overline{M}^0\rangle$, exhibiting the phenomenon of *flavour oscillations*.

Denoting by $|M^0(t)\rangle$ a state which *at time* $t = 0$ is a pure flavour eigenstate with eigenvalue $+1$, and analogously $|\overline{M}^0(t)\rangle$ (flavour-tagged states):

$$F|M^0(t = 0)\rangle = +|M^0(t = 0)\rangle \quad F|\overline{M}^0(t = 0)\rangle = -|\overline{M}^0(t = 0)\rangle \qquad (7.117)$$

its time evolution is obtained by expressing it in terms of the physical states which evolve freely according to eqn (7.24):

$$|M^0(t)\rangle = f_+(t)|M^0\rangle + \frac{q}{p}f_-(t)|\overline{M}^0\rangle \qquad (7.118)$$

$$|\overline{M}^0(t)\rangle = \frac{p}{q}f_-(t)|M^0\rangle + f_+(t)|\overline{M}^0\rangle \qquad (7.119)$$

where

$$f_\pm(t) = \frac{1}{2}e^{-im_a t}e^{-\Gamma_a t/2}\left[1 \pm e^{-i\Delta m t}e^{-\Delta\Gamma t/2}\right] \qquad (7.120)$$

with the definitions (7.32) and (7.33).

From the above equations the probability that at time t a state has the same flavour eigenvalue which it had at time $t = 0$ is the same for initial M^0 or $\overline{M}^0$:

$$P[M^0(t) \rightarrow M^0] = P[\overline{M}^0(t) \rightarrow \overline{M}^0] = |f_+(t)|^2 \qquad (7.121)$$

This is actually a consequence of *CPT* symmetry (for the case in which *CPT* is violated, see e.g. Branco, Lavoura, and Silva (1999)). Conversely, the probability that an initial M^0 becomes a $\overline{M}^0$ at time t is not necessarily the same as for a $\overline{M}^0$ becoming a M^0:

$$P[M^0(t) \to \overline{M}^0] = \left|\frac{q}{p}\right|^2 |f_-(t)|^2 \qquad P[\overline{M}^0(t) \to M^0] = \left|\frac{p}{q}\right|^2 |f_-(t)|^2 \quad (7.122)$$

which are different if there is *CPV* in the mixing (with their ratio being independent of time). Flavour oscillations exist independently from the validity of *CP* symmetry, which (as *T* symmetry as well) enforces the equality of the oscillation probabilities in the two directions.

Expressions (7.118), (7.119) also make evident how the time-dependent phase factor $e^{i\Delta mt}$ in the oscillation amplitude provides the complex term against which *CP*-violating phases in a decay amplitude can be detected (*CPV* in the interference of mixing and decay).

If *CPT* symmetry is not assumed, the evolution of the initial flavour eigenstates is modified with respect to eqns (7.118), (7.119) into

$$|M^0(t)\rangle = [f_+(t) + \delta_{CPT}f_-(t)]\,|M^0\rangle + \frac{q}{p}\sqrt{1 - \delta_{CPT}^2}\,f_-(t)|\overline{M}^0\rangle \qquad (7.123)$$

$$|\overline{M}^0(t)\rangle = \frac{p}{q}\sqrt{1 - \delta_{CPT}^2}\,f_-(t)|M^0\rangle + [f_+(t) - \delta_{CPT}f_-(t)]\,|\overline{M}^0\rangle \qquad (7.124)$$

where q/p was defined in eqn (7.34) also for the case of no *CPT* symmetry. When addressing such an exotic effect as *CPT* violation it is reasonable to question any other tacit assumption: considering semi-leptonic final states f, the λ_f parameters, which would be either zero or infinity if the $\Delta F = \Delta Q$ rule were exact (true flavour-specific decays), can be written in terms of the complex quantities $x^{(f)}, \overline{x}^{(f)}$ parametrizing possible violations of the above rule:

$$\lambda_f = \frac{q}{p}x^{(f)} \qquad\qquad \frac{1}{\lambda_{\overline{f}}} = \frac{p}{q}\overline{x}^{(\overline{f})*} \qquad (7.125)$$

$$x^{(f)} \equiv \frac{A(\overline{M}^0 \to f)}{A(M^0 \to f)} \qquad \overline{x}^{(\overline{f})} \equiv \frac{A(M^0 \to \overline{f})^*}{A(\overline{M}^0 \to \overline{f})^*} \qquad (7.126)$$

and the coefficients of the $f_-(t)$ terms in the time evolution of flavour-tagged mesons (which can be extracted from the analysis of their decay rates into, e.g. , the flavour-specific state f), are

$$\lambda_f\sqrt{1 - \delta_{CPT}^2} + \delta_{CPT} \simeq \lambda_f + \delta_{CPT} \qquad 1/\lambda_f\sqrt{1 - \delta_{CPT}^2} - \delta_{CPT} \simeq 1/\lambda_f$$

and the violations of CPT symmetry and of the $\Delta F = \Delta Q$ rule can be distinguished in principle (Lavoura and Silva, 1999).

Considering the decay rates of flavour-tagged mesons into a specific final state, CPT symmetry requires in general the oscillating term to have opposite phase (but possibly different amplitudes) for the case of initial $\overline{M}^0$ with respect to the case of initial M^0; conversely, T symmetry requires such terms to have the same amplitude (but possibly a phase difference).

From eqns (7.118), (7.119) the time-dependent decay rates[102] into a final state f (accessible to both M^0 and $\overline{M}^0$) for the states $|M^0(t)\rangle$, $|\overline{M}^0(t)\rangle$ can be expressed in terms of the decay amplitudes A_f and $\overline{A}_f$ for flavour eigenstates:

$$\Gamma[M^0(t) \to f] \propto \frac{1}{4}|A_f|^2 \left[|f_+(t)|^2 + |f_-(t)|^2|\lambda_f|^2 + 2\mathrm{Re}\left(f_+^*(t)f_-(t)\lambda_f\right) \right]$$
(7.127)

$$\Gamma[\overline{M}^0(t) \to f] \propto \frac{1}{4}|\overline{A}_f|^2 \left[|f_+(t)|^2 + |f_-(t)|^2\frac{1}{|\lambda_f|^2} + 2\mathrm{Re}\left(f_+^*(t)f_-(t)\frac{1}{\lambda_f}\right) \right]$$
(7.128)

The time dependence of the above expressions is contained in the functions $f_\pm(t)$; the explicit general formulæ are

$$\Gamma[M^0(t) \to f] \propto \frac{1}{4}\left[R_f^{(+)}e^{-\Gamma_a t} + R_f^{(-)}e^{-\Gamma_b t} + C_f e^{-\overline{\Gamma}t}\cos(\Delta m\, t) \right.$$
$$\left. +S_f e^{-\overline{\Gamma}t}\sin(\Delta m\, t) \right]$$
(7.129)

$$\Gamma[\overline{M}^0(t) \to f] \propto \frac{1}{4}\left[\overline{R}_f^{(+)}e^{-\Gamma_a t} + \overline{R}_f^{(-)}e^{-\Gamma_b t} + \overline{C}_f e^{-\overline{\Gamma}t}\cos(\Delta m\, t) \right.$$
$$\left. +\overline{S}_f e^{-\overline{\Gamma}t}\sin(\Delta m\, t) \right]$$
(7.130)

where

$$R_f^{(\pm)} = |A_f|^2 \left[|1 \pm \lambda_f|^2 \right] \qquad \overline{R}_f^{(\pm)} = |\overline{A}_f|^2 \left[|1 \pm 1/\lambda_f|^2 \right] \qquad (7.131)$$

$$C_f = 2|A_f|^2 \left(1 - |\lambda_f|^2 \right) \qquad \overline{C}_f = 2|\overline{A}_f|^2 \left(1 - 1/|\lambda_f|^2 \right) \qquad (7.132)$$

$$S_f = -4|A_f|^2 \,\mathrm{Im}(\lambda_f) \qquad \overline{S}_f = -4|\overline{A}_f|^2 \,\mathrm{Im}(1/\lambda_f) \qquad (7.133)$$

---

[102] We will often refer to these quantities as to decay rates, although it should be understood that phase space factors would have to be included in order to get a physical decay rate; these are really modulus-squared decay amplitudes. The difference is anyway irrelevant as long as we will consider only ratios of decay rates, which can be experimentally measured with good precision.

Note that

$$R_f^{(\pm)} = \left|\frac{q}{p}\right|^2 \overline{R}_f^{(\pm)} \qquad C_f = -\left|\frac{q}{p}\right|^2 \overline{C}_f \qquad S_f = -\left|\frac{q}{p}\right|^2 \overline{S}_f \qquad (7.134)$$

From the above expressions it is evident how CP violation manifests itself in non-exponential terms in the time evolution of flavour eigenstates: CPV in the mixing or in the decay induces a $\cos(\Delta m\,t)$ term, which is immediately relevant starting at $t = 0$ as it reflects a static property; conversely, CPV in the interference of mixing and decay introduces a $\sin(\Delta m\,t)$ term, which is not effective at $t = 0$, as this mechanism of CP violation requires flavour oscillations to start in order to become active.

Eqns (7.129) (7.130) can be rewritten in another form which is more convenient for heavy meson systems:

$$\Gamma[M^0(t) \to f] \propto (1/2)|A_f|^2(1 + |\lambda_f|^2)\,e^{-\overline{\Gamma}t}\,[\cosh(\Delta\Gamma t/2)$$
$$+A_f^{(\Delta)}\sinh(\Delta\Gamma t/2) + A_f^{(MD)}\cos(\Delta m\,t) - A_f^{(I)}\sin(\Delta m\,t)]$$

$$(7.135)$$

$$\Gamma[\overline{M}^0(t) \to f] \propto (1/2)|A_f|^2\left|\frac{p}{q}\right|^2(1 + |\lambda_f|^2)\,e^{-\overline{\Gamma}t}\,[\cosh(\Delta\Gamma t/2)$$
$$+A_f^{(\Delta)}\sinh(\Delta\Gamma t/2) - A_f^{(MD)}\cos(\Delta m\,t) + A_f^{(I)}\sin(\Delta m\,t)]$$

$$(7.136)$$

where

$$A_f^{(MD)} \equiv \frac{1 - |\lambda_f|^2}{1 + |\lambda_f|^2} \qquad A_f^{(I)} \equiv \frac{2\mathrm{Im}(\lambda_f)}{1 + |\lambda_f|^2} \qquad A_f^{(\Delta)} \equiv \frac{2\mathrm{Re}(\lambda_f)}{1 + |\lambda_f|^2} \qquad (7.137)$$

which are linked by

$$\left[A_f^{(MD)}\right]^2 + \left[A_f^{(I)}\right]^2 + \left[A_f^{(\Delta)}\right]^2 = 1 \qquad (7.138)$$

Consider now how the above expressions simplify in some limiting cases:

- If no flavour oscillations occur at all ($\Delta m = 0 = \Delta\Gamma$) the time evolution is exponential:

$$\Gamma[M^0(t) \to f] \propto |A_f|^2 e^{-\Gamma t} \qquad \Gamma[\overline{M}^0(t) \to f] \propto |\overline{A}_f|^2 e^{-\Gamma t}$$

- If $\Gamma_a \gg \Gamma_b$ (the case of K mesons), for short times such that $\Gamma_b t \ll 1$:

$$\Gamma[M^0(t) \to f] \propto R_f^{(-)} + e^{-\Gamma_a t} \left[R_f^{(+)} + C_f \cos(\Delta m\, t) + S_f \sin(\Delta m\, t) \right]$$
(7.139)

$$\Gamma[\overline{M}^0(t) \to f] \propto \overline{R}_f^{(-)} + e^{-\Gamma_a t} \left[\overline{R}_f^{(+)} + \overline{C}_f \cos(\Delta m\, t) + \overline{S}_f \sin(\Delta m\, t) \right]$$
(7.140)

and for long times $\Gamma_b t \gg 1$

$$\Gamma[M^0(t) \to f] \propto R_f^{(-)} e^{-\Gamma_b t} \qquad \Gamma[\overline{M}^0(t) \to f] \propto \overline{R}_f^{(-)} e^{-\Gamma_b t}$$

- In the limit $\Delta\Gamma = 0$ (a good approximation for B_d mesons, and definitely not for K):

$$\Gamma[M^0(t) \to f] \propto e^{-\overline{\Gamma} t} \left[2R_f + C_f \cos(\Delta m\, t) + S_f \sin(\Delta m\, t) \right] \quad (7.141)$$

$$\Gamma[\overline{M}^0(t) \to f] \propto e^{-\overline{\Gamma} t} \left[2\overline{R}_f + \overline{C}_f \cos(\Delta m\, t) + \overline{S}_f \sin(\Delta m\, t) \right] \quad (7.142)$$

having defined $R_f \equiv (R_f^{(+)} + R_f^{(-)})/2 = |A_f|^2(1 + |\lambda_f|^2)$ (and analogously $\overline{R}_f$).

- For a flavour-specific final state $\overline{A}_f = 0$ ($\lambda_f = 0$)

$$R_f^{(\pm)} = |A_f|^2 \qquad C_f = 2|A_f|^2 \qquad S_f = 0$$

$$\overline{R}_f^{(\pm)} = \left| \frac{p}{q} \right|^2 |A_f|^2 \qquad \overline{C}_f = -2 \left| \frac{p}{q} \right|^2 |A_f|^2 \qquad \overline{S}_f = 0$$

$$\Gamma[M^0(t) \to f] \propto |A_f|^2 \left[e^{-\Gamma_a t} + e^{-\Gamma_b t} + 2e^{-\overline{\Gamma} t} \cos(\Delta m\, t) \right] \quad (7.143)$$

$$\Gamma[\overline{M}^0(t) \to f] \propto \left| \frac{p}{q} \right|^2 |A_f|^2 \left[e^{-\Gamma_a t} + e^{-\Gamma_b t} - 2e^{-\overline{\Gamma} t} \cos(\Delta m\, t) \right] \quad (7.144)$$

which in absence of *CPV* in the mixing ($|q/p| = 1$) are just eqns (6.4) and (6.5). $\Gamma[M^0 \to \overline{f}]$ and $\Gamma[\overline{M}^0 \to f]$ vanish at $t = 0$ but not at later times, due to flavour oscillations.

- If *CP* symmetry holds ($\lambda_f = \pm 1$):

$$R_f^{(+)} = 0 = \overline{R}_f^{(+)} \quad \text{or} \quad R_f^{(-)} = 0 = \overline{R}_f^{(-)}$$

$$C_f = 0 = \overline{C}_f \qquad S_f = 0 = \overline{S}_f$$

and exponential decays into f are obtained:

$$\Gamma[M^0(t), \overline{M}^0(t) \to f] \propto e^{-\Gamma_a t} \quad \text{or} \quad \Gamma[M^0(t), \overline{M}^0(t) \to f] \propto e^{-\Gamma_b t}$$

- If CPV in the mixing can be neglected, $|q/p| = 1$ (a good approximation for B mesons):

$$R_f^{(\pm)} = |A_f|^2 + |\overline{A}_f|^2 \pm 2|A_f||\overline{A}_f|\cos(\overline{\theta}_f - \theta_f - 2\phi_M) = \overline{R}_f^{(\pm)} \qquad (7.145)$$

$$R_f = |A_f|^2 + |\overline{A}_f|^2 \qquad (7.146)$$

$$C_f = 2\left(|A_f|^2 - |\overline{A}_f|^2\right) = -\overline{C}_f \qquad (7.147)$$

$$S_f = -4|A_f||\overline{A}_f|\sin(\overline{\theta}_f - \theta_f - 2\phi_M) = -\overline{S}_f \qquad (7.148)$$

- If CPV in the decay can be neglected, $|A_f| = |\overline{A}_f|$ (as when the decay is dominated by a single amplitude):

$$R_f^{(\pm)} = |A_f|^2\left[1 + \left|\frac{q}{p}\right|^2 \pm 2\left|\frac{q}{p}\right|\cos(\overline{\theta}_f - \theta_f - 2\phi_M)\right]$$

$$\overline{R}_f^{(\pm)} = |A_f|^2\left[1 + \left|\frac{p}{q}\right|^2 \pm 2\left|\frac{p}{q}\right|\cos(\overline{\theta}_f - \theta_f - 2\phi_M)\right]$$

$$R_f = |A_f|^2\left[2 + \left|\frac{q}{p}\right|^2 + \left|\frac{p}{q}\right|^2\right]$$

$$C_f = 2|A_f|^2\left(1 - \left|\frac{q}{p}\right|^2\right)$$

$$\overline{C}_f = 2|A_f|^2\left(1 - \left|\frac{p}{q}\right|^2\right)$$

$$S_f = -4|A_f|^2\left|\frac{q}{p}\right|\sin(\overline{\theta}_f - \theta_f - 2\phi_M)$$

$$\overline{S}_f = +4|A_f|^2\left|\frac{p}{q}\right|\sin(\overline{\theta}_f - \theta_f - 2\phi_M)$$

- If CPV in both mixing and decay can be neglected:

$$R_f^{(\pm)} = 2|A_f|^2[1 \pm \cos(\overline{\theta}_f - \theta_f - 2\phi_M)] = \overline{R}_f^{(\pm)}$$

$$C_f = 0 = \overline{C}_f$$

$$S_f = -4|A_f|^2\sin(\overline{\theta}_f - \theta_f - 2\phi_M) = -\overline{S}_f$$

so that *CPV* in the interference of mixing and decay can still generate an interference term proportional to $\sin(\Delta m\, t)$, which is seen to violate *CP* as it has opposite sign for the two initial flavour states.

More information can be obtained by comparing the decays into the state f with those into the *CP*-conjugate state $\bar{f}$, and the situation is simpler when eqn (7.110) holds, as in that case

$$|A_{\bar{f}}| = |\bar{A}_f| \qquad |\bar{A}_{\bar{f}}| = |A_f|$$

and (besides not having *CPV* in the decay) the four expressions for the decays of initial $M^0, \overline{M}^0$ into $f, \bar{f}$ are all linked, e.g. in case of no *CPV*

$$R_f^{(\pm)} = \overline{R}_{\bar{f}}^{(\pm)} \qquad C_f = \overline{C}_{\bar{f}} \qquad S_f = \overline{S}_{\bar{f}}$$

Note also that for flavour-specific decays $1/\lambda_{\bar{f}} = 0$.

7.3.1 Integrated decay rates

In some cases experiments might not be able to measure the time dependence of the decay rates, e.g. if the time resolution is larger than the meson lifetime (lifetimes for B, D mesons are in the ps range). Decay rates integrated over time are then considered; even when the decay time can be measured, integrated rates clearly offer a large statistical advantage.

It is useful to introduce the two real dimensionless parameters

$$x \equiv \frac{\Delta m}{\Gamma} \qquad y \equiv \frac{\Delta\Gamma}{2\Gamma} \tag{7.149}$$

The range of y is between -1 and $+1$. For the K system $x_K \simeq 1$ and $y_K \simeq 1$, while for the D system $x_D \simeq 0 \simeq y_D$; for the B_d system $x_d \simeq 0.78$ but $y_d \ll 1$, while for the B_s system $x_s \gg 1$ and $y_s \sim O(0.2)$.

In terms of the parameters (7.149) the integrated decay rates for an initial M^0 or $\overline{M}^0$ are

$$\Gamma[M^0 \to f] \propto \frac{1}{4}\left[I_f^{(y)}\frac{1}{1-y^2} + I_f^{(x)}\frac{1}{1+x^2}\right] \tag{7.150}$$

$$\Gamma[\overline{M}^0 \to f] \propto \frac{1}{4}\left[\overline{I}_f^{(y)}\frac{1}{1-y^2} + \overline{I}_f^{(x)}\frac{1}{1+x^2}\right] \tag{7.151}$$

where

$$I_f^{(y)} = |A_f|^2\left[1 + |\lambda_f|^2 + 2y\,\text{Re}(\lambda_f)\right]$$

$$I_f^{(x)} = |A_f|^2\left[1 - |\lambda_f|^2 - 2x\,\text{Im}(\lambda_f)\right]$$

$$\overline{I}_f^{(y)} = |\overline{A}_f|^2 \left[1 + 1/|\lambda_f|^2 + 2y \, \text{Re}(1/\lambda_f) \right]$$

$$\overline{I}_f^{(x)} = |\overline{A}_f|^2 \left[1 - 1/|\lambda_f|^2 - 2x \, \text{Im}(1/\lambda_f) \right]$$

Again, the general expressions above simplify in some limits:

- For a flavour-specific final state ($\overline{A}_f = 0$):

$$I_f^{(y)} = |A_f|^2 = I_f^{(x)} \qquad \overline{I}_f^{(y)} = \left| \frac{p}{q} \right|^2 |A_f|^2 = -\overline{I}_f^{(x)}$$

$$\Gamma[M^0 \to f] \propto |A_f|^2 \frac{2 + x^2 - y^2}{(1 - y^2)(1 + x^2)} \tag{7.152}$$

$$\Gamma[\overline{M}^0 \to f] \propto \left| \frac{p}{q} \right|^2 |A_f|^2 \frac{x^2 + y^2}{(1 - y^2)(1 + x^2)} \tag{7.153}$$

Since in this case only M^0 can decay into f, these expressions are just proportional to the time-integrated probability $P[M^0 \to M^0]$ of a meson not to have oscillated, or respectively $P[\overline{M}^0 \to M^0]$ to have oscillated, into the opposite flavour eigenstate.

By writing analogous expressions for the decays into the CP-conjugate final state $\overline{f}$ (fed only by $\overline{M}^0$), the ratios of 'mixed' and 'unmixed' decays can be written for M^0 and $\overline{M}^0$ (Pais and Treiman, 1975):

$$\frac{\Gamma[M^0 \to \overline{f}]}{\Gamma[M^0 \to f]} = \left| \frac{\overline{A}_{\overline{f}}}{A_f} \right|^2 \frac{P[M^0 \to \overline{M}^0]}{P[M^0 \to M^0]} \tag{7.154}$$

$$\frac{\Gamma[\overline{M}^0 \to f]}{\Gamma[\overline{M}^0 \to \overline{f}]} = \left| \frac{A_f}{\overline{A}_{\overline{f}}} \right|^2 \frac{P[\overline{M}^0 \to M^0]}{P[\overline{M}^0 \to \overline{M}^0]} \tag{7.155}$$

with the ratio of oscillation to non-oscillation probabilities (Pais–Treiman parameters) being

$$r \equiv \frac{P[M^0 \to \overline{M}^0]}{P[M^0 \to M^0]} = \left| \frac{q}{p} \right|^2 \mathcal{F} \qquad \overline{r} \equiv \frac{P[\overline{M}^0 \to M^0]}{P[\overline{M}^0 \to \overline{M}^0]} = \left| \frac{p}{q} \right|^2 \mathcal{F} \tag{7.156}$$

$$\mathcal{F} \equiv \frac{\int_0^{+\infty} |f_-(t)|^2 \, dt}{\int_0^{+\infty} |f_+(t)|^2 \, dt} = \frac{x^2 + y^2}{2 + x^2 - y^2}$$

As is evident from (7.156), CP symmetry in the mixing would require $r = \bar{r}$:

$$\frac{\bar{r} - r}{\bar{r} + r} = \frac{P[\overline{M}^0 \to M^0] - P[M^0 \to \overline{M}^0]}{P[\overline{M}^0 \to M^0] + P[M^0 \to \overline{M}^0]} = \frac{1 - |q/p|^4}{1 + |q/p|^4} = \frac{2\delta_T}{1 + \delta_T^2} \quad (7.157)$$

and overall CP symmetry would require the ratios (7.154) and (7.155) to be equal.

Note that if $|A_f| = |\overline{A}_{\bar{f}}|$ the measurement of the ratio of mixed to unmixed events gives the parameters $r, \bar{r}$, which are related to CP violation in the mixing through the multiplicative factor $\mathcal{F}$ characterizing the amount of mixing, maximal ($\mathcal{F} = 1$) if either:

(a) the two physical states have very different lifetimes ($|y| \to 1$), when the state rapidly evolves into a mixture (in equal amounts if CP symmetry holds) of the two flavour eigenstates; or

(b) the mass difference is very large compared to the lifetimes ($|x| \to \infty$), when the state makes many flavour oscillations before decaying, appearing as an equal mixture of both flavour eigenstates.

On the contrary, if both x and y are small then $\mathcal{F}$ becomes small and sensitivity to CPV is lost.

The mixing probabilities are often expressed through the mixing parameter

$$\chi = \frac{P[M^0 \to \overline{M}^0]}{P[M^0 \to M^0] + P[M^0 \to \overline{M}^0]} \quad (7.158)$$

(and the corresponding $\overline{\chi}$ for the $\overline{M}^0 \to M^0$ transition); these just represent the probability that a tagged M^0 ($\overline{M}^0$) oscillates into the opposite flavour state. If $|A_f| = |\overline{A}_f|$ such parameters are

$$\chi(M^0 \to \overline{M}^0) = \frac{r}{1 + r} \qquad \chi(\overline{M}^0 \to M^0) = \frac{\bar{r}}{1 + \bar{r}} \quad (7.159)$$

and if CP symmetry holds

$$\chi(M^0 \to \overline{M}^0) = \chi(\overline{M}^0 \to M^0) = \frac{x^2 + y^2}{2(1 + x^2)} \quad (7.160)$$

Note that $\chi \simeq 1/2$ in the case of large mixing $x \gg 1$ (as for the B_s system).

• In the absence of oscillations ($x = 0 = y$) clearly

$$\Gamma[M^0 \to f] \propto |A_f|^2 \qquad \Gamma[\overline{M}^0 \to f] \propto |\overline{A}_f|^2$$

Since $M^0, \overline{M}^0$ are usually produced in flavour-conserving processes starting from $F = 0$ beams and targets, they are always accompanied by other flavoured particles (sometimes as a $M^0\overline{M}^0$ pair, Section 7.5): the observation of pairs of semi-leptonic decays with leptons of the *same* charge in an event provides evidence for flavour oscillations; any *asymmetry* in the number of pairs with two positive leptons and two negative leptons provides evidence for *CP* violation in the mixing.

• If *CP* symmetry holds:

$$\Gamma[M^0 \to f] \propto |A_f|^2 \frac{1}{1 \pm y} \qquad \Gamma[\overline{M}^0 \to f] \propto |\overline{A}_f|^2 \frac{1}{1 \pm y}$$

where the sign ambiguity reflects that of λ_f.

• If *CPV* in the mixing can be neglected:

$$I_f^{(y)} = |A_f|^2 + |\overline{A}_f|^2 + 2|A_f||\overline{A}_f|\cos(\overline{\theta}_f - \theta_f - 2\phi_M) = \overline{I}_f^{(y)}$$

$$I_f^{(x)} = |A_f|^2 - |\overline{A}_f|^2 - 2|A_f||\overline{A}_f|\cos(\overline{\theta}_f - \theta_f - 2\phi_M) = -\overline{I}_f^{(x)}$$

• If *CPV* in the decay can be neglected:

$$I_f^{(y)} = |A_f|^2 \left[1 + \left|\frac{q}{p}\right|^2 + 2y \left|\frac{q}{p}\right| \cos(\overline{\theta}_f - \theta_f - 2\phi_M) \right]$$

$$I_f^{(x)} = |A_f|^2 \left[1 - \left|\frac{q}{p}\right|^2 - 2x \left|\frac{q}{p}\right| \sin(\overline{\theta}_f - \theta_f - 2\phi_M) \right]$$

and $\overline{I}_f^{(y)}, \overline{I}_f^{(x)}$ are obtained from the above with the replacement $q/p \to p/q$.

• If *CPV* in both mixing and decay can be neglected:

$$I_f^{(y)} = 2|A_f|^2 \left[1 + y \cos(\overline{\theta}_f - \theta_f - 2\phi_M) \right] = \overline{I}_f^{(y)}$$

$$I_f^{(x)} = -2x|A_f|^2 \sin(\overline{\theta}_f - \theta_f - 2\phi_M) = -\overline{I}_f^{(x)}$$

7.4 Asymmetries

If the two physical states of a neutral flavoured meson system can be distinguished experimentally, *CP*-violating asymmetries can simply be formed by comparing their decays into *CP*-conjugated states, an example being eqn (7.112).

Other asymmetries can be formed by considering instead the decays of flavour eigenstates: this is the only possibility for the study of *CP* violation in heavy meson systems, for which experiments on only one of the physical states are not possible.

The expressions for time-dependent decay asymmetries for opposite initial flavour can be obtained by using the formulæ written in Section 7.3.

Knowledge of the initial flavour state of the neutral meson, called *tagging*, can be obtained from the identification of the flavour quantum number of the particles produced together with it, as the production process (strong or electromagnetic) is usually flavour-conserving, resulting in associate production starting from a $F = 0$ initial state. This can be done directly when the neutral meson is produced in association with a flavoured charged meson or a flavoured baryon: since the latter do not exhibit flavour oscillations (as in this case the flavour quantum number is linked to the charge or baryon number), their flavour tags the flavour eigenvalue of the neutral meson *at production time* as being opposite to it. If a second flavoured neutral meson is observed instead, its flavour (as determined when it decays into a flavour-specific state) usually gives no information on the flavour of its partner, as flavour oscillations can occur. Still, in the remarkable case of correlated meson pairs the flavour oscillations are completely coherent, and the knowledge of the flavour of one meson at a given time can tag the flavour of the other one *at the same instant of time*, as will be discussed in Section 7.5.2. In the case of uncorrelated neutral meson pairs, the flavour eigenvalues are related on statistical terms, and still (diluted) asymmetries can be formed.

In the case of a flavour-specific final state f, rather than considering the asymmetry for decays into f only, those obtained considering the decays into both CP-conjugate states f and $\bar{f}$ are more relevant; the decay rate into the state $\bar{f}$ is obtained from eqns (7.135), (7.136), which in this case are more conveniently written as

$$\Gamma[M^0(t) \to \bar{f}] \propto (1/2)|\bar{A}_{\bar{f}}|^2 \left|\frac{q}{p}\right|^2 \left(1 + \frac{1}{|\lambda_{\bar{f}}|^2}\right) e^{-\bar{\Gamma}t} \left[\cosh(\Delta\Gamma t/2)\right.$$
$$\left. + \bar{A}_f^{(\Delta)} \sinh(\Delta\Gamma t/2) - \bar{A}_f^{(MD)} \cos(\Delta m t) + \bar{A}_f^{(I)} \sin(\Delta m t)\right] \quad (7.161)$$

$$\Gamma[\overline{M}^0(t) \to \bar{f}] \propto (1/2)|\bar{A}_{\bar{f}}|^2 \left(1 + \frac{1}{|\lambda_{\bar{f}}|^2}\right) e^{-\bar{\Gamma}t} \left[\cosh(\Delta\Gamma t/2)\right.$$
$$\left. + \bar{A}_f^{(\Delta)} \sinh(\Delta\Gamma t/2) + \bar{A}_f^{(MD)} \cos(\Delta m t) - \bar{A}_f^{(I)} \sin(\Delta m t)\right] \quad (7.162)$$

where $\bar{A}_f^{(\Delta)}$, $\bar{A}_f^{(MD)}$ and $\bar{A}_f^{(I)}$ are obtained from (7.137) with the substitution $\lambda_f \to 1/\lambda_{\bar{f}}$. The advantage of the above expressions is evident for decays which are flavour-specific (or almost so), when A_f and $\bar{A}_{\bar{f}}$ are the 'large' amplitudes and $\lambda_f, 1/\lambda_{\bar{f}}$ are zero (or small).

The asymmetry for unmixed decays

$$\Delta_f^{(U)}(t) = \frac{\Gamma[\overline{M}^0(t) \to \bar{f}] - \Gamma[M^0(t) \to f]}{\Gamma[\overline{M}^0(t) \to \bar{f}] + \Gamma[M^0(t) \to f]} = \frac{|\bar{A}_{\bar{f}}|^2 - |A_f|^2}{|\bar{A}_{\bar{f}}|^2 + |A_f|^2} \equiv -\delta_A(f)$$

(7.163)

is independent of time and only sensitive to *CPV* in the decay. The one for mixed decays

$$\Delta_f^{(M)}(t) = \frac{\Gamma[\overline{M}^0(t) \to f] - \Gamma[M^0(t) \to \bar{f}]}{\Gamma[\overline{M}^0(t) \to f] + \Gamma[M^0(t) \to \bar{f}]} = \frac{|p/q|^2|A_f|^2 - |q/p|^2|\bar{A}_{\bar{f}}|^2}{|p/q|^2|A_f|^2 + |q/p|^2|\bar{A}_{\bar{f}}|^2}$$

(7.164)

is also independent of time and sensitive to *CPV* in either the mixing or in the decay, reducing to $\delta_A(f)$ if the former is absent, and to the oscillation asymmetry (7.157) if the latter is absent.

As these asymmetries are independent of time, they also result in a net difference in the number of f vs. $\bar{f}$ decays, as a consequence of the rate difference of either the decays $M^0 \to f$ and $\overline{M}^0 \to f$ or the oscillations $M^0 \to \overline{M}^0$ and $\overline{M}^0 \to M^0$.

In the general case the asymmetry for decays into a single (non flavour-specific) final state f can be considered

$$\Delta_f(t) = \frac{\Gamma[\overline{M}^0(t) \to f] - \Gamma[M^0(t) \to f]}{\Gamma[\overline{M}^0(t) \to f] + \Gamma[M^0(t) \to f]}$$

(7.165)

Using eqns (7.135) (7.136), if *CPV* in the mixing can be neglected (as for *B* mesons) (7.165) becomes

$$\Delta_f(t) = \frac{A_f^{(I)} \sin(\Delta m\, t) - A_f^{(MD)} \cos(\Delta m\, t)}{\cosh(\Delta\Gamma\, t/2) + A_f^{(\Delta)} \sinh(\Delta\Gamma\, t/2)}$$

(7.166)

in terms of the parameters defined in (7.137). Recalling the discussion of Section 7.2.1, $A_f^{(MD)}$ parametrizes *CPV* in the mixing and in the decay, reducing to $\delta_A(f)$ (7.163) if the former is absent, and to δ_T (7.57) if the latter is absent; $A^{(I)}$ parametrizes instead *CPV* in the interference of mixing and decay, and $A_f^{(\Delta)}$ is not a *CP*-violating parameter.

The asymmetries of the time-integrated rates can also give information on *CP* violation. For a flavour–specific final state, if there is no *CPV* in the decay the difference among the parameters r and $\bar{r}$ (7.157) which probes *CPV* in the mixing can be measured from the decay asymmetry

$$\Delta_f^{(M)} = \frac{\Gamma[\overline{M}^0 \to f] - \Gamma[M^0 \to \overline{f}]}{\Gamma[\overline{M}^0 \to f] + \Gamma[M^0 \to \overline{f}]} = \frac{\bar{r} - r}{\bar{r} + r}$$

which coincides with (7.164), as indeed in this case $\Delta_f^{(M)}$ just reflects the asymmetry in the oscillation probabilities for M^0 and $\overline{M}^0$.

In general, for a single (non flavour-specific) final state f the asymmetry

$$\Delta_f = \frac{\Gamma[\overline{M}^0 \to f] - \Gamma[M^0 \to f]}{\Gamma[\overline{M}^0 \to f] + \Gamma[M^0 \to f]} \tag{7.167}$$

can be considered; the expression for Δ_f is particularly simple if both *CPV* in the mixing and in the decay can be neglected, when

$$\Delta_f = \frac{1 - y^2}{1 + x^2} \frac{x \, \mathrm{Im}(\lambda_f)}{1 + y \mathrm{Re}(\lambda_f)}$$

Independently of the magnitude of *CP* violation, this asymmetry is suppressed for small x (D^0 mesons), large x (B_s^0 mesons) or $|y|$ close to 1 (K mesons). In the limit $|y| \ll 1$ (D^0 and B^0 mesons) this becomes

$$\Delta_f \simeq \frac{x}{1 + x^2} \mathrm{Im}(\lambda_f) \tag{7.168}$$

where the first factor is at most 1/2.

Note that, while the total integrated decay widths of M^0 and $\overline{M}^0$ are bound to be equal if *CPT* symmetry holds, such equality needs not hold at all times; indeed the time-dependent rate asymmetry for inclusive decays can be obtained by summing (7.129) and (7.130) over all final states f:

$$\Delta_f^{(\mathrm{incl})}(t) = \frac{\sum_f \Gamma[\overline{M}^0(t) \to f] - \sum_f \Gamma[M^0(t) \to f]}{\sum_f \Gamma[\overline{M}^0(t) \to f] + \sum_f \Gamma[M^0(t) \to f]}$$

$$= \delta_T \frac{-\cosh(\Delta\Gamma\, t/2) + y \sinh(\Delta\Gamma\, t/2) + \cos(\Delta m\, t) + x \sin(\Delta m\, t)}{\cosh(\Delta\Gamma\, t/2) - y \sinh(\Delta\Gamma\, t/2) - \delta_T^2 \cos(\Delta m\, t) - \delta_T^2 x \sin(\Delta m\, t)}$$

in the limit $|y| \to 0$ this asymmetry reduces to

$$\Delta_f^{(\mathrm{incl})}(t) \simeq \delta_T \left[x \sin(\Delta m\, t) - 2 \sin^2(\Delta m\, t/2) \right]$$

to first order in δ_T, from which Δm could in principle be measured.

Some information can also be extracted by considering the asymmetry for the decays to the *CP*-conjugate final states $f, \bar{f}$ irrespective of whether they originated from an initial M^0 or an initial $\overline{M}^0$: this is the case of an experiment with no flavour tagging of the initial state (which can benefit from a higher statistics). Summing eqns (7.135) and (7.136) the untagged decay rate is obtained

$$\Gamma_f(t) \propto |A_f|^2 e^{-\overline{\Gamma}t} \frac{1 + |\lambda_f|^2}{1 - \delta_T} \left[\cosh(\Delta\Gamma\, t/2) + A_f^{(\Delta)} \sinh(\Delta\Gamma\, t/2) \right.$$
$$\left. - \delta_T \left(A_f^{(MD)} \cos(\Delta m\, t) + A_f^{(I)} \sin(\Delta m\, t) \right) \right]$$
(7.169)

from which the untagged asymmetry can be formed:

$$A_f^{(\mathrm{untag})}(t) = \frac{\Gamma_{\overline{f}}(t) - \Gamma_f(t)}{\Gamma_{\overline{f}}(t) + \Gamma_f(t)}$$
(7.170)

For flavour-specific decays ($\lambda_f = 0 = 1/\lambda_{\overline{f}}$) this becomes

$$A_f^{(\mathrm{untag})}(t) = \frac{-[\delta_T + \delta_A(f)] \cosh(\Delta\Gamma\, t/2) + \delta_T[1 + \delta_T\delta_A(f)] \cos(\Delta m\, t)}{[1 + \delta_T\delta_A(f)] \cosh(\Delta\Gamma\, t/2) - \delta_T[\delta_T + \delta_A(f)] \cos(\Delta m\, t)}$$
(7.171)

and its time-integrated version is

$$A_f^{(\mathrm{untag})} = \frac{-\delta_A(f) \left[1 + x^2 - \delta_T^2(1 - y^2) \right] - \delta_T(x^2 + y^2)}{1 + x^2 - \delta_T^2(1 - y^2) + \delta_A(f)\delta_T(x^2 + y^2)}$$
(7.172)

where $\delta_A(f)$ was defined in eqn (7.163). Even in the case of no CPV in the decay, $\delta_A(f) = 0$, this integrated asymmetry can be suppressed by the factors containing the mixing parameters x, y, resulting in a reduced sensitivity to the parameter for CPV in the mixing δ_T. Note that (for flavour-specific decays), if CPV in the mixing can be neglected ($\delta_T = 0$) the oscillatory term of the untagged rate (7.169) provides a way to measure $\Delta\Gamma$.

Finally note that many final states exist with a particle content which is self-CP-conjugate, but which are not CP eigenstates because they contain states with different orbital angular momentum eigenvalues: in this case an angular analysis of the decay products allows to extract the contributions due to amplitudes with definite CP-parity.

7.5   Meson pairs

The interactions of the gluon and the photon are flavour-conserving, so quarks with non-zero flavour quantum number are always produced in pairs in the laboratory, starting from non-flavoured beams and targets. Sometimes this results in the production of a pair of CP-conjugate mesons $M^0\overline{M}^0$, and the decays of such pairs do exhibit a rich phenomenology. The two mesons can be produced in an uncorrelated way, or their wave functions can be linked due to some conservation law in the production process, forming an entangled state in which the probabilities of the two mesons to be found in any given state are correlated.

7.5.1 Meson pair states

For any pair of spinless mesons with the same intrinsic parity (considered in its centre of momentum frame) the parity eigenvalue is $\eta_P = (-1)^L$, where $L = J$ is the (orbital) angular momentum; the possible states are thus $J^P = 0^+, 1^-, 2^+, \ldots$; for identical mesons (e.g. $K^0 K^0$) Bose–Einstein symmetry only allows even values of L, so that $\eta_P = +1$. A neutral pair of flavoured mesons $M\overline{M}$ with no net flavour ($M^0\overline{M}^0$ or a pair of charged mesons $M^+ M^-$) has $\eta_{CP} = +1$ due to the generalized spin-statistics principle (Section 3.2.1), and is an eigenstate of charge conjugation with C-parity $\eta_C = (-1)^L$. In case of production by strong interactions the G-parity eigenvalue $\eta_G = (-1)^{L+I}$ (Section 3.2.4) is also relevant, with the isospin I being integer (0, 1 for K, D, B_d mesons, 0 for B_s).

$F = 0$ pairs of CP (and C) eigenstates (7.8) also have $\eta_{CP} = +1$: the C-parities are $\eta_C = +1$ for $M_+ M_+$ or $M_- M_-$, and $\eta_C = -1$ for $M_+ M_-$, so that the former are allowed for $J^P = 0^+, 2^+, \ldots$, and the latter for $J^P = 1^-, 3^-, \ldots$; moreover, the former states are necessarily symmetric under the exchange of the two particles, while the latter is antisymmetric, as can be explicitly verified by expressing it in terms of flavour eigenstates.

To study their time evolution, these states must be expressed in terms of physical eigenstates (7.25), as neither F nor C (or P) are absolutely conserved; the resulting freely evolving states are indeed not eigenstates of these quantities, nor of CP if CP violation in the mixing is present ($|q| \neq |p|$), but still they possess the same symmetry under particle exchange as the original states, because this property is conserved by the linear nature of time evolution. The effective Hamiltonian for the meson pair is assumed to be separable and symmetric: $\mathcal{H}(M\overline{M}) = \mathcal{H}(M) + \mathcal{H}(\overline{M})$, which amounts to neglecting the interactions among the two mesons, a good approximation for neutral particles separated by macroscopic distances. For this reason the states $M_+ M_-$ can only contain $M_a M_b$ pairs with *different* particles, at any time, while $M_+ M_+$ and $M_- M_-$ can also contain $M_a M_a$ or $M_a M_b$.

If a meson-antimeson pair is produced in a state with well-defined quantum numbers, constraints on the decays of the two particles can be obtained (Goldhaber et al., 1958a).Consider the exclusive production of an $M\overline{M}$ pair in a fermion-antifermion collision ($p\bar{p}$ or $e^+ e^-$): for a defined orbital angular momentum L_i of the initial fermion-antifermion pair the parity is $\eta_P = (-1)^{L_i+1}$, and since this is conserved in the (strong or EM) production process $L = L_i \pm 1$, and only the triplet state (spin $S_i = 1$) contributes, with $\eta_C = -1$; in this case only the antisymmetric $M\overline{M}$ state is formed from an initial state of even L_i ($S, D, \ldots$ waves), and only the symmetric state for odd L_i ($P, F, \ldots$ waves). The experimental observation that in $p\bar{p}$ annihilations at rest only $K_S K_L$ pairs are produced, but not $K_S K_S$ or $K_L K_L$, allowed the conclusion that the initial state is in an even L_i wave, thus providing evidence that $\bar{p}$ annihilations at rest occur in the S state (D'Espagnat, 1961).

The coherent symmetric $|C+\rangle$ and anti-symmetric $|C-\rangle$ states with $F = 0$, for which $C|C\pm\rangle = \pm|C\pm\rangle$, are written (in their centre of mass, with the convention that the first meson has momentum $\boldsymbol{p}$ and the second $-\boldsymbol{p}$) as

$$
\begin{aligned}
|C+\rangle &= \frac{1}{\sqrt{2}}[|M^0\overline{M}^0\rangle + |\overline{M}^0 M^0\rangle] = \frac{1}{\sqrt{2}}[|M_+ M_+\rangle - |M_- M_-\rangle] \\
&= \frac{1}{(p_a q_b + p_b q_a)\sqrt{2}} \left\{ \frac{2}{p_a/p_b + q_a/q_b} |M_a M_a\rangle \right. \\
&\quad \left. - \frac{2}{p_b/p_a + q_b/q_a} |M_b M_b\rangle - \delta_{CPT}[|M_a M_b\rangle + |M_b M_a\rangle] \right\}
\end{aligned} \tag{7.173}
$$

$$
\begin{aligned}
|C-\rangle &= \frac{1}{\sqrt{2}}\left[|M^0\overline{M}^0\rangle - |\overline{M}^0 M^0\rangle\right] = \frac{1}{\sqrt{2}}[|M_- M_+\rangle - |M_+ M_-\rangle] \\
&= \frac{1}{(p_a q_b + p_b q_a)\sqrt{2}}[|M_b M_a\rangle - |M_a M_b\rangle]
\end{aligned} \tag{7.174}
$$

in terms of flavour, CP eigenstates or physical states respectively. Note that while the composition of $|C\pm\rangle$ in terms of physical states $|M_{a,b}\rangle$ is constant in time, their strangeness content is not: the property of containing an equal mixture of K^0 and $\overline{K}^0$ is only true at the same instant of time.

As the above equations show, while the antisymmetric state always contains different physical particles, these can only appear in the symmetric state if CPT is violated.[103] If CPT symmetry holds, the above expressions in terms of physical states reduce (with appropriate choices of phases) to

$$
|C+\rangle = \frac{1}{2pq\sqrt{2}}[|M_a M_a\rangle - |M_b M_b\rangle] \qquad |C-\rangle = \frac{1}{2pq\sqrt{2}}[|M_b M_a\rangle - |M_a M_b\rangle] \tag{7.175}
$$

---

[103] This does not allow an easy test of CPT symmetry, as M_a and M_b cannot be individually distinguished on the basis of their decays because of CPV.

7.5.2 Coherent meson pairs

At an $e + e^-$ collider meson pairs can be produced in a correlated state $|C\pm\rangle$, by exploiting constraints arising from the nature of the initial state: this happens for instance when the centre of mass energy is at the production threshold for a meson pair. At a $\bar{p}p$ collider the centre of mass energy is instead not sharply determined by the machine parameters, since the elementary collision occurs between quarks, which carry a variable fraction of the hadron momentum.

Consider the production of a pair of flavoured mesons in a generic fermion-antifermion collision: if the initial state has orbital angular momentum $L_i = 0$ its parity is $\eta_P = (-1)^{L_i+1} = -1$ and a pair of spinless mesons formed by parity-conserving reactions must be in the antisymmetric state (7.174). In the exclusive production of a $M\overline{M}$ pair just above threshold the process proceeds through a virtual photon, and the pair has therefore quantum numbers $J^P = 1^-$, corresponding to the state $|C-\rangle$; if a $M^*\overline{M}$ state (or its charge conjugate) is produced, where M^* is a flavoured neutral vector meson, this decays rapidly into $M\overline{M}\gamma$, with the meson pair in a state with even orbital angular momentum, corresponding to the state $|C+\rangle$.

The correlation of the two mesons in a coherently produced pair is nothing peculiar for charged mesons: one of them has positive charge and the other negative. For *neutral* mesons instead, which exhibit an involved behaviour in time due to flavour oscillations, this leads to the remarkable correlation of their entire time evolution, as the pair forms an entangled quantum mechanical state in which the flavour oscillations are coherent (Day, 1961). Consider the antisymmetric $C = -1$ state: if one of the two mesons decays at a given time t into a flavour-specific state which identifies it as being a M^0, then at the same time t the other meson is known with certainty to be a $\overline{M}^0$. As the mesons can decay into states which tag their flavour or states which tag their CP eigenvalue (approximately, due to the small effect of CP violation), the correlation in their decays – even when they are separated by a space-like interval – is one example of the class of effects first emphasized by Einstein, Podolsky and Rosen (1935) as involving either the incompleteness of quantum theory of the presence of non-local interactions (Inglis, 1961). The celebrated inequalities first written by J.S. Bell (1964) in principle allow one to distinguish between these two possibilities, and indeed all experimental tests performed so far seem to confirm the validity of quantum theory (although the issue is still debated due to the presence of loopholes in the experimental tests); tests of the validity of quantum mechanics using correlated meson pairs are also possible (see e.g. Di Domenico (2006) for a review of this fascinating topic).

Coherent states have other interesting features related to the study of CP violation. From eqn (7.174) it is seen that $|C-\rangle$ always contains *different* states (M_+M_- or M_aM_b): if CP symmetry were exact the observation of one meson decaying into a CP-even state would forbid the other one to decay into the same state; conversely, the observation of one event with both mesons decaying into CP eigenstates with the same eigenvalue is definite evidence for CPV; indeed if CP symmetry held the physical eigenstates would also be CP eigenstates and thus when one meson is identified as having a given *CP*-parity the other one should have the opposite *CP*-parity, at all times. Such kind of events are easily observed in the decays of coherent antisymmetric $K^0\overline{K}^0$ pairs produced at a ϕ factory (Chapter 8), while for heavy mesons the smallness of the branching ratios makes this observation more difficult. Analogously, the symmetric state $|C+\rangle$ always contains (assuming *CPT* symmetry) *equal* states, and the observation of two decays into *CP* eigenstates with opposite eigenvalues indicates *CPV*.

In the presence of *CP* violation there is no practical way of distinguishing individually the decay of a M_a from that of a M_b, but still the antisymmetric nature of the state $|C-\rangle$ ensures that *at the time* in which one meson decays to any given final state f, the other one is in that specific linear superposition of states which is forbidden to decay into the same final state (Lipkin, 1968), due to the cancellation of the relevant decay amplitudes. In other words, B–E symmetry forbids a pair of bosons (as the final states from a $M_{a,b}$ decay are) to contain identical states with odd relative angular momentum L, as when the state is $|C-\rangle$. The situation changes with time though, as the linear combination of states above is not a physical eigenstate and thus at a later time the second meson can also decay into f: the decay rate $\Gamma^{(C-)}(f,t_1;f,t_2)$ of an antisymmetric coherent pair in which one meson decays into f_1 at time t_1 and the other into f_2 at time t_2 vanishes for $t_1 = t_2$ if $f_1 = f_2$ (which experimentally would appear as a dip in the event rate, smeared by the experimental time resolution). Further considerations on the use of antisymmetric coherent states will be made in Chapters 8 and 10.

This same important feature of the antisymmetric state offers an extremely important tool for studies of *CP* violation, as we now discuss. Starting from eqns (7.118), (7.119) the general expression for the decay rate from a coherent meson pair state $|C\pm\rangle$ can be written as

$$\Gamma^{(C\pm)}(f_1,t_1;f_2,t_2)$$

$$\propto e^{-\overline{\Gamma}(t_1+t_2)} \left[\frac{1}{8}|A^{(\pm)} + B^{(\pm)}|^2 e^{+\overline{\Gamma}y(t_1\pm t_2)} + \frac{1}{8}|A^{(\pm)} - B^{(\pm)}|^2 e^{-\overline{\Gamma}y(t_1\pm t_2)} \right.$$

$$+ \frac{1}{4}\left(|B^{(\pm)}|^2 - |A^{(\pm)}|^2\right)\cos[\overline{\Gamma}x(t_1\pm t_2)]$$

$$\left. - \frac{1}{2}\mathrm{Im}(A^{(\pm)}B^{(\pm)*})\sin[\overline{\Gamma}x(t_1\pm t_2)] \right] \tag{7.176}$$

where

$$A^{(\pm)} = A_{f1}\overline{A}_{f2}\left(\lambda_{f1} \pm \frac{1}{\lambda_{f2}}\right) \qquad B^{(\pm)} = A_{f1}\overline{A}_{f2}\left(1 \pm \frac{\lambda_{f1}}{\lambda_{f2}}\right)$$

Note that strictly speaking the times t_1, t_2 are the decay times as measured in the rest frame of the corresponding meson: such frames are different in general, but the difference is small if the two mesons move slowly in the laboratory, such as when they are produced close to threshold. Note also that for the symmetric state $|C+\rangle$ the expression (7.176) only depends on $t_1 + t_2$.

It can happen that the information available experimentally with good accuracy is not the decay time for each meson but rather the difference $\Delta t = t_1 - t_2$ of the two decay times, and the integral of expression (7.176) over $t_1 + t_2$ in the range $[\Delta t, +\infty]$ is more relevant (Bigi and Sanda, 1981).

Consider now a $M\overline{M}$ pair produced in the anti-symmetric state $|C-\rangle$: as mentioned, if one of the two mesons decays at a given time into a flavour-specific state, this determines the flavour of the other meson *at the same time*, resulting in a method for flavour tagging a neutral meson, at a variable time (not chosen at will) using another neutral meson. From that moment on, the undecayed meson evolves as a tagged meson of given flavour, and the situation discussed above for the decays of single tagged mesons is recovered, in which the decay time t must now be replaced by the time *difference* Δt with respect to the tagging time, when the flavour eigenvalue was determined. This also works for negative Δt, that is when the second meson decays *before* the one used for tagging: in this case the state of the first meson at the time of its decay must be that linear combination which, if it had not decayed, would have evolved to become the required flavour eigenstate (opposite to that of the tagging meson) at the tagging time. Note that this flavour tagging feature is specific to the antisymmetric coherent state, and it does not occur for the symmetric state $|C+\rangle$.

7.5.3 Uncorrelated meson pairs

If no specific constraint forces a correlation, meson pairs are produced in an incoherent state $|I\rangle$: this is usually the case at a hadronic collider or a fixed target experiment, but also at a e^+e^- collider with centre of mass energy much higher than the threshold energy for pair production (such as at LEP).

In this case the probability that one of the two mesons decays at time t_1 into the state f_1, and the other decays at time t_2 into the state f_2 (both non flavour-specific states) is given by an incoherent sum

$$\Gamma^{(I)}(f_1, t_1; f_2, t_2) \propto \frac{1}{2} \Big[\Gamma[M^0(t_1) \to f_1] \Gamma[\overline{M}^0(t_2) \to f_2]$$
$$+ \Gamma[\overline{M}^0(t_1) \to f_1] \Gamma[M^0(t_2) \to f_2] \Big]$$
$$= \frac{1}{2} \Big[\Gamma^{(C+)}(f_1, t_1; f_2, t_2) + \Gamma^{(C-)}(f_1, t_1; f_2, t_2) \Big]$$

If one meson is used as a time-independent flavour tag, with a decay into a flavour-specific state f or $\overline{f}$ at some time, the decay rate for the other meson to decay into

f_1 at time t_1 is

$$\Gamma^{(I)}(f_1, t_1; f) \propto |A_f|^2 \left\{ P[M^0 \to M^0] \Gamma[\overline{M}^0(t_1) \to f_1] \right.$$
$$\left. + P[\overline{M}^0 \to M^0] \Gamma[M^0(t_1) \to f_1] \right\}$$
$$\Gamma^{(I)}(f_1, t_1; \bar{f}) \propto |A_{\bar{f}}|^2 \left\{ P[\overline{M}^0 \to \overline{M}^0] \Gamma[M^0(t_1) \to f_1] \right.$$
$$\left. + P[M^0 \to \overline{M}^0] \Gamma[\overline{M}^0(t_1) \to f_1] \right\}$$

In the above expressions the first term represents the case in which the meson providing the tag did not oscillate and initially had the flavour corresponding to the observed decay ('right tag'); the second term represents the case in which the tagging meson did oscillate before decaying, so that initially it had opposite flavour ('wrong tag').

7.5.4 Meson pair asymmetries

As mentioned, a coherent state $|C-\rangle$ allows a time-dependent flavour tagging to be obtained from the flavour-dependent decay of the other meson in the pair, and asymmetries can be formed exploiting this information. Indeed, the antisymmetry of the state, which is preserved in time, guarantees in this case that the formulæ for time evolution and asymmetries depend on Δt, the difference of decay to tagging time, in the same way as those of Sections 7.3 and 7.4 depend on t, and one can just deal with the system as if it was that of a single meson tagged at $t = 0$ and decaying at $t = \Delta t$; this argument is independent of the validity of either CP or CPT symmetries.

In the case of one meson decaying into a flavour-specific state f and the other decaying into a CP eigenstate f_{CP}, the asymmetry for $|C-\rangle$ is indeed the same as for the single tagged meson case, eqn (7.166), with Δt replacing t.

This similarity does not continue to hold for decay rates integrated (over Δt, in this case), however: notably, when integrating over Δt in the range $[-\infty, +\infty]$ the term containing $A_f^{(I)}$, sensitive to CPV in the interference of mixing and decay, vanishes (the situation was different for single tagged mesons in which the time variable was positive definite: $t \in [0, +\infty]$). This fact is one of the main reasons why asymmetric B factories were built (Chapter 10); the time-integrated asymmetries in a symmetric B factory are only sensitive to CPV in mixing or in decay (Deshpande and He, 1996).

For incoherent meson pairs the asymmetry obtained by integrating over the flavour-tagging time is also given by (7.166), but a dilution factor appears due to the fact that the flavour of the tagging meson at the initial time (the only time at which it is known to be opposite to that of the other meson, in the uncorrelated case) is only partially determined from the one measured at the tagging time, due to flavour oscillations. Explicitly

$$\Delta_{f_{CP}}^{(I)}(t) = \frac{N(f;f_{CP},t) - N(\bar{f};f_{CP},t)}{N(f;f_{CP},t) + N(\bar{f};f_{CP},t)} = \frac{1 - y^2}{1 + x^2}\Delta_{f_{CP}}(t) \qquad (7.177)$$

and the same dilution factor appears in the time-integrated asymmetry.

As mentioned in Section 7.3.1, when the decays of both mesons into flavour-specific final states $f, \bar{f}$ are considered, CP violation can be probed by the time-integrated rate difference for events in which either meson has oscillated, so that both of them decay into the same final state. The fraction of 'mixed' events is

$$R = \frac{N(ff) + N(\bar{f}\bar{f})}{N(ff) + N(\bar{f}\bar{f}) + N(f\bar{f})} \qquad (7.178)$$

and its expression can be obtained from

$$\Gamma^{(C\pm)}(f;f) \propto |A_f|^4 \left|\frac{p}{q}\right|^2 \left[\mathcal{F}_y^{(\pm)} - \mathcal{F}_x^{(\pm)}\right] \qquad (7.179)$$

$$\Gamma^{(C\pm)}(\bar{f};\bar{f}) \propto |A_{\bar{f}}|^4 \left|\frac{q}{p}\right|^2 \left[\mathcal{F}_y^{(\pm)} - \mathcal{F}_x^{(\pm)}\right] \qquad (7.180)$$

$$\Gamma^{(C\pm)}(f;\bar{f}) \propto |A_f|^2|A_{\bar{f}}|^2 \left|\frac{q}{p}\right|^2 \left[\mathcal{F}_y^{(\pm)} + \mathcal{F}_x^{(\pm)}\right] \qquad (7.181)$$

$$\mathcal{F}_y^{(\pm)} = \frac{1 \pm y^2}{(1 - y^2)^2} \qquad \mathcal{F}_x^{(\pm)} = \frac{1 \mp x^2}{(1 + x^2)^2} \qquad (7.182)$$

In the case $|A_f| = |A_{\bar{f}}|$ the ratio R has a simple expression in terms of the mixing parameters (7.156) both for an anti-symmetric coherent state and for an incoherent state:

$$R^{(C-)} = \frac{r + \bar{r}}{2 + r + \bar{r}} = \chi + \bar{\chi} - 2\chi\bar{\chi} \qquad (7.183)$$

$$R^{(I)} = \frac{r + \bar{r}}{1 + r + \bar{r} + r\bar{r}} = \frac{\chi + \bar{\chi} - 2\chi\bar{\chi}}{2 - \chi - \bar{\chi}} \qquad (7.184)$$

The CP-violating asymmetry for two flavour-specific decays is, for any coherent or incoherent state:

$$\Delta_{FS} = \frac{N(ff) - N(\bar{f}\bar{f})}{N(ff) + N(\bar{f}\bar{f})} = \frac{1 - |q/p|^4|A_{\bar{f}}/A_f|^4}{1 + |q/p|^4|A_{\bar{f}}/A_f|^4} \qquad (7.185)$$

and if $|A_f| = |A_{\bar{f}}|$ reduces to (7.157), probing CPV in the mixing.

In the CP-conserving case ($r = \bar{r}$) the mixing ratio (7.178) reduces to

$$R^{(C-)} = \frac{r}{1+r} = \chi \qquad R^{(I)} = \frac{2r}{(1+r)^2} = 2\chi(1-\chi) \qquad (7.186)$$

and in the limit $\Delta\Gamma = 0$ one has $\chi = x^2/2(1+x^2)$.

Finally, if the decay of one meson is ignored, by integrating over all its decay modes and times, the asymmetry for the decay of the other one into either flavour-specific final state is just the untagged asymmetry (7.170), for any coherent or incoherent meson pair state.

Further reading

The phenomenology of flavoured neutral meson decays is discussed in many review articles and books, and is described in detail in Branco, Lavoura, and Silva (1999) and Bigi and Sanda (2000); the different types of CP violation in neutral meson systems are also discussed in Nir (1999).

Note that different parametrizations for flavoured neutral mesons can be found in the literature, which put the emphasis on rephasing-invariant quantities and minimize the number of parameters required: e.g. Palmer and Wu (1995), Kojima *et al.* (1997) or Kostelecký (2001); while some of these might be more appealing, they are not going to replace the standard formalism described here (and the relations between them are readily obtained).

Problems

1. Explain why the rich phenomenology of CP violation in flavoured meson decays is not available for the π^0.

2. With an arbitrary but fixed choice of the phase parameters, write the most general conditions imposed by CP symmetry on the ratio of mixing parameters q/p, and on the amplitudes A_f, $A_{\bar{f}}$, $\bar{A}_f$, $\bar{A}_{\bar{f}}$, on top of eqns (7.88) and (7.89) respectively, and derive the general condition (7.98) from these.

3. Verify that eqn (7.115) is invariant under a change of relative phase between $|M^0\rangle$ and $|\overline{M}^0\rangle$.

4. Derive the approximate relations (7.78), (7.79) in the limit of small ζ.

5. Consider a pair of (spinless neutral flavoured meson) CP eigenstates produced in a generic reaction, possibly together with other flavoured particles. What can be said concerning the quantum numbers of a M_+M_+ pair? And of a M_+M_- pair?

6. Write down the explicit time evolution of the $|C\pm\rangle$ coherent meson pair states (7.173) (7.174) and determine their strangeness content as a function of time.

8

CP VIOLATION IN THE *K* SYSTEM

There is no higher or lower knowledge,
but one only, flowing out of experimentation.
L. da Vinci

It was a <u>wonderful</u> *mess at that time.*
<u>Wonderful</u>*! Just great!*
It was so <u>confusing</u> *– physics at its best,*
when everything is confused and you know
something important lies just around the corner.
A. Pais

A baby I was when you took my hand
And the light of the night burned bright.
And the people all stared didn't understand
But you knew my name on sight.
B. H. May

In this chapter the system of *K* mesons is discussed in some detail. The phenomenology of *CP* violation in kaons is qualitatively rich, with most of its manifestations being present and experimentally accessible in several ways, but such richness is somewhat hidden by the limited amount of available final states and a strong suppression of some effects. For a long time *CP* violation was only observed in the neutral *K* system, and a formalism was developed which is mostly suited to it; this is introduced and connected to the more general one described in Chapter 7, highlighting the approximations specific to the case at hand, and describing some of the relevant experiments. Some decay modes are discussed in detail as they illustrate general approaches which can be applied to the decays of other mesons.

8.1 Neutral kaons: generalities

The two neutral *K* mesons K^0 ($\bar{s}d$ in terms of quarks) and its antiparticle $\overline{K}^0$ ($s\bar{d}$) are only distinguished by the strangeness quantum number S: $S(K^0) = +1$, $S(\overline{K}^0) = -1$; their mass of about $497 \, \text{MeV}/c^2$ is comparable to the typical hadronic scale of a few hundred MeV/c^2 (e.g. the pion mass), and this has several important (both beneficial and detrimental) consequences.

As the lightest neutral flavoured mesons, kaons are indeed the 'minimal flavour laboratory', exhibiting all the subtleties of a coupled system as described in Chapter 7 with just the minimum amount of complexity (e.g. number of available decay modes) required to support the physical phenomena discussed there. Unfortunately such simplicity does not carry over into a good control of the theoretical computations of their properties from first principles; this sets some limitation to the possibility of extracting precise quantitative information from measurements in the K system, with some remarkable exceptions.

The general formalism developed in Chapter 7 applies to the neutral K system, but some peculiarities of this system led historically to the development of a different notation, which is now standard; it mostly originated from the influential papers by Lee, Yang, Wolfenstein, and Wu (Wu and Yang, 1964; Lee and Wolfenstein, 1965; Lee and Wu, 1966b), written just after the discovery of CP violation; actually, most of the conclusions of the analysis by Wu and Yang in their August 1964 paper (one month after the CPV paper was published) stood for the following forty years.

As discussed in Chapter 6, the two physical neutral K states (eigenstates of the effective Hamiltonian $\mathcal{H}$) are characterized by a large lifetime difference, due to the relative smallness of the K mass with respect to the mass of the hadrons in which it can decay (pions); the physical states $|K_S\rangle$, $|K_L\rangle$ are thus labelled by their lifetime and related to (7.25) by $|K_S\rangle = |M_a\rangle$, $|K_L\rangle = |M_b\rangle$. Labelling the masses and decay widths of the physical states by subscripts, (Yao *et al.*, 2006)

$$m_L - m_S = (3.482 \pm 0.011) \cdot 10^{-6}\, \text{eV}/c^2 \tag{8.1}$$

$$\Gamma_S - \Gamma_L = (7.335 \pm 0.004) \cdot 10^{-6}\, \text{eV} \tag{8.2}$$

According to the definitions of Chapter 7 we set[104]

$$\Delta m \equiv m_L - m_S > 0 \qquad \Delta\Gamma \equiv \Gamma_L - \Gamma_S < 0 \tag{8.3}$$

so that the following empirical relation holds

$$\Delta m \simeq -\Delta\Gamma/2 \simeq \Gamma_S/2 \tag{8.4}$$

which allows flavour oscillations to be easily measurable, as mentioned.

It is useful to define the ratio

$$\tan(\phi_{SW}) \equiv \frac{-\Delta m}{\Delta\Gamma/2} \simeq \frac{\Delta m}{\Gamma_S/2} \qquad \phi_{SW} = (43.513 \pm 0.002)^\circ \tag{8.5}$$

---

[104] Note that for the K system it is usual to define mass and width differences as positive quantities $\Delta m_K \equiv m_L - m_S = |\Delta m| > 0$ and $\Delta\Gamma_K \equiv \Gamma_S - \Gamma_L = -\Delta\Gamma > 0$; to avoid confusion we stick to the definitions of Chapter 7.

The quantity ϕ_{SW} is called the *super-weak phase* for reasons which will become clear in the following: since $\tan(\phi_{SW}) \simeq 1$ its value is close to $45°$.

Note also that the Lee–Wolfenstein inequality (7.87), which follows from unitarity, gives the limit $|\langle K_S | K_L \rangle| \leq 0.06$, obtained from the values of the mass difference and decay widths, showing that (independently of measurements of *CP*-violating quantities) the physical states are almost orthogonal, and *CP* violation is known a priori to be a small effect.

The approximate validity of *CP* symmetry means that the physical states are not too far from being the *CP* eigenstates (7.8), which for the *K* system were named $|K_1\rangle = |M_+\rangle$ and $|K_2\rangle = |M_-\rangle$, eqn (6.2) with the phase convention $\eta_{CP} = 1$. Introducing two complex parameters $\epsilon_{S,L}$ (intended to be small, $|\epsilon_{S,L}|^2 \ll 1$), one writes therefore

$$|K_S\rangle = \frac{|K_1\rangle + \epsilon_S |K_2\rangle}{\sqrt{1 + |\epsilon_S|^2}} = \frac{1}{\sqrt{2(1 + |\epsilon_S|^2)}} \left[(1 + \epsilon_S)|K^0\rangle + (1 - \epsilon_S)|\overline{K}^0\rangle \right] \quad (8.6)$$

$$|K_L\rangle = e^{i\varphi} \frac{|K_2\rangle + \epsilon_L |K_1\rangle}{\sqrt{1 + |\epsilon_L|^2}} = \frac{e^{i\varphi}}{\sqrt{2(1 + |\epsilon_L|^2)}} \left[(1 + \epsilon_L)|K^0\rangle - (1 - \epsilon_L)|\overline{K}^0\rangle \right]$$
$$(8.7)$$

and (in view of the fact that *CPT* symmetry appears to be valid) the arbitrary choice is made

$$\langle K_S | K_L \rangle = e^{i\varphi} (\epsilon_L + \epsilon_S^*) \geq 0 \quad (8.8)$$

(a real quantity); in Section 8.2 it will be shown that $e^{i\varphi} = 1$.

Out of the four real parameters which are in principle required to describe K_S in terms of $K^0, \overline{K}^0$, one is not physically significant, as the relative phase of K^0 and $\overline{K}^0$ can be changed at will, and another is fixed by the normalization condition (8.8), so that a single complex parameter ϵ_S appears above; the same applies to K_L, reducing the total number of real parameters to four for both states. Moreover, the relative phase of K_S and K_L is also arbitrary, indicating that the number of independent real parameters is actually three. The *CP*-violating parameters $\epsilon_{S,L}$ are conveniently expressed in terms of two other quantities[105]

$$\overline{\epsilon} \equiv (\epsilon_S + \epsilon_L)/2 \qquad \delta \equiv (\epsilon_S - \epsilon_L)/2 \qquad (8.9)$$
$$\epsilon_S = \overline{\epsilon} + \delta \qquad \epsilon_L = \overline{\epsilon} - \delta \qquad (8.10)$$

---

[105] Again, different notations are found in the literature for these parameters; the one adopted here is rather straightforward, with 'epsilon bar' denoting the average of ϵ_S and ϵ_L.

The usefulness of the above definitions is understood when considering the conditions imposed by the CP, CPT, and T symmetries (Section 7.1.4):

- **CP symmetry:** $\epsilon_S = 0 = \epsilon_L, \bar{\epsilon} = 0 = \delta$.
  The physical states (8.6), (8.7) do not coincide with the CP eigenstates (6.2) if $\epsilon_{S,L} \neq 0$, indicating CP violation; this reflects into their non-orthogonality, expressed by the rephasing-invariant parameter ζ of eqn (7.28)

$$\zeta = \langle K_S | K_L \rangle = \frac{2\mathrm{Re}(\bar{\epsilon}) - 2i\mathrm{Im}(\delta)}{\sqrt{(1 + |\epsilon_S|^2)(1 + |\epsilon_L|^2)}} e^{i\varphi} \tag{8.11}$$

- **CPT symmetry:** $\epsilon_S = \epsilon_L = \bar{\epsilon}, \delta = 0$.
  The CP asymmetry in the strangeness content of the two physical states is the same for both physical states if $\epsilon_S = \epsilon_L$; the physical states are (setting $e^{i\varphi} = 1$):

$$|K_S\rangle = \frac{1}{\sqrt{2(1 + |\bar{\epsilon}|^2)}} \left[(1 + \bar{\epsilon})|K^0\rangle + (1 - \bar{\epsilon})|\overline{K}^0\rangle \right] \tag{8.12}$$

$$|K_L\rangle = \frac{1}{\sqrt{2(1 + |\bar{\epsilon}|^2)}} \left[(1 + \bar{\epsilon})|K^0\rangle - (1 - \bar{\epsilon})|\overline{K}^0\rangle \right] \tag{8.13}$$

- **T symmetry:** $\epsilon_S = \delta = -\epsilon_L, \bar{\epsilon} = 0$.
  If $\delta \neq 0$ CPT violation is present; the CP asymmetry in the composition of the two physical states is opposite.

The $\bar{\epsilon}, \delta$ parameters distinguish between CP violation also violating T symmetry and respecting CPT ($\bar{\epsilon} \neq 0$), and that also violating CPT while respecting T ($\delta \neq 0$). The relations with the parameters introduced in eqns (7.56), (7.43) are

$$\left(\frac{q}{p}\right)_{S,L} = \frac{1 - \epsilon_{S,L}}{1 + \epsilon_{S,L}} \qquad \epsilon_{S,L} = \frac{p_{S,L} - q_{S,L}}{p_{S,L} + q_{S,L}} \tag{8.14}$$

from which (eqns (7.56), (7.43))

$$\delta_T = \frac{2\mathrm{Re}(\bar{\epsilon})}{1 + |\bar{\epsilon}|^2} \qquad \delta_{CPT} = \frac{2\delta}{1 + \delta^2 - \bar{\epsilon}^2} \tag{8.15}$$

Note that $\bar{\epsilon}$ depends on the choice of phase between K^0 and $\overline{K}^0$, and only its real part (or more precisely δ_T) is a physical quantity; this is consistent with the fact that CPV in $\mathcal{H}$ is described (in case of CPT symmetry) by a single real parameter, namely the phase mismatch between M_{12} and Γ_{12} (7.75), rather than two ($\mathrm{Im}(M_{12})$ and $\mathrm{Im}(\Gamma_{12})$). The non rephasing invariant nature of $\bar{\epsilon}$ can be a nuisance, as this parameter can be chosen to assume different values in the limit of CP symmetry;

we note however that since *CPV* is known to be small, it is always *possible* to choose phase conventions in which all the quantities which parametrize *CPV* are numerically small.

$\bar{\epsilon}$ depends from the (arbitrary) choice of phase between K^0 and $\overline{K}^0$; in the case of *CPT* symmetry, the constraint

$$|\bar{\epsilon} - 1/\zeta|^2 = \frac{4|q/p|^2}{(1 - |q/p|^2)^2} = \text{constant}$$

shows that by varying the (non-physical) phase of q/p the quantity $|\bar{\epsilon}|$ takes extremal values for real q/p:

$$\bar{\epsilon}_{\min} = \frac{1 - |q/p|}{1 + |q/p|} \quad \text{(if } q/p = |q/p|) \qquad \bar{\epsilon}_{\max} = \frac{1 + |q/p|}{1 - |q/p|} \quad \text{(if } q/p = -|q/p|)$$

with $\bar{\epsilon}_{\min} + \bar{\epsilon}_{\max} = 2/\zeta$. The *proper phase convention* Wu (1989) is that in which $\bar{\epsilon} = \bar{\epsilon}_{\min}$, when

$$\text{Im}(\bar{\epsilon}) = 0 \quad |\bar{\epsilon}| < 1 \quad q/p = |q/p|$$

If $|\zeta| \ll 1$ (as it happens for small *CP* violation), a class of phase conventions in which $|\bar{\epsilon}| \ll 1$ is useful (*semi-proper phase conventions*), in which $\text{Im}(M_{12}) \simeq 0$ and $\text{Im}(\Gamma_{12}) \simeq 0$; the Wu–Yang phase convention (Section 8.3.4) is an example of this type. In this case $\zeta \simeq 2\text{Re}(\bar{\epsilon})$.

Introducing the eigenvalues $\lambda_{S,L}$ of $\mathcal{H}$, and expanding to first order in $\bar{\epsilon}, \delta$:

$$\mathcal{H}_{11} \simeq \frac{\lambda_S + \lambda_L}{2} + \delta(\lambda_S - \lambda_L) \qquad \mathcal{H}_{22} \simeq \frac{\lambda_S + \lambda_L}{2} - \delta(\lambda_S - \lambda_L) \qquad (8.16)$$

$$\mathcal{H}_{12} \simeq \frac{\lambda_S - \lambda_L}{2}(1 + 2\bar{\epsilon}) \qquad \mathcal{H}_{21} \simeq \frac{\lambda_S - \lambda_L}{2}(1 - 2\bar{\epsilon}) \qquad (8.17)$$

$$\lambda_{S,L} \simeq \frac{\mathcal{H}_{11} + \mathcal{H}_{22}}{2} \pm \frac{\mathcal{H}_{12} + \mathcal{H}_{21}}{2} \qquad (8.18)$$

$$\Delta\lambda \simeq -2\sqrt{\mathcal{H}_{12}\mathcal{H}_{21}} \simeq -(\mathcal{H}_{12} + \mathcal{H}_{21}) \qquad (8.19)$$

In this approximation

$$\bar{\epsilon} \simeq \frac{\mathcal{H}_{21} - \mathcal{H}_{12}}{2(\lambda_L - \lambda_S)} = \frac{\text{Im}(M_{12}) - i\text{Im}(\Gamma_{12})/2}{i\Delta m + \Delta\Gamma/2} \qquad (8.20)$$

$$\delta \simeq \frac{\mathcal{H}_{22} - \mathcal{H}_{11}}{2(\lambda_L - \lambda_S)} = \frac{\delta_{CPT}}{2} = \frac{(M_{11} - M_{22}) - i(\Gamma_{11} - \Gamma_{22})/2}{-2\Delta m + i\Delta\Gamma} \qquad (8.21)$$

The eigenvalues themselves are only affected by *CPV* at second order in $\bar{\epsilon}$ (since $\mathcal{H}_{12} + \mathcal{H}_{21}$ appears in eqn (8.18)) so that (assuming *CPT* symmetry, $\delta = 0$):

$$\lambda_{S,L} \simeq \mathcal{H}_{11} \pm \mathcal{H}_{12} \simeq M_{11} \pm \text{Re}(M_{12}) - i[\Gamma_{11} \pm \text{Re}(\Gamma_{12})]/2 \tag{8.22}$$

$$\Delta m \simeq -2\text{Re}(M_{12}) \qquad \Delta\Gamma \simeq -2\text{Re}(\Gamma_{12}) \tag{8.23}$$

$$\text{Im}(M_{12}) \simeq -\Delta m \, \text{Im}(\bar{\epsilon}) + (\Delta\Gamma/2)\text{Re}(\bar{\epsilon}) \tag{8.24}$$

$$\text{Im}(\Gamma_{12}) \simeq -\Delta\Gamma \, \text{Im}(\bar{\epsilon}) - 2\Delta m \, \text{Re}(\bar{\epsilon}) \tag{8.25}$$

$$M_{22} - M_{11} \simeq 2\Delta m \, \text{Re}(\delta) + \Delta\Gamma \, \text{Im}(\delta) \tag{8.26}$$

$$\Gamma_{22} - \Gamma_{11} \simeq -4\Delta m \, \text{Im}(\delta) + 2\Delta\Gamma \, \text{Re}(\delta) \tag{8.27}$$

and from (8.23) and (8.4)

$$\text{Re}(\Gamma_{12}) \simeq -2\text{Re}(M_{12}) \tag{8.28}$$

The mass difference Δm is driven by virtual $K^0 - \overline{K}^0$ transitions

$$\Delta m \equiv m_L - m_S \simeq m(K_2) - m(K_1) = \text{Re}\langle K_2|\mathcal{H}|K_2\rangle - \text{Re}\langle K_1|\mathcal{H}|K_1\rangle$$
$$= -\text{Re}\left(\langle K^0|\mathcal{H}|\overline{K}^0\rangle + \langle\overline{K}^0|\mathcal{H}|K^0\rangle\right) = -\text{Re}(\mathcal{H}_{12} + \mathcal{H}_{21}) \tag{8.29}$$

and its small value (with respect to the K mass) indicates that off-diagonal elements of $\mathcal{H}$ are indeed small with respect to diagonal ones. The hierarchy of the matrix elements of $\mathcal{H}$ is

$$M_{11} \simeq (m_L + m_S)/2 \sim 500 \, \text{MeV} \tag{8.30}$$

$$-\text{Re}(M_{12}) \simeq (m_L - m_S)/2 \sim 2 \cdot 10^{-6} \, \text{eV} \tag{8.31}$$

$$\Gamma_{11} + \text{Re}(\Gamma_{12}) \simeq \Gamma_S \sim 7 \cdot 10^{-6} \, \text{eV} \tag{8.32}$$

$$\Gamma_{11} - \text{Re}(\Gamma_{12}) \simeq \Gamma_L \sim 1 \cdot 10^{-8} \, \text{eV} \tag{8.33}$$

easily understood since M_{11} includes the effect of strong interactions while all the other terms are only induced by weak interactions.

The phenomenon of K_S regeneration in matter was briefly described in Chapter 6: besides representing a striking confirmation of the particle-mixture hypothesis of Gell-Mann and Pais, its importance stems from the fact that through it many experiments with 'controlled' coherent mixtures of K_S and K_L are possible (Whatley, 1962), and indeed a great wealth of information was obtained with this approach in early experiments and in more recent ones as well. A detailed discussion of the physics of regeneration is outside the scope of this book (see Kabir (1968) for details), and we just recall some concepts useful to understand the experiments.

Due to the different strong interactions of K^0 and $\overline{K}^0$ with matter, when a pure K_L beam traverses a material some K_S appear (the converse is also true, but the regeneration of K_L from K_S is not so interesting in absence of pure K_S beams and given the relatively long K_L lifetime). The regeneration due to the difference of forward scattering amplitudes (related by the optical theorem to the difference in total cross sections) is called *coherent* (or transmission) regeneration, and arises from the contributions of all the scattering centres of the material adding up coherently in the forward direction; this does not change the state of the regenerator material, as the recoil associated to the transition from K_L to K_S (with a different mass) is taken up by the macroscopic system as a whole, the energy transfer thus being practically zero. Regeneration can also arise from the contribution of individual scattering centres, due to the difference in the finite-angle scattering probability for K^0 and $\overline{K}^0$: as a consequence of such *incoherent* regeneration (*diffractive* at small angles) the state of the regenerator changes in a way that can in principle be observed.[106]

8.2 Semi-leptonic decays

Semi-leptonic decays of neutral kaons into the final states $\pi^\mp \ell^\pm \nu(\overline{\nu})$, where $\ell = e, \mu$ (K_{e3} and $K_{\mu3}$) are flavour-specific because of the $\Delta S = \Delta Q$ rule (Section 6.1.1), which is valid to a large extent within the SM (amplitudes with $\Delta S = -\Delta Q$ being $O(G_F^2)$, suppressed by a factor $\sim 10^{-6}$ with respect to the dominant ones). The allowed decays are thus $K^0 \to \pi^- \ell^+ \nu_\ell$ and $\overline{K}^0 \to \pi^+ \ell^- \overline{\nu}_\ell$, but not those with the final states interchanged; such decays tag the strangeness of the neutral kaon, a positive lepton indicating a K^0 decay and a negative lepton a $\overline{K}^0$ decay, and are very convenient for such purpose because of their large branching ratios: $BR(K_{e3}) \simeq 39\%$, $BR(K_{\mu3}) \simeq 27\%$. Conversely, since the final states are not *CP* eigenstates, both K_S, K_L can decay into them, with their interference resulting in a non-exponential decay rate, even if *CP* symmetry holds; the presence of *CP* violation manifests itself in the difference of decay rates into the two *CP*-conjugate states not averaging to zero.

No direct *CPV* is possible in semi-leptonic decays: with a single hadron in the final state f *CPT* symmetry enforces the equality (in modulus) of the decay amplitudes $K^0 \to f$ and $\overline{K}^0 \to \overline{f}$ (see Chapter 5), and the validity of the $\Delta S = \Delta Q$ rule forbids any interference of mixing and decay ($K^0 \to \overline{K}^0 \to f$); semi-leptonic

---

[106] A small coherent regeneration amplitude also originates from scattering on electrons.

decays allow one to probe the $K^0, \overline{K}^0$ content of the physical states, and thus to obtain information on CPV in the mixing only.

The decay amplitudes are usually written as (Section 8.5)

$$A(K^0 \to \pi^- \ell^+ \nu) = a + b \qquad A(K^0 \to \pi^+ \ell^- \overline{\nu}) = c + d \qquad (8.34)$$

$$A(\overline{K}^0 \to \pi^+ \ell^- \overline{\nu}) = a^* - b^* \qquad A(\overline{K}^0 \to \pi^- \ell^+ \nu) = c^* - d^* \qquad (8.35)$$

in which a possible violation of the $\Delta S = \Delta Q$ rule is present in the c, d amplitudes, parametrized as in eqn (7.126):

$$x^{(\ell)} \equiv \frac{A(\overline{K}^0 \to \pi^- \ell^+ \nu)}{A(K^0 \to \pi^- \ell^+ \nu)} = \frac{c^* - d^*}{a + b} \qquad (8.36)$$

$$\overline{x}^{(\ell)} \equiv \frac{A^*(K^0 \to \pi^+ \ell^- \overline{\nu})}{A^*(\overline{K}^0 \to \pi^+ \ell^- \overline{\nu})} = \frac{c^* + d^*}{a - b} \qquad (8.37)$$

Note that the imaginary parts of the amplitudes violate T symmetry, and the b, d amplitudes violate CPT symmetry, eqn (7.110), as parametrized by

$$y^{(\ell)} \equiv \frac{A^*(\overline{K}^0 \to \pi^+ \ell^- \overline{\nu}) - A(K^0 \to \pi^- \ell^+ \nu)}{A^*(\overline{K}^0 \to \pi^+ \ell^- \overline{\nu}) + A(K^0 \to \pi^- \ell^+ \nu)} = -\frac{b}{a} \qquad (8.38)$$

for the $\Delta S = \Delta Q$ amplitudes. Finally

$$x_+^{(\ell)} \equiv (x^{(\ell)} + \overline{x}^{(\ell)})/2 \qquad x_-^{(\ell)} \equiv (x^{(\ell)} - \overline{x}^{(\ell)})/2 \qquad (8.39)$$

respectively parametrize $\Delta S = \Delta Q$ violation in the decay amplitudes which respect or violate CPT symmetry.

Summarizing, the constraints imposed by the various symmetries on the above amplitudes are

- **CPT symmetry**: $b = d = 0$, $y^{(\ell)} = 0$, $x^{(\ell)} = \overline{x}^{(\ell)}$.
- **T symmetry**: $\text{Im}(a, b, c, d) = 0$, $\text{Im}(y^{(\ell)}, x^{(\ell)}, \overline{x}^{(\ell)}) = 0$.
- **CP symmetry**: $\text{Im}(a, c) = 0$, $\text{Re}(b, d) = 0$, $\text{Re}(y^{(\ell)}) = 0$, $x^{(\ell)} = \overline{x}^{(\ell)}$.
- **$\Delta S = \Delta Q$ rule**: $x^{(\ell)} = 0 = \overline{x}^{(\ell)}$.

As discussed in Chapter 7, several charge asymmetries can be formed from the decay rates into the two CP-conjugated semi-leptonic states; the general expressions for the time evolution of flavour eigenstates (7.129), (7.130) can be expanded in the small parameters $\overline{\epsilon}, \delta, x^{(\ell)}, \overline{x}^{(\ell)}, y$ to obtain approximate expressions for such asymmetries.

Assuming *CPT* symmetry, the charge asymmetry for semi-leptonic decays of states (7.117) initially tagged as K^0 ($\overline{K}^0$), is (at first order in $\bar\epsilon$)

$$\widehat{\Delta}_\ell(t) = \frac{\Gamma[K^0(\overline{K}^0)(t=0) \to \pi^-\ell^+\nu](t) - \Gamma[K^0(\overline{K}^0)(t=0) \to \pi^+\ell^-\overline{\nu}](t)}{\Gamma[K^0(\overline{K}^0)(t=0) \to \pi^-\ell^+\nu](t) + \Gamma[K^0(\overline{K}^0)(t=0) \to \pi^+\ell^-\overline{\nu}](t)}$$

$$\simeq \frac{X^{(-)}\left[\mathrm{Re}(\bar\epsilon)E^{(+)} \pm e^{-\overline{\Gamma}t}\cos(\Delta m\, t)\right]}{X^{(+)}E^{(+)}/2 + AE^{(-)} \pm 2X^{(+)}\mathrm{Re}(\bar\epsilon)e^{-\overline{\Gamma}t}\cos(\Delta m\, t) \mp 2Be^{-\overline{\Gamma}t}\sin(\Delta m\, t)}$$

$$(8.40)$$

where the upper sign is for initial K^0 and the lower for initial $\overline{K}^0$, and

$$X^{(\pm)} \equiv (1 \pm |x^{(\ell)}|^2) \qquad\qquad E^{(\pm)} \equiv \left(e^{-\Gamma_S t} \pm e^{-\Gamma_L t}\right)$$

$$A \equiv \mathrm{Re}(x^{(\ell)}) + 2\,\mathrm{Im}(\bar\epsilon)\,\mathrm{Im}(x^{(\ell)}) \qquad B \equiv \mathrm{Im}(x^{(\ell)}) - 2\,\mathrm{Im}(\bar\epsilon)\,\mathrm{Re}(x^{(\ell)})$$

Experiments at hadronic machines usually deal with initial states which are not pure flavour eigenstates but rather incoherent mixtures of those:[107] in this case the interference terms in the above asymmetries get multiplied by a kaon momentum-dependent *dilution factor* $\mathcal{D}(p)$ (replacing the $\pm$) which reflects the flavour production asymmetry

$$\mathcal{D}(p) \equiv \frac{\sigma(K^0, p) - \sigma(\overline{K}^0, p)}{\sigma(K^0, p) + \sigma(\overline{K}^0, p)} \qquad\qquad (8.41)$$

and due to the lower production threshold for K^0 usually $\mathcal{D}(p) > 0$. For $\mathrm{Im}(x^{(\ell)}) \ll 1$ and for proper times such that $\Gamma_S t \gg 1 \gg \Gamma_L t$ eqn (8.40) then becomes

$$\widehat{\Delta}_\ell(t \gg \tau_S) \simeq 2\frac{1 - |x^{(\ell)}|^2}{|1 - x^{(\ell)}|^2}\left[\mathrm{Re}(\bar\epsilon) + \mathcal{D}(p)\, e^{-\Gamma_S t/2}\cos(\Delta m\, t)\right]$$

which at large times reduces to the *CP* impurity in the mixing, as expected; note that the interference vanishes for $\mathcal{D}(p) = 0$ (an equal strangeness mixture).

The semi-leptonic charge asymmetries for physical states are defined as in eqn (7.112) and customarily denoted

$$\delta_{L,S}^{(\ell)} \equiv \frac{\Gamma(K_{L,S} \to \pi^-\ell^+\nu) - \Gamma(K_{L,S} \to \pi^+\ell^-\overline{\nu})}{\Gamma(K_{L,S} \to \pi^-\ell^+\nu) + \Gamma(K_{L,S} \to \pi^+\ell^-\overline{\nu})} \qquad\qquad (8.42)$$

---

[107] The mixture is indeed incoherent because the production process (by strong interactions) conserves strangeness, and by detecting the other strange particles produced it would be possible to determine for each event whether a K^0 or a $\overline{K}^0$ was produced in each interaction.

Note that δ_L is also the limit to which the untagged asymmetry (7.170) tends for large times. With no direct CPV, these charge asymmetries are expected to be equal for both K_{e3} and $K_{\mu3}$. Without any approximation (but assuming CPT symmetry in the decay), if the $\Delta S = \Delta Q$ rule holds exactly $(x^{(\ell)}, \bar{x}^{(\ell)} = 0)$ then

$$\delta_L = \frac{2\mathrm{Re}(\epsilon_L)}{1 + |\epsilon_L|^2} \tag{8.43}$$

which is just $\langle K_S | K_L \rangle$ if CPT symmetry also holds in the mixing ($\delta = 0$); in this case the experimental observation of $\delta_L > 0$ defines the relative phase between K_L and K_S introduced in (8.7): eqn (8.8) becomes $e^{i\varphi}2\,\mathrm{Re}(\bar{\epsilon}) \geq 0$, from which $e^{i\varphi} = 1$.

More generally, allowing for CPT violation, the asymmetry at first order in the symmetry violation parameters is

$$\delta_L \simeq 2\mathrm{Re}(\epsilon_L) - 2\mathrm{Re}(y^{(\ell)}) - 2\mathrm{Re}(x_-^{(\ell)}) \tag{8.44}$$

Using the decays of flavour-tagged neutral kaons, several other asymmetries can be formed for semi-leptonic decays, from which the various symmetry violating parameters can be extracted; an extensive discussion can be found in Hayakawa and Sanda (1993). As an example, the flavour-tagged asymmetries into the same semi-leptonic final state

$$\Delta_{\ell\pm}(t) = \frac{\Gamma[\overline{K}^0(t=0) \to \pi^{\mp}\ell^{\pm}\nu](t) - \Gamma[K^0(t=0) \to \pi^{\mp}\ell^{\pm}\nu](t)}{\Gamma[\overline{K}^0(t=0) \to \pi^{\mp}\ell^{\pm}\nu](t) + \Gamma[K^0(t=0) \to \pi^{\mp}\ell^{\pm}\nu](t)} \tag{8.45}$$

tend to $2\mathrm{Re}(\epsilon_S)$ for large times $t \gg \tau_S$, and this is true in general for any final state (note that ϵ_S appears and not ϵ_L, despite the fact that at large times only K_L are present); measurements of $\Delta_{\ell\pm}$ clearly fully depend on an accurate kaon flavour tagging.

It can be noted that, besides the usual flavour oscillations which can be observed through semi-leptonic decays, there are in general also oscillations in the total number of kaons present at a given time: starting from a state of defined strangeness at $t = 0$, the probabilities of having a neutral kaon of either strangeness are, at first order in $\bar{\epsilon}, \delta$:

$$P(K^0 \to K^0) + P(K^0 \to \overline{K}^0) \simeq (1/2)\Big\{[1 - 2\mathrm{Re}(\bar{\epsilon}) + 2\mathrm{Re}(\delta)]e^{-\Gamma_S t}$$

$$+ [1 - 2\mathrm{Re}(\bar{\epsilon}) - 2\mathrm{Re}(\delta)]e^{-\Gamma_L t} + [4\mathrm{Re}(\bar{\epsilon})\cos(\Delta mt) - 4\mathrm{Im}(\delta)\sin(\Delta mt)]e^{-\bar{\Gamma}t/2}\Big\}$$

$$P(\overline{K}^0 \to K^0) + P(\overline{K}^0 \to \overline{K}^0) \simeq (1/2)\Big\{[1 + 2\mathrm{Re}(\bar{\epsilon}) - 2\mathrm{Re}(\delta)]e^{-\Gamma_S t}$$

$$+ [1 + 2\mathrm{Re}(\bar{\epsilon}) + 2\mathrm{Re}(\delta)]e^{-\Gamma_L t} - [4\mathrm{Re}(\bar{\epsilon})\cos(\Delta mt) - 4\mathrm{Im}(\delta)\sin(\Delta mt)]e^{-\bar{\Gamma}t/2}\Big\}$$

The oscillations signal *CP* violation: terms containing $\cos(\Delta m\,t)$ respect *CPT* symmetry (and thus violate *T*), while those containing $\sin(\Delta m\,t)$ violate *CPT* (and respect *T*); the differences of the non-oscillating terms also violate *CP* but do not allow any distinction between *CPT* and *T* violation. In principle these oscillations could be measured from semi-leptonic decays if the relative tagging efficiencies were accurately known, or from the strong interactions of kaons in a beam which initially had the same number of K^0 and $\overline{K}^0$, but both approaches are challenging at best, at the required level of accuracy.

Semi-leptonic charge asymmetries

Early determinations of the charge asymmetry in semi-leptonic K_L decays (Bennett *et al.*, 1967; Dorfan *et al.*, 1967) were mentioned in Chapter 6; these experiments illustrate the general approach used in all following measurements of charge asymmetries: as magnetic analysis of charged particle tracks deflects oppositely charged tracks into different regions of the detector, the unavoidable spatial asymmetries of the apparatus are cancelled in first order by considering pairs of measurements with the magnetic field directed in opposite directions. When the magnetic field direction is reversed the particles of a given charge illuminate the region of the detector previously hit by their charge-conjugate, and any instrumental asymmetry reverses, while the physical asymmetry does not: the average of the two measurements is thus insensitive to imperfections of the apparatus in first order. The above is actually true for the arithmetic average of the two measurements, so that it is important to collect a comparable amount of data in the two configurations, to avoid worsening the statistical power with respect to a weighted average.

Particle identification (of $e^{\pm}$ using Čerenkov counters, and of $\mu^{\pm}$ using detectors behind thick absorbers) is an important issue, and the results had to be corrected for the charge-asymmetric interactions of leptons and pions into the detector: these are particularly relevant at low energies, and corrections were obtained from the analysis of data samples in which the amount of material crossed by the particles was varied in a controlled way. The time dependence of the charge asymmetry for semi-leptonic decays of neutral *K* was also measured (Gjesdal *et al.*, 1974), exhibiting the expected oscillatory behaviour due to the $K_S - K_L$ interference for early decay times, and the non-zero asymptotic limit for $t \gg \tau_S$ (see Fig. 8.1), which represents the effect of *CPV* corresponding to δ_L.

The K_{e3} charge asymmetries were remeasured with very high samples, $O(10^8)$ events, by KTeV at FNAL (Alavi-Harati *et al.*, 2002) and NA48 at CERN (Lazzeroni, 2004). The KTeV experiment only used the proper time region $t \geq 10.5\,\tau_S$ in order to reduce any effect due to the $K_L - K_S$ interference, which can be relevant for high energy kaons and has to be corrected for. Events were reconstructed from the two charged tracks, the missing neutrino introducing a twofold ambiguity in the solution for the kaon momentum: one solution was chosen and a small correction, determined from simulation, was applied afterwards. The critical issue for the measurement was to make sure that no systematic bias is introduced for oppositely charged particles, either in detector acceptance or in reconstruction efficiency; data were collected with opposite polarities of the spectrometer magnet to average out such

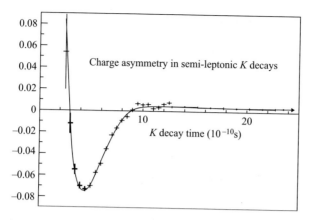

FIG. 8.1. Charge asymmetry as a function of proper decay time for K_{e3} decays, (Gjesdal *et al.*, 1974); the solid line indicates the best fit.

differences, and corrected for the small asymmetries in the magnetic field magnitudes. Intense K_L beams were used: KTeV had two alternating parallel beams, symmetrically shifted with respect to the axis of the detector, while NA48 used a single beam centred on the detector. Small corrections (up to $\sim 1.5 \cdot 10^{-4}$) were required to account for efficiency differences arising from the different interaction properties of π^+ and π^- (and at a lower level e^+ and e^-) in matter: isospin conservation in strong interactions implies that such corrections vanish for iso-scalar materials, residual effects arising from the excess of protons over neutrons. The efficiency asymmetries were measured using control data samples containing well identified $\pi^\pm$ and $e^\pm$. While backgrounds can be asymmetric (e.g. $\Lambda \to p\pi^-$ and $\overline{\Lambda} \to \overline{p}\pi^+$ have very different yields from the primary pp interaction), they were generally found to be negligible.

The results were $\delta_L^{(e)} = (3.322 \pm 0.058 \pm 0.047) \cdot 10^{-3}$ by KTeV and $\delta_L^{(e)} = (3.317 \pm 0.070 \pm 0.072) \cdot 10^{-3}$ by NA48. The size of the corrections which were applied to the raw results (those obtained by event counting) to obtain the final results gives an indication of the robustness of the method used and of the quality of the first-order cancellations of systematic biases: such correction was -95 ppm for KTeV (with the single largest contribution being (-160 ± 10) ppm due to the different energy deposition of $\pi^\pm$ in the calorimeter), and $+77$ ppm by NA48 (with the single largest contribution being $(+260 \pm 60)$ ppm, again due to the asymmetry of $\pi^\pm$ energy deposition in the formation of the trigger).

The average measured asymmetries are (Yao *et al.*, 2006)

$$\delta_L^{(e)} = (3.34 \pm 0.07) \cdot 10^{-3} \qquad \delta_L^{(\mu)} = (3.04 \pm 0.25) \cdot 10^{-3} \qquad (8.46)$$

Semi-leptonic decays of K_S were also observed for the first time by KLOE at Frascati (Aloisio *et al.*, 2002) with a branching ratio $BR(K_S \to \pi e\nu) \simeq 7 \cdot 10^{-4}$:

a first measurement of their charge asymmetry was performed (Ambrosino *et al.*, 2006)

$$\delta_S^{(e)} = (1.5 \pm 9.6 \pm 2.9) \cdot 10^{-3} \tag{8.47}$$

8.3   Two-pion decays and direct *CP* violation

Hadronic decays of neutral kaons, in which significant final state interactions can be present, allow for a richer phenomenology of *CP*-violating phenomena; decays into two or three pions are by far the most common ones.

The two-pion state obtained from the decay of a spinless particle was shown in Chapter 6 to be *CP* even; recalling the requirement imposed by *CP* symmetry (6.23) this implies $A(K^0 \to \pi\pi) = A(\overline{K}^0 \to \pi\pi)$ (phase convention $\eta_{CP} = +1$), so that the decay $K_L \to \pi\pi$ violates *CP* symmetry. In general, since *CP* violation is much more easily observed in the decays of K_L (for which dominant decays have much smaller widths) than in those of K_S, it is useful to introduce quantities describing *CPV* in the decays to final states f with $\eta_{CP}(f) = +1$ (forbidden for K_L if *CP* symmetry holds): these are the amplitude ratios

$$\eta_f \equiv \frac{A(K_L \to f)}{A(K_S \to f)} \frac{\langle K^0 | K_S \rangle}{\langle K^0 | K_L \rangle} \tag{8.48}$$

where the second factor (often omitted) has the purpose of making the above quantity invariant under rephasing of K^0 and $\overline{K}^0$. At first order in $\bar\epsilon, \delta$:

$$\eta_f \simeq \bar\epsilon - \delta + \frac{A(K_2 \to f)}{A(K_1 \to f)} = \epsilon_L + \epsilon_f \tag{8.49}$$

the last term describing *CPV* in the decay to the state f:

$$\epsilon_f \equiv \frac{A(K_2 \to f)}{A(K_1 \to f)} = \frac{A(K^0 \to f) - A(\overline{K}^0 \to f)}{A(K^0 \to f) + A(\overline{K}^0 \to f)} \tag{8.50}$$

(the second expression being valid for $\eta_{CP} = +1$).

If *CPT* symmetry holds ($\delta = 0$)

$$\Gamma(K^0 \to f) = \frac{1 + |\bar\epsilon|^2}{2|1 + \bar\epsilon|^2} |1 + \eta_f|^2 |A(K_S \to f)|^2$$

and the difference between the decay rates of $\overline{K}^0$ and K^0 into the state f is

$$\Gamma(\overline{K}^0 \to f) - \Gamma(K^0 \to f) = |A(K_S \to f)|^2 \frac{\langle K_S | K_L \rangle (1 + |\eta_f|^2) - 2\mathrm{Re}(\eta_f)}{1 - \langle K_S | K_L \rangle^2}$$

which for small $|\eta_f|$ and $\bar{\epsilon}$ becomes

$$\Gamma(\overline{K}^0 \to f) - \Gamma(K^0 \to f) \simeq [2\mathrm{Re}(\bar{\epsilon}) - 2\mathrm{Re}(\eta_f)]\,|A(K_S \to f)|^2 \qquad (8.51)$$

$$\Delta_f \equiv \frac{\Gamma(\overline{K}^0 \to f) - \Gamma(K^0 \to f)}{\Gamma(\overline{K}^0 \to f) + \Gamma(K^0 \to f)} \simeq -2\mathrm{Re}(\epsilon_f) \qquad (8.52)$$

so that the asymmetry (7.167) indeed vanishes in absence of direct *CPV*: in this case the composition of $K^0, \overline{K}^0$ in terms of *CP* eigenstates K_1, K_2 is the same (in modulus) and there can be no difference in the transition *rates*.

8.3.1 Isospin decomposition

Strong interactions induce transitions between $\pi^+\pi^-$ and $\pi^0\pi^0$, and the proper basis for the application of Fermi–Watson theorem (Chapter 5) to $\pi\pi$ states is that corresponding to a quantum number conserved by such interactions, namely isospin. $K \to 2\pi$ decay amplitudes can be written in terms of eigenstates of isospin I (with $I_3 = 0$, as the states are neutral) according to the Clebsch–Gordan decomposition. Recalling that for two pions in a state of zero angular momentum the antisymmetric $I = 1$ state is forbidden by Bose–Einstein symmetry[108] and introducing the symmetrized state

$$|(\pi^+\pi^-)_S\rangle \equiv \frac{1}{\sqrt{2}} \left[|\pi^+(\boldsymbol{p})\rangle|\pi^-(-\boldsymbol{p})\rangle + |\pi^+(-\boldsymbol{p})\rangle|\pi^-(\boldsymbol{p})\rangle \right] \qquad (8.53)$$

(where $\boldsymbol{p}, -\boldsymbol{p}$ are the pion momenta in the centre of mass) one has

$$|(\pi^+\pi^-)_S\rangle = \sqrt{\frac{2}{3}}|(\pi\pi)_{I=0}\rangle + \sqrt{\frac{1}{3}}|(\pi\pi)_{I=2}\rangle \qquad (8.54)$$

$$|\pi^0\pi^0\rangle = -\sqrt{\frac{1}{3}}|(\pi\pi)_{I=0}\rangle + \sqrt{\frac{2}{3}}|(\pi\pi)_{I=2}\rangle \qquad (8.55)$$

The partial decay rates into 2π states are

$$\Gamma = \frac{|\boldsymbol{p}|}{8\pi m_K^2} \frac{1}{2} |A_I|^2 \qquad (8.56)$$

where $A_I \equiv A(K^0 \to (\pi\pi)_I)$ denotes the amplitude for decays into an isospin eigenstate, and the factor $1/2$ is due to the final state with identical particles ($\pi^0\pi^0$ and symmetrized $\pi^+\pi^-$).[109] Isospin is not conserved by weak interactions, so that

---

[108] For $\pi^0\pi^0$ (identical pions) the antisymmetric state is identically zero and I must be even, independently of B–E symmetry.

[109] Sometimes in the literature isospin decompositions differing from the one above by a factor $\sqrt{2}$ are found, due to the $\pi^+\pi^-$ state not being symmetrized; in this case the relation between amplitudes and decay rates differs by a factor 2 between $\pi^0\pi^0$ and $\pi^+\pi^-$.

the transition matrix for the decays can be written as a sum of terms inducing an isospin change $\Delta I \neq 0$: since however $\Delta I_3 + \Delta S/2 = \Delta Q = 0$ (charge conservation), and $|\Delta S| = 1$ for decays of kaons into pions, all such terms must induce $|\Delta I_3| = 1/2$, and therefore $\Delta I = 1/2, 3/2, 5/2, \ldots$. Kaons have isospin $I = 1/2$, and it is found empirically that transitions with isospin change $\Delta I \leq 3/2$ dominate, so that $I = 0$ final states are reached by $\Delta I = 1/2$ transitions and those with $I = 2$ by $\Delta I = 3/2$ ones, found to be suppressed with respect to the former by a factor $\simeq 450$.

Decomposing the transition matrix in $\Delta I = 1/2, 3/2$ terms (neglecting possible amplitudes with $\Delta I > 3/2$) and recalling that antisymmetric states are not allowed by B–E symmetry,

$$A(K^0 \to \pi^+\pi^-) = \frac{1}{\sqrt{3}}(\sqrt{2}A_0 + A_2) \tag{8.57}$$

$$A(K^0 \to \pi^0\pi^0) = \frac{-1}{\sqrt{3}}(A_0 - \sqrt{2}A_2) \tag{8.58}$$

$$A(K^+ \to \pi^+\pi^0) = \sqrt{\frac{3}{2}}A_2 \tag{8.59}$$

Two comments are in order: first, if amplitudes with $\Delta I = 5/2$ are also allowed (as in the Standard Model with EM corrections) the above relations get modified (Gardner and Valencia, 2000). Second, in any case the above decomposition is not exact because isospin is just an approximate symmetry due to the smallness of the u, d quark masses with respect to those of other quarks: $m_d \neq m_u$ and EM effects (different electric charge of u and d quarks) break this symmetry and give rise to a mixing of π^0 with η, η' (see e.g. Donoghue *et al.* (1992)).

The two pions with isospin I in the final state interact, originating the strong phase shifts required for the observation of *CP* violation in decay rates

$$\langle (\pi\pi)_I | S_S | (\pi\pi)_I \rangle = e^{2i\delta_I} \tag{8.60}$$

The fact that the strong interaction just introduces phase shifts is due to the fact that elastic scattering is the only allowed process for pions at energies $\simeq m_K$ (elastic scattering in a given isospin eigenstate, which includes charge exchange). Recalling the Fermi–Watson theorem (5.18), *CPT* symmetry requires the decay amplitudes $K \to (\pi\pi)_I$ to have the form

$$A_I = a_I e^{i\delta_I} \qquad \bar{A}_I = a_I^* e^{i\delta_I} \tag{8.61}$$

The direct *CPV* parameter (8.50)

$$\epsilon_f = i\,\mathrm{Im}(a_f)/\mathrm{Re}(a_f) \tag{8.62}$$

would be just the phase of the weak amplitude in case the strong phases δ_I were absent.

As discussed in Chapter 5, *CPT* symmetry enforces the equality of the decay rates if all $\pi\pi$ states are considered (summing $\pi^+\pi^-$ and $\pi^0\pi^0$), independently of *CP* symmetry; from eqn (8.51) this means that the partial rate asymmetries for the two states have opposite sign:

$$|A(K_S \to \pi^+\pi^-)|^2 \operatorname{Re}(\epsilon_{\pi^+\pi^-}) = -|A(K_S \to \pi^0\pi^0)|^2 \operatorname{Re}(\epsilon_{\pi^0\pi^0}) \qquad (8.63)$$

and since $\Gamma(K \to \pi^+\pi^-) \simeq 2\Gamma(K \to \pi^0\pi^0)$ (eqn (8.57), (8.58) in the limit $A_0 \gg A_2$) these asymmetries (8.52) are related by[110]

$$\Delta_{\pi^+\pi^-} \simeq -\Delta_{\pi^0\pi^0}/2 \qquad (8.64)$$

Note that for the constraints enforced by unitarity the states $\pi\pi\gamma, \gamma\gamma$ and 3π should also be considered; however in the limit in which EM interactions are neglected the states with photons can be ignored, and transitions between 2π and 3π states for zero angular momentum are forbidden not only by *G*-parity (Chapter 3) but also by parity symmetry, and thus can only be induced by weak interactions. With these provisos the $\pi\pi$ states are effectively decoupled from the other states and the above constraints are valid.

8.3.2 Exact formulæ

For the decay into the $\pi\pi$ state with definite isospin I (a *CP* eigenstate) the asymmetry parameter (8.50) between *CP*-conjugate decay amplitudes is denoted by ϵ_I; recalling eqn (8.61) and writing

$$\xi_I \equiv \operatorname{Im}(a_I)/\operatorname{Re}(a_I) \qquad (8.65)$$

for the phase of the weak decay amplitude, *CPT* symmetry implies

$$\epsilon_I \equiv \frac{A_I - \bar{A}_I}{A_I + \bar{A}_I} = i\xi_I \qquad (8.66)$$

so that *CPV* in the decay is just given by the complex phase of the weak decay amplitude.

The overall *CP* violation in such states is parametrized by the amplitude ratio as defined[111] in eqn (8.48):

$$\eta_I \equiv \frac{A(K_L \to (\pi\pi)_I)}{A(K_S \to (\pi\pi)_I)} = \frac{\bar{\epsilon} + i\xi_I}{1 + i\bar{\epsilon}\,\xi_I} \qquad (8.67)$$

---

[110] Similar constraints hold for each isospin state $I = 0, 2$ individually, but are less relevant in practical terms.

[111] In the following we omit the explicit second factor in eqn (8.48), required to make η_f invariant under rephasing.

which for the dominant $(I = 0)$ $\pi\pi$ state is usually denoted as ϵ (or ϵ_K):

$$\epsilon = \eta_0$$

Note that ϵ includes contributions from all decay modes (through $\bar{\epsilon}$). This complex quantity is phase-convention independent, and as such it differs from the (unphysical) *CP* impurity parameter $\bar{\epsilon}$, which depends on an arbitrary choice of phase; note however that

$$\delta_T = \frac{\mathrm{Re}(\bar{\epsilon})}{1 + |\bar{\epsilon}|^2} = \frac{\mathrm{Re}(\epsilon)}{1 + |\epsilon|^2} = (3.32 \pm 0.06) \cdot 10^{-3} \qquad (8.68)$$

is a physical quantity, its value being determined by the average value of the semi-leptonic decay asymmetry, eqns (8.46), (8.68), assuming the validity of *CPT* symmetry and the $\Delta S = \Delta Q$ rule (8.43).

A second phase-convention independent quantity is

$$\epsilon' \equiv \frac{\eta_0}{\sqrt{2}} \left[\frac{A(K_L \to (\pi\pi)_{I=2})}{A(K_L \to (\pi\pi)_{I=0})} - \frac{A(K_S \to (\pi\pi)_{I=2})}{A(K_S \to (\pi\pi)_{I=0})} \right]$$

$$= \frac{i}{\sqrt{2}} \frac{\mathrm{Re}(a_2)}{\mathrm{Re}(a_0)} (1 - \bar{\epsilon}^2) e^{i(\delta_2 - \delta_0)} \frac{\xi_2 - \xi_0}{(1 + i\bar{\epsilon}\,\xi_0)^2} \qquad (8.69)$$

The above expressions clearly show that ϵ' is proportional to the phase difference $\xi_2 - \xi_0$ between two interfering decay amplitudes $a_{0,2}$, corresponding to decays into the $\pi\pi$ states with $I = 0, 2$.

A third parameter, not related to *CPV*, is also introduced

$$\omega \equiv \frac{A(K_S \to (\pi\pi)_{I=2})}{A(K_S \to (\pi\pi)_{I=0})} = e^{i(\delta_2 - \delta_0)} \frac{\mathrm{Re}(a_2)/\mathrm{Re}(a_0) + i\bar{\epsilon}\,\mathrm{Im}(a_2)/\mathrm{Re}(a_0)}{1 + i\bar{\epsilon}\,\xi_0}$$

$$(8.70)$$

Ignoring the small *CPV* (setting $K_S \simeq K_1$) and assuming *CPT* symmetry $|\omega| \simeq |a_2/a_0|$. ω parametrizes the suppression of the decays into the $I = 2$ $\pi\pi$ state (amplitudes with $\Delta I > 1/2$) with respect to those into the $I = 0$ state ($\Delta I = 1/2$), exemplified by the suppression of the $K^+ \to \pi^+\pi^0$ decay ($I = 2$ final state): $\Gamma(K^+ \to \pi^+\pi^0) \sim \Gamma(K_S \to \pi^+\pi^-)/450$. This empirical '$\Delta I = 1/2$' rule has been challenging theorists since the 1960s to receive a quantitative dynamical explanation.[112]

From the analysis of experimental data the value $\omega \simeq 1/22$ is obtained, as well as a value for the phase difference $\delta_2 - \delta_0 = (-47.8 \pm 2.8)°$ (Gatti, 2003). The

---

[112] If the rule held exactly $\Gamma(K_S \to \pi^+\pi^-)/\Gamma(K_S \to \pi^0\pi^0)$ should be 2 instead of the experimental value 2.19 ± 0.02 (it would be 1 if only $\Delta I = 3/2$ amplitudes existed instead).

difference of the elastic $\pi\pi$ scattering phase shifts for isospin $I = 2$ and $I = 0$, evaluated at a centre of mass energy corresponding to the neutral kaon mass, is $\delta_2 - \delta_0 = (-45.2 \pm 4.7)°$ (Ananthanarayan *et al.*, 2001), consistent with the above value as required by the Fermi–Watson theorem.

It should be noted that a proper treatment requires one to consider phase space differences between the decay modes and the effect of the EM interactions between pions: these effects introduce deviations from the exact isospin symmetry limit (Gardner *et al.*, 2001); moreover, if EM interactions are not neglected the application of the Fermi–Watson theorem requires more care, as $\pi\pi \to \pi\pi\gamma$ transitions must also be considered, and in general the strong phases of the $K \to \pi\pi$ decay amplitudes do not coincide exactly with the $\pi\pi$ elastic scattering phase shifts[113] (Cirigliano *et al.*, 2000).

The CP-violating amplitude ratios (8.48) for the $\pi^+\pi^-$ and $\pi^0\pi^0$ states are

$$\eta_{+-} \equiv |\eta_{+-}|e^{i\phi_{+-}} = \frac{\eta_0 + \eta_2\,\omega/\sqrt{2}}{1 + \omega/\sqrt{2}} = \eta_0 + \frac{\epsilon'}{1 + \omega/\sqrt{2}} \tag{8.71}$$

$$\eta_{00} \equiv |\eta_{00}|e^{i\phi_{00}} = \frac{\eta_0 - \sqrt{2}\,\eta_2\,\omega}{1 - \omega\sqrt{2}} = \eta_0 - \frac{2\epsilon'}{1 - \omega\sqrt{2}} \tag{8.72}$$

so that $$\epsilon' = \frac{\omega}{\sqrt{2}}(\eta_2 - \eta_0) = \frac{1}{3}(\eta_{+-} - \eta_{00})(1 - \omega/\sqrt{2} - \omega^2) \tag{8.73}$$

and those for isospin $I = 0, 2$ eigenstates

$$\eta_0 = \epsilon = \frac{2\eta_{+-} + \eta_{00}}{3} + \omega\sqrt{2}\,\frac{\eta_{+-} - \eta_{00}}{3}$$

$$= \frac{2}{3}\eta_{+-}(1 + \omega/\sqrt{2}) + \frac{1}{3}\eta_{00}(1 - \omega\sqrt{2}) \tag{8.74}$$

$$\eta_2 = \sqrt{2}\,\frac{\eta_{+-} - \eta_{00}}{3\omega} + \frac{\eta_{+-} + 2\eta_{00}}{3} \tag{8.75}$$

All the above formulæ are exact and independent of the choice of phase convention.[114]

---

[113] For quite some time the values of $\delta_2 - \delta_0$ extracted from kaon decay data were in significant disagreement with those obtained from $\pi\pi$ scattering data; the two were reconciled by a precise measurement of $\Gamma(K_S \to \pi^+\pi^-)/\Gamma(K_S \to \pi^0\pi^0)$ by KLOE (Aloisio *et al.*, 2002).
[114] The phases are unambiguously defined (observable) thanks to the choice in eqn (8.8).

8.3.3 Direct and indirect *CP* violation

Restricting now to phase conventions in which $|\bar{\epsilon}| \ll 1$, at first order in $\bar{\epsilon}$

$$\eta_2 - \eta_0 \simeq \epsilon_2 - \epsilon_0$$

Neglecting terms of order $\bar{\epsilon} \, \mathrm{Re}(a_2)/\mathrm{Re}(a_0)$ and $\bar{\epsilon} \, \mathrm{Im}(a_2)/\mathrm{Re}(a_0)$ according to the $\Delta I = 1/2$ rule, and if moreover[115] $|\bar{\epsilon} \, \mathrm{Im}(a_0)/\mathrm{Re}(a_0)| \ll 1$, the parameters defined in eqns (8.67), (8.3.2), (8.70) become

$$\eta_I \simeq \bar{\epsilon} + i\frac{\mathrm{Im}(a_I)}{\mathrm{Re}(a_I)} = \bar{\epsilon} + i\xi_I \tag{8.76}$$

$$\epsilon' \simeq \frac{i}{\sqrt{2}} e^{i(\delta_2 - \delta_0)} \frac{\mathrm{Re}(a_2)}{\mathrm{Re}(a_0)} \left[\frac{\mathrm{Im}(a_2)}{\mathrm{Re}(a_2)} - \frac{\mathrm{Im}(a_0)}{\mathrm{Re}(a_0)} \right] = \frac{i}{\sqrt{2}} \omega(\xi_2 - \xi_0) \tag{8.77}$$

$$\omega \simeq e^{i(\delta_2 - \delta_0)} \frac{\mathrm{Re}(a_2)}{\mathrm{Re}(a_0)} \tag{8.78}$$

Recalling eqn (8.66) and neglecting terms of order $\omega^2 \xi_I$:

$$\epsilon_{\pi^+\pi^-} \simeq i\xi_0 + \epsilon' \qquad \epsilon_{\pi^0\pi^0} \simeq i\xi_0 - 2\epsilon' \tag{8.79}$$

so that from eqns (8.76) and (8.49)

$$\eta_{+-} \simeq \epsilon + \epsilon' \qquad \eta_{00} \simeq \epsilon - 2\epsilon' \tag{8.80}$$

The above expressions clearly show that ϵ' cannot be ascribed to $K^0 - \overline{K}^0$ mixing, as it affects differently the $\pi^+\pi^-$ and $\pi^0\pi^0$ final states.

We note in passing that the parametrization of *CPV* in $K \to \pi\pi$ decays described above is redundant when *CPT* symmetry holds, as the partial decay rates must satisfy

$$\Gamma(K^0 \to \pi^+\pi^-) + \Gamma(K^0 \to \pi^0\pi^0) = \Gamma(\overline{K}^0 \to \pi^+\pi^-) + \Gamma(\overline{K}^0 \to \pi^0\pi^0) \tag{8.81}$$

resulting in a constraint (Sozzi, 2004) on the relations (8.71), (8.72), which at first order in the direct *CP*-violating parameters $\epsilon_{+-}, \epsilon_{00}$ of eqn (8.50) is

$$|1 + \omega/\sqrt{2}|^2 \, \mathrm{Re}(\epsilon_{+-}) + |1/\sqrt{2} - \omega|^2 \, \mathrm{Re}(\epsilon_{00}) = 0$$

(compare to eqn (8.63)). A consequence of this is that the approximate relations (8.80) cannot be consistently transformed into exact ones by invoking a suitable redefinition of ϵ and ϵ', as is sometimes done in the literature.

---

[115] This condition, defining a so-called *physical* phase convention, holds in the SM for most parametrizations of the CKM matrix, see Chapter 9.

The *CP*-violating quantities defined above can be written in terms of the phase-convention independent parameter λ_f introduced in eqn (7.91):

$$\lambda_f = \frac{1 - \bar{\epsilon}}{1 + \bar{\epsilon}} \frac{\overline{A}_f}{A_f} \tag{8.82}$$

In general (assuming *CPT* symmetry):

$$\eta_f = \frac{1 - \lambda_f}{1 + \lambda_f} \tag{8.83}$$

so that for the $I = 0 \, \pi\pi$ state

$$\epsilon = \frac{1 - \lambda_0}{1 + \lambda_0} \tag{8.84}$$

$$\mathrm{Re}(\epsilon) = \frac{1 - |\lambda_0|^2}{1 + 2\mathrm{Re}(\lambda_0) + |\lambda_0|^2} \qquad \mathrm{Im}(\epsilon) = \frac{-2\mathrm{Im}(\lambda_0)}{1 + 2\mathrm{Re}(\lambda_0) + |\lambda_0|^2} \tag{8.85}$$

Recalling the discussion of the different types of *CP* violation in Section 7.2.1, the above expressions show that $\mathrm{Re}(\epsilon) \neq 0$ represents *CPV* in the mixing $|(1 - \epsilon)/(1 + \epsilon)| \neq 1$, as in this case there is a single decay amplitude (in the isospin basis appropriate for strong interactions) and the Fermi–Watson theorem (5.18) requires $\overline{A}_0 = A_0 e^{-2i\delta_0}$, and thus $|\overline{A}_0/A_0| = 1$, even if *CP* is violated. $\mathrm{Im}(\epsilon) \neq 0$ instead represents *CPV* in the interference of decays with and without mixing. Experimentally the two contributions are found to be of comparable size, as the phase of ϵ is close to 45° (see Section 8.3.4). In the limit $|\bar{\epsilon}| \ll 1$ the former corresponds to $\mathrm{Re}(\bar{\epsilon})$ (8.68) and the latter to $\mathrm{Im}(\bar{\epsilon}) + \xi_0$, which is also a physical quantity split in two terms in phase convention dependent way.

For the charge eigenstates $(\pi^+\pi^-, \pi^0\pi^0)$ one defines instead

$$\eta_{+-} = \frac{1 - \lambda_{+-}}{1 + \lambda_{+-}} \qquad \eta_{00} = \frac{1 - \lambda_{00}}{1 + \lambda_{00}} \tag{8.86}$$

and if $|\bar{\epsilon}| \ll 1$

$$\lambda_{+-} \simeq \lambda_0 \left[1 - \left(\frac{A_2}{A_0} - \frac{\overline{A}_2}{\overline{A}_0} \right) / \sqrt{2} \right] \qquad \lambda_{00} \simeq \lambda_0 \left[1 + \sqrt{2} \left(\frac{A_2}{A_0} - \frac{\overline{A}_2}{\overline{A}_0} \right) \right] \tag{8.87}$$

and the relation

$$\frac{1}{3}\eta_{00} + \frac{2}{3}\eta_{+-} \simeq \epsilon \tag{8.88}$$

is valid including first-order terms in A_2/A_0 (compare with eqn (8.74)). Also, for $\lambda_{00} \simeq \lambda_{+-} \simeq 1$ (small CPV), neglecting terms of order $|\omega|$:

$$\frac{1}{3}(\eta_{+-} - \eta_{00}) \simeq \frac{1}{6}(\lambda_{00} - \lambda_{+-}) \simeq \epsilon' \tag{8.89}$$

and in this limit

$$\text{Re}(\epsilon') \simeq \frac{1}{3}\left[\frac{1 - |\lambda_{+-}|^2}{1 + 2\text{Re}(\lambda_{+-}) + |\lambda_{+-}|^2} - \frac{1 - |\lambda_{00}|^2}{1 + 2\text{Re}(\lambda_{00}) + |\lambda_{00}|^2}\right]$$

$$\text{Im}(\epsilon') \simeq \frac{1}{3}\left[\frac{\text{Im}(\lambda_{00})}{1 + 2\text{Re}(\lambda_{00}) + |\lambda_{00}|^2} - \frac{\text{Im}(\lambda_{+-})}{1 + 2\text{Re}(\lambda_{+-}) + |\lambda_{+-}|^2}\right]$$

showing that $\epsilon' \neq 0$ indicates direct CPV, as it requires $\arg(A_2) \neq \arg(A_0)$.

$\text{Re}(\epsilon') \neq 0$ requires $|\lambda_{+-}| \neq 1$ or $|\lambda_{00}| \neq 1$, and $\lambda_{+-} \neq \lambda_{00}$, so it cannot arise just from CPV in the mixing: CP violation in the decays ($|\bar{A}_f/A_f| \neq 1$) is necessary, as is also evident from the fact that $\text{Re}(\epsilon') \propto \sin(\delta_2 - \delta_0)$ also requires different strong phases in order not to vanish. Analogously $\text{Im}(\epsilon') \neq 0$ reflects CPV in the interference between decays with and without mixing, namely $\text{Im}(\lambda_{+-}) \neq 0$ or $\text{Im}(\lambda_{00}) \neq 0$ (and still $\lambda_{+-} \neq \lambda_{00}$). The phase $\phi(\epsilon') \simeq 45°$ indicates that the two contributions are comparable. Note that with no final state interactions (or equal strong phase shifts for the two isospin states) ϵ' could still be different from zero, but it would be a pure imaginary quantity (no CP violation 'in the decay').

All kinds of CP violation are thus present in the neutral K system, and the situation can be summarized in terms of the ϵ, ϵ' parameters as follows:

- CP violation in the mixing, arising from a relative phase between Γ_{12} and M_{12} in the effective Hamiltonian, is expressed by $\text{Re}(\epsilon) \neq 0$ (or more precisely $\delta_T \neq 0$). This is the only kind of CPV appearing in semi-leptonic decays (as long as the $\Delta S = \Delta Q$ rule holds).

- CP violation in the decays, arising from a relative phase between two interfering decay amplitudes, is present: for the $\pi\pi$ decay modes these are the amplitudes leading to the $I = 0, 2$ final states. This kind of CPV is expressed by $\text{Re}(\epsilon') \neq 0$.

- CP violation in the interference between decays with and without mixing, arising from a relative phase between the mixing amplitude and a decay amplitude is also present for the $\pi\pi$ decay modes, and expressed by $\text{Im}(\epsilon) \neq 0$ and $\text{Im}(\epsilon') \neq 0$.

It turns out that in the K system CP violation mainly manifests itself as a phase difference between M_{12} and the dominant part of Γ_{12}, that corresponding to decays into the $I = 0 \,\pi\pi$ state, while the contribution due to the phase difference between the $I = 0, 2$ decay amplitudes is much smaller:[116] *CP* violation can thus be said to be mostly of the indirect type in the neutral K system.

8.3.4 $\pi\pi$ dominance

Due to their relatively small mass, neutral kaons have few important decay modes available, and as mentioned not all of them support *CPV* in the decay amplitudes: this fact leads to some interesting relations.

Recalling the expression (8.20) for the mixing parameter $\bar{\epsilon}$ in the case of *CPT* symmetry, and using eqns (8.23) which are valid in any proper phase convention, together with (8.28) one obtains

$$\bar{\epsilon} \simeq \frac{e^{i\pi/4}}{2\sqrt{2}} \frac{\text{Im}(M_{12}) - i\text{Im}(\Gamma_{12})/2}{\text{Re}(M_{12})} \tag{8.90}$$

where $\phi_{SW} \simeq \pi/4$ was used. Introducing the (real) parameter

$$\bar{\epsilon}_M \equiv \frac{\text{Im}(M_{12})}{\text{Re}(M_{12})} \tag{8.91}$$

(which in an appropriate phase convention describes *CPV* in the mass matrix $\boldsymbol{M}$), the above expression is written as

$$\bar{\epsilon} \simeq \frac{e^{i\pi/4}}{2\sqrt{2}} \left[\bar{\epsilon}_M - \frac{i}{2} \frac{\text{Im}(\Gamma_{12})}{\text{Re}(M_{12})} \right] \simeq \frac{e^{i\pi/4}}{2\sqrt{2}} \left[\bar{\epsilon}_M + i \frac{\text{Im}(\Gamma_{12})}{\text{Re}(\Gamma_{12})} \right] \tag{8.92}$$

We recall that only physical (on mass shell) decay amplitudes contribute to the decay matrix $\boldsymbol{\Gamma}$, and semi-leptonic final states do not contribute to its off-diagonal element $\Gamma_{12} = \langle K^0 | \boldsymbol{\Gamma} | \overline{K}^0 \rangle \propto \sum_f A_f^* \overline{A}_f$, since the $\Delta S = \Delta Q$ rule implies that either K^0 or $\overline{K}^0$ (but not both) contribute to any such final state. The largest contribution to Γ_{12} therefore arises from $\pi\pi$ states, and mainly from the isospin $I = 0$ state ('$\Delta I = 1/2$ rule'), for which *CPT* symmetry (8.61) requires

$$(\Gamma_{12})_{\pi\pi(I=0)} \propto A_0^* \overline{A}_0 = (a_0^*)^2 = |a_0|^2 e^{-2i\phi_0}$$

---

[116] As the phases of ϵ and ϵ' are close to $\pi/4$ the amount of *CPV* in mixing, decay, and interference are of similar magnitude, but this same fact should warn the reader not to attribute a deep significance to the distinction between the third and the first two, since such phases are largely determined by *CPT* symmetry and the strong $\pi\pi$ scattering phases respectively, neither of which are linked to *CPV* itself.

(ϕ_0 being the phase of a_0). In any phase convention in which ϕ_0 is small: $\tan 2\phi_0 \simeq 2 \tan \phi_0 = 2\xi_0$, thus

$$\left(\frac{\text{Im}(\Gamma_{12})}{\text{Re}(\Gamma_{12})}\right)_{\pi\pi(I=0)} = -\tan 2\phi_0 \simeq -2\xi_0 \tag{8.93}$$

$$\bar{\epsilon} \simeq \frac{e^{i\pi/4}}{2\sqrt{2}}\left[\bar{\epsilon}_M - 2i\xi_0 - \frac{i}{2}\frac{\text{Im}(\widehat{\Gamma}_{12})}{\text{Re}(M_{12})}\right] \tag{8.94}$$

where the (small) term $\widehat{\Gamma}_{12} \equiv \Gamma_{12} - \Gamma_{12}^{\pi\pi(I=0)}$ represents the contribution to Γ_{12} of all decay modes but the dominant one. The phase of $\bar{\epsilon}$ is $\approx \pi/4$ in this limit.

In any such phase convention, in which the decay amplitude into the $(\pi\pi)_{I=0}$ state (the dominant one among those which can violate *CP*) is approximately real ($\xi_0 \simeq 0$), Γ_{12} is also approximately real, and thus[117] $\bar{\epsilon}$ is proportional to $\bar{\epsilon}_M$.

Within the same approximations the (phase-convention independent) parameter ϵ is, from (8.94)

$$\epsilon \simeq \bar{\epsilon} + i\xi_0 \simeq \frac{e^{i\pi/4}}{2\sqrt{2}}\left[\bar{\epsilon}_M + 2\xi_0 - \frac{i}{2}\frac{\text{Im}(\widehat{\Gamma}_{12})}{\text{Re}(M_{12})}\right] \tag{8.95}$$

and the smallness of the last term implies that the phase of ϵ is also close to $\pi/4$, due to the fact that *CP* violation mostly arises from a single decay mode and $\Delta m \simeq \Gamma_S/2$.

While this result was obtained assuming small ξ_0, it can be shown to be independent of phase conventions: the smallness of *CPV* is expressed by $\lambda_0 = (q/p)e^{-2i\phi_0} \simeq 1$, so that from eqn (8.84), using eqns (7.40) and (7.41):

$$\epsilon \simeq \frac{1 - \lambda_0^2}{4} \simeq \frac{i\,\text{Im}(e^{2i\phi_0}M_{12}) + \text{Im}(e^{2i\phi_0}\Gamma_{12})/2}{-\Delta m + i\Delta\Gamma/2}$$

The dominance of the $\pi\pi(I=0)$ state implies $\text{Im}(e^{2i\phi_0}\Gamma_{12}) \simeq 0$, and therefore

$$\epsilon \simeq -e^{i\pi/4}\frac{\text{Im}(e^{2i\phi_0}M_{12})}{\Delta m\sqrt{2}}$$

In the same approximation

$$\Delta m \simeq -2\,\text{Re}(\mathcal{H}_{12}e^{2i\phi_0}) \simeq -2\,\text{Re}(M_{12}e^{2i\phi_0})$$

which gives back (8.23) for phase choices in which ϕ_0 is small (indeed this is the case in which $|\bar{\epsilon}| \ll 1$).

---

[117] This justifies the loose statement, sometimes found in the literature, according to which $\bar{\epsilon}$ describes *CPV* in the mass matrix, as in this case the contribution from the decay matrix Γ is small. However this conclusion actually depends on the choice of phase convention: the physically meaningful quantity is always the phase difference between Γ_{12} and M_{12}.

The Bell–Steinberger unitarity relation (7.86) can be written in terms of the CP-violating parameters η_f (8.48) for all possible decay modes as

$$\sum_f \eta_f \Gamma(K_S \to f) = [(\Gamma_S + \Gamma_L)/2 + i\Delta m] \langle K_S | K_L \rangle$$

$$\simeq (\Gamma_S/2)(1 + i \tan \phi_{SW}) \langle K_S | K_L \rangle \qquad (8.96)$$

where the super-weak phase ϕ_{SW} was defined in (8.5) and eqn (7.85) was used. The above equations express the non-orthogonality of the physical states as a weighted sum over the CP-violating K_L decays. If CPV is assumed to be small in all decay modes except $\pi\pi$, from $\Gamma(K_S \to \pi^+\pi^-) + \Gamma(K_S \to \pi^0\pi^0) \simeq \Gamma_S \gg \Gamma_L$ and the experimental observation $\eta_{+-} \simeq \eta_{00} \simeq \eta_{\pi\pi} \simeq \epsilon$, the above relation reduces to

$$\Gamma_S(1 + i \tan \phi_{SW}) [\text{Re}(\bar{\epsilon}) - i\text{Im}(\delta)] \simeq \epsilon \Gamma_S$$

at first order in the CP- (and CPT-) violating parameters. Recalling eqn (8.49) one has, if CPT symmetry holds:

$$\eta_{\pi\pi} \simeq 2\text{Re}(\bar{\epsilon}) \left[\frac{1}{2} + \frac{i\Delta m}{\Gamma_S} \right] \qquad (8.97)$$

$$\phi_{\pi\pi} \simeq \arctan\left(\frac{2\Delta m}{\Gamma_S} \right) \simeq \phi_{SW} \qquad (8.98)$$

while, if T symmetry holds:

$$\eta_{\pi\pi} \simeq -2i\text{Im}(\delta) \left[\frac{1}{2} + \frac{i\Delta m}{\Gamma_S} \right] \qquad (8.99)$$

$$\phi_{\pi\pi} \simeq \arctan\left(\frac{2\Delta m}{\Gamma_S} \right) - \frac{\pi}{2} \simeq \phi_{SW} - \frac{\pi}{2} \qquad (8.100)$$

Thus if CPT symmetry holds (and CP violation is accompanied by T violation) the phase of $\eta_{\pi\pi} \simeq \epsilon$ is close to ϕ_{SW}, while if T symmetry holds (and CP violation comes together with CPT violation) this is orthogonal to it.

Allowing for the possibility of CPV in other decay modes, in the case of CPT symmetry the phase of ϵ is actually

$$\arg(\epsilon) \simeq \phi_{SW} - \arctan\left[\frac{\text{Im}(\Gamma_{12})/2}{\text{Im}(M_{12})} \right] \qquad (8.101)$$

and limits on the second term on the right-hand side of eqn (8.101) can be obtained by considering the partial decay widths for all decay modes:

$$\text{Im}(\Gamma_{12}) \leq \sum_f |\text{Im}(A_f^2)| = \sum_f |2 \text{Im}(A_f)\text{Re}(A_f)|$$

where the sum runs over all final states except the flavour-specific ones which are not accessible to both K^0 and $\overline{K}^0$. Since

$$\Gamma_L^f \equiv |A(K_L \to f)|^2 = |(pA_f - qA_f^*)|^2/(|p|^2 + |q|^2) \simeq 2\text{Im}(A_f)^2$$

$$\Gamma_S^f \equiv |A(K_S \to f)|^2 \simeq 2\text{Re}(A_f)^2$$

$$2\text{Im}(A_f)\text{Re}(A_f) \simeq \sqrt{\Gamma_L^f \Gamma_S^f}$$

$$\text{Im}(M_{12}) \simeq 2\sqrt{2}|\epsilon|\text{Re}(M_{12}) \simeq |\epsilon|\Gamma_S/\sqrt{2} \simeq \Gamma_S\text{Re}(\epsilon)$$

one gets

$$\left|\frac{\text{Im}(\Gamma_{12})/2}{\text{Im}(M_{12})}\right| \simeq \sum_f \sqrt{BR(K_S \to f)BR(K_L \to f)}\left(\frac{\Gamma_L}{\Gamma_S}\right)\frac{1}{2\text{Re}(\epsilon)} \qquad (8.102)$$

Recalling the discussion in Section 7.1.5, in terms of the phase difference (modulo π) $\Delta\xi$ between the off-diagonal terms Γ_{12} and M_{12} of the effective Hamiltonian, from eqn (7.84) one has

$$\Delta m = -2\text{Re}\left[\frac{q}{p}(M_{12} - i\Gamma_{12}/2)\right] \simeq \mp 2\overline{M}_{12} + O(\Delta\xi^2)$$

$$\Delta\Gamma = +4\text{Im}\left[\frac{q}{p}(M_{12} - i\Gamma_{12}/2)\right] \simeq \mp 2\overline{\Gamma}_{12} + O(\Delta\xi^2)$$

with the sign ambiguity corresponding to that of q/p in eqn (7.41).

For the K system $m_L > m_S$ and $\Gamma_L < \Gamma_S$ so that the relative phase of Γ_{12} and M_{12} is close to π (rather than 0) and independently of any convention

$$\Delta m \simeq 2|M_{12}| \qquad \Delta\Gamma \simeq -2|\Gamma_{12}|$$

Thanks to (8.4) the expression (7.83) for q/p reduces in this case to

$$\frac{q}{p} \simeq \pm e^{-i\xi_M}\left(1 + \Delta\xi\frac{\tan\phi_{SW} - i}{1 + \tan^2\phi_{SW}}\right) \qquad (8.103)$$

and the (convention dependent) phase of q/p is

$$\arg\left(\frac{q}{p}\right) \simeq -\xi_M - \frac{\Delta\xi/2}{1 + \tan^2\phi_{SW}}(+n\pi)$$

with $n = 0, 1$ for the choice of positive or negative sign in eqn (7.41).

The phase of $\epsilon \simeq (1 - \lambda_0)/2$ can be obtained from the above expression by noting that the smallness of *CPV* implies $\lambda_0 = (q/p)e^{-2i\phi_0} \simeq 1$ (*CPT* symmetry assumed), and the dominance of the $(\pi\pi)_{I=0}$ decay mode implies $\arg(\Gamma_{12}) = \xi_M + \Delta\xi \simeq -2\phi_0$; because of this not only ξ_M but also the $\Delta\xi$ cancels at first order, giving $\phi(\epsilon) \simeq \phi_{SW}$

Since a single final state dominates the decay amplitudes, a natural choice of phase convention is the one in which no *CP* violation appears at all in this decay mode: in the *Wu–Yang phase convention* (Wu and Yang, 1964) the decay amplitude a_0 corresponding to the $(\pi\pi)_{I=0}$ decay mode is chosen to be real and positive:[118]

$$a_0 = \mathrm{Re}(a_0) > 0 \tag{8.104}$$

so that $\bar{a}_0 = a_0$ and (*CPT* symmetry):

$$\epsilon \equiv \eta_0 = \bar{\epsilon} \qquad\qquad \epsilon' = \frac{i}{\sqrt{2}} e^{i(\delta_2 - \delta_0)} \frac{\mathrm{Im}(a_2)}{a_0}$$

$$\eta_{+-} = \bar{\epsilon} + \frac{\epsilon'}{1 + \omega/\sqrt{2}} \simeq \bar{\epsilon} + \epsilon' \qquad \eta_{00} = \bar{\epsilon} - \frac{2\epsilon'}{1 - \omega\sqrt{2}} \simeq \bar{\epsilon} - 2\epsilon'$$

which are just (8.80) with $\bar{\epsilon}$ in place of ϵ (showing that they are only valid in a given phase convention). In the Wu–Yang phase convention, if only the dominant decay mode is considered no direct *CPV* appears, as Γ_{12} is real in that limit (8.93), and

$$\bar{\epsilon} \simeq \frac{\mathrm{Im}(M_{12})}{i\Delta m - \Delta\Gamma/2} \simeq \frac{-\mathrm{Im}(M_{12})}{\Delta m \sqrt{2}} e^{i\phi_{SW}} \simeq \frac{e^{i\phi_{SW}}}{2\sqrt{2}} \frac{\mathrm{Im}(M_{12})}{\mathrm{Re}(M_{12})}$$

so that $\bar{\epsilon} \simeq \bar{\epsilon}_M$ and $\arg(\bar{\epsilon}) \simeq \phi_{SW}$, showing that (in this phase convention) the mixing parameter $\bar{\epsilon}$ is approximately real and originates in the mass matrix:

$$\epsilon = \bar{\epsilon} = -\frac{e^{i\pi/4}}{\Delta m \sqrt{2}} [\mathrm{Im}(M_{12}) + 2\mathrm{Re}(M_{12})(1 + i)\xi_0] \tag{8.105}$$

In the Wu–Yang convention and all the so-called *physical* ones in which $\bar{a}_0 \simeq a_0$, a_0 and η_{CP} are approximately real, so that *CP* violation is given by the imaginary parts of M_{12} and Γ_{12}: direct *CPV* is thus confined into a_2 (and possibly other small amplitudes) and indirect *CPV* in M_{12} (the contribution of a_2 to $\mathrm{Im}(\Gamma_{12})$ being smaller). In all such phase conventions several expressions get simpler: note that from eqn (8.67) and the smallness of $|\epsilon|$ this automatically implies $|\bar{\epsilon}| \ll 1$.

Figure 8.2 shows pictorially the different parameters describing the phenomenology of the neutral K system.

8.3.5   The super-weak model

Soon after *CP* violation was discovered in K decays, L. Wolfenstein observed that it could be ascribed entirely to a new very weak interaction hitherto undetected (Wolfenstein, 1964).

---

[118] In the usual parametrization of the CKM matrix (Chapter 9) a_0 is complex and a_2 is real.

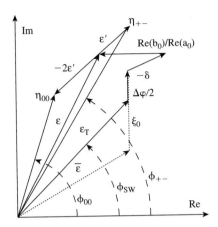

FIG. 8.2. The *CP*-violating parameters in the $K \to \pi\pi$ decays (in the limit $|\omega| \simeq 0$, not to scale). $\Delta\varphi/2 = \mathrm{Im}(a_0)/\mathrm{Re}(a_0) - \mathrm{Im}(\Gamma_{12})/\Delta\Gamma$. $\bar{\epsilon}$ and $\mathrm{Im}(a_0)/\mathrm{Re}(a_0)$ are phase-convention dependent, while $\mathrm{Re}(\bar{\epsilon})$ is not (for $|\bar{\epsilon}| \ll 1$).

Recalling the expression for the off-diagonal (*CP*-violating) terms of the effective Hamiltonian (7.19), an interaction changing strangeness by two units ($\Delta S = 2$) would have to compete in M_{12} with the usual weak interactions ($\Delta S = 1$) at *second order*.[119] Writing the effective Hamiltonian as in eqn (7.67) with $h_{CPT} = 0$ and $h_{CP} = ih_{SW}$ ($m_0, \delta m, h_{SW}$ real), *CP* violation is only due to the off-diagonal element h_{SW} in the mass matrix; recalling expression (8.20) for the *CP*-violating parameter $\bar{\epsilon}$ describing the *CP*-impurity of the physical states, this gives

$$\bar{\epsilon} \simeq \frac{-h_{SW}}{\Delta m} \frac{1+i}{2} \qquad (8.106)$$

where $\Delta m \simeq \Gamma_S/2$ was used. In order to obtain the measured effect $|\bar{\epsilon}| \simeq 2 \cdot 10^{-3}$ (Wu–Yang phase convention) the magnitude of h_{SW} has to be of order 10^{-3} of Δm, a quantity of second order in the usual weak interactions (see eqn (8.30)): $\Delta m \propto (G_F m_p^2)^2$, where the proton mass m_p was introduced as a typical scale to have a dimensionless constant. Now, if h_{SW} originates from some new interaction with coupling constant G_{SW} acting at *first* order: $h_{SW} \propto G_{SW} m_p^2$ (choosing G_{SW} with the same dimensions as G_F), so that $G_{SW} \sim 3 \cdot 10^{-8} G_F$ (since $G_F m_p^2 \sim 10^{-5}$). The new hypothetical interaction responsible for *CPV* can thus be so weak that it would go unnoticed everywhere, except in $K^0 - \overline{K}^0$ mixing, where a very high sensitivity is obtained due to the smallness of Δm.

---

[119] This means second order in G_F, which of course is actually fourth order in the electro-weak coupling constant, i.e. two gauge boson exchanges.

The basic feature of this model is that all *CP*-violating effects are due to the *CP* impurity of the physical states (in $\mathcal{H}$): *CP* symmetry is not violated in kaon decays ($\Delta S = 1$), and $K_2 \to \pi\pi$ decays do not occur: the super-weak interaction only produces indirect *CPV*. Explicit predictions (assuming *CPT* symmetry) are thus

$$\eta_{+-} = \eta_{00} = \bar{\epsilon} = \epsilon \qquad \epsilon' \simeq 0 \qquad \phi_{+-} \simeq \phi_{00} \simeq \phi_{SW} \qquad (8.107)$$

and this actually justifies the name 'super-weak phase' for ϕ_{SW}.

The super-weak model was never a dynamical model, but rather an *ansatz*, representing a class of models in which *CPV* effectively appears only in $\Delta F = 2$ mixing processes, leading to *CP*-violating effects which are the same in any kind of decay. Until a new manifestation of *CP* violation, different from the *CP*-impurity parameter ϵ, could be measured, the possibility that the elusive effect actually belonged to a different sector of the theory, almost completely decoupled from other observable phenomena, remained a viable possibility. Despite the rise of the Standard Model, which allowed for a mechanism for *CPV* (see Chapter 9) which is not super-weak (but accidentally almost so for the *K* system), this hypothesis could stand for thirty-five years, fuelled by the smallness of direct *CPV* and the fact that no other system appeared to exhibit any *CP*-violating effect at all.

In his 1980 Nobel speech, J. Cronin still noted that 'at present our experimental understanding of *CP* violation can be summarized by the statement of a single number' (ϵ) (Cronin, 1981). This state of affairs finally changed with the observation of direct *CP* violation in the *K* system, and later in the *B* system.

8.3.6 Measurements

Early measurements of *CPV* in $K \to \pi\pi$ decays were discussed in Chapter 8. After some initial confusion due to imprecise measurements, the scenario emerged in which indirect *CPV* dominates in such decays: $|\epsilon'|$ was known to be quite smaller than $|\epsilon|$, which from eqn (8.88) is (Yao *et al.*, 2006):

$$|\epsilon| = (2.232 \pm 0.007) \cdot 10^{-3} \qquad (8.108)$$

The moduli of the *CP*-violating parameters $\eta_{\pi\pi}$ (8.48) are extracted from the measurement of $\Gamma(K_L \to \pi\pi)/\Gamma(K_S \to \pi\pi) = |\eta_{\pi\pi}|^2$, and their phases $\phi_{\pi\pi}$ from the time-dependent decay rates of a known initial state (different from a pure K_L); the $K_L - K_S$ mass difference Δm can be extracted from the same data. The experimental values of the above quantities are (without assuming *CPT* symmetry): (Yao *et al.*, 2006)

$$|\eta_{+-}| = (2.236 \pm 0.018) \cdot 10^{-3} \qquad \phi_{+-} = (43.4 \pm 1.2)° \qquad (8.109)$$

$$|\eta_{00}| = (2.232 \pm 0.025) \cdot 10^{-3} \qquad \phi_{00} = (43.7 \pm 1.3)° \qquad (8.110)$$

(if *CPT* symmetry is assumed the error on the phases is reduced by an order of magnitude). In parallel with extensive tests of *CP* symmetry in other systems, experimental investigations with neutral kaons focused on the search for direct *CP* violation, starting what was to be a rather long quest.

We recall that *CPV* in the decay amplitudes (direct) is possible for $K_L \rightarrow \pi\pi$ decays, because in this case there are two interfering amplitudes of definite isospin, and the pions exhibit FSI. However by observing *CP* violation in a single decay mode of a neutral *K* one cannot distinguish this effect from *CPV* purely in the mixing: at least two measurements are necessary for such distinction, e.g. a comparison between *CPV* in the $\pi^+\pi^-$ and $\pi^0\pi^0$ modes, eqn (8.73). Again we see that the neutral kaon system turns out to be the 'minimal' one for this kind of investigation, because the constraint (8.81) enforced by *CPT* symmetry gives just two decay rates which must compensate against each other but can be different.

If *CP* violation were only due to an asymmetry in the $K^0 - \overline{K}^0$ mixing (which could be induced even by a super-weak interaction), all $\pi\pi$ decays of neutral kaons would be *CP*-conserving decays of the K_1 component, both for K_S and K_L: in this case all their properties, such as the ratio of $\pi^+\pi^-$ ('charged') to $\pi^0\pi^0$ ('neutral') decays, should be the same for both physical states. In other words, if *CP* violation only manifests itself as a 'constituent' property of the decaying meson itself, it should appear to be the same in any kind of decay process; the *CP*-violating amplitude ratios η_{+-} and η_{00} should be exactly equal in this case ($\epsilon_f = 0$).

The above amplitude ratios could be measured either ignoring the initial strangeness of the kaon, in which case the decay rate reduces to the sum of two exponentials from which $|\eta_f|^2$ can be extracted, or by fitting the interference term of the rate asymmetry in the proper decay time region around $\approx 14\tau_S$. The latter approach is more sensitive due to the small value of $|\eta_f|$, but the systematic uncertainties are very different in the two cases, leading in both cases to errors which are not sufficient for an accurate comparison and a measurement of ϵ'. Similarly, the extraction of ϵ'/ϵ from the comparison of η_{+-} (or η_{00}) with the mixing parameter ϵ, as obtained from the semi-leptonic charge asymmetry δ_L (8.43), does not provide a sufficient precision. Clearly, in searching for a difference between two quantities which are quite larger than their difference the comparison of separate measurements of the two terms quickly shows its limits, as the (uncorrelated) systematic errors involved in the two measurements become the dominating factor. A better approach is required in which at least part of such systematic effects do cancel in the difference of interest.

The parameter ϵ' representing direct *CPV* can be conveniently written in terms of experimentally measurable quantities in different ways (Mannelli, 1984). The close similarity of the phase of ϵ' ($\delta_2 - \delta_0 + \pi/2 = (47.8 \pm 2.8)°$) and that of η_{00} and η_{+-} ($\simeq \phi_{SW}$) is just an accidental fact (and would not hold if K_S were heavier than K_L, rather than the opposite); this however implies that only two real quantities have to be measured to obtain ϵ': $|\epsilon|$ and $|\epsilon'/\epsilon|$ (with a sign). Neglecting $|\omega| \ll 1$

$$\frac{3\epsilon'}{\eta_{+-}} \simeq 1 - \frac{\eta_{00}}{\eta_{+-}}$$

whose phase is $\phi \equiv \delta_2 - \delta_0 + \pi/2 - \phi_{+-}(\pm\pi)$ so that writing $\Delta\phi \equiv (\phi_{00} - \phi_{+-})$ one has

$$\pm\frac{3|\epsilon'|}{|\eta_{+-}|}e^{i\phi} = 1 - \frac{|\eta_{00}|}{|\eta_{+-}|}e^{i\Delta\phi}$$

and equating the real and imaginary parts gives (at first order in $\phi \ll 1$ rad):

$$\pm\frac{3|\epsilon'|}{|\eta_{+-}|} \simeq 1 - \frac{|\eta_{00}|}{|\eta_{+-}|} \qquad \pm\frac{3|\epsilon'|}{|\eta_{+-}|}\phi \simeq \left[1 - \frac{|\eta_{00}|}{|\eta_{+-}|}\right]\Delta\phi$$

and $|\epsilon'|$ can be obtained from $|\eta_{00}|$ without knowing ϕ. Moreover

$$\left|\frac{\eta_{00}}{\eta_{+-}}\right|^2 = \frac{|\epsilon|^2 - 4\mathrm{Re}(\epsilon^*\epsilon') + |\epsilon'|^2}{|\epsilon|^2 + 2\mathrm{Re}(\epsilon^*\epsilon') + |\epsilon'|^2} \simeq 1 - 6\frac{\mathrm{Re}(\epsilon^*\epsilon')}{|\epsilon|^2} = 1 - 6\,\mathrm{Re}(\epsilon'/\epsilon)$$

$$(8.111)$$

If $|\omega|$ is neglected

$$\mathrm{Re}(\epsilon'/\epsilon) \simeq \frac{1}{3}(1 - |\eta_{00}/\eta_{+-}|) \qquad \mathrm{Im}(\epsilon'/\epsilon) \simeq -\frac{1}{3}|\eta_{00}/\eta_{+-}|\Delta\phi \qquad (8.112)$$

Since the value of the phase of ϵ depends on the assumption of *CPT* symmetry (8.97) a large value of $\mathrm{Im}(\epsilon'/\epsilon)$ would indicate *CPT* violation. Recalling how the three types of *CP* violation are parametrized into ϵ and ϵ', the measured quantity $\mathrm{Re}(\epsilon'/\epsilon)$ is seen to contain all three of them.

All experimental measurements of ϵ' were obtained by exploiting relation (8.111) with a measurement of the double ratio of partial decay widths

$$R \equiv \frac{\Gamma(K_L \to \pi^0\pi^0)}{\Gamma(K_S \to \pi^0\pi^0)}\frac{\Gamma(K_S \to \pi^+\pi^-)}{\Gamma(K_L \to \pi^+\pi^-)} = \left|\frac{\eta_{00}}{\eta_{+-}}\right|^2 \qquad (8.113)$$

to look for its difference from 1. The story of this measurement is a long and fascinating one, involving several generations of dedicated precision measurements spanning more than twenty years (Sozzi and Mannelli, 2003).

Direct *CP* violation in neutral *K* decays

A round of dedicated experiments to measure $\mathrm{Re}(\epsilon'/\epsilon)$ was completed at the end of the 1990s, resulting in an unclear situation: the results of the NA31 experiment at CERN (Barr, *et al.*, 1993) showed some evidence of direct *CPV* : $\text{Re}(\epsilon'/\epsilon) = (2.30 \pm 0.65) \cdot 10^{-3}$, while the E731 experiment at FNAL (Gibbons *et al.*, 1993) was consistent with no effect: $\text{Re}(\epsilon'/\epsilon) = (0.74 \pm 0.60) \cdot 10^{-3}$. At the same time the theoretical situation concerning the value expected within the SM was somewhat confused.

The measurement of R (8.113) is basically a counting experiment: the number of K_S, K_L decays into $\pi^+\pi^-$ and $\pi^0\pi^0$ are counted and their ratio is formed: the difficulty lies in avoiding any normalization bias in the translation from event numbers to partial decay widths. In order to obtain a robust measurement, less sensitive to systematic biases, the experiments exploit cancellations of effects between pairs of decay modes: the beam intensity normalization is irrelevant if at least two decay modes are collected at the same time, as then it cancels in their ratio: for most of its data-taking period the E731 experiment measured alternately $|\eta_{00}|$ and $|\eta_{+-}|$ using two different detector arrangements; the NA31 experiment instead measured the $\pi^0\pi^0$ to $\pi^+\pi^-$ ratio by collecting both decay modes at the same time, alternating runs with K_L and (mainly) K_S decays. With these single-ratio measurements several sources of uncertainties cancelled to first order, and biases could only arise from e.g. time changes between runs in the beam intensity ratio (for E731) or in the relative detector inefficiencies (for NA31). Such biases can also be cancelled to first order if all four decay modes are collected simultaneously, which is what both following experiments KTeV at FNAL and NA48 at CERN did.

Both the new experiments (as their predecessors) produced intense neutral K_L beams by colliding protons (800 GeV/*c* at the FNAL Main Injector, 450 GeV/*c* at the CERN SPS) on targets, and letting neutral particles, collimated after sweeping magnets removed charged components, travel in vacuum for $\sim$ 100 m: only neutrons, photons, K_L and some long-lived neutral hyperons are left in such a beam.[120] The statistical error is dominated by the number of $K_L \to \pi^0\pi^0$ decays collected, so that intense primary beams and long decay regions are required. The use of relatively high energies gives large production cross sections and good detector resolutions for $\pi^0 \to \gamma\gamma$ decays, but also requires longer decay regions.

The techniques used to obtain K_S decays were very different in the two experiments: in KTeV a massive regenerator was set on one of two identical twin parallel K_L beams, thus producing some amount of K_S (Fig. 8.3); the regenerator was actually a fully active lead-scintillator detector, so that inelastically regenerated K_S (with different energy and angle with respect to the original K_L) could be identified from the energy deposit of the recoiling particles and discarded. In NA48 K_S were instead produced by steering (through crystal channeling) a small fraction of the primary protons onto a second target, placed just 6 m before the beginning of the decay region: shielding the fiducial region from the high flux of spurious particles required careful collimator and magnet arrangements (Fig. 8.4).

In both cases of course the secondary neutral beams ('K_S beams') actually contain only a small fraction of K_S, but these dominate in the $\pi\pi$ decay modes (and are nevertheless more abundant than $K_L \to \pi\pi$ decays from the K_L beams).

Both experiments had similar detectors, comprising large aperture magnetic spectrometers with multiple-view drift chambers to measure $\pi^+\pi^-$ decays, and most importantly very high performance state of the art electromagnetic calorimeters to measure $\pi^0\pi^0 \to 4\gamma$ decays. NA48 used a quasi-homogeneous liquid krypton ionization chamber with $\sim$13,000

---

[120] High-momentum K_S components were present at the per cent level and accounted for.

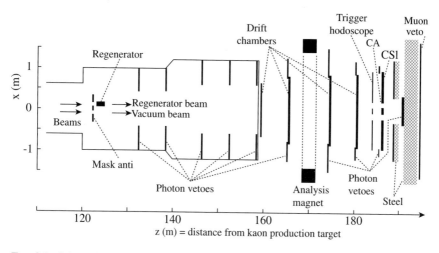

FIG. 8.3. Schematic drawing of the KTeV experimental setup (Alavi-Harati *et al.*, 2003)

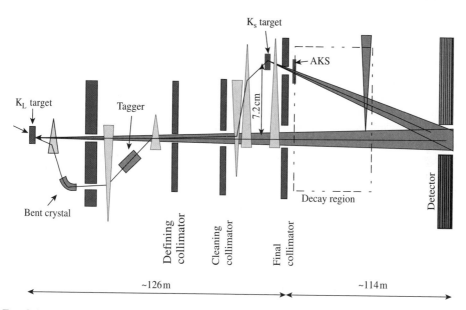

FIG. 8.4. Schematic drawing of the beam arrangement for the NA48 experiment (Lai *et al.*, 2001).

longitudinal cells arranged in a projective geometry, with very good energy, space and time resolution and excellent stability and uniformity; KTeV used a calorimeter made of ~3,000 pure CsI crystals with excellent energy resolution.

Other detectors, including muon counters behind thick iron walls and scintillation counters surrounding the fiducial volume were used to suppress the backgrounds, mostly arising from three-body K_L decays, and in KTeV from the regenerator. Backgrounds are a potentially dangerous source of bias, as each of them only contributes to one of the four measured counts with no cancellation being possible; for this reason very high resolution detectors, allowing good suppression by the use of kinematic constraints, were instrumental in reducing the backgrounds to such small levels (typically a fraction of percent) that the residual uncertainties on their size and subtraction did not affect the results in an important way.

In KTeV K_S and K_L events were distinguished by the position of the event centroid on the detector, since the two parallel beams were separated by $\sim$30 cm; this separation is potentially dangerous in view of possible differences in the local detector efficiencies and acceptances, and in order to cancel this effect the position of the regenerator was switched between the two beams every minute. In NA48 the two beams instead converged to the centre of the detector in order to minimize acceptance differences; the identification of K_S decays was based on the precise matching of the event time with the time of individually measured protons along the beam line leading to the K_S target. Events for which a fast scintillator detector arrangement indicated that a proton was present at the target within ± 2 ns of the event time, as measured by the downstream detectors, were labelled as K_S; accidental backgrounds could only give random mis-tagging diluting the measurement (in a correctable way), but could not induce a fake direct *CPV* signal in first order.

To avoid biases in the measurement it is crucial for the fiducial phase space volumes to be exactly the same at least for pairs of modes: since different detectors are involved in the measurement of the two decay modes, this in practice requires such volumes to be accurately equalized for K_L and K_S.

The kaon momentum spectra are necessarily different for K_S and K_L, (due to the differences in the production and regeneration cross sections), and this required the analyses to be performed in small momentum bins. The definition of the fiducial region along the direction of the beams requires some care: due to the unavoidable difference in lifetime, the coupling of the longitudinal decay distributions to resolution effects can introduce $K_S - K_L$ differences. For charged decays momenta are determined by the spectrometers, whose absolute scales were defined by their geometry and the magnitude of the magnetic field, fine tuned by adjusting the reconstructed kaon mass. For $K \to \pi^0\pi^0 \to 4\gamma$ decays the position of the kaon decay vertex can be reconstructed by using one of the three kinematical constraints available (the kaon and pion masses m_K, m_{π^0}), while the remaining two are used to suppress the background. Assuming zero transverse momentum of the decaying kaon with respect to the beam axis and denoting with E_i the photon energies and with r_{ij} their transverse separation ($i, j = 1, 4$) at the calorimeter, the longitudinal distance d of the decay vertex from the calorimeter was computed in NA48 assuming a $K \to 4\gamma$ decay as $d = (1/m_K)\sum_{i>j}\sqrt{E_i E_j r_{ij}^2}$, and similarly in KTeV in which the average position of the two best matching $\pi^0 \to 2\gamma$ vertices was used instead. The resolution on the longitudinal vertex position d was of order 40–50 cm in both experiments. This approach is seen to effectively link the distance scale to the absolute energy scale: any error in the latter shifts and expands (or shrinks) the fiducial region, differently for K_S and K_L, thus introducing a bias. In both experiments the absolute energy scales of the calorimeters were fixed from the data by adjusting the reconstructed position of a sharply defined detector edge to its nominal value.

The different longitudinal decay distribution of K_S and K_L induce a large acceptance difference: KTeV dealt with this by applying a sizeable correction factor ($\sim 85 \cdot 10^{-4}$) obtained from an accurate Monte Carlo simulation which was validated with large samples of auxiliary data (only 5% of the collected events were used for the measurement, the rest being used for systematic error evaluations). NA48, on the other hand, pursued an acceptance cancellation by weighting the K_L events in the analysis according to their decay proper time, so that the resulting distribution matched the one for K_S; such procedure eliminated the need for an acceptance correction at first order, but resulted in a non-negligible loss of statistics, as the number of $K_L \to 2\pi$ events was effectively reduced.

Among other factors to be kept under control were the detector and data acquisition dead times, for which any differential effect depending on the decay mode and/or the beam (through e.g. rate correlations) had to be reduced to negligible levels and continuously monitored.

KTeV collected data both in 1996–97 (3.3 million $K_L \to \pi^0\pi^0$ decays) and in 1999 (a sample of similar size); results from the analysis of the 1996–97 sample were published (Alavi-Harati *et al.*, 2003): $\mathrm{Re}(\epsilon'/\epsilon) = (2.071\pm0.148_{\text{stat}}\pm0.239_{\text{syst}})\cdot10^{-3} = (2.07\pm0.28)\cdot 10^{-3}$. NA48 collected data in 1997-1999 and 2001 (5 million $K_L \to \pi^0\pi^0$ decays in total), with the final result (Batley *et al.*, 2002): $\mathrm{Re}(\epsilon'/\epsilon) = (1.47\pm0.14_{\text{stat}}\pm0.09_{\text{stat/syst}}\pm0.15_{\text{syst}})\cdot 10^{-3} = (1.47 \pm 0.22)\cdot10^{-3}$ where the first quoted error is purely statistical, the second is the one induced by the finite statistic of the control samples used to study systematic effects, and the third is purely systematic. Both experiments thus proved the existence of direct *CP* violation.

The experiments considered many possible sources of systematic error and evaluated the related uncertainties (Alavi-Harati *et al.*, 2003; Lai *et al.*, 2003): in both cases the largest individual systematic error arose from the imperfect knowledge of the calorimeter reconstruction for $\pi^0\pi^0$ decays. In KTeV the error in the knowledge of the background in the neutral mode followed in importance, while in NA48 the uncertainty in the knowledge of the differential trigger efficiency for the charged mode was more relevant.

With the results from KTeV and NA48 (see Fig. 8.5) the world average

$$\mathrm{Re}(\epsilon'/\epsilon) = (16.3 \pm 1.6) \cdot 10^{-4}$$

is obtained (Sozzi, 2004), proving beyond any doubt[121] the existence of direct *CP* violation. The above measurement is best related to those performed on *B* mesons (Chapter 10) by writing it through eqn (8.52) as

$$\Delta_{\pi^+\pi^-} = \frac{\Gamma(\overline{K}^0 \to \pi^+\pi^-) - \Gamma(K^0 \to \pi^+\pi^-)}{\Gamma(\overline{K}^0 \to \pi^+\pi^-) + \Gamma(K^0 \to \pi^+\pi^-)} = (-5.2\pm0.5)\cdot10^{-6}$$

$$(8.114)$$

---

[121] With the inflated error figure used by the PDG (Yao *et al.*, 2006) to account for a non-satisfactory agreement of the experimental results, $2.3\cdot10^{-4}$, the significance is still above 7 standard deviations.

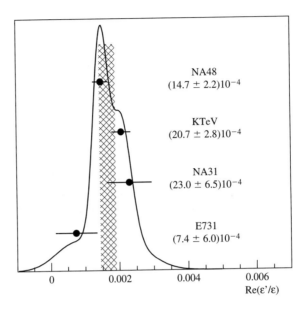

FIG. 8.5. Ideogram of recent experimental results for the direct *CP*-violating quantity Re(ϵ'/ϵ); the hatched band indicates the world average.

This long-awaited result finally invalidated the super-weak model (8.3.5), indicating that *CP* violation actually arises in the usual weak interactions responsible for flavour-changing decays, and is not restricted to the peculiar phenomenon of flavour mixing. This was actually the first significant test of the CKM mechanism by which *CPV* is incorporated into the SM (see Chapter 9).

In principle the ϵ' parameter could be also measured with other experimental approaches: one involves asymmetries of strangeness-tagged mesons, discussed in Section 8.3.7, and another one exploits correlated K pairs, discussed in Section 8.8.

It is interesting to note that the measured direct *CPV* can be turned into a limit on the possible amount by which Bose–Einstein symmetry is violated for pions (Greenberg and Mohapatra, 1989b). The absence of the antisymmetric $\pi^+\pi^-$ state with isospin $I = 1$ follows from the assumption that pions respect the above symmetry; as such state would be *CP*-odd, a K_L could decay into it without violating *CP*, with an amplitude of order βG_F, where β parametrizes the amount of B–E symmetry violation (Section 5.4). On the contrary, $K_L \rightarrow \pi^0\pi^0$ decay is not possible without *CP* violation (independently of the validity of B–E symmetry) as no antisymmetric state zero spin state can be formed with two π^0. Thus a violation of B–E statistics would contribute an apparent direct *CP*-violating effect, as a difference between η_{+-} and η_{00}: if the entire measured effect would be ascribed to such phenomenon (so that $\epsilon' = 0$) then $\beta \sim O(10^{-5})$ (a better limit on β could be obtained if a precisely predicted value of ϵ' could be subtracted).

The observation of direct *CPV* also invalidated other models attempting to maintain the overall validity of *CP* symmetry by ascribing the $K_L \rightarrow 2\pi$ effect to different causes, such as a matter-antimatter difference in the gravitational interaction (see Section 11.1).

8.3.7   Tagged kaon asymmetries

Several parameters of the neutral K system can be determined by studying decay asymmetries involving the complementary description in terms of strangeness-tagged kaons, as done by the CPLEAR experiment at CERN (Angelopoulos *et al.*, 2003): the η_{+-}, η_{00} *CP*-violating amplitude ratios and the corresponding ones for 3π decays, Δm, etc. Some information on direct *CPV* could also be obtained in principle in this way.

For *CP*-even final states f, which can be reached by both K_S and (through *CPV*) K_L ($|\eta_f| \ll 1$ in (8.48)), the decay asymmetry (7.165) with respect to the initial flavour reduces for $\mathrm{Re}(\bar{\epsilon}) \ll 1$ to

$$A_f(t) \simeq 2\mathrm{Re}(\bar{\epsilon}) - 2\frac{|\eta_f| e^{-\Delta\Gamma t/2} \cos(\Delta mt - \phi_f)}{1 + |\eta_f|^2 e^{-\Delta\Gamma t}} \tag{8.115}$$

If f is a *CP* eigenstate ($|\bar{f}\rangle = CP|f\rangle = \pm|f\rangle$) a study of *CPV* is possible using decays into a single final state.

It can be observed that any asymmetry in the tagged instantaneous partial decay rates into $\pi\pi$ states (either $\pi^+\pi^-$ or $\pi^0\pi^0$) could be used to extract a measurement of direct *CPV* (Backenstoss *et al.*, 1983): the initial value (at $t = 0$) of this asymmetry is

$$A_{\pi\pi}(0) \simeq 2\mathrm{Re}(\epsilon) - 2\mathrm{Re}(\eta_{\pi\pi}) \simeq \begin{cases} -2\mathrm{Re}(\epsilon') & (\text{for } \pi^+\pi^-) \\ +4\mathrm{Re}(\epsilon') & (\text{for } \pi^0\pi^0) \end{cases} \tag{8.116}$$

and ϵ' could be extracted from the measurement of the initial asymmetry in either mode: such a measurement would however require a very accurate control of K flavour tagging and the use of a very short decay time interval close to $t = 0$, which because it uses only a small fraction of the available decays makes the accumulation of a sufficient sample difficult.

Due to the finite experimental resolution on the proper decay time, in order to be able to measure an asymmetry with a statistically significant sample of events one has to resort to time-integrated measurements. The sensitivity to direct *CPV* effects is largest for short integration times: this is readily understood since any effect of *CP* violation due to $K^0 - \overline{K}^0$ mixing is relatively less important for $t \ll \tau_S \simeq \hbar/2\Delta m$, before strangeness oscillations can take place, while direct *CP* violation effects are independent of time.

Asymmetries integrated over a given range of proper decay time can also be conveniently used to extract the *CP* violation parameters (see e.g. Amelino-Camelia *et al.* (1992)). The ratio of such asymmetries for the two $\pi\pi$ states maintains some sensitivity to direct *CPV*, while at the same time being less sensitive to differential resolution effects; this would allow one to extract $\mathrm{Re}(\epsilon'/\epsilon)$ without the need to measure $\mathrm{Re}(\epsilon)$ to the (unattainable) level of precision necessary when using only one of the decay modes. The small size of the asymmetries (of order 10^{-3}) however implies that the number of events required to reach the same statistical accuracy as with the double ratio method comparing K_S and K_L decays (8.113) is many orders of magnitude larger, in practice again an insurmountable handicap. Other approaches can be devised by exploiting coherent mixtures of K_S and K_L, such as obtained by regeneration (Sozzi and Mannelli, 2003), but these are also somewhat poor from the point of view of statistics. Finally we mention that tagged kaon asymmetries also allow one to perform sensitive checks of the validity of quantum mechanics, and searches for the possible spontaneous decoherence of quantum states (Ellis *et al.*, 1996).

8.4   Time reversal violation

If *CPT* symmetry is valid the observed *CP* violation must necessarily be accompanied by *T* violation; as discussed in Section 8.3.4 the phase of the ϵ parameter carries information on whether the *CP* violation it describes also violates either the *T* or the *CPT* symmetries, with experiment indicating the former. Still, a direct measurement of time reversal violation could be obtained for the $K^0 - \overline{K}^0$ mixing process, providing an independent piece of evidence for the consistency of the picture.

Considering the time evolution of states which are initially strangeness eigenstates, *CPT* symmetry requires the probability P that a K^0 remains a K^0 at a later time to be equal to the probability that a $\overline{K}^0$ remains a $\overline{K}^0$, but the $K^0 \to \overline{K}^0$ transition probability can be different from the $\overline{K}^0 \to K^0$ one. The quantity which explicitly exhibits a violation of time reversal symmetry is the 'Kabir's asymmetry' (Kabir, 1970) (7.157) comparing the $K^0 - \overline{K}^0$ mixing transition occurring in opposite time directions; at first order in $\overline{\epsilon}$

$$A_T \equiv \frac{P(\overline{K}^0 \to K^0) - P(K^0 \to \overline{K}^0)}{P(\overline{K}^0 \to K^0) + P(K^0 \to \overline{K}^0)} \simeq 2\mathrm{Re}(\overline{\epsilon}) \simeq 2\mathrm{Re}\langle K_L | K_S \rangle \qquad (8.117)$$

With time reversal symmetry violation (and *CPT* symmetry) an incoherent mixture of K^0 and $\overline{K}^0$ with equal weights (zero strangeness) is seen to acquire a net non-zero strangeness of a given sign, having a maximum for $t \approx 5\tau_S$ before settling to an asymptotic value $\langle K_L | K_S \rangle$. It is interesting to note that A_T is not affected by the presence of matter (Dass et al., 1987).

The $K^0, \overline{K}^0$ states can be identified from the strangeness of the particles produced in association with them: since strangeness is not exactly conserved the above probabilities cannot be measured with arbitrary precision in principle, but in practice this fact has negligible effect. If the $\Delta S = \Delta Q$ rule holds for semi-leptonic decays, the net strangeness can be directly determined from the charge asymmetry. Considering the time-dependent decay rates $R(t)_\pm$, $\overline{R}(t)_\pm$ into states with positive or negative leptons originating from K^0 or $\overline{K}^0$:

$$A_T(t) \equiv \frac{\overline{R}_+(t) - R_-(t)}{\overline{R}_+(t) + R_-(t)} \tag{8.118}$$

which at first order in the CP-violating parameters is

$$A_T(t) \simeq 4\mathrm{Re}(\overline{\epsilon}) - 2\mathrm{Re}(x_-^{(\ell)}) - 2\mathrm{Re}(y^{(\ell)})$$

$$+ \frac{2\mathrm{Re}(x_-^{(\ell)}) \left[e^{\Delta \Gamma t/2} - \cos(\Delta mt) \right] + 2\mathrm{Im}(x_+^{(\ell)}) \sin(\Delta mt)}{\cosh(\Delta \Gamma t/2) - \cos(\Delta mt)} \tag{8.119}$$

where the parameters $x_\pm^{(\ell)}, y$ related to violations of the $\Delta S = \Delta Q$ rule and of CPT symmetry were defined in Section 8.2. The time dependence of the above asymmetry is due to $\Delta S \neq \Delta Q$ transitions, and vanishes even in presence of the $y^{(\ell)}$ term indicating CPT symmetry violation in the decay amplitudes, if these do not simultaneously also violate the $\Delta S = \Delta Q$ rule ($y^{(\ell)} \neq 0$ but $x_-^{(\ell)} = 0$). At large times ($t \gg \tau_S$) the asymmetry tends to a constant value

$$A_T(t \to \infty) \simeq 4\mathrm{Re}(\overline{\epsilon}) - 2\mathrm{Re}(x_-^{(\ell)}) - 2\mathrm{Re}(y^{(\ell)}) \tag{8.120}$$

and neglecting CPT violation in the decay amplitudes ($y^{(\ell)} = 0 = x_-^{(\ell)}$) only the first term is left (without making any assumption on the validity of the $\Delta S = \Delta Q$ rule, i.e. $x_+^{(\ell)}$ can be non-zero).

T violation in neutral K mixing

The CPLEAR experiment (Angelopoulos et al., 2003) at the CERN antiproton ring (LEAR) performed an extensive set of measurements in the neutral kaon system in the years 1992–5, using strangeness-tagged kaons. By exploiting the strangeness-conserving associate production reactions (driven by strong interactions)

$$\overline{p}p \text{ (at rest)} \to K^+ \pi^- K^0 \quad \text{or} \quad K^- \pi^+ \overline{K}^0$$

which amount to 0.4% of the total $\overline{p}p$ cross-section at rest, the initial strangeness of the produced neutral kaons could be known by detecting the charged particles $K^\pm \pi^\mp$ (only

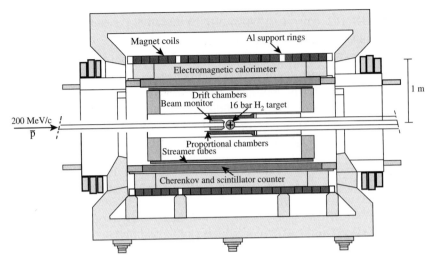

FIG. 8.6. Schematic longitudinal view of the CPLEAR detector (Angelopoulos *et al.*, 2003).

CPV in strong interactions could invalidate this procedure), and their time evolution and flavour-dependent asymmetries could be studied. An appealing feature of this approach is that equal amounts of K^0 and $\overline{K}^0$ are produced; moreover, by exploiting an initial state of precisely known energy and momentum, the neutral kaon direction and momentum could be determined from the production kinematics.

Intense fluxes of flavour-tagged K^0 and $\overline{K}^0$ were produced by stopping low-energy antiprotons (200 MeV/c, 10^6 $\overline{p}/s$) from LEAR in a low-density gaseous hydrogen target. The detector (Fig. 8.6) consisted of a cylindrical tracking chamber, a threshold Čerenkov counter to identify charged kaons, and an electromagnetic calorimeter made of lead plates interspaced with streamer tubes to detect photons from π^0 decay; everything was enclosed in a solenoidal magnet providing a constant 0.44 T field. Two layers of plastic scintillators provided further particle identification capabilities by specific ionization energy loss and time of flight measurements, resulting in an overall 4 standard deviation K/π separation for particle momenta above 350 MeV/c. The experimental acceptance extended to 20 τ_S on average (60 cm), covering the interesting $K_S - K_L$ interference region, with a resolution varying between 0.05 and 0.1 τ_S.

The measurement of the time asymmetry (8.118) requires the knowledge of the meson strangeness at two times: at production time (obtained as described) and at decay time (by exploiting the $\Delta S = \Delta Q$ rule in semi-leptonic decays); any error in the flavour determination at either time can introduce a bias in the measurement. Inefficiencies are less dangerous since they can only 'dilute' the measurement of an existing asymmetry but cannot induce a fake one by themselves.

The detection efficiencies common to the different states cancel to first order, and geometrical acceptance differences were averaged to zero by frequently reversing the magnetic field direction. Some care is required because the detection efficiencies for $K^\pm, \pi^\pm$ and $e^\pm$ depend on the charge sign. In particular the differences in the interactions of $K^+\pi^-$ and

$K^-\pi^+$ with the detector required all the observables to be independently normalized for K^0 and $\overline{K}^0$, to avoid systematic asymmetries; these can only be important close to the production target, before strangeness oscillations have taken place. Such normalization was performed by applying to each event a correction factor ξ, expressing the $\overline{K}^0/K^0$ efficiency ratio as a function of the $K^\pm$ and $\pi^\pm$ momenta. Such factors are clearly independent of the final state of decay and of decay time; they were obtained from the analysis of a large sample of $\pi^+\pi^-$ decays, by measuring the ratio of events tagged as $\overline{K}^0$ ($\overline{N}$) or K^0 (N) and fitting the expected time-dependence. Counting events up to n K_S lifetimes:

$$[1 + 4\text{Re}(\epsilon_L)]\,\xi(p_K, p_\pi) = \frac{\overline{N}(\pi^+\pi^-; n)}{N(\pi^+\pi^-; n)} \frac{1}{[1 - 4|\eta_{+-}|e^{n/2}\cos(\Delta m\, n\tau_S - \phi_{+-})]}$$

and the quntity $\text{Re}(\epsilon_L)$ was taken from the well measured semi-leptonic charge asymmetry δ_L (8.43) in the *CPT* symmetry limit. The measured asymmetry was thus

$$A_T^{(\exp)}(t) \equiv \frac{\overline{R}_+(t) - \frac{1+4\text{Re}(\epsilon_L)}{1+2\delta_L}R_-(t)}{\overline{R}_+(t) + \frac{1+4\text{Re}(\epsilon_L)}{1+2\delta_L}R_-(t)} = 4\text{Re}(\bar{\epsilon}) - 4\text{Re}(x_-^{(\ell)}) - 4\text{Re}(y^{(\ell)})$$

$$+ \frac{2\text{Re}(x_-^{(\ell)})\left[e^{\Delta\Gamma t/2} - \cos(\Delta mt)\right] + 2\text{Im}(x_+^{(\ell)})\sin(\Delta mt)}{\cosh(\Delta\Gamma t/2) - \cos(\Delta mt)} \qquad (8.121)$$

which becomes $\simeq 4\text{Re}(\bar{\epsilon}) - 4\text{Re}(y^{(\ell)} + x_-^{(\ell)})$ for $t \gg \tau_S$. The correction factors differed from unity by 12% on average. Since the final states are not charge symmetric, event-by-event momentum-dependent corrections were also applied to account for the small differences in detection efficiencies for oppositely charged final state particles, with average correction factors of order 1.5%, determined from pion and electron samples. Despite their relative smallness, these factors were the main source of systematic uncertainty for the final asymmetry value. Events were also corrected for regeneration in the detector, as a function of the neutral kaon momentum and the amount of traversed material: indeed regeneration could alter the free time evolution of the initial strangeness eigenstates; the amount of material in the detector was thus minimized and the resulting error on the asymmetry was kept below 10% of the statistical error. Backgrounds arose mainly from $K \to \pi^+\pi^-$ decays and their level was below 1% of the signal; the relative acceptances were determined by Monte Carlo simulation, and the corresponding uncertainties had a negligible effect on the asymmetry.

The total number of events measured was $6.4 \cdot 10^5$. Figure 8.7 shows the measured $A_T^{(\exp)}$ asymmetry in the decay time range $1 - 20\tau_S$, which exhibits a clear non-zero average $\langle A_T^{(\exp)} \rangle = (6.6 \pm 1.3 \pm 1.0) \cdot 10^{-3}$ indicating that a $\overline{K}^0$ develops into a K^0 with a slightly higher probability than a K^0 into a $\overline{K}^0$, violating time reversal symmetry.

By assuming *CPT* symmetry and fitting the asymmetry to the time-dependent expression (8.121) the parameters $\text{Im}(x_+^{(\ell)})$ and $\text{Re}(\bar{\epsilon})$ could be extracted, the former mostly from the value of the asymmetry at short decay times and the latter from that at long decay times. The results were (Angelopoulos *et al.*, 1998a): $\text{Re}(\bar{\epsilon}) = (1.55 \pm 0.35 \pm 0.25) \cdot 10^{-3}$ and $\text{Im}(x_+^{(\ell)}) = (1.2 \pm 1.9 \pm 0.9) \cdot 10^{-3}$, consistent with no T violation in amplitudes violating the

$\Delta S = \Delta Q$ rule (if they exist at all), and with Re($\bar\epsilon$) compatible with other determinations of the same parameter.

The above measurement represents the first direct determination of a quantity violating time reversal symmetry, independent from unitarity assumptions (Alvarez-Gaumé *et al.*, 1999).

Even without assuming any symmetry, limits on A_T can be obtained from the Bell–Steinberger unitarity relation (7.86) through the limits for $\langle K_L | K_S \rangle$ (see e.g. Angelopoulos *et al.* (2003)), using only the measured kaon branching fractions and assuming that the final state interactions respect *CPT* symmetry: a measurement of Re($y^{(\ell)}$) + Re($x_-^{(\ell)}$) can be obtained, from which it can be concluded that contributions to A_T due to *CPT* violation in the decay amplitudes are negligible.

8.5   *CPT* symmetry

Despite the difficulties in devising a fundamental theory in which *CPT* symmetry is not valid, the phenomenological description of the neutral *K* allows *CPT*-violating terms, whose size can be bound by experiment.

8.5.1   *CPT* and mixing

As discussed in Section 7.1.4, a difference in the diagonal elements of the effective $K^0 - \overline{K}^0$ Hamiltonian ($\delta \neq 0$) corresponds to *CPT* violation (called *indirect CPT violation*). The *CP*-violating parameters $\bar\epsilon, \delta$ are conveniently decomposed along two orthogonal axes, parallel and orthogonal to the direction defined by ϕ_{SW} (8.5): $\bar\epsilon = \bar\epsilon_\parallel + \bar\epsilon_\perp, \delta = \delta_\parallel + \delta_\perp$. The component $\delta_\perp$ orthogonal to the direction defined

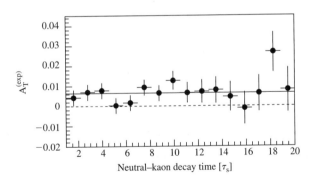

FIG. 8.7. The time reversal violating asymmetry $A_T^{(exp)}$ as a function of neutral *K* decay time (Angelopoulos *et al.*, 1998a); the solid line is the fitted average, with $\chi^2 = 0.84$ per degree of freedom.

by ϕ_{SW} is proportional to the mass difference between K^0 and $\overline{K}^0$, while the one parallel to this direction is proportional to the decay width (lifetime) difference, and better approximations for eqns (8.24)–(8.27) can be obtained:

$$\text{Im}(M_{12}) \simeq (\Delta\Gamma/2)\,[\text{Re}(\overline{\epsilon}) + \text{Im}(\overline{\epsilon})\tan(\phi_{SW})] \tag{8.122}$$

$$\text{Im}(\Gamma_{12}) \simeq \Delta\Gamma\,[\text{Re}(\overline{\epsilon})\tan(\phi_{SW}) - \text{Im}(\overline{\epsilon})] \tag{8.123}$$

$$M_{22} - M_{11} \simeq -\Delta\Gamma\,[\text{Re}(\delta)\tan(\phi_{SW}) - \text{Im}(\delta)] \tag{8.124}$$

$$\Gamma_{22} - \Gamma_{11} \simeq 2\Delta\Gamma\,[\text{Re}(\delta) + \text{Im}(\delta)\tan(\phi_{SW})] \tag{8.125}$$

Recalling that $M_{11} = m(K^0)$, $M_{22} = m(\overline{K}^0)$, from the above relations

$$m(\overline{K}^0) - m(K^0) \simeq \frac{-2\Delta m}{\sin(\phi_{SW})}|\delta_{\perp}| \simeq 2\Delta m[\text{Re}(\delta) - \text{Im}(\delta)] \tag{8.126}$$

Note that in general the effective Hamiltonian for the $K^0 - \overline{K}^0$ system can contain a term due to a possible (*CPT*-violating) mass difference between quarks and antiquarks, which would also contribute to the $K_L - K_S$ mass difference; to the extent that Δm is largely determined by the $K^0 - \overline{K}^0$ coupling term due to weak interactions, such quark-antiquark mass differences are therefore constrained (Fromerth and Rafelski, 2003).

The decay asymmetries of strangeness-tagged K allow the extraction of δ: without assuming *CPT* symmetry the time-dependent decay rate of a neutral K with definite strangeness at $t = 0$ into a state f reachable by both K^0 and $\overline{K}^0$ is (at first order in the symmetry-violating parameters)

$$\Gamma_f(t) \simeq \frac{1}{2}\left|A_f^{(S)}\right|^2 \left[Ae^{-\Gamma_S t} + 2C\,e^{-\overline{\Gamma}t}\cos(\Delta mt) + 2S\,e^{-\overline{\Gamma}t}\sin(\Delta mt) + D\,e^{-\Gamma_L t}\right] \tag{8.127}$$

$$A = 1 \mp [2\text{Re}(\overline{\epsilon}) - 2\text{Re}(\delta)] \tag{8.128}$$

$$C = \pm\{[1 \mp 2\text{Re}(\overline{\epsilon})]\,\text{Re}(\eta_f) \pm 2\text{Im}(\delta)\text{Im}(\eta_f)\} \tag{8.129}$$

$$S = \pm\{[1 \mp 2\text{Re}(\overline{\epsilon})]\,\text{Im}(\eta_f) \mp 2\text{Im}(\delta)\text{Re}(\eta_f)\} \tag{8.130}$$

$$D = |\eta_f|^2\,\{1 \mp [2\text{Re}(\overline{\epsilon}) + 2\text{Re}(\delta)]\} \tag{8.131}$$

with the upper (lower) sign is for an initial K^0 ($\overline{K}^0$).

With both tagged K^0 and $\overline{K}^0$ the comparison allows one to determine $\overline{\epsilon}$ and δ: for long times $t \gg \tau_S$

$$\frac{\Gamma(K^0 \to f; t \to \infty)}{\Gamma(\overline{K}^0 \to f; t \to \infty)} \simeq 1 - 4[\text{Re}(\overline{\epsilon}) + \text{Re}(\delta)] \tag{8.132}$$

which is difficult to measure for $\pi\pi$ states because of the smallness of $|\eta_{\pi\pi}|^2$; at the initial time instead

$$\frac{\Gamma(K^0 \to f; t = 0)}{\Gamma(\overline{K}^0 \to f; t = 0)} \simeq 1 - 4[\text{Re}(\bar{\epsilon}) - \text{Re}(\delta) - \text{Re}(\eta_f)] \tag{8.133}$$

and a measurement of this ratio would require a very good time resolution and a large sample.

For the measurement of parameters from tagged kaon asymmetries the semi-leptonic decays are a good choice, since all the coefficients in the time evolution (8.127) are of the same order of magnitude (Tanner and Dalitz, 1986) (this is not the case for $\pi\pi$ decays). Assuming the validity of the $\Delta S = \Delta Q$ rule ($x^{(\ell)} = \bar{x}^{(\ell)} = 0$)

$$\frac{\Gamma(K^0 \to \pi^{\mp}\ell^{\pm}\nu(\bar{\nu}); t \to \infty)}{\Gamma(K^0 \to \pi^{\mp}\ell^{\pm}\nu(\bar{\nu}); t = 0)} = 1 \mp 8\text{Re}(\delta) \tag{8.134}$$

and a comparison of the two also allows one to check this assumption. Considering the full expression (8.128) for the decay rate, $\text{Im}(\delta)$ can be obtained from the coefficients of the term containing $\sin(\Delta mt)$ for $K^0 \to \pi^-\ell^+\nu$ and $\overline{K}^0 \to \pi^+\ell^-\bar{\nu}$, $\text{Re}(y)$ from the ratio of such decays; finally $\text{Re}(\bar{\epsilon})$ can be extracted from the ratio of $\overline{K}^0 \to \pi^-\ell^+\nu$ and $K^0 \to \pi^+\ell^-\bar{\nu}$. If the $\Delta S = \Delta Q$ rule is not assumed $\text{Re}(x)$ can be obtained from the ratio of $\overline{K}^0 \to \pi^-\ell^+\nu$ decays for large and small times, and $\text{Im}(x)$ from the $\sin(\Delta mt)$ term in the same ratio; $\text{Re}(\bar{x}^{(\ell)})$ and $\text{Im}(\bar{x}^{(\ell)})$ are analogously extracted from $K^0 \to \pi^+\ell^-\bar{\nu}$ decays.

The parameter δ can be measured from the mixing asymmetry

$$A_{CPT}(t) = \frac{P[\overline{K}^0(t) \to \overline{K}^0] - P[K^0(t) \to K^0]}{P[\overline{K}^0(t) \to \overline{K}^0] + P[K^0(t) \to K^0]} \tag{8.135}$$

which for long decay times tends to $4\text{Re}(\delta)$. The CPLEAR experiment measured such asymmetry and by fitting it in the decay time range 0–20 τ_S obtained (Angelopoulos *et al.*, 2003)

$$\text{Re}(\delta) = (3.0 \pm 3.3) \cdot 10^{-4} \qquad \text{Im}(\delta) = (-1.5 \pm 2.3) \cdot 10^{-4} \tag{8.136}$$

Other ways of extracting *CPT*-violating parameters using tagged kaon asymmetries and semi-leptonic decays are discussed in Angelopoulos *et al.*, (2003).

8.5.2   *CPT* and decays

If *CPT* symmetry is not assumed for decays into the final state f (but is valid for strong interactions) then (Section 6.2.1)

$$\overline{A}_{\bar{f}} = e^{2i\delta_f} \left(A_f^{CPT}\right)^* \tag{8.137}$$

where with the notation introduced in eqn (8.61) $A_f^{CPT} \equiv \langle f|\mathcal{T}_{CPT}|K^0\rangle = a_f^{CPT} e^{i\delta_f}$, so that

$$\bar{a}_{\bar{f}} = \left(a_f^{CPT}\right)^* \tag{8.138}$$

and thus even for *CP*-eigenstates $\bar{f} = f$, a real a_f does not result in $\bar{a}_f$ being real (it is however always possible to exploit the arbitrariness of the phase convention to have $\bar{a}_f/a_{\bar{f}}$ real for at least one decay mode).

Semi-leptonic decays were discussed in Section 8.2; in principle a precise measurement of *CP* violation in K_S decays would allow to test *CPT* symmetry by comparing ϵ_S and ϵ_L: the difference between the K_S and K_L semi-leptonic charge asymmetries is

$$\delta_S - \delta_L = 4\text{Re}(\delta) - 4\text{Re}(d^*/a) \tag{8.139}$$

so that it signals *CPT* violation either in the effective Hamiltonian or in the $\Delta S = -\Delta Q$ amplitudes defined in eqns (8.34), (8.35). However, δ_S is difficult to measure with the high precision required for a significant test, due to the small semi-leptonic branching ratio for K_S.

Considering $\pi\pi$ decays, the decay amplitudes into an isospin I eigenstate can be generalized as

$$A_I = (a_I + b_I)e^{i\delta_I} \qquad \bar{A}_I = (a_I^* - b_I^*)e^{i\delta_I} \tag{8.140}$$

where the b_I terms are *CPT*-violating, so that

$$\epsilon_I \equiv \frac{A_I - \bar{A}_I}{A_I + \bar{A}_I} = \frac{\text{Re}(b_I) + i\text{Im}(a_I)}{\text{Re}(a_I) + i\text{Im}(b_I)} \tag{8.141}$$

and for small *CPT* violation ($|b_I| \ll |a_I|$)

$$\epsilon_I \simeq i\frac{\text{Im}(a_I)}{\text{Re}(a_I)} + \frac{\text{Re}(b_I)}{\text{Re}(a_I)} - i\frac{\text{Im}(b_I)}{\text{Re}(a_I)} \tag{8.142}$$

while in the limit of *CPT* symmetry ϵ_I is pure imaginary, eqn (8.66). $\text{Re}(\epsilon_I)$ is thus a measure of *CPT* violation in the decay amplitudes; note that since $\text{Im}(\epsilon_I) \propto \text{Im}(A_I\bar{A}_I^*)$, if A_I and $\bar{A}_I$ share the same phase (as it happens for the $I = 0$ mode in a physical phase convention, Section 8.3.4) *CPT* symmetry actually requires $\epsilon_I = 0$ (since $|a_I| = |\bar{a}_I|$).

The general expression for the *CP*-violating amplitude ratios

$$\eta_I = \sqrt{\frac{1 + |\epsilon_S|^2}{1 + |\epsilon_L|^2}} \frac{\text{Re}(b_I) + i\text{Im}(a_I) + \epsilon_L[\text{Re}(a_I) + i\text{Im}(b_I)]}{\text{Re}(a_I) + i\text{Im}(b_I) + \epsilon_S[\text{Re}(b_I) + i\text{Im}(a_I)]} \tag{8.143}$$

gives, at first order in the symmetry-violating parameters

$$\eta_I \simeq \epsilon_L + \epsilon_I = \bar{\epsilon} - \delta + \epsilon_I \tag{8.144}$$

In the limit $|\omega \ll 1|$ the general expressions for the *CP*-violating parameters ϵ, ϵ' are obtained (compare to (8.76), (8.77)):

$$\epsilon = \bar{\epsilon} + i \frac{\mathrm{Im}(a_0)}{\mathrm{Re}(a_0)} - \delta + \frac{\mathrm{Re}(b_0)}{\mathrm{Re}(a_0)} \tag{8.145}$$

$$\epsilon' = i \frac{\omega}{\sqrt{2}} \left[\frac{\mathrm{Im}(a_2)}{\mathrm{Re}(a_2)} - \frac{\mathrm{Im}(a_0)}{\mathrm{Re}(a_0)} - i \left(\frac{\mathrm{Re}(b_2)}{\mathrm{Re}(a_2)} - \frac{\mathrm{Re}(b_0)}{\mathrm{Re}(a_0)} \right) \right] \tag{8.146}$$

In eqns (8.145), (8.146) the first two terms respect *CPT* symmetry and the other two do not.

8.5.3 *CPT* and phases

As discussed in Section 8.3.4, *CPT* symmetry determines the phase of the $\bar{\epsilon}$ parameter in a given phase convention: when such symmetry holds the quantity expressing *CP* violation in the effective Hamiltonian is the difference between its off-diagonal elements

$$\mathcal{H}_{12} - \mathcal{H}_{21} = 2i\,\mathrm{Im}(M_{12}) + \mathrm{Im}(\Gamma_{12}) \tag{8.147}$$

where the first (dispersive) term receives contributions from all possible intermediate states in $K^0 - \overline{K}^0$ transitions, while the second (absorptive) only from those on the mass shell, actually reachable in a physical K decay. Except for the presence of common decay modes of K^0 and $\overline{K}^0$ (namely $2\pi, 3\pi, \pi\pi\gamma, \pi\ell\nu$ if the $\Delta S = \Delta Q$ rule is violated, etc.) the above difference would thus be pure imaginary, with the consequence that $\phi(\bar{\epsilon}) = \phi_{SW}$, eqn (8.20) (in the Wu–Yang phase convention this remains true also considering the dominant $(\pi\pi)_{I=0}$ decay mode). Equation (8.145) shows that, in the limit in which the decay amplitudes are saturated by the $\pi\pi$ mode, the *CPT*-conserving and *CPT*-violating components of ϵ are orthogonal:

$$\bar{\epsilon} + i\xi_0 \simeq \frac{1}{\sqrt{2}} \left[\frac{-\mathrm{Im}(M_{12})}{\Delta m} - \xi_0 \frac{2\Delta m}{\Delta\Gamma} \right] e^{i\phi_{SW}} \tag{8.148}$$

$$-\delta + \frac{\mathrm{Re}(b_0)}{\mathrm{Re}(a_0)} \simeq \frac{-i}{\sqrt{2}} \left[\frac{M_{11} - M_{22}}{2\Delta m} + \frac{\mathrm{Re}(b_0)}{\mathrm{Re}(a_0)} \right] e^{i\phi_{SW}} \tag{8.149}$$

It follows that the *CPT*-violating part can be obtained from the measurement of the phase difference between ϵ and ϕ_{SW}:

$$\left| -\delta + \frac{\mathrm{Re}(b_0)}{\mathrm{Re}(a_0)} \right| \simeq |\epsilon| \tan[\phi(\epsilon) - \phi_{SW}] \tag{8.150}$$

Experimentally $\phi_{+-} - \phi_{SW} = (0.61 \pm 1.19)^\circ$ (Alavi-Harati *et al.*, 2003).

The validity of the approximation $\phi(\epsilon) \simeq \phi_{SW}$ hinges on the smallness of *CPV* in decay modes different from $\pi\pi$. To better quantify the above, a (phase-convention independent) parameter ϵ_T can be introduced, which only violates *CP* and *T* but not *CPT* (and as such is proportional to δ_T of eqn (7.56)):

$$\epsilon_T \equiv |\epsilon_T|e^{i\phi_{SW}} = i\frac{|\mathcal{H}_{21}|^2 - |\mathcal{H}_{12}|^2}{\Delta\Gamma(\lambda_L - \lambda_S)}$$

$$= \frac{\mathrm{Re}(M_{12})\mathrm{Im}(\Gamma_{12}) - \mathrm{Re}(\Gamma_{12})\mathrm{Im}(M_{12})}{[(\Delta m)^2 + (\Delta\Gamma)^2/4]}\left(1 - i\frac{2\Delta m}{\Delta\Gamma}\right) \quad (8.151)$$

Note that ϵ_T lies *exactly* along the ϕ_{SW} direction. For small *CPV* and $\phi_{SW} \simeq \pi/4$ this becomes $\epsilon_T \simeq \bar{\epsilon} + i\frac{\mathrm{Im}(\Gamma_{12})}{\Delta\Gamma}$ so that $\mathrm{Re}(\bar{\epsilon}) = \mathrm{Re}(\epsilon_T)$ and

$$\epsilon = \epsilon_T - \delta + \frac{\mathrm{Re}(b_0)}{\mathrm{Re}(a_0)} + i\left[-\frac{\mathrm{Im}(\Gamma_{12})}{\Delta\Gamma} + \frac{\mathrm{Im}(a_0)}{\mathrm{Re}(a_0)}\right] \quad (8.152)$$

showing again that in case of *CPT* symmetry the phase of ϵ is approximately ϕ_{SW}. In any phase convention in which $\mathrm{Im}(a_0) \ll \mathrm{Re}(a_0)$ the term in square brackets is

$$\left[-\frac{\mathrm{Im}(\Gamma_{12})}{\Delta\Gamma} + \frac{\mathrm{Im}(a_0)}{\mathrm{Re}(a_0)}\right] \simeq \frac{1}{2}\left[\tan\arg(\Gamma_{12}) - \tan\arg(\bar{a}_0 a_0^*)\right] \simeq \frac{1}{2}\Delta\varphi$$

having set $\Delta\varphi \equiv \tan\arg(\Gamma_{12}a_0\bar{a}_0^*)$ for the phase difference between the non-diagonal terms in Γ_{12} corresponding to intermediate states different from $(\pi\pi)_{I=0}$. This term does not vanish in general, and to set experimental limits on *CPT* violation its value must be known. Factoring out the contribution from the dominant state $(\pi\pi)_{I=0}$ which practically saturates $\Gamma_S \gg \Gamma_L$:

$$\frac{\mathrm{Im}(\Gamma_{12})}{\Delta\Gamma} \simeq \frac{\mathrm{Im}(a_0)}{\mathrm{Re}(a_0)} + \frac{\mathrm{Im}(\widehat{\Gamma}_{12})}{\Delta\Gamma}$$

where the second term (which is practically $\Delta\varphi/2$) includes a sum over all intermediate states which can contribute *except* $\pi\pi(I = 0)$.

The $\delta_\perp$ component can be obtained from the measurements of *CP* violation in $\pi\pi$ decays. Writing the (measurable) *CP*-violating amplitude ratios for $\pi\pi$ decays (8.80) (neglecting terms of order $\omega\,\epsilon'$) with the help of (8.152) and eliminating ϵ' the *CPT*-violating parameter δ can be written in terms of quantities with well-defined phases

$$\delta = \epsilon_T - \frac{2}{3}\eta_{+-} - \frac{1}{3}\eta_{00} + \epsilon_0 + i\frac{\mathrm{Im}(\Gamma_{12})}{\Delta\Gamma} \quad (8.153)$$

Since $|\eta_{+-}| \simeq |\eta_{00}|$, at first order in $(\phi_{+-} - \phi_{SW})$ and $(\phi_{00} - \phi_{SW})$ one gets

$$|\delta_\perp| \simeq |\eta_{+-}| \left(\frac{2}{3}\phi_{+-} + \frac{1}{3}\phi_{00} - \phi_{SW} \right) + \frac{\Delta\varphi}{2} \cos(\phi_{SW}) - \frac{\text{Re}(b_0)}{\text{Re}(a_0)} \sin(\phi_{SW})$$

(8.154)

$$|\delta_\parallel| \simeq |\epsilon_T| - |\eta_{+-}| + \frac{\Delta\varphi}{2} \sin(\phi_{SW}) + \frac{\text{Re}(b_0)}{\text{Re}(a_0)} \cos(\phi_{SW})$$

(8.155)

and inserting in eqn (8.126) finally

$$m(\overline{K}^0) - m(K^0) \simeq \frac{2\Delta m |\eta_{+-}| [2\phi_{+-}/3 + \phi_{00}/3 - \phi_{SW}]}{\sin\phi_{SW}}$$

$$- \frac{\Delta m \Delta\varphi}{\tan\phi_{SW}} + 2\Delta m \frac{\text{Re}(b_0)}{\text{Re}(a_0)} \quad (8.156)$$

Neglecting the decay modes different from $\pi\pi$ ($\Delta\varphi = 0$) and the possibility of direct *CPT* violation ($b_0 = 0$) the above expression gives $m(\overline{K}^0) - m(K^0) \lesssim 2 \cdot 10^{-11}$ eV, rather impressive when related to the kaon mass

$$|m(\overline{K}^0) - m(K^0)| \lesssim 10^{-18} \, m_K$$

(8.157)

which however should not be interpreted to mean that *CPT* symmetry is tested at this level of precision, as m_K is due to strong interactions and is not a proper normalization factor (nor is Δm, which remains different from zero when $m(\overline{K}^0) = m(K^0)$ in the limit of *CP* symmetry).

Summarizing the above discussion, $\delta \neq 0$ ($\delta_{CPT} \neq 0$) in the effective Hamiltonian $\mathcal{H}$ parametrizes[122] a possible *CPT* violation in the amplitudes with $|\Delta S = 0|$; *CPT* violation in $|\Delta S = 1|$ decays is expressed by non-zero amplitudes b_I (for which $\overline{b} = -b^*$); finally *CPT* violation in $|\Delta S = 2|$ transitions shifts the phase of ϵ from its value ($\simeq \phi_{SW}$).

Indications of *CPT* symmetry violation in the neutral *K* system could be obtained from:

• $\delta \neq 0$ ($\epsilon_S \neq \epsilon_L$);

• $\phi(\epsilon) \neq \phi_{SW}$;

• a phase difference $\phi_{+-} - \phi_{00} \neq 0$;

---

[122] Note that since a violation of *CPT* symmetry would most likely be accompanied by a violation of Lorentz symmetry (Chapter 5), a momentum-dependence of the δ parameter might be expected in this case (Kostelecký, 2001).

- a difference between the semi-leptonic charge asymmetries δ_L and $2\text{Re}(\epsilon)$ obtained from $\pi\pi$ decays;

- a difference between the K_S and K_L semi-leptonic charge asymmetries δ_L and δ_S.

- evidence of $K_S - K_L$ interference terms in semi-leptonic decays;

The independent tests (in absence of violation of quantum mechanics) are the comparison of δ_L with $\text{Re}(\eta_{+-})$ and that of the phase of ϵ with the super-weak phase.

As discussed in Section 8.3.6, the values of the $\pi\pi$ scattering phase shifts are such that the phase of ϵ' is close to ϕ_{SW}, so that if *CPT* symmetry holds ϵ'/ϵ is approximately real (see eqn (8.97)). It follows that the imaginary part of ϵ'/ϵ is an (approximate) measure of *CPT* violation in the decay amplitudes (*direct CPT violation*): for small $|\epsilon'/\epsilon|$:

$$\phi_{00} - \phi_{+-} \simeq -3\,\text{Im}(\epsilon'/\epsilon) \tag{8.158}$$

A precise measurement by the KTeV experiment, obtained by fitting the time dependence of $K \to \pi\pi$ decays, gives $\phi_{00} - \phi_{+-} = (0.39 \pm 0.50)°$, consistent with *CPT* symmetry (Alavi-Harati *et al.*, 2003); this phase difference corresponds to $\text{Re}(b_2)/\text{Re}(a_2) = (-1 \pm 2)\cdot 10^{-4}$; the limitation of this approach lies in the approximations involved, so that it does not allow in principle to set arbitrarily small limits.

Also for ϵ' the *CPT*-conserving and -violating terms are orthogonal: decomposing this parameter along the direction of ϕ_{SW}

$$|\epsilon'_{\parallel}| \simeq \frac{1}{3}(|\eta_{+-}| - |\eta_{00}|) \qquad |\epsilon'_{\perp}| \simeq \frac{1}{3}|\eta_{00}|(\phi_{+-} - \phi_{00}) \tag{8.159}$$

and *CPT* symmetry requires the phase of ϵ' to be $\delta_2 - \delta_0 + \pi/2 \simeq \phi_{SW}$, so that

$$\arg(\epsilon'/\epsilon) = \phi_{CP} + \phi_{CPT} \tag{8.160}$$

$$\phi_{CP} = \pi/2 + \delta_2 - \delta_0 - \phi(\epsilon) \simeq 0 \tag{8.161}$$

$$\phi_{CPT} = -\frac{\text{Re}(b_2)/a_2 - \text{Re}(b_0)/a_0}{\text{Im}(a_2)/a_2} \tag{8.162}$$

and the smallness of $\text{Re}(\epsilon'/\epsilon)$ indicates that $\text{Im}(a_2)/a_2 \simeq 10^{-4}$, making the test quite sensitive. From eqn (8.146)

$$\text{Im}(\epsilon'/\epsilon) \simeq \frac{-\omega}{\epsilon\sqrt{2}}\left[\cos\phi_{CP}\left(\frac{\text{Re}(b_2)}{\text{Re}(a_2)} - \frac{\text{Re}(b_0)}{\text{Re}(a_0)}\right) - \sin\phi_{CP}\left(\frac{\text{Im}(a_2)}{\text{Re}(a_2)}\right)\right]$$

and as $\phi_{CP} \simeq 0$

$$\left(\frac{\mathrm{Re}(b_2)}{\mathrm{Re}(a_2)} - \frac{\mathrm{Re}(b_0)}{\mathrm{Re}(a_0)} \right) \simeq -2.4 \cdot 10^{-2} (\phi_{+-} - \phi_{00})$$

in which the strong phase shifts (and their errors) cancel. The deviations of ϕ_{+-} and ϕ_{00} from ϕ_{SW} due to direct *CP* violation are below $1°$: larger deviations would indicate either *CPT* violation or significant contributions to $K^0 - \overline{K}^0$ mixing from virtual states different from $\pi\pi$, to which however the phase difference $\phi_{+-} - \phi_{00}$ is insensitive.

Using only unitarity it is possible to link all the decay widths of neutral kaons to the symmetry violating parameters, exploiting the Bell–Steinberger relation (8.96), rewritten as

$$\mathrm{Re}(\bar{\epsilon}) - i\,\mathrm{Im}(\delta) \simeq \frac{(\epsilon + \kappa)\Gamma_S}{(\Gamma_L + \Gamma_S) + 2i\Delta m} \tag{8.163}$$

having used eqn (8.11), the dominance of the $(\pi\pi)_{I=0}$ state

$$A(K_L \to (\pi\pi)_{I=0}) A(K_S \to (\pi\pi)_{I=0})^* = \eta_0 \, |A[K_S \to (\pi\pi)_{I=0}]|^2 \simeq \epsilon\,\Gamma_S$$

$$\text{and} \quad \kappa \equiv (1/\Gamma_S) \sum_f A(K_L \to f) A(K_S \to f)^*$$

with the sum excluding the $\pi\pi_{(I=0)}$ state. The contributions of the various decay modes to κ are all suppressed by small quantities.

Separating the real and imaginary parts of (8.163)

$$\mathrm{Re}(\bar{\epsilon}) \simeq \mathrm{Re}\left(\frac{\epsilon + \kappa}{1 + 2i\Delta m/\Gamma_S} \right) \qquad \mathrm{Im}(\delta) \simeq \mathrm{Im}\left(\frac{\epsilon + \kappa}{1 + 2i\Delta m/\Gamma_S} \right) \tag{8.164}$$

Since $\pi\pi$ decays dominate in the sum, the determination of δ with this method strongly depends on knowledge of $\phi_{\pi\pi}$, but the b_I amplitudes do not enter in the expressions, so that no assumption about direct *CPT* violation has to be made (note however that something as dramatic as *CPT* violation might even be accompanied by a violation of unitarity (Kenny and Sachs, 1973), invalidating this approach). An example of this kind of analysis can be found in (Angelopoulos *et al.*, 2003).

8.6 Three-pion decays

Three-pion states with zero angular momentum have $J^{PG} = 0^{--}$, (*G* denoting the *G*-parity, Section 3.2.4), therefore their *CP*-parity is $\eta_{CP} = (-1)^I PG = (-1)^I$. The states can be classified in terms of their total isospin I (Grimus, 1988):

Isospin I	0	1	1	2	3
Number of states	1	1	2	2	1
Permutation symmetry	A	S	M	M	S
η_{CP} (state $I_3 = 0$)	+1	−1	−1	+1	−1
$I(\pi\pi)$	1	0	2	1	2
$L(\pi\pi)$	O	E	E	O	E

In the above table the states are labelled according to their properties under exchange of two pion momenta: S,A denote completely symmetric and antisymmetric states, and M states with mixed symmetry, which are symmetric under the exchange of the two identical pions (or of π^+ and π^- in the $\pi^+\pi^-\pi^0$ case); in the bottom part of the table the isospin I and angular momentum L of such pair are indicated (E,O denoting even and odd values respectively). For the state $\pi^0\pi^0\pi^0$ with three identical particles only the symmetric states with $I = 1, 3$ are possible, while all the states are admissible for $\pi^+\pi^-\pi^0$; for decays of $K^\pm$ the $I = 0$ states are not allowed (since $I_3 \neq 0$)

The three-pion decay amplitudes can be conveniently expressed as sums of terms in which the dependence on isospin (functions of isospin vectors), angular momentum and parity (functions of vector momenta) and energy (functions of the energies, namely form factors) are factorized (Fabri, 1954; Zemach, 1998).

As in the case of 2π decays, the decay amplitudes are related by isospin relations: if no transitions with $\Delta I > 3/2$ exist, the $I = 3$ state can be ignored, and the amplitudes are (D'Ambrosio and Isidori, 1998)

$$A(K^\pm \to \pi^\pm\pi^\pm\pi^\mp) = 2A_{1S}^{(\pm)} + A_{1M}^{(\pm)} + A_{2M} \tag{8.165}$$

$$A(K^\pm \to \pi^\pm\pi^0\pi^0) = A_{1S}^{(\pm)} - A_{1M}^{(\pm)} + A_{2M} \tag{8.166}$$

$$A(K^0 \to \pi^\pm\pi^-\pi^0) = \frac{1}{\sqrt{2}}[A_{1S}^{(0)} - A_{1M}^{(0)} + A_{0A} + (2/3)\widehat{A}_{2M}] \tag{8.167}$$

$$A(K^0 \to \pi^0\pi^0\pi^0) = \frac{3}{\sqrt{2}}A_{1S}^{(0)} \tag{8.168}$$

where subscripts indicate the isospin I of the final state and the symmetry under permutation, and $\widehat{A}_2 = A_2(p_3, p_2, p_1) - A_2(p_1, p_3, p_2)$.

For K decays the total angular momentum is zero and the angular part of the amplitude must be $\propto Y_0^0(\theta, \phi)$; since due to the (generalized) Bose–Einstein symmetry the amplitudes must be overall symmetric under the exchange of any two pions, the isospin part must have the same exchange symmetry of the form factor part, which can be written in terms of the s_i invariants (Appendix D). If the series expansion of the form factor is stopped at linear terms in the u, v variables (D.1) because of the limited phase space available, the $I = 0$ state is not possible (the totally antisymmetric function of lowest order is $(s_1 - s_2)(s_2 - s_3)(s_3 - s_1)$, of order 6 in the momenta); in this approximation the decay amplitudes are given by eqns (8.165)–(8.168) with the A_S being constants, the A_M linear in u and $\widehat{A}_M$ linear in v, all real if strong re-scattering is neglected and *CP* symmetry holds.

8.6.1 Neutral kaons

We recall from Section 6.1.2 that the $3\pi^0$ state must have total isospin $I = 1, 3$ so that $\eta_{CP}(3\pi^0) = -1$. The decay $K_S \rightarrow 3\pi^0$ thus violates *CP* (just as $K_L \rightarrow 2\pi$), but in this case the expected branching ratio is

$$BR(K_S \rightarrow 3\pi^0) \simeq |\epsilon_S|^2(\tau_S/\tau_L)\, BR(K_L \rightarrow 3\pi^0) \sim 10^{-9} \qquad (8.169)$$

This decay mode was searched in several ways: by measuring asymmetries in the decays of tagged mesons, eqn (8.115) (Angelopoulos *et al.*, 1998b), by looking for the $K_L - K_S$ interference term close to the production target of a $K^0, \overline{K}^0$ beam (the $\eta_{CP} = -1$ analogue of eqn (6.7)) (Lai *et al.*, 2005) and directly with K_S at 'K-factories' (Section 8.8) (Angelopoulos *et al.*, 2005); the latter approach gives at the best current limit $BR(K_S \rightarrow 3\pi^0) < 1.2 \cdot 10^{-7}$ (90% CL). Introducing the *CP*-violating amplitude ratio

$$\eta_{000} = \frac{A(K_S \rightarrow 3\pi^0)}{A(K_L \rightarrow 3\pi^0)} \qquad (8.170)$$

this limit corresponds to $|\eta_{000}| < 0.018$ (90% CL).

If *CPT* symmetry holds and transitions to the $I = 3$ state ($\Delta I \geq 5/2$) are neglected $\eta_{000} = \bar{\epsilon} + i\mathrm{Im}(a_1)/\mathrm{Re}(a_1)$, eqn (8.61), so that in the Wu–Yang phase convention the real part of η_{000} measures *CPV* in the mixing and its imaginary part *CPV* in the $I = 1$ decay amplitudes. Interference experiments allow to measure both the real and imaginary part of η_{000} (Lai *et al.*, 2005):

$$\mathrm{Re}(\eta_{000}) = -0.002 \pm 0.019 \qquad \mathrm{Im}(\eta_{000}) = -0.003 \pm 0.021 \qquad (8.171)$$

The $\pi^+\pi^-\pi^0$ state is not in general a *CP* eigenstate, although $\eta_{CP} = -1$ is kinematically favoured for K decays. The decay $K_S \rightarrow \pi^+\pi^-\pi^0$ therefore does not require *CP* violation, but the *CP*-conserving part of the amplitude vanishes at the centre of the Dalitz plot: the K_1 component of K_S (with $\eta_{CP} = +1$) can decay into $\pi^+\pi^-\pi^0$ by a *CP*-conserving $\Delta I = 3/2$ transition which however is antisymmetric under the exchange of π^+ and π^-, and thus vanishes when the corresponding momenta are equal; by integrating over the whole Dalitz plot (of course non-trivial from an experimental point of view) the contributions from $I = 0, 2$ final states ($\eta_{CP} = +1$) integrate to zero, and a pure CP-violating quantity is obtained:

$$\eta_{+-0} = \frac{A(K_S \to \pi^+\pi^-\pi^0; CP = -1)}{A(K_L \to \pi^+\pi^-\pi^0)} \qquad (8.172)$$

Indeed, of the four amplitudes for K_S, K_L decays to $\pi^+\pi^-\pi^0$ states of definite CP-parity the K_L decay into the CP-even state is suppressed both by angular momentum and CP symmetry and can be neglected. On the contrary the K_S decay to the CP-odd state violates CP, and that to the CP-even does not but is suppressed by the angular momentum barrier; however the CP-even state has the $\pi^+\pi^-$ pair in the antisymmetric isospin 1 state (while the K_L decay to the CP-odd state is unsuppressed). The decay amplitudes for tagged $K^0, \overline{K}^0$ are

$$A(K \to \pi^+\pi^-\pi^0) \simeq A_L(CP = -1) \pm [A_S(CP = +1) + A_S(CP = -1)]$$

where the $+$ $(-)$ sign is for K^0 ($\overline{K}^0$); the first and third terms correspond to even L (without and with CPV) and are symmetric in v, while the second corresponds to $L = 1$ (no CPV) and is antisymmetric in v ($A_L(CP = +1)$ neglected). This suggests that the η_{+-0} parameter can also be written as

$$\eta_{+-0} = \frac{\int A_L^*(u, v) A_S(u, v)\, du\, dv}{\int |A_L(u, v)|^2\, du\, dv}$$

with the integral extending over all the phase space.

The $A_{S,L}$ amplitudes can be decomposed in terms of isospin states, and the strong phase shifts are expected to be small in this case because of the limited phase space available; ignoring $I = 3$ then $\eta_{+-0} = \eta_{000}$ (that is, no direct CPV if only the $I = 1$ CP-violating amplitude is present).

8.6.2 Charged kaons

In the case of charged kaons (no mixing possible) the way to look for CP violation is to compare the decay properties of K^+ and K^-: any difference in partial decay widths or decay distributions indicates (direct) CPV.

For the most copious two-body hadronic (FSI needed) decays CPT symmetry requires (Chapter 5)

$$\Gamma(K^+ \to \pi^+\pi^0) = \Gamma(K^- \to \pi^-\pi^0)$$

when neglecting EM interactions, while if they are included the above constraint becomes[123]

$$\Gamma(K^+ \to \pi^+\pi^0) + \Gamma(K^+ \to \pi^+\pi^0\gamma) = \Gamma(K^- \to \pi^-\pi^0) + \Gamma(K^- \to \pi^-\pi^0\gamma)$$

The next most abundant hadronic decays are $K^\pm \to \pi^\pm\pi^\pm\pi^\mp$ (called $\tau^\pm$) and $K^\pm \to \pi^0\pi^0\pi^\pm$ (called $\tau'^\pm$) for which not only decay widths but also the Dalitz plot parameters $g_\pm, h_\pm, \ldots$ (see eqn (D.5)) can be compared. Rate differences are more difficult to measure experimentally at hadron machines, as they require accurate monitoring of the (different) beam intensities, while differences in the shape of the decay distributions are independent of the normalization and are only affected by acceptance differences.

Some experiments measured the difference in the linear Dalitz plot slopes for K^+ and K^- decays using alternating beams, with results at per cent precision affected by possible time variations of the detector properties. The NA48/2 experiment at CERN performed such a measurement using instead unique simultaneous and collinear K^+ and K^- beams: particles produced at zero angle with respect to the incident proton beam (thus with no first-order difference in the differential cross sections) were momentum selected and analysed in a two-stage achromatic magnet set-up which finally left the two beams collinear and superimposed to within 1 mm along the detector axis (see Fig. 8.8), the detector being mostly that of NA48 (Section 8.3.6). The K^+/K^- beam intensity ratio ($\sim$1.7) due to the different production cross section) was unimportant for the measurement.

With the above beam arrangement slow time drifts of detector properties become irrelevant, and the detector acceptances would be identical for $K^\pm$, except for the

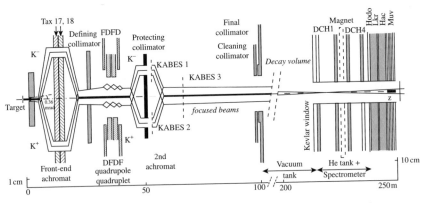

FIG. 8.8. Scheme of the beam arrangement for the NA48/2 experiment comparing $K^\pm$ decays (Batley *et al.*, 2006c).

---

[123] Note that the pure *bremsstrahlung* part of the radiative decay cannot by itself generate an asymmetry, as its rate is proportional to the non-radiative one.

fact that the spectrometer dipole magnet bends charged pions in opposite directions, thus illuminating different parts of the detector. This fact was compensated by frequent (6 to 24 h period) reversals of the direction of this field, which allowed the analysis to be performed on pairs of data sets corresponding to opposite field orientations, by forming double ratios of distributions in which the main acceptance asymmetries cancel at first order, and only second-order effects could introduce biases, due e.g. to time variations of acceptance asymmetries on a time scale shorter than that of field reversal. The directions of all the magnetic fields along the beam lines were also periodically reversed, and even more robust quadruple ratios were formed to obtain the final measurement. Results are (Batley *et al.*, 2006c; Batley *et al.*, 2006b; Batley *et al.* 2007)

$$\frac{g_+ - g_-}{g_+ + g_-} = (-1.7 \pm 2.1) \cdot 10^{-4} \qquad \frac{g_+ - g_-}{g_+ + g_-} = (1.8 \pm 1.8) \cdot 10^{-4} \qquad (8.173)$$

for the $K^\pm \to \pi^\pm \pi^+ \pi^-$ and $K^\pm \to \pi^\pm \pi^0 \pi^0$ decays respectively, consistent with no *CPV* (and with SM estimations).

8.7 Other decay modes

A $K_S - K_L$ interference term allows the observation of *CP* violation in decays to final states which are not necessarily *CP* eigenstates: for a non-leptonic state such as $\pi^+ \pi^- \pi^0$, $\gamma\gamma$ the presence of such a term is a clear indication of *CP* violation. In order to interfere after integration over angles, the final states (parity eigenstates) reached by K_S and K_L must have the same parity; if *CP* symmetry holds such states would have opposite *C*-parities: when dealing with states formed by particles (or sets of particles) which are *C* eigenstates, these would thus have different angular momenta, so that no interference would be possible after integration. Moreover this term is linear in the interfering amplitudes rather than quadratic line the decay rate: in the usual case of one amplitude being much smaller than the other it is thus a more sensitive quantity than the decay rate itself. Here we just briefly mention some other final states in which *CP* symmetry can be probed, as they exemplify some general approaches.

In the radiative decay $K \to \pi\pi\gamma$ the photon can be emitted either by QED 'inner' *bremsstrahlung* (IB) or from the weak decay vertex (*direct emission* DE): the former process dominates at low photon energies E_γ, its amplitude actually diverging for $E_\gamma \to 0$. As for all similar processes this is not a physical issue, since for low enough E_γ the radiative and non-radiative decays cannot be distinguished: what is actually measured experimentally is always the sum of the non-radiative process and the radiative one for photon energies up to a given value, below which the experimental resolution does not allow the two to be separated; while this is always a finite quantity, it is always necessary to indicate the actual cuts used when quoting results for radiative decays.

We recall that electric and magnetic multipoles have opposite transformation properties under *CP*: $CP(E_J) = (-1)^{J+1}$ and $CP(M_J) = (-1)^J$; as the photon polarization is usually not measured, these amplitudes do not interfere. *Bremsstrahlung* emission is essentially an electric dipole transition (Section 2.1.4): since the $K \to \pi^+\pi^-\gamma$ decay can occur via such a transition the corresponding amplitude dominates for K_S decay ($BR \simeq 1.8 \cdot 10^{-3}$); the corresponding K_L decay instead violates *CP* (as $K_L \to \pi\pi$ does), so that the smaller magnetic dipole amplitude (*CP*-conserving) has a chance to compete with it ($BR \simeq 4.4 \cdot 10^{-5}$)[124].

If a *CP*-violating E1 amplitude exists in the K_L decay, it can interfere with the E1 IB amplitude for K_S: a *CP*-violating parameter is defined as (D'Ambrosio and Isidori, 1998)

$$\eta_{+-\gamma} = \frac{A(K_L \to \pi^+\pi^-\gamma; E1)}{A(K_S \to \pi^+\pi^-\gamma; E1)} \tag{8.174}$$

This quantity was measured by studying the $K_L - K_S$ interference after a regenerator (Matthews *et al.*, 1995) and found to be compatible with ϵ (the expected value in absence of direct *CPV*, due to the K_1 component of K_L), being another manifestation of *CPV* in the mixing. If an E1 DE amplitude were present in the K_L decay, it could interfere with the IB one and shift $\eta_{+-\gamma}$ away from ϵ, exhibiting direct *CPV*.

If the photon polarization could be measured, an interference of the IB E1 amplitude and the DE M1 one (of comparable magnitude) could give rise to sizeable *CP*-violating effects; as discussed in Chapter 2, while polarization measurements at high energies are difficult, in $\gamma \to e^+e^-$ conversion the lepton plane orientation is correlated to the photon helicity, thus providing a handle to exploit the above interference by studying the $K_L \to \pi^+\pi^-e^+e^-$ decay. This rare decay was detected and a large ($\sim 14\%$) asymmetry in the relative orientation of the $\pi^+\pi^-$ and e^+e^- planes was observed by KTeV and NA48 (see Fig. 8.9) (Alavi-Harati *et al.*, 2000; Lai *et al.*, 2003). Assuming unitarity and small FSI such asymmetry is a *T*-violating effect, and its measured magnitude is in agreement with the expectations from *CPV* in the mixing.

Similarly, while $K \to \gamma\gamma$ decays are not *CP*-violating, a net circular polarization of the photons from a pure K_S or K_L would indicate *CP* violation: photons with parallel polarization from K_L or with perpendicular polarization from K_S are obtained with *CPV* (D'Ambrosio and Isidori, 1998). Again, the rare decays $K_S, K_L \to \mu^+\mu^-e^+e^-$ would allow this kind of measurement to be performed (Uy, 1991).

---

[124] The case of $K^\pm \to \pi^\pm\pi^0\gamma$ is similar, in this case the suppression being due to the $\Delta I = 3/2$ nature of the non-radiative parent process.

The observation of neutral kaon decays into $\ell^+\ell^-$ ($\ell = e, \mu$) does not violate CP symmetry, as the $\ell^+\ell^-$ state, even for zero total angular momentum, is not a CP eigenstate: $\eta_{CP}(\ell^+\ell^-) = (-1)^{S+1}$. The decays $K_L \to \mu^+\mu^-$ and $K_L \to e^+e^-$ have been observed, with branching ratios $\sim 10^{-8}$ and $\sim 10^{-11}$ respectively. A longitudinal polarization of the leptons in this decay would be a signal of CPV, since in this case the CP transformation is equivalent to the inversion of the helicities λ:

$$CP \, |\ell^+(\boldsymbol{p}, \lambda = +1) \, \ell^-(-\boldsymbol{p}, \lambda = +1)\rangle = |\ell^-(-\boldsymbol{p}, \lambda = -1) \, \ell^+(\boldsymbol{p}, \lambda = -1)\rangle$$

and analogously for the state with opposite helicity: if one of these states has a higher probability of emission (so that the emitted leptons have a net longitudinal polarization) CP is violated.

We also recall that T violation can be searched for by studying the transverse lepton polarization in $K_{\ell 3}$ decays, as discussed in detail in Section 4.4.1.

8.8 Coherent kaon pairs

Coherent kaon pairs (Section 7.5) can be produced in $p\bar{p}$ or e^+e^- annihilations: in this case the mesons are correlated in time and allow a different class of experiments to be performed, requiring a way to produce copious amounts of kaon pairs in a well-defined state. This was done in K-factories (or ϕ-factories), namely e^+e^- colliders with centre of mass energy 1020 MeV corresponding to the mass of the ϕ vector meson ($s\bar{s}$ 'strangeonium' state with $J^{PC} = 1^{--}$), which decays to $K_S K_L$ or

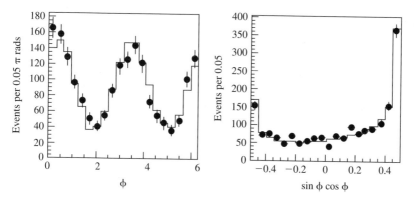

FIG. 8.9. The CP-violating decay plane asymmetry for $K_L \to \pi^+\pi^-e^+e^-$ decays (Alavi-Harati et al., 2000). The angle ϕ is defined in the K_L rest frame by $\sin\phi\cos\phi = (\hat{\boldsymbol{n}}_e + \hat{\boldsymbol{n}}_\pi) \cdot [\hat{\boldsymbol{n}}(\pi^+) + \hat{\boldsymbol{n}}(\pi^-)]$ where $\hat{\boldsymbol{n}}_e = \hat{\boldsymbol{n}}(e^-) \times \hat{\boldsymbol{n}}(e^+)$, $\hat{\boldsymbol{n}}_\pi = \hat{\boldsymbol{n}}(\pi^-) \times \hat{\boldsymbol{n}}(\pi^+)$ and $\hat{\boldsymbol{n}}$ are the unit vectors in the direction of the particles' momenta.

K^+K^- in 83% of the cases. Thanks to C-conservation in the (strong) decay the kaon pairs are necessarily produced in an antisymmetric coherent state $|C-\rangle$ (they could be in a symmetric $|C+\rangle$ state if produced by the radiative decay $\phi \rightarrow \gamma(K\bar{K})_{C=+1}$, which is however highly suppressed).

The formalism for dealing with such coherent pairs was developed in Section 7.5; we recall that the relevant variable is the difference Δt of the kaon decay times, and that a decay of the two correlated kaons into the same final state at the same time is forbidden not only by CP symmetry but also by Bose–Einstein symmetry.

We recall that the observation of one of the kaons decaying into a given final state f_1 constrains the other one to be (at that time) that particular combination of K_S and K_L which cannot decay into f_1: in particular if $f_1 = \pi^+\pi^-$ then the observation of a simultaneous decay of the other kaon into $\pi^0\pi^0$ indicates the existence of direct CP violation, as the second kaon must be a K_2, which only decays into $\pi\pi$ because of direct CPV.

Measurements at the 'same' decay time are obviously difficult in practice, while as soon as one kaon decays coherency is lost and decays (at different times) of the two kaons into the same final state become possible: such decays however can give much information on the parameters of the neutral K system.

First of all we note that contrary to the B system (Chapter 10) the branching ratios of K are mostly large enough that the study of exclusive decays of both kaons is practically feasible; the decay distributions are integrated over one of the decay times and expressed as a function of Δt:

$$I(f_1, f_2; \Delta t) = \int dt_1\, dt_2\, \delta(t_1 - t_2 - \Delta t)\, |A(f_1, t_1; f_2, t_2)|^2$$

When considering decays into the same final state f the time distribution of the rate is independent of the η_f factors:

$$I(f, f; \Delta t) = \frac{\Gamma_S(f)\Gamma_L(f)}{2(\Gamma_S + \Gamma_L)} \left[e^{-\Gamma_L|\Delta t|} + e^{-\Gamma_S|\Delta t|} - 2e^{-(\Gamma_S+\Gamma_L)|\Delta t|/2} \cos(\Delta m \Delta t) \right]$$

From the study of the decays into $\pi\pi$ the ratio ϵ'/ϵ can be extracted:

$$I(\pi^+\pi^-, \pi^0\pi^0; \Delta t > 0) = |\epsilon|^2 \frac{\Gamma(K_S \rightarrow \pi^+\pi^-)\Gamma(K_S \rightarrow \pi^0\pi^0)}{4\Gamma}$$

$$\left[[1 + 2\mathrm{Re}(\epsilon'/\epsilon)]e^{-\Gamma_L|\Delta t|} + [1 - 4\mathrm{Re}(\epsilon'/\epsilon)]e^{-\Gamma_S|\Delta t|} \right.$$

$$\left. -2e^{-\bar{\Gamma}|\Delta t|}[[1 - \mathrm{Re}(\epsilon'/\epsilon)]\cos(\Delta m \Delta t) + [1 + 3\mathrm{Im}(\epsilon'/\epsilon)]\sin(\Delta m \Delta t)] \right]$$

$$(8.175)$$

and similarly for $\Delta t < 0$ (with $\Gamma_S \leftrightarrow \Gamma_L$, $\mathrm{Im}(\epsilon'/\epsilon) \leftrightarrow -\mathrm{Im}(\epsilon'/\epsilon)$). Note that if $\epsilon' \neq 0$ the distribution is not symmetric around $\Delta t = 0$; the region of large $\Delta t \gg \tau_S$ is sensitive to $\mathrm{Re}(\epsilon'/\epsilon)$ while that of $|\Delta t| \sim (1 \div 5)\tau_S$ is sensitive to $\mathrm{Im}(\epsilon'/\epsilon)$. The asymptotic value ($|\Delta t| \gg \tau_S$) of the asymmetry between $I(\Delta t)$ and $I(-\Delta t)$ is just $3\mathrm{Re}(\epsilon'/\epsilon)$, which is also the value of the integrated asymmetry.

From the study of the decays into $\pi \ell \nu$ both the real and the imaginary parts of the CPT-violating parameter δ can be extracted, or actually the combination $\mathrm{Re}(\delta + x_+^{(\ell)})$, while for decays in which one of the kaons decays into $\pi \pi$ and the other semi-leptonically the charge asymmetries can be measured. More details can be found for example in Buchanan $et\ al.$ (1992).

The first K-factory (VEPP-2M) ran in Novosibirsk in the 1990s and collected $\sim 7 \cdot 10^6$ $K_S K_L$ pairs, then the DAΦNE machine in Frascati took over: this was built with two separate rings to minimize the beam-beam interactions, worrisome for low-energy colliders, for a design luminosity of $5 \cdot 10^{32}$ cm^{-2} s^{-1}. Kaon pairs are produced almost at rest and the $K^+ K^-$ ($1.5 \cdot 10^6$ pairs/pb^{-1}) and

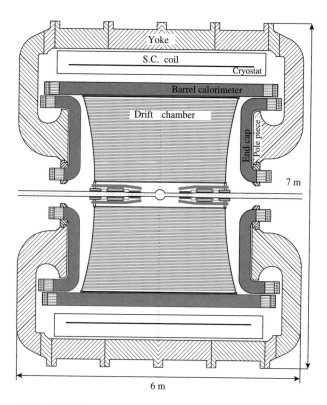

FIG. 8.10. Schematic longitudinal section of the KLOE detector (Aloisio $et\ al.$, 2002); the e^+ and e^- beams enter horizontally from left and right; the interaction region quadrupole magnets are also shown.

$K_L K_S$ ($1 \cdot 10^6$ pairs/pb^{-1}) have low momenta in the laboratory (127 and 110 MeV/c respectively). At DAΦNE the kaon decays are studied with the dedicated KLOE detector, which collected $\sim$2.4 pb^{-1} in 1999–2006: this essentially consists of a large drift chamber filled with helium-based gas to minimize multiple scattering, surrounded by an electromagnetic calorimeter made of a fine mesh of lead and scintillating fibres, all enclosed in a 0.52 T superconducting solenoid (see Fig. 8.10). The size of the detector (4 m diameter) is dictated by the need to have a sufficiently large acceptance for K_L (decay length 3.4 m), and that of the beam pipe (10 cm diameter) by the K_S (decay length 6 mm). The position of the EM photon shower centroid is obtained from the difference in arrival times of the light at the two ends of each fibre bundle in the calorimeter, and the absolute time is measured with respect to a reference given by particle bunch crossings (period 2.751 ns). The excellent time resolution obtained (well below 500 ps for the photon energies of interest) allows one to reconstruct the position of the π^0 decay vertices to $\sim$ 1.5 cm precision: this is obtained by knowing the kaon flight direction (from the decay of the other K) and determining the decay point along it from the photons' impact points and arrival times.

The properties of the coherent initial state allow the possibility of tagging the presence of a K_S by the identification of a K_L decay in the opposite hemisphere, to obtain a unique 'K_S beam' which is impossible to produce otherwise; for this reason these experiments are ideally suited to the study of rare K_S decays and the related *CP* violation. The identification of K_L is achieved not only by detecting its three-body decays but also by identifying its hadronic interaction with the detector, easily recognized by time-of-flight thanks to the very low K velocity ($\beta \simeq 0.2$). Some measurements performed by KLOE were already mentioned earlier; in principle by using both K_S and K_L a double-ratio measurement of direct *CPV* could be done, but this requires very high luminosities and a challenging control of systematics.

8.9   *C,P* properties of *CP* violation

Different *CP*-violating effects can also be distinguished in terms of their behaviour with respect to other discrete transformations, thus in principle probing different kinds of interactions.

As eigenstates of the strong interactions, K^0 and $\overline{K}^0$ are parity eigenstates, sharing the same eigenvalue (P is a multiplicative quantum number commuting with strangeness). Their usual parity assignment, consistent with those of the other particles participating in their (strong, parity-conserving) production reactions, is $\eta_P = -1$ (pseudo-scalar), but in this section we are going to keep η_P unspecified (but real) for a while. K^0 and $\overline{K}^0$ are clearly not C eigenstates, one being the antiparticle of the other, as in eqn (7.3); without loss of generality the phase factors

η_C can be chosen to be real (by setting $C^2 = 1$).[125] Charge conjugation eigenstates are defined as

$$C|K_1\rangle = -|K_1\rangle \qquad C|K_2\rangle = +|K_2\rangle \qquad (8.176)$$

Expressing the $K_{1,2}$ states in terms of $K^0, \overline{K}^0$ one obtains for these two orthogonal states

$$|K_1\rangle = \frac{e^{i\phi_1}}{\sqrt{2}} \left[|K^0\rangle - \eta_C |\overline{K}^0\rangle \right] \qquad |K_2\rangle = \frac{e^{i\phi_2}}{\sqrt{2}} \left[|K^0\rangle + \eta_C |\overline{K}^0\rangle \right] \qquad (8.177)$$

with arbitrary ϕ_1, ϕ_2. Since the relative phase between $|K^0\rangle$ and $|\overline{K}^0\rangle$ is also arbitrary, the choice $\eta_C = +1$ is always possible; the states $K_{1,2}$ are seen to be also eigenstates of CP:

$$CP|K_1\rangle = -\eta_P |K_1\rangle \qquad CP|K_2\rangle = +\eta_P |K_2\rangle \qquad (8.178)$$

and by construction the eigenvalues are independent from the arbitrary phase η_C (and ϕ_1, ϕ_2). The opposite CP-parity of the two states is due to their opposite C-parity (and equal parity). This shows that a CP-violating virtual transition $K_1 \leftrightarrow K_2$ violates C but not P, independently from any choice of phase factors. With the usual parity assignment $\eta_P = -1$ eqn (6.3) is obtained; in this case the state which decays into $\pi\pi$ with no suppression (CP even) is indeed K_1, while K_2 is CP-odd, and the CP-violating decay $K_2 \to \pi\pi$ (direct CPV) violates P but not C: the two kinds of CP violation (in mixing and in decay) have thus different C and P symmetry violating properties.

Other CP-violating observables can also be classified in terms of their parity and charge conjugation properties (Okun, 2003). For example, barring fake effects due to final state interactions, a T-odd (and P-even) transverse muon polarization in the semi-leptonic $K^+ \to \pi^0 \mu^+ \nu$ decay $\boldsymbol{S}_\mu \cdot (\boldsymbol{p}_\pi \times \boldsymbol{p}_\mu)$ would be required by CPT symmetry to be opposite for the corresponding K^- decay (thus a C-odd quantity); measuring it to be the same for both decays would rather indicate CPT violation (in absence of FSI effects). Conversely, static electric dipole moments are P-odd and T-odd, and CPT symmetry would require them to be the same for particle and antiparticle (C-even). The longitudinal polarization of muons in the $\eta \to \mu^+ \mu^-$ decay behaves similarly: $\boldsymbol{S}_\mu \cdot \boldsymbol{p}_\mu$ is P-odd, C-even and T-even.

Further reading

All books on CP violation deal extensively with the kaon system. Among the many review articles on CP-violating effects in flavoured neutral mesons a very

---

[125] With $C^2 = -1$ we would have $C|K^0\rangle = +|\overline{K}^0\rangle$ and $C|\overline{K}^0\rangle = -|K^0\rangle$, C would not be Hermitian and its eigenvalues would be $\pm i$.

comprehensive (if slightly outdated) one for *K* decays is D'Ambrosio and Isidori (1998). A great wealth of information, with emphasis on e^+e^- experiments, is contained in Maiani *et al.* (1995).

The phenomenology of neutral kaons is the subject of Belušević (1999). Several original papers concerned with the discovery of their peculiar behaviour are collected and briefly discussed in Goldhaber *et al.* (1957).

Early experiments are also discussed in the classic review papers by Lee and Wu (1966*a*, 1966*b*). A recent review of *CP* violation experiments with kaons can be found in Sozzi (2006) and Blucher (2006).

Searches for direct *CP* violation (not only in the *K* system) are reviewed in Sozzi and Mannelli (2003), while the phenomenology when *CPT* symmetry is not assumed is discussed in Barmin *et al.* (1984) and Tanner and Dalitz (1986). Possible measurements at *K*-factories are reviewed in Buchanan *et al.* (1992) and Hayakawa and Sanda (1993).

Problems

1. In presence of *CP* violation K_S and K_L can both decay into the same state, and $\langle K_S | K_L \rangle \neq 0$. In presence of direct *CP* violation both K_1 and K_2 can decay into the same state: does this imply $\langle K_1 | K_2 \rangle \neq 0$?

2. Can the copious leptonic decay $K^+ \to \mu^+ \nu$ (BR $\simeq 68\%$) and its charged conjugate be used to look for *CP*-violating effects? What about $K^+ \to \mu^+ \nu \gamma$? Or $K^+ \to \mu^+ \nu e^+ e^-$?

3. Prove the isospin decomposition for $K^0 \to 2\pi$ decays of eqns (8.57)–(8.59), and do the same for $K \to 3\pi$ decays, eqns (8.165)–(8.168).

4. Derive the expressions (8.122)–(8.125) for the elements of the effective Hamiltonian (recall that M_{ii} and Γ_{ii} are always real).

5. Which is the photon energy spectrum from $K_L \to \pi^0 \pi^0$ decay?

6. Explain why $K \to \pi^0 \pi^0 \gamma$ decay is much more suppressed than $K \to \pi^+ \pi^- \gamma$.

7. Discuss how *CP* violation might be studied with the decays $K^\pm \to \pi^\pm \pi^0 \gamma$. Which part of the radiative photon spectrum would be most sensitive to it?

8. Other strange mesons exist besides *K*, such as the vector K^* (mass $\sim$ 892 MeV). Do you expect a similar phenomenology (mixing, *CP* violation, etc.) for these?

9. Estimate the required level of photon detection efficiency in an experiment aiming at a measurement of the decays $K \to \pi \nu \bar{\nu}$.

9

THE LINK TO THEORY

The universe is simple;
it's the explanation that's complex.
W. Allen

There is a theory which states that
if ever anybody discovers exactly what the Universe is for
and why it is here, it will instantly disappear
and be replaced by something even more bizarre and inexplicable.
There is another theory which states that this has already happened.
D. Adams

In these days of cool reflection
You come to me and everything seems alright.
R. Taylor

This chapter provides a brief description of the way CP violation is accommodated in the Standard Model (SM) of particle physics, with which the reader is assumed to have some familiarity. Without entering in details, some SM predictions for CP-violating observables are mentioned, and conversely the informations which measurements of CP violation can provide about its parameters are discussed. Finally, we sketch how CP violation could arise in different ways, mostly as an invitation to further reading.

9.1 Early concepts

The C, P, T discrete symmetries are known to be violated only by weak interactions, and the history of the former is tightly intertwined to that of the latter: indeed the discovery of parity (and charge-conjugation) violation in weak interactions was instrumental in clarifying their theoretical structure and thus in shaping what we now call the Standard Model.[126] The story is well known and described in all textbooks of particle physics (see e.g. Nachtmann (1989) or Cahn and Goldhaber (1988) for a historical perspective): we just summarize here the main points, directing the reader elsewhere for a more detailed treatment and original references.

---

[126] It seems indeed it has come of age nowadays, and quite deserves the appellative of Standard Theory!

Briefly, the four fermion vector interaction written by Fermi to describe β decay was first generalized to the most general contact interaction of two fermion bilinears consistent with Lorentz invariance, see eqn (B.29), by including also scalar, tensor, axial vector, and pseudo-scalar terms using the complete set of matrices Γ (B.20). The study of the helicities of fermions and antifermions emitted in β decay indicated that only vector and axial interactions were actually involved: these could produce both Fermi and Gamow–Teller transitions (with or without nucleon spin flip). The product of two vector (V) or two axial-vector (A) currents is however a scalar quantity, which cannot describe parity violation; a pseudo-scalar term had to be added, obtained as the product of a V and an A current; the two terms had to have equal magnitude in order to get complete polarization for the neutrinos, as experiment showed, eventually leading to the emergence of the $V - A$ interaction

$$\mathcal{L} = -\frac{G_F}{\sqrt{2}} \sum_{i=V,A} \left[\overline{\psi}^{(p)} \gamma^\mu (c_V + c_A \gamma_5) \psi^{(n)} \right] \left[\overline{\psi}^{(e)} \Gamma_i (1 - \gamma_5) \psi^{(\nu)} \right] \qquad (9.1)$$

The factor $(1 - \gamma_5)$ selects the negative chirality component ψ_L of the lepton fields (3.34) as those participating in weak interactions, which in the massless limit correspond to purely left-handed (helicity -1) particles and right-handed (helicity $+1$) antiparticles (see the discussion in Section 3.1.3).

The form (9.1) of the Lagrangian was shown to be universal for weak interactions with the introduction of the Cabibbo angle describing the different magnitude of the coupling for strangeness-changing transitions as a rotation in flavour space of the charged quarks involved in weak interactions. Weak neutral currents, with no change of the fermion charge, were then discovered, and the fact that such currents did not change quark flavour at all (i.e. did not change s into d quarks) was explained by the introduction of a fourth quark, as we will discuss. Finally, the non-renormalizable four-fermion point interaction was shown to be the low-energy limit of a renormalizable interaction involving the exchange of vector bosons $W^\pm, Z$, culminating with the unification of weak and electromagnetic interactions in the Glashow–Salam–Weinberg model based on the $SU(2)_L \times U(1)$ gauge group, in which fermion masses are not due to explicit mass terms but rather to the spontaneous symmetry breaking of the electro-weak symmetry in the interactions with the scalar Higgs field. Adding the $SU(3)$ gauge interaction of coloured quarks and gluons we finally have the Standard Model in all its glory.

Note that the V and A interactions conserve chirality, by coupling only fields with the same chirality:

$$\overline{\psi}^{(1)} \gamma^\mu \psi^{(2)} = \overline{\psi}_R^{(1)} \gamma^\mu \psi_R^{(2)} + \overline{\psi}_L^{(1)} \gamma^\mu \psi_L^{(2)}$$

$$\overline{\psi}^{(1)} \gamma^\mu \gamma_5 \psi^{(2)} = \overline{\psi}_R^{(1)} \gamma^\mu \gamma_5 \psi_R^{(2)} + \overline{\psi}_L^{(1)} \gamma^\mu \gamma_5 \psi_L^{(2)}$$

This should be compared with the behaviour of non-derivative interactions with a scalar field ϕ (Yukawa interactions), which instead link the two chiralities just as mass terms do:

$$\phi \, \overline{\psi}^{(1)} \psi^{(2)} = \phi \, \overline{\psi}_R^{(1)} \psi_L^{(2)} + \phi \, \overline{\psi}_L^{(1)} \psi_R^{(2)}$$

In the massless limit a V or A gauge interaction with a *single* chiral fermion (and its antifermion) can thus be considered: this interaction indeed violates P (e.g. by involving only electrons of negative helicity, and not those of positive helicity) and C (e.g. by involving only electrons of negative helicity and positrons of positive helicity). CP, on the contrary, is seen to be quite the natural symmetry for this case, as interaction terms of the form

$$\overline{\psi}_L^{(1)} \gamma^\mu \psi_L^{(2)} \, W_\mu + \text{H.c.}$$

(where W^μ is the vector boson field) are CP symmetric, as can be verified explicitly.

The $V - A$ theory of weak interactions accommodated maximal P and C violation, but was automatically invariant under CP, as indeed it was required at the time it was developed. Also note that while purely left-handed neutrinos define maximal parity violation, an analogous definition of 'maximal' CP violation is not possible, since the existence of a right-handed antineutrino is already required by CPT symmetry.

9.2   The Standard Model

At present, we have no real 'explanation' of CP violation: in the last decades it was realized that the Standard Model, which so far has been able to describe all observed phenomena, has enough complexity to be able also to incorporate CP violation. Still, while the current theoretical framework has room for a single CP-violating phase and is consistent with all the observed manifestations of such phenomenon, the value of such phase is not better understood than, say, the value of the muon mass, and the physical origin of CP violation remains a mystery.[127]

As we saw in previous chapters, in general CP violation is linked to the appearance of complex factors in the Hamiltonian, and it is therefore useful to consider where such terms can appear in a non-trivial way in the Lagrangian of the SM.

---

[127] Of course, one could also adopt the point of view according to which there is nothing to explain, and the values of all the parameters in the SM are just what they are with no deeper explanation; or one might also follow the route leading to the conclusion that the observed set of parameters is the only one which is compatible with the existence of physicists measuring it. We don't dare to enter into these issues here, and stick to the view that a 'better' description of physical reality most likely exists.

9.2.1 Flavour mixing

Since *CP* violation has been observed only in weak processes, we start by considering the part of the Lagrangian density of the Standard Model which describes these, written as

$$\mathcal{L} = \mathcal{L}^{(G)} + \mathcal{L}^{(F)} + \mathcal{L}^{(QG)} + \mathcal{L}^{(Y)} \tag{9.2}$$

$\mathcal{L}^{(G)}$ contains the kinetic terms for the gauge vector bosons, of the form

$$\mathcal{L}^{(G)} = -\frac{1}{4} W^{(a)}_{\mu\nu} W^{(a)\mu\nu} - \frac{1}{4} B_{\mu\nu} B^{\mu\nu} \tag{9.3}$$

$$W^{(a)}_{\mu\nu} = \partial_\mu W^{(a)}_\nu - \partial_\nu W^{(a)}_\mu + g\,\epsilon_{abc} W^{(b)}_\mu W^{(c)}_\nu \tag{9.4}$$

$$B_{\mu\nu} = \partial_\mu B_\nu - \partial_\nu B_\mu \tag{9.5}$$

in which the field strengths $W^{(a)}_{\mu\nu}$ of the three $SU(2)$ gauge bosons $W^{(a)}$ ($a = 1, 2, 3$, coupling constant g) and that $B_{\mu\nu}$ of the $U(1)$ gauge boson B (coupling constant g') appear. These terms are necessarily real, and thus cannot induce *CP* violation.

The fermion kinetic terms and couplings to the gauge bosons appear in terms analogous to the minimal coupling of electromagnetism (Appendix B)

$$\mathcal{L}^{(EM)} = i\overline{\psi}\gamma^\mu(\partial_\mu + ieA_\mu)\psi \tag{9.6}$$

($e > 0$ being the electric charge of ψ) but quarks are arranged in left-handed $SU(2)$ doublets

$$q_L = \begin{pmatrix} u_L \\ d_L \end{pmatrix}$$

and right-handed $SU(2)$ singlets u_R, d_R (to streamline the notation we denote the field operator with the name of the particle, colour indexes are suppressed as all couplings are colour-diagonal). This pattern is replicated for the three quark generations, with (c, s) and (t, b) replacing (u, d); we first consider the (fictitious) case of a single generation, and later discuss the new features arising when more generations are added. In this case the kinetic terms have the form

$$\mathcal{L}^{(F)} = i\overline{q}_L\gamma^\mu\partial_\mu q_L + i\overline{u}_R\gamma^\mu\partial_\mu u_R + \cdots$$

and the interaction terms

$$\begin{aligned} \mathcal{L}^{(QG)} = &- g\,\overline{q}_L\gamma^\mu(\sigma_a/2)q_L W^{(a)}_\mu - g'\,Y(q_L)\overline{q}_L\gamma^\mu q_L B_\mu \\ &+ g'\,Y(u_R)\overline{u}_R\gamma^\mu u_R B_\mu + g'\,Y(d_R)\overline{d}_R\gamma^\mu d_R B_\mu \end{aligned}$$

where σ_a are the Pauli matrices and $Y(q_L) = 1/6$, $Y(u_R) = 2/3$, $Y(d_R) = -1/3$ is the weak hypercharge ($U(1)$ quantum number) of the quarks. The Y values for the different fermions are assigned in such a way as to obtain a pure vector neutral current interaction term corresponding to electromagnetism:

$$\mathcal{L}^{(QG)} = -\frac{g}{\sqrt{2}} \left(J^{\mu}_{(CC)} W^{(+)}_{\mu} + J^{\dagger\mu}_{(CC)} W^{(-)}_{\mu} \right)$$

$$- \frac{g}{\cos\theta_W} J^{\mu}_{(NC)} Z_{\mu} - g \sin\theta_W J^{\mu}_{(EM)} A_{\mu} \qquad (9.7)$$

where the field $W^{(+)} = (W^{(1)} - iW^{(2)})/\sqrt{2}$ annihilates a W^+ boson, $W^{(-)} = W^{(+)\dagger}$, and the neutral fields are

$$Z = \cos\theta_W W^{(3)} - \sin\theta_W B \qquad A = \sin\theta_W W^{(3)} + \cos\theta_W B \qquad (9.8)$$

where the weak mixing (Weinberg) angle is $\theta_W \equiv g'/g$, with $g \sin\theta_W = e$, the magnitude of the electron charge (A is the EM field). Here

$$J^{\mu}_{(CC)} = \bar{u}_L \gamma^{\mu} d_L \qquad (9.9)$$

$$J^{\mu}_{(NC)} = \frac{1}{2} \left(\bar{u}_L \gamma^{\mu} u_L - \bar{d}_L \gamma^{\mu} d_L \right) - \sin^2\theta_W J^{\mu}_{(EM)} \qquad (9.10)$$

$$J^{\mu}_{(EM)} = \frac{2}{3} \bar{u} \gamma^{\mu} u - \frac{1}{3} \bar{d} \gamma^{\mu} d \qquad (9.11)$$

are the weak charged, weak neutral, and EM (neutral) quark currents respectively. It can be checked explicitly that the Hermiticity of (9.7) requires the coupling constants g, g', e to be real quantities, so that also here no *CP* violation can be incorporated.[128]

Finally, the interaction of the quarks and leptons with the Higgs field, introduced to give masses to them and to the gauge bosons, are

$$\mathcal{L}^{(Y)} = -G^{(u)} \bar{q}_L \begin{pmatrix} \phi^0 \\ -\phi^- \end{pmatrix} u_R - G^{(d)} \bar{q}_L \begin{pmatrix} \phi^+ \\ \phi^0 \end{pmatrix} d_R + \text{H.c.} \qquad (9.12)$$

with $G^{(u,d)}$ being coupling constants. The spontaneous symmetry breaking due to the non-zero expectation value of the neutral Higgs field in the vacuum $\langle \phi^0 \rangle_{\text{vac}} = v$ gives fermion mass terms

$$-G^{(u)} v \, \bar{u}_L u_R - G^{(d)} v \, \bar{d}_L d_R + \text{H.c.}$$

Contrary to the gauge interaction terms (9.7), this part of the Lagrangian can support *CP* violation, as the coefficients $G^{(u,d)}$ are not necessarily real: consider a generic

---

[128] This follows from the fact that the coupling to gauge bosons is of the form (9.6) with $A_{\mu} \to A^{(a)}_{\mu} T^{(a)}$ where A_{μ} are Hermitian gauge fields and $T^{(a)}$ generators of the gauge group in the adjoint representation, which are pure imaginary.

Yukawa term in the Lagrangian

$$G\overline{\psi}_L^{(1)}\phi\,\psi_R^{(2)} \tag{9.13}$$

where the charge of the scalar field is chosen according to those of the spinor fields
to give charge conservation. Hermiticity requires that the Hermitian conjugate term

$$G^*\overline{\psi}_R^{(2)}\phi^\dagger\psi_L^{(1)}$$

also appears in $\mathcal{L}$. The CP transformation of (9.13) however gives

$$-G\,\eta_P(\phi)\eta_P^*(1)\eta_P(2)\eta_C(\phi)\eta_C^*(1)\eta_C(2)\,\overline{\psi}_R^{(2)}\phi^\dagger\psi_L^{(1)}$$

so that $\mathcal{L}$ will be invariant under CP only if the phase factors can be chosen such
that the coefficient of this term is equal to G^* (e.g. if the product of the phase factors
above were -1 a complex G coefficient in (9.13) would violate CP symmetry).

When more quark families are present a flavour index i ($i = 1, \ldots, n$; $n = 3$ in
the SM) is used to label them:

$$Q_{Li} = \begin{pmatrix} U_{Li} \\ D_{Li} \end{pmatrix} \qquad U_{Ri} \quad D_{Ri} \qquad U \equiv \{u,c,t\} \quad D \equiv \{d,s,b\}$$

and it should be noted that there is no reason why, e.g., U_1 (u) should be coupled
to D_1 (d) rather than to, e.g., D_2 (s), as all the U and all the D quarks are not
distinguished by the strong and EM interactions (they have the same charge and
are all massless at this stage). The most general Yukawa coupling is thus

$$\mathcal{L}^{(Y)} = -G_{ij}^{(U)}\overline{Q}_{Li}\begin{pmatrix} \phi^0 \\ -\phi^- \end{pmatrix}U_{Rj} - G_{ij}^{(D)}\overline{Q}_{Li}\begin{pmatrix} \phi^+ \\ \phi^0 \end{pmatrix}D_{Rj} + \text{H.c.}$$

where $G^{(U,D)}$ are now $n \times n$ matrices, which after spontaneous symmetry breaking
give the mass terms

$$-G_{ij}^{(U)}v\,\overline{U}_{Li}U_{Rj} - G_{ij}^{(D)}v\,\overline{D}_{Li}D_{Rj} + \text{H.c.} \tag{9.14}$$

The appearance of coupling constants which are now matrices in flavour (family)
space means that the down-type quark which couples to a particular up-type quark
in the interaction terms

$$J_{(CC)}^\mu = \overline{U}_{Li}\gamma^\mu D_{Li} \tag{9.15}$$

$$J_{(NC)}^\mu = \frac{1}{2}\left(\overline{U}_{Li}\gamma^\mu U_{Li} - \overline{D}_{Li}\gamma^\mu D_{Li}\right) - \sin^2\theta_W J_{(EM)}^\mu \tag{9.16}$$

$$J_{(EM)}^\mu = \frac{2}{3}\overline{U}_i\gamma^\mu U_i - \frac{1}{3}\overline{D}_i\gamma^\mu D_i \tag{9.17}$$

is in general not the same which couples to it in the mass terms. Passing to a matrix notation in flavour space ($\{U_i\} \rightarrow U$), the matrices $G^{(U,D)}$ are in general complex and non-diagonal.

As masses are usually more important in the dynamics of elementary particles than their weak interactions, the mass terms are made diagonal rather than the interaction terms. For each $G^{(U,D)}$ the matrices $GG^\dagger$ and $G^\dagger G$ are Hermitian and share the same (real) eigenvalues, and can be made diagonal with non-negative eigenvalues ($\langle\psi|G^\dagger G|\psi\rangle$ is the norm of $G|\psi\rangle$) with a similarity transformation using unitary matrices A, B, that is:

$$A(GG^\dagger)A^\dagger = D^2 \qquad B(G^\dagger G)B^\dagger = D^2$$

$$AGB^\dagger = D \qquad BG^\dagger A^\dagger = D$$

with D diagonal. The A, B matrices are defined by the above relations up to a number of phases equal to their dimension, since multiplying them with a diagonal unitary matrix the same diagonal matrix D is obtained; this fact will turn out to be useful later.

The freedom exists of rotating the basis for the quarks with the same quantum numbers, by writing

$$U_{L,R}^{(m)} = S_{L,R}^{(U)} U_{L,R} \qquad D_{L,R}^{(m)} = S_{L,R}^{(D)} D_{L,R} \tag{9.18}$$

with $S_{L,R}^{(U,D)}$ being unitary matrices. The $U^{(m)}, D^{(m)}$ fields are identified as the states of definite mass, and indeed by choosing the $S_L^{(U,D)}$ and $S_R^{(U,D)}$ matrices as A, B above the mass terms (9.14) become diagonal in this basis:

$$-v\overline{U}_L^{(m)} S_L^{(U)} G^{(U)} S_R^{(U)\dagger} U_R^{(m)} - v\overline{D}_L^{(m)} S_L^{(D)} G^{(D)} S_R^{(D)} D_R^{(m)\dagger} + \text{H.c.}$$

$$= m_u \bar{u}_L^{(m)} u_R^{(m)} + m_c \bar{c}_L^{(m)} c_R^{(m)} + \cdots$$

We note that (9.18) simultaneously diagonalizes the coupling of the quark fields to the Higgs field, while they have no effect on the neutral currents:

$$J_{(NC)}^\mu = \frac{1}{2}\left(\overline{U}_L \gamma^\mu U_L - \overline{D}_L \gamma^\mu D_L\right) - \sin^2\theta_W \frac{2}{3}\overline{U}\gamma^\mu U - \frac{1}{3}\overline{D}\gamma^\mu D$$

$$= \frac{1}{2}\left(\overline{U}_L^{(m)} \gamma^\mu U_L^{(m)} - \overline{D}_L^{(m)} \gamma^\mu D_L^{(m)}\right) - \sin^2\theta_W \frac{2}{3}\overline{U}^{(m)}\gamma^\mu U^{(m)}$$

$$- \frac{1}{3}\overline{D}^{(m)}\gamma^\mu D^{(m)} \tag{9.19}$$

which remain diagonal in the new fields due to the unitarity of $S_{L,R}^{(U,D)}$; processes in which quark flavour changes but charge does not, for example $s \rightarrow d$ transitions, only occur at second order in the weak interactions, as required by the experimental observations. As long as only one scalar field couples to U_R and one (possibly the same) couples to to D_R, it can be shown that also the interaction of neutral scalar fields remains diagonal in flavour.

Historically, the absence of flavour-changing neutral currents (FCNC) led Glashow, Iliopoulos, and Maiani (GIM) to note that this happens if the above matrices are unitary, which is possible if there are as many quarks of charge 2/3 as there are of charge $-1/3$, thus suggesting the existence of charm; indeed the absence of FCNC follows from having the same representation for all fermions ('weak universality') and the existence of a single neutral Higgs boson. In the expression for the charged weak currents, which couple U_L to D_L

$$J^\mu_{(CC)} = \overline{U}_L \gamma^\mu D_L = \overline{U}^{(m)}_L \gamma^\mu (S^{(U)}_L S^{(D)\dagger}_L) D^{(m)}_L \tag{9.20}$$

a non-trivial unitary matrix appears

$$V \equiv S^{(U)}_L S^{(D)\dagger}_L \tag{9.21}$$

called the Cabibbo–Kobayashi–Maskawa (CKM) matrix, which measures the mismatch between the matrices which diagonalize the U and D quark mass terms. The CKM matrix is responsible for the fact that charged weak interactions do change flavour, so that a u quark does not couple only to a d quark but also to s and b quarks; conversely, the flavour-conserving nature of strong and EM interactions follows from the absence of fundamental scalar fields in that sector, which allows the simultaneous diagonalization of the mass terms and the kinetic (plus gauge interaction) terms in flavour space.

9.2.2   CP violation

The appearance of the matrix V with complex elements opens the possibility of having CP violation; however this will only occur if the matrix cannot be made real with any choice of arbitrary phase factors, and in a now famous paper M. Kobayashi and T. Maskawa considered the general conditions under which this can happen (Kobayashi and Maskawa, 1973). First of all the unitarity of V

$$V^*_{ji} V_{jk} = \delta_{ik} \tag{9.22}$$

gives for n generations n^2 constraints on the matrix elements (not n^2 complex equations, which would be $2n^2$ constraints, since the complex conjugate of (9.22) gives the same equation): out of the $2n^2$ real parameters of a complex $n \times n$ matrix only n^2 are present for a unitary matrix.

The requirements that the mixing matrices $S^{(U,D)}_{L,R}$ diagonalize the mass terms do not completely determine V; in particular this matrix can be changed without any physical consequence by redefining arbitrarily the phases of the quark fields appearing in (9.20). In the absence of external sources interacting with the quark fields, the only way to measure the phases of such fields via the interactions present in the Lagrangian: as the SM contains many particles and not as many interactions linking them, a large reparametrization invariance is left, which can be exploited to remove irrelevant phases. Writing

$$U_i^{(m)} \rightarrow e^{i\phi_i^{(U)}} U_i^{(m)} \qquad D_i^{(m)} \rightarrow e^{i\phi_i^{(D)}} D_i^{(m)}$$

the CKM matrix gets multiplied from the left and from the right by two diagonal matrices whose elements are phase factors: $2n$ such phase factors appear, whose arbitrariness can be used to eliminate $2n - 1$ relative phases from V, since one overall phase is irrelevant (if all quark phases are changed in the same way V is unaffected).

Allowing for the rephasing of the quark fields $n^2 - (2n - 1) = (n - 1)^2$ real parameters are left. A unitary matrix is also orthogonal, and as such it contains $n(n - 1)/2$ parameters corresponding to the independent rotation angles between the n basis vectors; what is left must be complex phases, their number being

$$N(\text{phases}) = (n - 1)^2 - \frac{1}{2}n(n - 1) = \frac{1}{2}(n - 1)(n - 2)$$

For $n = 2$ quark families there is one rotation angle (the Cabibbo angle) and no phase; for $n = 3$ families there are three rotation angles (corresponding to the Euler angles) and one complex phase. At least three quark families are required for a non-trivial complex phase to appear in V and therefore to support CP violation via this mechanism.[129] Clearly, with more families the number of complex phases rapidly increases.

Note that if any two U-type or D-type quarks were degenerate in mass, they could not be distinguished by any means, and an arbitrary rotation among them would still be allowed (any linear combination of them would still be a mass eigenstate with the same mass): this would allow one to introduce an arbitrary 2×2 unitary rotation matrix among them (the pair of quarks would exhibit a $SU(2)$ symmetry rather than just a $U(1) \times U(1)$ rephasing symmetry), whose parameters (one angle and three phases) could be used to remove the single phase in V. Non-degenerate quarks are thus required to support CP violation (in practice this means non-zero masses, as any mass degeneracy between two massive quarks would be expected anyway to be lifted by radiative corrections). Also, if one of the mixing angles were zero, so that one family would be completely decoupled from the other two, the matrix V would become block-diagonal, with a 2×2 unitary sub-matrix which as seen would not support CP violation.

---

[129] Note that when Kobayashi and Maskawa made this remark only *three* quarks were known!

Summarizing, *CP* violation can arise in the weak charged current interactions of quarks through a single complex phase in the CKM mixing matrix if at least three generations of non-degenerate quarks are present which all mix with each other.

9.2.3 The CKM matrix

The CKM mixing matrix V can be parametrized in many equivalent ways by choosing the angles differently; a standard form (Yao *et al.*, 2006) is

$$
\begin{pmatrix}
V_{ud} & V_{us} & V_{ub} \\
V_{cd} & V_{cs} & V_{cb} \\
V_{td} & V_{ts} & V_{tb}
\end{pmatrix}
$$

$$
= \begin{pmatrix}
c_{12}c_{13} & s_{12}s_{13} & s_{13}e^{-i\delta_{CP}} \\
-s_{12}c_{23} - c_{12}s_{23}s_{13}e^{i\delta_{CP}} & c_{12}c_{23} - s_{12}s_{23}s_{13}e^{i\delta_{CP}} & s_{23}c_{13} \\
s_{12}s_{23} - c_{12}c_{23}s_{13}e^{i\delta_{CP}} & -c_{12}s_{23} - s_{12}c_{23}s_{13}e^{i\delta_{CP}} & c_{23}c_{13}
\end{pmatrix}
$$

$$(9.23)$$

where $s_{ij} \equiv \sin(\vartheta_{ij})$, $c_{ij} \equiv \cos(\vartheta_{ij})$ and δ_{CP} is the *CP*-violating phase. Each angle is here labelled with the indexes corresponding to the mixing of two families, so that $\vartheta_{ij} = 0$ would indicate that families i and j are decoupled; all these angles can always be chosen to lie in the first quadrant.

A discussion on the determination of the parameters of the CKM matrix from experimental measurements is beyond the scope of this book. We just mention that in general theoretical input is required to obtain the matrix elements V_{ij} from experimental data; semi-leptonic processes are generally used (including τ decays to hadrons), in which the complications arising from strong interactions are minimized, as they affect only half of the particles involved. From the experimental measurement of e.g. the shape of a decay spectrum one extracts the product of a CKM matrix element (squared) and a normalization factor, whose value must be computed from the theory. We briefly list some standard experimental measurements used to extract the moduli of CKM matrix elements (see, (Yao *et al.*, 2006) for more detail).

- $|\mathbf{V_{ud}}| \simeq 0.97$ can be obtained from nuclear β decays with $J^P = 0^+ \to 0^+$ (Fermi type), to which only the vector current contributes, by comparing to muon decay; in the decay of a free neutron the axial current also contributes, making this process less clean. Pion beta decay $\pi^{\pm} \to \pi^0 e^+ \nu$ is even cleaner but highly suppressed by phase space ($BR \sim 10^{-8}$). Hyperon β decays such as $\Sigma^+ \to \Lambda e^+ \nu$ are also sensitive to this matrix element.

- $|\mathbf{V_{us}}| \simeq 0.23$ (the Cabibbo angle) is determined from semi-leptonic K decays into electrons $K^+ \to \pi^0 e^+ \nu$ and $K_L \to \pi^{\pm} e^{\mp} \nu$, in which (neglecting the electron mass) a single form factor appears. Hyperon decays are also used, but the extraction of $|V_{us}|$ from the experimental data is usually more troublesome in this case.

- $|\mathbf{V_{ub}}| \simeq 0.004$ can be obtained from the semi-leptonic decays of B mesons above the kinematic limit for the $b \to c\ell\bar{\nu}$ process, using significant theoretical input.

- $|\mathbf{V_{cd}}| \simeq 0.23$ is determined from the inclusive production of *charm* in neutrino deep inelastic scattering on nucleons, $\nu d \to c\mu^-$ followed by $c \to d\mu^+\nu$ (the observation of μ^+ starting from a μ^- neutrino beam indicates charm production); again important theoretical input is required. Semi-leptonic decays of charmed quarks are usually more difficult to be handled theoretically.

- $|\mathbf{V_{cs}}| \simeq 0.96$ is also determined from inclusive charm production in neutrino scattering or semi-leptonic decays of D mesons into strange particles (with some more theoretical input) and from charm decays of W bosons.

- $|\mathbf{V_{cb}}| \simeq 0.04$ is extracted from exclusive and inclusive semi-leptonic decays of B mesons, which are heavy enough to allow different, more accurate, theoretical approaches to be used.

- $|\mathbf{V_{td}}| \simeq 0.007, |\mathbf{V_{ts}}| \simeq 0.04$: top quark decays are different from those of other quarks, as they can produce a real W (on the mass shell). These matrix elements are not extracted from top decays into d, s (which are highly suppressed) but can be obtained from processes in which the t quark is virtual, such as the longitudinal muon polarization in $K^+ \to \pi^+\mu^+\mu^-$ decays, or other rare K decays which are very clean from the theoretical point of view, such as $K \to \pi\nu\bar{\nu}$ with branching ratios $\sim O(10^{-10})$.

    This touches a general point: CKM matrix elements involving the top quark can be extracted from the study of FCNC processes, which are absent at tree level because of the GIM mechanism and arise from diagrams with intermediate quark loops, to which the heavy top quark can give the dominating contribution (this approach of course assumes no 'new physics' particles contributing to such loops).

- $|\mathbf{V_{tb}}| \simeq 1$ is roughly constrained by limits on top decays into particles not containing a b quark.

The use of unitarity relations (which assume the existence of three families only) usually constrains the small CKM matrix elements more tightly than the direct measurements. In some cases products or ratios of CKM matrix elements can be determined with better precision than individual terms: for example $|V_{tb}^* V_{td}|$ can be obtained from the mixing of neutral B mesons, and $|V_{td}|/|V_{ts}|$ from the ratio of the mass difference parameters Δm for B_s and B_d mesons.

In Nature the mixing angles are such that the matrix elements of V have very different magnitudes; an approximate parametrization in which this is evident was suggested by L. Wolfenstein (1983) and provides more insight than (9.23):

$$V = \begin{pmatrix} 1 - \lambda^2/2 & \lambda & A\lambda^3(\rho - i\eta + i\eta\lambda^2/2) \\ -\lambda & 1 - \lambda^2/2 - i\eta A^2\lambda^4 & A\lambda^2(1 + i\eta\lambda^2) \\ A\lambda^3(1 - \rho - i\eta) & -A\lambda^2 & 1 \end{pmatrix} + O(\lambda^4)$$

$$(9.24)$$

The four real parameters λ (the sine of the Cabibbo angle), A, ρ and η (the latter being the one responsible for the appearance of complex terms) are all of order unity ($\lambda \simeq 0.23$, $A \simeq 0.82$, $\rho \simeq 0.22$, $\eta \simeq 0.34$), so that the number of powers appearing in each element gives an indication of its relative size. The matrix (9.24) is unitary up to real terms $O(\lambda^3)$ and imaginary terms $O(\lambda^5)$, which is usually a sufficient approximation but can be improved when precision requires it (Buras et al., 1994).

The parametrization (9.24) immediately shows that the CKM has a 'hierarchical' structure, in which each up-type quark preferably couples to the down-type quark of its same family, and to other ones with couplings which are smaller the more the two families are distant; indeed the off-diagonal elements are of order λ between generations 1 and 2, λ^2 between 2 and 3 and λ^3 between 1 and 3: $|\theta_{12}| \gg |\theta_{23}| \gg |\theta_{13}|$ in the parametrization of (9.23). Another observation is that $\eta/\rho \sim O(1)$ shows that CP is not even an approximate symmetry of the SM: the smallness of CP-violating effects is just due to the small mixing angles which appear together with the complex phase δ_{CP} in the expression for specific observables, rather than a necessity. Conversely, it should be observed that the CP-violating parameter η can be determined also from experiments on CP-conserving processes, since the CKM matrix is uniquely determined by the moduli of its elements; for example if A is known the knowledge of $|V_{ub}|$ and $|V_{td}|$ gives ρ and η. The knowledge of the moduli of the CKM matrix elements is thus sufficient to put a limit on the amount of CP violation in the SM.

It is important to note that the position in which complex terms appear in the CKM matrix is not physically significant, as in different parametrizations the complex phase shifts to different matrix elements. Clearly, the physics should not depend on the choice of the parametrization, and quantities must be identified which are insensitive to such choice; again, CP violation will only be present if a complex term appears which cannot be made real with *any* choice of arbitrary phases, thus defining a rephasing-invariant complex quantity.

The above considerations can be formalized in terms of the CKM matrix elements by noting that only functions of V_{ij} which are invariant under rephasing of the quark fields can be physical: besides the moduli of the matrix elements, the other invariant quantities can be written in terms of those and the $V_{ij}V_{kl}V_{il}^*V_{kj}^*$ (no sum on i, j, k, l). In the case of three families, the unitarity of V constrains the imaginary parts of all such quantities to be equal within a sign, so that the following invariant (Jarlskog, 1985)

$$J_{CP} \equiv |\text{Im}(V_{ij}V_{kl}V_{il}^*V_{kj}^*)| = s_{12}s_{13}s_{23}c_{12}c_{13}^2 c_{23} \sin \delta_{CP} \simeq \lambda^6 A^2 \eta \qquad (9.25)$$

$$\text{Im}(V_{ij}V_{kl}V_{il}^*V_{kj}^*) = J_{CP} \sum_{m,n} \epsilon_{ikm}\epsilon_{ijn} \qquad (9.26)$$

is defined as the only CP violation term in the CKM matrix: any CP-violating quantity in the SM must be proportional to J_{CP}, reflecting the fact that a single complex phase appears in the 3×3 CKM matrix. The value extracted from experimental data is at present $J_{CP} = (3.08 \pm 0.17) \cdot 10^{-5}$ (Yao et al., 2006).

From the explicit expression (9.25) the fact of the disappearance of CP violation with the vanishing of any mixing angle ϑ_{ij} is manifest. On the other hand the product of all three mixing angles appears in J_{CP}, so that the hierarchical structure of the CKM matrix, with relatively small inter-family mixing, suppresses any CP-violating amplitude even for $\delta_{CP} = \pi/2$.

Equation (9.25) also shows that at least four different quarks (real or virtual) must be involved for a process to exhibit CP violation. Large effects are to be expected in transitions for which any competing CP-conserving amplitude is small, such as particle decays involving a change of flavour: K decays, Cabibbo-suppressed D decays ($c \rightarrow u$) and even more B decays involving the small element V_{ub}; virtual processes involving V_{td}, such as rare FCNC K decays, are also very good candidates.

The requirement of non-degeneracy of the quarks explicitly appears when a rephasing invariant quantity involving the mass matrices $G^{(U,D)}$ is considered: the necessary and sufficient condition to have CP symmetry in the SM is the vanishing of the quantity (Jarlskog, 1985)

$$\det\left[G^{(U)}G^{(U)\dagger}, G^{(D)}G^{(D)\dagger}\right] = 2i(m_t^2 - m_c^2)(m_t^2 - m_u^2)$$

$$(m_c^2 - m_u^2)(m_b^2 - m_s^2)(m_b^2 - m_d^2)(m_s^2 - m_d^2)J_{CP} \qquad (9.27)$$

and the case of more families is discussed in Branco, Lavoura, and Silva (1999).

It is important to remark that the SM implementation of CP violation is (in principle) a very predictive one, since all possible asymmetry measurements are correlated by their common origin from a single phase.

9.2.4 Unitarity triangles

The unitarity condition of the CKM matrix $V^\dagger V = 1$ gives six relations between the CKM matrix elements

$$V_{ud}^* V_{us} + V_{cd}^* V_{cs} + V_{td}^* V_{ts} = 0 \qquad [O(\lambda) + O(\lambda) + O(\lambda^5) = 0] \qquad (9.28)$$

$$V_{ud}^* V_{cd} + V_{us}^* V_{cs} + V_{ub}^* V_{cb} = 0 \qquad [O(\lambda) + O(\lambda) + O(\lambda^5) = 0] \qquad (9.29)$$

$$V_{us}^* V_{ub} + V_{cs}^* V_{cb} + V_{ts}^* V_{tb} = 0 \qquad [O(\lambda^4) + O(\lambda^2) + O(\lambda^2) = 0] \qquad (9.30)$$

$$V_{cd}^* V_{td} + V_{cs}^* V_{ts} + V_{cb}^* V_{tb} = 0 \qquad [O(\lambda^4) + O(\lambda^2) + O(\lambda^2) = 0] \qquad (9.31)$$

$$V_{td}^* V_{ud} + V_{ts}^* V_{us} + V_{tb}^* V_{ub} = 0 \qquad [O(\lambda^3) + O(\lambda^3) + O(\lambda^3) = 0] \qquad (9.32)$$

$$V_{ub}^* V_{ud} + V_{cb}^* V_{cd} + V_{tb}^* V_{td} = 0 \qquad [O(\lambda^3) + O(\lambda^3) + O(\lambda^3) = 0] \qquad (9.33)$$

As the above relations follow from the unitarity of the 3×3 CKM matrix, they are not true in general: if any of them fails to be verified by experiment then some new physics must be present; the same is true for the diagonal unitarity relations involving the sums of the squared moduli of the elements of one row or one column

$$\sum_i |V_{ij}|^2 = 1 \qquad (9.34)$$

which however (being real) bear no information on *CP* violation.

Each of the relations (9.28)–(9.33), requiring that the sum of three complex numbers vanishes can be visualized as a triangle in the complex plane. The shape of each triangle is independent of the choice of the quark field phases: by redefining any of those only the orientation of the triangles in the complex plane changes: this means that all their angles represent observable quantities. The fact that the triangles are not degenerate into a line represents *CP* violation, as in this case it is not possible to make real all three vectors describing the sides of the triangles.[130] Also, the area (clearly a rephasing-invariant quantity) is the same for all triangles and equal to $|J_{CP}|/2$, again as a consequence of the fact that there is a single source of *CP* violation.

The shapes of the six triangles are very different from each other, as indicated in eqns (9.28)–(9.33) by expressing each side in terms of the powers of λ which appear in the parametrization (9.24). The first two triangles ('ds' and '$\overline{u}c$'), relating the elements which appear in strange and charmed particle decays, are very squashed (despite having the same area as all the others) so that one of the angles representing the relative phases of the CKM matrix elements is tiny. The sides of the last two triangles ('$\overline{t}u$' and '$\overline{b}d$', which coincide at leading order in λ) are instead

---

[130] For two quark families the corresponding (single) relation would involve a sum of two terms only, an equality between two products of two matrix elements, required to be parallel in the complex plane.

all of comparable size, so that the angles (relative phases) are naturally large; the elements appearing in the last triangle (often called 'the' unitarity triangle) are those involved in B meson processes (respectively in Cabibbo-suppressed decays, Cabibbo-favoured decays and $B^0 - \overline{B}^0$ oscillations), so that relatively large CPV effects can be expected in that system.

As the parameters λ and A are better known, the unitarity triangles are usually plotted by fixing two of their vertices: focusing on the '$\overline{b}d$' triangle, a phase convention is chosen in which $V_{cb}^* V_{cd}$ is real, and each side is divided by its magnitude, so that the three vertices are in $(0, 0)$, $(1, 0)$ and (ρ, η). The two complex sides have lengths

$$R_u \equiv \sqrt{\rho^2 + \eta^2} = \frac{1}{\lambda} \left| \frac{V_{ub}}{V_{cb}} \right| \qquad R_t \equiv \sqrt{(1 - \rho)^2 + \eta^2} = \frac{1}{\lambda} \left| \frac{V_{td}}{V_{cb}} \right| \qquad (9.35)$$

and the three angles are (see Fig. 9.1).

$$\alpha \equiv \phi_2 \equiv \arg \left[-\frac{V_{tb}^* V_{td}}{V_{ub}^* V_{ud}} \right] \qquad (9.36)$$

$$\beta \equiv \phi_1 \equiv \arg \left[-\frac{V_{cb}^* V_{cd}}{V_{tb}^* V_{td}} \right] \qquad (9.37)$$

$$\gamma \equiv \phi_3 \equiv \arg \left[-\frac{V_{ub}^* V_{ud}}{V_{cb}^* V_{cd}} \right] \qquad (9.38)$$

(with $\alpha + \beta + \gamma = \pi$), while the area of the rescaled triangle is just $|\eta|/2$. Note that γ actually coincides with the CKM phase δ_{CP} of (9.23) to a good approximation.

In the Standard Model CP violation appears in a very specific way, related to the physics of quark flavours, and in particular all CP-violating phenomena

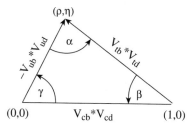

FIG. 9.1. The $\overline{b}d$ unitarity triangle.

are linked,[131] as they all descend from the appearance of a single phase. The value of this parameter is however completely arbitrary as far as we know, and is not understood at all: the physical origin of *CP* violation is still unknown, although it may be comforting to learn that (within the SM) it is deeply connected to the mystery of fermion family masses (one single big question rather than two).

All the observed *CP*-violating phenomena are in qualitative agreement with the SM implementation of *CP* violation, which is however neither unique nor completely tested: by enlarging the amount of experimental information available, and devising improved theoretical approaches to enable a more reliable connection between the observables and the underlying parameters, the model can be put to a more stringent quantitative test, by checking whether any observation does not fall in the tightly interconnected scheme of the SM.

It might also be noted that the fact that *CP* violation is observed at low energies, and that within the SM the only place in which this phenomenon can arise is in the coupling of fermions to a scalar field, is sometimes considered as indirect support for the existence of a Higgs boson.

9.2.5   The lepton sector

The above discussion for quarks can be repeated for the lepton sector: generally speaking *CP* violation might be expected to give large effects in the neutrino system, the intrinsic properties of which are completely determined by the (*CP*-violating) weak interactions. Note however that if neutrinos were truly massless (degenerate) *CP* violation would not be possible, as discussed in Section 9.2.3. This can also be seen from the fact that in this case no right-handed neutrino field ν_R would have to be introduced, as it would not interact with anything; a single mass term could then appear in the expression analogous to (9.14); spontaneous symmetry breaking would thus give mass only to the charged leptons (as desired), and the freedom to mix the three neutrino states would always be retained, allowing one to cancel any charged lepton mixing matrix appearing in the charged current, and leading to conserved lepton family numbers.

If neutrinos are non-degenerate in mass, as it now appears to be the case since neutrino oscillations show that lepton family numbers are not exactly conserved, a lepton mixing matrix which describes the relation of the physical (mass) eigenstates to the flavour eigenstates can be defined; as for quarks, complex matrix elements in such a Pontecorvo–Maki–Nakagawa–Sakata (PMNS) matrix (Pontecorvo, 1958; Maki, Nakagawa, and Sakata, 1962) could induce *CP* violation effects, and with three lepton flavours one complex phase remains which cannot be eliminated by field redefinitions.

---

[131] This unfortunately does not mean that their relative size can be computed in practice.

It should be recalled that *CP*-violating effects require all three mixing angles of a 3×3 mixing matrix to be non-zero, while at this time confirmed experimental indications of such a property only exist for two of them, as determined by solar and atmospheric neutrino experiments. In the three-family framework two of the mixing angles are known to be large (raising hopes for large *CP* violation effects) while for the third one only an upper limit (currently $\sim 13°$) exists; indeed, the main goal of the next generation of neutrino experiments is to find a *lower* bound on the third angle, thus hopefully proving that *CP* violation in neutrinos can be observed (see e.g. Gómez-Cadenas (2006)).

For two neutrino species $v_{i,j}$ with small masses (with respect to their average energy E) the oscillation phase difference is proportional to their squared mass difference $\Delta m_{ij}^2 = m_i^2 - m_j^2$, resulting in an oscillation probability over a distance L

$$P(v_i \rightarrow v_j) = \sin^2(2\vartheta_{ij}) \sin^2(\Delta m_{ij}^2 L/2E) \qquad (9.39)$$

where ϑ_{ij} is the $i-j$ neutrino family mixing angle. If the neutrino masses are such that, for a given experimental arrangement, one of them gives a negligible effect, $\Delta m_{ij}^2 L/2E \ll 1$, the *CP*-violating asymmetry is suppressed. It is often assumed therefore that only long-baseline experiments, sensitive to all three neutrino masses, can aim at directly measuring *CP* violation effects with neutrinos.

If neutrinos are Majorana (self-conjugate) particles (Section 3.1.3), two additional *CP*-violating phases appear in the mixing matrix (for three flavours); such phases do not produce any observable effect in oscillations, and only affect the rates for neutrino-less double-beta decay: the observation of this process, together with independent information on individual neutrino masses, could in principle allow one to extract information on such phases, although this might turn out to be difficult in practice (Barger *et al.*, 2002).

With massive neutrinos the leptonic sector can thus introduce a second source of *CP* violation, arising via the same kind of mechanism as for the quark sector. Still this is not yet the whole story for what concerns *CP* violation in the SM, as we will discuss in Section 11.2.

9.3   From theory to measurements

The poor theoretical control of strong interaction effects is such that the predictions for electro-weak decay processes cannot be reliably translated from the language of quarks to that of physical hadrons: theory is the weak link here, and the observables from which quantitative information on the SM parameters can be obtained are unfortunately not too many.

In general terms the computations of quantities involving hadrons are performed by separating the problem into two parts: the amplitude for the quark-level process is evaluated by using the theoretical techniques and approximations appropriate to the system under consideration, which usually involve considering all the effective four-quark operators contributing to the process and determining their coefficients (including QCD effects) by using the equations which describe the evolution from the high-energy regime, in which W, Z appear explicitly, to the effective theory in which such particle induce effective couplings. This part of the computation, despite being rather involved, is usually under good control and does not introduce large uncertainties. The second part is the translation from the quark-level process into that of the observed hadrons: this is a non-perturbative problem for which many approximate approaches have been developed, but the results usually have large uncertainties and model-dependencies; it is expected that lattice simulations will ultimately allow a better control of these aspects by the combination of clever techniques and brute computing power.

9.3.1   Neutral flavoured mesons

The GIM mechanism not only cancels FCNC at tree level, but it also suppresses their appearance at higher orders, as they would also vanish there if all the up-type and the down-type quarks were degenerate; as most quark masses (and thus mass differences) are small on the scale of the gauge vector bosons mediating weak interactions (except for the top quark), the above condition has important practical consequences.

The most important diagrams contributing to the $\Delta F = 2$ mixing between a neutral flavoured meson and its antiparticle are the so-called 'box' diagrams with two $W^\pm$ boson exchanges (Fig. 9.2); the intermediate states are clearly off the mass shell, as $m(W) \gg m(M^0)$.

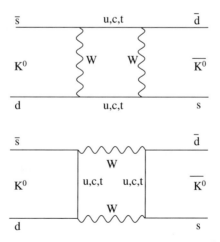

FIG. 9.2. The 'box' diagrams contributing to $K^0 - \overline{K}^0$ mixing.

In the comparison of neutral flavoured mesons M^0 built from up-type (U) quarks $(D^0$, as top quarks do not form hadrons since they decay before QCD hadronization can take place) to those built from down-type (D) quarks (K^0, B_d^0, B_s^0), some general features can be readily understood from simple considerations. According to their mass pattern, down-type quarks must decay outside of their family, so that their decays are suppressed by the small mixing angles; neglecting phase space considerations, this results in $\Gamma^{(U)} > \Gamma^{(D)}$. On the other hand, if the mixing angles are small with respect to the mass ratios of quarks in different families the contributions to the $M^0 - \overline{M}^0$ mixing process responsible for the mass difference between the two physical states are dominated by the exchange of the partner quark of the same family; as the mass difference is proportional to the squared mass of the virtual quark being exchanged, $(\Delta m)^{(U)} \propto m(D)^2 < (\Delta m)^{(D)} \propto m(U)^2$. Therefore in general $(\Delta m/\Gamma)^{(U)} < (\Delta m/\Gamma)^{(D)}$ and mixing in B^0 mesons is indeed large, as for K^0, while it is expected to be small (not observed so far) for D^0 mesons.

While all CP-violating effects are expected to be of the same order of magnitude in the SM (being all proportional to J_{CP}) the asymmetries, as ratios of CP-violating to CP-conserving quantities, are enhanced for suppressed decays, as all the decays of B mesons are (large B lifetime, small CKM couplings of the third family).

According to eqn (9.25), for a process to manifest a CP-violating phase it must involve at least four CKM matrix elements, with any three of them not belonging to the same row or column; this implies that below the charm threshold processes on the mass shell cannot violate CP, as only V_{us} and V_{ud} can be present: CP violation in the K meson sector only appears in virtual processes in which heavier quarks are involved, such as $K^0 - \overline{K}^0$ mixing or decays to which loop diagrams contribute.

Above the charm threshold CP violation in on-shell decays becomes possible, but it necessarily involves, using the Wolfenstein parametrization (9.24):

$$\arg\left(V_{cs}^* V_{cd}^* V_{ud}^* V_{us}\right) \simeq \arg\left(V_{cs}^*\right) \sim \eta A^2 \lambda^4$$

and is thus suppressed by powers of A and λ.

Above the beauty threshold other CKM matrix elements become available, and CP asymmetries in on-shell amplitudes can be proportional to e.g.

$$\arg\left(V_{ub}^* V_{cb} V_{cs}^* V_{us}\right) \simeq \arg\left(V_{ub}^*\right) = \eta/\rho$$

where the smallness of the mixing angles do not induce any suppression.

It should also be recalled that CPT symmetry imposes some constraints on the observability of CP violation in decay rate differences, by requiring the partial decay widths for some sets of final states to be equal for particle and antiparticle (see Section 5.1.3). For heavier meson decays such constraints become less effective, since the number of available decay modes increases and the situation in which CPT alone would enforce the decay rate equality for a specific channel becomes less common, as it becomes less likely that a subset of decay modes not mixed with others by strong and EM interactions only contains a single channel.

Finally we remark that semi-leptonic decay amplitudes are *CP*-conserving in the SM, and $\Delta S = \Delta Q$ holds with very high accuracy ($|x| \sim 10^{-7}$), and $x = 0 = y$ can be assumed. As discussed in Chapter 7 this means that charge asymmetries in these decays measure *CP* violation in $M^0 - \overline{M}^0$ mixing.

9.3.2 The strange sector

The $K^0 - \overline{K}^0$ mixing is described by the box diagrams (Fig. 9.2), in which u, c, t quarks appear in the loop. These diagrams are responsible for both Δm (the real part) and ϵ (the imaginary part), see Chapter 8. The fact that a diagram can be drawn does not mean that a reliable computation is possible, and indeed this is actually the case for the mentioned parameters. The uncertainties in the computation of the hadronic matrix elements (parametrized by $B_K \simeq 0.75 \div 1.10$) are the reason why the rather precise experimental measurement of ϵ translates into a rather wide band constraint in the (ρ, η) plane (see Fig. 10.1). Note that $|\epsilon|$ actually gives an overestimate of the size of *CP* violation, as (in the phase convention in which a_0 is real) in the denominator the small quantity Δm appears (8.105), which has nothing to do with *CP* violation and is small just because it is of fourth order in the weak coupling constant (and further suppressed by the GIM mechanism).

$K^0 \to 2\pi$ decays receive contributions from both tree diagrams and the so-called *penguin* diagrams[132] involving an electro-weak loop (Fig. 9.3), and the interference between the two can generate (direct) *CP* violation. Penguin diagrams are the basic ingredients for providing *CP*-violating partial rate asymmetries (Bander *et al.*, 1979), as they indeed provide the two necessary ingredients: the involvement of all three quark generations (in the loop), and the presence of absorptive parts (FSI) whenever real (on mass shell) particles are involved.

As in the case of ϵ, also ϵ' is zero at *tree* level in the SM: for example, all the tree diagrams corresponding to the quark transition $s \to d\bar{u}u$ share the same phase $\arg(V_{ud}^* V_{us})$. The penguin diagrams (Fig. 9.3) with an intermediate up-type quark q have a phase $\arg(V_{qd}^* V_{qs})$; those generated by strong interactions (with a gluon exchange) only contribute to the $I = 0$ state (the gluon has isospin 0), while those due to electro-weak interactions (γ, Z exchange) contribute to both $I = 0, 2$ states, thus providing a mechanism to have different weak phases in those modes (which also have different strong phases). The decay widths into a $\pi\pi$ state of definite isospin are equal for K^0 and $\overline{K}^0$, as there are no different strong phases available

---

[132] The name, supposedly due to some resemblance to the antarctic animals, originated from a lost bet by J. Ellis (Shifman, 1995): apparently when R. Feynman objected that penguin diagrams do not look like penguins at all, somebody remarked to him that Feynman diagrams do not bear any resemblance to him either.

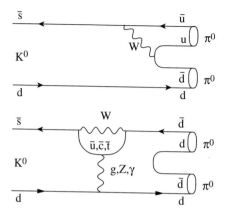

Fig. 9.3. Some diagrams contributing to $K^0 \rightarrow 2\pi^0$ decays. Top: tree diagram. Bottom: penguin diagram, with a Z, gluon or photon exchange in which a gluon creates a $d\bar{d}$ quark pair; all three quark families contribute to the loop, thus making the amplitude complex in general.

in this case. Besides the CKM factors, the ϵ' parameter is thus suppressed by the ratio of the A_2 and A_0 amplitudes ('$\Delta I = 1/2$ rule') and the ratio of penguin to tree amplitudes.

The computation of ϵ' in the SM is a large theoretical enterprise, which so far has not produced precise results; while the experimental result for ϵ'/ϵ (Chapter 8) represents a qualitative check of the SM, which in general predicts CP-violating effects also in $\Delta F = 1$ processes, despite the proportionality to $\mathrm{Im}(V_{td}V_{ts}^*)$ its value cannot be used yet to constrain the SM parameters of the CKM matrix because hadronic uncertainties plague the computation; the situation is worsened in this case by the presence of two contributions of comparable size, due to the gluonic and electro-weak penguin diagrams, which nearly cancel each other.

Note that in general the extraction of the weak phase difference (between the $I = 0$ and $I = 2$ final states in this case), which is related to the CKM matrix elements, requires knowledge of the strong phase difference; this is not a big issue for the K system in which only a few decay modes are present, but is a serious difficulty for heavy quark systems in which such phases, which cannot be computed from first principles, cannot be obtained from experiment either. For such systems a different approach is used to extract quantitative information on the weak amplitudes, as we later discuss.

It is important to remark that the mechanism by which CP violation is accommodated in the SM has several very specific features which were only experimentally verified recently: in particular the fact that CP violation is not compatible with a super-weak model (Section 8.3.5) was only established by the measurement of direct CP violation. The smallness of direct CP violation in the K system is accidental, due to the *top* quark mass having a value which enhances the electro-weak penguin contribution to ϵ' in such a way that it almost cancels the one due to the strong penguin diagrams. On the other hand, it was possible to conceive different mechanisms for CP violation, in which such phenomenon is mostly induced by virtual contributions of exotic high-mass particles in M_{12}, effectively of the super-weak type.

The importance of the measurement of direct CP violation is therefore independent of the actual value of the ϵ' parameter: $\epsilon' \neq 0$ shows that CP violation cannot be confined to the phenomenon of flavour mixing, and thus being consistent with the CKM model it is expected to be a ubiquitous phenomenon in all processes mediated by weak interactions. The experimental result is compatible with the SM predictions, which are however still not precise enough to allow a quantitative test of the CKM picture of CP violation; it is widely expected that progress in the numerical approach of lattice QCD will allow this number to be computed with good accuracy from first principles (actually an important test bench for the technique), thus allowing such a precisely measured quantity to also become a *quantitative* test of the SM.

For the CP-violating decays of K_S into three pions the situation is similar to that of K_L: indirect CP violation is expected to dominate, with direct CP violation representing a small correction; in this case however a CP-violating amplitude of comparable size to that of K_L results in a much smaller branching ratio (see Section 8.6.1).

For what concerns (direct) CP violation in charged K mesons, the necessary condition for CP violation in the decay, of having at least two weak amplitudes with different weak and strong phases, is not easily obtained in the SM. The $K^{\pm}$ decay modes with larger branching ratios are the leptonic and semi-leptonic ones, which lack strong phases, the $\pi^{\pm}\pi^0$ mode, which being purely $I = 2$ has a single amplitude, and the 3π states. The latter are in principle interesting, as in the isospin decomposition two different $\Delta I = 1/2$ amplitudes appear, of comparable magnitude, which could interfere among them resulting in larger effects than for ϵ' (in which one of the two interfering amplitudes is much smaller than the other). Unfortunately in first approximation these amplitudes have the same phase; the constraints obtained from direct CP violation in neutral K decays and the smallness of final state interactions are such that CP-violating asymmetries in charged K decays are expected to be very small $\sim O(10^{-5})$ (Gámiz *et al.*, 2002) in the SM and some of its extensions.

The contribution of the known CP violation to the transverse polarization of the muon in $K^+ \to \pi^0\mu^+\nu$ semi-leptonic decays is very small, as it cannot be generated by vector or axial-vector interactions only. Also, EM FSI only generate T-odd asymmetries of order 10^{-4} for the radiative $K_{\ell 3\gamma}$ decays (Braguta *et al.*, 2002).

The very rare decay $K_L \to \pi^0\nu\bar{\nu}$, with predicted branching ratio $3 \cdot 10^{-11}$ in the SM, is probably the cleanest weak decay from the theoretical point of view, for which theoretical uncertainties can be as small as 1%. Indeed, the presence of a single hadron in the final state accompanied by two particles which interact only weakly restricts the theoretical difficulties related to strong interactions to the evaluation of the matrix element for the $K \to \pi$ transition, which can however be obtained from the well-measured $K \to \pi e \nu$ decay. The ratio of decay rates for physical states into this state is

$$\frac{\Gamma(K_L \to \pi^0 \nu \bar{\nu})}{\Gamma(K_S \to \pi^0 \nu \bar{\nu})} = \frac{|1 - \lambda_{\pi \nu \bar{\nu}}|^2}{|1 + \lambda_{\pi \nu \bar{\nu}}|^2} \tag{9.40}$$

which vanishes if CP symmetry holds as this state is CP-even. CP violation in the mixing being small, and that in the decay negligible for this state, the above ratio measures CP violation in the interference of mixing and decay as the phase $2\theta_K$ of $\lambda_{\pi \nu \bar{\nu}}$, which is the mismatch between the phase of $K^0 - \overline{K}^0$ mixing and that of the $s \to d \nu \bar{\nu}$ decay amplitude. Exploiting the fact that $A(K^0 \to \pi^0 \nu \bar{\nu}) = A(K^+ \to \pi^+ \nu \bar{\nu})/\sqrt{2}$ from isospin, the ratio of two rare decay rates would give a measurement of the above phase

$$\frac{\Gamma(K_L \to \pi^0 \nu \bar{\nu})}{\Gamma(K^+ \to \pi^+ \nu \bar{\nu})} = \sin^2 \theta_K \tag{9.41}$$

with very small theoretical uncertainties (this relation is also not modified in most extensions of the SM). Clearly, the measurement of extremely rare final states containing two undetectable particles represents a formidable experimental challenge. The above K^+ mode was detected with very few events by the first generation of dedicated experiments, while the experimental limits for the K_L decay (Ahn *et al.*, 2006) are at present still higher than the SM prediction: these modes however hold a prominent place on the agenda of future experimental projects in flavour physics, because of their potential to provide very strong constraints on fundamental parameters, capable of putting in evidence even relatively small deviations from the SM (Buras, 2005).

9.3.3 The charm sector

In the SM larger CP violation effects are expected with heavy quarks, in which the complex phase of the CKM matrix can appear directly rather than only through virtual transitions; being a general feature of weak interactions, such effects should appear more easily as the number of relevant terms of the weak Hamiltonian which are probed by decays grows with the richer particle spectrum.

$D^0 - \overline{D}^0$ mixing is dominated by the first two quark families, and therefore large CP-violating effects are not expected: the top quark loops which provide the largest effects in K and B decays are absent for D. Moreover, many decay channels are possible for D mesons, which are not suppressed by small mixing angles ($c \to s$, same generation), resulting in large decay widths which make the observation of small effects more difficult.

The SM actually predicts very small mixing effects in the charm sector, $x, y \sim O(0.01)$; several intermediate states exist in the SM which can contribute to $D^0 - \overline{D}^0$ oscillations, all with branching ratios $O(10^{-3})$. For the above reason searches for $D^0 - \overline{D}^0$ oscillations can be considered as probes of new physics effects, to which x (but mostly not y) could be sensitive. Searches for CP-violating effects in the mixing are thus not practical for charmed mesons.

CP-violating effects in the interference of mixing and decay can be searched for both in decays into (almost) CP-eigenstates such as $\pi^+\pi^-$ and K^+K^-, and also in other flavour non-specific states such as $K^\pm\pi^\mp$: for example both $\overline{D}^0$ and D^0 can decay into $K^+\pi^-$, the latter being a doubly Cabibbo-suppressed (DCS) decay, and thus with a better chance to compete with the oscillation amplitude $D^0 \rightarrow \overline{D}^0 \rightarrow K^+\pi^-$. The SM predicts asymmetries at the $O(10^{-4})$ level due to the small mixing angles and slow oscillations. CP violation in the decays can be searched for in singly Cabibbo-suppressed (SCS) decays such as $D^0 \rightarrow \pi^+\pi^-, K^+K^-$ or $D^+ \rightarrow K_S K^+$; again, the SM predicts small asymmetries $O(10^{-3})$.

The fact that mixing and CP violation are highly suppressed in the SM makes the D system a good laboratory for searching for new physics effects: for example mixing could be enhanced by the presence of a new very heavy down-type quark disrupting the GIM cancellation, and large CP violation would indicate that some new particle contributes, as the effects of virtual b quarks are suppressed by the small CKM angles.

9.3.4 The beauty sector

The decays of B mesons can involve all three quark generations at tree level, and are thus a privileged system for the study of CP violation; asymmetries need not be suppressed by small CKM matrix elements. CP violation in the mixing is expected to be very small on rather general grounds, as we now discuss.

A rather general expectation for the B_d system is $|\Delta\Gamma| \ll |\Delta m|$: the decay width difference arises from the decay modes which are common to B^0 and $\overline{B}^0$ (recall eqn (7.22)), which all have branching ratios $O(10^{-3})$ (and in general can contribute to $\Delta\Gamma$ with different signs), so that $\Delta\Gamma/\Gamma \ll 1$; on the other hand $\Delta m/\Gamma \sim O(1)$ is known to be not small from experiment. In the case of B_s mesons $\Delta\Gamma/\Gamma$ can be larger, and this can be understood qualitatively from the quark composition: the dominant B_d decays are those in which the $\overline{b}$ quark decays to a $\overline{c}$ quark, and among the possible products of the emitted W^+ a (hadronic) final state common to B^0 and $\overline{B}^0$ can be obtained with $W^+ \rightarrow c\overline{s}$ thanks to $d\overline{s} \rightarrow \overline{d}s$ transitions ($K^0 - \overline{K}^0$ mixing). For the corresponding B_s decays the final state contains instead a $s\overline{s}$ pair

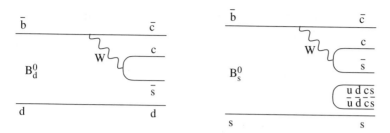

FIG. 9.4. Favoured tree diagrams for the decays of B_d (left) and B_s (right).

and has no net flavour, so that picking up a quark pair from a gluon (it's for free) more states which are common to $\bar{B}_s$ decays can be formed (see Fig. 9.4)

Within the SM the heavier neutral B meson is expected to have a (slightly) smaller decay width than the lighter one, so that the relative sign of Δm and $\Delta\Gamma$ would be opposite with respect to the case of K mesons.[133] From eqns (7.51) and (7.52), in the limit of small CP violation $|\Delta\Gamma/\Delta m|$ is given by $|\Gamma_{12}/M_{12}|$ so that for the B system(s)

$$|\Gamma_{12}| \ll |M_{12}| \tag{9.42}$$

The phase-convention independent measure of CP violation effects in the mixing, expressed by the phase mismatch between the absorptive and dispersive parts of the mixing amplitude, is thus suppressed by the large difference between the magnitudes of such terms:

$$\frac{\mathrm{Im}(M_{12}^*\Gamma_{12}/2)}{|M_{12}|^2 + |\Gamma_{12}|^2/4} \ll 1 \tag{9.43}$$

Note that Γ_{12}/M_{12} is approximately independent from CKM matrix elements, and thus comparable for B_d and B_s. The evaluation of M_{12} and Γ_{12} in terms of quark diagrams is expected to be much more reliable for B mesons than for K mesons, and the small CP violation in B mixing can be easily understood in terms of box diagrams (Carter and Sanda, 1981): the product of CKM matrix elements appearing in the box diagram with exchange of up-type quarks i and j ($i,j = u,c,t$) is

$$\Lambda_i^{(db)}\Lambda_j^{(db)} \equiv V_{ib}^*V_{id}V_{jb}^*V_{jd} \tag{9.44}$$

The part of M_{12} which is independent from quark masses cancels in the sum over the families due to the unitarity of the CKM matrix, eqn (9.22), while the largest remaining effect is due to the exchange of the top quark, which has the largest mass difference: $M_{12} \propto [\Lambda_t^{(db)}]^2 = (V_{tb}^*V_{ts})^2$ since, contrary to the case of K mesons, the

---

[133] Δm and $\Delta\Gamma$ are usually redefined to be both positive for K, and often also for B.

large top contribution is not suppressed by small CKM matrix elements ($V_{tb} \simeq 1$). Γ_{12} is obtained instead from physical decay amplitudes to which the top quark, way above in energy, cannot contribute; in the limit $(m_u^2 - m_c^2) \ll m_b^2$ all the other contributions become equal, apart from the CKM factors, so that by using the unitarity conditions again

$$\Gamma_{12} \propto [\Lambda_c^{(db)}]^2 + [\Lambda_u^{(db)}]^2 + \Lambda_u^{(db)}\Lambda_c^{(db)} + \Lambda_c^{(db)}\Lambda_u^{(db)} = [\Lambda_t^{(db)}]^2$$

and at leading order its phase coincides with that of M_{12}, so that CP violation (9.43) vanishes.

In the SM $|\Gamma_{12}/M_{12}| = O(m_b^2/m_t^2) \lesssim 10^{-2}$, and the additional suppression in mixing asymmetries (9.43) is of two more orders of magnitude. This result is rather model independent, as contributions from new physics, most likely linked to very heavy particles, are not expected to increase significantly Γ_{12}. The above considerations indicate that q/p is almost a pure phase

$$\frac{q}{p} \simeq \sqrt{\frac{M_{12}^*}{M_{12}}} = \frac{M_{12}^*}{|M_{12}|} = \frac{V_{tb}^* V_{td}}{V_{tb} V_{td}^*} \tag{9.45}$$

and $|q/p| \sim 1$ indeed indicates small CP violation in the mixing. The parameter q/p in (9.45) has a large phase in the Wolfenstein parametrization; we recall that the phase of q/p is not a physical quantity, as it depends on the choice of quark phase convention: what actually matters are the relative phases of q/p and ratios of decay amplitudes. Considering e.g. an amplitude in which the term $V_{cb}^* V_{cd}/V_{cb} V_{cd}^*$ appears, which is approximately real in the same parametrization, the phase difference β can result in significant CP-violating effects in the interference of mixing and decay.

For B_s mesons $\Delta\Gamma/\overline{\Gamma} \sim 0.3$ is larger, and so is $\Delta m/\overline{\Gamma} \sim 20$; moreover in this case the phase of the CKM factors appearing in the box diagrams is smaller than for B_d by two orders of magnitude: in this case

$$\frac{q}{p} \simeq \frac{V_{tb}^* V_{ts}}{V_{tb} V_{ts}^*} \tag{9.46}$$

and the phase difference which could be measured from the decay $B_s \to J/\psi\,\phi$ (requiring angular analysis into CP eigenstates) is

$$\beta_s \equiv \chi \equiv \arg\left[-\frac{V_{tb}^* V_{ts}}{V_{cb}^* V_{cs}}\right] \tag{9.47}$$

which is the small angle of the squashed triangle corresponding to the unitarity relation (9.30). The analogous quantity for the case of K mesons is

$$\beta_K \equiv \chi' \equiv \arg\left[-\frac{V_{cd}^* V_{cs}}{V_{ud}^* V_{us}}\right] \tag{9.48}$$

which is the small angle of the even more squashed triangle corresponding to (9.28).

Despite the small CP violation in the mixing, large CP-violating asymmetries can be present in the B system, as discussed. We recall that CP violation in the decay requires two interfering amplitudes with different weak and strong phases to be present; for a decay which can proceed via two different intermediate states different products of CKM matrix elements appear, providing different weak phases. In decays mediated by penguin diagrams strong phases are always present when the quark appearing in the loop can be put on mass shell (Bander *et al.*, 1979), so that a real intermediate state contributes an absorptive amplitude: this is possible for c and u quarks in the loops appearing in b quark decays (see Fig. 9.5). Since the two contributing amplitudes in general have different isospin structure the strong phases will be different and the conditions for observing CP violation in the decay are satisfied; as usual, CPT symmetry constrains some rate asymmetries to be equal (Gérard and Hou, 1991).

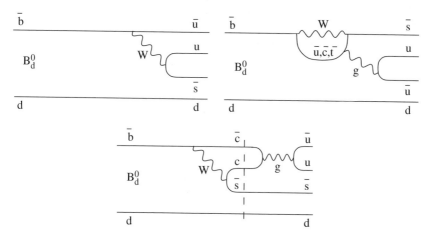

FIG. 9.5. Exemplification of the origin of weak and strong phases from penguin diagrams: in the case of $\bar{b} \to \bar{s}u\bar{u}$ decays. Top left: tree diagram; top right: penguin diagram; bottom: one penguin diagram redrawn in a way which makes more evident the presence of an intermediate on-shell state (here $c\bar{c}$) which re-scatters, generating an imaginary (absorptive) part of the amplitude.

9.3.5   Electric dipole moments

The SM implementation of *CP* violation, in weak charged currents, implies that *CP*-violating effects are expected to practically vanish in flavour-diagonal processes: this is a specific prediction of the SM which can be tested, as in principle *CP* violation could be quite independent from the flavour-changing nature of the processes.

The SM contribution to the neutron EDM is very small, because the one-loop corrections to the electromagnetic current from weak charged currents is a sum of terms in which each complex CKM matrix element (e.g. V_{cd} for the $d \to c$ transition) appears together with its complex conjugate (V_{cd}^* for the opposite $c \to d$ transition), and the total effect vanishes. The correction vanishes also at the two-loop level, and the first non-zero contribution appears with three loops. Predictions are of order 10^{-31} to 10^{-33} e cm, about 6 orders of magnitude below the experimental limits. Still, the large hadronic uncertainties at low energy prevent a reliable computation of the neutron EDM from the known *CP*-violating parameters. Modifications and extensions of the SM often lead to EDM contributions which are only first order in the weak interaction and therefore much larger. It should be noted that in principle a contribution could also arise from the *CP*-violating term of strong interactions (see Section 11.2), whose magnitude is actually bounded by the EDM limits.

Compared to the neutron, the electron EDM is free from hadronic uncertainties and can be computed more reliably, but it is also smaller (vanishing with massless neutrinos), and nothing is yet known on the *CP*-violating parameters in the leptonic sector. SM estimates for the electron EDM are usually many orders of magnitude smaller than for the neutron.

9.3.6   Other systems

Most predictions of *CP*-violating effects suffer from the above mentioned theoretical difficulties, but order of magnitude estimates are usually available for most possible signatures; we just mention a few for the systems which were discussed in previous chapters.

Contrary to $K \to \pi\pi$ decays, which are only sensitive to *CP* violation in the parity-violating amplitudes, hyperon decays are also sensitive to parity-conserving amplitudes. The predicted asymmetries in the SM are of order 10^{-5} at most (Tandean and Valencia, 2003).

Particle production asymmetries at colliders were mentioned in Chapter 6 as possible signatures of *CP* violation: these are negligible in the SM and thus good probes for new physics; the production of $\tau^+\tau^-$, $b\bar{b}$ and $t\bar{t}$ can be studied to maximize the sensitivity to *CP* violation mechanisms which grow of importance with the mass (such as those related to the Higgs coupling).

Anomalous couplings of the electro-weak gauge bosons, $W^+W^-Z\gamma$, $Z\gamma\gamma^*$, $ZZ^*\gamma$, ZZZ^*, were studied in e^+e^- and $p\bar{p}$ interactions with several final states, by detailed analysis of production cross sections and angular or energy distributions. Such couplings, which are zero at tree level in the SM, could in principle introduce CP violation effects, on which experimental limits were set (Yao $et\ al.$, 2006).

9.4 From measurements to theory

In a neutral flavoured meson system with sizeable mixing phases, asymmetries in the interference of mixing and decay can be large. Focusing on B mesons, the effective Hamiltonian contributing to $\Delta B = 2$ processes is written as

$$H_{\text{eff}}(\Delta B = 2) \propto e^{2i\theta_M} O(\Delta B = 2) + e^{-2i\theta_M} O^\dagger(\Delta B = 2)$$

with $O(\Delta B = 2)$ representing the sum of relevant operators: for $|M_{12}| \ll |\Gamma_{12}|$ one has $q/p \propto \exp[-2i\theta_M]$. Similarly the $\Delta B = 1$ effective Hamiltonian responsible for decays can be written in general as

$$H_{\text{eff}}(\Delta B = 1) \propto e^{i\theta_D} O(\Delta B = 1) + e^{-i\theta_D} O^\dagger(\Delta B = 1)$$

from which $\bar{A}_{\bar{f}}/A_f \propto \exp[-2i\theta_D]$, and for a final state f which is a CP eigenstate with eigenvalue $\eta_{CP}(f)$ the CPV parameter (7.91) is

$$\lambda_f = \eta_{CP}(f)\, e^{-2i(\theta_M + \theta_D)} \tag{9.49}$$

Little can be said in general about the phase θ_D, but for decays which are dominated by a single amplitude (no CP violation in the decay) this is just the weak phase of this amplitude, and the measurable asymmetries can be linked directly to CKM matrix elements: the measurement of the tagged meson asymmetry (7.145) allows the extraction of $\text{Im}(\lambda_f)$, which in this case is just the sum of the mixing and decay phases.

This kind of measurement in principle allows direct measurement the (relative) phases of CKM matrix elements from the experimental asymmetries: this programme was started with great success at the B factories. Note that the above in principle applies also to K decays, namely the measurement of CP violation in the interference of mixing and decays in $\text{Im}(\epsilon)$ (Chapter 8) provides a precise determination of the phase difference between M_{12} and Γ_{12} for that system, but unfortunately in this case this phase difference cannot be translated in a clean way into parameters of the Lagrangian. At this time therefore B mesons represent a very good laboratory for the quantitative exploration of the flavour sector of the SM.

Decays to which more amplitudes of comparable magnitude contribute are less useful from the above point of view: in general the relation between the measured asymmetries and the fundamental parameters of the SM is blurred by quantities which cannot be reliably computed or measured. In general $|\overline{A}_{\overline{f}}| \neq |A_f|$ for such decays (which for the B system for which $|q/p| \sim 1$ is equivalent to $|\lambda_f| \neq 1$) and CP violation in the decays is present; this is not a necessity, however, since in presence of two amplitudes with the same strong phase $|\overline{A}_{\overline{f}}| = |A_f|$: in this case the situation is cleaner, as the knowledge of the strong phase is not required, but still the extraction of the weak phase from the asymmetry requires the knowledge of the relative size of the two amplitudes involved.

If the two amplitudes share the same strong *and* weak phase however, such weak phase can be extracted from the asymmetry. This is the situation in the remarkable case of $B_d \to J/\psi K_S$ decays (Bigi and Sanda, 1981) (and in general $B_d \to$ charmonium K_S), driven by the $b \to c\bar{c}s$ quark transition. As mentioned above, this state is a CP eigenstate, as the $\bar{s}d$ and $s\bar{d}$ pairs resulting from the decays of B^0 and $\overline{B}^0$ mesons mix into physical K_S, K_L states[134] (Fig. 9.6).

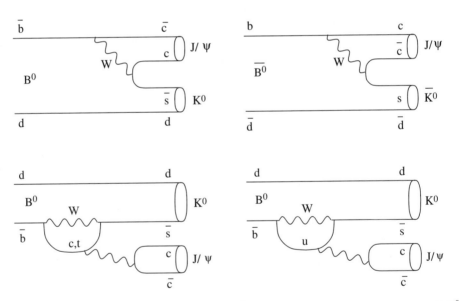

FIG. 9.6. Diagrams contributing to the $B_d \to J/\psi K^0$ decay. Top row: the tree decays $B_d \to J/\psi K^0$ and $\overline{B}_d \to J/\psi \overline{K}^0$ mixing into $B_d, \overline{B}_d \to J/\psi K_S$. Bottom row: the penguin decays, the left-hand one having the same phase as the tree decay.

[134] Strictly speaking the CP-eigenstates $K_{1,2}$ should be used instead of K_S, K_L, but the difference is quantitatively irrelevant.

The final state corresponds to a single isospin amplitude, and no direct CP violation is therefore expected (same strong phases); moreover the part which is difficult to compute, namely the hadronic matrix element linking the quark process to the observed hadronic process, drops out in the amplitude ratio $\overline{A}_{\overline{f}}/A_f$. The decay into $f = J/\psi K^0$ is dominated by the tree diagrams of Fig. 9.6, so that

$$\frac{\overline{A}_{J/\psi K_S}}{A_{J/\psi K_S}} = \eta_{CP}(J/\psi K_S) \frac{V_{cs}^* V_{cb}}{V_{cs} V_{cb}^*} \frac{V_{cd}^* V_{cs}}{V_{cd} V_{cs}^*}$$

where the last factor is the one due to $K^0 - \overline{K}^0$ mixing into K_S, and $\eta_{CP}(J/\psi K_S) = -1$. Multiplying this to (9.45) a large phase mismatch between the $B^0 - \overline{B}^0$ mixing amplitude and the decay amplitude is obtained for this decay:

$$\lambda_{J/\psi K_S} \simeq -\frac{V_{tb}^* V_{td}}{V_{tb} V_{td}^*} \frac{V_{cb}^* V_{cd}}{V_{cb} V_{cd}^*} = -e^{-2i\beta} \tag{9.50}$$

$$\mathrm{Im}\left(\lambda_{J/\psi K_S}\right) \simeq \sin(2\beta) \tag{9.51}$$

Besides the tree process, penguin diagrams containing quark loops also contribute to the decay (Fig. 9.6), but the dominant one (unsuppressed by CKM matrix elements) shares the same phase (9.51) as the tree amplitude, while the size of the other one is only 1% of the tree amplitude, so that the above conclusion is valid to a high degree of accuracy.[135] The above 'golden' decay mode was the first one in which CP violation was measured at B factories (Chapter 10), right after their start-up.

Generalizing to the case of final states which are not necessarily CP eigenstates, and recalling the discussion of Section 7.2.1, for decays into states dominated by a single amplitude, the decay phases are

$$\theta_f = \phi_f + \delta_f \qquad \theta_{\overline{f}} = \phi_{\overline{f}} + \delta_{\overline{f}} \qquad \overline{\theta}_{\overline{f}} = -\phi_f + \delta_f \qquad \overline{\theta}_f = -\phi_{\overline{f}} + \delta_{\overline{f}}$$

where $\phi_{f,\overline{f}}$ and $\delta_{f,\overline{f}}$ are respectively the weak and strong phases for the decays into the CP-conjugate states $f, \overline{f}$. In this case A_f and $\overline{A}_f$ are not related by CP (A_f and $\overline{A}_{\overline{f}}$ are), and the troublesome strong phases do not cancel in λ_f and $\lambda_{\overline{f}}$. If CP violation in the mixing and the decay width differences of the physical states are neglected (both good approximations for B mesons), the time-dependent rates of tagged decays (7.141), (7.142) are related by

$$R_f = R_{\overline{f}} \qquad C_f = -C_{\overline{f}} \qquad S_f = -S_{\overline{f}} \tag{9.52}$$

and the phase combinations appearing in $S_{f,\overline{f}}$ are

$$\arg(\lambda_f) = -2\theta_M - \phi_f - \phi_{\overline{f}} + \Delta\delta \qquad \arg(\lambda_{\overline{f}}) = -2\theta_M - \phi_f - \phi_{\overline{f}} - \Delta\delta$$

---

[135] For comparison, in decays mediated by $b \to c\overline{c}d$ transitions, such as $B^0 \to J/\psi\pi^0$, the argument holds only to the extent that penguin diagrams, with a different phase, can be neglected.

where $\Delta\delta = \delta_{\bar{f}} - \delta_f$; from the experimental measurement the quantity $\sin^2(2\theta_M + \phi_f + \phi_{\bar{f}})$ can be extracted with a twofold ambiguity, which can be removed by comparing with other final states with the same weak phases and different strong phases.

This discussion reduces to that presented earlier in the specific case of a CP eigenstate f; note that the asymmetries change sign according to the CP eigenvalue of f, so that when final states are considered in which CP eigenstates of opposite CP-parity are mixed together (such as when the final states f and $f\pi^0$ are not experimentally distinguished) the measured asymmetries are diluted.

The extraction of the SM parameters from B decay measurements is an ongoing activity, and it might well be that, as the number of channels being studied grows, some measurement will not fit in the framework briefly sketched here, pointing to the effect of physics beyond the SM. See e.g. Bigi and Sanda (2000), Branco, Lavoura, and Silva (1999) for a more detailed discussion of all the relevant decay modes and the parameters which can be extracted from them.

9.5   Other scenarios

The SM implementation of CP violation is by no means unique; in general in a gauge theory CP violation can arise either as an explicit symmetry breaking, due to the presence of CP-violating terms in the complete Lagrangian, or as a spontaneous symmetry breaking, when the Lagrangian is CP-symmetric but the physical vacuum is not. Explicit symmetry breaking requires that all the parameters in the Lagrangian which can be complex are complex, as even if they were set to be real at tree level they would receive complex renormalization corrections. As a consequence the number of CP-violating phases grows with the complexity of the theory, and the SM with three quark families is probably the simplest consistent model which can support explicit CP violation. In the following we briefly mention a few other possibilities, widely discussed in the literature.

The interactions of a single neutral Higgs boson are flavour-diagonal and CP symmetric, since the Hermiticity of the Lagrangian forces all the parameters of the Higgs potential to be real; with two $SU(2)$ Higgs doublets Φ_1, Φ_2, however, Hermiticity, renormalizability and gauge invariance do not forbid the appearance of complex terms in the scalar potential (Weinberg, 1976). For example

$$\mathcal{L} = k\,\Phi_1^\dagger\Phi_2 + h\,(\Phi_1^\dagger\Phi_2)^2 + \text{H.c.}$$

under a CP transformation

$$\Phi_{1,2} \rightarrow e^{i\xi_{1,2}}\,\Phi_{1,2}^\dagger$$

violates CP if $\mathrm{Im}(k^2 h^*) \neq 0$. With two Higgs doublets flavour-changing neutral currents arise at tree level, which have to be explicitly cancelled by imposing some discrete symmetry, to be in agreement with the observations; in doing so the only phase which is left is actually the usual CKM phase. With more than two Higgs doublets new CP-violating phases appear.

A neutral Higgs boson h^0 which is not a CP eigenstate could violate CP in its decay to a fermion-antifermion pair, and this would appear through the spin correlations of heavy fermions, such as $h^0 \to \tau^+\tau^-$ or $h^0 \to t\bar{t}$. When (if?) the Higgs boson is found, its quantum numbers will be measured and such possibility will also be checked.

An interesting possibility which has been considered in the literature is that CP symmetry is spontaneously broken, i.e. the Lagrangian respects such symmetry but the physical vacuum does not (Lee, 1973). The general features of models of this type are exposed by a simple example which is quite instructive. Consider a spin 1/2 field ψ coupled to a pseudo-scalar Hermitian massless field ϕ in the Lagrangian

$$\mathcal{L} = \frac{1}{2}\partial_\mu\phi\,\partial^\mu\phi - V(\phi) + \bar{\psi}(i\gamma^\mu\partial_\mu - m)\psi - ig\bar{\psi}\gamma_5\psi\phi$$

$$V(\phi) = -\frac{1}{2}\mu^2\phi^2 + \frac{1}{4}\lambda\phi^4$$

with real parameters m, g, μ, λ (so that $\mathcal{L}$ is Hermitian).[136] Note that this Lagrangian is explicitly CP-symmetric, with the appropriate choice of a field ϕ with $\eta_P(\phi) = -1$, $\eta_C(\phi) = +1$ and $\eta_T(\phi) = -1$ (recall eqns (2.52), (3.33), (4.52)). With $\lambda > 0$ spontaneous symmetry breaking occurs and ϕ acquires a non-zero vacuum expectation value $\langle\phi\rangle_{\mathrm{vac}} = v = \pm\sqrt{\mu^2/\lambda}$: the two possible choices for $\langle\phi\rangle_{\mathrm{vac}}$ are equivalent and related by a T transformation, but each of them is not T symmetric. Since the broken symmetry is not a continuous one, no massless Goldstone boson appears and the physical scalar field $\hat{\phi}$, defined by $\phi = v + \hat{\phi}$, has a mass $\hat{M} = \sqrt{2\mu^2}$.

Given the quantum numbers of the field ϕ, the vacuum is no longer an eigenstate of CP (nor of P and T); the breaking of T symmetry is made manifest by applying a unitary transformation on ψ of the form

$$\psi \to e^{-i(\alpha/2)\gamma_5}\psi$$

---

[136] Clearly this λ has nothing to do with the Cabibbo angle.

The kinetic term for ψ is unaffected, and choosing $\tan \alpha = g\sqrt{\mu^2/\lambda}/m$ the Lagrangian becomes

$$\mathcal{L} = \frac{1}{2}\partial_\mu \hat{\phi} \, \partial^\mu \hat{\phi} - V(\hat{\phi}) + \overline{\psi}(i\gamma^\mu \partial_\mu - M)\psi$$
$$- g \sin \alpha \, \overline{\psi}\psi\hat{\phi} - ig \cos \alpha \, \overline{\psi}\gamma_5 \psi\hat{\phi}$$
$$V(\hat{\phi}) = -\frac{1}{4}\frac{\mu^4}{\lambda} + \frac{1}{2}\hat{M}^2\hat{\phi}^2 + \sqrt{\lambda\mu^2}\,\hat{\phi}^3 + \frac{1}{4}\lambda\hat{\phi}^4$$

where $M = \sqrt{m^2 + g^2\mu^2/\lambda}$. In the interaction terms of ψ and $\hat{\phi}$, $\overline{\psi}\psi$ has $\eta_P = \eta_C = \eta_T = +1$, while $i\overline{\psi}\gamma_5\psi$ has $\eta_P = \eta_T = -1$ and $\eta_C = +1$; therefore in exchanges of $\hat{\phi}$ quanta the interference term between the two violates P, T, and CP. Realistic models of this type, which are consistent with the gauge symmetries of the SM, can be built by introducing more Higgs doublets. Simple models with spontaneous CP violation introduce cosmological problems related to the surfaces of contact between regions with opposite values of $\langle\phi\rangle_{\text{vac}}$ (see Chapter 11).

It may be aesthetically appealing to try restoring the broken left-right symmetry at a higher energy scale, by enlarging the gauge group to $SU(2)_L \times SU(2)_R \times U(1)$, and putting all fermions in left-handed and right-handed doublets; parity symmetry would then be broken spontaneously rather than explicitly in the Lagrangian. In such 'left-right symmetric' models the smallness of the (non-observed) interactions of right-handed quarks is explained by advocating large masses for the corresponding gauge bosons W_R (and for the additional neutral scalar fields which would otherwise mediate flavour-changing neutral currents of a too large magnitude). CP violation easily appears in this kind of models, due to the larger number of fields: even with a single quark generation uneliminable complex phases appear. In these models CP violation in the mixing can be enhanced, and the experimental limits on it actually give strong lower bounds on the mass of a possible W_R gauge boson.

While the SM gluon cannot contribute to the box diagrams describing neutral flavoured meson mixing, as its coupling is flavour-diagonal, in super-symmetric theories[137] the quark-squark-gluino vertex can change flavour, as the mass matrices of quarks and squarks are not necessarily simultaneously diagonalizable. A large number of uneliminable complex phases appear (43 new ones in a generic model!), and even in the minimal super-symmetric extension of the SM with restrictive hypothesis there are enough such phases to generate FCNC in general conflicting with observations. This is an example of a general issue: when extending the SM many possible additional sources of CP violation appear, which have to be somehow

---

[137] Super-symmetry is a theoretically appealing new kind of symmetry according to which each fermion (boson) has a boson (fermion) partner: spinless *squarks* for quarks, spin 1/2 *gluinos* for gluons, etc. Such symmetry (supposedly spontaneously broken) would solve several theoretical puzzles and is thus considered as a favoured possibility for physics beyond the SM, whose value is expected to be assessed by experimentation at the LHC; see e.g. Murayama (2000).

suppressed to agree with the smallness of the observed *CP* violation; this can be done by assuming that the new particles have very high masses so that their effects are barely noticeable, but in this way one often also forfeits the desired effects for which the new particles were introduced in the first place. There are of course ways out of this conundrum within specific models, but without entering into any detail we just note that the smallness of the measured *CP*-violating effects (and the limits on flavour-conserving *CP*-violating effects) represent a serious issue with which any extension of the SM must be confronted, going under the name of 'super-symmetric *CP* problem'.

While bets on the failure of *CPT* symmetry would incur rather high odds, the possibility that this symmetry is violated together with Lorentz symmetry has been actively considered. Violating *CPT* is somewhat like opening a Pandora's box, but a consistent general model in which this can happen without upsetting the known features of the SM exists (Colladay and Kostelecký, 1997); in particular, while the SM does not have the dynamics necessary for a spontaneous breaking of *CPT* and Lorentz symmetries, more general theories do. Within such a model a quantitative comparison of different tests of *CPT* becomes possible, and it is found that some specific *CPT* tests (such as the comparisons of g factors or $1S - 2S$ hydrogen transitions in matter and antimatter) are actually not sensitive to the possible effects (Bluhm, 2004).

Another speculative approach to possible *CPT* violation is that of apparent violations of unitarity arising in quantum gravity and leading to decoherence, or spontaneous transformation of a pure state into a mixed one (Mavromatos, 2005).

Further reading

The Standard Model is described in many textbooks at different levels of detail and we do not even try to give a partial listing. Branco, Lavoura, and Silva (1999), and Bigi and Sanda (2000) offer a thorough introduction to the computation of *CP*-violating quantities in the SM; these books also describe *CP* violation in possible extensions of the SM, which are also addressed in Nir (1999).

The advanced reader desiring to acquire a deep grasp on the link from the theoretical construct of the Standard Model to the observable quantities is directed to Donoghue *et al.* (1992), while the theoretically oriented reader might be interested in the more abstract approach of Froggatt and Nielsen (1991).

Problems

1. Show that the same flavour mixing matrix (9.21) which appears in the weak charged current also appears in the charged scalar interactions (before symmetry breaking).

2. Discuss the *CP* transformation of terms of the SM Lagrangian different from the Yukawa ones.

3. The quark flavour mixing is usually written in terms of down-type quarks. What about leptons? Why don't we measure charged lepton oscillations?

4. Show explicitly how any complex phase in a 2×2 unitary matrix can be eliminated by redefinition of the quark field phases.

5. Consider the decay $K^0 \rightarrow \pi^0 \pi^0$ in a world with only the first two quark generations: draw some possible diagrams for this decay and discuss their phases.

6. Show that if any two quarks were degenerate in mass the *CP*-violating phase appearing in the CKM matrix could be eliminated.

7. Explain qualitatively why the properties of $K \rightarrow \pi \nu \bar{\nu}$ decays can be predicted in a very reliable way.

10

CP VIOLATION IN HEAVY MESON SYSTEMS

There is no excellent beauty
that hath not some strangeness in the proportion.
F. Bacon

The more original a discovery,
the more obvious it seems afterwards.
A. Koestler

Just keep building on the ground
that's been won.
B. H. May

In this chapter the phenomenology of *CP* violation in heavy (*D* and *B*) flavoured mesons is discussed. After briefly reviewing the situation in the charm sector, we mostly focus on the *B* system, the only system besides that of *K* in which evidence of *CP* violation was found, with some added value: as a consequence of their heavier mass, *B* mesons are generally more sensitive to the mechanism of *CP* violation embodied in the Standard Model and also exhibit a much larger set of possible decays, some of them being more easily tractable from a theoretical point of view. While no *qualitatively* new features of *CP* violation were discovered in the *B* system, large effects were observed in many decay modes, some of them allowing significant *quantitative* test of the SM picture. Indeed for the *B* system the focus shifted to the accumulation of data with the purpose of probing the fundamental parameters of the underlying theory, and test of its consistency.

The large number of *CP* violation measurements and tests possible on the *B* system, by no means yet complete, precludes the possibility of dealing with them in an exhaustive way; a sampler will thus be provided, directing the interested reader to the growing literature on the subject for a more exhaustive treatment.

10.1 Flavour mixing

The peculiar features of a coupled system of neutral flavoured mesons discussed in Chapter 7 are also found in neutral mesons built from heavier quarks, which however are not quite simple higher-mass replicas of the kaon system but rather exhibit some specific features of their own. In contrast to K, in these heavier systems there is no single dominant decay mode, but rather many modes with comparable branching ratios, and this changes the picture considerably.

Flavour mixing is expected to be present for all families of neutral flavoured mesons. When dealing with a $M^0 - \overline{M}^0$ pair the mixing parameters can be extracted from the measurement of the fraction of mixed events, that is events in which the two mesons decay semi-leptonically producing leptons of the same charge, as discussed in Section 7.5.4; the leptons originating from semi-leptonic decays of a b quark are usually selected by imposing a minimum momentum threshold to distinguish them from those of different origin. In practice, the situation is often complicated by the fact that the available samples are not pure. In the case of a correlated state obtained from a resonance, one has $M^0 \overline{M}^0$ pairs but also $M^+ M^-$ pairs, and the latter tend to dilute the measurement, as they contribute to the number of 'unmixed' events: since semi-leptonic decays (with an undetected neutrino) are not completely reconstructed, but rather the number of leptons of a given charge are simply measured, opposite-sign leptons are also obtained from charged meson pairs. Labelling the measured number of pairs with subscripts indicating the charges of the parent M mesons, the observed number of lepton pairs are

$$N(\ell^\pm \ell^\pm) = N_0(\ell^\pm \ell^\pm) \qquad N(\ell^+ \ell^-) = N_0(\ell^+ \ell^-) + N_\pm(\ell^+ \ell^-)$$

The relative contributions of charged and neutral mesons are

$$\frac{N_\pm}{N_0} = \frac{F_\pm}{F_0} \left(\frac{BR_\pm}{BR_0} \right)^2$$

where $F_{\pm,0}$ are the charged and neutral meson production fractions, and $BR_{\pm,0}$ the respective branching ratios for semi-leptonic decay, so that the true mixing ratio R (7.178) for a $|C-\rangle$ state is obtained from the measured one $R^{(\mathrm{m})}$ as

$$R^{(C-)} = \left(1 + \frac{N_\pm}{N_0} \right) R^{(\mathrm{m})} \qquad (10.1)$$

and these are usually set equal to χ (7.158) by assuming the validity of the $\Delta F = \Delta Q$ rule ($|A_f| = |\overline{A}_{\bar{f}}|$) and the absence of *CP* violation in the mixing (Section 7.5.4).

In the case of uncorrelated meson production the situation is even more complicated, as many kinds of flavoured hadrons are produced, so that a knowledge of the production fractions is required to disentangle the mixing parameters from the observed mixing ratios. For example at high energy $e^+ e^-$ or hadron colliders both B_d and B_s are produced together, and (with the same assumptions indicated above) one measures an averaged mixing parameter (7.158)

$$\langle \chi \rangle = F_d \left(\frac{BR_d}{\langle BR \rangle} \right) \chi(B_d) + F_s \left(\frac{BR_s}{\langle BR \rangle} \right) \chi(B_s) \qquad (10.2)$$

where $F_{d,s}$ are the production fractions of B_d, B_s mesons, $BR_{d,s}$ their semi-leptonic branching ratios and $\langle BR \rangle$ the average semi-leptonic branching ratio of the sample, weighted by the relative production fractions.

10.2 *D* mesons

D mesons contain charm as the heavier quark, and as the only case (in the three family SM) in which such quark is of the up-type, they exhibit a behaviour some-what different from *K* and *B*, not too exciting for what concerns *CP* violation. Neutral *D* mesons contain *u* quarks ($D^0 = c\bar{u}$, mass ~ 1.9 GeV/c^2) and charged *D* mesons either *d* ($D^+ = c\bar{d}$, mass ~ 1.9 GeV/c^2) or *s* quarks ($D_s^+ = c\bar{s}$, mass ~ 2.0 GeV/c^2)[138]. The lifetimes are ~ 1 ps for $D^\pm$ and ~ 0.5 ps for D^0 and $D_s^\pm$, much shorter than those of *B* mesons when the quark mass scaling $\Gamma \propto m_q^5$ is taken into account, because *D* meson decays are not suppressed by small CKM angles. Such lifetimes are however long enough to allow direct experimental observation of the meson decay lengths.

Flavour mixing was observed recently in the $D^0 - \overline{D}^0$ system by the BABAR experiment (at 3.9 standard deviations significance) by studying $B^0, \overline{B}^0 \to K^\pm \pi^\mp$ decays (Aubert *et al.*, 2007*b*). Published limits (95% CL) on the mass difference of physical states are consistent with zero

$$\Delta m_D < 0.046 \cdot 10^{-3} \, \text{eV} \qquad (10.3)$$

and in terms of the parameters[139] x, y defined in (7.149) (Yao *et al.*, 2006)

$$x < 0.03 \ (95\% \ \text{CL}) \qquad y = (0.7 \pm 0.5) \cdot 10^{-4} \qquad (10.4)$$

This is consistent with the expectations in the SM, in which x, y are expected not to exceed 0.01, although precise predictions are difficult.

---

[138] The naming convention in terms of quark composition is opposite for mesons containing up-type quarks and down-type quarks: the 'unbarred' neutral state D^0, K^0, B^0 is always the isospin partner of the state with positive charge D^+, K^+, B^+, which means that it contains respectively a *c* quark or a $\bar{s}, \bar{b}$ quark (its flavour quantum number being always positive: $C, S, B = +1$).

[139] To extract x, y from the BABAR mixing result (Aubert *et al.*, 2007*b*) the value of an unmeasured strong phase difference is required: if the latter were zero one would obtain $y = (9.7 \pm 5.4) \cdot 10^{-3}$ and x consistent with zero.

In the case of truly flavour-specific decays ($A_{\bar{f}} = 0 = \bar{A}_f$), such as the semi-leptonic modes $D^0 \to K^- \ell^+ \nu$ and $\overline{D}^0 \to K^+ \ell^- \bar{\nu}$ ($BR \sim 3.3\%$), the 'wrong sign' decay rates entirely due to flavour mixing are (in the small mixing limit)

$$\Gamma[D^0 \to \bar{f}](t) \propto |\bar{A}_{\bar{f}}|^2 \left|\frac{q}{p}\right|^2 e^{-\bar{\Gamma}t}(\bar{\Gamma}^2 t^2)(x^2 + y^2) \tag{10.5}$$

$$\Gamma[\overline{D}^0 \to f](t) \propto |A_f|^2 \left|\frac{p}{q}\right|^2 e^{-\bar{\Gamma}t}(\bar{\Gamma}^2 t^2)(x^2 + y^2) \tag{10.6}$$

and these are usually normalized to the integrated number of 'right sign' decays

$$r_D(t) \simeq \frac{\Gamma[D^0 \to \bar{f}](t)}{N(\overline{D}^0 \to \bar{f})} \qquad \bar{r}_D(t) = \frac{\Gamma[\overline{D}^0 \to f](t)}{N(D^0 \to f)}$$

thus removing the squared amplitude factors from (10.5), (10.6). By fitting the experimental distributions to these expressions, limits on $D^0 - \overline{D}^0$ mixing can be obtained. In the case of *CP* conservation $r_D = \bar{r}_D$ and the time-integrated mixing rate is

$$R_M = \int_0^\infty r_D(t)\, dt \simeq \frac{1}{2}(x^2 + y^2) \tag{10.7}$$

The final states $f = K^- \pi^+$ and $\bar{f} = K^+ \pi^-$ are neither *CP* eigenstates nor truly flavour-specific ones, as they can be reached from both D^0 and $\overline{D}^0$: the decays $\overline{D}^0 \to f$ and $D^0 \to \bar{f}$ are not forbidden but rather doubly Cabibbo-suppressed ($BR \simeq 1.4 \cdot 10^{-4}$), as they involve $|V_{cd}^* V_{us}|^2 \sim \lambda^4$; on the contrary, the decays $D^0 \to f$ and $\overline{D}^0 \to \bar{f}$ ($BR \simeq 3.8\%$) involve $|V_{ud}^* V_{cs}|^2 \sim 1$. These are thus 'almost' flavour-specific decays, and for this reason in the decay e.g. $D^0 \to \bar{f}$ the mixed amplitude $D^0 \to \overline{D}^0 \to \bar{f}$ has a better chance of competing with the unmixed one to give observable effects. In this case the time dependence for the 'wrong sign' decays is given (for $x \ll 1, y \ll 1$ and $|\bar{A}_f| \ll |A_f|$, $|A_{\bar{f}}| \ll |\bar{A}_{\bar{f}}|$) by

$$\Gamma[D^0 \to \bar{f}](t) \propto 4|\bar{A}_{\bar{f}}|^2 \left|\frac{q}{p}\right|^2 e^{-\bar{\Gamma}t}$$

$$\left\{ \frac{1}{|\lambda_{\bar{f}}|^2} - \bar{\Gamma}t \left[x \operatorname{Im}\left(\frac{1}{\lambda_{\bar{f}}}\right) + y \operatorname{Re}\left(\frac{1}{\lambda_{\bar{f}}}\right) \right] + \frac{\bar{\Gamma}^2 t^2}{4}(x^2 + y^2) \right\} \tag{10.8}$$

$$\Gamma[\overline{D}^0 \to f](t) \propto 4|A_f|^2 \left|\frac{p}{q}\right|^2 e^{-\overline{\Gamma}t}$$

$$\left\{|\lambda_f|^2 - \overline{\Gamma}t\,[x\,\mathrm{Im}(\lambda_f) + y\,\mathrm{Re}(\lambda_f)] + \frac{\overline{\Gamma}^2 t^2}{4}(x^2 + y^2)\right\} \tag{10.9}$$

while for the Cabibbo-allowed ('right sign') decays

$$\Gamma[D^0 \to f](t) \propto 4e^{-\overline{\Gamma}t}|A_f|^2 \qquad \Gamma[\overline{D}^0 \to \bar{f}](t) \propto 4e^{-\overline{\Gamma}t}|A_{\bar{f}}|^2 \tag{10.10}$$

The expressions (10.8), (10.9) exhibit linear and quadratic terms in t: the doubly Cabibbo-suppressed amplitudes give the $e^{-\overline{\Gamma}t}$ terms, the mixing-induced amplitudes the $e^{-\overline{\Gamma}t}(\overline{\Gamma}t)^2$ terms, and the interference of the two the $e^{-\overline{\Gamma}t}\,\overline{\Gamma}t$ terms. Limits on mixing can also be obtained by fitting the time dependence of these decay rates.

All types of *CP* violation can be studied with non flavour-specific decays such as $K^{\pm}\pi^{\mp}$: assuming no *CP* violation in the Cabibbo-allowed decays, so that $|A_f| = |A_{\bar{f}}|$, a useful parametrization is obtained by defining

$$\frac{\overline{A}_f}{A_f} = -\sqrt{R_D}\,e^{-i\delta} \qquad \left|\frac{q}{p}\right| = 1 + A_M \tag{10.11}$$

$$\lambda_f = \frac{-\sqrt{R_D}\,(1 + A_M)}{1 + A_D}e^{-i(\delta+\phi)} \qquad \frac{1}{\lambda_{\bar{f}}} = \frac{-\sqrt{R_D}\,(1 + A_D)}{1 + A_M}e^{-i(\delta-\phi)} \tag{10.12}$$

Here R_D is the ratio of the doubly Cabibbo-suppressed to Cabibbo-allowed decay rate, measured as $(0.303 \pm 0.016 \pm 0.010)\%$ by the BABAR experiment (Aubert et al., 2007b), and δ is the strong phase difference between the two. At leading order:

$$r_D(t) = e^{-\overline{\Gamma}t}\left[R_D(1 + A_D)^2 + \sqrt{R_D}(1 + A_M)(1 + A_D)y'^{(+)}\overline{\Gamma}t\right.$$
$$\left. + \frac{R_M(1 + A_M)^2}{2}\overline{\Gamma}^2 t^2\right] \tag{10.13}$$

$$\bar{r}_D(t) = e^{-\overline{\Gamma}t}\left[\frac{R_D}{(1 + A_D)^2} + \frac{\sqrt{R_D}}{(1 + A_M)(1 + A_D)}y'^{(-)}\overline{\Gamma}t\right.$$
$$\left. + \frac{R_M}{2(1 + A_M)^2}\overline{\Gamma}^2 t^2\right] \tag{10.14}$$

where

$$y'^{(\pm)} \equiv y'\cos\phi \pm x'\sin\phi \tag{10.15}$$

$$x' \equiv x\cos\delta + y\sin\delta \qquad y' \equiv y\cos\delta - x\sin\delta \tag{10.16}$$

If *CP* symmetry holds A_M (*CPV* in the mixing), A_D (*CPV* in the decay), and ϕ (*CPV* in the interference of mixing and decay) all vanish, and the expressions (10.13), (10.14) reduce to

$$r_D = \bar{r}_D = e^{-\bar{\Gamma}t}\left(R_D + \sqrt{R_D}\, y'\,\bar{\Gamma}t + \frac{1}{2}R_M t^2\right) \qquad (10.17)$$

whose time-integrated form is

$$R = \int_0^\infty r_D(t)\, dt = R_D + \sqrt{R_D}\, y' + R_M \qquad (10.18)$$

which is the ratio of 'wrong sign' to 'right sign' decays, the most readily available experimental quantity.

A difference between D^0 and $\bar{D}^0$ in the $e^{-\bar{\Gamma}t} t^2$ term indicates *CP* violation in the mixing, while a difference in the $e^{-\bar{\Gamma}t}$ term indicates *CP* violation in the decay; the first measurement of D^0 mixing (Aubert *et al.*, 2007b) is consistent with no *CP* violation.

The normalized time-dependent decay rates into *CP* eigenstates $f_\pm$ (*CP*-parity ± 1), in the limit of small-mixing ($x, y \ll 1$) and small *CP* violation, are exponential with modified decay parameters:

$$r_D^{(\pm)} = \exp\left[-\bar{\Gamma}t\left(1 \pm \left|\frac{p}{q}\right|(y\cos\phi + x\sin\phi)\right)\right] \qquad (10.19)$$

$$\bar{r}_D^{(\pm)} = \exp\left[-\bar{\Gamma}t\left(1 \pm \left|\frac{q}{p}\right|(y\cos\phi - x\sin\phi)\right)\right] \qquad (10.20)$$

and a *CP*-violating quantity is

$$A_{CP} = \frac{r_D^{(\pm)}(t) - \bar{r}_D^{(\pm)}(t)}{r_D^{(\pm)}(t) + \bar{r}_D^{(\pm)}(t)} \simeq A_M y\cos\phi + x\sin\phi \qquad (10.21)$$

sensitive to *CP* violation in mixing and interference of mixing and decay. Measurements of A_{CP} in $D \to KK$ and $D \to \pi\pi$ decays are consistent with zero with errors below 1% range (Yao *et al.*, 2006).

The time-integrated decay asymmetries into *CP* eigenstates (7.167) are

$$\Delta_{f_\pm} = \frac{\Gamma(\bar{D}^0 \to f_\pm) - \Gamma(D^0 \to f_\pm)}{\Gamma(\bar{D}^0 \to f_\pm) + \Gamma(D^0 \to f_\pm)} = \delta_T - 2\mathrm{Re}(\eta_\pm) \qquad (10.22)$$

where the first factor is due to *CP* violation in the mixing and

$$\eta_\pm = \frac{1 \mp \lambda_{f_\pm}}{1 \pm \lambda_{f_\pm}} \qquad (10.23)$$

is sensitive to all types of *CP* violation (compare to eqn (8.51)).

Since in the neutral D system the *CP*-violating effects being looked for are tiny, it is important for the flavour tagging to be very accurate, as any significant mis-tagging would dilute small asymmetries. The charge of the slow pion in the $D^{*+} \rightarrow D^0 \pi^+$ and $D^{*-} \rightarrow \overline{D}^0 \pi^-$ is often used for this purpose, achieving mis-tagging rates $O(10^{-3})$.

D mesons are studied both at hadronic and at $e^+ e^-$ colliders. Large amounts of D meson pairs ($\sim 640 \cdot 10^3$ $c\bar{c}$ pairs per 100 pb^{-1}) can be produced at $e^+ e^-$ colliders running as 'charm factories' (CLEO-c at CESR, Cornell (Miller, 2006) and BES3 at BEPC, Beijing (Yuan, 2005)) at the $\psi(3770)$ resonance, which is just above the $D\overline{D}$ threshold but below the $D^*\overline{D}$ one, where the disadvantage of not being able to perform time-dependent analyses due to the low momenta is offset by the fact that the initial state is very clean (no fragmentation products accompanying the D mesons) and well constrained kinematically; average tagging fractions can be as large as 13% in this case. B-factories however also produce very large samples of D mesons ($\sim 120 \cdot 10^6$ $c\bar{c}$ pairs per 100 fb^{-1}), both from continuum and from B decays, and hadronic colliders currently provide the most precise results.

So far all measurements in the D system are consistent with *CP* symmetry.

The FOCUS experiment at Fermilab analysed $D^0, \overline{D}^0$ mesons produced in an uncorrelated way by the interaction of ~ 180 GeV photons on a target. Charmed particles were identified by their separated secondary vertices in silicon micro-strip detectors, their decay products being detected and identified using a double spectrometer, calorimeters, and Čerenkov counters. Tagging the D meson flavour at production by using the soft pion from $D^{*+} \rightarrow D^0 \pi^+$ decay (and its charge conjugate), T-violating triple products of momenta in about 800 $K^+ K^- \pi^+ \pi^-$ decays of D^0 and $\overline{D}^0$ were compared, thus allowing to obtain a limit on T violation independent from FSI effects (Chapter 4) (Link *et al.*, 2005):

$$\frac{1}{2}(A_T - \overline{A}_T) = 0.010 \pm 0.057 \pm 0.037 \tag{10.24}$$

where A_T ($\overline{A}_T$) is the asymmetry of $\boldsymbol{p}_{K^+}$ ($\boldsymbol{p}_{K^-}$) with respect to the $\boldsymbol{p}_{\pi^+} \times \boldsymbol{p}_{\pi^-}$ plane in D^0 ($\overline{D}^0$) decays.

Despite the absence of a mixing signal, tests of *CPT* violation in the mixing can be performed: if *CP* (and *CPT*) violation in the decays is neglected, but *CPT* violation in the mixing is allowed, the asymmetry for flavour-specific 'unmixed' decays (7.163) becomes sensitive to such *CPT* violation, and in the small mixing limit its expression becomes

$$\Delta_f^{(U)}(t) = \overline{\Gamma} t \left[x \operatorname{Im}(\delta_{CPT}) - y \operatorname{Re}(\delta_{CPT}) \right] \tag{10.25}$$

With a sample of about 35K neutral $D^0, \overline{D}^0$ mesons, the FOCUS experiment (Link *et al.*, 2003) compared the rates of the 'right sign' decays $D^0 \rightarrow K^- \pi^+$ and

$\overline{D}^0 \to K^+\pi^-$, obtaining

$$x \operatorname{Im}(\delta_{CPT}) - y \operatorname{Re}(\delta_{CPT}) = 0.0083 \pm 0.0065 \pm 0.0041 \qquad (10.26)$$

consistent with no *CPT* violation in the mixing.

10.3   *B* mesons phenomenology

The study of the invariant mass of muon pairs, which can be detected with relatively small background behind a shielding thick enough to absorb most other particles, was always a fruitful technique for the study of states produced in hadronic interactions. The *b* quark was discovered in 1977 by the group led by L. Lederman at Fermilab, studying the $\mu^+\mu^-$ invariant mass spectrum in 400 GeV proton nucleus collisions (Herb *et al.*, 1977): a two-arm muon spectrometer with 2% mass resolution was used, shielded from hadrons by 18 interaction lengths of beryllium. The statistically significant peak at 9.5 GeV/c^2 was later resolved as a superposition of two narrow resonances by experiments at the DORIS e^+e^- storage ring at DESY, which also showed their widths to be narrower than the resolution due to the beams' energy spread. As for the J/ψ a few years earlier, the partial decay width into e^+e^- was measured by the area under the resonance curve, and used to infer the charge of the new quark in what were identified as $b\overline{b}$ resonances; more 'bottomonium' states were later identified at the e^+e^- Cornell and DESY storage rings CESR and DORIS.

In 1983 the first measurements of the lifetime of *B* mesons, by the MAC (Fernandez *et al.*, 1983) and MARK II (Lockyer *et al.*, 1983) collaborations at the SLAC e^+e^- storage ring (29 GeV centre of mass energy), showed rather long lifetimes $\tau \sim 1.5$ ps. Vertex detectors developed for the measurement of charmed quark lifetimes (also in the ps range) were used, consisting of precision drift chambers with extrapolated resolutions of 100–200 μm on the transverse track position. Using *B*-enriched samples of semi-leptonic decays, the lifetimes were inferred from the distributions of the lepton track impact parameter with respect to the interaction point, averaging to 100–150 μm. The *B* samples were defined by cuts on the lepton momentum and angle with respect to the direction defined by all the charged particles in the event (the 'thrust axis'): heavier hadrons tend to follow the direction of the primary quark and to deliver larger transverse momentum to the daughter leptons. The measured lifetime was substantially larger than naive expectations from the quark mass scaling law, indicating that the couplings of the third family to the first two ($b \to c$ and $b \to u$) are rather small, and pointing to the 'hierarchical' structure of the CKM matrix. At the same time this discovery opened up the possibility of measuring *CP*-violating time-dependent asymmetries in *B* meson decays, by measuring the position of their decay vertex (into charged particles) using micro-vertex silicon detectors.

Two families of neutral mesons containing *b* quarks exist: B_d ($B_d^0 = d\bar{b}$, mass ~5.3 GeV/c^2) and B_s ($B_s^0 = s\bar{b}$, mass ~5.4 GeV/c^2); analogously, there are two kinds of charged mesons $B^\pm$ ($B^+ = u\bar{b}$, mass ~5.3 GeV/c^2) and $B_c^\pm$ ($B_c^+ = c\bar{b}$, mass ~6.3 GeV/c^2); the lifetimes are all around 1.5 ps, except for the heavier $B_c^\pm$ mesons for which $\tau \simeq 0.5$ ps. As discussed in Chapter 9 the two physical states have comparable decay widths (Yao *et al.*, 2006):

$$(\Delta\Gamma/\overline{\Gamma})_d = (0.009 \pm 0.037) \qquad (\Delta\Gamma/\overline{\Gamma})_s = (0.31^{+0.11}_{-0.13}) \qquad (10.27)$$

The two physical states are labelled instead by their mass, and called 'heavy' (B_H) and 'light' (B_L); the mass differences are (Yao *et al.*, 2006):

$$\Delta m_d = (3.34 \pm 0.03) \cdot 10^{-4}\,\text{eV} \qquad \Delta m_s = (115^{+2}_{-1}) \cdot 10^{-4}\,\text{eV} \qquad (10.28)$$

The above mass and lifetime differences clearly do not allow one to distinguish the two mesons in practice, and this precludes the possibility of experimenting with the physical states separately, as for kaons: all *B* meson experiments deal with the flavour eigenstates instead.

10.3.1 *B* mixing

Flavour mixing is large for *B* mesons: for the B_d and B_s mesons (Yao *et al.*, 2006)

$$x_d = 0.776 \pm 0.008 \qquad y_d = 0.005 \pm 0.019 \qquad (10.29)$$

$$x_s = 24.3^{+1.0}_{-0.9} \qquad y_s = 0.16^{+0.05}_{-0.06} \qquad (10.30)$$

Flavour mixing is studied using semi-leptonic decays (branching ratios ~10%), which in the SM are flavour-specific decays so that $B^0 \rightarrow X^-\ell^+\nu$ and $\overline{B}^0 \rightarrow X^+\ell^-\bar{\nu}$ are allowed, but not the decays with B^0 and $\overline{B}^0$ interchanged.

Indications of *B* flavour mixing were first obtained by the UA1 experiment at the CERN $p\bar{p}$ collider (Albajar *et al.*, 1987) and evidence was provided by the ARGUS experiment at the DESY DORIS II e^+e^- storage ring (Albrecht *et al.*, 1987). The first experiment observed an excess of like-sign di-muon events in collisions at ~500 GeV centre of mass energy, above what was expected due to the contribution of $b \rightarrow c \rightarrow \mu X$ decays; by focusing on high transverse momentum muons ($p_T > 3$ GeV/c), for which the contribution from $p\bar{p} \rightarrow c\bar{c}$ events is suppressed, the significance of the result obtained with a few hundred decays was 2.9 standard deviations. The second experiment detected the decays of $B^0\overline{B}^0$ pairs produced at the $\Upsilon(4S)$ resonance and identified a single fully reconstructed decay into $B^0 B^0$ (both B^0 decaying semi-leptonically), also observing a 4 standard deviation excess of like-sign di-muon events and a 3 standard deviation excess of events in which a fully reconstructed neutral *B* meson was accompanied by a 'wrong sign' high-energy lepton.

The measurement of the much faster B_s mixing required the higher sensitivity obtained by a time-dependent analysis, and was achieved only in 2006 by the CDF experiment at the TeVatron collider (Abulencia *et al.*, 2006).

Semi-leptonic decays are *CP*-conserving and dominated by a single amplitude in the SM, thus allowing measurements of *CP* violation in the mixing. As discussed in Chapter 9, this kind of *CP* violation is expected to be small for *B* mesons: for a system in which $\Delta\Gamma \ll \Delta m$, which for small *CP* violation implies $|\Gamma_{12}| \ll |M_{12}|$, from relation (7.81)

$$\left|\frac{q}{p}\right| \simeq 1 - \Delta\xi\,\frac{\overline{\Gamma}_{12}}{\overline{M}_{12}} \tag{10.31}$$

Contrary to what happens in the *K* system, eqn (8.103), in this case the effect of a non-zero relative phase $\Delta\xi$ between Γ_{12} and M_{12} on the amount of *CP* violation in the mixing is suppressed by the small factor $|\Gamma_{12}|/|M_{12}|$. The (phase-convention dependent) mixing parameter $\overline{\epsilon}_B$ is in this case (compare to the case of *K* in Section 8.3.4)

$$\epsilon_B \simeq \frac{\Delta\xi|\Gamma_{12}|/|M_{12}| + 2i\sin\xi_M}{2(1 + \cos\xi_M)} \tag{10.32}$$

which can be chosen to be almost purely imaginary: this means that the magnitude of *CP* violation in the mixing $\propto \mathrm{Re}(\overline{\epsilon}_B)$ is much smaller than $|\overline{\epsilon}_B|$ (while the phase of ϵ_K is close to $\pi/4$).

In the absence of *CP* violation in the decay, the asymmetry for wrong-sign flavour-specific decays (7.164) is

$$A_f^{(M)}(t) = \frac{|\Gamma_{12}|}{|M_{12}|}\Delta\xi \tag{10.33}$$

and predictions in the SM are $O(10^{-3})$ for B_d and $O(10^{-4})$ for B_s; measurements in the B_d system are consistent with these expectations.

Associate production of *B* mesons of different families is also possible in general; in this case there cannot be any interference between the two components: for $e^+e^- \to B_d\overline{B}_sX$ a coherent *C* eigenstate is

$$|B_d(t), \boldsymbol{p}_d; \overline{B}_s(t), \boldsymbol{p}_s; X, \boldsymbol{p}_X\rangle + (-1)^{n_C}|\overline{B}_d(t), \boldsymbol{p}_d; B_s(t), \boldsymbol{p}_s; \overline{X}, \boldsymbol{p}_X\rangle$$

The *X* state contains at least one strange particle: if it is a charged one then $X \neq \overline{X}$ and the two terms cannot interfere, but if it is a neutral *K* the large mixing between K^0 and $\overline{K}^0$ allows interference; however, if the *K* is not identified as either a K_S or a K_L and a sum over all states is considered instead, this interference cancels. Considering final states f_d (f_s) to which both B_d and $\overline{B}_d$ (B_s and $\overline{B}_s$) can contribute, the lepton charge asymmetry, in the case of no *CP* violation in the mixing nor in the decay (and $y = 0$), is given by (Bigi and Sanda, 1981):

$$\frac{\Gamma[\ell^+ f_i] - \Gamma[\ell^- f_i]}{\Gamma[\ell^+ f_i] + \Gamma[\ell^- f_i]} = \frac{-1}{1 + x_j^2} \frac{x_i}{1 + x_i^2} \sin(2\phi_i)$$

where $i, j = d, s$ (if $i = d$ then $j = s$ and vice versa) and $\lambda_{f_i} = -e^{2i\phi_i}$.

We recall that the measurement of the charge asymmetry for semi-leptonic decays of a physical state, the equivalent of (8.40) in the *K* system, is not possible for *B* mesons as the two physical states cannot be distinguished in practice; with correlated meson pairs one could think of using a decay into a *CP* eigenstate to tag the *CP*-parity of the partner meson and to measure its semi-leptonic charge asymmetry, but the tagging decay itself can be *CP*-violating, and thus such 'CP-tagging' procedure does not work.

10.3.2 Interference *CP* asymmetries

While the mixing of *B* mesons is largely *CP*-conserving, the fact that it is sizable is very important for the study of *CP* violation, as it can induce large *CP*-violating effects in the interference of mixing and decay (Chapter 6): indeed this kind of *CP* violation has been the main focus of experimental investigations, because of the rather clean handle it offers in some cases for the determination of fundamental parameters of the flavour sector of the SM. In the limit $|\Gamma_{12}|/|M_{12}| \to 0$, q/p is a pure phase $\exp[-i \arg(M_{12})]$ and the rephasing invariant quantity λ_f (7.91) is thus

$$\lambda_f = \exp[-i \arg(M_{12})] \frac{\overline{A}_f}{A_f} \tag{10.34}$$

For decays dominated by a single amplitude this is also a pure phase, which can be determined by measuring its imaginary part from interference *CP* violation. The study of decays into states which are not *CP* eigenstates can also provide useful information (Aleksan *et al.*, 1991).

In general *B* meson decay modes can be classified in terms of the expected SM contributions to their amplitude (Nir, 1999):

- Decays which have only contributions from tree diagrams. No *CP* violation is possible in the decays, and a single combination of CKM matrix elements appears: for example $b \to c\bar{u}d$ contains $V_{cb}V_{ud}^*$; these give final states which are not *CP* eigenstates. Examples are $B^+ \to \overline{D}^0\pi^+$, $B_d^0 \to D^-\pi^+$, $B_s^0 \to D^+K^-$.

- Decays in which both tree and penguin diagrams contribute, such as $b \rightarrow c\bar{c}s, u\bar{u}s, c\bar{c}d, u\bar{u}d$; the strong and weak phases appearing in the penguin diagram amplitudes are in general different for different quarks appearing in the loop, and only differences of penguin contributions are significant because the unitarity relations can be used to express one of the three (u, c, t) in terms of the other two. The ratio of penguin to tree amplitudes must be known in order to extract information on the CKM matrix elements; if the penguin contribution is relatively small there is little CP violation in the decay and the relation of the asymmetry to the CKM parameters is not spoiled too much (an example is $B_d \rightarrow \pi\pi$). Examples are $B^+ \rightarrow K^+\pi^0$, $B^0_d \rightarrow \pi^+\pi^-$, $B^0_s \rightarrow K_S\pi^0$. If the tree and penguin amplitudes share the same phase these are very 'clean' modes, as $B_d \rightarrow J/\psi K_S$.

- Decays to which only penguin diagrams contribute, such as $b \rightarrow s\bar{s}s, s\bar{s}d$: if only a single penguin amplitude appears the situation is similar to the first case and the decay is useful for the extraction of CKM parameters, as in the case of $B_d \rightarrow \phi K_S$. Moreover these decays are potentially sensitive to new physics contributions, since the SM amplitude is relatively small. Examples are $B^+ \rightarrow \phi\pi^+$, $B^0_d \rightarrow \phi K_S$, $B^0_s \rightarrow \phi\eta'$.

As discussed in Chapter 7, many tagged asymmetries for meson pairs lose most of the sensitivity to CP violation in the interference of mixing and decay when they are integrated over the decay time difference Δt in the range $[-\infty, +\infty]$, as necessarily happens at symmetric B-factories. However it should be recalled that the observation of decays to two CP eigenstates f_1, f_2 with opposite (equal) CP-parities, from a symmetric (anti-symmetric) coherent state is a signal of CP violation; in the absence of CP violation in both mixing and decay this rate is sensitive to CP violation in the interference of the two processes: if $\lambda_{f1} = \pm\lambda_{f2}$, as when the two states arise from the same diagram, for $|\lambda_{f1}| = |\lambda_{f2}| = 1$ the coefficients in eqn (7.176) are $A^{(\pm)} = A_{f1}\bar{A}_{f2} 2i \operatorname{Im}(\lambda_{f1})$, $B^{(\pm)} = 0$, and the integrated decay rate to (f_1, f_2) from a coherent pair state is

$$\Gamma^{(C\pm)}(f_1; f_2) \propto |A_{f1}|^2 |\bar{A}_{f1}|^2 \left[\mathcal{F}_y^{(\pm)} - \mathcal{F}_x^{(\pm)} \right] \operatorname{Im}^2(\lambda_f)$$

with $\mathcal{F}_{x,y}^{(\pm)}$ defined in eqn (7.182), allowing to determine $\operatorname{Im}(\lambda_f)$ to within a sign.

The statistical error in the measurement of an asymmetry Δ scales as $1/\sqrt{N}$ with the number N of collected events; observability requires Δ to be significantly larger than this error, thus $\Delta\sqrt{N} \gg 1$. The figure of merit for the measurability of an asymmetry in a particular decay mode is thus seen to be the product $BR \cdot \Delta^2$ of the branching ratio and the square of the asymmetry: this shows that it is more profitable to study decay modes with larger asymmetries, even if their branching ratios are smaller. On the other hand, the fact that larger asymmetries are usually present in rarer decay modes is just a consequence of the fact that (in the SM) all *CP*-violating effects arise from the same source (the single phase δ_{CP} in the CKM matrix) and are all proportional to the same quantity J_{CP} (9.25).

The decay $B_d \rightarrow J/\psi K_S$ is the 'golden' mode for *B*-factories, as it allows a clean extraction of the angle $\sin 2\beta$ of the $'\bar{b}d'$ unitarity triangle (Chapter 9). From the experimental point of view this mode, with an accessible branching ratio $\simeq 4 \cdot 10^{-4}$, has a very strong signature from the $J/\psi \rightarrow \mu^+\mu^-$ decay, and correspondingly low background. The value determined by *B*-factory experiments is $\sin 2\beta = 0.674 \pm 0.026$ (Aubert *et al.*, 2006a; Chen *et al.*, 2007), corresponding to $\beta = (21.2 \pm 1.0)°$ (or $\beta = (68.8 \pm 1.0)°$). Because of the unitarity relation $V_{tb}^* V_{ts} \simeq -V_{cb}^* V_{cs}$ the penguin-dominated decays induced by $b \rightarrow s\bar{q}q$ transitions ($B^0 \rightarrow \phi K^0$, $B^0 \rightarrow \eta' K^0$) have approximately the same phase as those arising from $b \rightarrow c\bar{c}s$, and as such they also measure $\sin(2\beta)$ in the SM; in this case, however, possible new physics contributions arguably have a better chance to compete and to drive the measured quantity away from the SM value.

For the measurement of $\sin 2\alpha$ decays driven by $b \rightarrow u\bar{u}d$ transitions can be used, such as $B_d \rightarrow \pi^+\pi^-$ (with $\eta_{CP} = +1$ and $BR \sim 5 \cdot 10^{-6}$). If only tree diagrams contribute then, along the lines of Section (9.4):

$$\frac{\bar{A}_{\pi^+\pi^-}}{A_{\pi^+\pi^-}} = \eta_{CP}(\pi^+\pi^-)\frac{V_{ud}^* V_{ub}}{V_{ud} V_{ub}^*} \qquad \lambda_{\pi^+\pi^-} = e^{2i\alpha} \qquad (10.35)$$

However penguin diagrams with different phases also contribute in a significant way ($\sim 25\%$), making the extraction of the angle α from the observed asymmetries more difficult ('penguin pollution'). The amount by which the measured angle is shifted from 2α could be estimated by exploiting isospin relations (Gronau and London, 1990) among amplitudes such as $A(B^0 \rightarrow \pi^+\pi^-)/\sqrt{2} + A(B^0 \rightarrow \pi^0\pi^0) = A(B^+ \rightarrow \pi^+\pi^0)$: since the pion pair must have isospin $I = 0, 2$ ($I = 2$ for $\pi^+\pi^0$) and (gluon) penguin amplitudes only contribute to $I = 0$ final states, accurate measurements of the rates and *CP* asymmetries for $\pi\pi$ decays would allow to disentangle the penguin contribution; unfortunately the experimental accuracy related to the tiny BR (of order 10^{-6}) does not allow this programme to be carried out with good precision (Aubert *et al.*, 2007c; Ishino *et al.*, 2007).

The analysis of the more complex vector-vector decay $B_d \rightarrow \rho^+\rho^-$ ($BR \sim 2.5 \cdot 10^{-5}$) is useful for extracting the angle α, as the final state turns out to be dominated (95–98%) by the *CP*-even longitudinally polarized component; the penguin contamination is also smaller in this case (Aubert *et al.*, 2006a; Somov *et al.*, 2006; Aubert *et al.*, 2007a).

Isospin analyses involving several decay modes give $\alpha = (93^{+11}_{-9})°$ (Charles *et al.*, 2006).

The angle γ does not depend on CKM matrix elements related to the top quark and thus can be measured from tree decays; measurements of $B \rightarrow DK, D\pi$ and related channels can be used for this purpose. The decays $B_d \rightarrow D^{\pm}\pi^{\mp}$ and $B_d \rightarrow D^{\pm}\rho^{\mp}$ are sensitive to γ because of the interference between the Cabibbo-favoured amplitude (such as $B_d \rightarrow D^-\pi^+$) and the doubly Cabibbo-suppressed amplitude (such as $B_d \rightarrow D^+\pi^-$), which have a relative phase $-\gamma$, and can give an interference between mixing and decay with phase $2\beta + \gamma$. In the decays $B^{\pm} \rightarrow DK^{\pm}$ (or those in which the excited states D^* and/or $K^{*\pm}$ appear) the neutral D meson is produced as an admixture of D^0 and $\overline{D}^0$, and by choosing a final state to which both can contribute an interference between two amplitudes is possible which is sensitive to γ, being the relative weak phase between them. Several approaches have been proposed for exploiting this interference, using D mesons reconstructed either in CP eigenstates such as $\pi^+\pi^-$, K^+K^-, $K_S\pi^0$) (Gronau and London, 1991; Gronau and Wyler, 1991) or in suppressed final states such as $K^+\pi^-$ (Atwood et al., 1997, 2001b) or in self-conjugate multi-body final states such as $K_S\pi^+\pi^-$ (Giri et al., 2003) by analysing their Dalitz plot distributions (Appendix D); in all these cases the rate ratios and asymmetries can be expressed in terms of amplitude ratios, strong phase differences and the angle γ. Experimental results (Aubert et al., 2006c,d; Abe et al., 2006a; Poluektov et al., 2006) indicate $\gamma = (60^{+38}_{-24})°$ (Charles et al., 2006).

Figure 10.1 shows a typical plot of the constraints on the standard unitarity triangle which can be obtained using some experimental input; the apex of the triangle is determined to be at[140] $(\bar{\rho}, \bar{\eta}) = (0.197 \pm 0.031, 0.351 \pm 0.020)$ (Bona et al., 2006; see also Charles et al., 2006).

It should be remarked that in case CP violation in the interference of mixing and decay could be entirely ascribed to the mixing process (indirect CP violation), then the λ_f parameters should be independent the decay mode: a measurement of $\mathrm{Im}(\lambda_f) \neq \pm\mathrm{Im}(\lambda_g)$ for any two CP eigenstates f, g indicates the presence of direct CP violation. Indeed a model of super-weak type confining all CP-violating phases to the mixing process would result in the same values for all asymmetries, while a quantity equivalent to ϵ' in the K system, invalidating kind of model, is the difference (apart from a sign) of any CP asymmetry with respect to the value $\sin 2\beta$ measured for the $J/\psi K_S$ mode.

As the decay rates into CP eigenstates are usually tiny, it is worth considering how final states can be combined in order to gain a statistical advantage.

When considering decays of neutral mesons into particles with spin, one has not CP eigenstates in general, as states with different values of orbital angular momentum L can be present, and for particles which are CP eigenstates even and odd L states have opposite CP-parities. If the final state contains at most one particle with non-zero spin its CP-parity is unique, since there is only one possible angular momentum state; on the contrary, for a decay into two particles with non-zero spin

---

[140] The $\bar{\rho}, \bar{\eta}$ parameters have slightly different definitions with respect to ρ, η of Section 9.2.3, providing improved unitarity to the parameterization.

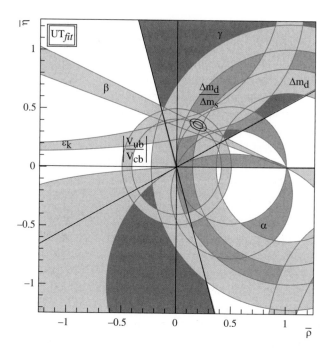

FIG. 10.1. Typical constraints on the vertex of the '$\bar{b}$d' unitarity triangle from the ϵ_K *CP* violation parameter of neutral *K* decays (Chapter 8), the angles α, β, γ from B_d decay asymmetries, the B_d and B_s mixing parameters and the ratio of $|V_{ub}/V_{cb}|$ (Bona *et al.*, 2006). The bands indicate 95% CL limits, and the 68% and 95% contours for the vertex position of the unitarity triangle are shown.

different angular momentum states can be present: an example is $B^0 \to J/\psi K^{0*} \to J/\psi K_S \pi^0$, which is *CP*-odd if $L = 0$ and *CP*-even if $L = 1$.

The above fact usually results in a dilution of the observed asymmetries: in the case of *CP* violation in the interference of mixing and decay the decay rates into a mixed *CP*-parity state can be written as

$$\Gamma[M^0 \to f] = (1 + \Delta)\Gamma^{(+)} + (1 - \Delta)\Gamma^{(-)} \tag{10.36}$$

$$\Gamma[\overline{M}^0 \to \bar{f}] = (1 - \Delta)\Gamma^{(+)} + (1 + \Delta)\Gamma^{(-)} \tag{10.37}$$

where $\Gamma^{(\pm)}$ are the decay widths into the two *CP* eigenstates and Δ would be the asymmetry if only one of them were present. The measured asymmetry into the mixture of states is

$$\Delta_f^{(U)} = \frac{\Gamma[\overline{M}^0(t) \to f] - \Gamma[M^0(t) \to \bar{f}]}{\Gamma[\overline{M}^0(t) \to f] + \Gamma[M^0(t) \to \bar{f}]} = \Delta \frac{\Gamma^{(-)} - \Gamma^{(+)}}{\Gamma^{(-)} + \Gamma^{(+)}} \tag{10.38}$$

which is diluted with respect to Δ by the presence of the two opposite *CP*-parities. In this case undiluted *CP*-violating asymmetries can be extracted by performing an angular analysis (Dunietz *et al.*, 1991). A simple method exploits the spin projections of a three-body intermediate state along the direction orthogonal to the plane defined by the momenta, in order to separate the two *CP* eigenstates. For a decay $M^0 \rightarrow AX \rightarrow AX_1X_2$, the *CP*-parity can be expressed in terms of the particles' spin projections t_j ($j = 1, 2, 3$ for A, X_1, X_2) along the normal to the (AX_1X_2) plane, in the M^0 or X rest frame, the so-called 'transversity axis' (this definition is invariant with respect to boosts between the X rest frame and the $X_{1,2}$ rest frames); the expectation values of these spin projections are related to the *CP* eigenvalues. Indeed a reflection with respect to the decay plane can be written as the product of a parity transformation P and a π rotation around the transversity axis (here chosen to be the y axis); the product of such a reflection and charge conjugation C gives

$$CPe^{i\pi J_y} = \eta_{CP}(A)\, \eta_{CP}(X_1)\, \eta_{CP}(X_2)e^{i\pi \sum_j t_j}$$

and for a state of total angular momentum zero (invariant under rotations, such as the one obtained in the decay of a B meson), this is just CP. The different polar angle distributions for the decays of the three particles with respect to the transversity axis can be used to separate the contributions with different t_i, and for each of those undiluted asymmetries can be formed. Note that the above analysis applies equally to time-dependent and integrated rates, as it can be performed for each time bin (the angular dependence and the time dependence factorize).

10.3.3 *CP* violation in decays

CP violation in the decays of B mesons can be large in some cases, but in most cases it cannot be reliably predicted in terms of fundamental parameters of the SM, both because the relative size of two contributing amplitudes is not well known and, even more important, because of the necessary presence of strong phases which cannot be computed from first principles. While at low energies (on the typical hadronic scale) final state interactions are mostly just elastic scattering, at higher energies inelastic interactions with many open channels dominate, so that the decay amplitude contains in general a sum of many terms with different matrix elements: think to the interactions of two pions at a centre of mass energy of 0.5 GeV, relevant for $K \rightarrow 2\pi$ and at 5 GeV, relevant for $B \rightarrow 2\pi$.

    CP violation in decays of B mesons can be measured either as a time-integrated partial rate asymmetry between two *CP*-conjugated final states, or as an asymmetry between the time-dependent decay rates of tagged mesons which is not induced by flavour oscillations, e.g. for short times after flavour tagging.

10.4 *B* meson experiments

The study of *B* mesons has been carried out both at high-energy e^+e^- and hadronic colliders, where large *B* samples can be obtained with high backgrounds, and since 2000 at dedicated *B*-factories (Section 10.4.1), which despite lower production cross sections present several important advantages due to the coherent nature of meson pairs.

In the large background environment of a high-energy collider the triggering of events is the most delicate issue in the collection of *B* meson samples; this is most efficiently accomplished by identifying decay vertices which are displaced from the production vertex thanks to the relatively long lifetime of *B* mesons and their Lorentz boost in the laboratory; trigger efficiencies however depend strongly on the proper decay time, effectively going to zero for small values of this quantity.

A crucial issue for the study of *B* meson decays is thus the use of high-resolution devices to detect their decay vertices, both to identify the *B* meson and to measure its decay time. The proper decay time is obtained from the measurement of the spatial separation L between the decay vertex and the production vertex: $t = (m/p)L$, where m is the *B* meson mass and p its measured momentum. The resolution on this quantity

$$\sigma_t \sim \frac{m}{\langle p \rangle} \sigma_L \oplus t \frac{\sigma_p}{p}$$

has a time-independent term proportional to the position resolution σ_L of the vertex tracking device (improving with the particle boost, of order ~ 0.1 ps at high energy colliders and ~ 1 ps at *B*-factories) and a term which increases with the decay time due to the relative resolution σ_p/p of meson momentum; if the *B* momentum is known by other means (as it happens at *B*-factories) the first term dominates.

The techniques used for flavour tagging differ depending on the nature of the meson pair: at *B*-factories the correlated nature of the pair can be exploited (Section 7.5.2) by identifying the flavour of the neutral meson partner on the opposite side, but this possibility is not available in the case of incoherent production, e.g. at hadron colliders, as in this case the flavour of a neutral meson does not give information on that of its partner at an arbitrary time, because of (incoherent) flavour oscillations. Semi-leptonic decays of charged *B* mesons are instead always useful for identifying the flavour of the partner meson at production, since charged mesons do not oscillate.

Tagging methods at high-energy colliders are generally classified as 'same-side' tagging and 'opposite-side' tagging: in the first case the correlation of the meson flavour with the charge of other particles which are produced together with it is exploited; in the second case one tries to identify the flavour of the partner of the meson under study, for example by detecting the lepton charge from its semi-leptonic decay, or the charge of a kaon obtained from the $b \rightarrow c \rightarrow s$ decay chain.

When the charge of the lepton in a semi-leptonic decay is used mis-tagging can be due to a cascade decay in which the b quark decays to c, which in turn decays semi-leptonically resulting in the same lepton charge as would be obtained from a $\bar{b}$. Any error in the flavour tagging dilutes the observed mixing ratio: with mis-tagging probabilities $w_{i,f}$ for the initial state (at production) and the final state (at decay) the measurement gets degraded by a factor $(1 - 2w_i)(1 - 2w_f)$.

For each tagging procedure an effective tagging efficiency is defined as $Q_k = f_k(1-2w_k)^2$, where f_k is the tagging efficiency of procedure k (the fraction of events to which it can be applied), and w_k the corresponding mis-tagging probability: the quality of a particular signature with a higher mis-tagging probability can be higher than that of a cleaner one, if it can be applied to a larger fraction of events (untagged events are not used at all): the statistical error on the result indeed scales with $1/\sqrt{\sum_k Q_k}$, so this factor describes the effective loss of statistics in a given measurement.

Mis-tagging probabilities vary from a few per cent to about 40% depending on the method, with asymmetries between B^0 and $\overline{B}^0$ of a few percent; different tagging signatures are often combined. The overall effective tagging efficiencies $\sum_k Q_k$ are $\sim$30% at B-factories (and can be comparable at high-energy e^+e^- colliders), while the corresponding figures at hadron colliders are often significantly lower (1–4%).

10.4.1 *B*-factories

For the study of *CP*-violating asymmetries in the B system large samples of B mesons are required: a relatively high signal-to-background ratio for the production of B mesons is obtained at the first $b\bar{b}$ resonance above the $B\overline{B}$ production threshold, the $\Upsilon(4S)$ (the fourth radially excited $^3S_1$ $b\bar{b}$ bound state) with mass 10.58 GeV/c^2, which decays almost exclusively into $B\overline{B}$ mesons (roughly equally into B^+B^- and $B^0-\overline{B}^0$) with momentum 0.34 GeV/c in the CM. An e^+e^- collider with its precisely tunable centre of mass energy set at this value (a B-factory) is thus an excellent choice for such investigations.

At a symmetric B-factory (colliding beams with the same energy), such as CESR at Cornell, the centre of momentum (CM) frame is at rest in the laboratory system, and the two B mesons obtained from the decay of a Υ produced at rest fly off in opposite directions with momenta of 335 MeV/c ($\beta\gamma = 0.06$); the average decay length of a B^0 meson at this energy is only $\sim$30 μm, which cannot be easily resolved with vertex detectors. At higher energies (such as at LEP) the measurement would become possible but the background would be higher and the important advantages of correlated production would be lost.

Apart from resolution issues, the measurement of the proper decay time of a B meson in an $e^+e^- \to B^0\overline{B}^0$ event is not possible because the position of its production point (the e^+e^- collision point, as the Υ is produced at rest and anyway has a lifetime $O(10^{-23})$ s) cannot be determined with sufficient accuracy: it is different from event to event and not sufficiently constrained due to the spatial extension of the beam bunches, and its position as measured from the visible tracks in the event is typically poorly determined (of order ~10 mm in the longitudinal direction). As the two *B* mesons fly away back-to-back, the measurement of the sum of their decay times does not require the knowledge of the production point, but this is not a useful quantity for tagging, as discussed in Chapter 7.

It was noticed however (Oddone, 1987) that with an *asymmetric B*-factory, with different energies for the two beams, two advantages would arise: the boost would result in increased decay lengths, thus allowing their measurement with micro-vertex silicon detectors; moreover the CM frame in which the Υ is produced would be moving in the laboratory, and the two *B* mesons would fly off in the same direction (for a boost larger than the *B* momentum in the CM), so that the measurement of their decay positions would give information directly on the decay time *difference* of interest, without requiring an accurate knowledge of the production point.

Two double-ring asymmetric *B*-factories were built in the '90s: PEP-II at SLAC and KEK-B at KEK, and both proved to be extremely successful machines, exceeding their design luminosities (an account of the history of these projects can be found in Hitlin (2006)). Both machines obtain high luminosities by storing a large number of bunches in two rings and using continuous beam injection. PEP-II collides 3.1 GeV e^+ and 9 GeV e^- head-on, resulting in a $\beta\gamma = 0.56$ boost of the $\Upsilon(4S)$ in the direction of the e^- beam, which corresponds to an average separation of 260 μm between the decay vertices of the two *B* mesons; it reached an instantaneous luminosity $\mathcal{L} = 1.2 \cdot 10^{34}$ cm^{-2} s^{-1}, delivering an integrated luminosity of more than 410 fb^{-1} to the BABAR experiment (as of March 2007). KEK-B collides 3.5 GeV e^+ and 8 GeV e^- at $\pm$ 11 mrad crossing angle, for a boost $\beta\gamma = 0.43$ corresponding to a 200 μm separation of *B* decay vertices; it reached $\mathcal{L} = 1.7 \cdot 10^{34}$ cm^{-2} s^{-1}, delivering an integrated luminosity of more than 710 fb^{-1} to the Belle experiment (as of March 2007).

A useful conversion figure is one million $B\bar{B}$ pairs produced per year at $1 \cdot 10^{32}$ cm^{-2} s^{-1} luminosity. A small non-zero crossing angle helps in reducing the parasitic collisions between the two beams as they travel away from the interaction point at the price of a smaller effective luminosity for the same machine parameters; head-on collisions require separation of the two beams with magnets close to the interaction point, which lead to somewhat higher machine background.

The experiments at *B*-factories exploit the features of production at threshold and the small *Q*-value in the $\Upsilon(4S) \to B\bar{B}$ decay to suppress the backgrounds: the two Lorentz-invariant quantities traditionally used are the 'beam-constrained' (or 'energy-substituted') mass M_{bc} and the energy difference ΔE, defined as

$$M_{bc} = \sqrt{\frac{(s/2 + \boldsymbol{p}_\Upsilon \cdot \boldsymbol{p}_B)^2}{E_\Upsilon^2} - \boldsymbol{p}_B^2} = \sqrt{E_{\text{beam}}^{*2} - \boldsymbol{p}_B^{*2}} \qquad (10.39)$$

$$\Delta E \equiv \frac{(\boldsymbol{p}_B \boldsymbol{p}_\Upsilon) - s/2}{\sqrt{s}} = E_B^* - E_{\text{beam}}^* \qquad (10.40)$$

where E_B, p_B are the energy and momentum of the B meson candidate (starred quantities refer to the CM frame), and $E^*_{\text{beam}} = \sqrt{s}/2$ is the 'beam energy' (half of the CM energy). The two variables are largely uncorrelated: M_{bc} peaks at $m(B)$ for the signal, and its resolution ($\sim$2.5 MeV/c^2) is dominated by the spread in the beam energies, thus one order of magnitude smaller than the resolution in the invariant mass computed from the decay products, and independent of the B meson decay mode; ΔE peaks at 0 for the signal and its resolution ($\sim$15–30 MeV) depends on the detector resolution. Note that if a particle is mis-identified the value of the variable ΔE changes but M_{bc} remains the same if the first expression in eqn (10.39) is used. In most of the cases the backgrounds do not exhibit peaks in the above variables, and the corresponding shapes are parametrized with empirical formulæ.

With the beam energy spread of B-factories, the $\Upsilon(4S)$ has an effective peak production cross section $\simeq$ 1.1 nb (the physical peak cross section being 3.6 nb, the difference is due to the beam energy spread and radiation), but it sits on top of a large ($\sim$3 nb) non-resonant background due to $e^+e^- \rightarrow q\bar{q}$ ($q = u, d, s, c$) production, which must be measured off the resonance energy and subtracted (for comparison, the $B\bar{B}$ production cross section is 6.3 nb at the Z peak). The optimal fraction f of data to be collected off-resonance, in order to minimize the total error on a measurement, depends on the expected background to signal ratio N_B/N_S (BABAR collaboration, 1999): from a given number of events $N_{\text{on}}, N_{\text{off}}$ taken on- and off-resonance out of an integrated luminosity L

$$N_{\text{on}} = (1-f)L(1+N_B/N_S)N_S \qquad N_{\text{off}} = fL(N_B/N_S)N_S$$

the number of signal events N_S is obtained as

$$N_S = (N_S + N_B) - N_B = \frac{N_{\text{on}}}{(1-f)L} - \frac{N_{\text{off}}}{fL}$$

with an error

$$\sigma_S \simeq \sqrt{\frac{f + N_B/N_S}{f(1-f)}} \sqrt{\frac{N_S}{L}}$$

The minimum error is obtained with $f \simeq 0.15$ for a 10% background, rising to $f \simeq 0.35$ for a 50% background: B-factory experiments routinely collect roughly 10% of the data at about 40 MeV below resonance.

The techniques to suppress the hadronic background have evolved to be rather sophisticated, and are based on the combination of several shape variables which discriminate continuum events (low mass states, collimated and back to back) with respect to $B\bar{B}$ events which tend to have a more isotropic distribution of the decay particles, due to the small energy available. The variables used are modifications of

Fox–Wolfram moments (Fox and Wolfram, 1978): these are rotationally invariant sums over each pair of particles, defined as

$$H_l = \sum_{i,j} \frac{|\boldsymbol{p}_i||\boldsymbol{p}_j|}{s} P_l(\cos \theta_{ij})$$

where $\boldsymbol{p}_{i,j}$ are the particles' momenta, $\sqrt{s}$ the centre of mass energy, and P_l Legendre polynomials of order l in terms of the angle θ_{ij} between the i and j particle directions. Neglecting particle masses, energy and momentum conservation require $H_0 = 1$ and $H_1 = 0$, but moments of higher order can be used to distinguish the signal from the background: in particular the ratio H_2/H_0 is close to 0 for 'spherical' events (signal) and close to 1 for collimated 'jet-like' distributions (background).

These and other variables are usually combined into Fisher discriminants (Fisher, 1936), which are linear combinations with optimized weights: considering a set of measurements x_i ($i = 1, 2, \ldots, n$) the linear combination $F = \sum_i K_i x_i$ is searched which maximizes the ratio of the difference $D = \langle F \rangle_S - \langle F \rangle_B$ between the average values for the two samples (signal and background) over the standard deviation σ_F of F, due to the fluctuations within each sample. The optimal weights K_i are determined by minimizing the above ratio, and their expressions are obtained by solving the system of equations defined by $(1/\sigma_F)\partial\sigma_F/\partial K = (1/D)\partial D/\partial K$. In this way an overall discriminating shape variable is formed, which can be used in a Likelihood ratio test for the signal vs. background hypothesis, usually together with the cosine of the angle between the B flight direction and the beam axis.

The residual background can be modelled and subtracted by measuring it outside the signal region (in a 'sideband') and obtaining in this way an estimate of the contamination, if the scaling factor from the sideband to the signal region is known (e.g. from simulation).

Particle identification is the other crucial issue for flavour physics: $K - \pi$ separation in particular is both important and challenging. Early experiments only used the energy losses in the active detectors, while modern ones also exploit Čerenkov radiation and time-of-flight measurements.

The measurement of time differences Δt between meson decays is obtained from the longitudinal spatial separation Δz of the vertices and the relativistic factors (β_B^*, γ_B^*) and polar angle (θ_B^*) of the reconstructed B meson in the $\Upsilon(4S)$ rest frame:

$$\Delta z = \beta_\Upsilon \, \gamma_\Upsilon \, \gamma_B^* \, c\Delta t + \gamma_\Upsilon \, \beta_B^* \, \gamma_B^* \, \cos\theta_B^* \, c(\tau_B + |\Delta t|)$$

where $\beta_\Upsilon, \gamma_\Upsilon$ are the centre of mass (Υ) relativistic factors. Typical figures for longitudinal decay point resolutions are $\sim$70 μm for a fully reconstructed B meson and $\sim$180 μm for the tagging B meson.

B-factory experiments

The detectors for the two *B*-factory experiments, BABAR at SLAC and Belle at KEK, have rather similar designs dictated by the physics requirements: they were required to have a large acceptance and a high efficiency for *B* decays to maximize the number of collected events, good momentum and energy resolutions to separate the small signals from the backgrounds, very good vertex position resolution, efficient and clean particle identification capabilities for hadrons. Since the average charged particle momentum is below 1 GeV/*c*, the minimization of the amount of material producing multiple scattering is very important, and the same concern is present for the amount of dead material in front of the electromagnetic calorimeters, which have to detect showers with energies as low as 20 MeV.

The detectors are offset with respect to the interaction point, to adapt to the asymmetric energy configuration, and comprise: silicon vertex detectors close to the beam pipe for secondary vertex measurement and to provide the first tracking points, drift chambers with helium-based gas for charged particle tracking and particle identification through ionization energy loss (*dE*/*dx*) with minimal multiple scattering, Čerenkov detectors for particle identification, and CsI(Tl) crystal calorimeters for the measurement of neutral particles. The high light yield and small Molière radius of CsI allow excellent energy and angular resolutions; moreover, the high yield permits the use of silicon photodiodes, which can work in magnetic fields: all the above detectors are inside 1.5 T cryogenic (4.5 K) superconducting solenoids. The magnet flux return yokes are used to absorb hadrons and contain single-gap resistive plate chambers to perform as muon and neutral hadron detectors. Two-level hardware-software trigger systems cope with the full rate of e^+e^- interactions.

The BABAR detector (Fig. 10.2) covers polar angles from about 15° in the forward (e^- beam) direction to about 160°; permanent dipole and quadrupole magnets are placed inside it, to allow head-on collisions.

The thin water-cooled beam pipe has a gold-plated inner surface to attenuate synchrotron radiation, and is surrounded by the silicon vertex tracker (SVT), consisting of five double-sided silicon strip layers arranged in a cylindrical geometry, covering 90% of the solid angle; this detector is also used for stand-alone tracking of low momentum tracks and provides ionization energy loss information (*dE*/*dx*) with 14% resolution for a minimum ionizing particle. The limit for the point resolution required is dictated by the multiple scattering, and is of order 10–40 μm per layer; each layer has longitudinal strips on one side (to measure the azimuthal angle) and transverse strips on the other (to measure the longitudinal coordinate along the beam direction).

The drift chamber (DCH) has 40 layers and ~7,000 cells with wires at a small angle with respect to the detector axis providing also some longitudinal position information; it is filled with a helium-based gas to minimize multiple scattering and also provides *dE*/*dx* information from charge collection, useful for particle identification in the forward region not covered by other devices. SVT and DCH together provide a single track position resolution of 25 μm and 40 μm in the transverse and longitudinal directions at the interaction point.

Particle identification is mostly provided by a novel Čerenkov detector (DIRC) made of 144 long bars of fused silica with rectangular cross section placed parallel to the beam direction. The light produced inside the bars is internally reflected, preserving its angle while

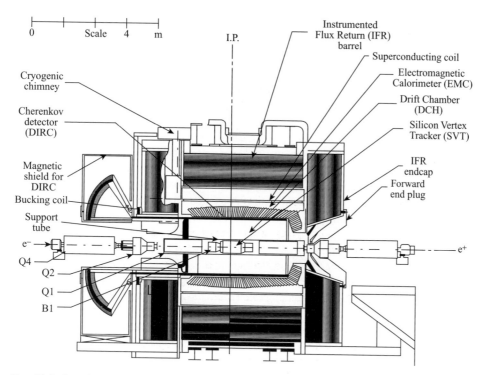

FIG. 10.2. Longitudinal cross section of the BABAR detector at SLAC (Aubert *et al.*, 2002) (see text). B1: dipole magnet; Q1,Q2,Q4: quadrupole magnets.

propagating towards an array of ~11,000 photomultipliers (PM), placed behind a water-filled expansion tank at the backward end of the bars (the forward end having mirrors): as the index of refraction of purified water is close to that of silica, total reflection at the interface is minimized with this arrangement. The radial thickness of the detector, placed between the drift chamber and the electromagnetic calorimeter, is only ~8 cm, thus allowing a relatively large chamber without requiring a too large (and expensive) calorimeter. Knowing the track direction, the positions of the PM which are hit with respect to the position of the active bar provide the information on the Čerenkov angle and thus the particle velocity, also complemented by the hit time information: the detector provides good K/π separation for momenta of 500 MeV/c and above (more than 4 standard deviations at 3 GeV).

The EM calorimeter with ≃6,500 crystals is used to identify photons, electrons and neutral hadrons, using energy and shower shape information, and provides a ~6.5 MeV/c^2 π^0 mass resolution. Electron/hadron separation is based on EM shower shapes and the ratio of measured calorimeter energy to track momentum: an electron efficiency of ~95% with 0.3% pion misidentification probability was achieved.

The flux return (IFR) was instrumented with single-gap planar and cylindrical resistive plate chambers interspersed with steel, which provided a muon detection efficiency of ≃90% with a fake pion rate around 7% but exhibited a marked dependence of the efficiency on

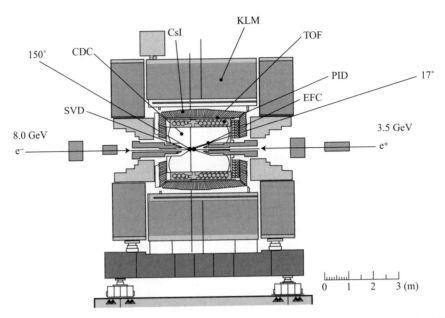

Fɪɢ. 10.3. Longitudinal cross section of the Belle detector at KEK (Abashian *et al.*, 2002) (see text).

environmental conditions and were replaced by limited streamer tubes. Neutral hadrons such as K_L can also be identified in the IFR as clusters not associated to charged tracks, with an efficiency of order 30%.

The hardware trigger is based on track information from DCH and IFR and coarse-grained EM showers in the calorimeter; its rate is about 1 kHz with efficiency above 99.9% for $B\bar{B}$ pairs; a software trigger performs more accurate track finding and fitting and EM cluster identification, reducing the rate to 100 Hz without degrading the efficiency.

The Belle detector (Fig. 10.3) covers polar angles from 17° to 150°, with smaller angles being covered by forward calorimeters.

The silicon vertex detector (SVD) originally had three double-sided silicon layers providing a ∼80 μm vertex position resolution, and was later upgraded adding a fourth layer. The drift chamber (CDC) consists of ∼8,500 cells in 50 layers, with small-angle wires, providing a ∼130 μm transverse position resolution, as well as dE/dx information.

Particle identification is mostly obtained with two detectors: the first is an array of ∼1,200 silica aerogel threshold Čerenkov detectors (PID) pointing to the interaction region and read by fine-mesh photomultipliers which can operate in the magnetic field, providing a good K/π separation for momenta above 1.2 GeV/c. The second detector, used for lower momenta, is a time-of-flight (TOF) system based on fast plastic scintillator slabs with no light guides and large PMs; by using two-sided light detection the crossing time can be determined independently of the time delay due to the distance of the particle impact point from the PM (mean timing), allowing one to obtain a 100 ps time resolution and a precise velocity discrimination on the basis of the flight time over a 1.2 m distance with respect to a RF clock synchronized with the beam collisions. With a *K* identification efficiency above 80% a π fake rate below 10% is achieved.

The highly-segmented EM calorimeter, with $\sim$9,000 crystals, provides a $\sim$4.9 MeV/c^2 π^0 mass resolution. The electron identification efficiency obtained combining the EM calorimeter information with the other detectors can be above 90% with a 0.3% fake rate. A forward calorimeter (EFC) made of radiation resistant bismuth germanate (BGO) crystals extends the acceptance in the forward and backward regions to 6.5° and 170°; silicon photodiodes are used for reading it.

Outside the solenoid, the muon and K_L detector (KLM) consists of alternating iron and glass RPC layers, and achieves a 90% muon efficiency with a fake hadron rate below 5% for momenta above 1.5 GeV/c.

The hardware trigger is based on the drift chamber and TOF for charged particles and on EM cluster counting and energy for neutral ones, plus the information from the KLM and forward calorimeter. The combined trigger efficiency is above 99.5% and the final rate after the software trigger is $\sim$200 Hz.

As the first dedicated *B* physics experiments, the BABAR and Belle experiments are providing a wealth of data, and we cannot attempt a full coverage of all the measurements related to *CP* violation: we will discuss some of the most significant results, dealing with the three types of *CP* violation, which provide an illustration of the general techniques used.

10.4.2 Interference *CP* violation

The first *CP*-violating effect detected with *B* mesons (actually the first manifestation of this phenomenon outside the *K* system) was the decay asymmetry into the 'golden' mode $B^0 \rightarrow J/\psi K_S$, arising in the interference of mixing and decay amplitudes (Section 9.4).

Interference *CP* violation at *B*-factories

The first results were published in 2001 by both BABAR and Belle (see Aubert *et al.* (2001) and Abe *et al.* (2001)), for decays into two-body modes containing charmonium and a neutral strange meson (Browder and Faccini, 2003).

In these analyses the decay of one meson into the *CP* eigenstate was fully reconstructed, with the J/ψ being identified from its decay into e^+e^- or $\mu^+\mu^-$ pairs. The background fractions were determined by extrapolating the observed number of events in sidebands of M_{bc} and ΔE. Backgrounds due to final states which are similar to the one being studied except for a low-momentum particle have the same M_{bc} distribution as the signal ('peaking backgrounds') but a different ΔE distribution.

The meson flavour was tagged using one of several different signatures based on the charge of some particles in the rest of the event, such as a high momentum lepton from a semi-leptonic decay or a secondary lepton, a kaon from the $b \to c \to s$ decay chain, slow pions from $D^* \to D\pi$ decays, or combinations of the above (either obtained analytically or by using neural networks). Each signature had a non-zero probability of providing a wrong tag: the mis-tagging parameters w were separately measured for each tagging category by fully reconstructing one meson decay into a flavour-eigenstate and studying the fraction F_{mix} of mixed events, i.e. those for which the tagging procedure identifies the flavour of the meson as being opposite: the observed value for such fraction is

$$F_{\text{mix}}(\Delta t) = 1/2 - (1 - 2w)\cos(\Delta m\Delta t)/2 \qquad (10.41)$$

where Δt is the time difference between the decay times of the fully reconstructed meson and of its partner (compare to eqn (7.144) in the limit of no *CP* violation in the mixing and $\Delta\Gamma = 0$); for very small decay time differences the observed mixing fraction is just the mis-tagging rate, while for larger $|\Delta t|$ the flavour oscillation probability sets in and F_{mix} is different from zero for perfect tagging also. By fitting the observed time-dependent mixing fraction to eqn (10.41) the mis-tagging fractions are obtained with small errors due to the background contaminations.

In another approach the integrated fraction of mixed events over a given range of Δt, which are expected to be (compare to eqn (7.158))

$$\int_0^{\Delta T} F_{\text{mix}}(\Delta t)\, d\Delta t = \chi(\Delta T) + [1 - 2\chi(\Delta T)]w$$

is used to determine the w parameters. Over the full range of time differences $\chi(\Delta T) \to \chi(B_d)$, but all the statistical power is in the small Δt region, and a limited range ΔT optimizes the error; this method is less sensitive to the details of the Δt distribution but more heavily affected by the presence of backgrounds.

The momentum of the tagging meson was obtained from that of the fully reconstructed one and from the knowledge of the average position of the interaction point: tracks were fitted to a common vertex and outliers were removed until a good fit was obtained. The presence of charmed mesons, with a lifetime comparable to that of B, introduced a bias in the measurement which had to be taken into account.

For each event the error on Δt was obtained from the vertex fits taking into account the finite resolution of detectors, multiple scattering, beam spot size, and the B flight path transverse to the beam axis; other sources of error are absorbed in a Δt resolution function, parametrized as a sum of two or more Gaussian distributions, whose standard deviations are linked to the above error by fitted scale factors, which account for any underestimation or overestimation of the errors. The Δt RMS resolution was of order 1 ps ($\sim$160 µm), dominated by the resolution of the partially reconstructed tagging meson vertex, and thus independent of the specific *CP* eigenstate considered (with a small dependence on the tagging category). Note that wrong assumptions on the origin of the tracks used for vertexing can introduce biases, as when tracks originating from charmed meson decays are included.

The observed asymmetries of the proper time distributions for the two flavours were fitted by the maximum likelihood method, each event contributing as an average of different time distributions, obtained as convolutions of 'true' time distributions with the appropriate time resolution function, weighted with the probability that the event belongs to that signal (or background) category.

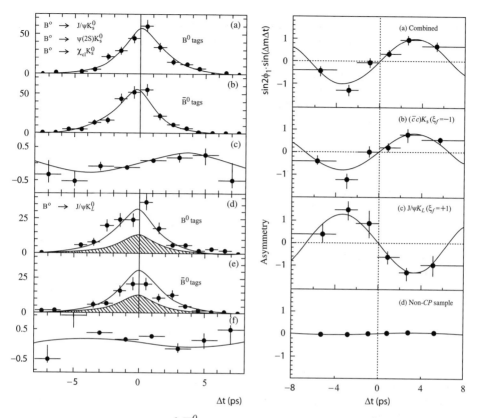

FIG. 10.4. Decay distributions for $B^0, \overline{B}^0 \to J/\psi\, K_S$ and related modes as a function of the difference Δt between decay time and tagging time. Left (BABAR experiment): decays of (a) $\overline{B}^0$ and (b) B^0 into *CP*-odd final states and (c) their asymmetry; (d,e,f) are the same for the *CP*-even final states; the shaded area is the background contribution (Aubert *et al.*, 2001). Right (Belle experiment): (a) combined decay asymmetry and separate ones for (b) *CP*-odd and (c) *CP*-even states; (d) is the asymmetry of a control sample of self-tagged modes such as $B^0 \to D^{*-}\ell^+\nu$, which is consistent with zero as expected (Abe *et al.*, 2001).

The final state $J/\psi\, K_L$ was also reconstructed, with the K_L identified by their interactions in the electromagnetic calorimeters (exploiting the hadronic shower shape) or muon detectors.

The *CP*-parities of the different final states considered are different, e.g. $\eta_{CP} = -1$ for $J/\psi\, K_S$ and $\eta_{CP} = +1$ for $J/\psi\, K_L$, resulting in asymmetries of opposite sign (recall eqn (7.93)); with no *CP* violation in mixing or decay the asymmetry is (eqn (7.166)):

$$A_{CP}(\Delta t) = \frac{\Gamma[\overline{B}^0 \to f](\Delta t) - \Gamma[B^0 \to f](\Delta t)}{\Gamma[\overline{B}^0 \to f](\Delta t) + \Gamma[B^0 \to f](\Delta t)} = \eta_{CP}(f)\,\mathrm{Im}(\lambda_f)\,\sin(\Delta m\,\Delta t)$$

as shown in Fig. 10.4.

States which are mixtures of *CP* eigenstates were also considered, e.g. $J/\psi K^{0*}$ (with $K^{0*} \rightarrow K_S\pi^0$) is *CP*-even (odd) for even (odd) orbital angular momentum: ignoring the angular information, the contribution of such decays to the asymmetry is diluted by the presence of the different components, whose disproportion can be measured as discussed in Section 10.3.2, corresponding to a state with effective fractional *CP*-parity (of order +0.6 in the example considered, depending on the experimental acceptance).

The results published in 2001 were based on about 1,000 signal events per experiment, extracted from samples with backgrounds varying from 5 to 40%, corresponding to 32 million $\Upsilon(4S) \rightarrow B\bar{B}$ decays analysed. *CP*-violating asymmetries with 4–6 standard deviation significance were measured as a non-zero $\sin 2\beta$ parameter: $\sin 2\beta = 0.59 \pm 0.14 \pm 0.05$ (BABAR), $\sin 2\beta = 0.99 \pm 0.14 \pm 0.06$ (Belle). The analyses were performed using a blind approach, in which the physical result was hidden from the experimenters until the end of the adjustments, to avoid the possibility of human bias. Systematic errors mainly originated from the uncertainties related to the tails of the vertex distributions, backgrounds and mis-tagging factors; the $B^0 - \bar{B}^0$ asymmetries in the reconstruction or tagging efficiencies, also extracted from the data, were also included in the determination of the errors.

The above analyses have been much improved since, leading to the value of $\sin 2\beta$ quoted in Section 10.3.2, with samples corresponding to hundreds of million $B\bar{B}$ decays.

10.4.3   *CP* violation in mixing

The asymmetry (7.157) of the $B^0 - \bar{B}^0$ oscillation probabilities was measured using inclusive events with two leptons in the final state, with $BR(B_d \rightarrow \ell^+ X \simeq 10\%)$. The lepton charge was used to tag the neutral meson flavour; by assuming the semi-leptonic decays to be flavour-specific ($\Delta B = \Delta Q$ rule) and *CP*-conserving, the asymmetry between like-sign lepton pairs is the quantity of interest.

Leptons not originating from semi-leptonic decays of B mesons should not be used in forming the asymmetry: electrons from photon conversions and leptons from J/ψ decays were rejected; to reduce the contamination from leptons ℓ originating in $b \rightarrow c \rightarrow \ell$ decays (resulting in flavour mis-tagging) several discriminating event variables were used. Moreover, as the background appears predominantly at small proper decay time differences, a cut was made on this difference, as determined from the separation $|\Delta z|$ of the estimated decay vertex positions; although the asymmetry is time-independent, any effect gets diluted as a function of $|\Delta z|$ by the varying background fraction:

$$\frac{N(\ell^+\ell^+) - N(\ell^-\ell^-)}{N(\ell^+\ell^+) + N(\ell^-\ell^-)} = \frac{N_S(\ell^+\ell^+) - N_S(\ell^-\ell^-)}{N_S(\ell^+\ell^+) + N_S(\ell^-\ell^-)} \frac{N_S}{N_S + N_B(|\Delta z|)}$$

where the total number of events $N = N_S + N_B$ is the sum of the signal and background contributions, the latter assumed to have zero asymmetry. The fitting

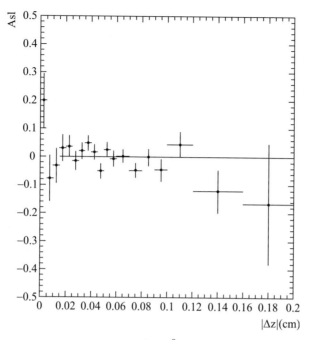

FIG. 10.5. Charge asymmetry in the di-lepton $B^0 - \bar{B}^0$ events in the Belle experiment as a function of the vertex separation $|\Delta z|$ (Nakano *et al.*, 2006).

of the asymmetry as a function of $|\Delta z|$, including the background contribution with the shape predicted by simulation, was performed above a minimum $|\Delta z|$ of 150–200 μm (see Fig. 10.5). The assumption of null asymmetry for the background could be confirmed by studying off-resonance data and the background-dominated small-$|\Delta z|$ events.

Lepton identification efficiencies were measured on independent samples such as $e^+e^- \rightarrow e^+e^-e^+e^-$, and ranged from 75% to 90%, with hadron fake rates of order 1%. Possible charge asymmetries induced by the lepton detection or reconstruction (and tagging) were studied: effects of $\sim$0.5–1% were measured and used for the determination of the systematic errors.

Results of the measurements from BABAR (Aubert *et al.*, 2006*f*) and Belle (Nakano *et al.*, 2006) were

$$|q/p| = 0.9992 \pm 0.0027 \pm 0.0019 \quad \text{(BABAR)} \tag{10.42}$$

$$|q/p| = 1.0005 \pm 0.0040 \pm 0.0043 \quad \text{(Belle)} \tag{10.43}$$

respectively, on samples of different size corresponding to 230 million (BABAR) and 85 million (Belle) $\Upsilon \rightarrow B\bar{B}$ decays. The main contributions to the systematic errors arise from the uncertainties in the subtraction of continuum events and the limits on asymmetries in the event selection.

The above results confirmed that *CP* violation in the mixing (and *T* violation, assuming *CPT*) is small for *B* mesons.[141]

Analyses of B_d mixing parameters not assuming *CPT* symmetry were also performed. At BABAR (Aubert *et al.*, 2004*b*), out of 88 million $\Upsilon(4S)$ decays, events were selected in which one *B* meson was fully reconstructed into either a *CP* or flavour eigenstate, and the other particles in the event were used for flavour tagging; the *CP* eigenstates considered were those consisting of a charmonium state and a neutral kaon, for which no direct *CP* violation is expected. A global fit to the Δt distributions gave

$$(\mathrm{Re}(\lambda_f)/|\lambda_f|)\,\mathrm{Re}(\delta_{CPT}) = 0.014 \pm 0.035 \pm 0.034 \qquad (10.44)$$

$$\mathrm{Im}(\delta_{CPT}) = 0.038 \pm 0.029 \pm 0.025 \qquad (10.45)$$

with no signal of *CPT* violation. Effects due to backgrounds were checked by adding background events to simulated data in the fit, and did not introduce any bias; the procedure was also validated by applying it to samples of $B^{\pm}$ decays (in which mixing is not present). Other systematic effects were estimated by varying several of the parameters in the fitted functions; note that charge asymmetries in the reconstruction and tagging efficiencies are themselves extracted from the data fit. The largest systematic errors were due to the finite statistics of simulated data and to the fact that a fully reconstructed flavour eigenstate cannot always be unambiguously tagged due to the presence of doubly Cabibbo-suppressed decays, which allow the same final state to be reached from both B^0 and $\overline{B}^0$. From a higher statistics (230 million decays) di-lepton sample BABAR also obtained (Aubert *et al.*, 2006*f*)

$$\mathrm{Im}(\delta_{CPT}) = -0.0139 \pm 0.0073 \pm 0.0032$$

still consistent with *CPT* symmetry.

Analysing di-lepton events from 32 million $\Upsilon(4S)$ decays, and assuming *CP* symmetry and the validity of the $\Delta B = \Delta Q$ rule in the semi-leptonic decays, the Belle collaboration obtained (Hastings *et al.*, 2003) results compatible with *CPT* symmetry corresponding to

$$\mathrm{Re}(\delta_{CPT}) = 0.00 \pm 0.12 \pm 0.01 \qquad \mathrm{Im}(\delta_{CPT}) = -0.03 \pm 0.01 \pm 0.03$$

---

[141] Although the present results, taken at face value, would not actually rule out mixing *CP* violation of the same size as that of the *K* system.

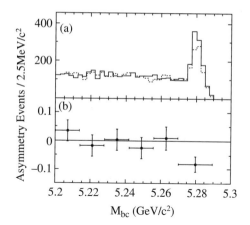

FIG. 10.6. $B^0 - \bar{B}^0 \rightarrow K^{\pm}\pi^{\mp}$ events in the BABAR experiment: (a) distribution of $K^+\pi^-$ (solid line) and $K^-\pi^+$ (dashed line) in the M_{bc} variable; (b) *CP*-violating asymmetry for intervals of M_{bc} (Aubert *et al.*, 2004*a*). No asymmetry is seen in the left part of the plot in which background dominates.

10.4.4 *CP* violation in decays

In 2004 both *B*-factory experiments published the first evidence for direct *CP* violation in neutral *B* meson decays[142] to $K^{\pm}\pi^{\mp}$ (BR $\sim 1.8 \cdot 10^{-5}$).

The (time-integrated) partial decay rate asymmetry was measured:

$$A_{CP}(K\pi) = \frac{N(K^-\pi^+) - N(K^+\pi^-)}{N(K^-\pi^+) + N(K^+\pi^-)}$$

In the SM the $B^0 \rightarrow K^+\pi^-$ decay has both tree and penguin amplitudes with different weak (and most likely strong) phases, so that *CP* violation in the decay is indeed possible; the kaon charge identifies the meson flavour, apart from second-order weak transitions which could be neglected at the present level of precision; *CP* violation in the mixing might contribute in principle, but its effect is known to be small, as discussed.

Backgrounds, mostly due to light quark pair production, were rather high, and the good particle identification in the range of momenta 2–4 GeV/*c* was essential; they were suppressed by exploiting their jet-like pattern and flavour tagging quality. With about 1,600 $K\pi$ signal events out of 230 million $\Upsilon(4S) \rightarrow B\bar{B}$ decays (BABAR, see Fig. 10.6) (Aubert *et al.*, 2004*a*), and about 3,000 $K\pi$ events out of 390 million decays (Belle) (Abe *et al.*, 2005*b*), the *CP*-violating asymmetry was

---

[142] It only took three years to get from the first *CP* violation measurements to direct *CP* violation, as compared to thirty-five years in the case of kaons: the *B* system appears to be much less 'super-weak' than the *K* one, but Nature and technological advances helped too!

measured to be

$$A_{CP}(K^\pm \pi^\mp) = -0.133 \pm 0.030 \pm 0.009 \quad \text{(BABAR)}$$

$$A_{CP}(K^\pm \pi^\mp) = -0.113 \pm 0.022 \pm 0.008 \quad \text{(Belle)}$$

thus establishing the existence of *CP* violation in the decay of *B* mesons with a significance exceeding 5 standard deviations. It is maybe interesting to note that the corresponding asymmetries for charged *B* meson decays into $K^\pm \pi^0$ or $K^0 \pi^\pm$ (with slightly larger errors) are so far consistent with zero (Yao *et al.*, 2006; Aubert *et al.*, 2006*b*, *e*; Lin *et al.*, 2006):

$$
\begin{aligned}
A_{CP}(K^\pm \pi^0) &= 0.016 \pm 0.041 \pm 0.012 && \text{(BABAR)} \\
A_{CP}(K^\pm \pi^0) &= 0.04 \pm 0.05 \pm 0.02 && \text{(Belle)} \\
A_{CP}(K^0 \pi^\pm) &= -0.029 \pm 0.039 \pm 0.010 && \text{(BABAR)} \\
A_{CP}(K^0 \pi^\pm) &= 0.03 \pm 0.03 \pm 0.01 && \text{(Belle)}
\end{aligned}
$$

Direct *CP* violation was also observed in the decay asymmetry of neutral *B* mesons to $\rho\pi$ states which, not being *CP* eigenstates, require more complex analysis procedures (Aubert *et al.*, 2006*c*; Wang *et al.*, 2005).

Radiative decays induced by $b \rightarrow s\gamma$ (e.g. $B \rightarrow K^{0*}\gamma$) are also very interesting as the emitted photon acts as a probe to study the dynamics of the *b* quark inside the *B* meson; the performances of the electromagnetic calorimeters are crucial for these measurements, which are performed both with inclusive (no reconstruction of the recoiling state) and exclusive final states; in the latter case somewhat higher systematics effects are a trade-off for the higher resolutions (and lower model dependence when comparing with theory) which can be achieved. *CP*-violating asymmetries, expected to be very small in the SM, were measured to be compatible with zero.

Evidence for *CP* violation in $\pi^+ \pi^-$ decays (BR $\simeq 5 \cdot 10^{-6}$) was also obtained, originally by Belle (Abe *et al.*, 2005*a*), from time-dependent asymmetry analyses. The tagged asymmetry (7.166) for correlated meson pairs was measured (in the approximation $\Delta\Gamma = 0$): as *CP* violation in the mixing is small, the parameter $A^{(MD)}$ defined in (7.137), which is the coefficient of the $\cos(\Delta m \, \Delta t)$ term in the asymmetry is mostly sensitive to *CP* violation in the decay.

In BABAR (Belle) out of 383 (535) million $\Upsilon(4S) \rightarrow B\bar{B}$ decays about 1,100 (1,500) $\pi^+\pi^-$ signal events are extracted, and their measured asymmetry, taking into account the effect due to misidentification and mis-tagging probability differences, corresponds to (Aubert *et al.*, 2007c; Ishino *et al.*, 2007)

$$A^{(MD)}_{\pi^+\pi^-} = -0.21 \pm 0.09 \pm 0.02 \qquad A^{(I)}_{\pi^+\pi^-} = -0.60 \pm 0.11 \pm 0.03 \quad \text{(BABAR)}$$

$$A^{(MD)}_{\pi^+\pi^-} = -0.55 \pm 0.08 \pm 0.05 \qquad A^{(I)}_{\pi^+\pi^-} = -0.61 \pm 0.10 \pm 0.04 \quad \text{(Belle)}$$

showing direct *CP* violation at more than 5 standard deviation significance, with the main contribution to the systematic error arising from uncertainties in the vertex reconstruction and signal event fraction.

Finally it should be recalled that with *CP* violation in the interference of mixing and decay being now measured in several decay modes besides $J/\psi K_S$ (Yao *et al.*, 2006; Barberio *et al.*, 2006):

$$A^{(I)}_{J/\psi K_S} = +0.69 \pm 0.03$$

$$A^{(I)}_{K^+K^-K_S} = -0.45 \pm 0.13 \qquad A^{(I)}_{D^{*+}D^{*-}} = -0.75 \pm 0.23$$

$$A^{(I)}_{\eta'K_S} = +0.50 \pm 0.09 \qquad A^{(I)}_{f_0 K_S} = -0.75 \pm 0.24$$

the differences in the values of $|A^{(I)}_f|$ also indicate that direct *CP* violation is present in *B* mesons.

10.4.5 Future colliders

In a *B* meson experiment the collected statistics depends on the *B* production cross section and the efficiency, but the capability of extracting the signal from the background is an equally important issue. Hadron colliders and e^+e^- colliders are complementary in this sense: the *B* production cross section is $\sim$100 μb at the Fermilab TeVatron $\bar{p}p$ collider (2 TeV CM energy) and $\sim$500 μb at the upcoming CERN LHC pp collider, several 10^5 times higher than at *B*-factories. On the other hand the background is also much higher in a hadronic environment, as the *B* cross section is only a fraction of per cent of the total; as a consequence efficiencies are also generally low: the control of backgrounds and efficiencies are key issues for *B* physics studies at hadron colliders.

The different production mechanisms at hadron colliders result in a strong dependence of the *B* momentum from the production angle, so that at small angles decay lengths are relatively long; in this angular region moreover the *B* mesons are strongly correlated, and they tend to go in the same direction (both forward or both backward), so that a detector on just on side of the collision point can have a high acceptance for both the meson being studied and its tagging partner. The positions of the primary (production) and secondary (decay) vertices can be measured with silicon detectors, from which also the spatial separation of primary and secondary vertices is obtained, to study the time evolution of *B* mesons.

Besides the possible competition with *B*-factories on the modes which can be accessed by both, experiments at hadron machines have the unique possibility of producing B_s mesons, whose pair production threshold is above the $\Upsilon(4S)$ mass; the CLEO and Belle experiments ran shortly at the $\Upsilon(5S)$ centre of mass energy (10.86 GeV), at which B_s meson pairs can also be produced, measuring a $b\bar{b}$ production cross-section about 1/20 of that at the $\Upsilon(4S)$ (Acquines *et al.*, 2006).

General purpose detectors at hadron colliders, such as CDF and D0 at the TeVatron (and CMS and ATLAS at the LHC) are mostly designed for searches of high mass particles, and thus focus on the central rapidity (polar angles around $\pi/2$ with respect to the beams) and high transverse momentum regions, which are optimal for that purpose. Despite their lack of excellent particle identification capabilities, they can perform *B* physics studies thanks to their good vertexing capabilities, if dedicated trigger systems are implemented. Indeed the trigger system is a crucial element in the high-rate environment of a hadron collider experiment, as selecting the few events containing *B* mesons among the plethora of more mundane events in a very short time (so that they can be recorded for further analysis) is one of the biggest challenges. As an example the CDF experiment has successfully implemented a displaced vertex trigger to select events containing tracks with large impact parameters with respect to the interaction point, which identify long-lived particles decaying into charged tracks; this works as a second-stage processing applied to events containing muons or high transverse momentum tracks, and is able to identify tracks with an impact parameter resolution of ∼50 μm above a few GeV. In this way large samples of *B* decaying exclusively into hadrons can be collected.

The general purpose LHC experiments ATLAS and CMS will be able to perform some limited *B* physics studies despite their lack of particle identification capability and their need to use muons for triggering on *B* mesons: events with muons of transverse momentum above ∼6 GeV/*c* are indeed dominated by *B* decays. The use of radiation-resistant silicon pixel detectors allow one to identify displaced vertices and large impact parameter tracks. Again the issue will be triggering, as the very high rates will require compromises and competition between physics programmes: *B* physics could probably be performed at the early stages of these experiments, when the limited luminosity will allow a less aggressive sharing of the available trigger bandwidth.

LHCb is a dedicated experiment for *B* physics in preparation at the LHC: its detector is on one side only of the interaction region and has a large angular acceptance (from 15 to 300 mr polar angles), resulting in a large effective $b\bar{b}$ cross section ∼230 μb (which compensates the lower luminosity at which the experiment is expected to work with respect to other LHC experiments). Vertices are reconstructed by a silicon micro-strip detector consisting of twenty-one stations with alternating $r - \phi$ strips. An aerogel ring imaging Čerenkov counter (RICH) is used for large-angle particle identification, and is followed by more silicon micro-strip detectors. The spectrometer consists of a 4 T m dipole field integral, followed by straw tracking chambers and a second small angle RICH. Additional particle identification capability over the large range of track momenta (1-100 GeV/c) is provided by a lead-scintillator electromagnetic calorimeter used to reconstruct photons with the help of a pre-radiator, and to identify electrons; a hadron calorimeter and muon chambers complete the set-up. As with other LHC experiments, the very high rates required most of the detectors and front-end electronics to be radiation tolerant.

The LHCb trigger will select high transverse momentum muons and events with high transverse energy in the calorimeters, while rejecting events with multiple identified interaction vertices; the vertex tracker will also be used in the early stages of the trigger, providing a primary vertex resolution of about 60 μm along the beam axis and 20 μm transverse to it with a fast tracking algorithm; tracks originating from secondary vertices displaced by 0.1–3 mm from the primary one will then be selected as *B* meson candidates. Running at a design luminosity of $2 \cdot 10^{32}$ cm^{-2} s^{-1}, smaller than the typical figure for LHC, about 10^{12} $b\bar{b}$ pairs will be produced per year, and many decay modes of both B_d and B_s mesons could be studied, complementing the information which can be obtained from *B*-factories (see e.g. Uwer (2004)).

Apart from some specific channels, theoretical predictions for *CP*-violating asymmetries are plagued by large hadronic uncertainties, and if these are not drastically reduced precise measurements cannot be exploited to test the SM picture. For some modes which are considered 'clean' the uncertainties can be kept under reasonable control (of order 10%, say): it would therefore be useful to reduce the experimental error below this level, and to reach this on a large set of decay modes a large increase in statistics is required, from the presently available $\sim$1 ab^{-1} to 15–40 ab^{-1}. This cannot be achieved in a reasonable time by upgrading the present *B*-factories, and would rather require new e^+e^- colliders with luminosities in the 10^{36} cm^{-2} s^{-1} range, dubbed 'super *B*-factories' (Hewett and Hitlin, 2004). Such machines would require high beam currents, which in conventional schemes also imply very high rates and backgrounds, not sustainable by the present detectors, thus requiring new developments also in this field. If such projects are pursued the required integrated luminosity could be available around 2020.

Further reading

Charm physics is reviewed in e.g. Bianco *et al.* (2003) and Briere (2006).

Many reviews of *B* physics exist, but as the subject is in its full blossoming and new results appear at a rather high rate, any of them (including the present one) are doomed to become obsolete rather quickly; up-to-date information can be obtained from the proceedings of topical conferences and schools; a recent one is

Lanceri (2006), while theoretical aspects are discussed in some detail for example in Buras (2005). As of 2007 online compilations of results can be found at Barberio *et al.* (2006), Charles (2006) and Bona (2006).

A nice introduction to *B*-factory experiments (written before they started) is in BABAR collaboration (1999). Prospects for experiments at hadron machines are discussed in Butler (2006).

Problems

1. The μ lifetime is approximately given by $\tau_\mu^{-1} = (G_F^2 m_\mu^5)/(192\pi^3)$. Use the same formula to estimate the *B* meson lifetime.

2. Which is the phase-convention independent version of the statement that the mixing parameter $\bar{\epsilon}_B$ is almost purely imaginary?

3. Consider the decay of a correlated *B* meson pair with odd *C*-parity into two *CP* eigenstates with the same eigenvalue. Would this be evidence of *CP* violation? Of which kind? Why are no such measurements reported?

4. *B* mesons often decay into charmed mesons: discuss the effects of these daughter particles for the flavour tagging and for the decay time determination.

5. Why are the branching ratios of purely leptonic *B* decays, such as $B^0 \to \tau^+\tau^-$ or $B^\pm \to \tau^\pm \nu$ relatively small? What physical information might be extracted from these decay modes?

6. In what respects would a *B*-factory running at the $\Upsilon(5S)$ resonance (capable of producing $B_s\bar{B}_s$ meson pairs) differ from those running at the $\Upsilon(4S)$?

11

WIDENING THE PICTURE

No science is more pretentious than physics,
for the physicist lays claim to the whole universe
as his subject matter.
J. R. Brown, P. C. W. Davies

'After all, we only outnumber them
about a thousand million to one.'
Rugon laughed at his captain's little joke.
Twenty years afterward, the remark didn't seem funny.
A. C. Clarke

Made in heaven, made in heaven
it was all meant to be.
F. Mercury

In this chapter some additional issues which are somehow linked to the violation of discrete symmetries are briefly outlined. We cannot give the reader more than a fleeting glimpse of a few diverse topics, with the goal of setting the importance of the physics issues addressed in previous chapters in a wider context. Somewhat differently from the rest of the book the discussion will thus be mostly qualitative, and the reader should be warned that the focus is shifted from rather solid facts to topics which range from the speculative to the highly speculative; we feel nevertheless that it is useful to provide a broader view with a snapshot on some important open questions at the edge of current research, which have a close connection to discrete symmetries.

11.1 Symmetries of gravity

Gravity is by far the weakest interaction, when applied to 'elementary particles', and as such its microscopic properties have not been really tested experimentally. A comparison of the gravitational and electromagnetic forces can be given by the ratio of the fine structure constant $\alpha = e^2/4\pi\hbar c \simeq 1/137$ to the dimensionless quantity $G_N m^2/4\pi\hbar c$ (G_N being Newton's constant and m an appropriate mass): for a subatomic particle such as the proton the latter quantity has the value $\simeq 5 \cdot 10^{-40}$. A feeling for the relative weakness of gravity on the microscopic scale[143] can be obtained by considering that the radius of an electron-proton system bound only by gravitational interactions would be comparable to the size of the known universe.

From the point of view of particle physics, gravity is described by the exchange of a tensor (spin 2) boson (the graviton), coupling to mass-energy; in general Lorentz-invariant interactions mediated by the exchange of even-spin particles are always attractive (under similar assumptions as those required by the *CPT* theorem) (Deser and Pirani, 1967), contrary to what happens for those mediated by vector fields, such as electrodynamics in which charges can either attract or repel each other. Actually, theories of gravitation including small non-tensorial components have been proposed, and limits on the magnitude of specific terms in the gravitational potential can be obtained by considering their expected contribution to accurately known energy differences: for example the hyperfine splitting of hydrogen allows to constrain a possible spin-dependent ($S \cdot r$) term (*P*- and *T*-violating) in the gravitational potential of the Earth.

On the other hand a microscopic (quantum) theory of gravitation does not yet exist, and the classical theory (general relativity) leads one to think that this interaction respects the C, P, and T symmetries.[144] It has been suggested that quantum gravity effects might violate *CPT* symmetry, and the Planck mass[145] $M_P c^2 = \sqrt{\hbar c^5 / G_N} = 1.2 \cdot 10^{19}$ GeV sets the energy scale at which such effects are expected to become relevant.

The gravitational interaction of antimatter has not been observed so far, but is widely assumed to be the same (attractive) as for matter. Experimental support for this comes from the observation of neutrinos from the explosion of supernova 1987A: if, as it appears to be very likely, both neutrinos and antineutrinos were observed by the three detectors which were active at that time around the world, the simultaneity of the observations puts rather stringent bounds on a different gravitational interaction for antiparticles.

Laboratory experiments on gravitational interaction of elementary particles are notably difficult, since electromagnetic interactions are much stronger. Moreover, the scarcity of antiparticles and the difficulties in producing (often at high energies relative to their rest masses) and handling them make such

---

[143] Of course gravity is macroscopically much more relevant than electromagnetism, the other long-range force, because it is always attractive and matter is electrically neutral. Clearly the relative weakness of gravity can be rephrased in terms of the small mass to charge ratio of the fundamental constituents of matter as we know them.

[144] Strictly speaking, the coupling of purely left-handed neutrinos to gravity would spoil the validity of P, C, although by a very tiny amount.

[145] The natural unit for mass, corresponding to the value for which the Schwartzschild radius equals the Compton length divided by π.

experiments very challenging. Cold, electrically neutral systems such as anti-hydrogen atoms, are the most promising for future direct experimental tests (Holzscheiter *et al.*, 2004).

As discussed in Chapter 5, *CPT* symmetry requires the equality of *inertial* masses for particles and antiparticles, but it says nothing about *gravitational* masses, which are the 'charges' of gravitation theory (whatever that is). The equivalence principle of general relativity states that all bodies undergo the same acceleration in the same gravitational field[146]: this is equivalent to the statement that gravitational mass coincides with inertial mass; together with *CPT* symmetry this principle requires the same gravitational properties for matter and antimatter. Note also that the *CPT* theorem is valid for a Lorentz-invariant field theory, but when dealing with gravitational effects it is general relativity which must be considered, and thus the issue of the gravitational interaction of antimatter is connected to the great unsolved issue of a quantum theory of gravity.

A repulsive interaction between antimatter particles would violate *CPT* symmetry, while a repulsive interaction between matter and antimatter would require the 'passive' gravitational mass of an antiparticle (expressing the force exerted on it by an external gravitational field, possibly in contrast to the 'active' gravitational mass expressing the force exerted by the particle itself) to be negative, thus violating the equivalence principle.[147]

Arguments against the possibility of 'anti-gravity' (a generic term for the case of antiparticles behaving oppositely to particles in gravitational interactions) were put forward by Schiff (1959): in case positrons had a gravitational interaction equal and opposite to that of electrons their contribution to the masses of atoms, through virtual effects, might exceed the limits on the difference of inertial and gravitational masses as set by the experiments pioneered by Eötvös (Eötvös *et al.*, 1922). Such experiments use torsion balances to search for a composition-dependence of the free-fall acceleration of macroscopic samples, reaching precisions of order 10^{-12} (Su *et al.*, 1994); while the experiments are of course performed only on matter, implications for antimatter arise from the fact that the size of the induced effect would be expected to be different for various atomic species, due to the different atomic sizes (and average distance of the virtual pairs from the nuclei). Due to the difficulties in evaluating the expected effect, however, the argument above was considered to be inconclusive (Nieto and Goldman, 1991).

On the other hand the equality in magnitude and sign of the inertial masses of particles and antiparticles (by *CPT*) is related to the finite amount of energy required

---

[146] This is properly called the *weak equivalence principle*, to distinguish it from more general formulations extended to non-mechanical phenomena.

[147] *CPT* symmetry tells us how an anti-apple would fall towards an anti-Earth, but in itself it has nothing to say on how an anti-apple would fall towards the Earth (Nieto and Goldman, 1991).

for pair creation: equal and opposite inertial masses would result in a vanishing rest energy associated to a particle antiparticle pair.

A violation of the equivalence principle was claimed to be inconsistent with energy conservation (Morrison, 1958): with opposite gravitational coupling for antiparticles, a particle-antiparticle pair would experience no net gravitational attraction, while the photon from which such a pair can be produced (and into which it can annihilate) does, so that by having pair production and annihilation occurring at different heights in a gravitational field an cyclic energy-producing engine appears to be possible; a careful analysis shows however that this is not the case (Charlton *et al.*, 1994).

Before the discovery of *CP* violation it was argued that differences in the gravitational interaction of particles and antiparticles would be detectable in the $K^0 - \overline{K}^0$ system (Chapter 6), which behaves as a very sensitive interferometric system due to the very small mass difference between the physical states (Good, 1961); if matter and antimatter had opposite couplings to gravity a gravitational potential Φ would induce a relative phase between K^0 and $\overline{K}^0$ equal to $2m_K\Phi t$, as a consequence of unequal diagonal terms in the effective Hamiltonian (7.67). As the potential energy of a kaon in the gravitational field of the Earth is ~ 0.4 eV, this would give a $K^0 - \overline{K}^0$ oscillation period of $\sim 10^{-15}$ s, resulting in unobservably fast strangeness oscillations, contrary to experimental observation; in this way an opposite gravitational coupling for antiparticles was claimed to be excluded.

The above argument was criticized for what concerns the estimated magnitude of the potential, and actually anti-gravity was claimed to be a possible origin of the $K_L \to 2\pi$ decay (Chapter 6) usually ascribed to *CP* violation (Leitner and Okubo, 1964; Chardin and Rax, 1992). While a K_L propagates in vacuum, weak interactions continuously transform quarks into antiquarks and vice versa ($K^0 - \overline{K}^0$ oscillations): it could be assumed that if quarks and antiquarks get separated by more than the 'size' of the kaon, e.g. by a different gravitational interaction, the state would become either a K^0 or a $\overline{K}^0$, thus regenerating a K_S component which would decay to 2π, apparently violating *CP* symmetry: the order of magnitude which a hypothetical anti-gravity would provide matches the size of the ϵ parameter.

With the discovery of direct *CP* violation (Section 8.3.3) the above speculative argument lost its appeal, since a difference in the gravitational interaction of $\overline{K}^0$ and K^0, as any property of the decaying system, cannot account for any decay mode difference: while this clearly cannot disprove the anti-gravity hypothesis it makes it unnecessary (in the context of neutral kaons) through Occam's razor argument.[148]

---

[148] '*Pluralitas non est ponenda sine necessitate*': plurality shouldn't be posited without necessity, William of Occam (1285–1349).

11.2   The strong *CP* problem

The symmetry properties of the Standard Model discussed so far (*C* symmetry of strong interactions, *P* violation of weak interactions, etc.) can be shown
to be necessary properties of the most general renormalizable Lagrangian density consistent with Poincaré invariance and gauge invariance under the group
$S(U(2) \times U(3))$, when the matter fields are arranged in the irreducible representations discussed in Chapter 9 (Froggatt and Nielsen, 1991); such a Lagrangian
density is not complicated enough to violate the other symmetries exhibited by the
Standard Model.[149]

Following this line of reasoning there is, however, one potential source of trouble: in the SM Lagrangian the above conditions do not rule out the presence of the
following term

$$\mathcal{L}^{(SCP)} = \theta \frac{g^2}{32\pi^2} F^{(a)}_{\mu\nu} \tilde{F}^{\mu\nu(a)} \tag{11.1}$$

where θ is an arbitrary coefficient and $F^{(a)}_{\mu\nu} = -F^{(a)}_{\nu\mu}$ is the field strength tensor
of a non-Abelian gauge group, either $SU(3)$ of colour (strong interactions, $a =
1, \ldots, 8$) or $SU(2)$ of weak isospin (electro-weak interactions, $a = 1, \ldots, 3$, see eqn
(9.4)),[150] g being the corresponding coupling constant; the dual tensor is $\tilde{F}^{\mu\nu(a)} =
\epsilon^{\mu\nu\rho\sigma} F^{(a)}_{\rho\sigma}/2$.

From the transformation properties of the field strength tensor under C, P, T (see
Section 4.3.4) it is seen that the term in eqn (11.1) violates both P and T symmetries
(and thus CP if CPT symmetry is taken for granted); while this might not appear to
be a big concern for electro-weak interactions, it certainly is for strong interactions,
which experimentally are not seen to violate such symmetries.

One could simply postulate $\theta = 0$ to eliminate the troublesome term, but things
are not so simple because when the quark fields are rotated to diagonalize the
corresponding flavour matrices $G^{(U,D)}$ (Section 9.2.1) a term like (11.1) arises due
to a quantum anomaly: this is a symmetry which is valid classically (resulting in
a conservation law) but is violated after quantization[151] (see e.g. Holstein (1993)).

---

[149] It is probably unnecessary to stress the fact that this kind of 'derivation' of the symmetry
properties of the SM follows from the choices of gauge group and matter multiplets, made to reproduce
its experimentally observed properties.

[150] The explicit expression in terms of the gauge fields is $F^{(a)}_{\mu\nu} = \partial_\mu A^a_\nu - \partial_\nu A^a_\mu + g f_{abc} A^b_\mu A^c_\nu$,
where f_{abc} are the structure constants of the group, defined by the commutator of the generators T^a:
$[T^a, T^b] = i f_{abc} T^c$ ($f_{abc} = \epsilon_{abc}$ for $SU(2)$).

[151] A simple example is scale invariance of massless QED: with the electron mass set to zero
the theory is scale invariant and the coupling constant is independent of distance, but regularization
introduces a mass scale in the form of an ultraviolet cut-off, which ultimately results in a running
coupling constant.

The coefficient of the above term becomes in this way

$$\bar{\theta} = \theta + \arg \det \left[G^{(U)} G^{(D)} \right] \tag{11.2}$$

which is invariant under chiral transformations. In order to have *CP* violation in weak interactions the *G* matrices are in general complex, so that they indeed give a non-zero contribution to $\bar{\theta}$ when they are rotated to give real masses.

The term (11.1) is diagonal in flavour, so that it could contribute e.g. to the electric dipole moment of the neutron (Section 4.3.4), whose size is estimated to be of order $d_E(n) \sim 10^{-16}\bar{\theta}$ *e* cm: the current experimental limits on such quantity thus imply $\bar{\theta} \lesssim 10^{-9}$. The so-called 'strong *CP* problem' is the lack of an 'explanation' for such an unnaturally small value of an otherwise arbitrary parameter. In any case, even starting from $\bar{\theta} = 0$ in the SM Lagrangian (which requires some peculiar matching of the strong and the electro-weak contributions), this value would not be stable under quantum corrections, and in general the problem remains.

While the strong *CP* problem might be called a 'theoretical' problem, the falsification of its possible solutions lies on experiment. The simplest solution would require one of the quark masses to be exactly zero: in this case $\det[G^{(U)}G^{(D)}] = 0$ and its argument is arbitrary and unphysical, so that any value of θ can be removed to 0 by a redefinition of the quark fields. Indirect determinations based on the effects of quark masses on the properties of hadrons, using theoretical input (since quarks are confined and unobservable as physical particles), strongly suggest that none of the quarks is massless, the lightest one being the *u* quark with a mass of a few MeV (Yao *et al.*, 2006).

Another possible solution which received a lot of attention requires the introduction of a new spontaneously broken global symmetry; $\bar{\theta}$ then becomes a dynamical variable, which in some conditions relaxes spontaneously to zero (Peccei and Quinn, 1977). These models require the existence of an hypothetical boson, the *axion*, associated with the new degree of freedom and coupling to photons. Such a particle has been actively searched for, so far without any success; nevertheless, since the axion mass is not fixed by the model, some varieties of axion models are not excluded.

11.3   The cosmic connection

The issue of *CP* violation, the smallest symmetry violation detected so far, is not only relevant for the understanding of a small set of relatively rare weak processes; on the contrary it actually bears on one of the most intriguing mysteries of cosmology, namely the fact that the universe is almost empty but not exactly so, allowing our own existence.

11.3.1 The matter-antimatter asymmetry

The fundamental and deep connection between particles and antiparticles which is expressed by *CPT* symmetry, the only discrete symmetry which (so far) appears to be exactly valid in Nature, and also by the *C* or *CP* symmetries valid for the dominant interactions of Nature, could lead to the expectation that matter and antimatter, being on equal footing, are equally abundant in the universe. In his 1933 Nobel lecture, P. A. M. Dirac said that 'we must regard it rather as an accident that the Earth (and presumably the whole solar system), contains a preponderance of negative electrons and positive protons. It is quite possible that for some of the stars it is the other way about [...] In fact, there may be half of the stars of each kind'.

This appears, however, not to be the case and only one of the two, which we call matter, is present around us (Steigman, 1976). Antiparticles are indeed found in Nature in tiny amounts, created together with particles in (natural or artificial) high energy processes, but they do not appear to be present in significant amounts in the form of stable macroscopic lumps of antimatter.[152]

The presence of antimatter within the solar system is excluded because the annihilations of the solar wind protons on antimatter bodies would be intense sources of γ rays, which are not observed. Moreover, any amount of antimatter present in a pre-solar gas cloud would have annihilated well before solid bodies were formed (the typical annihilation time being much shorter than the characteristic time for free gravitational fall).

Information on the presence of antimatter in the universe can be obtained through cosmic rays: those with energies greater than O(100) MeV are believed not to be of solar origin, and thus provide samples of matter originating elsewhere in the galaxy or beyond. Unfortunately, except for those of extremely high energies, cosmic rays cannot be traced to their sources, because their trajectories are highly randomized by the presence of poorly known magnetic fields.

The nature and composition of primary cosmic rays have been a subject of investigation for a long time, mostly by using detectors mounted on balloons launched in the higher regions of the atmosphere, in order to reduce the probability of observing a cascade of secondary particles which limits the amount of information which can be obtained on the primary. Measurements with detectors on satellites have great advantages in terms of sensitivity and statistics. The presence of heavy nuclei in cosmic rays indicates that they originate in stars, and they probably provide a representative information on the matter content of the whole galaxy.

The only antiparticles which have been identified in cosmic rays to date are positrons and antiprotons (see e.g. Galaktionov (2002) for a review). The positron

---

[152] While antimatter is very scarce in the universe, antiparticles are present even in the human body: out of about 100 g of potassium, 0.02% is in the form of the radioactive isotope ^{40}K, which has a β^+ decay with a half-life of 1.3 billion years.

fraction in cosmic rays is of order 0.1, but they are not very significant as far as the presence of antimatter in the universe is concerned, because e^+ are easily produced in the interactions of matter particles and photons with the interstellar gas or the dense atmosphere around stars:

$$pp \to \pi X \qquad \pi \to \mu^{\pm} \nu_{\mu} \qquad \mu^{\pm} \to e^{\pm} \nu_e \nu_{\mu} \qquad \gamma \to e^+ e^-$$

Other sources of positrons are the β^+ decays of nuclei produced in novæ or supernovæ, and pair production from high energy photons, such as those emitted when matter falls within a black hole. A source of 511 MeV positron annihilation photons was identified close to the centre of our galaxy (Jean *et al.*, 2003) (see also Johnson *et al.* (1972)), not corresponding to any known structure.

A measurement of the asymmetry between the electron and positron densities within our galaxy is also provided by the Faraday rotation of the plane of polarized light (from radio pulsars) as it travels through regions of the galaxy containing magnetic fields (Steigman, 1976).

The relative abundance of antiprotons in cosmic rays is $\bar{p}/p \sim 10^{-4}$. Interactions of primary protons can easily produce secondary antiprotons: the threshold is 5.6 GeV, corresponding to the reaction $pp \to \bar{p} + 3p$ (baryon number conservation). The shape of the secondary antiproton spectrum produced in this way can be modelled: it falls steeply below ~ 2 GeV because of the production kinematics, and its ratio with the primary proton spectrum has a maximum around 10 GeV, falling to 10^{-5} at some hundred GeV. Primary antiprotons can thus be searched either below 2 GeV or at high energies.

The separation of primary and secondary components can also be obtained by exploiting the anti-correlation of the low-energy primary component with the solar activity (with a period of 22 years) originated by diffusion effects due to the solar wind, with a variation up to a factor 20 (Mitsui *et al.*, 1996); the secondary component would be much less affected by solar modulation. Spectrum measurements performed in different phases of the solar cycle show that the antiproton component is largely of secondary origin, with no evidence of a significant primary component.

Contrary to the case of antiprotons, the production of secondary antinuclei $\overline{N}$ with mass m_N in particle collisions is highly suppressed:

$$n(\overline{N})/n(\bar{p}) \approx \exp[-2(m_N - m_p)/T_0] \tag{11.3}$$

where $T_0 \approx 160$ MeV. The above ratio ranges from $\sim 10^{-9}$ for anti-^{3}He (Chardonnet *et al.*, 1997) down to 10^{-56} for anti-C. The observation of even a single antinucleus in cosmic rays would thus rather indicate the existence of antistars, in which such nuclei could be produced (note that only antinuclei of relatively high-energy (10–100 GeV/nucleon) could have reached us from other galaxies, corresponding to a very suppressed region of the spectrum). Limits of order $\sim 10^{-6}$ have been obtained for the ratio of anti-He to He (Galaktionov, 2002); note that each experimental limit corresponds to a given maximum rigidity (momentum per unit charge) essentially dictated by the magnetic field integral of the spectrometer used: for high energy particles which are barely deflected the determination of the charge becomes ambiguous. The next frontier for these experimental investigations is based on detectors on satellites, such as AMS (Pereira, 2005).

Hypothetical antimatter galaxies cannot be detected from their electromagnetic radiation because the photon is its own antiparticle, and in this respect such galaxies would look exactly the same as those made of matter. While we do not have direct measurements on the presence of antimatter in distant galaxies, indirect limits on its presence can be obtained from the analysis of its possible annihilation products.

The energy released in e^+e^- annihilations is rather small, while that from $p\bar{p}$ annihilation is significantly larger, so the latter processes are more useful for setting limits on the presence of antimatter, which would produce energetic γ rays from the decay of π^0 produced in $\bar{p}p$ annihilations. Stable particles produced in $\bar{p}p$ annihilations are on average: (a) $\sim$3.8 $e^\pm$ of $\langle E \rangle \sim$90 MeV, (b) $\sim$3 ν_e of $\langle E \rangle \sim$100 MeV, (c) $\sim$6 ν_μ of $\langle E \rangle \sim$100 MeV (d) $\sim$3 γ of $\langle E \rangle \sim$180 MeV (with an energy peak at 70 MeV from π^0 decay), which because of the *red-shift* are observed as $\sim$1 MeV photons.

The $e^\pm$ component is difficult to detect because of the background from secondary production and the fact that it does not travel very far from the production site due to the presence of magnetic fields and the rapid energy loss by Compton scattering. The electron neutrino flux is hard to measure because of the background from the Sun. One third of the muon neutrinos produced in annihilations can produce muons with a spectrum below a few hundred MeV, but the background from neutrinos produced by cosmic ray interactions in the atmosphere hinders their detection. It follows that annihilation γ photons are the best candidates for observing the presence of antimatter. As 0.5 MeV annihilation photons can also be produced by secondary e^+, more stringent limits are obtained from higher-energy photons (the $\gamma\gamma$ component from $\bar{p}p$ annihilations is not significant).

If antimatter indeed exists in the universe it must be well separated from matter, otherwise the amount of annihilation radiation produced in the regions where matter and antimatter come into contact will be much larger than observed. Any massive antimatter body would be a very intense source of annihilation γ due to its motion within the interstellar (matter) medium. The absence of significant annihilation radiation implies that all observed galaxies contain only a single type of matter (matter or antimatter), and the observation of galaxy collisions or galaxies in the same cloud of intergalactic gas gives no evidence of annihilation radiation, allowing one to set stringent limits on the presence of antimatter in the universe: it is generally concluded that the nearest region rich in antimatter (if any) should be at least 10 Mpc away from us (1 parsec (pc) $\simeq$3.09 $\cdot$ 10^{16} m).

A strong bound was obtained by an analysis of the expected diffuse γ-ray spectrum from annihilations in the case of a matter-antimatter symmetric universe

(Cohen *et al.*, 1998): the spectrum in the ~ 1 MeV energy range would be affected by the red-shifted photons from π^0 produced in baryon-antibaryon annihilations at large distances. Measurements are performed by satellite-borne equipment detecting Compton scattered photons or $e^{\pm}$ pairs. Under mild assumptions, compatibility with measurements requires that matter and antimatter regions be separated from each other by distances of at least 1 Gpc, comparable to the observable part of the universe (~ 3 Gpc).

11.3.2 Evolution of the universe

A description of the cosmological picture which has emerged in the last decades and which satisfactorily describes several observed features of the evolution of the universe after the Big Bang (the Standard Model of cosmology, see e.g. Olive and Peacock (2006)) cannot be attempted here, but we briefly recall some relevant facts required for a better appreciation of the following discussion.

On a large scale the universe is assumed to be homogeneous and isotropic, as also indicated by the properties of the background radiation. The expansion of the universe, or the evolution of its characteristic scale $R(t)$ is described by the Hubble parameter $H \equiv \dot{R}/R$, whose value at present times is $H_0 \simeq 72$ km s^{-1} Mpc^{-1}; this can be obtained by measuring the red-shift (and thus the velocity) of distant objects (not affected by local deviations from the ideal Hubble expansion) whose distance is reliably known.

If the universe is dominated by non-relativistic particles the corresponding energy density ρ decreases as $\rho \propto R^{-3}$ because of the decreasing number density due to the expanding volume. For radiation (or relativistic particles with masses much below the thermal energy: $m \ll T$) an extra factor $1/R$ accounts for the red-shift decreasing the energy E (frequency) of each particle, so that $\rho \propto R^{-4}$.

In early epochs the universe was hotter and denser, and dominated by radiation; the rapid rate of particle interactions allowed thermodynamic equilibrium to be established. Indeed at any temperature T the universe contains a corresponding black-body distribution of photons (the equilibrium of massless particles is not disturbed by the expansion of the universe), which have sufficient energy to produce particles of mass well below T, together with their antiparticles, in large numbers; in this case such particles are in equilibrium with the radiation and share the same temperature.

In conditions of thermodynamic equilibrium, in which the universe has been for most of its history, the number densities in phase space for particle type i are given by

$$f_i^{(\mathrm{eq})}(\boldsymbol{x}, \boldsymbol{p}) = f_i^{(\mathrm{eq})}(\boldsymbol{p}) = f_i^{(\mathrm{eq})}(|\boldsymbol{p}|) = \frac{g_i}{e^{(E-\mu_i)/kT} \pm 1} \tag{11.4}$$

where the first two equalities express the homogeneity and isotropy of the universe; g_i counts the number of degrees of freedom for the particle type (e.g. two polarization states for a photon), $E = \sqrt{p^2 + m_i^2}$ is the particle energy, μ_i its chemical potential and T the absolute temperature; the plus sign is for bosons and the minus sign for fermions.

We recall that the chemical potential represents the internal energy change arising from a variation of the number of particles in a non-isolated system. In a relativistic system particle creation and annihilation is possible, and the number of particles can vary even in absence of any flux; this requires the chemical potential to be zero for (singly-created) bosons, and opposite for particles and antiparticles. A particle species is in *chemical equilibrium* (required for thermodynamical equilibrium) when the inelastic scattering processes which can change its number are balanced; in this case for a reaction $A + B \rightarrow C + D$ the chemical potentials are related by $\mu_A + \mu_B = \mu_C + \mu_D$.

In a relativistic regime $T \gg m_i, \mu_i$ in which particles are effectively massless their equilibrium number density is

$$n_i = g_i \int \frac{d\boldsymbol{p}}{(2\pi)^3} f_i^{(\mathrm{eq})}(|\boldsymbol{p}|) \propto T^3 \tag{11.5}$$

and the corresponding equilibrium energy density is $\rho \propto T^4$. In the non-relativistic limit ($T \ll m_i$) the equilibrium number density is instead

$$n_i \propto g_i (m_i T)^{3/2} e^{-(m_i - \mu_i)/T} \tag{11.6}$$

and since the corresponding energy density is exponentially smaller than in the relativistic case, the total energy density of the universe is well approximated by the part due to the relativistic species only.

From general relativity the expansion rate of the universe is linked to its average energy density ρ by

$$H = \sqrt{\frac{8\pi G_N \rho}{3}} = \sqrt{\frac{8\pi \rho}{3 M_P^2}} \tag{11.7}$$

For a radiation-dominated universe $\rho \propto R^{-4} \propto T^4$, so that the temperature diminishes as $T \propto 1/R$ and $H \sim T^2/M_P$. If the scale factor evolves in time as a power law $R(t) \propto t^\alpha$ (with constant α), then $H \propto 1/t$, and from the above $\alpha = 1/2$. As long as the interactions necessary for the adjustment of the particle distribution functions are rapid compared to the expansion rate, the universe will evolve through a succession of states very close to thermodynamic equilibrium; the explicit condition for this to happen is that such reactions responsible for the equilibrium have a rate $\Gamma \gtrsim H$, so they can quickly react by adjusting the distributions on the time scale $\sim 1/H$. If the rates of all interactions involving a particle species are instead much smaller than H then they cannot 'keep up' with the expansion of the universe to maintain thermodynamic equilibrium, and the species is effectively decoupled.

The present universe contains mainly radiation, mostly in the form of the Cosmic Microwave Background Radiation (CMBR): this is a relic of the time (about 10^5 years after the Big Bang) when the temperature dropped to a fraction of eV, and electrons and nuclei joined to form neutral atoms (recombination); around this time matter and radiation decoupled, not being any more in thermodynamic equilibrium. Nevertheless, the contribution to the energy density of the universe by non-relativistic matter is today some orders of magnitude greater than that due to relativistic particles. Baryonic matter (in the form of protons and neutrons), is known to account only for about 5% of the total energy density of the universe, with most of the rest being of unknown origin.

As the baryon number density evolves with the expansion of the universe, the excess of baryons over antibaryons is conveniently expressed by the dimensionless ratio

$$\eta = \frac{n_B - n_{\bar{B}}}{n_\gamma} \simeq \frac{n_B}{n_\gamma} \approx 6 \cdot 10^{-10} \tag{11.8}$$

where $n_B, n_{\bar{B}}$ are the number densities of baryons and antibaryons and $n_\gamma \simeq$ 411 cm^{-3} the density of photons, practically all in the CMBR, which is determined with high accuracy from the corresponding black-body spectrum at $T = 2.725$ K. The value of η is experimentally determined in two independent ways: from Big-Bang nucleosynthesis and from the angular distribution of the CMBR.

The synthesis of light elements ('Big-Bang nucleosynthesis') is a non-equilibrium process which occurred when the temperature of the universe dropped below the binding energies (2–30 MeV) of the first four light nuclei: ^{2}H (D), ^{3}H, ^{3}He, ^{4}He. About 1 s after the Big Bang the temperature was $T \sim 1$ MeV and the weak interactions converting neutrons to protons and vice versa froze out, leaving them in a ratio $n_n/n_p \sim \exp[-(m_n - m_p)/T] \sim 1/6$. In the following 3 minutes growing amounts of light nuclei formed, in amounts dictated by nuclear statistical equilibrium, eventually ending up mostly in ^{4}He (the nucleus with highest binding energy), with a $\sim$24% mass fraction, essentially determined by n_n/n_p. Coulomb barriers and the lack of stable nuclei with mass number 5 and 8 prevent significant nucleosynthesis beyond ^{4}He.

The abundance of ^{4}He in the universe grows because stars also produce it, so that the primordial abundance is inferred from measurements of its ratio to hydrogen in metal-poor[153] (low contamination from stars) regions of hot ionized gas in other galaxies.

---

[153] In astronomy all elements heavier than ^{4}He are 'metals'.

The primordial abundances of other light elements depend primarily on the baryon to photon ratio η, and thus allow a measurement of such quantity: the number densities of D and ^{3}H diminish according to the reaction rates for ^{4}He formation: $\Gamma \propto X_A n_B \langle \sigma v \rangle$, where X_A is the mass fraction of the nuclear species with mass number A, $n_B = n_p + n_n \simeq \eta \, n_\gamma$ the baryon density, and $\langle \sigma v \rangle$ the average product of reaction cross section times relative velocity. When these reactions also freeze out the amount of D and ^{3}He left, of order 10^{-5} with respect to hydrogen, decreases with η. Deuterium is destroyed by most astrophysical processes, and therefore its abundance has declined since the Big Bang; the D/H ratio is measured by absorption lines in distant hydrogen clouds against even more distant quasars: the Lyman-α ('forest') lines of hydrogen (shifted for D), are in the visible part of the spectrum for red-shifts $z \geq 3$, and can thus be observed from Earth. The information obtained from ^{3}He is less useful since details of the processes involving it are uncertain. ^{7}Li cannot be produced by reactions of free protons and neutrons on nuclei; yet a small amount of ^{7}Li is formed by fusion of lighter nuclei, in a fraction $\sim 10^{-10}$ of hydrogen: its primeval abundance is measured in the atmosphere of the oldest stars.

The experimentally allowed ranges for the primordial abundances of light elements (spanning 9 orders of magnitude) are compatible with each other only for a restricted range of η, thus constraining the value of this parameter on rather solid grounds: quantitative predictions are based on the values of nuclear cross sections at the relevant energies (which, contrary to the case of stellar nucleosynthesis, are experimentally known), and of the neutron lifetime.

The angular power spectrum of the CMBR is

$$C(\theta) = \left\langle \frac{\Delta T(\hat{m})}{T_0} \frac{\Delta T(\hat{n})}{T_0} \right\rangle$$

where $\Delta T(\hat{n})/T_0$ is the fractional temperature fluctuation in direction $\hat{n}$ with respect to the average temperature $T_0 \simeq 2.73$ K (with $\hat{m} \cdot \hat{n} = \cos\theta$) and the average is taken over all directions in the sky. This spectrum is conveniently expressed in terms of multipole moments, the coefficients in an expansion in Legendre polynomials.

The CMBR has a prominent (10^{-3}) dipole ($\ell = 1$) anisotropy, interpreted as due to the peculiar motion of the solar system which red-shifts half of the sky; other smaller structures in the power spectrum at the 10^{-5} level (up to index $\ell \sim 900$) are interpreted as due to the matter density fluctuations at the time of recombination: in the early universe such local fluctuations induce compressions of the primordial plasma, which are stopped by the rise in photon pressure causing rarefactions and so on, thus inducing acoustic oscillations. The compression of the plasma creates spots with higher temperature, and after decoupling these remain visible in the CMBR. Prominent peaks at $\ell \simeq 200, 500$ etc., corresponding to angular structures below $1°$ are correlated to cosmological parameters (Scott and Smoot, 2006), and in particular the ratio of the size of these 'acoustic' peaks provides a measurement of the baryon to photon ratio η, consistent with the value obtained from nucleosynthesis constraints (Jungman et al., 1996).

Note that the two measurements (from light element abundance and from CMBR power spectrum) relate to very different epochs in the evolution of the universe, and standard cosmology does not predict any change in η between the two. The smallness of η indicates that we live in a universe almost completely void of matter, with an average baryon density of 1 proton in 4 m^3 (most of them being invisible).

The quantity η is usually called the Baryon Asymmetry of the Universe (BAU), which corresponds to the matter-antimatter asymmetry at early times: when the temperature T of the universe was very high the baryons (or actually the quarks), with masses $m \ll T$ were highly relativistic and their density was comparable to that of photons

$$n_B^{(\text{then})} \sim n_{\bar{B}}^{(\text{then})} \sim n_\gamma^{(\text{then})} \propto T^3$$

As long as baryonic charge is conserved pair annihilation does not change $n_B - n_{\bar{B}}$, and it can somewhat increase n_γ but without changing its order of magnitude (the relatively few photons emitted by stars likewise do not alter the argument), so that the number densities just decrease because of the increase in the scale factor R:

$$[(n_B - n_{\bar{B}})R^3]^{(\text{then})} = [(n_B - n_{\bar{B}})R^3]^{(\text{now})} \simeq [n_B R^3]^{(\text{now})}$$

$$\text{and similarly} \quad [n_\gamma R^3]^{(\text{then})} = [n_\gamma R^3]^{(\text{now})}$$

$$\text{so that} \quad \left(\frac{n_B - n_{\bar{B}}}{n_B}\right)^{(\text{then})} \simeq \left(\frac{n_B - n_{\bar{B}}}{n_\gamma}\right)^{(\text{now})} \simeq \left(\frac{n_B}{n_\gamma}\right)^{(\text{now})} = \eta$$

This means that while today the baryon asymmetry is relatively large (as no antibaryons are found), in the early universe it was very small: there was an excess of a single baryon for each $\sim 10^9$ baryon-antibaryon pairs; since then essentially all the pairs annihilated, leaving a fraction of order η of baryons to build the universe of today. The tiny value of this asymmetry makes it indeed quite puzzling.

11.3.3   A symmetric universe?

The universe, at least in the neighbourhood that is accessible to our investigation, appears to be highly charge asymmetric: only particles appear while antiparticles are practically absent. It is hard to obtain information on the matter-antimatter asymmetry at scales larger than galaxy clusters by observation, because of ignorance about the properties of the medium between them, but the possibility that the universe consists of 'patches' of matter and antimatter above such scales presents other difficulties.

One could make the hypothesis that maybe matter and antimatter were actually formed in equal amounts and then somehow separated from each other, so that no annihilation should be observed. No viable mechanism capable of providing such a separation has been found, though. A hypothetical repulsive gravitational interaction between antiparticles and particles was sometimes considered as a possible mechanism for separating matter and antimatter on a cosmological scale in a symmetric universe; as discussed in Section 11.1, this possibility would imply a violation of the equivalence principle.

The measured uniformity of the CMBR (within $\sim 10^{-5}$) also supports the smallness of the antimatter density: at the time matter and radiation decoupled domains containing antimatter must have been separated from those containing matter by regions of empty space which today cannot be larger than 15 Mpc, not to contradict experimental observations; on the other hand the diffusion of baryons would have canceled empty regions smaller than that. Matter and antimatter domains, if they existed, should therefore be in contact, and thus annihilation must be present.

In a universe which was at an early time matter-antimatter symmetric the expansion and cooling would have led to almost complete annihilation: the amount of matter and antimatter left can be estimated from the knowledge of the proton-antiproton annihilation cross section.

At temperatures below ~ 1 GeV, when protons become non-relativistic, their equilibrium abundance ratio to photons is obtained from eqn (11.6) as

$$\frac{n_B}{n_\gamma} \simeq \frac{n_{\overline{B}}}{n_\gamma} \simeq \left(\frac{m_p}{T}\right)^{3/2} e^{-m_p/T} \tag{11.9}$$

(m_p is the proton mass) and as long as their annihilation rate Γ_a is larger than the expansion rate of the universe H the number of nucleons and antinucleons continues to decrease; at $T \sim 20$ MeV the annihilation rate (proportional to nucleon density) becomes $\Gamma_a \simeq H$, and 'freezes out' as nucleons and antinucleons become so rare that they cannot annihilate any more (being difficult for an antinucleon to find a nucleon to annihilate with). At this time however $n_B/n_\gamma \simeq n_{\overline{B}}/n_\gamma \simeq 10^{-18}$, much smaller than η which is required to explain the subsequent nucleosynthesis (the 'annihilation catastrophe') (Chiu, 1966).

One might assume that some mechanism separated matter and antimatter before $T \sim 40$ MeV ($t \simeq 10^{-3}$ s), when $n_B/n_\gamma \sim 10^{-10}$; at that time, however, the size of the causally connected regions was small and contained only an amount of matter corresponding to $\sim 10^{-7}$ solar masses, and one would thus expect regions with only one kind of matter of such size, much smaller than observed.

Another attempt to explain the observed asymmetry within a baryon symmetric universe, which also fails, invokes just the statistical fluctuations in the baryon and antibaryon densities (Alpher *et al.*, 1953): the comoving volume V encompassing our galaxy contains today about 10^{79} photons and 10^{69} baryons, but at times when

$T \gtrsim 1$ GeV it also contained $\sim 10^{79}$ baryons and antibaryons on average. One could therefore expect from (Poisson) statistical fluctuations that $(n_B - n_{\bar{B}})/n_B \sim 1/\sqrt{n_B V} \sim 10^{-79/2}$, again much smaller than the observed η.

Even if the laws of Nature were *exactly* symmetric between matter and antimatter (no C or CP violation, which is quite a good approximation in any case), and the preponderance of matter were not just a local phenomenon on some (large) scale, one might perhaps adopt the position that the observed asymmetry between matter and antimatter was just a result of arbitrary 'initial conditions' in the universe, *ex nihilo*. While this possibility might not appear 'philosophically' appealing, as it seems to violate the 'principle' according to which any effect should be at least as symmetric as its cause (Curie, 1894), Nature is definitely not bound to satisfy our innate desire for symmetry[154] or our expectations. In this sense the matter-antimatter asymmetry of the universe would just be a fact, an 'external' initial condition not calling for any deeper explanation: by starting with asymmetric initial conditions, involving more matter than antimatter at the very beginning of the universe (a time which is clearly beyond our present understanding), symmetric laws of Nature would conserve such asymmetry forever.

Remarkably enough, recent developments in cosmology appear to make the above position untenable, leading to the much more interesting situation in which, irrespective of the 'initial conditions', a dynamical mechanism is required to generate the matter-antimatter asymmetry which we seem to observe. The reason is that it is now believed that the universe underwent a period of 'inflation', during which its size increased exponentially. While this is not proved beyond any doubt, there are very strong indications that it was indeed the case: inflationary theory is the only known mechanism to explain simultaneously several long-standing cosmological problems concerning the observed flatness, homogeneity and isotropy of the universe, as well as being consistent with the measured features of the background radiation. A period of inflation, during which the scale factor $R(t)$ of the universe increased by $\approx e^{70}$ is not compatible with the conservation of baryonic charge (see e.g. Dolgov (1992)): eqn (11.7) shows that exponential inflation $R(t) = \exp(Ht)$ is obtained if the energy density ρ is constant in time. As the energy density of any conserved (baryonic) charge scales as $\rho \sim 1/R^3 \sim \exp(-3Ht)$ for non-relativistic matter (and $\rho \sim 1/R^4 \sim \exp(-4Ht)$ for relativistic matter), given the present value of η in eqn (11.8), going backward in time during the inflation period by $Ht \sim 4 \div 5 \ll 70$ the baryon energy density becomes the dominating one in the past, and being not at all constant it is not compatible with an exponential inflation.

The above means that any initial net baryon number present before inflation would be diluted to a negligible level by such process, returning to the case of

---

[154] Actually an interesting subject of investigation for evolutionary biology.

symmetric 'initial' conditions: generation of baryonic asymmetry must take place after inflation.

11.3.4 Generation of the asymmetry

In 1967 Andrei Sakharov wrote a paper (Sakharov, 1967) which suggested that the baryon density of the universe might be understandable in terms of physical laws, and indicated the three requirements which must be simultaneously present in order for a baryon asymmetry to evolve from an initial symmetric state:

- **Baryon number violation** must be present in some process. This is rather obvious, as otherwise any baryon asymmetry could only be introduced as an initial condition, and a state of initial zero baryon number (which modern cosmology suggests) would always remain symmetric. Baryon number conservation indeed never appeared to have a deeper explanation, in contrast to e.g. electric charge conservation which is connected with the gauge symmetry of electrodynamics and the fact that the photon is massless, and which manifests itself in the presence of the long-range electromagnetic force. No corresponding long-range force coupled to baryon number has ever been detected, and the equality of gravitational and inertial mass, experimentally verified to high precision for materials with different baryon-to-mass ratios, poses strong constraints on any such force (Lee and Yang, 1955). There are thus no strong theoretical grounds for a global symmetry such as baryon (and lepton) number conservation. Still, to date there is no direct evidence of the violation of baryon number conservation, such as proton decay (lifetime $> 10^{29}$ years) or neutron-antineutron oscillations (period $> 10^8$ s) (Yao *et al.*, 2006).

  At the time Sakharov wrote his article this point was highly speculative, and was probably one of the reasons why the work did not receive much attention initially; the situation is quite different today[155] and several mechanisms for baryon number violation are conjectured.

  Note that baryon number violation could also originate from a non-trivial topology of space-time, even if it was not present in the Lagrangian: since baryon number is not linked to any long-range field, there is no way of knowing the baryonic charge contained inside a black hole (contrary to what happens for its mass and electric charge); as black holes can evaporate into charge symmetric radiation, matter can actually disappear into them leaving no trace of its baryonic charge.

---

[155] The very existence of our matter universe was thought to provide evidence of the stability of baryonic matter: 'We exist, therefore the proton must be stable'. Today our conclusions are exactly opposite, as according to Sakharov's conditions the observed matter-antimatter asymmetry requires the violation of baryon number conservation: 'We exist, therefore the proton must be unstable'.

▪ **Violation of C and CP symmetries** must be present in Nature. If this were not the case then any reaction (maybe violating baryon number conservation) producing a particle would have as a counterpart a reaction producing the charge-conjugate particle, occurring at exactly the same rate, so that no net baryon number would be generated. We recall that CPT symmetry alone does not guarantee the rate equality for charge-conjugate processes (Chapter 6), as enforced by C symmetry. Note also that if C symmetry is violated but CP holds, the decay widths of particles and antiparticles with different helicities λ are related by

$$\Gamma[i(\lambda) \to f] = \Gamma[\bar{i}(-\lambda) \to \bar{f}] \tag{11.10}$$

and by summing over helicities the equality of decay widths results: both C and CP must be broken to have a decay width difference.

This ingredient was known to be present in Nature after the discovery of CP violation in weak interactions, although today the actual mechanism at play is believed to be most likely a different one.

▪ **Departure from thermodynamic equilibrium** is required to define an arrow of time and obtain a particle-antiparticle asymmetry. Recalling the form of particle distribution functions in thermodynamic equilibrium (11.4), CPT symmetry requires particles and antiparticles to have the same masses, so that E in such distributions is also the same for them (for a given $\boldsymbol{p}$); since in thermodynamic equilibrium the chemical potential for a non-conserved charge must be zero the particle and antiparticle number densities in space are just equal: $n_i = n_{\bar{i}}$ in this case, and no BAU is possible.

In other words, baryon number B is a C-odd (and CP-odd) quantity

$$P^\dagger B(t) P = B(t) \qquad C^\dagger B(t) C = -B(t) \qquad T^\dagger B(t) T = B(-t)$$

as can be verified from its explicit expression in terms of quark fields $q(t, \boldsymbol{x})$

$$B = \frac{1}{3} \sum_q \int d\boldsymbol{x} : q^\dagger(t, \boldsymbol{x}) q(t, \boldsymbol{x}) :$$

and the transformation rules written in Chapters 2, 3, 4. The above implies

$$(CPT)^\dagger B(0)(CPT) = -B(0) \tag{11.11}$$

The average value of B in thermodynamic equilibrium at temperature T is obtained from the density matrix operator $\exp(-H/T)$ (where H is the Hamiltonian) as

$$\langle B(t) \rangle_T = \mathrm{Tr}\left[e^{-H/T} e^{iHt} B(0) e^{-iHt} \right] = \langle B(0) \rangle_T$$

but from (11.11)

$$\langle B(0)\rangle_T = \text{Tr}\left[(CPT)(CPT)^\dagger e^{-H/T} B(0)\right]$$

$$= \text{Tr}\left[(CPT)^\dagger e^{-H/T} B(0)(CPT)\right] = -\langle B(0)\rangle_T$$

and the baryon number must be zero.

Note that CPT symmetry was assumed to hold in the above, and if CP is violated this implies that time reversal symmetry is also violated, so that detailed balance (Chapter 4) cannot be invoked to justify the above result by an inverse reaction having the same rate as the direct one. While detailed balance is the assumption normally used to derive the form of the equilibrium distributions (11.4), it can be shown that the unitarity of the S matrix is sufficient to obtain it. If CPT symmetry is violated an asymmetry between particle and antiparticle densities can also arise in thermodynamic equilibrium, but conservation of probability is enough to ensure that the equilibrium distributions still have the form of (11.4) (Dolgov, 2006).

We note in passing that the above properties of B also prove explicitly that C (and CP) violation are required to generate a BAU: a charge-symmetric universe U is an eigenstate of C

$$C|U\rangle = \eta_C(U)|U\rangle$$

and if C is conserved $[H, C] = 0$, from which

$$\langle U|B(t)|U\rangle = \langle U|e^{iHt} B(0)e^{-iHt}|U\rangle$$

$$= \langle U|e^{iHt} CC^\dagger B(0)CC^\dagger e^{-iHt}|U\rangle = -\langle U|B(t)|U\rangle = 0$$

Note that, conversely, in presence of baryon number violating interactions, any baryon asymmetry gets canceled in thermodynamic equilibrium, so that when a BAU has been generated and equilibrium is established again, if such interactions remain effective the asymmetry gets washed out to zero.

Whatever the mechanism for generating a baryon asymmetry from initially symmetric conditions, all three of the Sakharov conditions have to be *simultaneously* satisfied at some time.[156]

11.3.5 Mechanisms for baryogenesis

Many possible ways of generating the baryon asymmetry ('baryogenesis') of the universe have been proposed, and this is indeed a very active field of research: a

---

[156] This is not strictly true, and rather exotic (weird) models exist in which a BAU can be generated with some of the above conditions not being fulfilled; see (Dolgov, 1992) for a review.

discussion of those would be beyond the scope of this book, and we just briefly mention some of the important possibilities.

Grand-Unified Theories (GUT), in which the strong, weak, and electromagnetic gauge interactions are unified into a single gauge group with a single coupling constant at high energy, violate baryon number conservation and easily accommodate CP-violating phases (indeed they should at least contain the CKM phase). Such theories involve very heavy particles of mass $m_X \sim 10^{16}$ GeV, whose decays provide a straightforward mechanism for departure from thermodynamic equilibrium.

At early times, with high temperatures $T \gg m_X$, the X particles (and their antiparticles) are effectively relativistic and as abundant as photons, eqn (11.5). As the temperature of the universe drops below m_X these particles become non-relativistic and their abundance, eqn (11.6), must decrease rapidly with T to remain in thermodynamic equilibrium; whether they manage to do so depends on their decay rate (other processes, such as pair annihilation, are of higher order in the coupling constant and thus less important), which for such particles can be estimated on dimensional grounds as $\Gamma_X \sim \alpha_X m_X$, where $\alpha_X \sim 10^{-2} \div 10^{-1}$ is a coupling constant (note that for $T \lesssim m_X$ inverse decay and annihilation are less important than decay).

The parameter determining the effectiveness of the decays in maintaining equilibrium is thus the ratio of the expansion and decay rates at the temperature corresponding to m_X:

$$\frac{H(T \simeq m_X)}{\Gamma_X} \sim \frac{m_X}{\alpha_X M_P} \tag{11.12}$$

(indeed as the temperature drops below m_X the deviation from equilibrium can be larger, but the number of X particles becomes exponentially suppressed, so that they do not play an important role). For very heavy particles (or very lightly coupled ones) the above quantity can be larger than 1, so that the X and $\overline{X}$ particles cannot 'keep up' with the expansion and remain as abundant as photons for $T \lesssim m_X$, out of equilibrium. Numerically it is found that this can indeed occur for $m_X \gtrsim 10^{16}$ GeV, as expected in GUTs (corresponding to very early times $t \lesssim 10^{-34}$ s).

An explicit example helps to clarify the features of GUT baryogenesis (Nanopoulos and Weinberg, 1979). A heavy particle X violates baryon number conservation, as it can decay into two channels with different baryon number: $X \to f_1$ (baryon number B_1) and $X \to f_2$ (baryon number $B_2 \neq B_1$); the two states could be e.g. $f_1 = qq$ ($B_1 = +2/3$) and $f_2 = q\bar{\ell}$ ($B_2 = +1/3$), where q is a quark and ℓ a lepton. The corresponding antiparticle $\overline{X}$ decays to $\overline{X} \to \bar{f}_1$ and $\overline{X} \to \bar{f}_2$ (baryon number $-B_1$ and $-B_2$). CPT symmetry requires the total decay widths of X and $\overline{X}$ to be equal, but since (at least) two channels are present, if CP (and C) symmetry is violated the partial decay widths for charge-conjugate channels can be different. This can only occur at higher orders in perturbation theory, when intermediate states are present which can introduce re-scattering phases (final state interactions, see Chapter 6). One has therefore

$$\Gamma(X \to f_1) = (1 + \Delta_1)\Gamma_1 \qquad \Gamma(\overline{X} \to \overline{f}_1) = (1 - \Delta_1)\Gamma_1 \qquad (11.13)$$

$$\Gamma(X \to f_2) = (1 - \Delta_2)\Gamma_2 \qquad \Gamma(\overline{X} \to \overline{f}_2) = (1 + \Delta_2)\Gamma_2 \qquad (11.14)$$

Assuming no other decay modes are present, CPT symmetry requires

$$\Delta_1\Gamma_1 = \Delta_2\Gamma_2 \qquad (11.15)$$

Note that the time-reversed reactions, such as $f_1 \to X$, are not those responsible for the fact that in thermodynamic equilibrium no net asymmetry is generated: T symmetry is violated, while the CPT transformation gives

$$\Gamma(\overline{f}_1 \to \overline{X}) = (1 + \Delta_1)\Gamma_1 \qquad \Gamma(f_1 \to X) = (1 - \Delta_1)\Gamma_1 \qquad (11.16)$$

$$\Gamma(\overline{f}_2 \to \overline{X}) = (1 - \Delta_2)\Gamma_2 \qquad \Gamma(f_2 \to X) = (1 + \Delta_2)\Gamma_2 \qquad (11.17)$$

so that the direct and inverse reactions actually give the same sign for the asymmetry. The reactions which (in thermodynamic equilibrium) compensate the ones above are actually baryon number non-conserving inelastic scattering $f_{1,2} \leftrightarrow f_{1,2}$, with exchange of X and $\overline{X}$.

The average baryon number produced in X and $\overline{X}$ decays are respectively

$$B_1(1 + \Delta_1) \frac{\Gamma_1}{\Gamma_X} + B_2(1 - \Delta_2) \frac{\Gamma_2}{\Gamma_X}$$

$$-B_1(1 - \Delta_1) \frac{\Gamma_1}{\Gamma_X} - B_2(1 + \Delta_2) \frac{\Gamma_2}{\Gamma_X}$$

The average net baryon number is therefore

$$\Delta B = 2(B_1 - B_2) \Delta_1 \frac{\Gamma_1}{\Gamma_X} \qquad (11.18)$$

using eqn (11.15). This vanishes if baryon number is conserved ($B_1 = B_2$, which would be taken as the baryon number of X) or if C (or CP) symmetry holds ($\Delta_1 = 0$).[157]

When the universe becomes as old as the lifetime of the X particles, at time $t \sim 1/\Gamma_X$ these start to decay; this happens when $H \sim 1/t \sim \Gamma_X$, at a temperature $T_D < m_X$, as the condition for having out-of-equilibrium decay is $H(T = m_X) > \Gamma_X$. At temperatures above T_D the density of the X particles is $n_X \sim n_{\overline{X}} \sim n_\gamma$, so that the net baryon number density produced is $n_B - n_{\overline{B}} = \Delta B n_X$, and the quantity ΔB is equal to η of eqn (11.8). Note that below T_D both inverse decays and baryon

---

[157] If C symmetry were violated but CP held, then the rate of the decay $X \to f_1$ with some particle being emitted in direction $\hat{n}$ would be equal to that of $\overline{X} \to \overline{f}_1$ with the corresponding antiparticle being emitted in direction $-\hat{n}$, and still $\Delta_1 = 0$.

number violating scatterings can be ignored, so that the above net baryon number is not cancelled.

It is instructive to consider further the conditions which have to be satisfied in this kind of model for an asymmetry to arise (Harvey *et al.*, 1982). It was noted that in order to avoid the *CPT* constraint on the decay rates at least two decay modes must be present for the heavy particle X (the final states $f_{1,2}$ above), now explicitly written as $X \rightarrow \bar{a}_1 b_1$ and $X \rightarrow \bar{a}_2 b_2$, see Fig. 11.1 (a,b). It should be recalled that at tree level no *CP* asymmetry can be obtained, as re-scattering phases are necessary: one has to consider then the one-loop diagram with the exchange of another boson Y with decay modes $Y \rightarrow \bar{a}_2 a_1$ and $Y \rightarrow \bar{b}_2 b_1$ (Fig. 11.1 (c,d)), which can provide an interfering diagram.

The Lagrangian describing this system is

$$\mathcal{L} = g_1 X b_1^\dagger a_1 + g_2 X b_2^\dagger a_2 + h_1 Y a_1^\dagger a_2 + h_2 Y b_1^\dagger b_2 + \text{H.c.}$$

(where the particles stand for the field operators and $g_{1,2}, h_{1,2}$ are coupling constants), and at two-loops order the decay rate for $X \rightarrow \bar{a}_1 b_1$ is

$$\Gamma(X \rightarrow \bar{a}_1 b_1) = |g_1|^2 F_X + (g_1 g_2^* h_1 h_2^*) F_{XY} + [(g_1 g_2^* h_1 h_2^*) F_{XY}]^*$$

where the first term is the contribution of the tree diagram (Fig. 11.1 a) and the other two are the interference terms with the Y-exchange diagram (Fig. 11.1 (e)); here F_X and F_{XY} are kinematic factors (phase space integrals, etc.): F_{XY} can be complex if the intermediate particles in the loop can be on mass shell, which requires $m_X > m(a_2) + m(b_2)$. One can work out the corresponding expression for the conjugate decay $\overline{X} \rightarrow a_1 \bar{b}_1$, and the difference between the two is

$$\Gamma(X \rightarrow \bar{a}_1 b_1) - \Gamma(\overline{X} \rightarrow a_1 \bar{b}_1) = 4 \operatorname{Im}(F_{XY}) \operatorname{Im}(g_1 g_2^* h_1 h_2^*)$$

The diagrams with an X boson exchange (Fig. 11.1 (g,h)) do not contribute, as the product of coupling constants is real in this case.

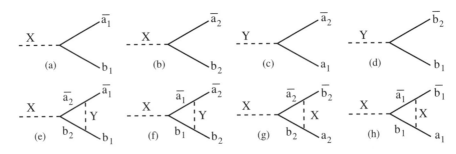

FIG. 11.1. Some diagrams in a simple baryogenesis model involving two heavy bosons X, Y (see text). (a,b): tree diagrams for X decays; (c,d): tree diagrams for Y decay; (e,f): 1-loop diagrams for X decay with Y exchange and (g,h): X exchange.

Considering also the other decay mode of X, and weighting the rate asymmetries with the baryon number of the respective final states, the net baryon number produced in the decay of X and $\overline{X}$ is found to be

$$\Delta B = \frac{4}{\Gamma_X} \operatorname{Im}(F_{XY}) \operatorname{Im}(g_1 g_2^* h_1 h_2^*) \left[[B(b_2) - B(a_2)] - [B(b_1) - B(a_1)] \right]$$

showing that the particle exchanged in the loop must also violate baryon number conservation (Nanopoulos and Weinberg, 1979). This requirement is actually a general one, which follows from the discussion in Chapter 5: when summing over a set of states which is complete for what concerns the interaction responsible for re-scattering, CPT symmetry guarantees the equality of the rates, and if this interaction conserves baryon number the ensemble I_B of states with baryon number B is indeed such a set. Explicitly, at one-loop order the decay amplitude can be written as

$$\mathcal{M}(i \to f) = \sum_n \mathcal{M}_0(i \to n) \mathcal{M}_1(n \to f)$$

where $\mathcal{M}_0$ is a tree (CP-conserving) amplitude violating baryon number conservation and $\mathcal{M}_1$ the re-scattering correction, and a complete sum over intermediate states has been introduced; if $\mathcal{M}_1$ conserves baryon number then unitarity implies

$$\sum_{f \in I_B} \mathcal{M}_1(n \to f) \mathcal{M}_1^*(f \to n') = \delta_{nn'}$$

for the sum over all states with baryon number B, and this implies

$$\sum_{f \in I_B} |\mathcal{M}(i \to f)|^2 = \left| \sum_{n \in I_B} \mathcal{M}_0(i \to n) \right|^2$$

$$= \left| \sum_{\bar{n} \in I_{\bar{B}}} \mathcal{M}_0(\bar{i} \to \bar{n}) \right|^2 = \sum_{\bar{f} \in I_{\bar{B}}} |\mathcal{M}(\bar{i} \to \bar{f})|^2 \qquad (11.19)$$

where the CP symmetry of $\mathcal{M}_0$ has been used (Kolb and Wolfram, 1980).

Finally it should be noted that in the above model if the X, Y bosons have the same mass, then the asymmetry produced in the decay of X (and $\overline{X}$) is exactly cancelled by the one produced in the decay of Y (and $\overline{Y}$).

The kind of mechanism for baryogenesis discussed above was the first one considered, but is now disfavoured because it appears that after inflation (which erased any pre-existing BAU) the universe never reached temperatures corresponding to the mass scales $T \sim m_X$ of grand-unified theories.

Remarkably enough, in principle all three of the Sakharov requirements could be present within the SM, leading to the model of electro-weak baryogenesis. The Standard Model actually includes baryon number violating processes due to a quantum anomaly ('t Hooft, 1976). SM baryon number violating processes are non-perturbative and extremely suppressed in normal conditions, occurring with probabilities of order $\exp(-4\pi/\alpha_W) \sim 10^{-150}$ (α_W is the weak gauge coupling constant): with such rates these processes would never have occurred in the entire history of the universe. In the extreme conditions of the early universe, however, such processes can be dramatically enhanced by coherent effects which involve the so-called 'sphaleron' (for 'ready to fall') solutions of the equations of motion (Klinkhamer and Manton, 1984) corresponding to quantum tunnelling transitions between two topologically distinct vacua with different baryon numbers. Such processes indeed become significant at temperatures close to the W boson mass ($T_{EW} \sim 100$ GeV), which the universe had at the time of the electro-weak phase transition ($t \sim 10^{-9}$ s), when it become energetically favourable for the Higgs scalar field to acquire a non-zero expectation value.

The second Sakharov condition is also satisfied in principle, because CP violation is present in the SM.

Finally, while the W and Z boson masses are far too small to provide significant deviations from thermodynamic equilibrium according to eqn (11.12), such a departure could be present if the electro-weak phase transition were of first order.[158] A first-order phase transition is not an adiabatic process: as the temperature decreases super-cooling can occur, and the transition proceeds through the formation of growing bubbles, inside which the new phase (the true vacuum, in which EW symmetry is broken) is present: the bubble walls are regions in which the Higgs fields are changing and conditions there are far from thermodynamic equilibrium, so that baryogenesis could in principle occur (Shaposhnikov, 1987). One problem with this model is that after the bubble wall has passed a region, all baryon number violating processes should stop, otherwise the asymmetry which was produced would be washed away (as thermodynamic equilibrium is now present): this in turn can be shown to require the Higgs mass to be relatively small ($m_H \lesssim 70$ GeV); the present experimental lower limits for this mass from direct searches at accelerators rule out the possibility of electro-weak baryogenesis within the context of the SM. Similar difficulties arise also in minimal extensions of the SM.

Moreover, it has been shown that the amount of CP violation provided by the CKM matrix cannot generate the required size of the baryon asymmetry, because its contribution to baryon number violation only occurs at the three-loop level (Barr *et al.*, 1979). We recall from Chapter 9 that the amount of CP violation in

---

[158] A phase transition of order n is one in which the n-th derivative of a thermodynamic potential is discontinuous, while all derivatives of order $n - 1$ are continuous; in a first-order phase transition two phases can coexist at the critical point.

the CKM model is proportional to the quantity in eqn (9.27) involving the mass differences between all down and up quarks, and the order of magnitude of η which can be obtained with this mechanism is given by such quantity divided by T_c^{12}, where $T_c \sim O(100 \text{ GeV})$ is the critical temperature for the EW phase transition; this results in $\eta \sim O(10^{-19})$, much smaller than required. This means that other sources of CP violation *must* be present beyond the SM.

Even if electro-weak baryogenesis within the SM seems not to work, the presence of SM processes violating baryon number has some important consequences: these processes actually violate both baryon and lepton number, but conserve their difference $B - L$. The rate of the processes which violate the sum $B + L$ is much larger than the expansion rate of the universe H for a large range of temperatures, between T_{EW} and $\sim 10^{12}$ GeV; such processes tend to reduce $B + L$ to its equilibrium value, which is zero unless some of the conserved charges are non vanishing. This actually poses an important constraint on any baryogenesis mechanism operating at temperatures above T_{EW}: the baryon number is $B = (B+L)/2 + (B-L)/2$, and SM processes can erase $B + L$ but conserve $B - L$; if the processes which generate B conserve $B - L$ (so that this quantity remains zero), then any baryon or lepton number excess generated will be washed out by SM processes as soon as the temperature drops below $\sim 10^{12}$ GeV (Kuzmin *et al.*, 1985).

The origin of the BAU might have to be searched for in the lepton sector: in theories in which heavy ($M \sim 10^{10}$ GeV) Majorana neutrinos appear, their out-of-equilibrium decays can establish a lepton asymmetry in the universe at temperatures $T \sim M$, with the same mechanism described above for GUTs. The fact that SM processes do not change $B - L$ but erase $B + L$ allows such leptonic asymmetry to be translated into a baryon asymmetry $B \simeq -L$ (Fukugita and Yanagida, 1986).

Another appealing model of baryogenesis involves the coherent motion of condensates of scalar fields (Affleck and Dine, 1985), such as the hypothetical scalar partners of quarks and leptons which are predicted by super-symmetry.

It should be noted that the sign of the baryon asymmetry generated depends on the sign of the CP asymmetry: this implies that in a scenario in which CP symmetry is broken spontaneously (Section 9.5), causally disconnected regions would be dominated by either matter or antimatter, randomly. In three dimensions, for a distribution of cells with the value of an attribute chosen randomly between two equiprobable values, results from percolation theory show that almost all cells are part of infinite connected domains, so that large regions of matter and antimatter are expected in general. This is a potential problem, as it would result in large energy densities at the boundary surfaces between those, which would have destroyed the observed homogeneity of the universe (Kobsarev *et al.*, 1974), unless appropriate mechanisms are devised to destroy those boundaries at early times.

Other possible mechanisms for baryogenesis can be found in the literature (see e.g. Dolgov (2006)).

11.4   Mirror matter

Perhaps an 'explanation' of the left-right asymmetry of nature might be looked for in the anthropic principle (Foot and Silagadze, 2005): as fundamental ingredients in the stellar cycle driven by weak interactions, neutrinos play a crucial role in allowing our existence, but if they had right-handed partners one could expect them to acquire Dirac masses comparable to those of charged leptons, and in this case their mass density would overclose the universe.

As discussed in Chapter 6, left-right symmetry cannot be restored with the aid of charge conjugation, and it seems the only way to maintain it would be to further double the spectrum of particles (besides the doubling corresponding to the existence of antiparticles), by introducing a 'mirror' partner for each type of ordinary particle, which is identical to it but with right-handed weak interactions (Foot *et al.*, 1991). The violation of parity symmetry would then be just apparent and due to the fact that we failed so far to observe the mirror particles, exhibiting an opposite symmetry violation. With this hypothesis it could be assumed that the operator corresponding to coordinate inversion $x \rightarrow -x$ is actually MP (rather than CP), where M is the operator which transforms ordinary particles in mirror particles and vice versa: such an operator could then be conserved, even if P is not.

The apparent left-right asymmetry of the laws of Nature could thus be interpreted as just due to a local preponderance of ordinary matter in our environment, while mirror matter with the opposite asymmetry existed elsewhere. If such mirror matter did not interact at all with ordinary matter, thus remaining undetectable, this possibility would lie outside the domain of science proper. Ordinary matter and mirror matter should at least interact through gravity, however, and possibly also (weakly) in other ways, opening up a possibility for detection; thinking to such a possible weak interaction, mirror matter was suggested as a candidate for the dark matter of the universe, which supposedly comprises a large part of its matter density. Particles common to both worlds should be neutral, but mirror photons would also exist, so that electromagnetic interactions between the particles belonging to different (normal and mirror) worlds would either be absent or at least highly suppressed (Kobzarev *et al.*, 1966). If particles existed which interact strongly with both kinds of matter, then a mixing between photons and mirror photons would give small (fractional) electric charges to mirror electric charges (Glashow, 1986).

The idea of the possible existence of 'mirror particles' was originally put forward by Lee and Yang in their 1956 paper on parity symmetry (Lee and Yang, 1956c), and resurfaces periodically in the literature as a possible explanation for several puzzles: oscillations into a mirror neutrino were proposed, for example, to explain neutrino anomalies, and mirror supernovæ were claimed to be viable candidates for the positron source at the centre of our galaxy (see Section 11.3.1). However it is fair to say that, although appealing, the somewhat *ad hoc* idea of mirror matter has not received significant support from experimental observations, so far. Perhaps for the left-right asymmetry, as R. Feynman remarked: 'The paradox is only a conflict between reality and your feeling what reality ought to be'.

11.5 Macroscopic chirality

Several complex molecules exist in two chiral versions, called *enantiomers*. As discovered by L. Pasteur in 1848, chiral molecules often exhibit optical activity (of opposite sign) and are thus also called optical isomers. The properties of enantiomers for ordinary chemical reactions are identical, but their interactions with other chiral molecules can differ: these include for example smell differences (right-handed limonene smells like orange, while the left-handed one smells like lemon) and the selective action of penicillin.[159]

Most organic molecules are chiral, and this can be linked to the tetravalent chemical nature of the carbon atom: when four different groups are attached to it a chiral structure emerges (in three dimensions a triangle with numbered vertices can always be superimposed to its mirror image by a rotation, while a tetrahedron cannot). A striking fact is that most biological molecules share the same chirality: only the left-handed version of amino-acids[160] and the right-handed version of sugars are involved in life; all biological processes maintain such chirality, and in particular RNA replication cannot work on achiral mixtures (which indicates that the development of biochirality cannot originate in natural selection).

Any link to the fundamental chirality of weak interactions is so far speculative: tiny energy differences between enantiomers, linked to the small chiral contribution of the neutral weak current on top of the electromagnetic interaction which controls chemistry could lead to a chiral excess of about 1 in 10^{17}, but it is not at all clear whether this can be amplified to have a significant effect, or whether other properties of the environment in which life originated, such as light polarization, could have had an influence (Bailey, 2000).

The human body exhibits a clear chirality in the disposition of the internal organs, but this does not appear to be related to any functional reason; indeed about 1 out of 8,500 people have *situs inversus*, a complete reversal of the left-right pattern, which does not give any problem to the individual (and incidentally does not change the probability of being left-handed). The origin of such macroscopic chirality is not entirely clear yet, but has been traced to microscopic chiral structures, and might eventually be linked to the chirality of biological molecules. It is thus tempting to think that the origin of all biological handedness, and ultimately of the

---

[159] Another dramatic case is that of thalidomide, prescribed to pregnant women in the 1960s as a sedative: it was soon discovered that only one enantiomer was effective for that purpose, while the other one caused child malformations at birth: the synthesized compound was a 50%–50% (*racemic*) mixture.

[160] Of the twenty amino-acids of life only one (glycine) is non-chiral.

design of spiral staircases in medieval castles and the choice of driving side on roads, lies in microscopic physics processes, although no proof exists at this time.

Further reading

Brief and up-to-date introductions to some of the topics discussed here can be found in the short review articles in (Yao *et al.*, 2006), which also provide references to more extensive treatments.

Many reviews exist on the strong *CP* problem and axion searches; an introductory one is Dine (2000), in which other references can be found.

An excellent introduction to anomalous symmetries and baryon number violation is Arnold (1991). Many excellent reviews of baryogenesis mechanisms exist: a recent one is Dine and Kusenko (2004), while more details can be found e.g. in Riotto (1998). The evolution of the early universe is described in Kolb and Turner (1990); for an introduction to nucleosynthesis see e.g. Schramm and Turner (1998), and for an alternative account of the early history of the universe see Tolkien (1977).

Much information on the issue of the origin of the chiral asymmetry in living systems can be found in Cline (1997).

CODA

*The most important fundamental laws and facts
of physical science have all been discovered,
and these are now so firmly established
that the possibility of their ever being supplemented
in consequence of new discoveries is exceedingly remote.*
A. A. Michelson (1903)

*Well I tell you my friend
this might seem like the end
But the continuation
is yours for the making.*
B. May

The study of symmetries has always been one of the most fruitful approaches to physics, even more so in the twentieth century, starting with the work of giants like Einstein and Wigner, when the systematic investigation of symmetry properties provided both powerful tools for the understanding of the puzzling and complex microscopic phenomena, and the possibility for new fundamental discoveries. Such studies opened new fields of research and led to giant steps in the understanding of Nature together with a deeper appreciation of the rôle that symmetry plays in Her realm. Most of the interest and the usefulness of symmetries actually lies in their violation, and in this context discrete symmetries have had a central place in the developments of our understanding of the basic laws of Nature since the 1950s.

The discrete symmetries addressed in this book lie at the heart of our understanding of Nature, and bear deep and wide connections to very diverse fields of physics, some of which we tried to describe. *CP* violation in particular appears to be a very fundamental, and still quite obscure, property of Nature, linking as it does to deep questions such as quantum mechanical interference, microscopic time reversibility, and the matter antimatter asymmetry of the universe which allows our very existence.

Moreover, the Sakharov conditions (Section 11.3.4) and the fact that the *CP* violation in the Standard Model seems too small to explain the observed asymmetry indicate that such new physics must necessarily involve new sources of *CP* violation. Such sources can also be reasonably expected to have some observable effect on the low-energy processes which are observed in laboratories, and this is one of the reasons why the study of *CP*-violating phenomena maintains a high potential for further discoveries.

The fact that the SM has just enough complexity to be able to accommodate *CP* violation does not shed any light on the true nature of this phenomenon, and we feel the conclusion of J. Cronin's 1980 Nobel speech still stands: 'We must continue to seek the origin of the *CP* symmetry violation by all means at our disposal. [. . .] We are hopeful, then, that at some epoch, perhaps distant, this cryptic message from nature will be deciphered.' (Cronin, 1981)

The study of symmetry violations is a very stimulating field, both for the development of new theories and models and for the experimental measurements which put such theories to the test or discover new and unexpected phenomena. Such experimental studies often require, besides the effort and dedication which are the basis of any scientific endeavour, a good deal of ingenuity and the technical developments which go together with it. While such aspects are very evident to the practitioner and represent an important part of the intellectual pleasure provided by research activity, they are often not fully appreciated by the student, as indeed the formal beauty of a theory, which usually enjoys the privilege of being allowed to focus only on some chosen aspects of a problem, is often easier to understand and appreciate (and to convey as well!).

Our aim with this work was to try giving the reader a feeling, as partial and limited as dictated by our space and ability, of the intellectual content of this kind of experimental research. If a single reader is stimulated by this work to delve deeper into the fascinating issues which underlie the design of a successful experiment, we will consider our goal as having been fulfilled.

We conclude with the fitting words of a great physicist:

Even the best [mathematical physicists] have a tendency to treat physics as purely a matter of equations. I think this is shown by the poverty of the theoretical communications on the problems which face the experimenter. I quite recognize that the experimenter is inclined to drop his mathematics also... As a matter of fact it is extremely difficult to keep up the latter when all your energies are absorbed in experimentation.

Ernest Rutherford (Letter to Sir Arthur Schuster, 1907).

APPENDIX A

NOTATION

Latin indices usually run over space coordinates ($i = 1, 2, 3$) while Greek indices run over space-time coordinates; with boldface letters indicating (three-) vectors, four-vectors are $x^\mu = (ct, \boldsymbol{x})$, $x_\mu = (ct, -\boldsymbol{x})$. The space-time metric is $g_{\mu\nu} = \mathrm{diag}(1, -1, -1, -1)$, so that for a four-vector $a^2 = a_0^2 - \boldsymbol{a}^2$ and $\Box = \partial_\mu \partial^\mu = (\partial/\partial t)^2 - \boldsymbol{\nabla}^2$ (repeated indexes indicate summation: $A_i B_i = \sum_i A_i B_i$). A hat indicates a versor (unit vector): $\hat{\boldsymbol{a}} = \boldsymbol{a}/|\boldsymbol{a}|$. The same notation is used for classical observables and the corresponding quantum-mechanical operators, as the context makes the distinction clear.

ϵ_{ijk} and $\epsilon_{\alpha\beta\gamma\delta}$ are the totally antisymmetric tensors in three and four dimensions respectively, which are $+1$ (-1) when their indexes are an even (odd) permutation of $(1, 2, 3)$ or $(0, 1, 2, 3)$ respectively, and 0 otherwise; note that the former is invariant for cyclic permutation of its indexes, while the latter is not.

Square brackets indicate commutators and curly ones anti-commutators: $[A, B] \equiv AB - BA$, $\{A, B\} \equiv AB + BA$.

The step function is

$$\theta(t) = \begin{cases} +1 & (t > 0) \\ 0 & (t < 0) \end{cases} \tag{A.1}$$

A T superscript indicates the transpose M^T of a matrix M (not to be confused with the T subscript used to indicate time-reversed quantities), and the Hermitian conjugate (H.c.) is $M^\dagger = M^{*T}$ where the star denotes complex conjugation.

Dirac matrices γ^μ are defined by

$$\{\gamma^\mu, \gamma^\nu\} = g^{\mu\nu} \tag{A.2}$$

so that $\gamma_0^2 = 1$, $\gamma_i^2 = -1$. Moreover, $\gamma_0^\dagger = \gamma_0$, $\gamma_i^\dagger = -\gamma_i$ (thus $\gamma_\mu^\dagger = \gamma^\mu$), and $\gamma_\mu^\dagger = \gamma_0 \gamma_\mu \gamma_0$. For Dirac spinors the bar indicates $\bar{u} \equiv u^\dagger \gamma_0$.

The matrix $\gamma_5 = \gamma^5 \equiv i\gamma^0\gamma^1\gamma^2\gamma^3$ satisfies $\{\gamma_5, \gamma^\mu\} = 0$ and $\gamma_5^2 = 1, \gamma_5^\dagger = \gamma_5$. With γ_5 (and the unit matrix) together with $\{\gamma_\mu\}$ any complex 4×4 matrix can be expressed as a sum of terms, each being a product of at most 2 such matrices.

In the Dirac–Pauli representation the γ matrices are (in terms of 2×2 sub-matrices)

$$\gamma^0 = \begin{pmatrix} 1 & 0 \\ 0 & -1 \end{pmatrix} \quad \gamma^k = \begin{pmatrix} 0 & \sigma_k \\ -\sigma_k & 0 \end{pmatrix} \quad \gamma^5 = \begin{pmatrix} 0 & 1 \\ 1 & 0 \end{pmatrix}$$

where the usual Pauli matrices are

$$\sigma_1 = \begin{pmatrix} 0 & 1 \\ 1 & 0 \end{pmatrix} \quad \sigma_2 = \begin{pmatrix} 0 & -i \\ i & 0 \end{pmatrix} \quad \sigma_3 = \begin{pmatrix} 1 & 0 \\ 0 & -1 \end{pmatrix} \tag{A.3}$$

and the notation $\boldsymbol{\sigma} = (\sigma_1, \sigma_2, \sigma_3)$ is sometimes used. Other useful representations are the Weyl (or chirality) one

$$\gamma^0 = \begin{pmatrix} 0 & -1 \\ -1 & 0 \end{pmatrix} \quad \gamma^k = \begin{pmatrix} 0 & \sigma_k \\ -\sigma_k & 0 \end{pmatrix} \quad \gamma^5 = \begin{pmatrix} 1 & 0 \\ 0 & -1 \end{pmatrix}$$

in which γ_5 is diagonal, and the Majorana one

$$\gamma^1 = \begin{pmatrix} i\sigma_3 & 0 \\ 0 & i\sigma_3 \end{pmatrix} \quad \gamma^2 = \begin{pmatrix} 0 & -\sigma_2 \\ \sigma_2 & 0 \end{pmatrix} \quad \gamma^3 = \begin{pmatrix} -i\sigma_1 & 0 \\ 0 & -i\sigma_1 \end{pmatrix}$$

$$\gamma^0 = \begin{pmatrix} 0 & \sigma_2 \\ \sigma_2 & 0 \end{pmatrix} \quad \gamma^5 = \begin{pmatrix} \sigma_2 & 0 \\ 0 & -\sigma_2 \end{pmatrix}$$

in which all the matrices are pure imaginary and the Dirac equation is real.

According to Pauli's fundamental theorem Pauli (1936) if two sets of 4×4 matrices $\{\gamma_\mu\}$, $\{\gamma'_\mu\}$ exist which both satisfy (A.2) then there exists a non-singular 4×4 matrix S, unique up to a multiplicative constant, such that $\gamma'_\mu = S\gamma_\mu S^{-1}$. The matrices S_W, S_M to obtain respectively the Weyl or the Majorana representations from the Dirac–Pauli one are

$$S_W = \frac{1}{\sqrt{2}} \begin{pmatrix} 1 & -1 \\ 1 & 1 \end{pmatrix} \quad S_M = \frac{1}{\sqrt{2}} \begin{pmatrix} 1 & \sigma_2 \\ \sigma_2 & -1 \end{pmatrix}$$

BR often stands for Branching Ratio. When quoting experimental results, the first error figure is the statistical error and the second the systematic one, unless indicated otherwise.

$e > 0$ is the positron charge, $\alpha = e^2/(4\pi) \simeq 1/137$ is the fine structure constant. Natural units ($\hbar = c = 1$) are mostly used, except where they are not.

APPENDIX B

QUANTUM FIELD THEORY

It is never hard to find trouble in field theory.
J. D. Bjorken, S. D. Drell

In this appendix a brief glimpse of quantum field theory (QFT) is provided, mostly as a collection of useful formulæ and results. The reader who knows nothing about quantum field theory won't learn it here, but might just get a sufficient idea of some of its basic features to go through the book; the reader who just needs a refresher will probably profit most from this appendix; the advanced reader might just be driven to tears by the lack of rigour and depth, and is directed to Itzykson and Zuber (1980) and Weinberg (1995).

B.1 Scalar fields

Quantum field theory is the formalism used to describe relativistic quantum phenomena and therefore particle physics, generalizing quantum mechanics to systems with uncountably infinite degrees of freedom, corresponding to the values of a field at every point of space, such as the classical electromagnetic fields $E(t, x)$ and $B(t, x)$.

The simplest case of a scalar field (one component) is considered first. According to Bohr's correspondence principle a real classical scalar field $\phi_c(x)$ can be considered as the expectation value of a Hermitian quantum-mechanical field operator $\phi(x)$ for some appropriate quantum state $|a\rangle$: $\phi_c(x) = \langle a|\phi(x)|a\rangle$, with both ϕ_c and ϕ satisfying the same (Klein-Gordon) equation

$$(\Box + m^2)\, \phi(x) = 0$$

m being the mass of the particle associated with the field.

By requiring that such field operator ϕ satisfies the relativistic generalization of the equation for time evolution in the Heisenberg description: $\partial_\mu \phi(x) = (i/\hbar)[P_\mu, \phi(x)]$ (P_μ being the four-momentum operator), if the field is expanded in a series of complete solutions such as the plane waves[161]

$$\phi(x) = \int \frac{d\boldsymbol{p}}{(2\pi)^3 2E} \left[a(\boldsymbol{p}) e^{-ipx} + a^\dagger(\boldsymbol{p}) e^{ipx} \right] \tag{B.1}$$

then the following relations hold

$$[P_\mu, a^\dagger(\boldsymbol{p})] = p_\mu a^\dagger(\boldsymbol{p}) \qquad [P_\mu, a(\boldsymbol{p})] = -p_\mu a(\boldsymbol{p}) \tag{B.2}$$

where $a(\boldsymbol{p}), a^\dagger(\boldsymbol{p})$ are operators in Hilbert space (an infinite set of them, labelled by the three-times infinite 'indices' $\boldsymbol{p}$) and $p^0 = E = +\sqrt{\boldsymbol{p}^2 + m^2}$. Relations (B.2) lead to the particle interpretation: by applying $a^\dagger(\boldsymbol{p})$ to the vacuum state $|0\rangle$, defined as the one with no particles present (and normalized to $\langle 0|0\rangle = 1$) an eigenstate of P_μ with eigenvalue p_μ is obtained:

$$P_\mu |\boldsymbol{p}\rangle \equiv P_\mu a^\dagger(\boldsymbol{p})|0\rangle = p_\mu |\boldsymbol{p}\rangle \tag{B.3}$$

so that this state indeed corresponds to a single free particle of mass m and momentum $\boldsymbol{p}$; $a^\dagger$ is accordingly called a creation operator. By applying several creation operators in sequence multi-particle free states are obtained. Since $a(\boldsymbol{p})$ applied to $|0\rangle$ generates a state of negative energy which is not observed in Nature, it is required that $a(\boldsymbol{p})|0\rangle = 0$ for any $\boldsymbol{p}$ (this can also be considered as the definition of the vacuum state).

Consistency with special relativity, demanding that two measurements at space-time points x, y separated by a space-like distance $((x - y)^2 < 0)$ cannot influence each other, leads to the postulate of micro-causality for the fields

$$[\phi(x), \phi(y)] = 0 \qquad \text{for} \qquad (x - y)^2 < 0$$

which can be shown to require

$$[a(\boldsymbol{p}_1), a(\boldsymbol{p}_2)] = 0 \qquad [a^\dagger(\boldsymbol{p}_1), a^\dagger(\boldsymbol{p}_2)] = 0 \tag{B.4}$$

for any $\boldsymbol{p}_{1,2}$. The second of the commutation relations (B.4) shows in particular that a two-particle state is not affected by an interchange:

$$|\boldsymbol{p}_1 \boldsymbol{p}_2\rangle = |\boldsymbol{p}_2 \boldsymbol{p}_1\rangle \tag{B.5}$$

corresponding to the case of Bose–Einstein (B–E) statistics for the particles.

Note also that a single-particle state with sharply defined momentum $|\boldsymbol{p}\rangle$ is not normalizable (and physically unrealizable due to the uncertainty principle, just as

---

[161] Note that a variety of conventions exist for the factors appearing in front of the integral, the normalization of states, the phase-space factors and the commutation relations between operators, which are all linked; see Donoghue *et al.* (1992) for a discussion.

a plane wave), and one actually deals with superpositions of states (wave-packets) in which each momentum component is weighted by some function $\varphi(\boldsymbol{p})$:

$$|\varphi\rangle = \int \frac{d\boldsymbol{p}}{(2\pi)^3\sqrt{2E}}\, \varphi(\boldsymbol{p})\, a^\dagger(\boldsymbol{p})|0\rangle \quad \text{with} \quad \int \frac{d\boldsymbol{p}}{(2\pi)^3}\, |\varphi(\boldsymbol{p})|^2 = 1 \qquad \text{(B.6)}$$

Given eqns (B.4) the simplest assumption needed to give a non-trivial theory is that not all a commutes with any $a^\dagger$, but rather

$$\left[a(\boldsymbol{p}_1), a^\dagger(\boldsymbol{p}_2)\right] = 2E(2\pi)^3 \delta(\boldsymbol{p}_1 - \boldsymbol{p}_2) \qquad \text{(B.7)}$$

The commutation relations (B.7) in turn show that the operator $a(\boldsymbol{p})$ eliminates a particle of momentum $\boldsymbol{p}$ from the state to which it is applied (this can be seen by evaluating the norm of the state (B.6)), and is thus called an annihilation operator. By comparing the time dependence in the exponential which accompanies $a, a^\dagger$ in (B.1) with the usual e^{-iEt} term familiar from quantum mechanical time evolution the two terms in this expression are also called positive-energy and negative-energy parts respectively.

If a time-dependent state is defined with $\varphi(t, \boldsymbol{p}) \equiv \varphi(\boldsymbol{p})\, e^{+iEt}$, the function $\varphi^*(t, \boldsymbol{p})$ is seen to be the ordinary wavefunction in momentum space, whose square modulus gives the probability of finding the particle at time t with momentum $\boldsymbol{p}$. In quantum field theory the wavefunction of a state is thus just the (complex, $c - number$) matrix element of the field $\phi(x)$ between a given one-particle state $|\varphi\rangle$ and the vacuum:

$$\varphi(t, \boldsymbol{x}) = \langle 0|\phi(t, \boldsymbol{x})|\varphi\rangle \qquad \text{(B.8)}$$

which transforms just as in quantum mechanics[162].

Classically the vacuum is the state of lowest energy in which any field vanishes at every point; the quantum mechanical field is an operator which does not commute with the Hamiltonian, whose expectation value is indeed zero in the vacuum state, but whose mean square deviation does not vanish and actually diverges

$$\langle 0|\phi^2(x)|0\rangle = \int \frac{d\boldsymbol{p}}{(2\pi^3)2E}$$

becoming finite (but non-zero) when considering the field over a finite volume; this reflects the fact that the vacuum is actually a very complex system, in which virtual particle-antiparticle pairs are continuously created and annihilated.

---

[162] In the relativistic case such wavefunction cannot be consistently interpreted as the probability distribution function for the particle, however.

B.2   Fields with spin

The above discussion can be generalized to non-Hermitian fields (describing charged particles) with spin degrees of freedom. In the general case a quantum field (operator) can be expressed in terms of Hilbert-space creation and annihilation operators $a^\dagger(p, s)$, $a(p, s)$ for momentum p and spin projection s (along a given quantization direction); for non-Hermitian fields ϕ and $\phi^\dagger$ are independent, and corresponding antiparticle annihilation and creation operators $b^\dagger(p, s), b(p, s)$ also appear. The tensorial or spinorial nature of the field is contained in appropriate factors $u_i(p, s)$, $v_i(p, s)$ for particles and antiparticles respectively, where i stands for a set of spinorial or four-vector indexes:

$$\phi_i(x) = \sum_s \int \frac{dp}{(2\pi)^3 2E} \left[a(p, s)u_i(p, s)e^{-ipx} + b^\dagger(p, s)v_i(p, s)e^{ipx} \right] \quad \text{(B.9)}$$

$$\phi_i^\dagger(x) = \sum_s \int \frac{dp}{(2\pi)^3 2E} \left[b(p, s)v_i^\dagger(p, s)e^{-ipx} + a^\dagger(p, s)u_i^\dagger(p, s)e^{ipx} \right] \quad \text{(B.10)}$$

Note that this plane-wave expansion can be rewritten in a form which is explicitly relativistically covariant (for orthochronous transformations which do not change the sign of p^0) using

$$\frac{dp}{(2\pi)^3 2E} = \frac{d^4p}{(2\pi)^4} 2\pi \delta(p^2 - m^2)\theta(p^0) \quad \text{(B.11)}$$

For a Hermitian field (describing particles identical to their antiparticles) $a = b$ and $u = v$.

As an example, the Hermitian electromagnetic field can be written as

$$A_\mu(x) = \sum_{s=1,2} \int \frac{dp}{(2\pi)^3 2E} \left[a^{(s)}(p)\epsilon_\mu^{(s)}(p)e^{-ipx} + a^{(s)\dagger}(p)\epsilon_\mu^{(s)*}(p)e^{ipx} \right] \quad \text{(B.12)}$$

where only the two transverse polarization states $s = 1, 2$ appear.

For the spin-1/2 (Dirac) field ψ and $\overline{\psi} = \psi\gamma^0$ are used (the reason being that $\overline{\psi}\psi$, rather than $\psi^\dagger\psi$ behaves as a scalar under Lorentz transformations)

$$\psi(x) = \sum_{s=\pm 1/2} \int \frac{dp}{(2\pi)^3 2E} \left[a(p, s)u(p, s)e^{-ipx} + b^\dagger(p, s)v(p, s)e^{ipx} \right] \quad \text{(B.13)}$$

$$\overline{\psi}(x) = \sum_{s=\pm 1/2} \int \frac{dp}{(2\pi)^3 2E} \left[b(p, s)\overline{v}(p, s)e^{-ipx} + a^\dagger(p, s)\overline{u}(p, s)e^{ipx} \right] \quad \text{(B.14)}$$

and the u, v are four-component spinors, being in the free case

$$u(\boldsymbol{p}, s) = \sqrt{\frac{E+m}{2m}} \begin{pmatrix} \varphi^{(s)} \\ \frac{\boldsymbol{\sigma} \cdot \boldsymbol{p}}{E+m} \varphi^{(s)} \end{pmatrix} \qquad v(\boldsymbol{p}, s) = \sqrt{\frac{E+m}{2m}} \begin{pmatrix} \frac{\boldsymbol{\sigma} \cdot \boldsymbol{p}}{E+m} \chi^{(s)} \\ \chi^{(s)} \end{pmatrix}$$

with $\varphi^{(s)}, \chi^{(s)}$ two-component spinors. The Dirac equations for the fields are

$$(i\gamma^{\mu}\partial_{\mu} - m)\psi = 0 \qquad (i\gamma^{\mu}\partial_{\mu} + m)\overline{\psi} = 0 \tag{B.15}$$

$$\gamma_{\mu}p^{\mu}u(\boldsymbol{p}, s) = mu(\boldsymbol{p}, s) \qquad \gamma_{\mu}p^{\mu}v(\boldsymbol{p}, s) = -mv(\boldsymbol{p}, s) \tag{B.16}$$

From eqns (B.13) the field operator ψ is seen to annihilate particles and create antiparticles, while $\overline{\psi}$ has the opposite effect.

For an integer-spin field the commutation relations of creation and annihilation operators are

$$\left[a(\boldsymbol{p}_1, s_1), a^{\dagger}(\boldsymbol{p}_2, s_2)\right] = 2E(2\pi)^3 \delta_{s_1, s_2}\delta(\boldsymbol{p}_1 - \boldsymbol{p}_2) = \left[b(\boldsymbol{p}_1, s_1), b^{\dagger}(\boldsymbol{p}_2, s_2)\right] \tag{B.17}$$

with all other commutators between $a, a^{\dagger}, b$ and $b^{\dagger}$ being zero.

For half-integer spin fields the commutation relations (B.17) can be shown to lead unavoidably to negative energies; however, since such fields are not observables, it is possible to require them to *anti-commute* instead

$$\left\{a(\boldsymbol{p}_1, s_1), a^{\dagger}(\boldsymbol{p}_2, s_2)\right\} = 2E(2\pi)^3 \delta_{s_1, s_2}\delta(\boldsymbol{p}_1 - \boldsymbol{p}_2) = \left\{b(\boldsymbol{p}_1, s_1), b^{\dagger}(\boldsymbol{p}_2, s_2)\right\} \tag{B.18}$$

with all other anti-commutators being zero. In this way a consistent theory can be obtained, in which the measurable quantities built out of the field operators themselves actually commute at space-like separations; the use of anti-commuting fields leads to particles obeying Fermi–Dirac (F–D) statistics. This link between spin and the sign of commutation rules is the *spin-statistics connection* (Chapter 5).

The Hamiltonian for the free field ϕ can be written in terms of creation and annihilation operators as

$$H =: \int dp \sum_s E \left[a^{\dagger}(\boldsymbol{p}, s)a(\boldsymbol{p}, s) + b^{\dagger}(\boldsymbol{p}, s)b(\boldsymbol{p}, s)\right] : \tag{B.19}$$

where the colons denote normal ordering of the fields, meaning that in the above expression all annihilation operators are moved to the right of all creation operators, taking into account the commutativity or anti-commutativity of the fields (but disregarding any delta function); this prescription is equivalent to subtracting an infinite (but unmeasurable) contribution to the vacuum energy.

From the above expression quantum field theory is thus seen to be intrinsically a many body theory, as particles can be created and destroyed by the Hamiltonian. Indeed the classical limit of the theory corresponds to the case of a very large number of excitations (quantum electrodynamics reducing to classical electrodynamics for very large number of photons), while the classical limit of quantum mechanics is the classical mechanics of a single mass point: while both electron and photon exhibit particle-like and wave-like behaviour, it is thus not surprising that only the particle nature of the electron and the wave nature of light were initially apparent.

Generalizing (B.19) the four-momentum is

$$P^\mu =: \int d\mathbf{p} \sum_s p^\mu \left[a^\dagger(\mathbf{p}, s) a(\mathbf{p}, s) + b^\dagger(\mathbf{p}, s) b(\mathbf{p}, s) \right] :$$

and these four operators are the generators of the group of space-time translations.

The generators of the Lorentz group $J_{\rho\sigma}$ can be split into the generators of three-dimensional rotations $J^i = -(1/2)\epsilon_{ijk} J_{jk}$ ($i = 1, 2, 3$) and the generators of the Lorentz boosts $K^i = M^{0i}$. While the J^i form a sub-group, the K^i do not: $[K^i, K^j] = -i\epsilon^{ijk} J_k$. By forming (non-Hermitian) linear combinations of the two generators, the Lorentz group can be split in two distinct pieces with generators:

$$J_i^{(\pm)} = \frac{1}{2}(J_i \pm iK_i)$$

for which $[J_i^{(+)}, J_j^{(-)}] = 0$, each having the algebra of $SU(2)$, i.e. of the usual rotation group: $[J_i^{(\pm)}, J_j^{(\pm)}] = i\epsilon_{ijk} J_k^{(\pm)}$. Representations of the Lorentz group can thus be obtained formally by combining two representations of angular momentum, labelled by eigenvalues $(j^{(+)}, j^{(-)})$ where $j^{(+)}, j^{(-)} = 0, 1/2, 1, \ldots$. If $j^{(+)} + j^{(-)}$ is half-integer, the representation is spinorial, otherwise it is tensorial:

$(j^{(+)}, j^{(-)})$	
(0,0)	spin 0 (scalar)
(1/2,0) and (0,1/2)	spin 1/2 (spinor)
(1/2,1/2)	spin 1 (vector)
(1,1/2) and (1/2,1)	spin 3/2
(1,1)	spin 2

Spinorial representations are always double, the two components being connected by a parity transformation.

Continuous Lorentz transformations form the *proper, orthochronous* (or *restricted*) Lorentz group, which includes all homogeneous Lorentz transformations with determinant $+1$. The full Lorentz group also contains two other space-time operations which preserve the Minkowski interval $s^2 = t^2 - \boldsymbol{x}^2$: space inversion (parity), transforming $(t, \boldsymbol{x}) \rightarrow (t, -\boldsymbol{x})$ (improper transformation), and time inversion, transforming $(t, \boldsymbol{x}) \rightarrow (-t, \boldsymbol{x})$ (non-orthochronous transformation); these cannot be reached by continuous transformations starting from the identity[163] and have determinant -1.

The measurability of field operators is subject to limitations: in particular, for a half-integer spin field these limitations are important, since physically measurable quantities can be scalars, vectors, etc. but never spinors. Covariant quantities bilinear in the fermion fields have the form $\overline{\psi} \Gamma \psi$, where the generic set of 4×4 matrices

$$\Gamma \equiv \{1, \gamma_\mu, \sigma_{\mu\nu}/2, \gamma_\mu \gamma_5, i\gamma_5\} \tag{B.20}$$

accounts for all possible tensor structures built with Dirac matrices:

$$S(x) \equiv: \overline{\psi}(x)\psi(x) : \tag{B.21}$$

$$V^\mu(x) \equiv: \overline{\psi}(x)\gamma^\mu \psi(x) : \tag{B.22}$$

$$T^{\mu\nu}(x) \equiv: \overline{\psi}(x)\sigma^{\mu\nu}\psi(x) : \tag{B.23}$$

$$A^\mu(x) \equiv: \overline{\psi}(x)\gamma^\mu \gamma_5 \psi(x) : \tag{B.24}$$

$$P(x) \equiv: i\overline{\psi}(x)\gamma_5 \psi(x) : \tag{B.25}$$

(here A^μ should not be confused with the electromagnetic field, nor P with the parity operator!), where $\sigma^{\mu\nu} \equiv (i/2)[\gamma^\mu, \gamma^\nu]$; the factor i in $P(x)$ is inserted so that all bilinears are Hermitian: $\overline{\Gamma} \equiv \gamma_0 \Gamma^\dagger \gamma_0 = \Gamma$.

In the non-relativistic limit the above quantities reduce to

$$S(x) \sim V^0(x) \sim \psi^\dagger(x)\psi(x) \tag{B.26}$$

$$T^{ij}(x) \sim \epsilon_{ijk} \psi^\dagger(x)\sigma^k \psi(x) \tag{B.27}$$

$$A^k(x) \sim \overline{\psi}(x)\sigma^k \psi(x) \tag{B.28}$$

while $V^i(x), T^{0i}(x), A^0(x)$ and $P(x)$ vanish in the same limit.

B.3 Interactions

A quantum field theory is usually defined by providing its Lagrangian (density) function, which is an explicitly covariant function of the fields and its derivatives from which the equations of motion (and the Hamiltonian) can be derived. The most

---

[163] Note that improper or non-orthochronous transformations do not form a group, since they miss the identity element.

general Lorentz-invariant interaction Lagrangian for fermion fields $\psi_{1,2}$ interacting with other fields $\phi_S, \phi_V^\mu, \ldots$ (scalar, vector, ...) can be written as

$$
\begin{aligned}
\mathcal{L} =: & \phi_S \, \overline{\psi}_1 (c_S + c_S' \gamma_5) \psi_2 + \phi_V^\mu \, \overline{\psi}_1 \gamma_\mu (c_V + c_V' \gamma_5) \psi_2 \\
& + \phi_T^{\mu\nu} \, \overline{\psi}_1 \sigma_{\mu\nu} (c_T + c_T' \gamma_5) \psi_2 + \phi_A^\mu \, \overline{\psi}_1 (c_A + c_A' \gamma_5) \psi_2 \\
& + \phi_P \, \overline{\psi}_1 i \gamma_5 (c_P + c_P' \gamma_5) \psi_2 : + \text{H.c.}
\end{aligned}
\tag{B.29}
$$

The interaction of a spin 1/2 field ψ of charge Q with the electromagnetic field A_μ occurs through its current $j^\mu(x)$ and is described by the interaction Hamiltonian

$$
H_{\text{int}} = j^\mu(x) A_\mu(x) = Q : \overline{\psi}(x) \gamma^\mu \psi(x) A_\mu(x) :
\tag{B.30}
$$

which is called *minimal coupling*, equivalent to substituting the momentum of the spin 1/2 particle with $p_\mu \to p_\mu - Q A_\mu$, just as in classical electrodynamics. The appearance of a product of fields at the very same space-time point is what leads to some mathematical difficulties in the form of infinite integrals, requiring some reinterpretation of ill-defined physical quantities obtain finite results (the renormalization procedure).

The interaction term (B.30) is gauge-invariant, that is unchanged by the transformations

$$
\psi(x) \to e^{iq\alpha(x)} \psi(x)
\tag{B.31}
$$

$$
\overline{\psi}(x) \to e^{-iq\alpha(x)} \overline{\psi}(x)
\tag{B.32}
$$

$$
A_\mu(x) \to A_\mu + [\partial_\mu \alpha(x)] A_\mu(x)
\tag{B.33}
$$

with an arbitrary (space-time dependent) $\alpha(x)$; this is now understood to be a fundamental and powerful unifying principle for fundamental interactions.

APPENDIX C

SCATTERING IN QUANTUM THEORY

You don't understand anything
until you learn it more than one way.
M. Minsky

Here is a brief review of the description of scattering in quantum field theory. The purpose is to collect some useful formulæ and derive some results in detail to help avoid the confusion which can sometimes arise due to the alternative descriptions adopted in the literature to deal with this vast subject; the reader is directed to standard textbooks such as Newton (1982) or Weinberg (1995) for details.

C.1 Scattering

Because of the impossibility of following the details of a microscopic interaction, all measurements in particle physics are performed by scattering experiments, so that the quantities of interest are the probability amplitudes for a given transition. In order to deal with interacting particles the Hamiltonian can be split in the form $H = H_0 + H_{int}$, where H_0 describes the free motion and H_{int} describes all the interactions among the particles, assumed to vanish in the remote past and future, when they are sufficiently far apart.

It is useful to work with states that have simple transformations under the Lorentz group: this is the case for the (rather uninteresting) single-particle states and for states consisting of several non-interacting particles, which are direct products of single-particle states. One therefore introduces the so-called 'in' and 'out' states, which are eigenstates of the *full* Hamiltonian

$$H|\alpha_{(in,out)}\rangle = E_\alpha|\alpha_{(in,out)}\rangle \tag{C.1}$$

(where α stands for the full specification of number and type of particles, momenta, spins, etc.). For these states the particle content is specified at time $t = -\infty$ and $t = +\infty$ respectively: e.g. the state $|\alpha_{(in)}\rangle$ contains the set of particles specified by the quantum numbers α at $t = -\infty$, and not necessarily at other times. Such specification in the far past and future corresponds to *free* states at such time $|\alpha\rangle = |i,f\rangle$, since $H_{\text{int}} \to 0$ for $t \to \pm\infty$:

$$H_0|i\rangle = E_i|i\rangle \qquad H_0|f\rangle = E_f|f\rangle \tag{C.2}$$

For a scattering experiment the particle content at those asymptotic times can be conveniently specified plane waves (actually wave packets of finite extension in time and space)[164]. The connection between the free states and the 'in' and 'out' states which is appropriate for wave packets (superpositions of energy eigenstates) is given by the conditions

$$e^{-iHt}|\alpha_{(\text{in})}\rangle \xrightarrow{t\to-\infty} e^{-iH_0t}|\alpha\rangle \qquad e^{-iHt}|\alpha_{(\text{out})}\rangle \xrightarrow{t\to+\infty} e^{-iH_0t}|\alpha\rangle$$

so that

$$|\alpha_{(\text{in})}\rangle = \lim_{t\to-\infty} e^{iHt}e^{-iH_0t}|\alpha\rangle \qquad |\alpha_{(\text{out})}\rangle = \lim_{t\to+\infty} e^{iHt}e^{-iH_0t}|\alpha\rangle \tag{C.3}$$

The amplitudes for finding in the remote future the system in the (free) final state $|f\rangle$, when it was prepared to be in the (free) initial state $|i\rangle$ in the remote past are the set of numbers of interest in scattering theory

$$S_{fi} \equiv \langle f_{(\text{out})}|i_{(\text{in})}\rangle \tag{C.4}$$

which define the S (scattering) matrix. In other words, the 'in' and 'out' states form two (complete) sets of basis states spanning the Hilbert space, and the coefficients S_{fi} describe the change of basis among them, or how a given 'in' state is expanded in terms of 'out' states:

$$|i_{(\text{in})}\rangle = \sum_f S_{fi}|f_{(\text{out})}\rangle \tag{C.5}$$

In absence of interactions the 'in' and 'out' states coincide, and $S_{fi} = \delta_{fi}$.
    The S matrix is unitary, since[165]

$$\sum_\gamma S_{\gamma\beta}^* S_{\gamma\alpha} = \sum_\gamma \langle\gamma_{(\text{out})}|\beta_{(\text{in})}\rangle^* \langle\gamma_{(\text{out})}|\alpha_{(\text{in})}\rangle = \sum_\gamma \langle\beta_{(\text{in})}|\gamma_{(\text{out})}\rangle \langle\gamma_{(\text{out})}|\alpha_{(\text{in})}\rangle$$

$$= \langle\beta_{(\text{in})}|\alpha_{(\text{in})}\rangle = \delta_{\beta\alpha}$$

[164] The above is also valid in the Heisenberg picture of motion, in which the states do not depend explicitly on time but contain the whole 'history' of time evolution. The 'in' states are those containing incoming plane waves and outgoing spherical waves, while the 'out' states contain outgoing plane waves and incoming spherical waves.
    [165] This proof requires the completeness of the 'in' and 'out' states: in the case at hand in which *all* interactions vanish at times $t \to \pm\infty$ these can be the scattering states, specified at those times as containing free particles, since no bound states can exist; in the general case in which bound states are possible they should also be included in the complete set.

The analogous relation $\sum_\gamma S_{\gamma\alpha}S^*_{\gamma\beta} = \delta_{\beta\alpha}$ (not equivalent to the previous one for infinite matrices) also holds. The unitarity of the S matrix is the expression of probability conservation: *any* given state $|i_{(\text{in})}\rangle$ whose content is specified in the remote past will evolve into *some* of the states $|f_{(\text{out})}\rangle$ whose content is specified in the remote future with unity probability:

$$\sum_f |S_{fi}|^2 = \sum_f S^*_{fi}S_{fi} = 1 \qquad (\text{C.6})$$

and analogously *any* given state $|f_{(\text{out})}\rangle$ should be obtainable from *some* $|i_{(\text{in})}\rangle$ state. This also requires

$$\sum_f |S_{fi}|^2 = \sum_f |S_{if}|^2 \qquad (\text{C.7})$$

A scattering *operator* S is also introduced, which is defined to have matrix elements between *free* states corresponding to the elements of the S matrix[166]:

$$\langle f|S|i\rangle \equiv S_{fi} \qquad (\text{C.8})$$

The unitarity of the S matrix implies that of the S operator:

$$\delta_{\beta\alpha} = \sum_\gamma \langle\gamma|S|\beta\rangle^*\langle\gamma|S|\alpha\rangle = \langle\beta|S^\dagger|\gamma\rangle\langle\gamma|S|\alpha\rangle = \langle\beta|S^\dagger S|\alpha\rangle$$

due to the completeness of the free states, so that

$$SS^\dagger = S^\dagger S = 1 \qquad (\text{C.9})$$

We recall that in the quantum-mechanical interaction picture of time evolution (see e.g. Sakurai (1985)), only the evolution of the state vectors due to H_0 is factored out, so that their time evolution only depends on H_{int}, while the operators O evolve according to H_0:

$$i\hbar\frac{\partial}{\partial t}|\psi(t)\rangle^{(I)} = H^{(I)}_{\text{int}}|\psi(t)\rangle^{(I)}$$

$$\frac{dO^{(I)}(t)}{dt} = \frac{\partial O^{(I)}(t)}{\partial t} + \frac{i}{\hbar}\left[H^{(I)}_0, O^{(I)}(t)\right]$$

---

[166] Note that this is not the operator $\widehat{S}$ which is sometimes defined to map an 'in' scattering state with a given plane-wave particle content at $t = -\infty$ into the 'out' state with the same particle content at $t = +\infty$: $|\alpha_{(\text{in})}\rangle = \widehat{S}|\alpha_{(\text{out})}\rangle$; such an operator is guaranteed to exist (and be unitary) by Wigner's theorem (Chapter 1) if the set of scattering states is complete, i.e. in absence of bound states.

(the superscript (I) denotes quantities in the interaction picture). The relations with Schrödinger and Heisenberg pictures are

$$|\psi(t)\rangle^{(I)} = e^{iH_0t} |\psi(t)\rangle^{(S)} = e^{iH_0t} e^{-iHt} |\psi\rangle^{(H)}$$

$$O^{(I)}(t) = e^{iH_0t} O^{(S)} e^{-iH_0t} = e^{iH_0t} e^{-iHt} O^{(H)}(t) e^{iHt} e^{-iH_0t}$$

The operator $U^{(I)}(t, t_0)$ describing time evolution from time t_0 to t in the interaction picture satisfies the equation

$$i\hbar \frac{\partial}{\partial t} U^{(I)}(t, t_0) = H^{(I)}_{\text{int}}(t) U^{(I)}(t, t_0) \qquad (C.10)$$

with the condition $U^{(I)}(t_0, t_0) = 1$, and its action on free states is

$$|i_{(\text{in})}\rangle^{(I)} = U^{(I)}(0, -\infty)|i\rangle^{(I)} \qquad |f_{(\text{out})}\rangle^{(I)} = U^{(I)\dagger}(\infty, 0)|f\rangle^{(I)}$$

i.e. the time evolution operator $U^{(I)}$ transforms the free $|i, f\rangle$ (eigenstates of H_0) into $|i_{(\text{in})}\rangle, |f_{(\text{out})}\rangle$ (eigenstates of H) so that

$$S_{fi} = \lim_{t_2 \to \infty} \lim_{t_1 \to -\infty} {}^{(I)}\langle f | U^{(I)}(t_2, 0) U^{(I)}(0, t_1)|i\rangle^{(I)}$$

and a formal expression for the S operator is obtained

$$S \equiv U^{(I)}(\infty, -\infty) \qquad (C.11)$$

which is not of much practical help, since knowing $U^{(I)}$ and the time evolution of the states amounts to having already solved the scattering problem.

The explicit form of the time evolution operator in the interaction picture is

$$U^{(I)}(t_2, t_1) = e^{iH_0t_2} e^{-iH(t_2 - t_1)} e^{iH_0t_1} \qquad (C.12)$$

which is consistent with eqns (C.3).

Another expression for the scattering operator is obtained (omitting the superscripts in the following) from the Lippmann–Schwinger integral equation for the 'in' states (Sakurai, 1985)

$$|i_{(\text{in})}\rangle = |i\rangle + \lim_{\epsilon \to 0} \frac{1}{E_i - H_0 + i\epsilon} H_{\text{int}} |i_{(\text{in})}\rangle \qquad (C.13)$$

from which, using eqn (C.12) and (C.2)

$$S_{fi} = \lim_{t_2 \to \infty} \langle f | e^{iH_0t_2} e^{-iHt_2} |i_{(\text{in})}\rangle = \lim_{t_2 \to \infty} e^{i(E_f - E_i)t_2} \langle f | i_{(\text{in})}\rangle$$

$$= \delta_{fi} + \lim_{t_2 \to \infty} \frac{e^{i(E_f - E_i)t_2}}{E_f - E_a + i\epsilon} \langle f | H_{\text{int}} | i_{(\text{in})}\rangle = \delta_{fi} - i(2\pi)\delta(E_f - E_i) T_{fi} \qquad (C.14)$$

thus defining the transition matrix, with elements

$$T_{fi} \equiv \langle f | H_{int} | i_{(in)} \rangle = \langle f_{(out)} | H_{int} | i \rangle \tag{C.15}$$

and a corresponding operator T, in analogy with (C.8), such that

$$\langle f | T | i \rangle \equiv T_{fi} \tag{C.16}$$

In a shorthand notation

$$S = 1 - i(2\pi)\delta(E)T \tag{C.17}$$

In a relativistic theory, energy and momentum are treated on the same footing and one writes instead[167]

$$S_{fi} = \delta_{fi} - (2\pi)^4 i \delta^4 (p_f - p_i) \mathcal{M}_{fi}$$

where $\mathcal{M}_{fi}$ is a Lorentz-invariant amplitude.

The transition matrix elements give the differential transition rate (probability per unit time, normalized to the number of particles) as

$$d\Gamma(i \rightarrow f) = (2\pi)^4 \delta^4 (p_f - p_i) \prod_{j=1,N_i} \left(\frac{1}{2E_j} \right) V^{(1-N_i)} |\mathcal{M}_{fi}|^2 \prod_{k=1,N_f} \left[\frac{d\mathbf{p}_k}{(2\pi)^3 2E_k} \right]$$

where N_i and N_f are the number of particles in the initial and final states, V the normalization volume and the last factor is the Lorentz-Invariant Phase Space volume element for the final state particles, denoted as d(LIPS). For the decay of a particle of mass M ($N_i = 1$) the above expression gives the relativistic generalization of Fermi's golden rule

$$\Gamma(i \rightarrow f) = \frac{(2\pi)^4}{2M} \int \delta^4 (p_f - p_i) |\mathcal{M}_{fi}|^2 \, d(\text{LIPS}) \tag{C.18}$$

The cross section (transition rate normalized to relative flux) for the collision of two particles of masses m_1, m_2 ($N_i = 2$) is

$$\sigma(i \rightarrow f) = (2\pi)^4 \int \delta^4 (p_f - p_i) \frac{1}{4\sqrt{(p_1 p_2)^2 - m_1 m_2}} |\mathcal{M}_{fi}|^2 \, d(\text{LIPS}) \tag{C.19}$$

The unitarity of the S matrix (C.9) gives the following relation for the transition matrix:

$$T_{if}^* - T_{fi} = 2\pi i \delta(E_n - E_i) T_{nf}^* T_{ni} \tag{C.20}$$

---

[167] Care should be taken as several alternative definitions of the transition matrix can be found in the literature, differing by 2π factors, the inclusion of delta functions or energy factors (see e.g. Donoghue *et al.* (1992) for a discussion); the one in eqn (C.15) is not Lorentz-invariant and keeps a close connection to the non-relativistic theory. Of course the final expressions for physical quantities such as cross sections or decay rates are the same in every notation.

(valid for $E_i = E_f$), sometimes written formally as

$$T^\dagger - T = 2\pi i \delta(E) T^\dagger T \tag{C.21}$$

The introduction of the T operator defined in eqn (C.16) trades the complication of the 'in' and 'out' states for that of having to deal with an operator which does not share the simple properties of the Hamiltonian, such as Hermiticity. Indeed the description of scattering can be done entirely in terms of free states, the 'in' and 'out' states being required only in the connection of the transition matrix elements with the Hamiltonian.

C.2 Approximations

If the interaction is weak, i.e. the matrix elements of H_{int} are much smaller than those of H_0, then the 'in' states are not too different from free states, so that instead of the exact equation (C.15) a good approximation is

$$\mathcal{T}_{fi} \simeq \langle f | H_{int} | i \rangle \tag{C.22}$$

This is the *Born approximation*: in this case, by comparing with eqn (C.16) the operator T is seen to coincide with the interaction Hamiltonian H_{int} and is therefore Hermitian; this can also be seen from the unitarity relation (C.20) in the limit in which the right-hand side, of second order in T, can be neglected. In this approximation therefore the following relation holds for the transition amplitudes

$$\mathcal{T}_{fi} \simeq \mathcal{T}_{if}^* \quad \text{(Born approximation)} \tag{C.23}$$

Higher-order approximations are obtained considering that eqn (C.10) for the time evolution operator in the interaction picture can be written as an integral equation

$$U(t, t_0) = 1 - i \int_{t_0}^{t} dt' \, H_{int}(t') \, U(t', t_0) \tag{C.24}$$

which when iterated gives

$$U(t, t_0) = \sum_{n=0}^{\infty} (-i)^n \int_{t_0}^{t} dt_1 \int_{t_0}^{t_1} dt_2 \cdots \int_{t_0}^{t_{n-1}} dt_n \, H_{int}(t_1) H_{int}(t_2) \ldots H_{int}(t_n)$$

$$= \sum_{n=0}^{\infty} \frac{(-i)^n}{n!} \int_{t_0}^{t} dt_1 \int_{t_0}^{t} dt_2 \cdots \int_{t_0}^{t} dt_n \, \theta(t, t_1) \theta(t_1 - t_2) \ldots \theta(t_{n-1} - t_n)$$

$$H_{int}(t_1) H_{int}(t_2) \ldots H_{int}(t_n) \tag{C.25}$$

where the step function orders the interaction terms in time (note that $H_{int}(t)$ and $H_{int}(t')$ need not commute if $t \neq t'$). The above expression is often written as

$$U(t, t_0) = \sum_{n=0}^{\infty} \frac{(-i)^n}{n!} \int_{t_0}^{t} dt_1 \int_{t_0}^{t} dt_2 \cdots \int_{t_0}^{t} dt_n \, T\,[H_{int}(t_1) H_{int}(t_2) \ldots H_{int}(t_n)] \tag{C.26}$$

by introducing the time-ordering operator[168] T

$$T\,[O(t_1) \ldots O(t_n)] \equiv \sum_{P} \theta(t_{\alpha_1}, t_{\alpha_2}) \ldots \theta(t_{\alpha_{n-1}} - t_{\alpha_n})\, O(t_1) \ldots O(t_n)$$

where the sum is over all permutations of indexes $\alpha_i = 1, 2, \ldots, n$; T reorders a product of time-labelled operators according to their times, so that the latest one occurs first:

$$T\,[O(t_1) \ldots O(t_n)] = O(t_i) \ldots O(t_j) \ldots O(t_k) \qquad (t_i > \ldots t_j \ldots > t_k)$$

allowing one to write formally

$$U(t, t_0) = T \exp\left(-i \int_{t_0}^{t} dt' H_{int}(t')\right) \tag{C.27}$$

The usefulness of the above series expansion is when H_{int} is small enough so that the first few terms suffice. While this is often not the case, (e.g. when dealing with weak transitions among particles which also interact through strong or electromagnetic interactions) the formalism can be adapted to deal with the case in which the interaction contains two terms, one strong and one weak:

$$H_{int} = H_S + H_W \tag{C.28}$$

The above discussion can be repeated, introducing new states $|\alpha_{S(in,out)}\rangle$ defined as what the 'in' and 'out' states would be if H_W were absent, and obtaining

$$T_{fi} = \langle f_{S(out)} | H_W | i_{(in)} \rangle + \langle f | H_S | i_{S(in)} \rangle \tag{C.29}$$

For a transition which cannot be driven by H_S due to some selection rule (which makes the existence of H_W relevant) the second term in eqn (C.29) vanishes and

$$T_{fi} = \langle f_{S(out)} | H_W | i_{(in)} \rangle \tag{C.30}$$

This is analogous to eqn (C.15), but now the state appearing on the left is not a free state but rather one in which the particles do interact according to H_S, the strong but 'uninteresting' part of the interaction. This leads to a picture in which, although

---

[168] Not to be confused with the time reversal operator T!

the complete interaction H_{int} is not weak, the first-order approximation in which the effect of H_W on the states can be neglected in eqn (C.30) might still be valid:

$$T_{fi} \simeq \langle f_{S(\text{out})} | H_W | i_{S(\text{in})} \rangle \tag{C.31}$$

This is the *distorted Born approximation*, valid at first order in H_W but to all orders in H_S; it is the same as the previous case except that eigenstates of $H_0 + H_S$ are used instead of free states (eigenstates of H_0). The difference is due to the presence of interactions induced by H_S in the initial and final states, among the particles participating in the transition induced by H_W; note however that for single-particle states (e.g. the initial state for a decay process) there are no interactions and $| i_{S(\text{in})} \rangle = | i \rangle$.

Introducing operators T_S, T_W for the strong and weak parts of the interaction:

$$T = T_S + T_W \tag{C.32}$$

the scattering operator can be written as

$$S = S_S - i(2\pi)\delta(E)T_W \tag{C.33}$$

$$\text{where} \tag{C.34}$$

$$S_S = 1 - i(2\pi)\delta(E)T_S \tag{C.35}$$

The unitarity condition (C.9) can now be written as

$$S_S^{\dagger} S_S + i\left(T_W^{\dagger} S_S - S_S^{\dagger} T_W \right) + (2\pi)\delta(E)T_W^{\dagger} T_W = 1$$

which at zeroth order in T_W gives $S_S^{\dagger} S_S = 1$ and at first order

$$T_W^{\dagger} = S_S^{\dagger} T_W S_S^{\dagger} \tag{C.36}$$

instead of the simple Hermiticity of T_W in the Born approximation.

In the case of strongly interacting particles the first-order approximation in the weak interaction which induces a transition is not simply equivalent to taking the (Hermitian) weak Hamiltonian as the transition matrix.

APPENDIX D

THE DALITZ–FABRI PLOT

In this appendix a widely used construction for analysing three-body decays is discussed.

The Lorentz-invariant phase space element $d(\text{LIPS})$ for the three-body decay of a particle of mass M into three particles of masses m_i $(i = 1, 2, 3)$ is

$$d(\text{LIPS}) = \frac{1}{(2\pi)^9} \frac{d\boldsymbol{p}_1}{2E_1} \frac{d\boldsymbol{p}_2}{2E_2} \frac{d\boldsymbol{p}_3}{2E_3} \delta(E - E_1 - E_2 - E_3) \delta^3\left(\boldsymbol{P} - \sum_i \boldsymbol{p}_i\right)$$

$$d\boldsymbol{p}_i = \boldsymbol{p}_i^2 \, d|\boldsymbol{p}_i| \, d\Omega_i = \boldsymbol{p}_i^2 \, d|\boldsymbol{p}_i| \, d\phi_i \, d\cos\theta_i$$

where $\boldsymbol{p}_i$, E_i are the momenta and energies of the three particles, θ_i, ϕ_i their polar and azimuthal angles, and $\boldsymbol{P}$, E the momentum and energy of the decaying one. Due to energy-momentum conservation this expression actually depends on only two independent variables. The integration on one of the momenta is cancelled by the momentum delta function; the uniformly distributed variables Ω_1 and ϕ_2 can be integrated over: the first gives the overall orientation of the decay in space and the second one the azimuthal orientation of particle 2 with respect to particle 1 (having defined the z axis along the direction of particle 1) uniform for spinless particles. Therefore in this case

$$d(\text{LIPS}) = \frac{\pi^2}{(2\pi)^9} \frac{\boldsymbol{p}_1^2 \, \boldsymbol{p}_2^2}{E_1 \, E_2 \, E_3} \frac{1}{} \, d|\boldsymbol{p}_1| \, d|\boldsymbol{p}_2| \, d\cos\theta_2 \, \delta(E - E_1 - E_2 - E_3)$$

where with the chosen set of axis θ_2 is the angle between particles 1 and 2. Working in the centre of mass (but the same is true for any fixed $\boldsymbol{P}$) $E_3^2 = m_3^2 + |\boldsymbol{p}_1|^2 + |\boldsymbol{p}_2|^2 - 2|\boldsymbol{p}_1||\boldsymbol{p}_2|\cos\theta_2$ so that for fixed $|\boldsymbol{p}_1|$, $|\boldsymbol{p}_2|$ one has $E_3 dE_3 = -|\boldsymbol{p}_1||\boldsymbol{p}_2| d\cos\theta_2$ and the integration over the angular variable is canceled by the energy delta function, resulting in

$$\int d(\text{LIPS}) = \frac{\pi^2}{(2\pi)^9} \int dE_1 dE_2$$

which indicates a uniform bi-dimensional distribution in the plane (E_1, E_2).

The two independent variables describing the decay configuration can be chosen conveniently as the kinetic energies in the centre of mass frame $T_i^* = E_i^* - m_i$

(where the star indicates variables in this frame), or some linear combination of them: a standard choice (explicitly Lorentz-invariant) is[169]

$$u \equiv \frac{s_3 - s_0}{m^2} \qquad v \equiv \frac{s_1 - s_2}{m^2} \tag{D.1}$$

where m is one of the m_i and

$$s_i \equiv (P - p_i)^2 = (M - m_i)^2 - 2MT_i^* \tag{D.2}$$

$$s_0 \equiv \sum_i s_i/3 = \left(M^2 + \sum_i m_i^2\right)/3 \tag{D.3}$$

In case of equal masses $m_i = m$ the s_i range between $(s_i)_{\min} = 4m^2$ and $(s_i)_{\max} = (M - m_i)^2$.

A common application (for which this was invented) is to $K \rightarrow 3\pi$ decays: conventionally, in this case the subscript 3 is used for the 'odd' pion, i.e. the one with charge opposite to that of the decaying particle for $K^{\pm} \rightarrow \pi^{\pm}\pi^{\pm}\pi^{\mp}$, the charged one for $K^{\pm} \rightarrow \pi^0\pi^0\pi^{\pm}$, the neutral one for $K^0 \rightarrow \pi^+\pi^-\pi^0$.

Plotting two independent variables on a plane (the 'Dalitz–Fabri plane'), each decay configuration is represented by a point, constrained by energy-momentum conservation to lie within a given boundary. Given (D.3) a convenient construction (Dalitz, 1953; Fabri, 1954) is that in which each $s_i - (s_i)_{\min}$ represents the distance of the point from one side of an equilateral triangle (see Fig. D.1) of side $2\sqrt{3}(M^2/3 - 5m^2)$; the centre of the Dalitz plane corresponds to a decay configuration symmetric in the three particles, with $s_i = s_0$ and $u = 0 = v$.

In the non-relativistic limit ($Q \equiv M - \sum_i m_i \rightarrow 0$) the physical boundary is the circumference inscribed in the triangle; using relativistic kinematics (which is a minor correction in the case of $K \rightarrow 3\pi$ decays for which $Q \simeq 80$ MeV $< m_i$) the actual physical boundary is distorted from a circumference, while still being tangential to the triangle at the centre of each side. The exact equation for the physical boundary is (in case of equal masses)

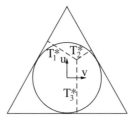

FIG. D.1. The Dalitz–Fabri plane; the circular non-relativistic boundary is shown.

---

[169] Sometimes in the literature u and v are denoted y and x respectively.

$$\frac{1}{M^2}(s_0 - s_1)(s_0 - s_2)(s_0 - s_3) + \frac{s_0^2}{2M^2}\left(\sum_i s_i^2 - 3s_0^2\right) = \frac{1}{27}\left(M + \sum_i m_i\right)^2 Q^2$$

$$\text{(D.4)}$$

and the limits of the u, v ranges are (assuming $m_1 = m_2$).

$$u_{\text{min}} = 4 - s_0/m_1^2$$

$$u_{\text{max}} = 2(s_0/m_1^2 - 1) - \sqrt{(3s_0/m_1^2 - 2)^2 - (M^2 - m_3^2)^2/m_1^4}$$

$$v_{\text{max}}^2(u) = u^2 - 4u\, s_0/m_1^2 - (s_0/m_1^2 - 4)(5s_0/m_1^2 - 4)$$

$$+ [1 - 4/(u + s_0/m_1^2)](M^2 - m_3^2)^2/m_1^4$$

In the non-relativistic limit the boundary (D.4) reduces to

$$0 \le u^2 + v^2 \le Q^2/3$$

All the points on the boundary correspond to collinear decay configurations, those on the sides of the triangle correspond to events in which one of the particles is at rest in the centre of mass frame, and those diametrically opposite to two particles with equal and parallel (minimum) momentum.

The Dalitz plane also has a uniform measure in the u, v variables: $du\, dv = 8(M^2/m_\pi^4)\, T_1^* T_2^*$. Since the phase space measure element is proportional to the area in the plane, any departure from uniformity in the distribution reflects the non-trivial dynamics of the process; this is the usefulness of this representation, which factorizes the trivial phase space dependence. For example different spin-parity assignments for the decaying particle lead to different distributions in the (u, v) plane, and if e.g. particles 1 and 2 form a resonance an enhancement is expected for $s_3 = (p_1 + p_2)^2$ corresponding to its mass.

The decay distribution in the Dalitz plot has traditionally been parametrized in a generic way with the first few terms of a series expansion around its centre:

$$|A|^2 \propto 1 + gu + hu^2 + jv + kv^2 + fuv + \cdots \tag{D.5}$$

Note that if two of the final particles ($i = 1, 2$) are identical (as for $K^\pm \to \pi^\pm\pi^\pm\pi^0, \pi^0\pi^0\pi^\pm$) odd terms in v cannot appear, and in case of three identical particles (as for $K^0 \to 3\pi^0$) g, j, f are not defined and a single quadratic term is present since $u^2 = v^2/3$. With particles $i = 1, 2$ being CP-conjugate (as for $K^0 \to \pi^+\pi^-\pi^0$) a non-zero value for the parameters j or f indicates the presence of CP violation.

The above parametrization is of course just an approximation, which works reasonably well for $K \to 3\pi$ in which the available phase space is small and no $\pi\pi$ resonances with mass below m_K exist, although even in this case accurate measurement shows that the actual matrix element is more complicated and cannot be written in such a form (Batley *et al.*, 2006a).

APPENDIX E

SPIN AND POLARIZATION

Nothing shocks me. I'm a scientist.
Indiana Jones

You've never seen nothing like it,
no never in your life.
F. Mercury

Spin is a fundamental intrinsic property of elementary particles which, despite the analogies with the mechanical phenomenon, exhibits its non-classical nature in several ways, sometimes puzzling for the uninitiated (see Tomonaga (1997) for a history with great insight). The relativistically covariant description of spin was developed in the early times of quantum field theory, but is not often presented in modern introductory particle physics textbooks; being very useful for the understanding of several experiments, in which one rarely works in the particle rest system, we briefly review it here.

E.1 Polarization

The polarization of a beam of massive particles is defined in their rest frame: a polarized beam has a net angular momentum in such a frame, which can only be due to a non-uniform and asymmetric population of the possible spin states for the particles. The *polarization vector* is defined as the expectation value of the spin operator S in units of the spin j itself:

$$\mathcal{P} = \langle S \rangle / j\hbar \tag{E.1}$$

e.g. the x component of the polarization is the normalized expectation value of S_x, the net angular momentum along the direction x in the particle rest frame. When dealing with non-pure states, i.e. a statistical ensemble, the polarization is the expectation value averaged over the ensemble.

The modulus of the polarization vector $\mathcal{P} = |\mathcal{P}|$ (usually called the *degree of polarization*, or simply the polarization) ranges between 0 (unpolarized state) and 1 (fully polarized state); intermediate values correspond to partial polarization.

Note that a spin 1/2 particle in a pure state is always completely polarized in some direction ($|\mathcal{P}| = 1$): for example a particle in the pure state $|S, S_z\rangle = |1/2, +1/2\rangle$ has a polarization vector $\mathcal{P} = \hat{z}$; the state $(1/\sqrt{2})[\,|1/2, +1/2\rangle + |1/2, -1/2\rangle\,]$, with equal weights for the two spin states along $\hat{z}$, has no net polarization along such axis but its polarization vector is $\mathcal{P} = \hat{x}$, again of modulus 1. On the contrary an ensemble of spin 1/2 particles, half of which are in the state $|1/2, +1/2\rangle$ and half in the state $|1/2, -1/2\rangle$ has zero polarization.

In general, equal population of the $2j+1$ spin states implies a spatially isotropic distribution, while an unequal population singles out a preferred direction in space and can give rise to anisotropies; if the (unequal) relative population of the sub-states is independent of the sign of S_z the system is said to be *aligned* (determining a privileged axis but not a privileged direction, i.e. possessing a cylindrical symmetry around the spin quantization axis and also a reflection symmetry with respect to a plane orthogonal to such axis), resulting in no net magnetic moment, while if this is not the case the system is said to be *polarized*, a privileged direction being also defined (and reflection symmetry being lost).

Particles with integer spin can have zero polarization also for pure states: a spin 1 particle in the state $|S, S_z\rangle = |1, 0\rangle$ is unpolarized, as is one in the state $\alpha[\,|1, +1\rangle + |1, -1\rangle\,] + \beta|1, 0\rangle$; in the latter case if $\alpha > \beta$ the state has alignment along the z axis.

In the case of particles with zero rest mass no rest frame exists, and the polarization must be defined differently: we consider for the moment particles with non-zero mass and come back to massless ones in Section E.4.

Note that for a fully polarized spin j state the direction of polarization is the one along which every spin measurement would yield the maximal value $\hbar j$, rather than the direction along which all the spins of an ensemble are pointing (which is actually not defined), as the magnitude of the spin vector is not $\hbar j$ but $\hbar\sqrt{j(j+1)}$ (the difference being related to the uncertainty principle). Also, the polarization is a property of a state or an ensemble of particles, rather than an observable: for a completely unpolarized beam of spin 1/2 particles individual spin measurements along any given direction would always yield either $+\hbar/2$ or $-\hbar/2$, and the same is true even for a beam which is fully polarized in a direction orthogonal to that. If it is possible to determine (e.g. by repeated measurements on an ensemble of identical particles) the probabilities $P(\pm; k)$ of measuring either a positive or a negative value for the spin component along direction k then the corresponding component of the polarization is $\mathcal{P}_k = [P(+; k) - P(-; k)]/[P(+; k) + P(-; k)]$. The complete information on the polarization of the ensemble is then obtained by performing such measurement along three mutually orthogonal directions to determine $\mathcal{P}$.

For the description of mixed (non-pure) spin states the formalism of the density matrix is used (see e.g. Messiah (1961)): in this case the matrix has dimension $(2j + 1) \times (2j + 1)$ and involves $(2j + 1)^2 - 1$ real parameters. As an example, in the case of spin 1/2 particles the 2×2 density matrix can be written in general with the help of the Pauli matrices σ_i as[170]

$$\rho = \frac{1}{2}[1 + \mathcal{P} \cdot \boldsymbol{\sigma}] \qquad\qquad (E.2)$$

which has $\text{Tr}(\rho) = 1$, as can be verified using eqn (E.1) and $\langle S \rangle = \text{Tr}(S\rho)$.

For scattering reactions in which one of the particles has non-zero spin many spin-dependent observables exist. The simplest is the *analysing power* $\mathcal{A}$: in the scattering of a spin 0 beam particle from a spin 1/2 target particle this is defined as

$$\mathcal{A}(\theta) = \frac{1}{\mathcal{P}_T} \frac{N_L - N_R}{N_L + N_R} \qquad\qquad (E.3)$$

where $N_{L,R}$ denote the number of events in which the particle is scattered at an angle θ to the left or to the right, and $\mathcal{P}_T$ is the polarization of the target. Instead of using two spectrometers placed at symmetric angles with respect to the forward direction, $\mathcal{A}$ can be measured by using a single one at angle θ and comparing the number of events $N_\pm$ with the target polarization in opposite directions (e.g. up and down):

$$\mathcal{A}(\theta) = \frac{1}{\mathcal{P}_T} \frac{N_+ - N_-}{N_+ + N_-} \qquad\qquad (E.4)$$

which is much more cost effective; errors related to the geometry of the two spectrometers in the first approach are traded for errors related to the equality of the target polarization in the two opposite directions, and possible time differences (e.g. efficiency, normalization) between the two measurements.

A difficulty arising in the analysis of polarized target data is due to the fact that only the protons in hydrogen atoms contained in the target material are polarized, while any remaining nuclei are not.

More observables can be measured by using both polarized beams and targets.

E.2 Covariant polarization

In order to know the polarization in other frames of reference, the transformation properties of $\mathcal{P}$ must be known. According to Ehrenfest's theorem (see e.g. Messiah (1961)) the expectation value of a quantum-mechanical observable follows a classical equation of motion: in the case of spin S, for a particle with magnetic moment $\boldsymbol{\mu} = g\mu_0 S$ ($\mu_0 = Q/2m$ for a charged particle with charge Q, see eqn (4.91)) in

---

[170] The set of four numbers $(I, \mathcal{P})$, with I the intensity, are called Stokes parameters.

presence of a magnetic field $\boldsymbol{B}$ this gives (in the particle rest frame)

$$\frac{d\langle S\rangle}{dt} = \langle\boldsymbol{\mu}\rangle \times \boldsymbol{B} = g\mu_0 \langle S\rangle \times \boldsymbol{B} \tag{E.5}$$

$$\frac{d\mathcal{P}}{dt} = g\mu_0 \mathcal{P} \times \boldsymbol{B} \tag{E.6}$$

see (4.112).

Consider an ensemble of spin 1/2 particles described by a density matrix (E.2) in a region with a magnetic field $\boldsymbol{B}$; the time evolution of the polarization vector, which in this case is $\mathcal{P} = \langle\boldsymbol{\sigma}\rangle$, can be obtained from the time evolution of the density matrix $i\hbar\,\partial\rho/\partial t = [H(t), \rho(t)]$ with the Hamiltonian being $H = -\boldsymbol{\mu}\cdot\boldsymbol{B}$ (see eqn (4.93)):

$$i\hbar\frac{d\mathcal{P}_k}{dt} = i\hbar\frac{d\langle\sigma_k\rangle}{dt} = i\hbar\frac{\partial}{\partial t}\,\mathrm{Tr}(\rho\sigma_k) = i\hbar\,\mathrm{Tr}\left(\frac{\partial\rho}{\partial t}\sigma_k\right) = \mathrm{Tr}([H,\rho]\,\sigma_k)$$

$$= \mathrm{Tr}([\sigma_k, H]\,\rho) = -g\mu_0\frac{\hbar}{2}B_j\,\mathrm{Tr}([\sigma_k, \sigma_j]\,\rho)$$

$$= -g\mu_0\frac{\hbar}{4}B_j\left[\mathrm{Tr}([\sigma_k, \sigma_j]) + \mathcal{P}_l\mathrm{Tr}([\sigma_k, \sigma_j]\sigma_l)\right] = -g\mu_0\hbar\,i\epsilon_{kjl}B_j\mathcal{P}_l$$

which is just eqn (E.6) in vector form.

The polarization vector can be considered as the space component of a covariant polarization vector $s^\mu \equiv (s^0, s)$, defined by

$$s_R^\mu = (0, \mathcal{P}) \quad \text{in the particle rest frame} \tag{E.7}$$

$$s^2 = -\mathcal{P}^2 \tag{E.8}$$

(one also needs to check that the quantity s^μ as defined actually transforms as a four-vector, which indeed it does). Note that the Lorentz-invariant product of s^μ with the four-velocity $v^\mu = (\gamma, \gamma\boldsymbol{\beta})$ vanishes

$$s^\mu v_\mu = s^0 v^0 - s\cdot v = 0 \tag{E.9}$$

as can be verified by evaluating it in the rest frame; this allows one to compute the time derivative of the s^0 component in this frame:

$$\frac{ds^\mu}{dt}v_\mu = -s^\mu\frac{dv_\mu}{dt} = \left(\frac{ds^0}{dt}\right)_R \tag{E.10}$$

where the last equality holds in the rest frame.

The unique covariant generalization of the equations of motion for $\mathcal{P}$ (E.6) and s^0 (E.10) is (Bargmann et al., 1959)

$$\frac{ds^\mu}{d\tau} = g\mu_0\left[s_\nu F^{\nu\mu} - (s_\nu v_\rho F^{\nu\rho})\,v^\mu\right] - \left(s_\rho\frac{dv^\rho}{d\tau}\right)v^\mu \tag{E.11}$$

where $\tau \equiv t/\gamma$ denotes proper time and $F^{\mu\nu}$ is the electromagnetic field strength tensor.[171] Indeed since the above equation is manifestly covariant and reduces in the rest frame to the equations (E.6) and (E.10) it is the correct generalization of those.

In a homogeneous field the equation of motion for a charged particle with charge Q and mass m is $dv^\mu/d\tau = -(Q/m)F^{\mu\rho}v_\rho$, so that setting $\mu_0 = Q/2m$ eqn (E.11) reduces to

$$\frac{ds^\mu}{d\tau} = \frac{Q}{2m}\left[gS_\nu F^{\nu\mu} - (g-2)(S_\nu v_\rho F^{\nu\rho})v^\mu\right] \tag{E.12}$$

For a neutral particle instead

$$\frac{ds^\mu}{d\tau} = g\mu_0\left[S_\nu F^{\nu\mu} - (S_\nu v_\rho F^{\nu\rho})v^\mu\right] \tag{E.13}$$

The expression of s^μ in any frame is now easily obtained: in the laboratory frame in which the particle moves with velocity $\boldsymbol{\beta}c$ one has

$$s_L^\mu = \left(\gamma\boldsymbol{\beta}\cdot\boldsymbol{\mathcal{P}}, \boldsymbol{\mathcal{P}} + \boldsymbol{\beta}\frac{\gamma^2}{\gamma+1}\boldsymbol{\beta}\cdot\boldsymbol{\mathcal{P}}\right) \tag{E.14}$$

The three-vector s_L in the laboratory frame does not have any immediate interpretation: both its magnitude and direction depend on β:

$$s_L^2 = \mathcal{P}^2\left[1 + \beta^2\gamma^2\cos^2\theta_R\right] \tag{E.15}$$

where θ_R is the angle between the polarization direction (in the rest frame) and the boost direction $\boldsymbol{\beta}$; eqn (E.14) indicates that at relativistic velocities ($\beta \to 1$) s_L becomes either parallel or anti-parallel to $\boldsymbol{\beta}$, their relative angle being

$$\cos\theta_L = \sqrt{\frac{\gamma^2\cos^2\theta_R}{1 + \beta^2\gamma^2\cos^2\theta_R}} \tag{E.16}$$

so that the helicity is

$$\lambda = \gamma\,\mathcal{P}\cos\theta_R \tag{E.17}$$

Note that the transverse component of the polarization is not affected by the Lorentz boost: $s_L = s_R = \mathcal{P}$ if $\cos\theta_R = 0$.

Taking as an example the $\pi \to \mu\bar{\nu}$ decay discussed in Chapter 2, in the rest frame of the pion the muon has a longitudinal polarization (due to the violation of parity symmetry); if the pion is moving in the laboratory the muon will also have a transverse polarization component there.

---

[171] It was implicitly assumed that the particle has no electric moments nor magnetic moments higher than the dipole one (see Bargmann *et al.* (1959) and Section 4.4.3 for the case of an electric dipole moment).

E.3 Time evolution

The degree of polarization $\mathcal{P}$ is a Lorentz-invariant quantity (E.8) which is conserved in interactions with electromagnetic fields,[172] since using the equation of motion (E.11) and recalling eqn (E.9)

$$\frac{ds^2}{d\tau} = 2s_\mu \frac{ds^\mu}{d\tau} = g\mu_0 \left[s_\mu s_\nu F^{\nu\mu} - (s_\nu v_\rho F^{\nu\rho}) s_\mu v^\mu \right] - \left(s_\rho \frac{dv^\rho}{d\tau} \right) s_\mu v^\mu = 0$$

What is most relevant is therefore the change in direction of the polarization vector, and in particular how the angle between the polarization (in the rest frame) and the direction of motion changes.

Two unit vectors $\hat{\ell}$ and $\hat{n}$ are introduced in the laboratory frame: the first parallel to the direction of motion of the particle $\boldsymbol{\beta} = \beta\,\hat{\ell}$, and the second orthogonal to it in the plane containing $\hat{\ell}$ and the vector s_L. The covariant polarization vector in the laboratory frame (E.14) can be written as

$$s_L^\mu = \mathcal{P} \left(L_L^\mu \cos\theta_R + N_L^\mu \sin\theta_R \right) \tag{E.18}$$

having defined the two four-vectors L^μ, N^μ which in the laboratory frame have components

$$L_L^\mu = \gamma\,(\beta, \hat{\ell}) \qquad N_L^\mu = (0, \hat{n})$$

Note that $L^2 = N^2 = -1$ and $L^\mu N_\mu = L^\mu v_\mu = N^\mu v_\mu = 0$; note also that the expression (E.18) is also valid for the covariant polarization vector in the rest frame, where

$$L_R^\mu = (0, \hat{\ell}) \qquad N_R^\mu = (0, \hat{n})$$

and in particular the 'angle' between s^μ and L^μ from (E.18) is just θ_R, the angle between $\mathcal{P}$ and the direction of motion $\hat{\ell}$ in the rest frame.

Inserting (E.18) into the equation of motion (E.12) for the case of homogeneous fields and evaluating the derivatives of L^μ and N^μ explicitly one finally obtains

$$\frac{d\theta_R}{dt} = \left(g\mu_0\beta - \frac{Q}{m\beta} \right) \boldsymbol{E} \cdot \hat{n} + \left(g\mu_0 - \frac{Q}{m} \right) \hat{\ell} \cdot \boldsymbol{B} \times \hat{n} \tag{E.19}$$

in terms of the electric and magnetic fields $\boldsymbol{E}, \boldsymbol{B}$ in the laboratory frame; for $Q \neq 0$ the above equation can be written as

$$\frac{d\theta_R}{dt} = \frac{Q}{2m} \left[\frac{(g-2) - g/\gamma^2}{\beta} \boldsymbol{E} \cdot \hat{n} + (g-2)\,\hat{\ell} \cdot \boldsymbol{B} \times \hat{n} \right] \tag{E.20}$$

---

[172] In presence of inhomogeneous fields the trajectory of a particle depends on its polarization, so that beam particles might be deflected differently and experience different fields, thus affecting the overall degree of polarization (Good, 1962).

The above equations simplify in specific cases: an interesting one is that of a magnetic field orthogonal to the particle trajectory, $\boldsymbol{E} = 0$ and $\boldsymbol{B} = |\boldsymbol{B}|\,\hat{\boldsymbol{n}} \times \hat{\boldsymbol{\ell}}$, when

$$\frac{d\theta_R}{dt} = \left(g\mu_0 - \frac{Q}{m}\right)|\boldsymbol{B}| = \frac{Q|\boldsymbol{B}|}{2m}(g - 2) \qquad \text{(E.21)}$$

so that the angle between the polarization vector and the direction of motion varies at a constant rate if the particle has an anomalous magnetic moment ($g \neq 2$), independently of the particle velocity: this is actually what is exploited for the measurement of $g - 2$.

The component of the polarization along the direction of motion is called *longitudinal*, and the one orthogonal to it *transverse*.

Longitudinal polarization cannot be produced by scattering (if parity symmetry holds) nor measured from scattering asymmetries; however longitudinal and transverse polarization can often be transformed into each other: if the Larmor frequency (precession of a magnetic dipole) and the cyclotron frequency (revolution of a charged particle) are different (as is the case for the proton) such a transformation can be performed by having the particle passing through a suitably oriented magnetic field, in which the momentum and the spin rotate at different rates (the momentum does not rotate at all for a neutral particle). For an electron such frequencies are almost the same so that the above technique is not usable, but one can rather rotate only the momentum (in an electric field, at low energies) or only the spin (in crossed electric and magnetic fields).

E.4 Massless particles

For particles with zero mass ($m = 0$) the construction of a covariant polarization vector presented above is not possible, since no rest frame is defined; indeed for $\gamma \to \infty$ the four-vector of eqn (E.14) diverges. In this case it is possible to consider instead the four-vector $W^\mu = ms^\mu$, which has a finite limit for $m \to 0$:

$$W^\mu \to s_R \cos\theta_R\,(E, E\hat{\boldsymbol{\ell}}) = \lambda P^\mu$$

where P^μ is the four-momentum of the massless particle. The quantity λ is obviously Lorentz-invariant and can be called the degree of polarization for the particle in this case.

The direction of polarization is always parallel ($\lambda > 0$) or anti-parallel ($\lambda < 0$) to the direction of motion; in this sense the properties of the massless spin 1 photon are closer to those of a massive spin 1/2 particle, in that there are only two possible polarization states. In this case the degree of polarization is the same in any reference frame, i.e. the helicity is always $\pm\lambda$, and there is no need for an equation of motion for the polarization.

The above fact can also be seen from eqn (E.16): as the particle speed increases the longitudinal component of s_L also increases in modulus; for a massless particle, which has no rest frame and moves at the speed of light, the angle between s_L and the direction of motion becomes either 0 or π. Thus the fact that a massless particle has only two possible polarization states is not due to some property of spin but is just a consequence of Lorentz transformations (Wigner, 1939). Actually, since for a massless particle the helicity is a Lorentz invariant quantity, only one polarization state would be required by special relativity (either $\lambda > 0$ or $\lambda < 0$): the second one is actually necessary only because of parity symmetry, which implies that left-handed polarized light must exist if right-handed one does.

For mixed states an expression similar to (E.2) can be used for photons, which also have two possible polarization states (e.g. linear horizontal and vertical, or circular right and left):

$$\rho = \frac{1}{2}[1 + \boldsymbol{\xi} \cdot \boldsymbol{\sigma}] \qquad \text{(E.22)}$$

although in this case $\boldsymbol{\xi}$ is not the photon polarization vector but rather a vector in what is called 'Poincaré space':[173] if the two states defining the basis of the 2×2 space are chosen to be those of vertical and horizontal polarization respectively, then the components of $\boldsymbol{\xi}$ correspond to:

- $\xi_1 = +1(-1)$: full vertical (horizontal) linear plane polarization;

- $\xi_2 = +1(-1)$: full right (left) circular polarization (helicity or longitudinal polarization);

- $\xi_3 = +1(-1)$: full $45°$ ($135°$) linear plane polarization;

while if $|\boldsymbol{\xi}| < 1$ the photon beam is partially polarized. This description of photon polarization is invariant with respect to Lorentz transformations between reference systems with relative velocities along the direction of the photon momentum, so that there is no need to work in any special frame.

More details can be found e.g. in Hagedorn (1963).

[173] The four numbers $(I, \boldsymbol{\xi})$, with I being the intensity, are the Stokes' parameters for the photon.

APPENDIX F

SYSTEMATIC ERRORS

The aim of science is not to open
the door to infinite wisdom,
but to set a limit on infinite error.
B. Brecht

Statistics: The only science that enables
different experts using the same figures
to draw different conclusions.
E. Esar

There are three kinds of lies:
lies, damned lies, and statistics.
B. Disraeli

A student once came to the author for advice, unhappily remarking about his inability to find any good textbook on the subject of systematic errors. Any experimentalist would probably agree on the fact that the handling of systematic errors amounts to the most important and most challenging part of her or his work, and one which cannot be reduced to a well defined set of rules. While the reader of this book has had the opportunity to appreciate several examples of how systematic errors are treated in successful experiments, this appendix briefly discusses a few general issues and tools of the trade, without having the presumptuous goal of 'teaching' something which is really only learned by personal experience.

F.1 Of errors, mistakes and experiments

The result of an experiment is ultimately a number (or a set of numbers) with an attached error. The importance of the latter is not at all secondary with respect to the former, and the error might actually be the more important figure; while an experimental result might sometimes be just an error (or a range of variation of a physical quantity), a result in the form of a numerical value with no error estimate at all is of no use whatsoever. The error is often quoted as two distinct parts, the first one (by convention) usually being the statistical part and the second the systematic part: the comparison of the two already gives some information on the limitations of an experiment.

While statistical research produces an impressive amount of highly sophisticated mathematical results, found in scholarly books and journals, and also a fair amount of useful techniques for treating the data, with which the physicist should be familiar, most of these deal with errors of statistical nature, introduced by the finite sampling of an idealized parent probability distribution. Recipes on how to handle systematic errors, which might be loosely defined as all kinds of error which are not statistical, are much scarcer, for several rather obvious reasons.

First of all many widely different effects go under the name 'systematic': backgrounds, detector and data analysis efficiencies, measurement biases, resolution effects, time-variations of instrument properties, dead times, detector calibration, etc. Clearly, each of these effects has its own peculiarities and usually cannot be treated in the same way as another. Second, even in well-designed experiments some effects can fall in the systematic category which might be called 'mistakes'; there is clearly no such thing as a 'theory' of mistakes.[174]

The uncertainty in the estimation of a systematic effect is called systematic error, and often requires an amount of experimental effort which can even exceed that needed to perform the measurement in the first place, as several experiments described in this book show.

The best way to handle systematic effects is not to have any: indeed a well-designed experiment is one in which, by careful design, the largest number of potentially important sources of perturbation do not influence the result, for example because their effect cancels in the ratio (or the difference) of two physical quantities. Even if such cancellations are never perfect in practice, good experimental design leaves only 'second-order' (smaller) effects to be cared for; this helps because, while there is no strict connection between the size of a systematic effect and the corresponding systematic error, generally speaking it is clear that smaller effects give smaller errors, as it is very difficult to estimate a (large) correction with very high relative precision. As an example, an experiment in which the signal sits on a very large background might not be affected by it in an important way if the size of the background can be very accurately estimated by a large data sample: in practice however one would have to care also about the uncertainty in the shape of the background in the region below the signal, where it cannot be measured separately, possible differences between the effect of perturbations on the signal region and on the region from which the background is estimated, etc.; all such uncertainties are clearly proportional to the amount of background to be subtracted, and thus diminish with it.

---

[174] On the other hand the neglect of a systematic effect is also a mistake.

Whenever a systematic effect cannot be cancelled or reduced by design, the experimental set-up should provide the means of estimating it to an adequate accuracy: a somewhat larger but measurable systematic effect is much better than a smaller one whose size is completely unknown and can only be estimated a priori on paper.

In some cases a systematic effect is estimated in a separate experiment, specifically designed for such purpose, focusing on a single source of disturbance which might affect the result, while all the rest of the experimental apparatus and procedure is left untouched, in the same conditions used for the main measurement. Since usually the possibility that disturbing effects change with time cannot be completely dismissed, such ancillary experiments must often be performed *simultaneously* with the the main measurement: this calls for the interleaving of the data collection for the two purposes, so that the systematic effect measured in the control experiment is as faithful a representation of the one affecting the main measurement as possible. This also implies that such a procedure has to be foreseen right from the start in the design of the experiment; this point leads again to the rather trivial but important point that the quality of an experiment mostly depends on its design.

Finally, there are obviously some systematic effects which neither cancel in the measurement nor are continuously monitored during the data taking: this happens for example when the presence of some source of disturbance only becomes apparent a posteriori after the experiment has been performed. In such cases, if ancillary measurements cannot be performed at a later time, the experimentalist has to resort to the analysis of the collected data: redundant information, not strictly necessary for the main measurement, then becomes precious, and one can try to make a model of the perturbing mechanism and estimate its effect on the measured quantity by measuring the effect on one or more other quantities, whose values are known by other means (maybe because they are fundamental constants, or because they have been measured with sufficiently high accuracy in other experiments). When no other weapon can be used, the last resort is hard cuts: a part of the data, which could possibly be affected by an uncontrollable systematic effect, is simply rejected and not used for the measurement; the quality of the result lies in the size (and reliability) of its error, so that it is often fruitful to trade some data (larger statistical error) for better control (smaller systematic error), resulting in an overall smaller uncertainty.

F.2   Treatment of systematic errors

Usually an experiment will be affected by more than one systematic effect, with the corresponding errors being often of the same order of magnitude (in a case of good design). The issue therefore arises of how the individual contributions to the systematic error should be combined: there are many empirical approaches to this problem, none of which seems to lie on solid arguments. One problem arises from the fact that by definition the statistical variations between two measurements of the same quantity are uncorrelated, while systematic effects might be correlated: if the correlation coefficient were known it could be used to provide a proper combination of those, but this is usually not the case and for lack of a better procedure the different systematic errors are often treated as being independent (resulting in an error which can be either larger or smaller than what it ought to be). The above assumption amounts to adding the errors in quadrature, but sometimes another procedure is adopted in which the errors are added linearly: this amounts to assuming total (positive) correlation between the individual errors, and usually results in an overestimate of the total systematic error.

In the case of asymmetric errors the positive and negative contributions are often added separately to obtain the overall error: this procedure is however inconsistent with the assumed non-linearity (implicit in the use of asymmetric errors), and a better one was suggested in Barlow (2003), based on the use of quantities which sum algebraically under convolution. Using the simplest (quadratic) asymmetric continuous dependence of the measured quantity x on the unknown perturbing factors, assumed (in a Bayesian sense) to have Gaussian distributions, for a central value x_0 and n asymmetric systematic errors σ_i^+, σ_i^- ($i = 1, \ldots, n$), the overall bias δ, variance V, and skew γ are formed as

$$\delta = \sum_i (\sigma_i^+ - \sigma_i^-)/2$$

$$V = \sum_i \left[(\sigma_i^+ + \sigma_i^-)^2/4 + (\sigma_i^+ - \sigma_i^-)^2/2 \right]$$

$$\gamma = \sum_i \left[3/4(\sigma_i^+ + \sigma_i^-)^2(\sigma_i^+ - \sigma_i^-) + (\sigma_i^+ - \sigma_i^-)^3 \right]$$

and then the equation

$$\alpha = \frac{\gamma}{6V - 4\alpha^2}$$

is solved numerically for α; the final result is

$$x_0 + \sum_i (\sigma_i^+ - \sigma_i^-)/2 - \alpha$$

with asymmetric errors

$$\sigma^\pm = \sqrt{V - 2\alpha^2} \pm \alpha \qquad \text{(F.1)}$$

Note that the asymmetric errors have the effect of *shifting* the result from the central value x_0: this is unavoidable if one really believes in asymmetric errors.

There is some tendency to overestimate the systematic errors, in order to provide a 'conservative' estimate: while this kind of caution might be dictated by wisdom (and should anyway be explicitly indicated, if adopted), it cannot be considered as good practice in general: the quoted error should actually reflect the best honest estimate of the experimenter corresponding to the canonical 68% confidence level interval.

In a generic experiment one has a set of observations x_i $(i = 1, \ldots, n)$ of the quantity x, from which an inference on an unknown parameter θ is desired, within the context of a model providing a probability distribution function $p(x_i|\theta)$; systematic effects are included as some other unknown parameter ξ introducing a disturbance in the measurement. With such ingredients a likelihood function can be written as

$$\mathcal{L}(\theta, \xi) = \prod_i p(x_i|\theta, \xi)$$

The formal way to proceed actually depends on whether the Bayesian or the frequentist conceptual approach is adopted: we cannot discuss the good and the bad of the two approaches (see e.g. Cousins and Highland (1992) for an introduction), nor are we going to indicate a preference for either of them.[175]

In a Bayesian approach one assumes a (somewhat arbitrary) prior probability distribution $\pi(\xi)$ for ξ, reflecting the subjective degree of belief that ξ actually has a given value (e.g. 'it can have any value in the range $[\xi_0 - \Delta\xi/2, \xi_0 + \Delta\xi/2]$, with all values being equiprobable'); then a probability ('belief') distribution for θ can be written

$$\int \mathcal{L}(\theta, \xi)\pi(\xi)\, d\xi$$

from which confidence ('credibility') intervals can be formed (Cousins and Highland, 1992). The dependence of the conclusions on the choice of $\pi(\xi)$ should be checked: for example it is often not at all evident in which variable the prior probability should be chosen to be flat, and what is equiprobable in one variable is usually not in another one (an unambiguous – if not necessarily better – choice is provided by the Jeffreys' prior, which is invariant under reparametrization (Jeffreys, 1946)).

In a frequentist approach one does not speak of probability for a parameter to take any value, and for example a likelihood function can be used which is obtained by maximizing $\mathcal{L}(\theta, \xi)$ with respect to ξ for each possible value of θ. Extensive discussion can be found in the literature concerning the best procedure

---

[175] We just dare to state that we know of no significant scientific result whose validity or influence on subsequent developments is affected by the choice of Bayesian or frequentist approach for its analysis: if the two led to different conclusions on some issue then the result would just be considered not to be solid enough.

to incorporate systematic uncertainties into upper limits for the determination of the rate of a rare process, see e.g. Lyons *et al.* (2003).

When a result R depends on the value of a quantity ξ ('nuisance parameter') whose value is not exactly known, the effect of an uncertainty in ξ is usually evaluated using the standard error propagation formulas, according to which, in the general case of several ξ_i ($i = 1, \ldots, n$), with uncertainties σ_i

$$\langle R \rangle \simeq R(\langle \xi_i \rangle)$$

$$\sigma_R^2 \simeq \sum_i \left(\frac{\partial R}{\partial \xi_i} \right)^2 \sigma_i^2 + \left[2 \sum_{j<k} \frac{\partial R}{\partial \xi_j} \frac{\partial R}{\partial \xi_k} \rho_{jk} \sigma_j \sigma_k \right]$$

with $\rho_{jk} = \text{cov}(\xi_j, \xi_k)/(\sigma_j \sigma_k)$ being the correlation coefficients.

The derivatives can be estimated from the results obtained for the values $\xi_i = \langle \xi_i \rangle \pm \sigma_i$: from these the quantities

$$\frac{\partial R}{\partial \xi_i} \simeq \frac{R(\langle \xi_i \rangle + \sigma_i) - R(\langle \xi_i \rangle - \sigma_i)}{2\sigma_i}$$

are evaluated in the linear approximation; more than two points can be used to estimate the derivative in case non-linearities are important (the error on the knowledge of such derivative can be ignored).

A common technique used to estimate systematic errors is that of determining the 'maximum' range ΔR by which the result can be changed by the systematic effect under consideration (while still being consistent with observations): this range is then transformed into a systematic error of size $\Delta R / \sqrt{12}$, under the assumption that the actual effect might be anywhere within the range ΔR.

Sometimes the size of a systematic error is actually linked to the amount of data available to estimate the effect: in this case the error is of statistical nature, as with a larger control data sample it would be reduced, at least down to some limit where the validity of the assumptions in the procedure used for the estimation starts to be questioned and enters as a separate source of error. It is good practice to quote this kind of error separately, since, for example, its scaling properties with respect to further data-taking are different with respect to other systematic errors.

Proper experimental design would require that the amount of ancillary data collected for the estimation of systematic effects is adequate, so that the resulting systematic error is not overall dominant, e.g. with respect to the statistical error: this is however not always possible, either because the size of the required sample is not known a priori, or because by educated judgement the presence of a systematic effect was estimated to be so unlikely that forfeiting the collection of a significant amount of main-measurement data for the accumulation of a larger control sample was not considered adequate.

As an example we consider the common case of an experiment in which the collection of events is driven by a trigger system which is only partially efficient. Suppose an experiment is designed to count a number N of signal events (to measure e.g. a branching ratio after proper normalization), and that the data is recorded only when some trigger condition $T0$ is satisfied: the measured number of events will be $N_0 = \epsilon_0 N$, where $\epsilon_0 \lesssim 1$ is the trigger efficiency. The error on $N = N_0/\epsilon_0$ will thus be

$$\frac{\sigma(N)}{N} = \frac{\sigma(N_0)}{N_0} \oplus \frac{\sigma(\epsilon_0)}{\epsilon_0}$$

and it is desirable that the error due to the knowledge of the trigger efficiency is smaller than the statistical error $\sigma(N_0)/N_0 = 1/\sqrt{N_0}$ due to the size of the sample. In order to measure ϵ_0, data is also collected when a second independent trigger condition $T1$ (with efficiency ϵ_1) is satisfied, irrespective of $T0$; if $T1$ and $T0$ are indeed independent, (because their sources of inefficiency are uncorrelated), an estimate of ϵ_0 is given by its value as measured on the sample triggered by $T1$: $\epsilon_0 \simeq N_1(T0)/N_1$, where $N_1(T0)$ indicates the number of events of the $T1$-triggered sample for which condition $T0$ is also seen to be satisfied. The corresponding error (for a binomial distribution) is

$$\frac{\sigma(\epsilon_0)}{\epsilon_0} = \frac{\sqrt{\epsilon_0(1 - \epsilon_0)N_1}}{N_1\epsilon_0}$$

Frequently the data recording bandwidth is limited, so that it is necessary to collect only a fraction $1/D$ of the data triggered by $T1$; such *downscaling*, being done in a pseudo-random way, does not affect the efficiency estimate, but it reduces the statistical power of the auxiliary sample thus increasing the error of this estimate. Observing that the true number of signal events is

$$N = N_0/\epsilon_0 = DN_1/\epsilon_1$$

in order for the efficiency error not to exceed the statistical error the condition on the downscaling factor D is

$$D \lesssim \frac{\epsilon_1}{1 - \epsilon_0} \simeq \frac{1}{1 - \epsilon_0}$$

For example, if the trigger efficiency is estimated a priori not to be less than 95%, some auxiliary trigger downscaled by a factor not exceeding 20 should be included in the data-taking. Note that as long as $\epsilon_1 \simeq 1$ the condition above is practically independent of the choice of auxiliary trigger $T1$: *any* (efficient) trigger condition can be chosen and downscaled by a factor D to achieve the same error on the ϵ_0. This appears surprising at first sight but is actually easy to understand: if a looser ('minimum bias') trigger is chosen for $T1$ the corresponding rate (for the same downscaling factor D) will be higher (and harder to accommodate in the limited bandwidth), but the fraction of useful signal events in it ('purity') will be smaller; on the contrary a tighter (but still highly efficient) trigger will have a lower rate and a higher purity[176].

Another class of systematic errors are those due to the propagation of the uncertainty of some quantity which enters the result but is not directly measured by the experiment (such as the value of the fine structure constant, or a cross section or the mass of a particle). Such 'external' errors should also be quoted separately from the others, since they are mostly unrelated to the quality of the measurement itself and would be reduced when a better knowledge of the relevant parameters is available. Still, while the sensitivity of a measurement to the precise value of external parameters might be in some cases unavoidable, it can often be reduced by good experiment design, in particular when some poorly known quantity is involved.

It should also be mentioned that a rather subtle source of bias can come from the experimenter, as influences on the measurement which push it in the direction of previous results or theoretical expectations can be introduced unintentionally. For example, if selection cuts are chosen while knowing the effect these have on the result, the experimenter might tend to tune them in order to eliminate residual background events or to increase the statistical significance of the signal. More subtly, the analysis effort aimed at trying to identify additional possible sources of systematic bias might be stopped when the result reaches a value which meets prior expectations.

Since the experimenter's bias cannot be estimated, techniques have been devised and used in experiments to avoid it completely: such techniques usually go under the name of *blind analysis* (Klein and Roodman, 2005), and consist in artificially hiding the result from the analyser until all the criteria for selection have been established (including those for deciding whether an observation or an upper limit will be claimed in searches for rare events, depending on the number of measured events). In practice this is done by either ignoring (masking) the signal region while the analysis is being developed, or by artificially biasing the data in a deterministic way (unknown to the analysers); the use of only a sub-sample of the data to define the analysis criteria is also possible, although in this case the reduced statistical power might hide some other potential systematic effect which would be evident with the full sample.

---

[176] Clearly, if a highly efficient trigger condition were identified, with such a high purity (low rate) that no downscaling ($D=1$) could be chosen, it would be a valid alternative to $T0$ itself!

F.3 Of errors and cross checks

A case which frequently arises in practice is that in which the stability of a result is checked against the variation of some parameter, to which the result itself is assumed to be insensitive. One might, for example, repeat a data analysis after changing some criteria or cut values which have some degree of arbitrariness: in this way a new (in general different) result (possibly with a different error) is obtained, and the interest lies in assessing whether the new result is fully consistent with the original one, thus confirming the initial insensitivity assumption, or not, in which case a previously unsuspected source of trouble has to be investigated.

Any comparison between the two results will be limited by statistics to some level, and moreover it is important to note that when the comparison is performed on the same data a fraction of the events will be common to the two samples, so that it is not sufficient for the discrepancy to be small with respect to the statistical error for the check to be considered as satisfied: the two results must agree much better than dictated by their respective statistical errors, which partly reflect their (common) statistical fluctuations, as the common part of the samples is obviously fully correlated.

Consider a stability check performed by tightening some cuts, thus reducing from the total sample T of N_T events to a sub-sample $S \subset T$ with $N_S < N_T$ events. Assuming the measured quantity x can be evaluated for each event, the results obtained from T and S with their statistical errors are

$$x_T = \frac{1}{N_T} \sum_{i \in T} x_i \qquad \sigma_T = \sqrt{\langle x_T^2 \rangle - \langle x_T \rangle^2} = \frac{\sigma}{\sqrt{N_T}}$$

$$x_S = \frac{1}{N_S} \sum_{i \in S} x_i \qquad \sigma_S = \sqrt{\langle x_S^2 \rangle - \langle x_S \rangle^2} = \frac{\sigma}{\sqrt{N_S}}$$

(angle brackets indicating averaging over repeated experiments), where σ is the standard deviation of x, which is the same for both samples under the hypothesis that the cut variation has no effect on the x distribution, so that any difference between x_S and x_T is of pure statistical origin. For a meaningful test the difference $\Delta = x_S - x_T$ must be compared to its error, which is

$$\sigma_\Delta = \sqrt{\sigma_S^2 + \sigma_T^2 - 2\,\mathrm{cov}(x_S, x_T)} \tag{F.2}$$

where the covariance of the two results is

$$\mathrm{cov}(x_S, x_T) = \langle x_S\, x_T \rangle - \langle x_S \rangle \langle x_T \rangle = \rho_{ST}\, \sigma_S\, \sigma_T$$

Using the definition of x_T, separating the contribution from S and exploiting the fact that the remaining part of the sample is uncorrelated to it

$$\langle x_S \, x_T \rangle = \left\langle x_S \left(N_S \, x_S + \sum_{i \in T-S} x_i \right) \middle/ N_T \right\rangle = (N_S/N_T)\langle x_S^2 \rangle + (1/N_T)\left\langle x_S \left(\sum_{i \in T-S} x_i \right) \right\rangle$$

$$= (N_S/N_T)\left(\sigma_S^2 + \langle x_S \rangle^2\right) + (1/N_T)\langle x_S \rangle (N_T \langle x_T \rangle - N_S \langle x_S \rangle)$$

$$= (N_S/N_T)\sigma_S^2 + \langle x_S \rangle \, \langle x_T \rangle$$

so that

$$\mathrm{cov}(x_S, x_T) = \frac{N_S}{N_T} \sigma_S^2 = \frac{\sigma^2}{N_T}$$

and $\rho_{ST} = \sigma_T/\sigma_S$. Finally one obtains

$$\sigma_\Delta = \sqrt{\sigma_S^2 - \sigma_T^2} \qquad \text{(F.3)}$$

for the error on the difference between the two results, corresponding to the standard deviation of the uncorrelated part of the sample $T - S$.

In more complicated cases the correlation coefficient ρ_{AB} between the two samples A, B involved in the comparison might not be known: it is presumably positive if the two measurements share some of the data, and in any case an upper limit for it can be obtained by using the Cramer–Rao bound on the variance of an estimator (see e.g. Frodesen *et al.* (1979)): two estimates $x_{A,B}$ of a quantity which are independent to some level can be combined in a weighted average

$$x_w = w \, x_A + (1 - w) \, x_B$$

with an error which is smaller than those of the individual measurements

$$\sigma_w = \sqrt{w^2 \sigma_A^2 + (1 - w)^2 \sigma_B^2 + 2w(1 - w)\rho_{AB}\sigma_A\sigma_B}$$

The minimum of this error with respect to the choice of weight w

$$\sigma_w(\min) = \sqrt{\frac{\sigma_A^2 \sigma_B^2 (1 - \rho_{AB}^2)}{\sigma_A^2 + \sigma_B^2 - 2\rho_{AB}\sigma_A\sigma_B}}$$

must exceed the minimum value

$$\sigma_0 = \left[N \int \left(\frac{d \ln P}{d x_t} \right)^2 P \, dx \right]^{-1/2}$$

where $P(x; x_t)$ is the probability density function for measuring x given the true value x_t, and N is the number of events used (in evaluating σ_0 the measured value

x is substituted for x_t if needed). From the above condition limits on the allowed range of ρ_{AB} can be obtained, and the corresponding bounds on the error σ_Δ of the difference between the results are (Barlow, 1994)

$$\sigma_\Delta^{(\pm)} = \left| \sqrt{(\sigma_A - \sigma_0)^2} \pm \sqrt{(\sigma_B - \sigma_0)^2} \right| \qquad (F.4)$$

One can therefore state that the two results agree within *at least* $(x_B - x_A)/\sigma_\Delta^{(-)}$ standard deviations.

It is important always to maintain a clear distinction between *known systematic effects* and *crosschecks*: often the experimenter assumes that the result R does not depend on the value of some specific quantity ξ (an analysis cut, the supply voltage of a detector, the temperature of the laboratory, etc.), and checks this hypothesis by evaluating the result for different values of ξ. Except in the case discussed above of an analysis cut, when (almost) the entire data sample is used for each evaluation of R (and correlations must be taken into account), this will usually require splitting the data in smaller sub-samples by binning it with respect to ξ: the statistical error of R for each ξ bin will obviously be larger than for the unbinned data, and the sensitivity to observe a systematic effect will be reduced accordingly. From the plot of R as a function of ξ the validity of the original assumption will be assessed with some statistical hypothesis test: in practice the points representing the different (and this time uncorrelated) results will be likely scattered around an horizontal line (assuming the assumption was correct), and the result of the test will be that, any slope of $R(\xi)$ is less than some given value $(dR/d\xi)_0$ at some confidence level[177].

Since such a test cannot exclude the possibility that R depends on ξ with a slope smaller than the above value, it might be tempting to attach a systematic error $\sigma_R = (dR/d\xi)_0 \, \Delta\xi/\sqrt{12}$ to the result, where $\Delta\xi$ is the maximum estimated range of variability for ξ. This procedure would be incorrect, however, and even in the case that σ_R is small compared to other contributions to the error, it would inflate it unnecessarily. Indeed one can always check the stability of the result with respect to *any* quantity (say, the phase of the moon); and one will only be able to exclude a dependence up to a certain level: considering the effect which cannot be excluded as a systematic error would thus lead to an indefinite increase of the total error. An a priori judgement is required here: if there is no reason to suspect a dependence on some quantity the above test should be only considered as a cross check, and treated accordingly thereafter. If the test is passed (e.g. no dependence within say 1 standard deviation) nothing more needs to be done, while if the test is failed (a 4 standard deviation effect, say) one should worry about the origin of a

---

[177] A likelihood-ratio test might be used, or a comparison of the goodness-of-fit χ^2 variable for a linear (for example) fit and a constant fit, etc. Note that if several points are available any trend can be detected with higher statistical significance than provided by a simple χ^2 test, by using e.g. a runs test, a Kolmogorov–Smirnov test or by computing the auto-covariance function of the residuals, as discussed in statistics textbooks.

previously ignored effect, and go back to the analysis to look for ways of estimating the consequences on the result and the related uncertainty. In no case is one justified to simply use the resulting limit as a systematic error.

To conclude, there is at least one general rule which can be given, namely that a result should be presented together with all the information allowing the reader to understand and judge the procedure used for estimating the errors, which includes the size of all the individual corrections which were applied and the way these were estimated.

As a last remark, the scientist should always keep in mind what can sometimes be found explicitly stated in experimental papers: 'This [result] does not imply that there are no other errors; it is only a reflection of the fact that we are ignorant of any substantial systematic errors.' (Bennett *et al.*, 1967)

BIBLIOGRAPHY

Abashian, A. *et al.* (1957). Angular distributions of positrons from $\pi^+ - \mu^+ - e^+$ decays observed in a liquid hydrogen bubble chamber. *Phys. Rev.*, **105**, 1927.

Abashian, A. *et al.* (1964). Search for *CP* nonconservation in K_2^0 decay. *Phys. Rev. Lett.*, **13**, 243.

Abashian, A. *et al.* (Belle collaboration) (2002). The Belle detector. *Nucl. Instr. Meth. Phys. Res. A*, **479**, 117.

Abe, K. *et al.* (Belle collaboration) (2001). Observation of large *CP* violation in the neutral *B* meson system. *Phys. Rev. Lett.*, **87**, 091802.

Abe, K. *et al.* (Belle collaboration) (2005a). Improved evidence for direct *CP* violation in $B^0 \to \pi^+\pi^-$ decays and model-independent constraints on ϕ_2. *Phys. Rev. Lett.*, **95**, 101801.

Abe, K. *et al.* (Belle collaboration) (2005b). Improved measurements of direct *CP* violation in $B \to K^+\pi^-, K^+\pi^0$ and $\pi^+\pi^0$ decays. Technical Report BELLE-CONF-0523, BELLE. Preprint hep-ex/0507045.

Abe, K. *et al.* (Belle collaboration) (2006a). Study of $B^\pm \to D_{CP}K^\pm$ and $D_{CP}^*K^\pm$ decays. *Phys. Rev. D*, **73**, 051106.

Abe, M. *et al.* (KEK E246 collaboration) (2006). Search for *T*-violating transverse muon polarization in the $K^+ \to \pi^0\mu^+\nu$ decay. *Phys. Rev. D*, **73**, 072005.

Abov, Y. G. *et al.* (1964). On the existence of an internucleon potential not conserving spacial parity. *Phys. Lett.*, **12**, 25.

Abulencia, A. *et al.* (CDF collaboration) (2006). Observation of $B_{(s)}^0 - \overline{B}_{(s)}^0$ oscillations. *Phys. Rev. Lett.*, **97**, 242003.

Acquines, O. *et al.* (CLEO collaboration) (2006). Measurements of the exclusive decays of the $\Upsilon(5S)$ to *B* meson final states and improved B_s^* mass measurement. *Phys. Rev. Lett.*, **96**, 152001.

Adelberger, E. G. and Haxton, W. C. (1985). Parity violation in the nucleon-nucleon interaction. *Ann. Rev. Nucl. Part. Sci.*, **35**, 501.

Affleck, I. and Dine, M. (1985). A new mechanism for baryogenesis. *Nucl. Phys. B*, **249**, 4244.

Ahn, J. K. *et al.* (KEK-E391a collaboration) (2006). New limit on the $K_L^0 \to \pi^0\nu\overline{\nu}$ decay rate. *Phys. Rev. D*, **74**, 051105(R).

Aizu, K. (1954). In *Proc. Int. Conf. Theoretical Physics, 1953 (Kyoto)*, pp. 200. Kyoto-Tokyo Science Council.

Akhmetshin, R. R. *et al.* (CMD-2 collaboration) (1999a). First observation of the $\phi \to \pi^+\pi^-\gamma$ decay. *Phys. Lett. B*, **462**, 371.

Akhmetshin, R. R. *et al.* (CMD-2 collaboration) (1999b). Study of the ϕ decays into $\pi^0\pi^0\gamma$ and $\eta\pi^0\gamma$ final states. *Phys. Lett. B*, **462**, 380.

Alavi-Harati, A. *et al.* (KTeV collaboration) (2000). Observation of *CP* violation in $K_L \to \pi^+\pi^-e^+e^-$ decays. *Phys. Rev. Lett.*, **84**, 408.

Alavi-Harati, A. *et al.* (KTeV collaboration) (2002). Measurement of the K_L charge asymmetry. *Phys. Rev. Lett.*, **88**, 181601.

Alavi-Harati, A. *et al.* (KTeV collaboration) (2003). Measurements of direct *CP* violation, *CPT* symmetry, and other parameters in the neutral kaon system. *Phys. Rev. D*, **67**, 012005.

Albajar, C. *et al.* (UA1 collaboration) (1987). Search for $B^0 - \overline{B}^0$ oscillations at the CERN proton-antiproton collider. *Phys. Lett. B*, **186**, 247.

Albrecht, H. *et al.* (ARGUS collaboration) (1987). Observation of $B^0 - \overline{B}^0$ mixing. *Phys. Lett. B*, **192**, 245.

Aleksan, R. *et al.* (1991). *CP* violation using non-*CP* eigenstate decays of neutral *B* mesons. *Nucl. Phys. B*, **361**, 141.

Alff-Steinberger, C. *et al.* (1966). K_S and K_L interference in the $\pi^+\pi^-$ decay mode, *CP* invariance and the $K_S - K_L$ mass difference. *Phys. Lett.*, **20**, 207.

Aloisio, A. *et al.*(KLOE collaboration) (2002). Measurement of the branching fraction for the decay $K_S \rightarrow \pi e \nu$. *Phys. Lett. B*, **535**, 37.

Aloisio, A. *et al.* (KLOE collaboration) (2002). Measurement of $\Gamma(K_S \rightarrow \pi^+\pi^-(\gamma))/\Gamma(K_S \rightarrow \pi^0\pi^0)$. *Phys. Lett. B*, **538**, 21.

Alpher, R. A. *et al.* (1953). Physical conditions in the initial stages of the expanding universe. *Phys. Rev.*, **92**, 1347.

Alvarez-Gaumé, L. *et al.* (1999). Violation of time-reversal invariance and CPLEAR measurements. *Phys. Lett. B*, **458**, 347.

Amado, R. D. and Primakoff, H. (1980). Comments on testing the Pauli principle. *Phys. Rev. C*, **22**, 1338.

Amati, D. and Vitale, B. (1955). On conservation laws in production and annihilation of antinucleons. *Nuovo Cim.*, **2**, 719.

Ambrosino, F. *et al.* (KLOE collaboration) (2005). A direct search for the *CP*-violating decay $K_S \rightarrow 3\pi^0$ with the KLOE detector at DAΦNE. *Phys. Lett. B*, **619**, 61.

Ambrosino, F. *et al.* (KLOE collaboration) (2006). Study of the branching ratio and charge asymmetry for the decay $K_S \rightarrow \pi e \nu$ with the KLOE detector. *Phys. Lett. B*, **636**, 173.

Amelino-Camelia, G. *et al.* (1992). *CP* violating asymmetries in the neutral kaon decays. *Z. Phys. C*, **55**, 63.

Amoretti, M. *et al.* (ATHENA collaboration) (2002). Production and detection of cold antihydrogen atoms. *Nature*, **419**, 456.

Ananthanarayan, B. *et al.* (2001). Roy equation analysis of $\pi\pi$ scattering. *Phys. Rep.*, **353**, 207.

Anderson, C. D. (1933). The positive electron. *Phys. Rev.*, **43**, 491.

Angelopoulos, A. *et al.* (CPLEAR collaboration) (1998a). First direct observation of time-reversal non-invariance in the neutral-kaon system. *Phys. Lett. B*, **444**, 43.

Angelopoulos, A. *et al.* (CPLEAR collaboration) (1998b). Search for *CP* violation in the decay of tagged $\overline{K}^0$ and K^0 to $\pi^0\pi^0\pi^0$. *Phys. Lett. B*, **425**, 391.

Angelopoulos, A. *et al.* (CPLEAR collaboration) (2001). *T*-violation and *CPT*-invariance measurements in the CPLEAR experiment: a detailed description of the analysis of neutral-kaon decays to $e\pi\nu$. *Eur. Phys. J. C*, **22**, 55.

Angelopoulos, A. *et al.* (CPLEAR collaboration) (2003). Physics at CPLEAR. *Phys. Rep.*, **374**, 165.

Anisimovsky, V. V. *et al.* (KEK E246 collaboration) (2003). First measurement of the *T*-violating muon polarization in the decay $K^+ \rightarrow \mu^+\nu\gamma$. *Phys. Lett. B*, **562**, 166.

Anthony, P. L. *et al.* (2004). Observation of parity nonconservation in Møller scattering. *Phys. Rev. Lett.*, **92**, 181602.

Antipov, Yu. M. *et al.* (1971). Observation of antihelium-3. *Nucl. Phys.*, **31**, 235.

Arash, F. *et al.* (1985). Dynamics-independent null experiment for testing time-reversal invariance. *Phys. Rev. Lett.*, **54**, 2649.

Arbic, B. K. *et al.* (1988). Angular-correlation test of *CPT* in polarized positronium. *Phys. Rev. A*, **37**, 3189.

Arnold, P. B. (1991). An introduction to baryon violation in standard electroweak theory. In M. Cvetič *et al.* (eds), *Proc. 1990 TASI in Elementary Particle Physics, 1990 (Boulder)*, Theoretical Advanced Study Institute, Singapore.

Atwood, D. and Soni, A. (1992). Analysis for magnetic moment and electric dipole moment form factors of the top quark via $e^+e^- \rightarrow t\bar{t}$. *Phys. Rev. D*, **45**, 2405.

Atwood, D. *et al.* (1997). Enhanced *CP* violation with $B \rightarrow KD^0(\overline{D}^0)$ modes and extraction of the Cabibbo-Kobayashi-Maskawa angle γ. *Phys. Rev. Lett.*, **78**, 3257.

Atwood, D. *et al.* (2001a). *CP* violation in top physics. *Phys. Rep.*, **347**, 1.

Atwood, D. *et al.* (2001b). Improved methods for observing *CP* violation in $B^\pm \rightarrow KD$ and measuring the *CKM* phase γ. *Phys. Rev. D*, **63**, 036005.

Aubert, B. *et al.* (BABAR collaboration) (2001). Observation of *CP* violation in the B^0 meson system. *Phys. Rev. Lett.*, **87**, 091801.

Aubert, B. *et al.* (BABAR collaboration) (2002). The BABAR detector. *Nucl. Instr. Meth. Phys. Res. A*, **479**, 1.

Aubert, B. *et al.* (BABAR collaboration) (2004a). Direct *CP* violating asymmetry in $B^0 \to K^+\pi^-$ decays. *Phys. Rev. Lett.*, **93**, 131801.

Aubert, B. *et al.* (BABAR collaboration) (2004b). Limits on the decay-rate difference of neutral *B* mesons and on *CP*, *T*, and *CPT* violation in $B^0\overline{B}^0$ oscillations. *Phys. Rev. D*, **70**, 012007.

Aubert, B. *et al.* (BABAR collaboration) (2004c). Measurement of *CP*-violating asymmetries in $B^0 \to (\rho\pi)^0$ using a time-dependent Dalitz plot analysis. Preprint hep-ex/0408099.

Aubert, B. *et al.* (BABAR collaboration) (2006a). Improved measurement of *CP* asymmetries in $B^0 \to (c\bar{c})K^{(*)0}$ decays. Technical Report SLAC-PUB-11986, SLAC. Preprint hep-ex/0607107.

Aubert, B. *et al.* (BABAR collaboration) (2006b). Measurement of *CP* asymmetries and branching fractions in $b \to \pi\pi$ and $b \to K\pi$ decays. Technical Report SLAC-PUB-12032, SLAC. Preprint hep-ex/0607106.

Aubert, B. *et al.* (BABAR collaboration) (2006c). Measurement of *CP* observables for the decays $B^\pm \to D^0_{CP}K^\pm$. *Phys. Rev. D*, **73**, 051105.

Aubert, B. *et al.* (BABAR collaboration) (2006d). Measurement of the CKM angle γ in $B^\mp \to D^{(*)}K^\mp$ decays with a Dalitz analysis of $D^0 \to K^0_s\pi^-\pi^+$. Technical Report SLAC-PUB-12028, SLAC. Preprint hep-ex/0607104.

Aubert, B. *et al.* (BABAR collaboration) (2006e). Observation of $B^+ \to \overline{K}^0K^+$ and $B^0 \to K^0\overline{K}^0$. *Phys. Rev. Lett.*, **97**, 171805.

Aubert, B. *et al.* (BABAR collaboration) (2006f). Search for *T*, *CP* and *CPT* violation in $B^0 - \overline{B}^0$ mixing with inclusive dilepton events. *Phys. Rev. Lett.*, **96**, 251802.

Aubert, B. *et al.* (BABAR collaboration) (2006g). Updated measurement of the CKM angle α using $B^0 \to \rho^+\rho^-$ decays. Technical Report SLAC-PUB-12012, SLAC. Preprint hep-ex/0607098.

Aubert, B. *et al.* (BABAR collaboration) (2007a). Evidence for $B^0 \to \rho^0\rho^0$ decays and implications for the Cabibbo-Kobayashi-Maskawa angle α. *Phys. Rev. Lett.*, **98**, 111801.

Aubert, B. *et al.* (BABAR collaboration) (2007b). Evidence for $D^0 - \overline{D}^0$ mixing. Technical Report SLAC-PUB-07/019, SLAC. Preprint hep-ex/0703020.

Aubert, B. *et al.* (BABAR collaboration) (2007c). Observation of *CP* violation in $B^0 \to k^+\pi^-$ and $B^0 \to \pi^+\pi^-$. *Phys. Rev. Lett.* **99**, 021603.

Ayres, D. S. *et al.* (1971). Measurements of the lifetimes of positive and negative pions. *Phys. Rev. D*, **3**, 1051.

BABAR collaboration (1999). *The BABAR physics book*. Technical Report SLAC-R-504, SLAC.

Backenstoss, G. *et al.* (1983). Letter of intent: test of symmetries with K^0 and $\overline{K}^0$ beams. Technical Report CERN/PSCC/83-28, CERN, Geneva.

Bailey, J. (2000). Chirality and the origin of life. *Acta Astronautica*, **46**, 627.

Bailey, J. *et al.* (1978). New limits on the electric dipole moment of positive and negative muons. *Jou. Phys. G*, **4**, 345.

Bailey, J. *et al.* (1979). Final report on the CERN muon storage ring including the anomalous magnetic moment and the electric dipole moment of the muon, and a direct test of relativistic time dilation. *Nucl. Phys. B*, **150**, 1.

Baker, C. A. *et al.* (2006). Improved experimental limit on the electric dipole moment of the neutron. *Phys. Rev. Lett.*, **97**, 131801.

Bander, M. *et al.* (1979). *CP* noninvariance in the decays of heavy charged quark systems. *Phys. Rev. Lett.*, **43**, 242.

Barberio, E. (Heavy Flavours Averaging Group) *et al.* (2006). Averages of *b*-hadron properties at the end of 2005. On the web http://www.slac.stanford.edu/xorg/hfag/.

Bardin, G. *et al.* (1984). A new measurement of the positive muon lifetime. *Phys. Lett. B*, **137**, 135.

Barger, V. *et al.* (2002). No-go for detecting *CP* violation via neutrinoless double beta decay. *Phys. Lett. B*, **540**, 247.

Bargmann, V. *et al.* (1959). Precession of the polarization of particles moving in a homogeneous electromagnetic field. *Phys. Rev. Lett.*, **2**, 435.

Barlow, R. (1994). Evaluating systematic errors. Technical Report MAN/HEP/93/9, Manchester University.

Barlow, R. (2003). Asymmetric systematic errors. Technical Report MAN/HEP/03/02, Manchester University. Preprint physics/0306138.

Barmin, V. *et al.* (1984). *CPT* symmetry and neutral kaons. *Nucl. Phys. B*, **247**, 293.

Barnes, P. D. *et al.* (CERN PS185 collaboration) (1996). Observables in high-statistics measurements of the reaction $\bar{p}p \to \bar{\Lambda}\Lambda$. *Phys. Rev. C*, **54**, 1877.

Barr, G. *et al.* (NA31 collaboration) (1993). A new measurement of direct *CP* violation in the neutral kaon system. *Phys. Lett. B*, **317**, 233.

Barr, S. *et al.* (1979). Magnitude of the cosmological baryon asymmetry. *Phys. Rev. D*, **20**, 2494.

Bartlett, D. F. (1968). $K_2^0 \to 2\pi^0$ decay rate. *Phys. Rev. Lett.*, **21**, 558.

Batley, J. R. *et al.* (NA48 collaboration) (2002). A precision measurement of the direct *CP* violation in the decay of neutral kaons into two pions. *Phys. Lett. B*, **544**, 97.

Batley, J. R. *et al.* (NA48/2 collaboration) (2006*a*). Observation of a cusp-like structure in the $\pi^0\pi^0$ invariant mass distribution from $K^{\pm} \to \pi^{\pm}\pi^0\pi^0$ decay and determination of the $\pi\pi$ scattering lenghts. *Phys. Lett. B*, **633**, 173.

Batley, J. R. *et al.* (NA48/2 collaboration) (2006*b*). Search for direct *CP*-violation in $K^{\pm} \to \pi^{\pm}\pi^0\pi^0$ decays. *Phys. Lett. B*, **638**, 22.

Batley, J. R. *et al.* (NA48/2 collaboration) (2006*c*). Search for direct *CP* violation in the decays $K^{\pm} \to 3\pi^{\pm}$. *Phys. Lett. B*, **634**, 474.

Batley, J. R. *et al.* (NA48/2 collaboration) (2007). Search for direct CP violating charge asymmetries in $K^{\pm} \to \pi^{\pm}\pi^+\pi^-$ and $K^{\pm} \to \pi^{\pm}\pi^0\pi^0$ decays. To appear in *Eur. Phys. J. C*. Preprint ∂rXIV:0707.0697 [hep-ex], July 2007.

Bell, J. S. (1955). Time reversal in field theory. *Proc. Roy. Soc. (London) A*, **231**, 479.

Bell, J. S. (1962). Electromagnetic properties of unstable particles. *Nuovo Cim.*, **24**, 452.

Bell, J. S. (1964). On the Einstein Podolsky Rosen paradox. *Physics*, **1**, 195.

Bell, J. S. and Mandl, F. (1958). The polarization-asymmetry equality. *Proc. Phys. Soc. (London)*, **71**, 272.

Bell, J. S. and Steinberger, J. (1966). Weak interactions of kaons. In *Proc. Oxford International Conference on Elementary Particles, 1965*, Chilton, pp. 195. Rutherford Laboratory.

Belušević, R. (1999). *Neutral kaons*. Springer, Berlin.

Bennett, S. *et al.* (1967). Measurement of the charge asymmetry in the decay $K_L^0 \to \pi^{\pm} + e^{\mp} + \nu$. *Phys. Rev. Lett.*, **19**, 993.

Berestetskiǐ, V. B. *et al.* (1978). *Relativistic quantum theory*. Pergamon Press, Oxford.

Bernreuther, W. and Nachtmann, O. (1981). Weak interaction effects in positronium. *Z. Phys. C*, **11**, 235.

Bernreuther, W. *et al.* (1988). How to test CP, T and CPT invariance in the three photon decay of polarized $^3S_1$ positronium. *Z. Phys. C*, **41**, 143.

Bernreuther, W. *et al.* (1993). *CP*-violating electric and weak dipole moments of the τ lepton from threshold to 500 GeV. *Phys. Rev. D*, **48**, 78.

Bernstein, J. and Michel, L. (1960). T, P, C symmetries of π^0 decay. *Phys. Rev.*, **118**, 871.

Bernstein, J. *et al.* (1965). Possible C, T noninvariance in the electromagnetic interaction. *Phys. Rev.*, **139**, B1650.

Beyer, M. (ed.) (2002). *CP violation in particle, nuclear and astrophysics*. Springer, Berlin.

Bianco, S. *et al.* (2003). A Cicerone for the physics of charm. *Nuovo Cim.*, **26**, 1.

Bigi, I. I. and Sanda, A. I. (1981). Notes on the observability of *CP* violations in *B* decays. *Nucl. Phys. B*, **193**, 85.

Bigi, I. I. and Sanda, A. I. (2000). *CP violation*. Cambridge University Press, Cambridge.

Birge, R. W. *et al.* (1960). Proton helicity from Λ decay. *Phys. Rev. Lett.*, **5**, 254.

Birss, R. R. (1963). Macroscopic symmetry in space-time. *Rep. Prog. Phys.*, **26**, 307.

Bjorken, J. D. and Drell, S. D. (1965). *Relativistic quantum fields*. McGraw-Hill, New York.

Blanke, E. *et al.* (1983). Improved experimental test of detailed balance and time reversibility in the reactions ^{27}Al+p $\leftrightarrow$^{24}Mg+α. *Phys. Rev. Lett.*, **51**, 355.

Blucher, E. (2006). Experiments on *CP* violation with *K* mesons – part II. In M. Giorgi *et al.* (eds), *CP* violation: from quark to leptons. SIF and IOS Press.

Bludman, S. A. (1964). $K_2 \rightarrow \pi^+\pi^-$ and the question of Bose statistics for pions. *Phys. Rev.*, **138**, B213.

Bluhm, R. (2004). Lorentz and *CPT* tests in matter and antimatter. *Nucl. Instr. Meth. Phys. Res. B*, **221**, 6.

Boehm, F. *et al.* (1957). Positron polarization in a mirror transition. *Phys. Rev.*, **108**, 1497.

Böhm, A. *et al.* (1969). The phase difference between $K_L \rightarrow \pi^+\pi^-$ and $K_S \rightarrow \pi^+\pi^-$ decay amplitudes. *Nucl. Phys. B*, **9**, 605.

Bohr, A. (1959). Relation between intrinsic parities and polarizations in collision and decay processes. *Nucl. Phys.*, **10**, 486.

Boldt, E. *et al.* (1958). Helicity of the proton from Λ^0 decay. *Phys. Rev. Lett.*, **1**, 256.

Bolotov, V. N. *et al.* (ISTRA collaboration) (2005). Study of $K^- \rightarrow \pi^0 e^- \bar{\nu}_e \gamma$ decay with ISTRA+ setup. Technical Report INR-1150/2005, Institute for Nuclear Research (Moscow). Preprint `hep-ex/0510064`.

Boltzmann, L. (1895). On certain questions of the theory of gases. *Nature*, **51**, 413.

Bona, M. *et al.* (UTfit collaboration) (2006). On the web `http://www.utfit.org`.

Bott-Bodenhausen, M. *et al.* (1966). Time-dependent interference effects in two-pion decays of neutral kaons. *Phys. Lett.*, **20**, 212.

Bouchiat, M. A. and Bouchiat, C. C. (1974). Weak neutral currents in atomic physics. *Phys. Lett. B*, **48**, 111.

Bouchiat, M. A. *et al.* (1984). New observation of a parity violation in cesium. *Phys. Lett. B*, **134**, 463.

Braguta, V. V. *et al.* (2002). *T*-odd correlation in the $K_{l3\gamma}$ decay. *Phys. Rev. D*, **65**, 054038.

Branco, G. C., Lavoura, L., and Silva, J. P. (1999). *CP violation*. Oxford University Press, Oxford.

Briere, R. A. (2006). A review of charm physics. *AIP Conf. Proc.*, **815**, 169.

Browder, T. E. and Faccini, R. (2003). Establishment of *CP* violation in *B* decays. *Ann. Rev. Nucl. Part. Sci.*, **53**, 353.

Brown, L. M. *et al.* (eds) (1989). *Pions to quarks*. Cambridge University Press, Cambridge.

Brown, L. S. and Gabrielse, G. (1986). Geonium theory: physics of a single electron or ion in a Penning trap. *Rev. Mod. Phys.*, **58**, 233.

Buchanan, C. D. *et al.* (1992). Testing *CP* and *CPT* violation in the neutral kaon system at a ϕ factory. *Phys. Rev. D*, **45**, 4088.

Buhler, A. *et al.* (1963). A measurement of the e^+ polarisation in muon decay: the e^+ annihilation method. *Phys. Lett.*, **7**, 368.

Buras, A. J. (2005). Flavour physics and *CP* violation. Technical Report TUM-HEP-590/05, Munich Technical University. Preprint `hep-ph/0505175`.

Buras, A. J. *et al.* (1994). Waiting for the top quark mass, $K^+ \rightarrow \pi^+\nu\bar{\nu}$, $B_s^0 - \bar{B}_s^0$ mixing, and *CP* asymmetries in *B* decays. *Phys. Rev. D*, **50**, 3433.

Burgy, M. T. *et al.* (1958). Test of time-reversal invariance of the beta interaction in the decay of free polarized neutrons. *Phys. Rev. Lett.*, **1**, 324.

Burgy, M. T. *et al.* (1960). Measurements of spatial asymmetries in the decay of polarized neutrons. *Phys. Rev.*, **120**, 1829.

Burkard, H. *et al.* (1985). Muon decay: measurement of the transverse positron polarization and general analysis. *Phys. Lett. B*, **160**, 343.

Butler, J. (2006). *B* physics at hadron colliders. In M. Giorgi *et al.* (eds), *CP* violation: from quark to leptons. SIF and IOS Press.

Cahn, R. N. and Goldhaber, G. (1988). *The experimental foundations of particle physics*. Cambridge University Press, Cambridge.

Carruthers, P. (1968). Isospin symmetry, *TCP*, and local field theory. *Phys. Lett. B*, **26**, 158.

Carter, A. B. and Sanda, A. I. (1980). *CP* nonconservation in cascade decays of *B* mesons. *Phys. Rev. Lett.*, **45**, 952.

Carter, A. B. and Sanda, A. I. (1981). *CP* violation in B-meson decays. *Phys. Rev. D*, **23**, 1567.

Chaichian, M. and Hagedorn, R. (1998). *Symmetries in quantum mechanics*. Institute of Physics Publishing, Bristol.

Chamberlain, O. *et al.* (1955). Observation of antiprotons. *Phys. Rev.*, **100**, 947.

Chardin, G. and Rax, J.-M. (1992). *CP* violation: a matter of (anti)gravity? *Phys. Lett. B*, **282**, 256.

Chardonnet, P. *et al.* (1997). The production of anti-matter in our galaxy. *Phys. Lett. B*, **409**, 313.

Charles, J. *et al.* (CKMfitter group) (2006). On the web http://ckmfitter.in2p3.fr/.

Charlton, M. *et al.* (1994). Antihydrogen physics. *Phys. Rep.*, **241**, 6.

Chase, C. T. (1930). The scattering of fast electrons by metals. II. Polarization by double scattering at right angles. *Phys. Rev.*, **36**, 1060.

Chauvat, P. *et al.* (CERN R608 collaboration) (1985). Test of *CP* invariance in Λ^0 decay. *Phys. Lett. B*, **163**, 273.

Chen, K. F. *et al.* (Belle collaboration) (2005). Measurement of polarization and triple-product correlations in $B \to \phi K^*$ decays. *Phys. Rev. Lett.*, **94**, 221804.

Chen, K. F. *et al.* (Belle collaboration) (2007). Observation of time-dependent *CP* violation in $B^0 \to \eta' K^0$ decays and improved measurements of *CP* asymmetries in $B^0 \to \phi K^0$, $K_S^0 K_S^0 K_S^0$ and $B^0 \to J/\psi K^0$ decays. *Phys. Rev. Lett.*, **98**, 031802.

Chinowsky, W. and Steinberger, J. (1954). Absorption of negative pions in deuterium: parity of the pion. *Phys. Rev.*, **95**, 1561.

Chiu, H.-Y. (1966). Symmetry between particle and antiparticle populations in the universe. *Phys. Rev. Lett.*, **17**, 712.

Chollet, J. C. *et al.* (1970). Observation of the interference between K_L^0 and K_S^0 in the $\pi^0 \pi^0$ decay mode. *Phys. Lett. B*, **31**, 658.

Christenson, J. H. *et al.* (1964). Evidence for the 2π decay of the K_2^0 meson. *Phys. Rev. Lett.*, **13**, 138.

Cirigliano, V. *et al.* (2000). $K \to \pi\pi$ phenomenology in the presence of electromagnetism. *Eur. Phys. J. C*, **18**, 83.

Cline, D. B. (ed.) (1997). *Physical origin of homochirality in life*. Oxford University Press, Oxford.

Coester, F. (1957). Weak decay interactions. *Phys. Rev.*, **107**, 299.

Cohen, A. G. *et al.* (1998). A matter-antimatter universe? *Astrophys. J.*, **495**, 539.

Colladay, D. and Kostelecký, V. A. (1997). *CPT* violation and the standard model. *Phys. Rev. D*, **55**, 6760.

Conti, R. S. *et al.* (1986). Tests of the discrete symmetries *C*, *P* and *T* in one-photon transitions of positronium. *Phys. Rev. A*, **33**, 3495.

Conzett, H. E. (1993). Null tests of time-reversal invariance. *Phys. Rev. C*, **48**, 423.

Coombes, C. A. *et al.* (1957). Polarization of μ^+ mesons from the decay of K^+ mesons. *Phys. Rev.*, **108**, 1348.

Cork, B. *et al.* (1957). Antineutrons produced from antiprotons in charge-exchange collisions. *Phys. Rev.*, **104**, 1193.

Costantini, F. *et al.* (eds) (2002). *Proc. KAON2001 international conference on CP violation, 2001 (Pisa)*. INFN Frascati.

Cousins, R. D. and Highland, V. L. (1992). Incorporating systematic uncertainties into an upper limit. *Nucl. Instr. Meth. Phys. Res. A*, **320**, 331.

Cox, R. T. *et al.* (1928). Apparent evidence of polarization in a beam of β rays. *Proc. Nat. Acad. Sci. USA*, **14**, 544.

Crawford, F. S. *et al.* (1957). Detection of parity nonconservation in Λ decay. *Phys. Rev.*, **108**, 1102.

Cronin, J. W. *et al.* (1967). Measurement of the decay rate of $K_2^0 \to \pi^0 \pi^0$. *Phys. Rev. Lett.*, **18**, 25.

Cronin, J. W. (1981). *CP* symmetry violation – the search for its origin. *Rev. Mod. Phys.*, **53**, 373.

Culligan, G. et al. (1957). Longitudinal polarization of the positrons from the decay of unpolarized positive muons. *Nature*, **180**, 751.

Curie, P. (1894). Sur la symétrie dans les phénomènes physiques, symétrie d'un champ électrique et d'un champ magnétique. *J. Phys. (Paris)*, **3**, 393.

Dalitz, R. H. (1953). On the analysis of τ-meson data and the nature of the τ-meson. *Phil. Mag.*, **44**, 1068.

Dalitz, R. H. (1989). In: L. M. Brown et al. (eds). *Pions to Quarks*. Cambridge University Press, Cambridge.

D'Ambrosio, G. and Isidori, G. (1998). *CP* violation in kaon decays. *Int. Jou. Mod. Phys. A*, **13**, 1.

Dass, G. V. et al. (1987). $K^0 \leftrightarrow \overline{K}^0$ transitions in matter. *Phys. Rev. D*, **35**, 1730.

Davy, H. (1839). *Elements of Chemical Philosophy*. Smith, Elder & Co., London.

Day, T. B. (1961). Demonstration of quantum mechanics in the large. *Phys. Rev.*, **121**, 1204.

Dehmelt, H. et al. (1999). Past electron-positron g-2 experiments yielded sharpest bound on *CPT* violation for point particles. *Phys. Rev. Lett.*, **83**, 4694.

Deilamian, K. et al. (1995). Search for small violations of the symmetrization postulate in an excited state of helium. *Phys. Rev. Lett.*, **74**, 4787.

DeMille, D. et al. (1999). Search for exchange-antisymmetric two-photon states. *Phys. Rev. Lett.*, **83**, 3978.

Deser, S. and Pirani, F. A. E. (1967). The sign of the gravitational force. *Ann. Phys. (NY)*, **43**, 436.

Deshpande, N. G. and He, X.-G. (1996). *CP* asymmetry in the neutral *B* system at symmetric colliders. *Phys. Rev. Lett.*, **76**, 360.

D'Espagnat, B. (1961). A proposed experimental test of the Day, Snow and Sucher argument based on antiproton annihilation into kaons. *Nuovo Cim.*, **20**, 1217.

de Shalit, A. et al. (1957). Detection of electron polarization by double scattering. *Phys. Rev.*, **107**, 1459.

Dine, M. (2000). TASI lectures on the strong *CP* problem. Technical Report SCIPP-00/30, Theoretical Advanced Study Institute. Preprint hep-ph/0011376.

Dine, M. and Kusenko, A. (2004). Origin of the matter-antimatter asymmetry. *Rev. Mod. Science*, **76**, 1.

Dirac, P. A. M. (1931). Quantized singularities in the electromagnetic field. *Proc. Roy. Soc. (London) A*, **133**, 60.

Dirac, P. A. M. (1933). Theory of electrons and positrons. In *Nobel Lectures, Physics 1922–1941*. Elsevier, Amsterdam, 1965.

Dirac, P. A. M. (1949). Forms of relativistic dynamics. *Rev. Mod. Phys.*, **21**, 392.

Dobrzynski, L. et al. (1966). Test of *CP* and *C* invariances in $\overline{p}p$ annihilations at 1.2 GeV/c involving strange particles. *Phys. Lett.*, **22**, 105.

Dolgov, A. D. (1992). Non-GUT baryogenesis. *Phys. Rep.*, **222**, 309.

Dolgov, A. D. (2006). *CP* violation in cosmology. In M. Giorgi et al. (eds), *CP violation: from quark to leptons*. SIF and IOS Press.

Di Domenico, A. (2006). Correlations in ϕ decays into $K^0 - \overline{K}^0$. In M. Giorgi et al. (eds), *CP violation: from quark to leptons*. SIF and IOS Press.

Donoghue, J. F. et al. (1986a). *CP* violation in low energy $p\overline{p}$ reactions. *Phys. Lett. B*, **178**, 319.

Donoghue, J. F. et al. (1986b). Hyperon decays and *CP* nonconservation. *Phys. Rev. D*, **34**, 833.

Donoghue, J. F. et al. (1992). *Dynamics of the standard model*. Cambridge University Press, Cambridge.

Dorfan, J. et al. (1967). Charge asymmetry in the muonic decay of the K_2^0. *Phys. Rev. Lett.*, **19**, 987.

Dress, W. B. et al. (1977). Search for an electric dipole moment of the neutron. *Phys. Rev. D*, **15**, 9.

Duck, I. and Sudarshan, E. C. G. (eds) (1997). *Pauli and the spin-statistics theorem*. Singapore.

Dunietz, I. et al. (1991). How to extract *CP*-violating asymmetries from angular correlations. *Phys. Rev. D*, **43**, 2193.

Eades, J. and Hartmann, F. J. (1999). Forty years of antiprotons. *Rev. Mod. Phys.*, **71**, 373.

Ehrenfest, P. and Oppenheimer, J. R. (1931). Note on the statistics of nuclei. *Phys. Rev.*, **37**, 333.

Einstein, A., Podolsky, B., and Rosen, N. (1935). Can quantum-mechanical description of physical reality be considered complete? *Phys. Rev.*, **47**, 777.

Eisler, F. *et al.* (1957). Demonstration of parity nonconservation in hyperon decay. *Phys. Rev.*, **108**, 1102.

Ellis, J. *et al.* (1996). Precision tests of *CPT* symmetry and quantum mechanics in the neutral kaon system. *Phys. Rev. D*, **53**, 3846.

Enz, C. P. and Lewis, R. R. (1965). On the phenomenological description of *CP* violation for *K*-mesons and its consequences. *Helv. Physica Acta*, **38**, 860.

Eötvös, R. von *et al.* (1922). Beiträge zum Gesetz der Proportionalität von Trägheit und Gravität. *Ann. Physik (Leipzig)*, **68**, 11.

Ericson, T. E. O. (1966). Nuclear enhancement of *T* violation effects. *Phys. Lett.*, **23**, 97.

Escribano, R. and Massó, E. (1997). Improved bounds on the electromagnetic dipole moments of the τ lepton. *Phys. Lett. B*, **395**, 369.

Eversheim, P. D. *et al.* (1991). Parity violation in proton-proton scattering at 13.6 MeV. *Phys. Lett. B*, **256**, 11.

Fabri, E. (1954). A study of τ-meson decay. *Nuovo Cim.*, **XI**, 479.

Farley, F. J. M. *et al.* (2004). New method of measuring electric dipole moments in storage rings. *Phys. Rev. Lett.*, **93**, 052001.

Fee, M. S. *et al.* (1993). Measurement of the positronium $1^3S_1 - 2^3S_1$ interval by continuous-wave two-photon excitation. *Phys. Rev. A*, **48**, 192.

Feinberg, G. (1957). Selection rules imposed by *CP* invariance. *Phys. Rev.*, **108**, 878.

Feinberg, G. (1960). Invariance under antiunitary operators. *Phys. Rev.*, **120**, 640.

Feinberg, G. and Weinberg, S. (1959). On the phase factors in inversions. *Nuovo Cim.*, **14**, 571.

Fermi, E. (1955). Lectures on pions and nucleons. *Suppl. Nuovo Cim.*, **2**, 17.

Fermi, E. and Marshall, L. (1947). Interference phenomena of slow neutrons. *Phys. Rev.*, **71**, 666.

Fernandez, E. *et al.* (MAC collaboration) (1983). Lifetime of particles containing *b* quarks. *Phys. Rev. Lett.*, **51**, 1022.

Ferretti, B. (1946). The absorption of slow mesons by an atomic nucleus. In *Report of an International Conference on Low Temperatures and Fundamental Particles (Cambridge)*, The Royal Society, London.

Feynman, R. P. (1948). A relativistic cut-off for classical electrodynamics. *Phys. Rev.*, **74**, 939.

Feynmann, R. P. (1961). *Theory of Fundamental Process*. Benjamin, New York.

Feynman, R. P. (1965). *The character of physical law*. MIT Press, Cambridge (USA).

Feynman, R. P. (1970). *The Feynman Lectures on Physics*. Addison-Wesley. Readings, MA.

Feynman, R. P. (1987). The reason for antiparticles. In *Elementary particles and the laws of physics*, Cambridge University Press, Cambridge.

Feynman, R. P. and Gell-Mann, M. (1958). Theory of the Fermi interaction. *Phys. Rev.*, **109**, 193.

Fierz, M. (1939). Über die relativistische Theorie kräftefreier Teilchen mit beliebigem Spin. *Helv. Physica Acta*, **12**, 3.

Fisher, R. A. (1936). The use of multiple measurements in taxonomic problems. *Ann. Eugenics*, **7**, 179.

Fitch, V. L. (1981). The discovery of charge-conjugation parity asymmetry. *Rev. Mod. Phys.*, **53**, 367.

Fitch, V. L. *et al.* (1965). Evidence for constructive interference between coherently regenerated and *CP*-nonconserving amplitudes. *Phys. Rev. Lett.*, **15**, 73.

Fitch, V. L. *et al.* (1967). Studies of $K_2^0 \to \pi^+\pi^-$ decay and interference. *Phys. Rev.*, **164**, 1711.

Fonda, L. and Ghirardi, G. C. (1970). *Symmetry principles in quantum physics*. Marcel Dekker Inc., New York.

Foot, R. *et al.* (1991). A model with fundamental improper spacetime symmetries. *Phys. Lett. B*, **272**, 67.

Foot, R. and Silagadze, Z. K. (2005). Supernova explosions, 511 keV photons, gamma ray bursts and mirror matter. *Int. Jou. Mod. Phys. D*, **14**, 143.

Forte, M. *et al.* (1980). First measurement of parity-nonconserving neutron-spin rotation: the tin isotopes. *Phys. Rev. Lett.*, **45**, 2088.

Fortson, E. N. and Lewis, L. L. (1984). Atomic parity nonconservation experiments. *Phys. Rep.*, **113**, 289.

Fox, G. C. and Wolfram, S. (1978). Observables for the analysis of event shapes in e^+e^- annihilation and other processes. *Phys. Rev. Lett.*, **41**, 1581.

Frauenfelder, H. *et al.* (1957a). Parity and electron polarization: Møller scattering. *Phys. Rev.*, **107**, 643.

Frauenfelder, H. *et al.* (1957b). Parity and the polarization of electrons from Co^{60}. *Phys. Rev.*, **106**, 386.

Friedman, J. I. and Telegdi, V. L. (1957a). Nuclear emulsion evidence for parity nonconservation in the decay chain $\pi^+ \to \mu^+ \to e^+$. *Phys. Rev.*, **105**, 1681.

Friedman, J. I. and Telegdi, V. L. (1957b). Nuclear emulsion evidence for parity nonconservation in the decay chain $\pi^+ \to \mu^+ \to e^+$. *Phys. Rev.*, **106**, 1290.

Frodesen, A. G. *et al.* (1979). *Probability and statistics in particle physics*. Universitetsforlaget, Bergen.

Froggatt, C. D. and Nielsen, H. B. (1991). *Origin of symmetries*. World scientific, Singapore.

Fromerth, M. J. and Rafelski, J. (2003). Limit on quark-antiquark mass difference from the neutral kaon system. *Acta Physica Polonica B*, **34**, 4151.

Fukugita, M. and Yanagida, T. (1986). Baryogenesis without grand unification. *Phys. Lett. B*, **174**, 45.

Fumi, F. G. and Wolfenstein, L. (1953). Allowed final states for annihilation into three photons. *Phys. Rev.*, **90**, 498.

Furry, W. H. (1937). A symmetry theorem in the positron theory. *Phys. Rev.*, **51**, 125.

Gabrielse, G. *et al.* (1999). Precision mass spectroscopy of the antiproton and proton using simultaneously trapped particles. *Phys. Rev. Lett.*, **82**, 3198.

Gabrielse, G. *et al.* (ATRAP collaboration) (2002). Background-free observation of cold antihydrogen with field-ionization analysis of its states. *Phys. Rev. Lett.*, **89**, 213401.

Gaillard, J.-M. *et al.* (1967). Measurement of the decay of the long-lived neutral K meson into two neutral pions. *Phys. Rev. Lett.*, **18**, 20.

Galaktionov, Yu. V. (2002). Antimatter in cosmic rays. *Rep. Prog. Phys.*, **65**, 1243.

Gámiz, E. *et al.* (2002). Charged kaon $K \to 3\pi$ CP violating asymmetries at NLO in CHPT. *J. High Energy Phys.*, **0310**, 042.

Gardner, M. (1990). *The new ambidextrous universe*. W. H. Freeman and Co., New York.

Gardner, S. and Valencia, G. (2000). Impact of $|\Delta I| = 5/2$ transitions in $K \to \pi\pi$ decays. *Phys. Rev. D*, **62**, 094024.

Gardner, S. *et al.* (2001). Watson's theorem and electromagnetism in $K \to \pi\pi$ decay. *Phys. Lett. B*, **508**, 44.

Garwin, R. L. *et al.* (1957). Observation of the failure of conservation of parity and charge conjugation in meson decays: the magnetic moment of the free muon. *Phys. Rev.*, **105**, 1415.

Garwin, R. L. and Lederman, L. M. (1959). The electric dipole moment of elementary particles. *Nuovo Cim.*, **11**, 776.

Garwin, R. L. (1974). One researcher's personal account. *Adventures in experimental physics*, Vol. γ, 124.

Gasiorowicz, S. (1966). *Elementary particle physics*. John Wiley & Sons, New York.

Gatti, C. (KLOE collaboration) (2003). Scalar mesons and $\delta_0 - \delta_2$ at KLOE.

Gatto, R. (1957). Possible experiments on the behaviour of the weak hyperon decay interactions under P, C, and T. *Nucl. Phys.*, **5**, 183.

Gell-Mann, M. (1953). Isotopic spin and new unstable particles. *Phys. Rev.*, **92**, 833.

Gell-Mann, M. and Pais, A. (1955). Behavior of neutral particles under charge conjugation. *Phys. Rev.*, **97**, 1387.

Gentile, G. (1940). Osservazioni sopra le statistiche intermedie. *Nuovo Cim.*, **17**, 493.

Gérard, J.-M. and Hou, W.-S. (1991). CP violation in inclusive and exclusive charmless B decays. *Phys. Rev. D*, **43**, 2909.

Gervais, J.-L. *et al.* (1966). On a test of electromagnetic T-violation in charged k decays. *Phys. Lett.*, **20**, 432.

Geweniger, C. *et al.* (1974). A new determination of the $K^0 \to \pi^+\pi^-$ decay parameters. *Phys. Lett. B*, **48**, 487.

Gibbons, L. K. *et al.* (E731 collaboration) (1993). Measurement of the *CP*-violation parameter Re(ϵ'/ϵ). *Phys. Rev. Lett.*, **70**, 1203.

Gibson, W. M. and Pollard, B.R. (1976). *Symmetry principles in elementary particle physics*. Cambridge University Press, Cambridge.

Giorgi, M. *et al.* (eds.) (2006). *CP violation: from quark to leptons*. SIF and IOS Press. Proc. International School of Physics "Enrico Fermi", Course CLXIII, 2005 (Varenna), vol. 163.

Giri, A. *et al.* (2003). Determining γ using $B^\pm \to DK^\pm$ with multibody D decays. *Phys. Rev. D*, **68**, 054018.

Gjesdal, S. *et al.* (1974). A measurement of the $K_L - K_S$ mass difference from the charge asymmetry in semi-leptonic kaon decays. *Phys. Lett. B*, **52**, 113.

Glashow, S. L. (1986). Positronium versus the mirror universe. *Phys. Lett. B*, **167**, 35.

Goebel, C. (1956). Selection rules for $N\overline{N}$ annihilation. *Phys. Rev.*, **103**, 258.

Goldhaber, M. and Scharff-Goldhaber, G. (1948). Identification of beta-rays with atomic electrons. *Phys. Rev.*, **73**, 1472.

Goldhaber, M. *et al.* (1957). Evidence for circular polarization of bremsstrahlung produced by beta rays. *Phys. Rev.*, **106**, 826.

Goldhaber, M. *et al.* (1958a). Decay modes of a $(\theta + \overline{\theta})$ system. *Phys. Rev.*, **112**, 1796.

Goldhaber, M. *et al.* (1958b). Helicity of neutrinos. *Phys. Rev.*, **109**, 1015.

Golub, R. and Lamoreaux, S. K. (1994). Neutron electric-dipole moment, ultracold neutrons and polarized ^{3}He. *Phys. Rep.*, **237**, 1.

Gómez-Cadenas, J. J. (2006). Measuring leptonic *CP* violation in future neutrino facilities. In M. Giorgi *et al.* (eds), *CP violation: from quark to leptons*. SIF and IOS Press.

Good, M. L. (1961). K_2^0 and the equivalence principle. *Phys. Rev.*, **121**, 311.

Good, R. H., Jr. (1962). Classical equation of motion for a polarized particle in an electromagnetic field. *Phys. Rev.*, **125**, 2112.

Good, R. H. *et al.* (1961). Regeneration of neutral K mesons and their mass difference. *Phys. Rev.*, **124**, 1223.

Gottfried, K. (1966). *Quantum mechanics – Vol. I: Fundamentals*. Benjamin, New York.

Gould, C. R. and Davis, E. D. (2002). Time reversal invariance in nuclear physics: from neutrons to stochastic systems. In M. Beyer (ed.), *CP violation in particle, nuclear and astrophysics*, Berlin. Springer.

Greenberg, O. W. (2002). *CPT* violation implies violation of lorentz invariance. *Phys. Rev. Lett.*, **89**, 231602.

Greenberg, O. W. (2003). Why is *CPT* fundamental? Technical Report PP-04004, Univ. of Maryland.

Greenberg, O. W. and Mohapatra, R. N. (1989a). Difficulties with a local quantum field theory of possible violation of the Pauli principle. *Phys. Rev. Lett.*, **62**, 712.

Greenberg, O. W. and Mohapatra, R. N. (1989b). Phenomenology of small violations of Fermi and Bose statistics. *Phys. Rev. D*, **39**, 2032.

Grimus, W. (1988). *CP* violating phenomena and theoretical results. *Fortschr. Phys.*, **36**, 201.

Gronau, M. and London, D. (1990). Isospin analysis of *CP* asymmetries in B decays. *Phys. Rev. Lett.*, **65**, 3381.

Gronau, M. and London, D. (1991). How to determine all the angles of the unitarity triangle from $B_d^0 \to DK_S$ and $B_s^0 \to D\phi$. *Phys. Lett. B*, **253**, 483.

Gronau, M. and Wyler, D. (1991). On determining a weak phase from charged B decay asymmetries. *Phys. Lett. B*, **265**, 172.

Gudkov, V. P. *et al.* (1992). On *CP* violation in nuclear reactions. *Phys. Rep.*, **212**, 77.

Haas, R. *et al.* (1959). Conservation of parity in strong interactions. *Phys. Rev.*, **116**, 1221.

Hagedorn, R. (1963). *Relativistic kinematics*. Benjamin, New York.

Hamilton, J. (1959). *The theory of elementary particles*. Clarendon Press, Oxford.

Harvey, J. A. *et al.* (1982). Calculation of cosmological baryon asymmetry in grand unified gauge models. *Nucl. Phys. B*, **201**, 16.

Hastings, N. C. *et al.* (Belle collaboration) (2003). Studies of $B^0 - \bar{B}^0$ mixing properties with inclusive dilepton events. *Phys. Rev. D*, **67**, 052004.

Hayakawa, M. and Sanda, A. I. (1993). Searching for T, CP, CPT, and $\Delta S = \Delta Q$ rule violations in the neutral K meson system: a guide. *Phys. Rev. D*, **48**, 1150.

Henley, E. M. (1969). Parity and time-reversal invariance in nuclear physics. *Ann. Rev. Nucl. Sci.*, **19**, 367.

Henley, E. M. and Jacobsohn, B. A. (1959). Time reversal in nuclear interactions. *Phys. Rev.*, **113**, 225.

Herb, S. W. *et al.* (1977). Observation of a dimuon resonance at 9.5 GeV in 400-GeV proton-nucleus collisions. *Phys. Rev. Lett.*, **39**, 252.

Hewett, J. L. and Hitlin, D. G. (eds) (2004). The discovery potential of a super B factory. Technical Report SLAC-R-709, SLAC.

Hilborn, R. C. and Tino, G. M. (eds) (2000). *Proc. Spin-statistics connection and commutation relations, 2000 (Anacapri)*. Am. Inst. of Physics.

Hillman, P. *et al.* (1958). Time-reversal invariance in nuclear scattering. *Phys. Rev.*, **110**, 1218.

Hitlin, D. G. (2006). Asymmetric B factories. In M. Giorgi *et al.* (eds), *CP* violation: from quark to leptons. SIF and IOS Press.

Hollister, J. H. *et al.* (1981). Measurement of parity nonconservation in atomic bismuth. *Phys. Rev. Lett.*, **46**, 643.

Holmstrom, T. *et al.* (HyperCP collaboration) (2004). Search for *CP* violation in charged-Ξ and Λ hyperon decays. *Phys. Rev. Lett.*, **93**, 262001.

Holstein, B. (1993). Anomalies for pedestrians. *Am. J. Phys.*, **61**, 142.

Holzscheiter, M. H. *et al.* (2004). The route to ultra-low energy antihydrogen. *Phys. Rep.*, **402**, 1.

Hori, M. *et al.* (2003). Direct measurement of transition frequencies in isolated $\bar{p}\,\mathrm{He}^+$ atoms, and new *CPT*-violation limits on the antiproton charge and mass. *Phys. Rev. Lett.*, **91**, 123401.

Huang, K. (1992). *Quarks, leptons & gauge fields (2nd ed.)*. World Scientific, Singapore.

Huffman, P. R. *et al.* (1997). Test of parity-conserving time-reversal invariance using polarized neutrons and nuclear spin aligned holmium. *Phys. Rev. C*, **55**, 2684.

Hughes, R. J. and Deutch, B. I. (1992). Electric charges of positrons and antiprotons. *Phys. Rev. Lett.*, **69**, 578.

Hurley, J. (1981). Time-asymmetry paradox. *Phys. Rev. A*, **23**, 268.

Ignatiev, A. Yu. *et al.* (1996). The search for the decay of the Z boson into two photons as a test of Bose statistics. *Mod. Phys. Lett. A*, **11**, 871.

Inami, K. *et al.* (Belle collaboration) (2003). Search for the electric dipole moment of the τ lepton. *Phys. Lett. B*, **551**, 16.

Inglis, D. R. (1961). Completeness of quantum mechanics and charge-conjugation correlations of theta particles. *Rev. Mod. Phys.*, **33**, 1.

Ioffe, B. L. *et al.* (1957). The problem of parity non-conservation in weak interactions. *Soviet Phys. JETP*, **5**, 328.

Ishino, H. *et al.* (Belle collaboration) (2007). Observation of direct *CP*-violation in $B^0 \to \pi^+\pi^-$ decays and model-independent constraints on ϕ_2. *Phys. Rev. Lett.* **98**, 211801.

Itzykson, C. and Zuber, J.-B. (1980). *Quantum field theory*. McGraw Hill, New York.

Jackson, J. D. *et al.* (1957a). Coulomb corrections in allowed beta transitions. *Nucl. Phys.*, **4**, 206.

Jackson, J. D. *et al.* (1957b). Possible tests of time reversal invariance in beta decay. *Phys. Rev.*, **106**, 517.

Jacobsohn, B. A. and Henley, E. M. (1959). Gamma-ray angular correlation tests for time-reversal invariance in nuclear forces. *Phys. Rev.*, **113**, 234.

Jane, M. R. *et al.* (1974). A measurement of the charge asymmetry in the decay $\eta \to \pi^+\pi^-\pi^0$. *Phys. Lett. B*, **48**, 260.

Jarlskog, C. (1985). Commutator of the quark mass matrices in the standard electroweak model and a measure of maximal *CP* nonconservation. *Phys. Rev. Lett.*, **55**, 1039.

Jean, P. *et al.* (INTEGRAL/SPI collaboration) (2003). Early SPI/INTEGRAL measurements of galactic 511 keV line emission from positron annihilation. *Astron. Astrophys.*, **407**, L55.

Jeffreys, H. (1946). An invariant form for the prior probability in estimation problems. *J. R. Statist. Soc. A*, **186**, 453.

Johnson, W.N., III *et al.* (1972). The spectrum of low-energy gamma radiation from the galactic-center region. *Astrophys. J.*, **172**, L1.

Jost, R. (1957). Eine Bemerkung zum *CTP* Theorem. *Helv. Phys. Acta*, **30**, 409.

Jungman, G. *et al.* (1996). Cosmological-parameter determination with microwave background maps. *Phys. Rev. D*, **54**, 1332.

Kabir, P. K. (1968). *The CP puzzle*. Academic press, London.

Kabir, P. K. (1970). What is not invariant under time reversal? *Phys. Rev. D*, **2**, 540.

Kabir, P. K. (1988*a*). Polarization-asymmetry relations in neutron optics. *Phys. Rev. Lett.*, **60**, 686.

Kabir, P. K. (1988*b*). Transformation of neutron polarization in polarized media and tests of *T* invariance. *Phys. Rev. D*, **37**, 1856.

Kaplan, D. M. (1988). Remarks on muon $g-2$ experiments and possible *CP* violation in $\pi \to \mu \to e$ decay. *Phys. Rev. D*, **57**, 3827.

Kemmer, N. *et al.* (1959). Invariance in elementary particle physics. *Rep. Prog. Phys.*, **22**, 368.

Kenny, B. G. and Sachs, R. G. (1973). Non-Hermitian interactions and the evidence for violation of *T* invariance. *Phys. Rev. D*, **8**, 1605.

Khriplovich, I. B. (1991). *Parity nonconservation in atomic phenomena*. Gordon and Breach, Philadelphia.

Khriplovich, I. and Lamoreaux, S. (1997). *CP violation without strangeness*. Springer, Berlin.

Khriplovich, I. B. and Pospelov, M. E. (1990). Anapole moment of a chiral molecule. *Z. Phys. D*, **17**, 81.

Klein, J. R. and Roodman, A. (2005). Blind analysis in nuclear and particle physics. *Ann. Rev. Nucl. Part. Sci.*, **55**, 141.

Kleinknecht, K. (2003). *Uncovering CP violation*. Springer, Berlin.

Klinkhamer, F. R. and Manton, N. S. (1984). A saddle-point solution in the Weinberg-Salam theory. *Phys. Rev. D*, **30**, 2212.

Kobayashi, M. and Maskawa, T. (1973). *CP*-violation in the renormalizable theory of weak interaction. *Prog. Theor. Phys.*, **49**, 652.

Kobzarev, I. Yu. *et al.* (1966). On the possibility of experimental observation of mirror particles. *Sov. Jou. Nucl. Phys.*, **3**, 837.

Kobsarev, I. Yu. *et al.* (1974). Spontaneous *CP* violation and cosmology. *Phys. Lett. B*, **50**, 340.

Kojima, K. *et al.* (1997). Experimental constraints to rephasing-invariant *CP, T* and *CPT* violating parameters in the neutral kaon system. *Prog. Theor. Phys.*, **97**, 103.

Kolb, E. and Turner, M. (1990). *The early universe*. Addison Wesley, New York.

Kolb, E. W. and Wolfram, S. (1980). Baryon number generation in the early universe. *Nucl. Phys. B*, **172**, 224.

Kostelecký, V. A. (2001). *CPT, T* and Lorentz violation in neutral-meson oscillations. *Phys. Rev. D*, **64**, 076001.

Kostelecký, V. A. (ed.) (2005). *Proc. CPT and Lorentz symmetry III, 2004 (Bloomington)*. World Scientific, Singapore.

Kramers, H. A. (1930). Théorie générale de la rotation paramagnétique du plan de polarisation dans les cristaux. *Proc. Acad. Sci. Amsterdam*, **33**, 959.

Kramers, H. A. (1937). The use of charge-conjugated wave-functions in the hole-theory of the electron. *Proc. K. Ned. Acad. Wet. Amsterdam*, **40**, 814.

Kuzmin, V. A. *et al.* (1985). On anomalous electroweak baryon-number non-conservation in the early universe. *Phys. Lett. B*, **155**, 36.

Lai, A. *et al.* (NA48 collaboration) (2001). A precise measurement of the direct *CP* violation parameter $\mathrm{Re}(\epsilon'/\epsilon)$. *Eur. Phys. J. C*, **22**, 231.

Lai, A. *et al.* (NA48 collaboration) (2003). Investigation of $K_{L,S} \rightarrow \pi^+\pi^- e^+ e^-$ decays. *Eur. Phys. J. C*, **30**, 33.

Lai, A. *et al.* (NA48 collaboration) (2005). Search for *CP* violation in $K^0 \rightarrow 3\pi^0$ decays. *Phys. Lett. B*, **610**, 165.

Lanceri, L. (2006). Experiments on *CP* violation with *B* mesons. In M. Giorgi et al. (eds), *CP violation: from quark to leptons*. SIF and IOS Press.

Landau, L. D. (1948). On the angular momentum of a system of two photons. *Doklady Akad. Nauk. (USSR)*, **60**, 207.

Landau, L. D. (1957*a*). Conservation laws in weak interactions. *Soviet Phys. JETP*, **5**, 336.

Landau, L. D. (1957*b*). On the conservation laws for weak interactions. *Nucl. Phys.*, **3**, 127.

Landau, L. D. and Lifshitz, E. M. (1965). *Quantum mechanics: non-relativistic theory*. Pergamon Press, Reading.

Lande, K. *et al.* (1956). Observation of long-lived neutral V particles. *Phys. Rev.*, **103**, 1901.

Lande, K. *et al.* (1957). Report on long-lived K^0 mesons. *Phys. Rev.*, **104**, 1925.

Laporte, O. (1924). The structure of the iron spectrum. *Zeit. Physik*, **23**, 135.

Lavoura, L. and Silva, J. P. (1999). Disentangling violations of *CPT* from other new-physics effects. *Phys. Rev. D*, **60**, 056003.

Lazzeroni, C. (NA48 collaboration) (2004). New NA48 results on *CP* violation. *Eur. Phys. J. C*, **33**, s330.

Lee, T. D. (1965). Possible *C* noninvariant effects in the 3π decay modes of η^0 and ω^0. *Phys. Rev.*, **139**, B1415.

Lee, T. D. (1973). A theory of spontaneous *T* violation. *Phys. Rev. D*, **8**, 1226.

Lee, T. D. (1988). *Particle physics and introduction to quantum field theory*. Harwood Academic Publishers, Chur.

Lee, T. D. and Orear, J. (1955). Speculations on heavy mesons. *Phys. Rev.*, **100**, 932.

Lee, T. D. and Wick, G. C. (1966). Space inversion, time reversal and other discrete symmetries in local field theories. *Phys. Rev.*, **148**, 1385.

Lee, T. D. and Wolfenstein, L. (1965). Analysis of *CP*-noninvariant interactions and the K_1^0, K_2^0 system. *Phys. Rev.*, **138**, B1490.

Lee, T. D. and Wu, C. (1966*a*). Decays of charged *K* mesons. *Ann. Rev. Nucl. Sci.*, **16**, 471.

Lee, T. D. and Wu, C. (1966*b*). Decays of neutral *K* mesons. *Ann. Rev. Nucl. Sci.*, **16**, 511.

Lee, T. D. and Yang, C. N. (1955). Conservation of heavy particles and generalized gauge transformations. *Phys. Rev.*, **98**, 932.

Lee, T. D. and Yang, C. N. (1956*a*). Charge conjugation, a new quantum number *G*, and selection rules concerning a nucleon-antinucleon system. *Nuovo Cim.*, **3**, 749.

Lee, T. D. and Yang, C. N. (1956*b*). Mass degeneracy of the heavy mesons. *Phys. Rev.*, **102**, 290.

Lee, T. D. and Yang, C. N. (1956*c*). Question of parity conservation in weak interactions. *Phys. Rev.*, **104**, 254.

Lee, T. D. and Yang, C. N. (1957). General partial wave analysis of the decay of a hyperon of spin 1/2. *Phys. Rev.*, **108**, 1645.

Lee, T. D. *et al.* (1957*a*). Possible detection of parity nonconservation in hyperon decay. *Phys. Rev.*, **106**, 1367.

Lee, T. D., Oehme, R., and Yang, C. N. (1957*b*). Remarks on possible noninvariance under time reversal and charge conjugation. *Phys. Rev.*, **106**, 340.

Leipuner, L. B. *et al.* (1963). Anomalous regeneration of K_1^0 mesons from K_2^0 mesons. *Phys. Rev.*, **132**, 2285.

Leitner, J. and Okubo, S. (1964). Parity, charge conjugation, and time reversal in the gravitational interaction. *Phys. Rev.*, **136**, B1542.

Leprince-Ringuet, L. and L'Héritier, M. (1944). Existence probable d'une particule de masse 990 m_0 dans le rayonnment cosmique. *Comptes Rendus Acad. Sci. Paris*, **219**, 618.

Limon, P. *et al.* (1968). Proton-proton triple scattering at 430 MeV. *Phys. Rev.*, **169**, 1026.

Lin, S. W. *et al.* (Belle collaboration) (2006). Observation of *b* decays to two kaons. Technical Report BELLE-CONF-0633, BELLE. Preprint hep-ex/0608049.

Link, J. M. *et al.* (FOCUS collaboration) (2003). Charm system tests of *CPT* and Lorentz invariance with FOCUS. *Phys. Lett. B*, **556**, 7.

Link, J. M. *et al.* (FOCUS collaboration) (2005). Search for *T* violation in charm meson decays. *Phys. Lett. B*, **622**, 239.

Lipari, P. (2001). *CP* violation effects and high energy neutrinos. *Phys. Rev. D*, **64**, 033002.

Lipkin, H. J. (1968). *CP* violation and coherent decays of kaon pairs. *Phys. Rev.*, **176**, 1715.

Lloyd, S. P. (1951). 2^{L-1}-magnetic 2^L-electric interference terms in $\gamma - \gamma$ angular correlations. *Phys. Rev.*, **81**, 161.

Lobashov, V. M. *et al.* (1967). Parity non-conservation in the gamma decay of ^{181}Ta. *Phys. Lett. B*, **25**, 104.

Lobkowicz, F. *et al.* (1966). Precise measurement of the K^+/K^- and π^+/π^- lifetime ratios. *Phys. Rev. Lett.*, **17**, 548.

Lockyer, N. S. *et al.* (Mark II collaboration) (1983). Measurement of the lifetime of bottom hadrons. *Phys. Rev. Lett.*, **51**, 1316.

Leite Lopes, J. (1969). *Lectures on symmetries*. Gordon and Breach, New York.

Low, F. E. (1967). *Symmetries and elementary particles*. Gordon and Breach, New York.

Lüders, G. (1957). Proof of the *TCP* theorem. *Ann. Phys. (NY)*, **2**, 1.

Lüders, G. and Zumino, B. (1957). Some consequences of *TCP*-invariance. *Phys. Rev.*, **106**, 385.

Lyons, L. *et al.* (eds) (2003). Proc. statistical problems in particle physics, astrophysics and cosmology (PHYSTAT2003), 2003 (stanford). Technical Report SLAC-R-703, SLAC. eConf C030908.

Mach, E. (1893). *Science of Mechanics*. Open Court Publ. Co., Chicago.

Macq, P. C. *et al.* (1958). Helicity of the electron and positron in muon decay. *Phys. Rev.*, **112**, 2061.

Magritte, R. (1937). *La reproduction interdite* (Portrait d'Edward James). Oil on canvas, 79×65.5 cm (Museum Boymans-van Beuningen, Rotterdam).

Maiani, L. *et al.* (eds) (1995). *Second DAΦNE physics handbook*. INFN, Frascati.

Majorana, E. (1937). Teoria simmetrica dell'elettrone e del positrone. *Nuovo Cim.*, **14**, 171.

Maki, Z., Nakagawa, M., and Sakata, S. (1962). Remarks on the unified model of elementary particles. *Prog. Theor. Phys.*, **28**, 870.

Mannelli, I. (1984). Neutral $K^0 - \overline{K}^0$ mixing and *CP* violation. In A. Bertin et al. (eds), *50 anni di fisica delle interazioni deboli*, Bologna. SIF.

Matthews, J. N. *et al.* (E731 collaboration) (1995). New measurement of the *CP* violation parameter $\eta_{+-\gamma}$. *Phys. Rev. Lett.*, **75**, 2803.

Mavromatos, N. E. (2005). *CPT* violation: theory and phenomenology. In *Proc. EXA05, Int. Conf. on Exotic Atoms, 2005 (Vienna)*. Austrian Academy of Sciences Press.

Mazzotti, D. *et al.* (2001). Search for exchange-antisymmetric sttes for spin-0 particles at the 10^{-11} level. *Phys. Rev. Lett.*, **86**, 1919.

McDonough, J. *et al.* (1988). New searches for the *C*-noninvariant decay $\pi^0 \rightarrow 3\gamma$ and the rare decay $\pi^0 \rightarrow 4\gamma$. *Phys. Rev. D*, **38**, 2121.

McNabb, R. (BNL muon $g-2$ collaboration) (2005). An improved limit on the electric dipole moment of the muon. In J. Tran Thanh Van (ed.), *Proc. XXXIXth Rencontres de Moriond on Electroweak Interactions and Unified Theories, 2004 (La Thuile)*, The Gioi Publishers, Hanoi.

Messiah, A. (1961). *Quantum mechanics – Vol. 2*. North Holland, Amsterdam.

Messiah, A. M. L. and Greenberg, O. W. (1964). Symmetrization postulate and its experimental foundation. *Phys. Rev.*, **136**, B 248.

Michel, L. (1953). Selection rules imposed by charge conjugation. *Nuovo Cim.*, **10**, 319.

Michel, L. (1987). Charge conjugation. In M. G. Doncel et al. (eds), *Symmetries in Physics (1600–1980)*. Barcelona University.

Miller, D. H. (2006). Charmonium at CLEO-c. In T. Akesson et al. (eds), *Proc. HEP2005 Int. Europhysics Conf. on High Energy Physics, 2005 (Lisboa)*. Proceedings of Science.

Mills, A. P. and Berko, S. (1967). Search for *C* nonconservation in electron-positron annihilation. *Phys. Rev. Lett.*, **18**, 420.

Mitsui, T. *et al.* (1996). Expected enhancement of the primary antiproton flux at the solar minimum. *Phys. Lett. B*, **389**, 169.

Morpurgo, G. *et al.* (1954). On time reversal. *Nuovo Cim.*, **12**, 677.

Morrison, P. (1958). Approximate nature of physical symmetries. *Am. J. Phys.*, **26**, 358.

Morse, W. M. *et al.* (1980). Search for the violation of time-reversal invariance in $K^0_{\mu3}$ decays. *Phys. Rev. D*, **21**, 1750.

Muller, F. *et al.* (1960). Regeneration and mass difference of neutral K mesons. *Phys. Rev. Lett.*, **4**, 418.

Murayama, H. (2000). Supersymmetry phenomenology. Technical Report UCB-PTH-00/05, UC Berkeley. Preprint hep-ph/0002232.

Nachtmann, O. (1989). *Elementary particle physics*. Springer, Berlin.

Nakano, E. *et al.* (Belle collaboration) (2006). Charge asymmetry of same-sign dileptons in $B^0 - \overline{B}^0$ mixing. *Phys. Rev. D*, **73**, 112002.

Nakano, T. and Nishijima, G. (1953). Charge independence for V-particles. *Prog. Theor. Phys.*, **10**, 581.

Nanopoulos, D. V. and Weinberg, S. (1979). Mechanism for cosmological baryon production. *Phys. Rev. D*, **20**, 2484.

Neubeck, K. *et al.* (1974). Parity non-conservation in the alpha particle decay of the 8.87 MeV 2^- state of ^{16}O. *Phys. Rev. C*, **10**, 320.

Newton, R. G. (1982). *Scattering theory of waves and particles (2nd ed.)*. Springer, New York.

Niebergall, F. *et al.* (1974). Experimental study of the $\Delta S/\Delta Q$ rule in the time-dependent rate of $K^0 \to \pi e \nu$. *Phys. Lett. B*, **49**, 103.

Nieto, M. M. and Goldman, T. (1991). The arguments against 'antigravity' and the gravitational acceleration of antimatter. *Phys. Rep.*, **205**, 221. See also: PRP, **216**, 343 (1992).

Nir, Y. (1999). *CP* violation in and beyond the standard model. In *XXVII SLAC Summer Institute on Particle Physics*, Stanford. SLAC.

Noether, E. (1918). Invariante Variationsprobleme. *Nachr. König. Gesellsch. Wiss. Göttingen, Math.-Physik.*, 235.

Oddone, P. (1987). Detector considerations. In D. Stork (ed.), *Proc. Workshop on linear collider $B\overline{B}$ factory conceptual design, 1987 (Los Angeles)*, pp. 243. World Scientific, Singapore.

Okubo, S. (1958). Decay of the Σ^+ hyperon and its antiparticle. *Phys. Rev.*, **109**, 984.

Okun, L. B. (1982). *Leptons and Quarks*. North Holland, Amsterdam.

Okun, L. B. (1989). Tests of electric charge conservation and the Pauli principle. *Sov. Phys. Usp.*, **32**, 543.

Okun, L. B. (2003). C, P, T are broken, why not *CPT*? In J. Tran Thanh Van (ed.), *Proc. XIV*[th] *Rencontres de Blois (2002)*, The Gioi Publishers, Hanoi.

Olive, K. A. and Peacock, J. A. (2006). Big-bang cosmology. *Jou. Phys. G*, **33**, 210.

Orear, J. *et al.* (1956). Spin and parity analysis of Bevatron τ mesons. *Phys. Rev.*, **102**, 1676.

Overseth, O. E. and Roth, R. F. (1967). Time reversal invariance in Λ^0 decay. *Phys. Rev. Lett.*, **19**, 391.

Page, L. A. (1957). Polarization effects in the two quantum annihilation of positrons. *Phys. Rev.*, **106**, 394.

Pais, A. (1952). Some remarks on the V-particles. *Phys. Rev.*, **86**, 663.

Pais, A. (1959). Notes on antibaryon interactions. *Phys. Rev. Lett.*, **3**, 242.

Pais, A. (1986). *Inward bound*. Clarendon Press, Oxford.

Pais, A. and Jost, R. (1952). Selection rules imposed by charge conjugation and charge symmetry. *Phys. Rev.*, **87**, 871.

Pais, A. and Piccioni, O. (1955). Note on the decay and absorption of the θ^0. *Phys. Rev.*, **100**, 1487.

Pais, A. and Treiman, S. B. (1975). *CP* violation in charmed-particle decays. *Phys. Rev. D*, **12**, 2744.

Palmer, W. F. and Wu, Y. L. (1995). Rephase-invariant *CP*-violating observables and mixings in the B^0-, D^0-, and K^0- systems. *Phys. Lett. B*, **350**, 245.

Pauli, W. (1925). Über den Zusammenhang des Abschlusses der Elektronengruppen im Atom mit der Komplexstruktur der Spektren. *Zeit. Physik*, **31**, 761.

Pauli, W. (1936). Contributions mathématique à la théorie des matrices de Dirac. *Ann. Inst. H. Poincaré*, **6**, 109.

Pauli, W. (1940). The connection between spin and statistics. *Phys. Rev.*, **58**, 716.

Pauli, W. (1955). Exclusion principle, Lorentz group and reflection of space-time and charge. In W. Pauli, L. Rosenfeld (eds), *Niels Bohr and the development of physics*, McGraw-Hill, New York.

Pauli, W. (1964) 'Letter to V. Weisskopf'. In: *Collected Scientific Works*, Wiley, New York.

Peccei, R. D. and Quinn, H. R. (1977). *CP* conservation in the presence of pseudoparticles. *Phys. Rev. Lett.*, **38**, 1440.

Pereira, R. (2005). Astrophysics with the AMS-02 experiment. In G. Barreira et al. (eds), *Proc. Int. Europhysics Conf. on High Energy Physics, 2005 (Lisboa)*. Proceedings of Science.

Perkins, D. H. (2000). *Introduction to high energy physics (4th ed.)*. Cambridge University Press, Cambridge.

Perret, P. (ed.) (2003). *Proc. second international conference on flavor physics and CP violation (2003)*, Paris.

Piper, D. (1977). The oscillating universe. *Observatory Magazine*, **97**, 10P. Reproduced in: J.D. Barrow, *The artful universe*, Oxford University Press, New York (1995).

Poluektov, A. *et al.* (Belle collaboration) (2006). Measurement of ϕ_3 with a dalitz plot analysis of $B^+ \to D^{(*)} K^{(*)+}$ decay. *Phys. Rev. D*, **73**, 112009.

Pondrom, L. *et al.* (1981). New limit on the electric dipole moment of the Λ hyperon. *Phys. Rev. D*, **23**, 814.

Pontecorvo, B. (1958). Inverse beta processes and nonconservation of lepton charge. *Soviet Phys. JETP*, **7**, 172.

Postma, H. *et al.* (1957). Asymmetry of the positron emission by polarized ^{58}Co-nuclei. *Physica*, **23**, 259.

Prakhov, S. *et al.* (Crystal Ball collaboration) (2000). Search for the *CP* forbidden decay $\eta \to 4\pi^0$. *Phys. Rev. Lett.*, **84**, 4802.

Prescott, C. Y. *et al.* (1978). Parity non-conservation in inelastic electron scattering. *Phys. Lett. B*, **77**, 347.

Purell, E. M. and Ramsey, N. F. (1950). On the possibility of electric dipole moments for elementary particles and nuclei. *Phys. Rev.* **78**, 807.

Rabi, I. I. (1957). *The New York Times*, 16 January.

Rabi, I. I. *et al.* (1954). Use of rotating coordinates in magnetic resonance problems. *Rev. Mod. Phys.*, **26**, 167.

Raffelt, G. (1994). Pulsar bound on the photon electric charge reexamined. *Phys. Rev. D*, **50**, 7729.

Ramberg, E. and Snow, G. A. (1990). Experimental limit on a small violation of the Pauli principle. *Phys. Lett. B*, **238**, 438.

Ramsey, N. F. (1949). A new molecular beam resonance method. *Phys. Rev.*, **76**, 996.

Ramsey, N. F. (1956). *Molecular beams*. Oxford University Press, Oxford.

Ramsey, N. F. (1958). Time reversal, charge conjugation, magnetic pole conjugation and parity. *Phys. Rev.*, **109**, 225.

Regan, B. C. *et al.* (2002). New limit on the electron electric dipole moment. *Phys. Rev. Lett.*, **88**, 071805.

Riotto, A. (1998). Theories of baryogenesis. In *Lectures of Summer School in High Energy Physics and Cosmology, 1988 (Miramare)*. Report CERN-TH/98-204, CERN (Geneva).

Rochester, G. D. and Butler, C. C. (1947). Evidence for the existence of new unstable elementary particles. *Nature*, **160**, 855.

Rosner, J. and Slezak, S. A. (2001). Classical illustrations of *CP* violation in kaon decays. *Am. J. Phys.*, **69**, 44.

Sachs, R. G. (1955). Classification of the fundamental particles. *Phys. Rev.*, **99**, 1573.

Sachs, R. G. (1987). *The physics of time reversal*. University of Chicago Press, Chicago.

Sachs, R. G. and Treiman, S. B. (1962). Test of *CP* conservation in neutral *K*-meson decay. *Phys. Rev. Lett.*, **8**, 137.

Sakharov, A. D. (1967). Violation of *CP* invariance, *C* asymmetry, and baryon asymmetry of the universe. *JETP Lett.*, **5**, 24.

Sakurai, J. J. (1964). *Invariance principles and elementary particles*. Princeton University Press, Princeton.

Sakurai, J. J. (1967). *Advanced quantum mechanics*. Addison Wesley, Reading, MA.

Sakurai, J. J. (1985). *Modern quantum mechanics*. Benjamin Cummings, Menlo Park, CA.

Sakurai, J. J. and Wattenberg, A. (1967). Absolute distinction between particles and antiparticles and *CP* violation in $K_{L,S} \rightarrow 2\pi$. *Phys. Rev.*, **161**, 1449.

Samios, N. P. *et al.* (1962). Parity of the neutral pion and the decay $\pi^0 \rightarrow 2e^+ + 2e^-$. *Phys. Rev.*, **126**, 1844.

Scadron, M. D. (1991). *Advanced quantum theory*. Springer, Berlin.

Schäfer, A. and Adelberger, E. G. (1991). *T*-violation experiments using Mössbauer transitions. *Z. Phys. A*, **339**, 305.

Schiff, L. I. (1959). Gravitational properties of antimatter. *Proc. Nat. Acad. Sci*, **45**, 69.

Schiff, L. I. (1963). Measurability of nuclear electric dipole moments. *Phys. Rev.*, **132**, 2194.

Schopper, H. (1957). Circular polarization of γ rays: further proof for parity failure in β decay. *Phil. Mag.*, **2**, 710.

Schramm, D. N. and Turner, M. S. (1998). Big-bang nucleosynthesis enters the precision era. *Rev. Mod. Phys.*, **70**, 303.

Schwinger, J. (1951). The theory of quantized fields I. *Phys. Rev. D*, **82**, 914.

Scott, D. and Smoot, G. F. (2006). Cosmic microwave background. *Jou. Phys. G*, **33**, 238.

Shaposhnikov, M. E. (1987). Baryon asymmetry of the universe in standard electroweak theory. *Nucl. Phys. B*, **287**, 757.

Shifman, M. A. (1995). *ITEP lectures on particle physics and field theory*. World Scientific, Singapore.

Smith, J. H. *et al.* (1957). Experimental limit to the electric dipole moment of the neutron. *Phys. Rev.*, **108**, 120.

Somov, A. *et al.* (Belle collaboration) (2006). Measurement of the branching fraction, polarization, and *CP* asymmetry for $B^0 \rightarrow \rho^+ \rho^-$ decays, and determination of the CKM phase ϕ_2. *Phys. Rev. Lett.*, **96**, 171801.

Sozzi, M. S. (2004). On the direct *CP*-violating parameter ϵ'. *Eur. Phys. J. C*, **36**, 37.

Sozzi, M. S. (2006). Experiments on *CP* violation with *K* mesons – part I. In M. Giorgi et al. (eds), *CP violation: from quark to leptons*. SIF and IOS Press.

Sozzi, M. S. and Mannelli, I. (2003). Measurements of direct *CP* violation. *Riv. Nuovo Cim.*, **26**(3), 1.

Sromicki, J. *et al.* (1996). Study of time reversal violation in β decay of polarized ^{8}Li. *Phys. Rev. C*, **53**, 932.

Stapp, H. P. (1957). Some experimental tests of the Lüders theorem. *Phys. Rev.*, **107**, 634.

Stedman, G. E. (1983). Space-time symmetries and photon selection rules. *Am. J. Phys.*, **51**, 750.

Steigman, G. (1976). Observational tests of antimatter cosmologies. *Ann. Rev. Astron. Astrophys.*, **14**, 336.

Streater, R. F. and Wightman, A. S. (1964). *PCT, spin statistics, and all that*. Benjamin Cummings, Reading, MA.

Stueckelberg, E. C. G. (1942). La mécanique du point matériel en théorie de relativité et en théorie des quanta. *Helv. Phys. Acta*, **15**, 23.

Su, Y. *et al.* (1994). New test of the universality of free fall. *Phys. Rev. D*, **50**, 3614.

Sudarshan, E. C. G. (1975). Relation between spin and statistics. *Stat. Phys. Suppl.: J. Indian Inst. Sci.*, **June**, 123.

Suzuki, Y. *et al.* (Kamiokande collaboration) (1993). Study of invisible nucleon decay $n \rightarrow \nu\nu\bar{\nu}$ and a forbidden nuclear transition in the kamiokande detector. *Phys. Lett. B*, **311**, 357.

't Hooft, G. (1976). Symmetry breaking through Bell-Jackiw anomalies. *Phys. Rev. Lett.*, **37**, 8.

Tandean, J. and Valencia, G. (2003). *CP* violation in hyperon nonleptonic decays within the standard model. *Phys. Rev. D*, **67**, 056001.

Tanner, N. W. and Dalitz, R. H. (1986). The determination of *T*- and *CPT*- violations for the $(K^0, \overline{K}^0)$ complex by $\overline{K}^0/K^0$ comparisons. *Ann. Phys. (NY)*, **171**, 463.

Telegdi, V. L. (1973). In J. Mehra (eds). *The Physicist's Conception of Nature*. Reidel, Dordrecht.

Telegdi, V. L. (1987). Parity violation. In M. G. Doncel et al. (ed.) *Symmetries in Physics (1600-1980)*. Barcelona University.

Telegdi, V. L. (1990). Mind over matter – the intellectual content of experimental physics. Technical Report CERN 90-09, CERN.

Thomson, W. (1st Baron Kelvin) (1904). *Baltimore Lectures on Molecular Dynamics and the Wave Theory of Light*. C. J. Clay & Sons, London.

Tolkien, J. R. R. (1977). *The Silmarillion*. George Allen & Unwin, London.

Tomonaga, S. I. (1997). *The story of spin*. Chicago University Press, Chicago.

Tripp, R. D. (1965). Spin and parity of elementary particles. *Ann. Rev. Nucl. Sci.*, **15**, 325.

Uwer, U. (2004). LHCb physics performance. In M. Paulini, S. Erhan (eds), *Proc. 9th International Conference on B-Physics at Hadron Machines, 2003 (Pittsburgh)*, pp. 231. AIP Conference Proceedings 722.

Uy, Z. E. S. (1991). Determining the *CP*-violating and *CP*-conserving form factors in the decay modes $K_S \to \gamma\gamma$ and $K_L \to \gamma\gamma$. *Phys. Rev. D*, **43**, 802.

Van Dyck, R. S. Jr. *et al.* (1986). Electron magnetic moment from geonium spectra: early experiments and background concepts *Phys. Rev. D*, **34**, 722.

Van Dyck, R. S. Jr. *et al.* (1987). New high-precision comparison of electron and positron *g* factors. *Phys. Rev. Lett.*, **59**, 26.

Vetter, P. A. and Freedman, S. J. (2002). Branching-ratio measurements of multiphoton decays of positronium. *Phys. Rev. A*, **66**, 052505.

Vetter, P. A. and Freedman, S. J. (2003). Search for *CPT*-odd decays of positronium. *Phys. Rev. Lett.*, **91**, 263401.

Wang, C. C. *et al.* (Belle collaboration) (2005). Study of $B^0 \to \rho^{\pm}\pi^{\mp}$ time-dependent *CP* violation at belle. *Phys. Rev. Lett.*, **94**, 121801.

Watson, K. M. (1954). Some general relations between the photoproduction and scattering of π mesons. *Phys. Rev.*, **95**, 228.

Weinberg, S. (1976). Gauge theory of *CP* nonconservation. *Phys. Rev. Lett.*, **37**, 657.

Weinberg, S. (1995). *The quantum theory of fields – Vol. I: Foundations*. Cambridge University Press, Cambridge.

Weisskopf, V. F. and Wigner, E. (1930*a*). Berechnung der natürlichen Linienbreite auf Grund der Diracschen Lichttheorie. *Zeit. Physik*, **63**, 54.

Weisskopf, V. F. and Wigner, E. (1930*b*). Über die natürliche Linienbreite in der Strahlung des harmonischen Oszillators. *Zeit. Physik*, **65**, 18.

Weyl, H. (1929). Elektron und Gravitation. *Zeit. Physik*, **56**, 330.

Weyl, H. (1952). *Symmetry*. Princeton University Press, Princeton.

Whatley, M. C. (1962). *CP* invariance and K_1^0 regeneration. *Phys. Rev. Lett.*, **9**, 317.

Wick, G. *et al.* (1952). The intrinsic parity of elementary particles. *Phys. Rev.*, **88**, 101.

Wigner, E. P. (1927). Some consequences of the Schrödinger theory for term structure. *Zeit. Physik*, **43**, 624.

Wigner, E. P. (1932). Über die Operation der Zeitumkehr in der Quantenmechanik. *Nachr. Akad. Ges. Wiss. Göttingen, Math.-Physik*, **31**, 546.

Wigner, E. P. (1939). On unitary representations of the inhomogeneous Lorentz group. *Ann. Math.*, **40**, 149.

Wigner, E. P. (1959). *Group theory and its applications to the quantum mechanics of atomic spectra*. Academic Press, New York.

Wilkin, C. (1980). On the parity of the positive pion. *Jou. Phys. G*, **6**, 25.

Wilkinson, D. H. (1958). Parity conservation in strong interactions: introduction and the reaction $\mathrm{He}^4(d, \gamma)\mathrm{Li}^6$. *Phys. Rev.*, **109**, 1603.

Wolfenstein, L. (1964). Violation of *CP* invariance and the possibility of very weak interactions. *Phys. Rev. Lett.*, **13**, 562.

Wolfenstein, L. (1983). Parametrization of the Kobayashi-Maskawa matrix. *Phys. Rev. Lett.*, **51**, 1945.

Wolfenstein, L. and Ashkin, J. (1952). Invariance conditions on the scattering amplitudes for spin 1/2 particles. *Phys. Rev.*, **85**, 947.

Wolfenstein, L. and Ravenhall, D. G. (1952). Some consequences of invariance under charge conjugation. *Phys. Rev.*, **88**, 279.

Wu, C. S. (1988). The discovery of nonconservation of parity in beta decay. In R. Novick (ed.), *Thirty years since parity nonconservation*, Boston. Birkhauser.

Wu, C. S. and Shaknov, I. (1950). The angular correlation of scattered annihilation radiation. *Phys. Rev.*, **77**, 136.

Wu, C. S. *et al.* (1957). Experimental test of parity conservation in beta decay. *Phys. Rev.*, **105**, 1413.

Wu, D. D. (1989). Signs of Δm and $\Delta \gamma$ of the $B_d - \bar{B}_d$ system. *Phys. Rev. D*, **40**, 806.

Wu, T. T. and Yang, C. N. (1964). Phenomenological analysis of violation of *CP* invariance in decay of K^0 and $\bar{K}^0$. *Phys. Rev. Lett.*, **13**, 380.

Yamazaki, T. *et al.* (2002). Antiprotonic helium. *Phys. Rep.*, **366**, 183.

Yang, C. N. (1950). Selection rules for the dematerialization of a particle into two photons. *Phys. Rev.*, **77**, 242.

Yang, C. N. and Tiomno, J. (1950). Reflection properties of spin 1/2 fields and a universal Fermi-type interaction. *Phys. Rev.*, **79**, 495.

Yang, J. *et al.* (1996). Four-photon decay of orthopositronium: a test of charge-conjugation invariance. *Phys. Rev. A*, **54**, 1952.

Yao, W.-M. *et al.* (Particle Data Group) (2006). Review of particle physics. *Jou. Phys. G*, **33**, 1.

Yuan, C. (2005). Recent BES results on charmonium decays. In *Proc. XXXIX Rencontres de Moriond 2004 – QCD and hadronic interactions, 2004 (La Thuile)*, Gif-sur-Yvette. Editions Frontières.

Zel'dovich, Ya.B. (1959). Parity nonconservation in the first order in the weak-interaction constant in electron scattering and other effects. *Soviet Phys. JETP*, **9**, 682.

Zemach, C. (1964). Three-pion decays of unstable particles. *Phys. Rev.*, **133**, B1201.

Zocher, H. and Török, C. (1953). About space-time asymmetry in the realm of classical general and crystal physics. *Proc. Nat. Acad. Sci. Wash.*, **39**, 681.

Zumino, B. (1966). Geometries and space-time symmetries. *Suppl. Nuovo Cim.*, **IV**, 384.

LIST OF EXPERIMENTS

INDEX